SECOND EDITION

PROCESS MODELING IN COMPOSITES MANUFACTURING

SECOND EDITION

PROCESS MODELING IN COMPOSITES MANUFACTURING

SURESH G. ADVANI • E. MURAT SOZER

CRC Press
Taylor & Francis Group
Boca Raton London New York

CRC Press is an imprint of the
Taylor & Francis Group, an **informa** business

CRC Press
Taylor & Francis Group
6000 Broken Sound Parkway NW, Suite 300
Boca Raton, FL 33487-2742

International Standard Book Number: 978-1-4200-9082-6 (Hardback)

Library of Congress Cataloging-in-Publication Data

Advani, Suresh G.
Process modeling in composites manufacturing / Suresh G. Advani, E. Murat Sozer. -- 2nd ed.
p. cm.
Includes bibliographical references and index.
ISBN 978-1-4200-9082-6 (hardcover : alk. paper)
1. Composite materials. 2. Manufacturing processes. I. Sozer, E. Murat. II. Title.

TA418.9.C6A325 2010
620.1'18--dc22 2010017665

Visit the Taylor & Francis Web site at
http://www.taylorandfrancis.com

and the CRC Press Web site at
http://www.crcpress.com

Dedication

to our families:

Yolanda Chetwynd, Madhu and Diana Advani;
and
Hanife, Zehra and Eray Sozer.

Contents

Preface

When the publisher asked us to write a second edition of this book, we wanted to bring something new into the mix in addition to (i) correcting the errors and typos of the first edition and (ii) updating the process physics and modeling state in the field. The new material we have introduced is the computer-based solutions using MATLAB code and also flow simulation-based analysis. This should complement closed form solution discussed in the book to understand the role of different material, geometric and process parameters. The book now is 614 pages instead of the 436 pages of the first edition. We have also developed a solution manual for all the problems and questions at the end of each chapter and introduced many of the latest references that update the state of modeling in composite manufacturing processes. This book is about how to formulate and use science-based models in polymer composite manufacturing processes. Three decades ago our understanding of mass, momentum, and energy transfer during composites processing was nonexistent. As a result, almost all manufacturing was based on experience, intuition, and trial and error. We have come a long way since then. Many researchers did delve into this difficult and poorly understood area to uncover the physics and chemistry of processing and to develop the fundamental and constitutive laws to describe them.

There is currently a wealth of literature on modeling and simulation of polymer composite manufacturing processes. However, we felt that there was a need to systematically introduce how one would go about modeling a composite manufacturing process. Hence, we focused on developing a textbook instead of a researcher's reference book to provide an introduction to modeling of composite manufacturing processes for seniors and first-year graduate students in material science and engineering, industrial, mechanical, and chemical engineering. We have explained the basic principles, provided a primer in fluid mechanics and heat transfer, and created a self-contained text. Many example problems have been solved to facilitate the use of back-of-the-envelope calculations to introduce a scientific basis to manufacturing. The end of each chapter has questions and problems that reinforce the content and help the instructor. "Fill in the Blanks" sections were created by E. Murat Sozer to add to the qualitative knowledge of process modeling of composites manufacturing that will develop the "experience base" of the manufacturing, materials, and design engineer or scientist.

A project of this magnitude obviously cannot be realized without the help of others. First, we would like to thank Dr. Ali Gokce who created many of the graphics and figures you see in this book. Dawn Fiore and Yolanda Chetwynd were responsible for the artwork on the cover, which depicts the concept of process modeling. Diane Kukich helped in technical editing of the first edition. Of course we thank all the graduate students in our research group who over the years have helped create the research and the science base to develop models of composite manufacturing processes. We would especially like to mention Petri Hepola, Steve Shuler, Terry Creasy, Krishna Pillai, Sylvia Kueh, Simon Bickerton, Hubert Stadtfeld, Pavel Nedanov, Pavel Simacek, Kuang-Ting Hsiao, Gonzalo Estrada,

Jeffery Lawrence, Roopesh Mathur, Justin Alms, Valentin Neascu, Claire Steggall, and Justin Merotte. Some of the examples and figures used in the book were first developed with their help.

The book contains eight chapters. The first two introduce the composite materials and manufacturing processes. Chapters 3–5 provide the tools needed to model the processes, and Chapters 6–8 apply these tools to some of the well-known manufacturing processes.

The appendix files can be downloaded from the following web sites:

Appendix A, MATLAB files:

`http://home.ku.edu.tr/~msozer/Appendix_A_Matlab_files.zip`

Appendix C, LIMS input files:

`http://www.ccm.udel.edu/Tech/LIMS/index.html`

About the Authors

Suresh G. Advani is the George W. Laird Professor of the Mechanical Engineering Department and Associate Director of the Center for Composite Materials, University of Delaware. His research interests are in rheology, fluid mechanics, and heat transfer as applied to manufacturing processes especially for polymer composites and in fuel cells. He has co-authored over 350 journal and conference proceeding articles and has contributed chapters and co-edited over twenty books. Advani is a Fellow of the American Society of Mechanical Engineers and is the North American Editor for the journal *Composites A: Applied Science and Manufacturing.* Professor Advani serves also on the Scientific Advisory Committee of *Journal of Forming Processes and International Conference on Flow Processes in Composites Manufacturing.* He has been or is a member of the American Society of Mechanical Engineers, Society of Plastic Engineers, Society of Rheology, and Society for Advancement of Materials and Processes.

E. Murat Sozer is Associate Professor of the Mechanical Engineering Department, Koc University, Istanbul, Turkey. His research interests are in manufacturing of composite materials and fluid dynamics. He and his co-authors S. Bickerton and S. G. Advani received the Outstanding Technical Paper Award by the Composites Manufacturing Association (CMA) of the Society of Manufacturing Engineers (SME) in recognition of their outstanding contribution to the composites manufacturing body of knowledge for the technical paper at the Composites Manufacturing and Tooling 2000 Conference, Newport Beach, California, February 23-25, 2000. Before joining Koc University in 2000, he worked as a post-doctoral researcher at the Center for Composite Materials, University of Delaware, between 1997 and 2000, and as a technical editor for Prentice-Hall Publishers between 1996 and 1997.

Chapter 1

Introduction

1.1 Motivation and Contents

Polymer composites have been in use for a few decades now. Their advantages over other materials for high-performance, lightweight applications have attracted many industries such as aerospace, automobile, infrastructure, sports and marine to explore and increase their usage. The path to the design and manufacturing of composite structures was pursued in evolutionary as well as revolutionary ways. They ranged from using hand layup with labor and cost intensive autoclave processing to the use of automated processes such as injection molding and extrusion, traditionally employed by the polymer processing industry. Many new manufacturing techniques were invented and introduced during the last two decades, and some of them were incrementally improved to increase the yield of manufactured composite parts. The process engineer has relied on experience and "trial and error" approaches to improve the manufacturability of a *prototype*. Very little analysis of process physics and back-of-the-envelope calculations were done to approach a prototype development of a composite structure. Even the choice of the process was dictated by familiarity and experience rather than appropriateness and methodology. This has proved to be very expensive. However, in the last few decades the composites manufacturing industry has come under intense pressure to be cost-effective and focus on *cost avoidance* in prototype development. Design and manufacturing engineers have resorted to use of process modeling and simulations to address some of these concerns. The virtues of virtual manufacturing are becoming more obvious to the manufacturing engineer when formulating the guidelines and methodology for the design and manufacturing of composites.

As a result, many books on composites manufacturing have been written in the last few years. They have served as good research references for the composites manufacturing engineers and personnel. The underlying science for many of these manufacturing processes is described by a process model and incorporated into simulations to allow one to perform trial and error experiments in virtual space instead of the laboratory space. Currently, the available books are either chapters written by multiple authors on different processes which summarize the state of the art in the field and are excellent research reference materials [7–9] or they paint with a broad brush on the qualitative aspects of manufacturing of composite materials with polymer, ceramics and metal matrices [10]. Thomas Astrom's book [11] is an excellent book to get practical information about the manufacturing processes as well as a great resource for material property data. The book gives a very detailed qualitative insight into the materials and processes addressing the issues encountered from designing to shop floor manufacturing.

Our book will have two complementary focuses as compared to the recent books written on composites manufacturing. First, this book will introduce to the reader the approach to model the processing operation during composite manufacturing using fundamental principles. The most important aspect will be to identify the key transport phenomena that surface during the manufacturing process and the approach to incorporate them in a process model. Thus, it will reduce the reliance on the trial and error methodology used to achieve an acceptable composite part and will increase the use of science base in the manufacturing process. Second, as the reader or practitioner understands more about the physics of the process and the transport phenomena that drive the process, he or she will be in a position to invent a novel composite processing method that can improve upon the existing manufacturing methods. Selected examples of this will be presented in later chapters. This could attract many more industries to accelerate insertion of composite materials into their products.

In this book, we will restrict ourselves to modeling the processing step of the polymer composite fabrication process, although the modeling philosophy could be extended to manufacturing processes with other materials. The book is written with the undergraduate senior and the first-year graduate student in mind who has some understanding of the basics of fluid mechanics and heat transfer and ordinary and partial differential equations. A brief introduction of useful equations in fluid mechanics and heat transfer will be presented as a primer for those unfamiliar with the subject and should serve as a refresher for those who can't quite recall the details.

1.2 Preliminaries

Composite materials generically consist of two different materials that are combined. In engineering, the definition can be narrowed down to a combination of two or more distinct materials into one with the intent of suppressing undesirable constituent properties in favor of the desirable ones. Atomic level combinations such as metal alloys and polymer blends are excluded from these definitions [11]. Usually composites which contain fibers held together with a polymer, metal or ceramic matrix are referred to as polymer matrix composites, metal matrix composites or ceramic matrix composites, respectively. This text is entirely devoted to modeling polymer composite manufacturing processes although some of the modeling techniques could be extended to metal and ceramic composites.

In polymer composites, the individual constituents are polymer resin and fibers as shown in Figure 1.1. The role of the polymer resin, which is also called the matrix phase of the composite, is to primarily bind the fibers together, give the composite a nice surface appearance in addition to environmental tolerance and provide overall durability. The matrix also provides the strength in compression and shear. Inclusion of nano materials such as nano clays and nanotubes in the matrix phase in recent years has helped improve the shear properties of such composites. The fibers, also known as the reinforcing phase, carry the structural load, reduce thermal stresses and provide macroscopic stiffness and strength [10, 11]. The polymer matrix is either a thermoset or a thermoplastic material. The reinforcing fibers are made from glass, carbon or polymer. Some of the fiber forms are shown in Figure 1.2.

From the processing and manufacturing viewpoint, the type of matrix plays an important role. Thermoset materials are only 50 to 500 times more viscous than water and can impregnate the empty spaces between the fibers readily. They do require an additional processing step which involves chemical reaction of cross-linking the polymer chains known as curing. This is schematically shown in Figure 1.3. On the other hand, thermoplastic

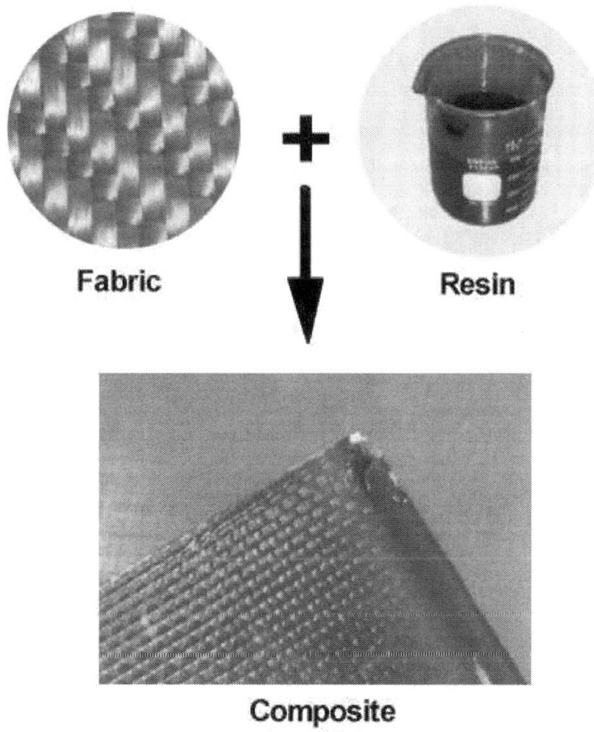

Figure 1.1: Fiber composite made from fibers and resin.

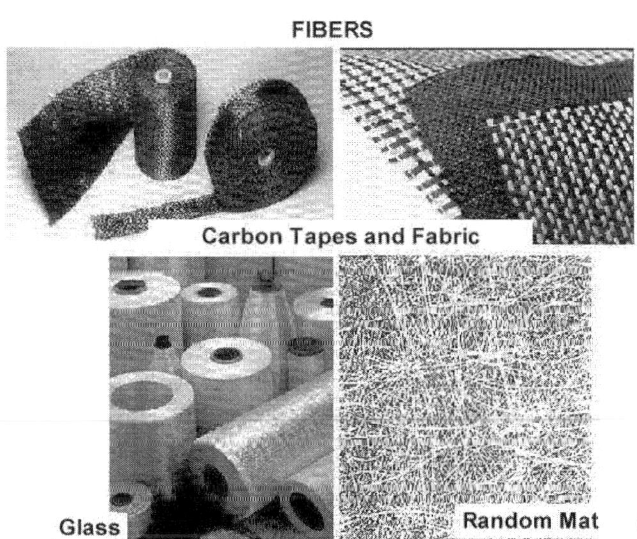

Figure 1.2: Different fiber forms.

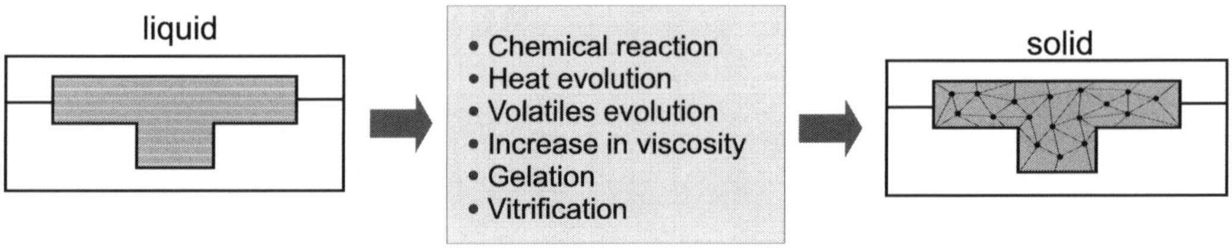

Figure 1.3: Curing is the process of cross-linking a thermoset polymer.

materials do not require this step but are highly viscous. Their viscosity can be as high as a million times more than that of water. Hence, it is difficult to make them flow and fill the tiny empty spaces between the reinforcing fibers. The advantage of thermoplastic matrices are their recyclable nature and higher toughness than thermosets which tend to be brittle and more prone to microcracking. Figure 1.4 displays the important differences between thermoplastics and thermosets. The constitutive equations that describe the chemorheology of the matrix materials such as the influence of temperature, shear rate and degree of cure on the viscosity will play an important role in the processing step during the composite manufacturing process.

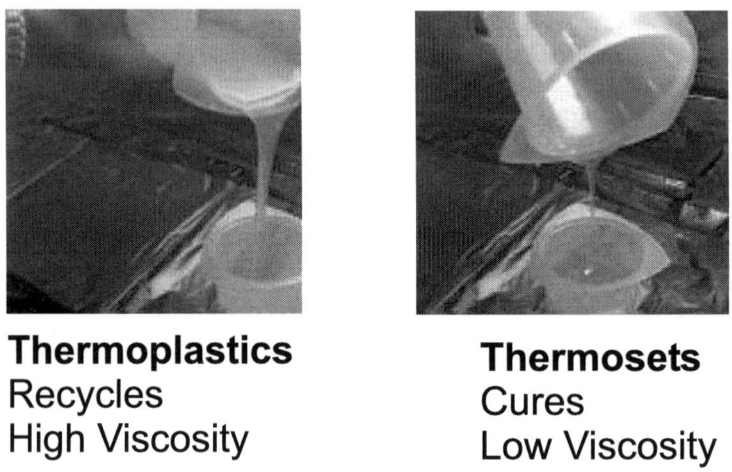

Figure 1.4: Polymer resin types used to make composites.

Glass and carbon are the most common materials used for the fibers. The fiber material usually will not influence the modeling of the manufacturing process in a very significant way. However, whether the fibers are discontinuous or continuous will influence the modeling approach. Also, if the fibers are discontinuous, their aspect ratio (the ratio of fiber length to its diameter) will be important during processing, and if the fibers are continuous then their fiber architecture will play a major role in the manufacturing process. The continuous fibers can be introduced into the polymer matrix as unidirectional fiber arrays or by utilizing appropriate fabrication methods, e.g., weaving, braiding, knitting, or stitching, shaped into 2-D or 3-D reinforcing fabrics before being embedded. Some of these structures are displayed in Figures 1.5 and 1.6. In general, the ease of processing decreases as we move from discontinuous short fibers to continuous fiber preforms that are woven or stitched as schematically shown in Figure 1.7.

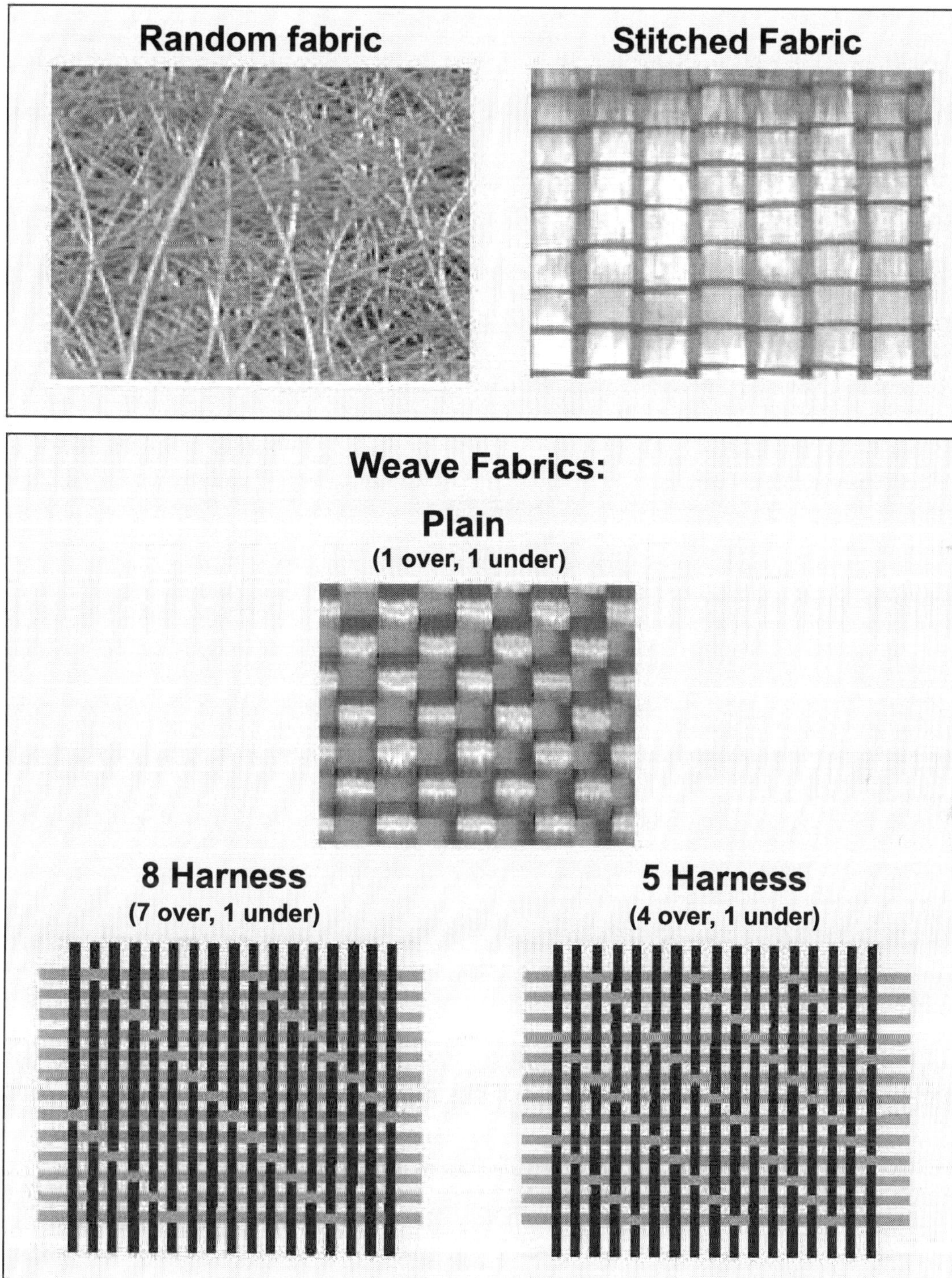

Figure 1.5: Typical fabrics and schematics with continuous fibers.

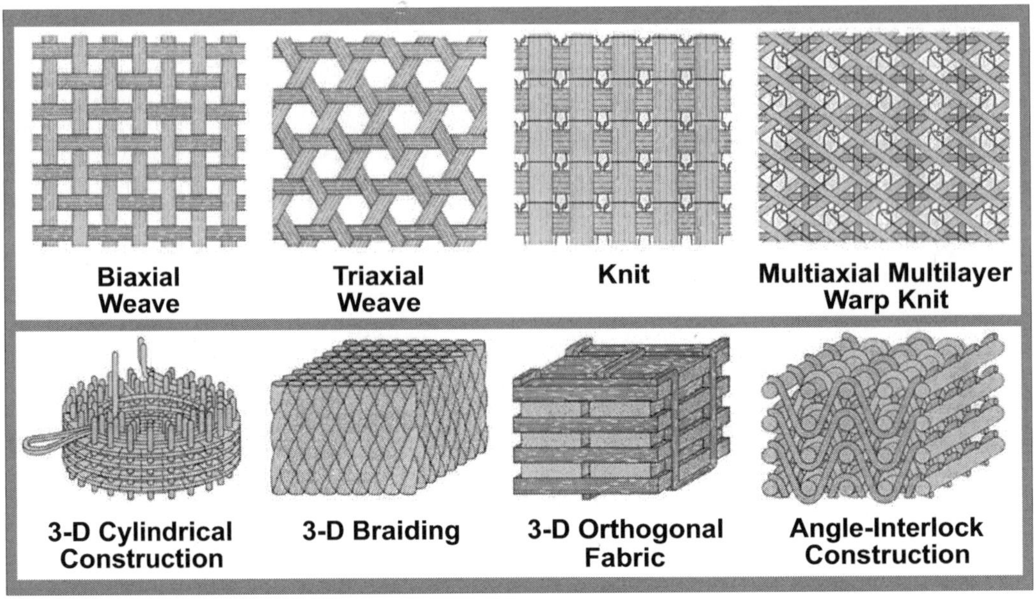

Figure 1.6: More detailed architectures of reinforcing fiber preforms [12].

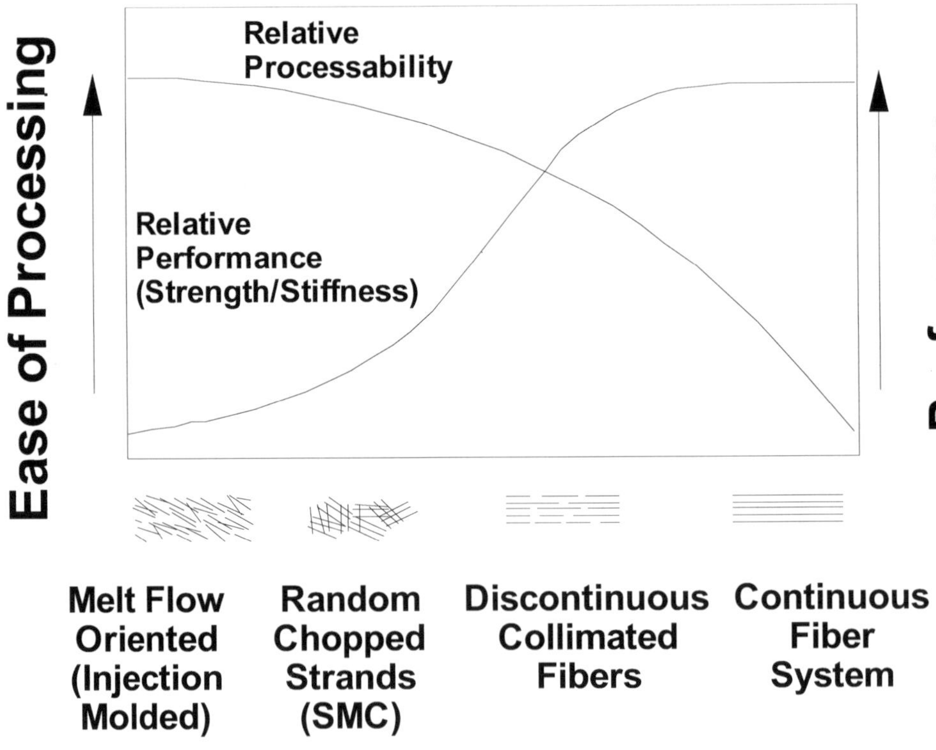

Figure 1.7: Schematic of role of fiber form on processing and performance [13].

1.3 Polymer Matrices for Composites

The polymer matrix in a composite consists mainly of a thermoplastic or thermoset resin. In addition, it may contain small quantities of additives, inert fillers and adhesives.

1.3.1 Polymer Resins

Polymers or plastics are high molecular weight (MW) compounds consisting of many (poly) repeated small segments (mers). They possess the characteristic of a chemical reaction utilizing the smallest building block, the *monomer*, or oligomer (a few monomers together) and assembling approximately 10^3 up to 10^6 of these blocks into a *polymer*. There are two different types of reactions that can be used to create most modern polymers: poly-condensation and poly-addition. They can all produce linear or branched polymers. The primary bonds formed between the molecules are the strong covalent bonds. However, the molecules also form secondary bonds which are an order of magnitude weaker than the covalent bonds. These bonds are due to van der Waals' forces.

Polymers are classified as thermoplastics or cross-linked polymers such as elastomers or thermosets. Thermoplastic polymers can be either amorphous (without regular structure), see Figure 1.8(a), or semicrystalline (amorphous base structure with embedded regular substructures), displayed in Figure 1.8(b).

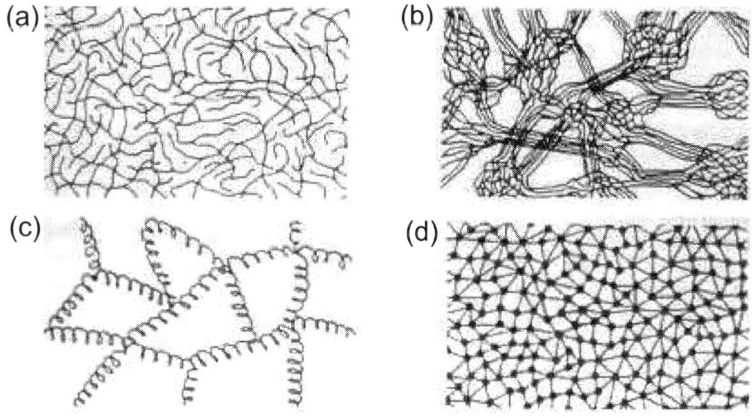

Figure 1.8: Schematic of the molecular structure: (a) amorphous thermoplastic, (b) semi-crystalline thermoplastic, (c) elastomer and (d) thermoset [14].

The thermoplastic properties are determined by the resulting microstructure, which is strongly influenced by the cooling dynamics. In general, an amorphous polymer is transparent, has lower mechanical properties and is less resistant to other chemicals than the semicrystalline thermoplastics. Thermoplastic polymers can be best compared to a plate of spaghetti. Each noodle can be thought of as a long chain of repetitive molecules. When it is heated, the molecules can move around. If cooled quickly, the current arrangement of the noodles or molecule chains can be frozen due to the van der Waals' forces. If cooled slowly, the molecules align themselves in regular crystal formation which is the lowest possible energy state for the arrangement of molecules. As there are only secondary bonds between the molecules, thermoplastics can be melted and reformed. If the molecules are in an orderly form they are known as semicrystalline. If they are randomly organized they are known as amorphous. Hence thermoplastics can be cooled in amorphous form or with various degrees of crystallinity. Crystalline state packs better than the amorphous state. However, it is not possible to get 100% crystalline state. The resulting degree of crystallinity depends on pressure, molecular weight, temperature and most importantly cooling rate, as shown in Figure 1.9. Note that T_g in the figure refers to the glass transition temperature. When the polymer is a solid, the atoms vibrate around their fixed positions. As heat is supplied to it, and its temperature is increased, the atoms' vibration amplitude increases and above glass transition temperature the atoms are no longer fixed and the atoms start to move laterally but their movement is restricted due to the entanglements. Thus between the glass transition and the melting temperature the material becomes soft but is still a solid [15]. At T_g, one expects the secondary bonds to initiate breaking. The molecules continue to slide over one another with relative ease, and flow is possible. The softening occurs over a temperature range of 50-80°C. As a shear force is applied to the bulk material, the long chain molecules will start to slide relative to each other, in response to the applied force. Thus most thermoplastic resins exhibit shear thinning behavior, which is an important property to consider when one develops the process model. This causes the thermoplastic resin viscosity to decrease with temperature and applied shear as shown in Figure 1.10.

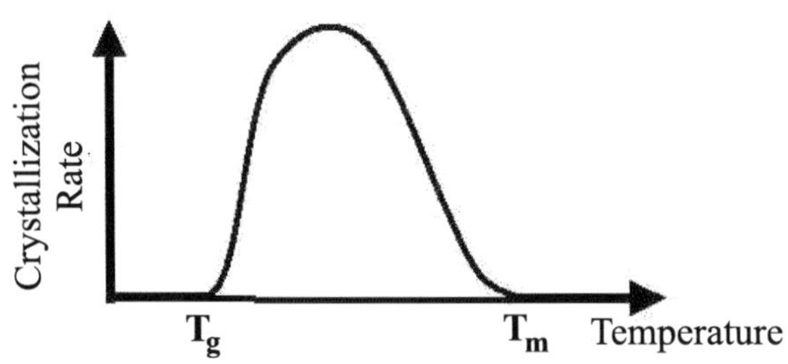

Figure 1.9: Change in crystallization rate as a function of temperature. T_g is the glass transition temperature and T_m is the melting temperature.

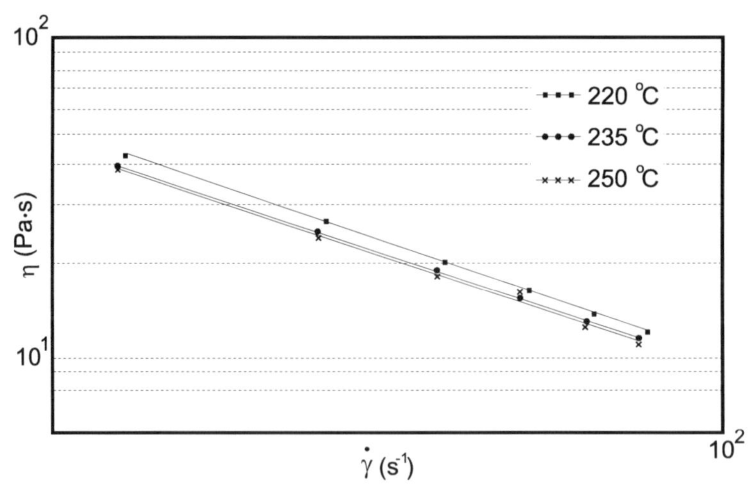

Figure 1.10: Change in viscosity as a function of temperature and applied shear for polypropylene resin [11].

In Figure 1.8(c), one can notice the slightly cross-linked structure of an elastomer. Its structure can be modeled as a network of springs to capture their rubber like character. Figure 1.8(d) shows the cross-linked structure of the family of thermosets. Both thermosets and elastomers always cure to form an amorphous structure. Thermosets initially consist of long chain molecules with weak bonds. Chemical reactions can initiate covalent bonds between atoms in the chain that cross-link. The process of cross-linking is called curing. As more and more atoms cross-link, the glass transition and the melting temperature of the polymer increases as the atoms are no longer free to move laterally. Thus, the melting temperature of the polymer becomes higher than the decomposition temperature and hence when thermosets are heated they will decompose before they can melt. The high density of cross-linking within the thermoset structure is responsible for the superior thermal stability and the mechanical properties, as can be seen in Table 1.1 [16]. The materials selected in Table 1.1 reflect the most commonly used thermoplastics and thermosets. Thermoplastics are listed first followed by the thermosets.

1.3.2 Comparison between Thermoplastic and Thermoset Polymers

Thermosets form cross-link between atoms which are as strong as the covalent bonds within the chain whereas thermoplastics do not. A key difference between thermoplastics and thermosets is that one needs to apply heat to melt a thermoplastic and hence initiate a phase change from solid to liquid before or during processing, whereas thermosets are generally provided by the manufacturer in the liquid phase. However, some thermosets such as phenolics and unsaturated polyesters are solid at room temperature and need to be heated to convert them into liquids. A diluent/solvent is added to thermosets to lower their viscosity. After processing, the thermoplastic melt must be cooled down to solidify it, whereas the thermoset will turn into a solid after it is chemically and thermally activated to form the cross-linked network which is also referred to as curing. If the material viscosity is low, it is easy to get it to flow in the empty spaces between the fibers and/or into a mold. Thermoset viscosity is usually between 0.050 to 0.500 Pa.s. Thermoplastics at room temperature are solid, but their viscosity around the processing temperature range is between 10^2 and 10^6 Pa.s. Also, thermoplastics exhibit non-Newtonian behavior such as shear thinning of the viscosity with applied stress, whereas thermosets are relatively insensitive to shear. All polymers exhibit reduction in viscosity with temperature, although thermoplastics can exhibit a steeper reduction than thermosets. A variety of different material viscosities are compared in Table 1.2. In the last two decades there has been a concerted effort in development of reactive thermoplastics. The main motivation is to reduce their viscosity so one can process them at lower temperatures and be able to impregnate fibers at the microlevel. In the emerging field of reactive processing of thermoplastics, a low viscosity mono- or oligomeric precursor of low molecular weight is injected to impregnate all the fibers and in-situ polymerization is initiated with heat or ultraviolet radiation. The polymerization is of the addition type in which the double bonds between carbon atoms in a monomer are broken into single bonds. This generates two free electrons which can join the monomers to form a long chain of carbon atoms with single bond between them. For cyclic or ring-shaped molecules, the rings are opened into monomers or oligomers which are then collected into long chain molecules. Additional advantages of processing with reactive thermoplastics instead of traditional thermoplastics include good bond between fibers and resin as the polymerization takes place around the bond and one can include nano particles to the monomer before the polymerization step. The disadvantages include that not all polymer can be reacted in this fashion especially the high-performance thermoplastics due to the requirement of high temperature. Figure 1.11 summarizes this in a graph [18].

Table 1.1: Properties and Typical Applications for Commonly Used Polymers [17]

Name / Abbreviation	**Material Family**	**Application Temperature Range (^oC)**	**Important Properties**	**Application Examples**
Polyethylene / PE	Thermoplastic	100	Low strength, high ductility, resists most chemicals	Bottles, fuel tanks, sealing material, tubing, plastic films
Polypropylene / PP	Thermoplastic	110	In general, better properties than PE	Suitcases, tubing, enclosures, bottles
Polyvinylchloride / PVC	Thermoplastic	60	Good chemical resistance	Flooring material, plastic films, tubing
Polytetrafluorethylene / PTFE	Thermoplastic	-200 to +270	Highest chemical resistance, strongest anti-adhesive	Lab. equipment, coatings (pans, etc.)
Poliamide / PA	Thermoplastic	-40 to +120	High strength and ductility	Ropes, bearings, gears, dowels
Polyethylentherephthalate / PET	Thermoplastic	110	Low creeping tendency, clear	Most soda bottles, miniature parts, parts with small tolerances
Polycarbonate / PC	Thermoplastic	-100 to +130	High strength and ductility, clear	Visors for helmets, safety glasses, quality flatware
Polyacrylate / PMMA	Thermoplastic	70	Excellent optical properties (organic glass), easy bonding	Magnifying glasses, lenses of all kind, showcases
Polystyrole / PS	Thermoplastic	60	High strength, brittle, glass clear, low chem. resistance	Wrapping film, low quality cutlery and plastic cups
Unsaturated Polyester / UP	Thermoset	100 to 180	High tensile strength (close to steel), good chemical resistance	Structural parts for boats and cars, fishing rods
Epoxy / EP	Thermoset	80 to 200	Structural parts in airplanes with high demands for stiffness and strength	Helicopter rotor blades, fuselage, commonly used as adhesive
Phenolics / PF	Thermoset	150	High strength and stiffness, brittle	Enclosures, printed circuit boards
Vinylesters / VE	Thermoset	200	Room curing, high chemical resistance, good strength and ductility	Applications in marine industry, corrosion resistant tanks and pipes

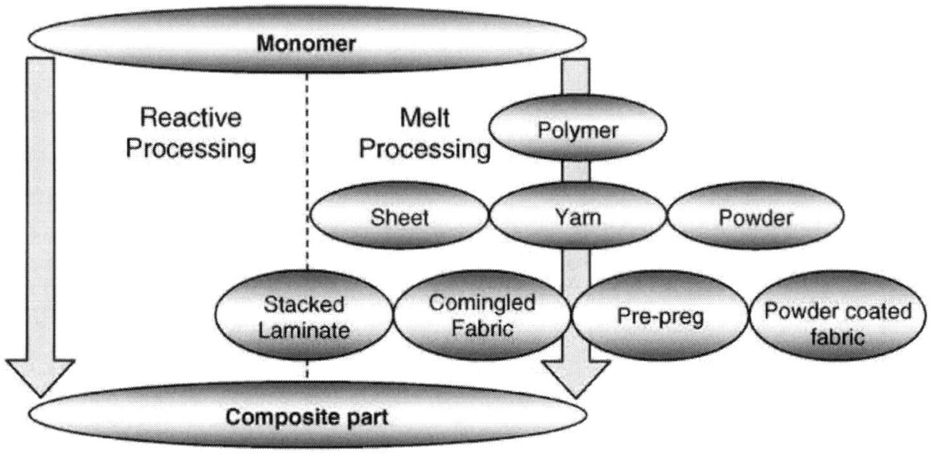

Figure 1.11: Reactive processing of thermoplastic composites (reprinted with permission from [18]).

Table 1.2: Comparison of Different Material Viscosities [16]

Material	Viscosity [Pa.s]	Consistency
Air	10^{-5}	Gaseous
Water	10^{-3}	Thin-bodied
Glycerin	1	Liquid
Thermosets	0.05 to 0.5	Liquid
Molten thermoplastics at processing temperature	10^2 to 10^6	Thick-flowing
Glass	10^{21}	Solid

A general comparison between thermoplastic and thermoset matrices is depicted in Table 1.3.

1.3.3 Additives and Inert Fillers

Additives are added to the matrix in small quantities to change specific properties such as processibility, shrinkage, mechanical or optical properties. For thermosets, small amounts of additives are added which are cross-linking agents that can initiate, inhibit or accelerate the reaction. For thermoplastics, plasticizers are added which are low molecular weight compounds to lower the viscosities. To toughen the resin system, nano particles or elastomers may be added in small quantities to the resin. However, this action usually will increase the viscosity of the resin.

Inert fillers are solid materials usually added to the resin to reduce cost and sometimes to improve stiffness, electrical properties, decrease shrinkage and provide resistance to ultraviolet radiation and fire or provide benefits as colorants. Fillers are usually added to the molten resin and mixed with it. This results in increase in viscosity of the resin, hence the need to control the volume fraction of fillers that are added to the resin.

Table 1.3: Summary of Differences between Thermoplastics and Thermosets from Processing Viewpoint

Characteristic	Thermoplastics	Thermosets
Viscosity	High	Low
Initial state	Usually solid	Usually liquid
Post processing	None	Heat necessary
Reversibility	Can be remelted and reformed	Once formed, virgin state cannot be recovered
Heat transfer requirement	Heat needed to melt it	Heat may be needed to initiate cure
Processing temperature	Usually high	Can be at room temperature
Usage	Large volumes in injection molding	Mainly used in advanced composites
Solidification	Cooling for change of phase	Extraction of exothermic heat during curing

1.4 Fibers

Reinforcements carry structural loads and provide stiffness and strength to the composite. They can be in the form of particles, whiskers or fibers. Particles or flakes have a low aspect ratio of the diameter to the length, whiskers are usually 0.1 micron in diameter and made from a single crystal. Particles do not provide a substantial change in mechanical properties and whiskers are too expensive to manufacture.

Fibers are usually spun from a solution or a melt which orients the molecules of the material. They are made from either glass, carbon or polymer or natural materials such as jute or hemp. Their diameter is usually less than 10 μm. Fibers come in various forms, shapes and materials, and are primarily used for reinforcements. Composites containing continuous fibers made from glass, carbon or polymeric materials are known as advanced composites. Composites containing discontinuous fibers are called short fiber or long fiber reinforced plastics. Composites containing hemp, flax or jute fibers are called natural composites. These natural fibers are already woven into fabrics as shown in Figure 1.12. The fabric has fiber tows in warp and weft directions. The warp tows are in the lengthwise direction where the weft has horizontal yarns that intertwine at right angles with the vertical warp threads. Jute, which is also known as burlap, is widely grown in India and Bangladesh, is one of the strongest and cheapest natural fibers and is second only to cotton in worldwide production. Studies have shown great properties of several natural bast fibers especially flax [19]. These natural fibers are not as strong as synthetic fibers such as high performance carbon but they are comparable to fibers such as glass commonly used for lower end high volume production in composites. Since these natural bast fibers are already spun into continuous fibers and woven in fabrics for various products, they are well suited and can be easily integrated into many composite manufacturing processes such as Resin Transfer Molding (RTM) and Vacuum Assisted RTM (VARTM).

For advanced composites, the fibers are used in the form of rovings, yarns, strands and

Figure 1.12: Woven fabrics containing hemp (left), flax (middle) and jute (right) fibers [20–22].

tows. These yarns or tows can be combined in various forms to create a preform. Some of these structures are shown in Figures 1.5 and 1.6.

Many preforms are formed using fabric reinforcements. Fabrics are formed from a network of continuous fibers. One large class of fabrics is manufactured by either weaving or stitching together bundles ("tows") of fibers. These tows are generally elliptical in cross-section, and may contain from 100 to 48000 single fibers as can be seen in Figure 1.5. The cross-sectional width and thickness of tows are of the order of millimeters, as can be seen in Figure 1.13. Another large class of preform fabrics include "chopped" and "continuous strand" random mat also shown in Figure 1.5. These fabrics are typically formed from low cost E-glass fibers, cheaper than woven and stitched fabrics, and used for low-strength applications. These fabrics are formed using smaller tows (approximately 100 fibers), and

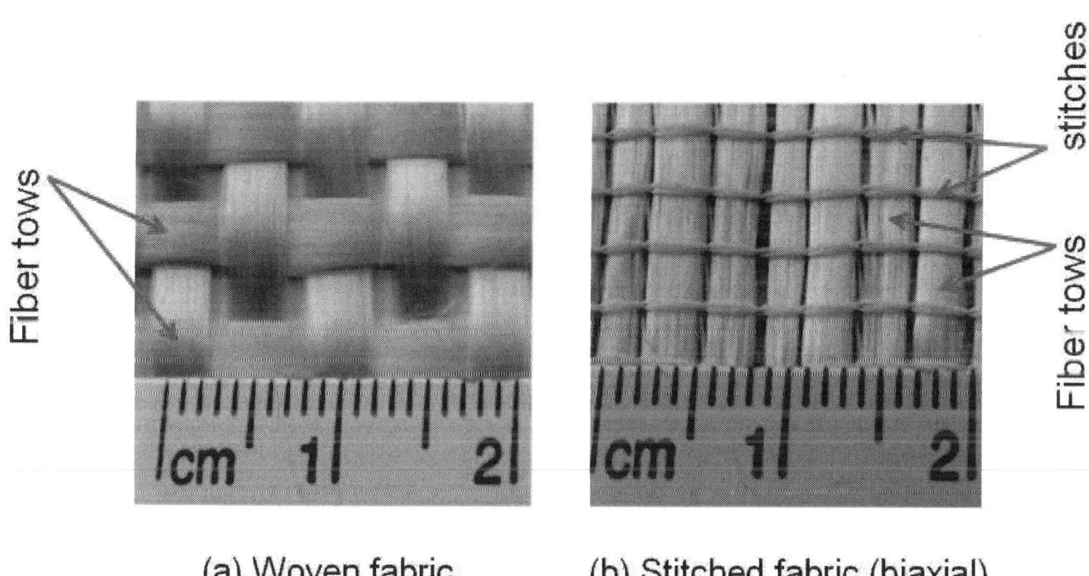

Figure 1.13: A network of fiber tows containing 1000 to 2000 fibers in each tow (a) woven or (b) stitched together to form the fabric.

have a structure which is more random in nature than woven or stitched fabrics. Random mats are typically isotropic in their structural and flow properties in in-plane directions, which is often untrue for woven or stitched fabrics.

Fiber tows can be formed into three-dimensional shapes using braiding and weaving techniques. Tows can be oriented at different angles in three-dimensional space, providing structural support in a multitude of directions, depending upon the application. A fiber preform is an assembly of fabrics. Figure 1.14 shows a complex preform constructed from woven fabric that forms the skeleton of the shape of the composite part. Such preforms are typically used for advanced composites. From the mold filling process viewpoint, the architecture of the preforms dictates the resistance to the flow of resins.

Figure 1.14: Fiber preform constructed from woven fabrics and placed in a mold that forms a skeleton of the composite part.

It is difficult to stack the layers in a desired orientation when draping multi-layers of a dry fabric preform over a tool surface. Use of preimpregnated fabrics (prepregs) or a tackifier (a binder that holds various layers together) eases the preforming and draping process for various net-shape structures [23]. This also reduces the chance of fiber wash (movement of the fibers) during the resin injection.

1.4.1 Fiber-Matrix Interface

A strong bond between the fiber and the matrix will improve the interlaminar shear strength, delamination resistance, fatigue properties and corrosion resistance. However, a weak bond is useful for damage tolerance and energy absorption. Per unit volume, the interface area between the fibers and the resin is given approximately by four times the volume fraction of the fibers divided by the diameter of the fiber. The interface area for a small volume of 1 meter by 1 meter by 2 centimeter composite part containing 50% fibers can be the size of a baseball field (about 10,000 square meters) if the fiber diameter is a few microns [8].

A good bond between the fiber and the matrix is created by wetting of the fiber by the resin. The thermodynamics of wetting states that low energy liquids wet high energy solids. The surface energy for glass is around 500 dynes/cm^2 and for the polymers it is around 30-40 dynes/cm^2. Hence, glass is a great wetting agent. Carbon has a surface energy of around 50 dynes/cm^2 and can still be easily wetted by the resin. Addition of sizing[1] to the fibers promotes handling but can inhibit wetting. Thermodynamics can tell you if the resin will wet the fibers, but does not tell you the rate of wetting and if we can get the resin to reach the fiber surface. For this we have to understand how the resin flows and impregnates between the fiber surfaces.

1.5 Classification

Thermosets have been around much longer than thermoplastic materials; hence, almost all manufacturing techniques developed for thermoplastics today were originally derived from the processes that used thermoset matrices, and the most important ones are listed in Table 1.4. The practice of choosing an appropriate manufacturing method is usually based on the actual part size and geometry, the unit count, the precursor material (initial state of the composite material), the selected components of the composite, i.e., the reinforcement and the matrix, and the cost.

1.5.1 Short Fiber Composites

The fibers can be cut or chopped and compounded in an extruder with any polymer to form a pellet consisting of short fibers or could be pultruded consisting of aligned fibers as shown in Figure 1.15.

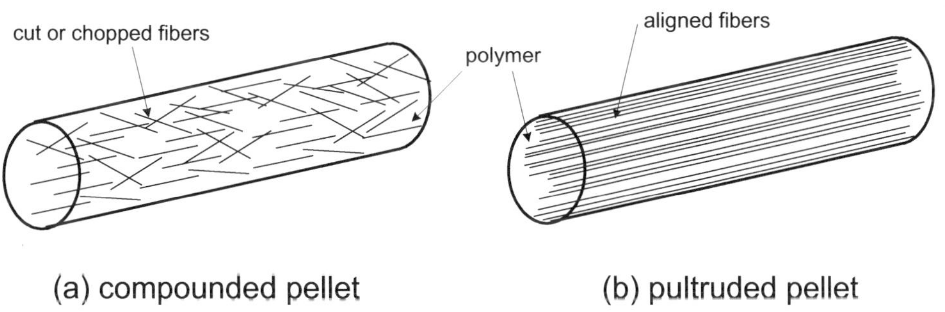

Figure 1.15: Schematic of pellets for injection molding.

The first generation of composites used chopped or short fibers preimpregnated with the thermoplastic polymer matrix in the form of pellets. The pellets are usually a few centimeters in length and a few millimeters in diameter.

The composite types can be broadly divided into composites made from short fibers (aspect ratio less than 100) and continuous fibers. The three most common mass produc-

[1]Sizing is a chemical coating that is applied on the surface of the fibers, sometimes by grafting the molecules on the surface in order to improve the adhesion between the resin and the fiber.

Table 1.4: Examples of Composite Manufacturing Processes

Process	**Precursor Materials**	**Tooling**	**Production**
Injection Molding	Pellets with short fibers	Aluminum or steel molds	Small complex shape parts, high volume
Extrusion	Pellets with short fibers	Dies for continuous operation	Tubes, T-sections, any cross-sectionally long part
Compression Molding	Laminates with long discontinuous fibers or continuous fibers	Aluminum or steel molds	Large net shape parts such as wind deflectors
Wet layup and Tape Layup	Unidirectional continuous fiber prepregs	One-sided steel mold	Small low curvature parts
Autoclave	Unidirectional continuous fiber prepregs	One-sided steel mold	Part size limited by autoclave size
Filament Winding	Prepregs or fiber rovings	Open mold process over a mandrel (steel/aluminum)	Usually axisymmetric parts with hollow cores
Liquid Molding	Random, woven, knitted fabric preforms in any form	Closed mold process with aluminum or steel mold	Near-net shape parts
Pultrusion	Unidirectional tape or fabric rovings	Closed die made of steel	Continuous cross-section parts
Sheet Forming	Preimpregnated sheets	Metal mold	Parts with slight curvatures

tion processes for short fiber composites manufacturing are injection molding, compression molding and extrusion. These processes were adopted from the polymer processing industry that had developed the equipment to produce parts in high volumes with polymers.

These pellets are fed into the hopper of either an injection molding machine or an extruder. The polymer contained in the pellets melts inside the barrel of the machine because of the heating and viscous dissipation due to the shearing of the polymer against the barrel and the screw. The short fibers suspended in the molten polymer are forced either (i) into a mold cavity along with the polymer in the case of discontinuous operations, such as injection molding, or (ii) through a die in the case of continuous operations, such as extrusion. Thus, the existing machines used for polymer processing could be easily adopted for fiber reinforced polymers. The advantages are stiffer and stronger components with lower degrees of warpage and shrinkage. The disadvantage is that there is no control on the fiber orientation in the component and fiber attrition.

1.5.2 Advanced Composites

From the analysis, it was evident that composites with continuous fibers could enhance the mechanical properties by one to two orders of magnitude as compared to short fiber composites. These composites are referred to as advanced composites. There are several primary steps that are common in manufacturing of advanced composites. First, all advanced composites require a skeleton structure of the fibers or fiber network that is tailored for the particular part geometry and the property requirements. Second, this fiber structure must be covered and impregnated by the liquid resin in some way. Finally, the part should be supported by a rigid tool to allow the resin to solidify or cross-link, permanently freezing the microstructure created by the fiber network.

Many fiber structures are available as seen from Figures 1.2, 1.5 and 1.6. Broadly, they can be divided into two groups. The first group consists of continuous fibers in sheets, tapes or tows aligned in one direction. These fibers may be prewet with the resin and then laid in different directions by hand or by a machine to construct the desirable structure. Aligned fibers allow for creation of very high fiber volume fraction and hence high specific in-plane strengths and stiffnesses. The second group uses fiber interlacing to create two-dimensional and three-dimensional interlocked textile structures as shown in Figure 1.6. This allows the composite to achieve higher stiffness and strength both in and out of plane directions, and potentially allow the designer to tailor the mechanical or physical properties to the desired application. However, as shown in Figure 1.16, the degree of complexity that can be handled with short fiber composites cannot be duplicated with advanced composites.

Thus, the material manufacturers started to make resin impregnated prepregs with continuous aligned fibers that could be bonded and fused together to make an advanced composite. Textile preforms were being made where the fiber tows could be woven, stitched or braided together to create the underlying microstructure to provide strength to the composite. Various techniques were invented to induce the resin to wet the fibers and infiltrate the empty spaces between the fibers to retain the integral structure of the composite. This started the evolution of new composite manufacturing processes to make advanced composites. The details of all the processes are explained very well in Astrom [11]. Here, we will briefly introduce the general approach to model the processing step of composites manufacturing, and in the next chapter we will briefly introduce some of the important manufacturing processes and identify the underlying physics of transport of mass, momentum and energy in these processes.

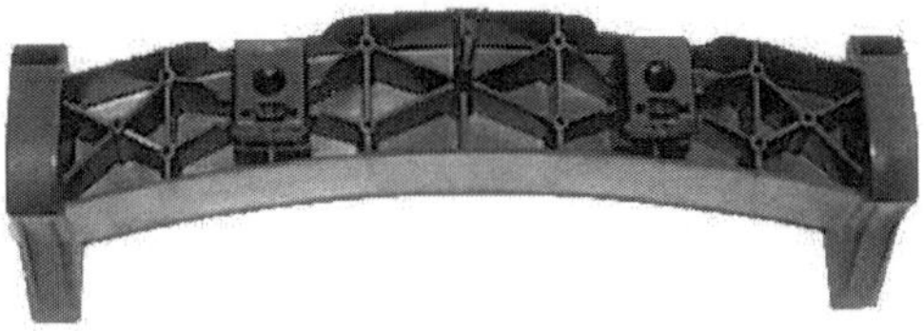

(a)

(b)

Figure 1.16: (a) Injection molded short fiber component with very high degree of geometric complexity [24]. (b) Resin transfer molded advanced composite with modest level of geometric complexity [25].

1.6 General Approach to Modeling

The manufacturing process physics and modeling are greatly influenced by the type of fibers being used: short or long, continuous or discontinuous, aligned or interlaced, etc. The type of resins being used, thermoplastic or thermoset, also influences the process. Thus, the fiber form and the matrix type play a key role in the selection of the manufacturing process. The geometry of the part to be manufactured influences the decision and also the case of whether the process is carried out in an open mold or a closed mold. These choices influence the physics of mold filling.

The process modeling step in composite manufacturing is generally approached by researchers on two scales. The macroscale is usually the order of the smallest dimension of the composite being manufactured (millimeters). The microscale is more on the order of a fiber or tow diameter (microns). At the macroscale, the modeler is generally interested in the overall relationship between the process parameters (such as pressure, flow rate and temperature) and global deformation of the composite material that is being formed. One can use a continuum mechanics approximation to describe this physics. However, as composite materials are heterogeneous materials by definition, macrolevel physics cannot capture phenomena that occur on the scale of a fiber diameter (usually a few microns). Hence one may need to model this physics separately and find an approach to couple it with the macroscale physics.

Broadly, one can divide the manufacturing processes into three different categories on the macroscale. The first category deals with materials that contain either thermoplastic or thermoset resin but in which the precursor material is formed into pellets with short or chopped fibers. In such cases, once the resin is melted, the fibers and the resin flow together as a suspension either into a mold or through a die. Hence, the key physics here is the flow physics of fiber suspensions and the rheology of such suspensions as they deform under the applied pressure to occupy the mold or the die. Rheology is usually defined as the deformation science of materials. It becomes important to model the transport, attrition and orientation of the fibers during the flow as the final microstructure is decided by the resulting frozen-in fiber distribution and orientation. Processes such as extrusion, injection molding and compression molding fall under this category.

The second category involves long, discontinuous fibers or continuous fibers preimpregnated with viscous thermoplastic resin. In such cases, heat and pressure are applied to form and consolidate the composite part. Here the physics of squeeze flow with anisotropic viscosity of the composite is used to model the flow process. The heat transfer during heating and cooling of the material and the mold are also important aspects. The non-Newtonian and shear thinning nature of the composite complicates the rheology but needs to be addressed. Thermoplastic sheet forming, thermoplastic pultrusion and fiber tow placement are some examples of processes that can be modeled in this manner.

The third category involves thermosets of low viscosity resins and continuous fibers in the form of aligned, woven, or stitched fibers. In almost all processes involving thermoset resins, the approach to modeling uses resin infiltration into a porous network of fibers. Here the flow through porous media physics allows one to model the impregnation process. Heat transfer is also important along with the cure kinetics which cross-links the resin, rapidly changes the viscosity and also introduces heat into the composite. Resin transfer molding, thermoset pultrusion, thermoset filament winding and autoclave processing are examples of such manufacturing processes.

Thus, depending on the category, the modeling physics on the macroscale will differ, and in this book we will introduce approaches and philosophy to model these processes. On the macroscale, our goal is to find processing conditions such as flow rates, pressures and rates of heating and cooling to manufacture a successful part. However, as composites are quite heterogeneous materials on a macroscale, one is forced to address microlevel issues such as creation of microvoids due to volatiles or air entrapped in the resin that does not escape from the mold cavity. Interface adhesion between the fibers and resin is decided by the type of sizing on the fibers and their compatibility with the resin. Changes in cure kinetics and rheology may also occur due to the presence of the sizing on the fibers. In this book we also discuss how to address some of these microscale issues.

The important challenge in modeling is how one can couple what occurs at the microscale to the physics that gets influenced at the macroscale. This may not be intuitively obvious in all cases, but we will present a few examples of how one can endeavor to approach it. Usually constitutive equations are formulated to bridge this gap. This is a fertile area of research, especially in areas of material processing in which the modeler tends to use continuum mechanics principles to describe the macroscale physics for heterogeneous materials and ignores the associated microlevel physics.

1.7 Organization of the Book

Chapter 2 will discuss briefly grouping the composite manufacturing processes with the underlying theme of processing. The philosophy in modeling these processes will be outlined, along with a brief introduction to the important manufacturing methods. Issues important for modeling the transport processes will be highlighted at the micro- and macro-scales. Chapter 3 will review the basics of fluid mechanics and heat transfer as required for processing of polymers and polymer composites. Fundamental principles involved in modeling and the approach that couples the physical laws and the constitutive laws to describe the physics with the help of appropriate boundary conditions will be outlined. Thus, it will introduce the transport equations necessary for modeling along with boundary conditions and examples. Chapter 4 will delve into details of constitutive laws and relationships based on phenomenological behavior. Chapter 5 will discuss details on the tools necessary for modeling these processes. The usefulness of tools such as dimensionless analysis and simple back-of-the-envelope calculations will be illustrated with the help of examples. As there are different scales involved in polymer composite manufacturing and as the material behavior at the microscale can influence the issues at the macroscale, coupling of microscale physics with the phenomena at the macroscale will be discussed. Different phenomenology involved in characterization of material parameters required in modeling will be introduced, and their usefulness, challenges and uncertainties will be unveiled. Chapters 6-8 will apply the tools and the fundamental principles studied in earlier chapters to model composite manufacturing processes. We will illustrate how the modeling principles can be incorporated in some of the composite manufacturing processes to reveal some of the understanding based on scientific principles rather than trial and error approaches. Solved examples are presented in all chapters to enhance the physical understanding of processing these complex heterogeneous materials. The questions, fill in the blanks and some problems are formulated and solved analytically or with MATLAB to reinforce qualitative understanding of transport phenomena in various processes and also understanding the key manufacturing issues in composites manufacturing with seniors and general practitioners of composites in mind. The analysis and some problems are introduced for graduate and advanced students who would like to delve further into understanding and modeling of such processes.

1.8 Exercises

1.8.1 Questions

1. What are the advantages of polymer composites over other materials?

2. List a few industries that use polymer composites.

3. When polymer composites were used a few decades ago, did the process engineers rely on (i) experience and trial and error approaches, or (ii) accurate mathematical modeling of process physics, in order to improve the manufacturability of a certain prototype?

4. What are the two major ingredients of a composite material? How do they enhance the properties of the composite?

5. What are the two types of polymer resins used in composites processing? What are major differences between them?

6. What are the three different types of materials used to manufacture fibers?

7. Is it easier to inject a thermoplastic or thermoset resin through a tightly knit fiber preform?

8. What are the typical viscosities of thermoplastic and thermoset resins relative to the viscosity of water?

9. Which one of the following has more influence on the mathematical modeling of the manufacturing process: (a) the fiber material (e.g., glass or carbon) or (b) whether the fibers are discontinuous or continuous? Why?

10. Although in general the ease of processing decreases as we move from discontinuous short fibers to continuous fiber preforms that are woven or stitched, why are continuous fiber preforms preferred in some composite parts instead of discontinuous short fibers?

11. List various manufacturing processes that use continuous fibers.

12. What is the smallest building block in plastics? How many of them are there in a polymer?

13. What are the different types of reactions that can be used to create all modern polymers?

14. What are the main differences between thermoplastic and thermoset resins in terms of processing?

15. What are the criteria used to choose an appropriate composite manufacturing method?

16. What is a pellet?

17. What are the advantages and disadvantages of short fiber composites?

18. What composites are referred to as advanced composites? Why?

19. What is a prepreg fabric?

20. What are the two scales used to model composite manufacturing processes? Why do we need to couple them?

21. List the modeling approaches for manufacturing processes on the macroscale. Why do we need to model them differently?

22. What are fiber sizings and why are they necessary?

1.8.2 Fill in the Blanks

1. The thermoplastic properties are determined by their resulting ------------, which is strongly influenced by the ------------ dynamics.

2. In general, an amorphous polymer is transparent, has ------------ mechanical properties and is ------------ resistant to other chemicals than the ------------------------ thermoplastics.

3. When a shear force is applied to thermoplastic polymers, the long chain molecules will start to ------------ relative to each other, in response to the applied force. Thus, most of them exhibit shear ------------ behavior, which is an important property to consider when one develops the process model.

4. The high ___________ of cross-linking points within a thermoset structure is responsible for the superior ___________ stability and ___________ properties.

5. ___________ is applied to ___________ a thermoplastic and hence initiates a ___________ change from ___________ to ___________ before or during processing.

6. Thermosets are generally provided by the manufacturer in the ___________ phase.

7. After processing, the thermoplastic melt must be cooled down to ___________ it, whereas the thermoset will turn into a ___________ after it is chemically and thermally activated to form the cross-linked network, which is also referred to as ___________.

8. Fiber diameter is usually less than ___________. Fibers are cut or chopped and compounded in an extruder with any ___________ to form ___________ consisting of ___________ fibers, or could be pultruded consisting of ___________ fibers.

9. Fabrics are laminate structures having fibers aligned in the ___________ directions. One large class of fabrics is manufactured by either ___________ or ___________ together bundles, or "___________", of fibers. These bundles are generally ___________ in cross-section, and may contain from ___________ to ___________ single fibers.

10. Random mats are typically ___________ in their structural and flow properties, which is often untrue for woven or stitched fabrics.

11. Fiber tows can be formed into three-dimensional shapes using ___________ and weaving techniques.

12. Use of preimpregnated fabrics (prepregs) enables these two: (i) it eases the ___________ process during placement over a tool surface, (ii) it reduces the chance of fiber ___________ during the resin injection.

13. The existing machines used for polymer processing could be adopted easily for fiber reinforced polymers. The advantages would be ___________ and ___________ components with low degrees of ___________ and ___________. The disadvantage would be no ___________ on the fiber ___________ in the component and fiber attrition.

14. Composites with ___________ fibers could enhance the mechanical properties by one to two orders of magnitude as compared to ___________ fiber composites. These composites are referred to as ______________________.

15. There are several primary steps that are common in manufacturing of advanced composites. First, all advanced composites require a skeleton structure of ___________ that is tailored for the particular part ___________ and the property requirements. Second, this structure must be _____________ by the liquid ___________ in some way. Finally, the part should be supported by a rigid ___________ to allow the ___________ to solidify or cross-link, permanently freezing the microstructure created by the fiber network.

16. Aligned fibers allow for creation of very high fiber __________________ and hence high stiffness and strength in ___________ directions.

17. Two- and three-dimensional interlocked textile structures can allow the composite to achieve high stiffness and strength in ___________ and ___________ directions and potentially allow the designer to tailor the mechanical or physical properties to the desired application.

18. The manufacturing process physics and modeling are greatly influenced by the types of ____________ and ____________.

19. The geometry of the part to be manufactured influences the decision if the process is carried out in an open or closed ____________.

20. The process modeling step in composite manufacturing is generally approached by researchers on two scales. The ____________ scale is usually the order of the smallest dimension of the composite part being manufactured which is in ____________. The ____________ scale is more on the order of a ____________ diameter which is in microns.

21. Broadly, one can divide the manufacturing processes into three different categories on the ____________ scale from the flow viewpoint. (i) Once the solid pellets are heated, ____________ and short or chopped ____________, which are the two ingredients of pellets, flow together as a ____________. (ii) Long ____________ fibers or ____________ fibers are preimpregnated with viscous ____________ resin by applying heat and ____________. (iii) The flow of low viscosity ____________ resins through ____________ fibers is modeled as infiltration of resin into a ____________ network of fibers.

22. On the macroscale, a modeler's goal is to find processing conditions such as ____________, ____________ and rates of ____________ and ____________ to manufacture a successful part. However, as composites are heterogeneous materials on a macroscale, one is forced to address microlevel isssues such as creation of micro ____________ due to ____________ or ____________ entrapped in the resin that does not escape from the mold cavity.

23. Interface adhesion between the fibers and resin is decided by the type of ____________, which is a chemical ____________ that is applied on the surface of the fibers, sometimes by grafting the molecules on the surface in order to improve the ____________ between the resin and the fiber.

Chapter 2

Overview of Manufacturing Processes

2.1 Background

Several industries have been using fiber reinforced composite materials for a few decades now. Glass fibers were available commercially in the 1940s. Within a decade, composites were being used by several industries; for example, the automobile industry was producing polyester panels with approximatcly 25% glass fibers [26].

Manufacturing with composite materials is very different from metals. This is because when making a metal part, the properties of the virgin material and the finished part are fundamentally unchanged. For composites, the manufacturing process plays a key role. During composite processing, one makes not only the part of the desired shape, but also the material itself with specific properties. In addition, the quality of the composite material and the part fabricated depends on the manufacturing process, because it is during the manufacturing process that the matrix material and the fiber reinforcement are combined and consolidated to form the composite.

In early stages of development, the cost of composite materials was very high and only selected industries, for which the importance of the property of the material greatly outweighed the cost factor, were willing to use them. These industries were primarily the aerospace and the aeronautical industries. They valued the properties of the composites greatly and could justify the higher costs because of the weight savings. Both industries took advantage of the light weight and, in the case of defense oriented projects, the stealth properties. The lack of automated and repeatable manufacturing processes drove the cost of composite parts up and limited the number of potential users. Another industry that has been using composites since the 1970s is the marine industry. It could deal with low production volume and relatively high costs while taking advantage of the corrosion resistance property of composites. The majority of the manufacturing work done in these industries in the 1970s was very labor intensive and not very cost effective as the manufacturing modus operandi was "experience" and trial and error.

It was imperative that in order for composites to be widely used, especially by the consumer goods industry, such as the automotive and sports industry, two major goals had to be achieved. First of all, the cost of raw materials had to go south. Secondly, and most importantly, manufacturing methods had to be developed to achieve high-volume production that relied more on the fundamental understanding of the physics of the process rather than the accepted trial and error practice ingrained on the shop floor of the manufacturing sites. Over the last two decades, composites research has moved in this direction, and the development and improvement of several manufacturing methods has been achieved.

The objective of this chapter is to briefly outline the different composites manufacturing techniques that are commonly used today. Details are described in other books mentioned in the introduction chapter. As the emphasis of this book is process modeling, the manufacturing methods will be categorized in groups that involve similar transport process physics. The dominant transport processes will be identified in each process to help clarify the physics. In addition, a brief evaluation of each method, emphasizing its advantages and disadvantages, will be listed to help the users identify the best possible method for their applications.

2.2 Classification Based on Dominant Flow Process

Transport processes encompass the physics of the phenomena of mass, momentum and energy transfer on all scales. As composites are heterogeneous materials, there is simultaneous transfer of heat, mass and momentum at micro-, meso- and macroscales, often along with chemical reaction, in a multiphase system with time-dependent material properties and boundary conditions.

Composite manufacturing processes are generally grouped into two general classes: open mold and closed mold. Open mold are those processes in which the part is not inside the mold during the complete duration of the manufacturing process such as pultrusion or filament winding. In closed mold processes, the preform is placed in a mold, the mold is closed, and when it is reopened the part is fabricated. However, as the focus of this book is transport phenomena in composites processing, instead of classifying processes as open and closed mold, we will use the mechanisms of transport processes as the yardstick to group them.

One may broadly group composites manufacturing processes into three categories. The first category consists of manufacturing processes that involve the transport of fibers and resin as a suspension into a mold or through a die to form the composite. In such processes, the fibers in the molten deforming resin can travel large distances and are usually free to rotate and undergo breakage; thus the microstructure of the final product is linked with the processing method and the flow of the suspension in the mold. We will describe injection molding, compression molding and extrusion processes in this category. The reinforcements are usually discontinuous glass, kevlar or carbon fibers, and the resin may be either thermoset or thermoplastic. We will call this category *short fiber suspension manufacturing methods*. In recent years, nano particles such as carbon nano tubes or nano clays are being added into the resins to toughen the resin or impart other properties such as fire retardancy. Such suspensions can also be addressed under this category; however, the challenge for such processes is obtaining good dispersion.

The second category, which we will refer to as *squeeze flow manufacturing methods* or *advanced thermoplastic composites manufacturing methods*, usually involves continuous or long aligned discontinuous fibers preimpregnated with thermoplastic resin either partially or completely. In these processes, the fibers and the resin deform together like the dough containing strands of continuous wires or wire screens under applied stress to form the composite shape. However, the presence of fibers creates anisotropic resistance to the applied load, and the viscosity can be over a million times that of water, preventing large bulk movements of the composite. Thermoplastic sheet forming, thermoplastic pultrusion and fiber tape laying methods can be described by this physics. The precursor materials can be in various forms such as the thermoplastic tapes impregnated with aligned and continuous or long discontinuous fibers (for example, APC2 and LDF materials, respectively) [27–29]. The other popular form is to weave a preform of the polymer fibers commingled with glass or carbon fibers. Thus under applied heat and pressure, the polymer fibers melt and occupy the space in between the reinforcing fibers [30]. The polymer could also be in powder form attached to the fibers during the initial stages, which the heat will melt and the pressure will help fuse and consolidate the fiber assembly [31, 32]. The reality of thermoplastic resins is that they cannot travel and infiltrate large distances due to their very high viscosity. Thus the precursor material form has to accommodate a distributed resin percolation among the fiber architectural network.

The third and final group, which we will term *porous media manufacturing methods* or *advanced thermoset composite manufacturing methods*, involves usually continuous and nearly stationary fiber networks into which the resin will impregnate and displace the air, forming the composite in an open or a closed mold. The resin in such processes is almost always a thermoset due to its low viscosity. But one does have to account for complex chemical reactions that are prevalent in these methods. We will discuss liquid composite molding, filament winding and autoclave processing under this category. The precursor material here can take various forms from partially impregnated prepregs to applying the liquid thermoset resin to the fibers during the process or impregnating a stationary network of fibers with the resin. The low viscosity of thermoset resin allows for this versatility. However, the disadvantage is the introduction of complex chemical reaction, gelling and curing phenomena. In addition, thermosets are environmentally unfriendly, and it is difficult to repair and recycle them.

Figures 2.1 and 2.2 schematically show the type of flows one expects in these categories. In the section below, we will describe the commonly used composite manufacturing methods under the categories created.

2.3 Short Fiber Suspension Manufacturing Methods

The underlying transport process here during manufacturing is the flow of the resin along with the discontinuous fibers. Mechanical and physical property control is a primary issue in this process. Denton [33] showed that tensile strength and the elastic modulus of the samples made by compression molding of short fiber composites in a carefully controlled laboratory environment exhibited a standard deviation of about 50% around their mean value. The properties vary due to two main reasons: precursor material variability and fiber orientation variability. It is well known that flow will change the orientation of fibers, which, in turn, will influence the properties. In this book, in addition to developing process parameters which influence manufacturing, we will try to quantify this effect in an attempt to tailor the properties of such materials.

The three processes, injection molding, extrusion and compression molding, which evolved as composite processes from polymer processing techniques and the transport issues associated with them, will be discussed here. The precursor materials (material form used as the initial input to the process) in injection molding and extrusion are polymer pellets containing short fibers, and for compression molding it is usually a charge of material containing resin, fillers and fibers.

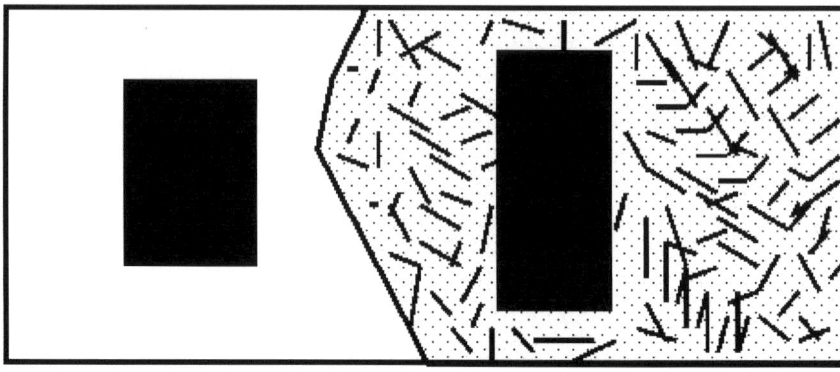

a) Short Fiber Suspension Flow

b) Squeeze Flow in Advanced Thermoplastic Manufacturing Methods

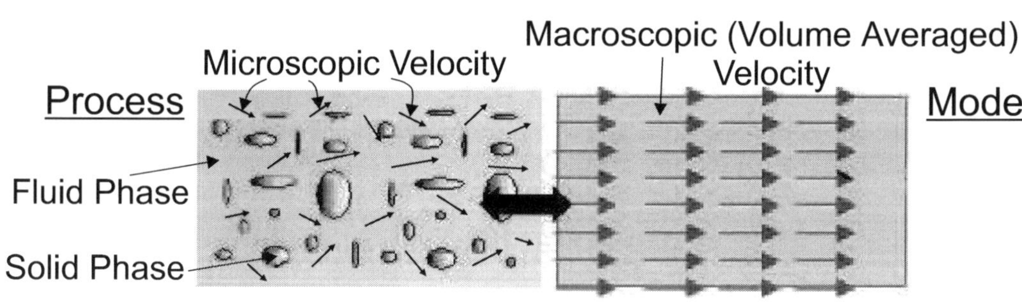

c) Darcy Flow for Advanced Thermoset Manufacturing Methods

Figure 2.1: Schematic of types of flows expected in composites processing: (a) short fiber suspension flow, (b) advanced thermoplastic flow, and (c) flow through porous media.

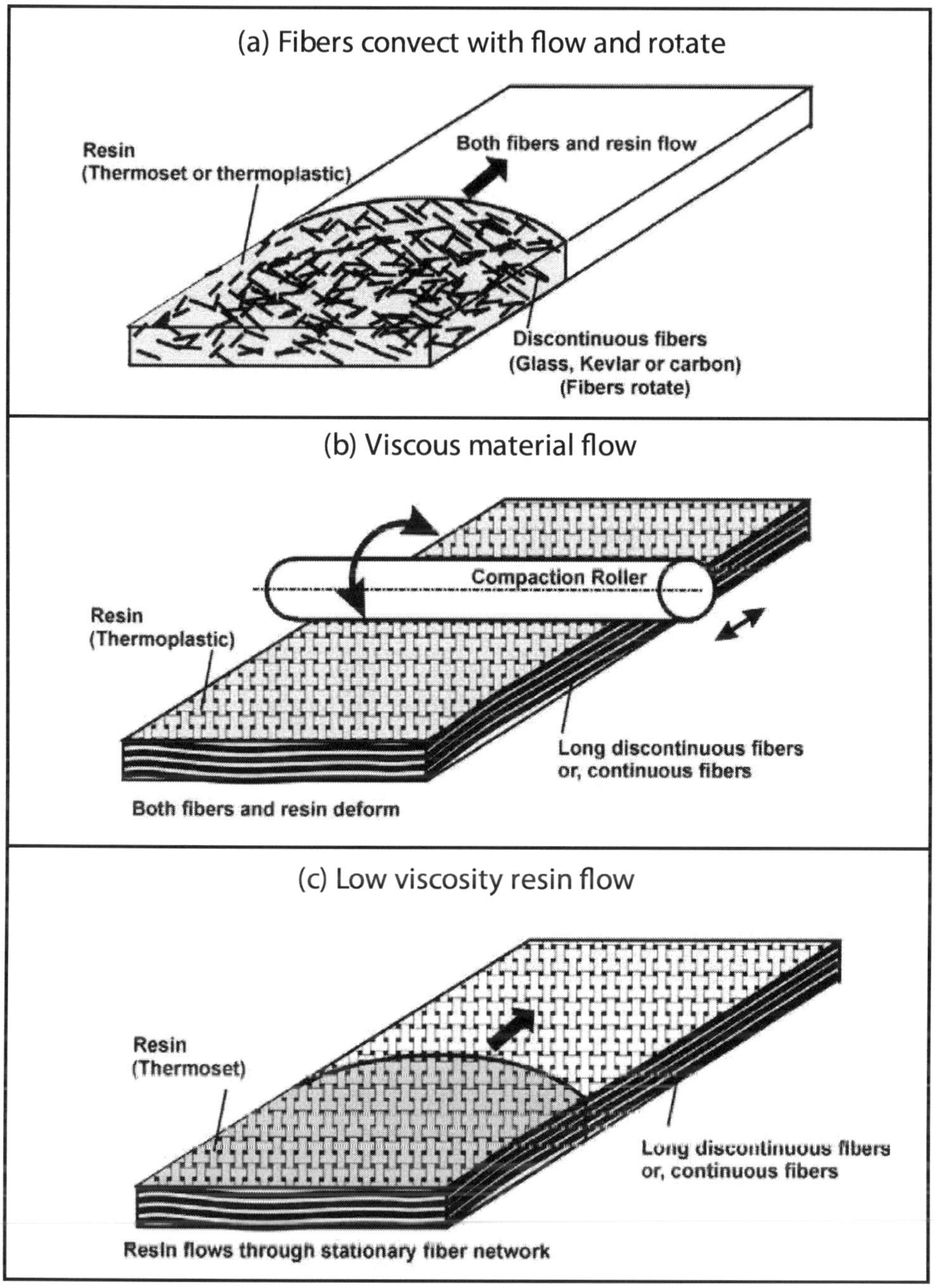

Figure 2.2: Flow types in composite manufacturing processes.

2.3.1 Injection Molding

Process

Injection molding is the most common and widely used manufacturing process for high-volume production of thermoplastic resin parts, reinforced with fibers or otherwise. Solid pellets of resin (usually the size of a small piece of chalk) containing the fibers and sometimes fillers are fed through a hopper into a heated barrel with a rotating screw. The function of the screw is to mix the reinforcements and the resin and also to generate heat by viscous shearing against the barrel. This melts the resin. The screw then acts as a piston and forces the mixture of fibers, fillers and molten resin under high pressure through sprues and runners into a matched-metal mold where the polymer solidifies, freezing the orientation and distribution of fibers and fillers in the part. The mold cavity is then opened and the composite part is ejected. As the fabrication of metal molds can run into thousands of dollars, one can justify the use only for high-volume production parts such as laser disks, etc. Recently, researchers have begun to explore the use of plastic molds reinforced with metal powder for small-volume production or for prototype development with the use of rapid prototype technology [34].

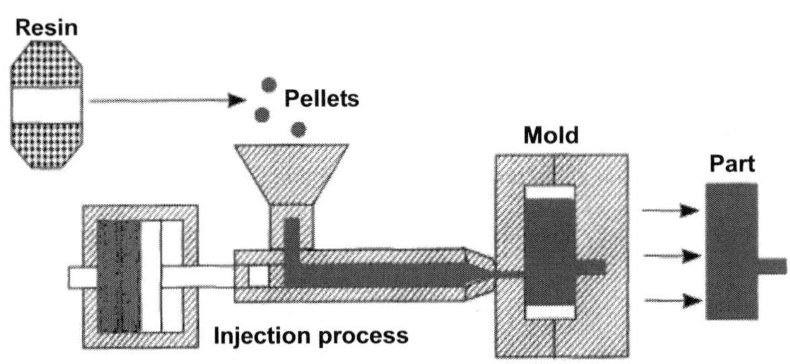

Figure 2.3: Schematic of injection molding process.

The injection molding is a cyclic process in which a molten polymer along with reinforcements and fillers is injected into a closed mold cavity where it takes the shape of the mold cavity and solidifies because the mold is usually cold with high thermal inertia (see Figures 2.3 and 2.4). The molding time is usually of the order of a few seconds, and parts can be complex and precise as shown in Figure 1.16(a). The mold is then opened, the part is ejected with ejector pins and the cycle is repeated. Since the first hand operated injection molding machine was introduced over seven decades ago, it has evolved into a complex, sophisticated and automated process that can produce many thousands to millions of parts.

The injection machines can be as simple as a plunger and a torpedo unit to a reciprocating single or double screw injection unit. Description of complex machines can be found in various other books on injection molding [35–37]. As the shear rates experienced by the material can be of the order of 10^4 to 10^6 s^{-1}, the thermoplastic viscosities can decrease rapidly due to their shear thinning nature, allowing low power requirement and rapid molding of such materials. Some of the injection machines can be modified to handle thermoset materials with reinforcement as well.

The important process parameters that can be controlled on the injection units are the melt temperature, injection and screw speed, injection pressure and in some instances, the mold temperature. The material parameters that will influence the manufacturing process

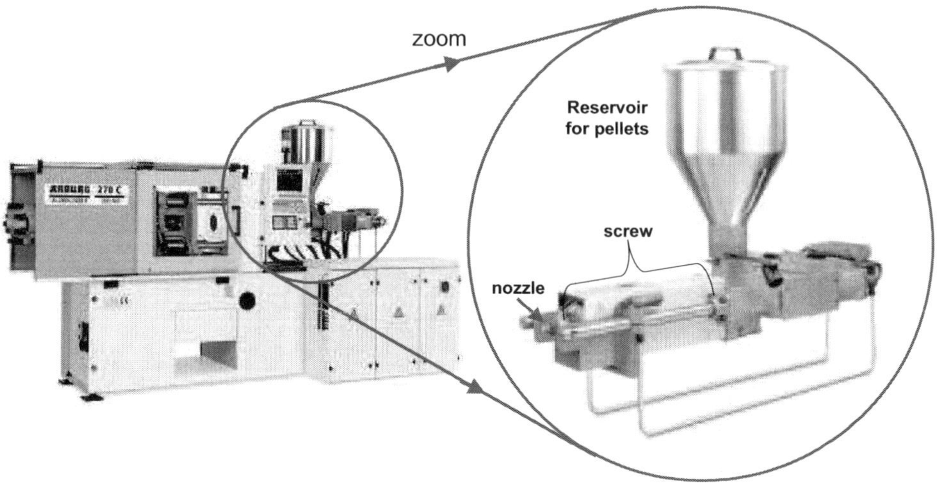

Figure 2.4: Details of injection molding machine [24].

and the final properties of the part are the resin rheology and the filler type and content. The geometric parameters that will play a key role are the mold cavity shape and size, and the locations of injection gates, through which the resin enters the cavity, and the vents that allow the air to escape.

Precursor Materials

The filled thermoplastic pellets usually contain a second, discontinuous, usually more rigid phase blended into the polymer. When the aspect ratio (ratio of largest to smallest dimension) of the second component is around one, it is referred to as a *filler*. If the aspect ratio is one to two orders of magnitude larger, then it is called a reinforcement. The most commonly used reinforcements are *particles*, *whiskers* and *short fibers* usually less than one inch in length. The parts obtained usually have a fiber volume fraction between 30% and 40%. Filled or reinforced materials provide much different properties than the base resin. For example, reinforced polypropylene provides higher rigidity and lower warpage characteristics than neat polypropylene. Also, the viscosity of the filled resin will be different in magnitude and sometimes anisotropic as compared to the neat resin. In practice, fibrous reinforcements made of glass fibers dominate the market although the carbon and aramid fibers provide higher stiffness and strength, but are seldom used due to the high cost of raw materials.

The traditional injection molding process limits the fiber length that solidifies in the final part since the high shear rates in the barrel and the passage of fibers through narrow gates and openings in the mold cause significant fiber attrition. Usually, the fiber diameter is of the order of a few microns, and the final length distribution, irrespective of the starting fiber length, is of the order of 50 to 500 μm. The starting length of these fibers in the log like pellets is usually of the order of 1 to 3 mm. As a result, new methods to produce pellets containing longer fibers were developed in which the fibers were pultruded and stayed bundled together and were not dispersed in the pellet by the action of compounding. These pellets (see Figure 2.5) produced final parts that retained a higher percentage of longer fibers and consequently showed a significant increase in modulus and impact toughness.

The thermoplastic matrix material selected also plays a role in the final physical and

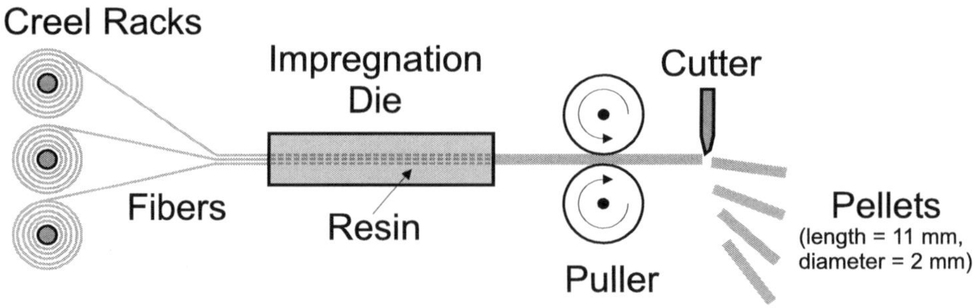

Figure 2.5: Schematic to make pultruded pellets.

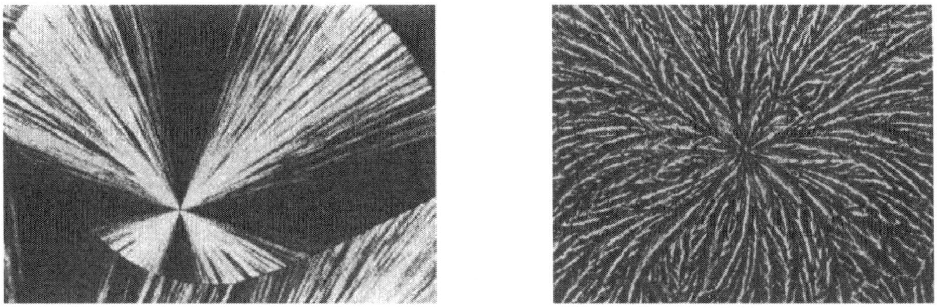

(a) (b)

Figure 2.6: Crystalline and amorphous materials. (a) Spherulite in malonamide containing 10% d-tartaric acid (low crystallinity). (b) Spherulite grown in mixture of isotactic and atactic polypropylene (high crystallinity) (redrawn from Keith and Padden [38]).

optical properties. Most thermoplastic materials when they solidify do so as an amorphous matrix or exhibit various degrees of crystalline behavior depending on the thermal history the material undergoes during the injection molding process. Samples of crystalline and amorphous materials are shown in Figure 2.6. More details on crystallization of thermoplastics will be discussed later.

Transport Issues

The issues that relate to transport phenomena in this process are the flow of fiber suspensions as they occupy the closed mold, the orientation of the fibers during flow, fiber length distribution, fiber breakage and the heat transfer that changes the microstructure of the resin. So, if we consider mass conservation, we have to account for the mass balance of the suspension which can be treated as a continuum material at least for the short fiber materials as the fiber length is of much smaller scale as compared to the domain dimensions. One also needs to characterize and describe the orientation of the fibers in a flowing suspension. The physical concept that one may have to invoke here is the conservation of the orientation field, which simply states that if the orientation of the fibers disappears in one direction, it should reappear in some other direction. One also needs to conserve the momentum to describe the flow and the pressure field during the flow process. This requires one to describe the constitutive equation between the stress applied and the strain rate experienced by the material. For Newtonian fluids, this is usually constant and the constant of proportionality is called viscosity. However, as the thermoplastic melts are shear thinning, the viscosity is known to decrease with shear rate, and the addition of fibers can change the stress-strain rate behavior and even makes it anisotropic. One needs a rheological equation to describe this behavior. The energy conservation allows one to describe the temperature history of the melt in the channel between the screw and the barrel, where it gets its heat input from the heaters on the barrel and due to viscous dissipation caused by the shearing of the suspension. It also allows one to keep track of the cooling history in the closed mold as the suspension enters in it. The cooling dynamics play a crucial role in the resulting microstructure.

The transport phenomena modeling also requires one to consider initial and boundary conditions; hence the physical laws of mass, momentum and energy balance need to be applied at the boundaries along with imposed external conditions which we refer to as process parameters. The material parameters enter into the modeling through the description of the constitutive equations.

There are other microscale phenomena occurring simultaneously such as the molecular orientation and spherulitic growth of polymers during solidification and fiber breakage due to shearing action in the screw that results in a length distribution. For long fiber suspensions, the constitutive equation may change, and issues such as fiber clustering may also need to be addressed, as seen in Figure 2.7 [39].

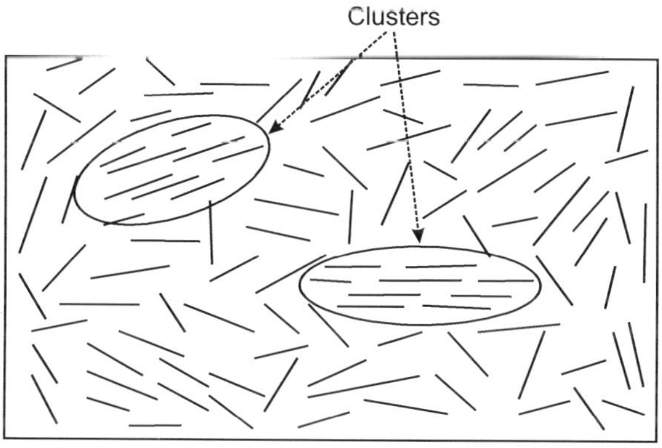

Figure 2.7: Schematic of fiber clustering for long fiber suspensions [39].

The coupling between the energy transport and the momentum creates a *fountain flow* mechanism in injection molding. As the walls are cooler than the core, the suspension viscosity is higher near the walls (polymer viscosity increases as temperature decreases) as compared to the core. Hence, under the same pressure, the suspension in the core moves ahead of the suspension near the walls, spreading from the center outwards like a fountain, as shown in Figure 2.8.

The fibers align in the direction of shearing and also in the direction of stretching as shown in Figure 2.9. The shear flow near the mold walls aligns the fibers in the direction of the flow and is called the skin. Next to this layer, the suspension continues to experience shear and fibers orient along the shear lines; this layer is known as the shear layer. Finally, the core has fibers that are influenced by the bulk deformation of the flow in the mold which usually has an elongated component, causing the material to stretch in and out of the paper direction aligning the fibers. This skin-core structure is a common microstructural observation, and it is the study of the coupling between flow, heat transfer and fiber orientation which will allow us to understand this phenomenon.

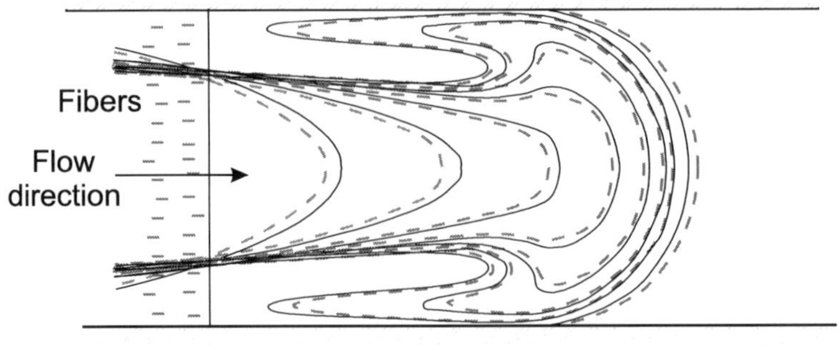

Figure 2.8: Schematic of fountain flow effect encountered during filling (redrawn from Coyle et al. [40]).

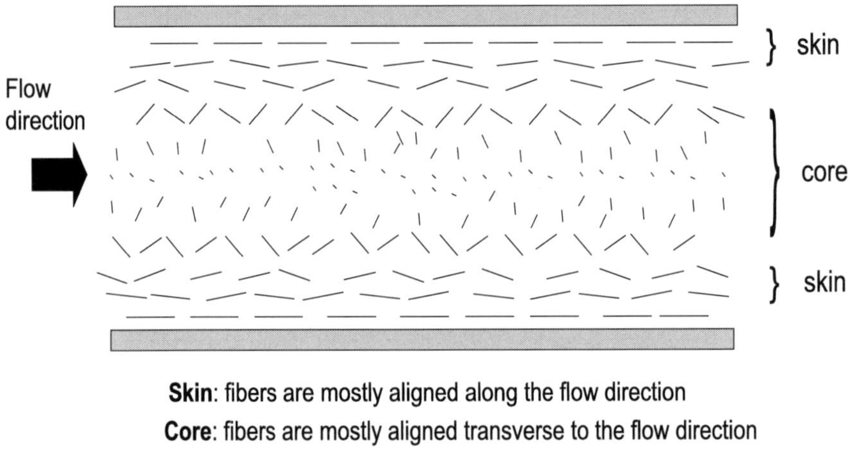

Figure 2.9: Influence of flow on fiber orientation.

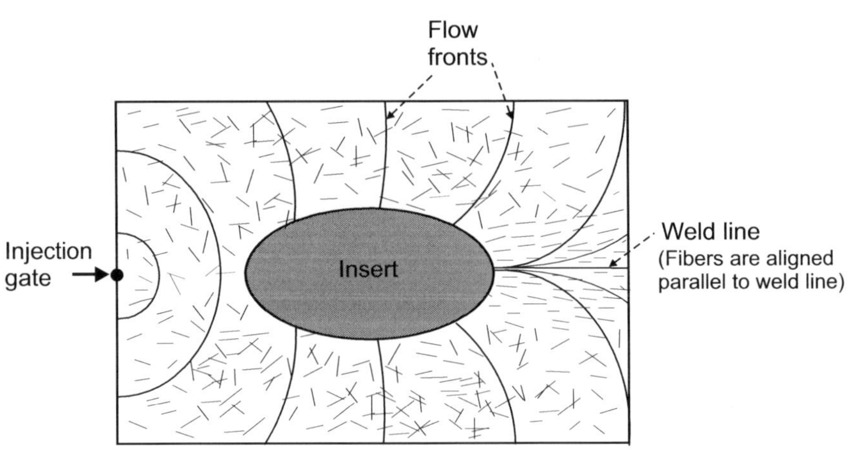

Figure 2.10: Schematic of flow front locations and weld line with fibers aligning along them during injection molding.

Another phenomenon associated with flow and fibers at the microlevel has been that the fibers at the flow front tend to align along the flow boundary and not across it. This may have to do with the surface tension phenomenon between a solid, polymer and air. However, because of this phenomenon, whenever two flow fronts meet, the boundary where they meet is called a weld line and is usually the weak link for the mechanical properties because the fibers align along the weld line and not across it, as shown in Figure 2.10. Thus, understanding of this issue could help modelers address and develop flow management techniques to create strength along the weld lines. Many modeling aspects in fiber suspensions such as fiber orientation, shear thinning, fiber attrition are being adopted to model flow of resins containing carbon nanotubes [41–48].

Applications

Nearly 20% of the goods manufactured nowadays use injection molding due to its versatility and low cost. However, as short fiber composites can improve the desired physical, optical and mechanical properties, structural integrity and dimensional stability, injection molding machines and the screw geometry were modified to handle fibers along with the polymer. Many applications such as housing for electric tools, automotive parts under the hood, plastic drawers, metal inserts and attachments, seats in airplanes, etc. are routinely manufactured using injection molding with thermoplastic pellets containing discontinuous fibers.

The fundamental advantage of injection molding is the ease of automating the process and the short cycle times, which together allow for the possibility of high volume production. In addition, molds can also be constructed to make more than one part at a time. The major disadvantages are the high initial costs of the capital equipment and the molds and the variation in part properties due to lack of control of fiber orientation and distribution.

2.3.2 Extrusion

Extrusion resembles injection molding because it contains a screw. The main difference is that there is no closed mold in extrusion. Instead a die is used to shape the polymer suspension into specific cross sections. This process is used to plasticize and compound polymer pellets containing short fibers and also for manufacturing continuous parts with different cross sections. As in the case of injection molding, the screw melts the polymer and acts as a piston to push the suspension into the die geometry. Inside the die, the suspension takes the form of the die cross-section and exits from the other side of the die and can be continuously pulled to make long tubes, I-beams and reinforced pipes.

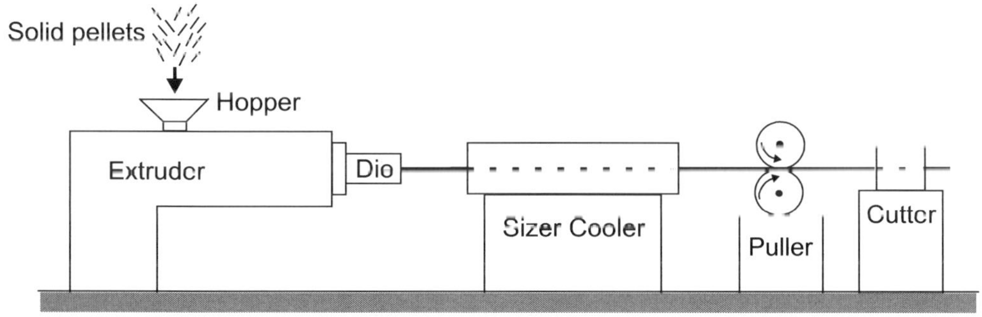

Figure 2.11: Schematic of extrusion process line [49].

Process

Figure 2.11 shows a simple sketch of an extrusion line. The process starts with a hopper into which one pours solid pellets. The extruder melts the plastic (resin or polymer) and may seriously cause fiber attrition. It pumps the fiber suspension through a die hole of the desired shape. It then enters a sizer and cooler trough where the correct size and shape are developed. Next, the formed product enters a puller which pulls it through the sizer. At the end of the line, either a cutter cuts the product at the desired longitudinal length, or a coiler winds it on a drum. Thus, it is a continuous process, and as long as the operator makes sure that the hopper is filled with pellets and the final product is moved away for storage and shipment, it can operate for days without much attention, which makes it very cost effective.

The heart of the extruder is a barrel and the screw that turns in it. A sketch of the extruder is given in Figure 2.12. The screw channel connects the hopper at the rear end to the die at the front of the extruder. The screw is the moving part that melts and pumps the plastic. The extruder screw is turned in the barrel with the power supplied by the motor operating through a gear reducer. The screw is usually machined from a solid steel rod and fits within the barrel with less than a millimeter clearance. It is hardened and chrome plated to resist corrosive action of some resins. To pump a suspension through a die, the screw is designed to generate over 100 to 200 atmospheres of pressure in the suspension. The barrel of the extruder resembles the barrel of a cannon. It is made of steel and has thick walls to withstand very high internal pressures. The inside of the extruder is made of a hard steel alloy for corrosion resistance, and the inside dimension of the barrel determines the extruder size. The power to turn the screw comes from an electric motor. Usually, the power requirements are large due to the pumping pressure and rate required. The pellets are usually fed through the hopper by gravity. The outside of the barrel is equipped with heating and cooling systems to maintain the barrel at a desired temperature. We will consider the details of the working of the screw later, but for a simple and crude explanation, one may think of a bolt as the screw, the nut as the pellet and the wrench as the barrel. Now if one turns the bolt (screw) and holds the wrench (barrel) in place, the nut (pellet) will move forward. The heat from the heaters and the mechanical frictional work will melt the polymer and push it forward along with the short fibers. This action is usually called plasticating. Most of the heat needed for softening and melting comes from the viscous dissipation due to the turning of the screw inside a stationary barrel. This heat generated from mechanical work is sometimes more than sufficient to operate an extruder with its heaters turned off, thus the cooling system has to be turned on to avoid overheating of the resin.

Some important characteristics of the injection molding and extrusion processes, and their similarities and differences are highlighted below [51]:

Extrusion:

- Raw material (in the form of pellets or granules) is placed into the hopper and fed into the barrel.
- The barrel is equipped with a rotating screw which blends the pellets and conveys them along the barrel.
- Heat is generated by (a) the shearing of the plastic due to friction and (b) the external heaters.
- The polymer resin melts creating a fiber suspension which builds up pressure as the screw rotates and reaches the maximum pressure at the die.

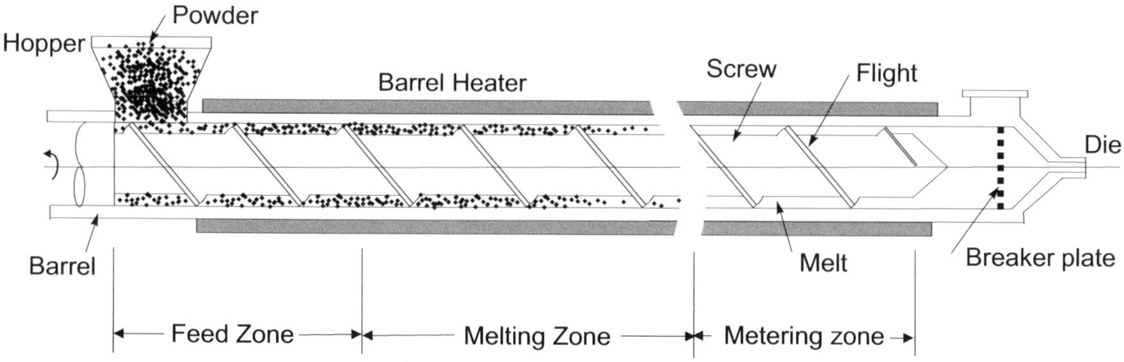

Figure 2.12: A schematic of the extruder (redrawn from [50]).

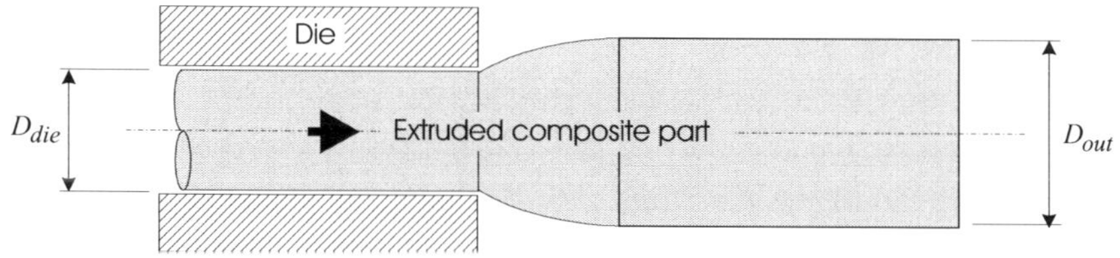

Figure 2.13: Schematic of die swelling.

- The extruded part is then cooled either by (a) exposing it to air or (b) passing it through water-channels.

Injection Molding:

- In injection molding, the suspension is pumped into a closed mold instead of a die. So, extrusion is a continuous process whereas injection molding is a batch process, as one must demold the part before one can inject the material again.
- Since the pressure of molten material is high when injected into the mold, complex shapes (such as thin sections and ribs) and good dimensional accuracy can be achieved. (However, the design engineer has to adjust the dimensions of the mold cavity considering the shrinkage due to cooling of the part.)
- The entrapped air escapes from the vents.
- Vents are closed using valves.
- The mold might be cooled especially for large parts to shorten the solidification time and thus shorten the cycle time.
- Ejectors are used to remove the part after cooling.
- There are 3 basic types of molds:

  a) The "cold runner 2 plate mold" (with usual gates, runners, and sprue which are trimmed later).

  b) The "cold-runner 3-plate mold" (the runner system is separated from the part automatically when the mold opens).

  c) The "hot-runner mold" (no gates, runners, or sprue).

- Before letting polymer enter, metallic components (such as pins) are placed in the mold cavity and are a permanent part of the injected product (this process is known as insert molding).
- Typical cycle time is 5-60 seconds for thermoplastics, and of the order of minutes for most thermosets.
- Typical injection molding machines have a clamping force of the order of Mega Newtons (MN).
- The clamping force on the mold is generally supplied by a hydraulic system.

Transport Issues

The action of the screw as a pump is a complex process and involves drag flow and pressure driven flow acting against each other. Hence, the important transport issues are to understand the interaction between the drag and pressure driven flow and their role in calculation of the pumping rate and the power requirement for turning of the screw. Another transport phenomenon is the interaction of the viscous dissipation in the momentum transport with the heat transport which helps understand and control the heating and cooling system. As the suspension is non-Newtonian when it exits the die, the normal stresses can distort the cross-section by causing a phenomenon known as die swelling as shown in Figure 2.13. Just after the material exits the die, the longitudinal tension on it will be relaxed, and thus it contracts in the longitudinal direction. Assuming incompressible flow (and thus conservation of volume during this plastic deformation), the material expands in the transverse directions. This phenomenon is called die-swelling.

Die swelling and melt fracture are significant processing issues for extruded structures and with the addition of fibers can distort the final cross-section. Transport processes and their interaction with surface tension and normal stresses will allow one to study and hopefully understand this relationship.

Applications

Extrusion is considered as the positive displacement pump for producing over six billion polymer products each year. A partial list of extruded products includes films, pipes, tubing, insulated wire, filaments for brush bristles, profiles for home siding, storm windows and gaskets, etc. [52]. The process of making these products is termed extrusion. The processing physics of flow and heat transfer of plastic melts in an extruder have been studied in detail by many researchers [53, 54]. In the last few years, the process has been slightly modified to allow extrusion of polymers containing reinforcements.

2.3.3 Compression Molding

The principle in compression molding is very simple. The material (called the charge) is placed inside the mold cavity. The mold is closed by applying pressure. The material deforms to take the shape of the cavity. The mold is opened and the part is ejected. Although conceptually simple, there are many critical issues that need to be resolved before one can produce a part without defects.

Compression molding has been around for decades and was used for a long time as a standard method for molding phenolics and similar thermosets. Injection molding partially replaced it primarily because of its ease in material handling and automation. However, compression molding offers a distinct advantage when processing composites. As compared to injection molding, in compression molding the material undergoes a very modest amount of deformation and there are no regions of very high stress, unlike a gate of an injection mold. Also as there is no gate through which the fibers have to enter the mold, the reinforcing fibers are not damaged by the flow during mold filling, and longer fibers and higher fiber volume fractions can be achieved. The other advantages of compression molding are that it is fairly simple, cycle times can be relatively fast, repeatability is excellent, high-volume production is easily obtained and parts with tight tolerances can be produced. In addition, mixing the resin and fibers before the compression allows for good control over the chemistry and the mix of the final product. The major disadvantages are that a large initial investment in molds and presses is necessary, the material must be stored under certain environmental conditions and must be used within a certain time for thermosets, and flow related problems can arise during the process so mold design is difficult and tedious. Several minor defects can arise in the part such as residual stresses, delamination, warpage, and significant variation in fiber orientation; all these give the process a large number of variables that can influence the process and hence the part quality [55].

Precusor Material Forms

A wide variety of composite materials are compression molded. The most commonly used material for compression molding of composites is sheet molding compound (SMC). This material consists of a thermoset matrix with short fibers dispersed through it. This may involve compounding a resin, combining it with fillers and fibers or impregnating a fiber preform with a resin. Subsequently, thickening of the resin or B-staging of the resin is carried out for proper bonding between the fibers and the resin. Figure 2.14 shows schematically how SMC is made.

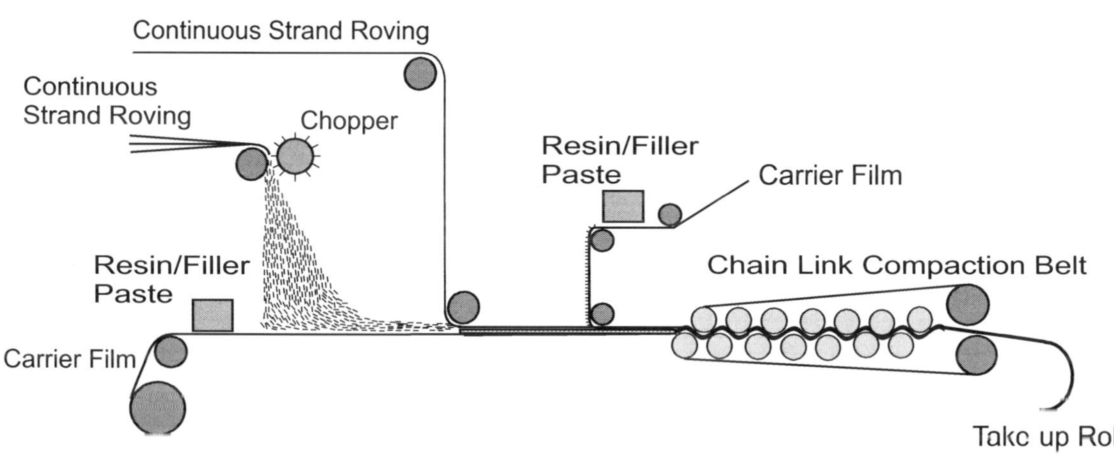

Figure 2.14: Schematic of sheet molding compound (SMC) production (redrawn from Tucker [55]).

To form the sheets of SMC material a specific procedure is used as shown in Figure 2.14. A thin layer of resin is placed on a sheet of nonporous material, such as nylon. As the nylon moves along the production line, fibers are added to it; these can be in random, unidirectional, or other orientations. Next a layer of resin, placed on a cover sheet, is applied onto the fibers so that the resin is in contact with the fibers. This sheet, enclosed between the nylon sheet and the cover sheet, is then passed through several compaction rollers. These serve two main purposes: they mix the resin and fibers together, and they compact the sheets. The resin is now in a continuously changing state (i.e., it is slowly curing); it is left to thicken for approximately 5 days, after which the SMC is ready. At this point the SMC sheet must be used within a certain period of time, which can be up to several weeks, and must be stored under certain environmental conditions, such as low humidity. Several types of SMC are currently used in industry:

- SMC-R (reinforced with fibers oriented randomly),
- SMC-C (reinforced with unidirectional continuous fibers),
- SMC-C/R (reinforced with both randomly oriented and continuous unidirectional fibers),
- SMC-D (reinforced with directional but discontinuous fibers).

It is possible to use both thermoplastics and thermosets in SMC, but the majority of SMC is done using thermosets [56].

Automobiles have been using SMC in many primary and secondary structures. For example, Ford Explorers and Rangers used SMC beams in 1995, and their number increased to 1.75 million by the end of 1996 [57] and has continued to increase despite the fact that one cannot recycle this material.

Other composite materials that can be compression molded are thermoset matrices with continuous reinforcing fibers and thermoplastic matrices with continuous random or aligned fiber reinforcement. When thermoplastic resins are used instead of thermoset resins, the material is called Glass Mat Thermoplastics (GMT). They consist of a glass mat preimpregnated with a thermoplastic resin. Some of the thermoplastic resins are PP, PE, PC and PPS [58–60]. GMT is prepared by using E-glass or S-glass fiber in the form of chopped fibers, randomly chopped fibers or continuous mats and impregnating them with the polymer powder using a solvent medium. The GMT sheet is then dried and consolidated by heating it in an oven and compressing it in a press. Traditional laminates usually with woven or stitched fiber preforms are also compression molded. The processing of such materials will be discussed under squeeze flow manufacturing methods for thermoplastic materials and porous media manufacturing methods for thermoset matrices as categorized by the dominant transport processes. Compression molding of short fiber reinforced materials such as SMC is interesting because the properties of the finished product are strongly affected by the processing.

Process

The SMC or the GMT is stacked into the mold cavity. This is referred to as the initial charge. This charge may be preheated by infrared or other means before it is placed in the mold. The initial charge shape and its placement location in the mold are crucial parameters as they influence the final properties of the product. Sometimes this charge is preheated using dielectric sensors before closing the mold to initiate the flow. The temperature field that results from this stage is part of the initial condition for the mold filling stage. Mold filling begins when the polymer begins to flow and ends when the mold cavity is filled. The heated top and bottom platens containing the two halves of the mold cavity are brought together, generating heat and pressure to initiate the flow. Temperature range is between 135 and 160°C, pressure is between 3.5 and 15 MPa and cycle time is between 1 and 6 minutes. The amount of flow in compression molding is small but critical to the properties and the quality of the part because the flow controls the orientation of the short fibers and the final orientation pattern is what will determine the physical and mechanical properties of the composite.

In-mold curing is the next stage in compression molding of thermosets, where the liquid resin starts to gel and cross-link and forms a solid part. The curing may be initiated during mold filling stage but the bulk of the curing takes place after the mold is filled. The part is removed from the mold as soon as it is solid-like and may be placed in an oven for post-cure to complete the curing process. For GMT one must wait for the thermoplastic material to cool below its glass transition temperature before demolding it.

The mold is usually made of steel and it is hardened in key areas where the mold can wear out more easily. This is important because the mold is subject to high pressures and temperatures and also undergoes many cycles continuously. For these reasons mold design is very important, and the overall cost of the molds is usually high. Several different resin systems can be used in SMC. Vinyl ester and polyester are the most used systems in the automotive industry, while epoxy resins are widely used in the aerospace industry [61]. There are several variations and modifications that different industries have developed over time, in order to improve the process and to tailor them to their own needs. The automotive industry, for example, has a specific need for parts with excellent surface finish. For this, a technique known as in-mold coating was developed.

In in-mold coating, after the part is partially cured inside the mold, the mold is opened slightly and a resin, such as a urethane, is injected in the mold. Subsequently the mold is closed again, causing the resin to coat the outside of the part, filling any voids on it. This greatly improves the surface finish on the part and can save several stages in the painting process.

Transport Issues

Heat transfer by conduction, interaction between cure and heat transfer, cure and viscosity, temperature and viscosity, velocity, deformation gradient and fiber orientation, fiber orientation and viscosity are some of the transport phenomena that are important to understand and to model the flow behavior of SMC like materials by compression molding. Part cool down is the final phase which plays an important role in the distortion of the part and the development of residual stresses. A difference in thermal shrinkage through the thickness and in different sections of the part is one of the sources that gives rise to residual stresses inducing distortions as the part cools down to room temperature. Thus, the temperature distribution and rate of cooling are important in determining how these stresses can be relaxed during cool down.

Another issue related to the transport processes is the cycle time it takes to produce a part from a single mold and press. This is important in high-volume applications. Closing speed and placement of the charge in the mold can influence the time to fill the mold with the material. Considerable efforts on trial and error methods have been made to reduce the cycle time. Understanding of the cure kinetics and its interaction with heat transfer should provide fruitful avenues for enlightening experiments and methods to improve the cycle time.

Mold and part design is related to flow and heat transfer and indirectly controls the quality of the part and the cycle time. For example, thermal design of the mold will influence the cycle time. Fabrication of molds is an expensive and time consuming task, so modeling the process, which can virtually verify the design efficiency before the mold or the tool is built, will be extremely handy.

Process automation is also a critical issue as it will make this process more competitive with injection molding. Currently the initial charge shape is cut and placed by hand in the mold cavity. Use of devices to load and unload the part, along with cutting and placement of the charge as shown in Figure 2.15 will allow this process to handle material in bulk and for high-volume production.

Performance of the manufactured part is tied to the flow, heat transfer and chemical reactions which occur in the mold. For example, initial charge shape and location will change the flow pattern, which in turn will influence the fiber orientation. The fiber orientation will influence the physical and mechanical properties.

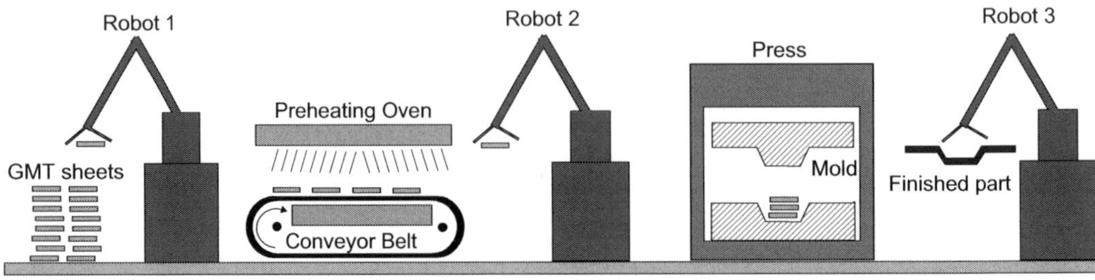

Figure 2.15: Schematic of automated compression molding [11].

2.3.4 Structural Foam Molding

Structural foam molding is a modification of the injection molding process. This process can be carried out at low injection pressures and can make thick-walled parts due to the addition of a blowing agent. Any injection molded plastic can be foam molded. The pellets may contain fibers and some nitrogen or a small blowing agent. The amount of blowing agent controls the cellular microstructure and the density. The fibers provide additional stiffness and strength for load bearing applications. One may have fibers in the pellets or inject the resin into a dry preform in which the resin is allowed to foam. The use of low pressures makes this process attractive as the tooling can be made from materials such as aluminum instead of high grade steel that is required for injection molding. When a solid outer skin and a cellular inner structure is required, this process is used. A blowing agent (a gas such as nitrogen) is added to the thermoplastics and this causes the material to expand. The overall density could be as low as 40% of the neat plastic [51]. Typical products are furniture components, TV cabinets, large instrument housing, in-ground housing for water systems, storage cabins and trash cans. Glass-filled foams can be used for car bumpers as the presence of voids helps with energy absorption. Due to their large wall thickness and insulation properties, the cooling time is much longer than in injection molding. Also the entrapped gas escapes over the course of three to four months from the part which could cause discoloration of the part.

Transport Issues

This process in addition to momentum and heat transfer also has to address mass transfer. In addition to non-isothermal flow of a viscous liquid, during structural foam molding, there is nucleation and growth of bubbles, interaction between bubbles to coalescence and form a larger bubble. The bubble growth process involves simultaneous mass, momentum and heat transfer between an expanding bubble and the fluid surrounding the bubble. It is governed by many competing transport and material parameters such as the diffusion rate of the gas into the polymer, surface tension of the polymer and the concentration of the gas into the viscous or viscoelastic resin. Also, it is important to describe growth of many bubbles in close proximity separated by a thin film of the polymer as is experienced during foam molding instead of growth of a bubble in an infinite fluid medium [62,63].

2.3.5 Rotational Molding

This process is used to manufacture hollow parts such as tanks of various sizes (trash cans, buckets, toys and footballs). A pre-measured quantity of powdered plastic mixed with chopped fibers or a resin mixed with fibers is placed inside a mold. The mold is heated in a large oven while it is being rotated about the multiaxes. The stages of the process are schematically shown in Figure 2.16. The part is heated to approximately 250°C until the material melts and sticks to the inner surface of the mold [64]. After the molten material forms almost a uniform thickness due to the centrifugal effect, it is transferred to a cooling chamber while being continuously rotated. Finally, the part is demolded after it achieves a green strength (sufficient rigidity to remove it from the mold). Both thermoplastics and some thermosets can be formed using rotational molding. Cycle times are usually longer than in other molding processes. As a major advantage of this process, the equipment cost is lower than the other alternative processes because usually thin-walled molds are used [51,64].

2.4 Advanced Thermoplastic Manufacturing Methods

Thermoplastic resins are usually solid at room temperature and exhibit some softening when heated above their crystallization temperature, which is usually above 100°C. Even at their melt temperature, thermoplastic resins are very viscous. It is not uncommon to have a resin with a viscosity about a million times that of water (10,000 Poise) at its melt temperature. It is extremely difficult to process such resins if they have to be forced to occupy the empty spaces between fibers to form an advanced thermoplastic composite. In the last decade, there has been emergence of reactive thermoplastics as introduced in Chapter 1. In reactive thermoplastic processing as opposed to melt processing, the fibers are impregnated with low viscosity monomer or oligomeric precursor and the polymerization is carried out in-situ with heat or radiation. As the monomer viscosity is of the order of thermosets, one can use thermoset manufacturing methods to process reactive thermoplastics. In this section, we will address methods only applicable to melt processing of advanced thermoplastic composites. Unlike short fiber suspensions, advanced thermoplastics contain continuous fibers or nearly continuous fibers. However, two mechanisms can help reduce the viscosity of the resin. The viscosity reduces rapidly with temperature for most thermoplastics. Secondly almost all thermoplastics are shear-thinning materials, which implies that under high shear rate their viscosity reduces. Nevertheless, high processing temperatures translate into higher costs in cycle times and processing equipment. As the polymer can degrade at higher temperatures, the window to raise the temperature much higher than the crystallization temperature is narrow. Also, it is difficult to shear the resin to occupy the spaces between the fibers at such high viscosity levels. To circumvent this problem, the resin is preimpregnated into the fiber bundles or is sprayed as commingled powder or resin fibers are woven with the reinforcing fibers. This ensures that resin does not have to travel far when heat and pressure are applied to move the resin to occupy the empty spaces between the fibers.

When heat is applied the resin melts, and the applied pressure makes the resin flow and redistribute. As the resin is viscous, the fibers and the resin on the macroscale will deform together and consolidate into the shape of the structure being manufactured. Hence,

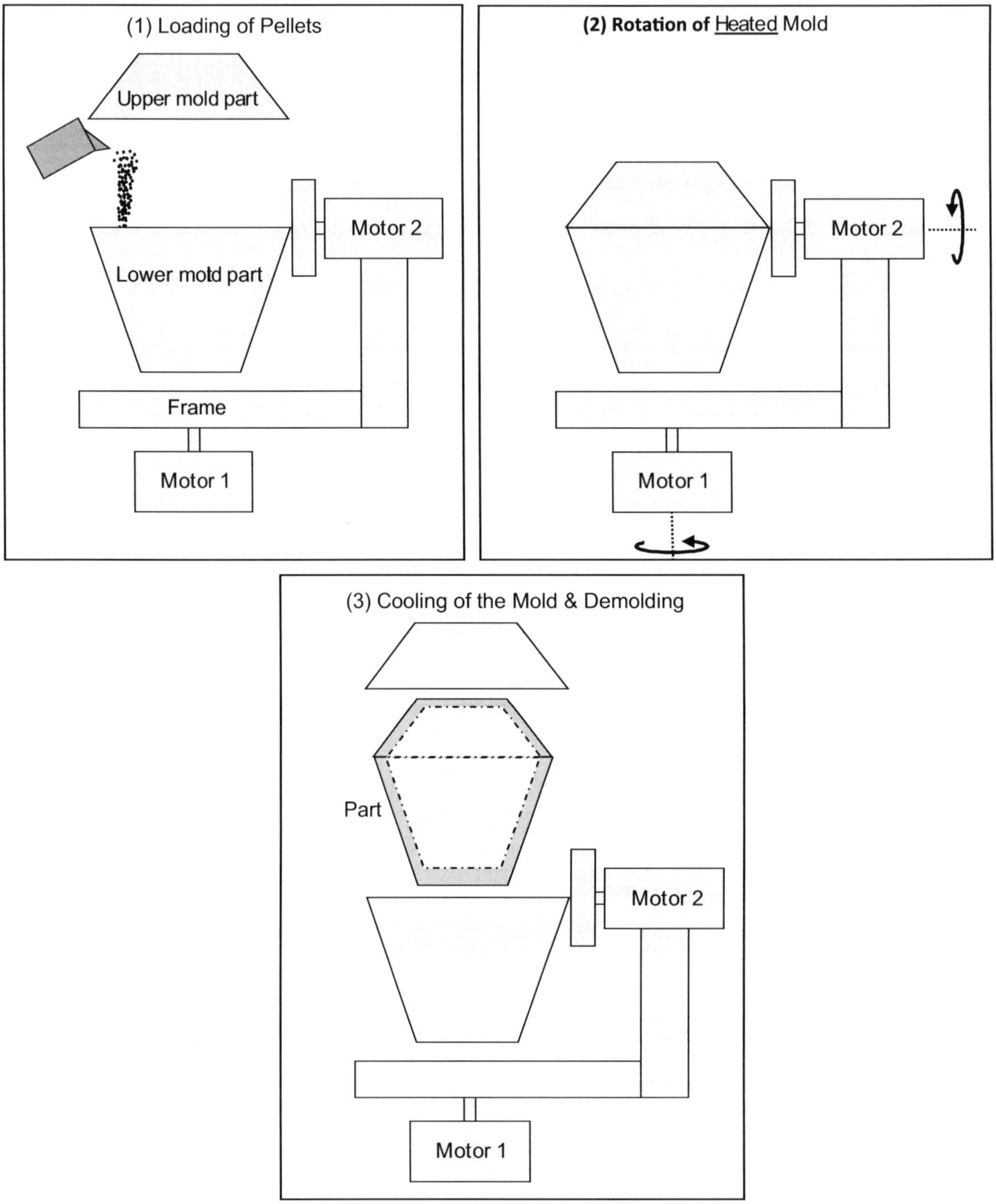

Figure 2.16: Stages in the Rotational Molding process.

for such materials practice has converged, from the modeling viewpoint, to treat them as one material with a viscosity that is modified due to the presence of the fibers. The three manufacturing methods that may be modeled with this philosophy and will be discussed in this book are (1) sheet forming, (2) tape laying or advanced fiber placement and (3) thermoplastic pultrusion. We will give a brief description of these processes below.

2.4.1 Sheet Forming

The processing science for long and short fiber reinforced thermoplastic sheets grew out of a need for large parts with higher strength and stiffness over non-reinforced sheets and monolithic materials with faster processing cycle times than those for thermoset-matrix composites. In addition to faster forming cycles, thermoplastics have more flexible processing parameters since the viscosity is only a function of temperature — not of cross-linking or cure as in thermosets. Sheet formed parts also have the potential to reduce the total part count in a structure by forming into "net shape" with reinforcements in desired areas.

Process and Precursor Materials

Composite sheet forming is a process well suited for the forming and shaping of thermoplastic matrix short- and long-fiber reinforced composites. The material preform may be in unidirectional or multi-axial sheets either in stacked or preconsolidated form. The basic sheet forming sequence starts with heating the preform to its forming temperature, defined as the temperature where the viscosity of the resin is soft enough to allow the reinforcing fibers to slide relative to one another and permit easy shaping of the sheet. Then, using either mechanical or hydraulic press, the sheet is formed over a curvilinear tool surface. The forming step is analogous to several common sheet metal bending and forming operations and includes deformation of the sheet both in and out of the plane. This material flow is viscous and is characterized by the flow of both resin and fibers together. After the forming step is complete, consolidation pressure is maintained on the part until the part cools below the matrix melt temperature. Once sufficiently cooled, the part is removed from the tool surface and, if required, an edge trimming step is performed. If necessary, the reversible solid-liquid phase change characteristic of thermoplastics enables the once formed part to be transferred to another tool surface and repeatedly re-formed or incrementally formed until the final desired geometry is attained.

The oldest form of sheet forming exists with the processing of non-reinforced thermoplastic sheets. This process is also known as "thermoforming." Isotropic in nature, the sheets are usually held in place along the edges over a tool surface and brought up to their material softening temperature. This is usually somewhere slightly below the actual melt temperature in order to work the material while in a compliant but not liquid state. The most common forming methods are "hot stamping," where the sheet is pressed between matching dies, and "vacuum forming," where a vacuum is pulled through small holes in the tool face, pulling and spreading the sheet down over the surface. Because thermoforming is a combination of drawing and stretching operations, the thermoplastic material should exhibit high uniform elongation, otherwise it may neck or fail.

The sheet forming process also known as fabric thermoforming holds several unique advantages over other thermoplastic processing methods such as injection molding, pultrusion and tape laying methods. The nature of injection molding combined with the high viscosity of the thermoplastic precludes the use of high aspect ratio fibers (>1000) which provide the necessary mechanical properties desired in high performance applications. Additionally, it is difficult to make large parts by injection molding. Sheet forming can potentially make large parts and provide a much greater control over and ability to predict the final fiber architecture. However, thermoplastic sheets containing fabrics allow very little stretching in the fiber direction. Therefore if large movement of crossing fibers during forming of compound curvature parts from flat sheets is not permitted by some sort of shear mechanism, the final part will not be wrinkle free. Hence it is important to address the shear mechanism between fibers and resin when modeling the forming process [60].

Sheet Forming Methods

The major composite sheet forming processing methods can be broadly classified as hot stamping, diaphragm forming and incremental processing. Composite sheet stamping or matched-die press forming imitates the stamping methods employed in the field of sheet metal forming as a high-volume, low-cost manufacturing process as shown in Figure 2.17(b). The composite blank is heated to the forming temperature and then stamped against the tool surface. A variation on this is rubber tool stamping wherein one or both sides of the die are made compliant. This helps maintain an even consolidation pressure across the part in case of any tool misalignment.

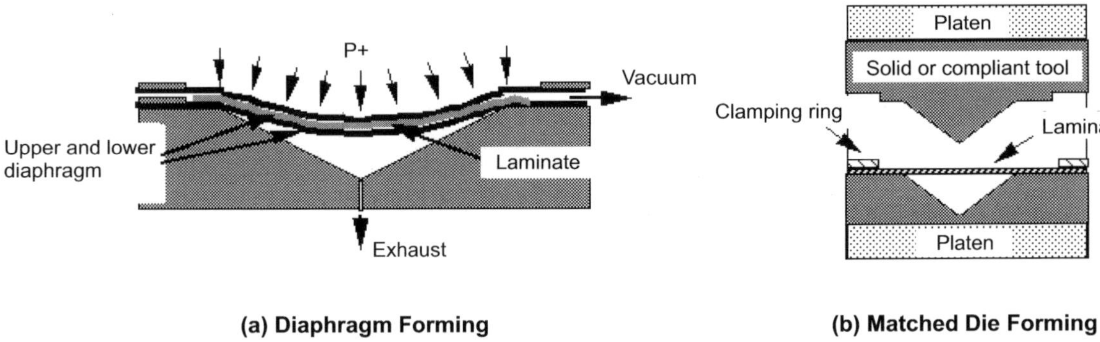

Figure 2.17: Composite sheet forming processing methods: (a) diaphragm forming and (b) matched die forming [65].

In diaphragm forming as shown in Figure 2.17(a), the blank is held between two disposable, plastically deformable diaphragms of either super-plastic aluminum or polyimide polymer. During the forming cycle, the diaphragm edges are clamped, heated along with the blank and deformed through the use of air pressure to the tool surface. The diaphragms serve to hold the blank in tension and prevent fiber buckling that can occur under compressive stresses. When forming parts containing continuous fiber reinforcement, the diaphragms are clamped but the blank cannot be. This is due to the inextensibility of the fiber reinforcement. Hydroforming is a process similar to diaphragm forming; however, hydraulic fluid is used to provide the pressure behind a permanent rubber diaphragm.

Incremental processing enables the forming of large structures using smaller, lower cost fabricating equipment. After the usual heating and forming steps, there is an additional part transfer step added to the forming cycle. Incremental forming also provides the ability to construct final shapes that would be very difficult otherwise. Stock shapes can be produced and later incrementally formed into custom configurations. One promising incremental forming method is stretch forming. Stretch forming makes exclusive use of aligned, long, discontinuous fiber reinforcement technology and as the name implies, an extensional mode of deformation in the fiber direction is enlisted during the forming process. Since the fibers are discontinuous, the composite sheets may be locally heated and deformed. This allows certain incremental forming techniques not possible with continuous fiber reinforced materials. For example, linear beams can be stretch formed into curved sections with favorable mechanical properties since the fibers follow the curvature of the beam [65] as shown in Figure 2.18.

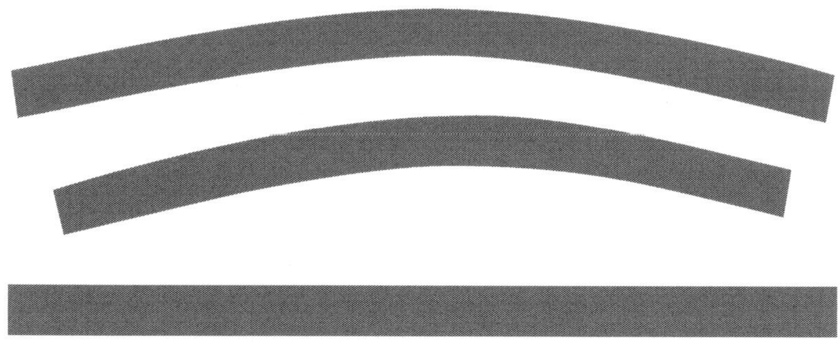

Figure 2.18: Linear beams can be stretch formed into curved sections with favorable mechanical properties since the fibers follow the curvature of the beam [65].

The key to successful stretch forming is precise control over the final fiber placement. This is achieved by clamping both ends of the unformed part, heating up the portion between the clamps and then carefully forming the stock shape to the desired curvature [65]. Another process which can be included in this category is roll forming in which preheated thermoplastic sheets are pulled through a series of rollers which bend the sheet to the required form [66]. This process can produce different profiles at rates of 10 meters per minute but high precision in sheet thickness is required.

Transport and Other Issues

The major disadvantage of this process is that due to the presence of fibers, the viscosity of the material is highly anisotropic and hence forming compound curvature shapes leads to wrinkles and folds during manufacturing as shown in Figure 2.18. During forming of textile or woven fabrics thermoplastic sheets into complex structures, the fibers need to rearrange due to the local deformation. However, there is a limit to which the fibers can slide and change the unit cell architecture until the interlacing fiber lock which represents the maximum deformation. Further deformation causes fold and wrinkles which are the most common flaws of this process. Long discontinuous fiber (LDF) material alleviates this problem to some extent, but nevertheless this has been a big hurdle in the lack of interest in using this process for more complex curvature parts. Modeling of the sheet forming process is quite a challenge. To describe the transformation of a stack of flat sheets into a complicated three-dimensional shape involves movement of a free surface and contact with a tool. The deformation can be large, and one must be able to describe the deformation physics of such highly viscous anisotropic materials. The anisotropy can be very severe, and the material can have multiple inextensible directions. The layers may slip over one another during forming.

Qualitatively, one can describe the micro- and macromechanisms the material undergoes before the part is formed. First, when one places various sheets of thermoplastic tapes containing continuous or discontinuous fibers and applies heat and pressure, at the micro-level, one can expect the resin to melt and percolate to form a more uniform fiber-resin mixture. Further pressure causes squeeze flow of this material. However, due to the presence of the fibers, the material has anisotropic viscosity and will only flow in the transverse direction to the fibers at the macroscale. If the layers of sheets containing unidirectional fibers were arranged in different directions, one could expect the layers to shear based on the deformation process it is undergoing.

No comprehensive model for sheet forming exists. There are many geometric mapping issues as well as transport issues in this process. The modeling should be able to provide information about the thickness of the composite after deformation, location of the edges of the blank and the occurrence of defects, such as laminate wrinkling and/or fiber buckling. The important parameters that do influence this process are part geometry, tool geometry, initial blank shape (fiber composite sheet shape), initial thickness, fiber orientation in the initial blank, material properties and processing conditions such as forming speeds, applied pressure and temperature.

2.4.2 Thermoplastic Pultrusion

As it has been around since the late 1940s, pultrusion is one of the oldest composite manufacturing processes. It was originally designed to manufacture fishing rods [67]. This process can be used for both thermosets and thermoplastics. However, the thermosets have dominated the composites industry.

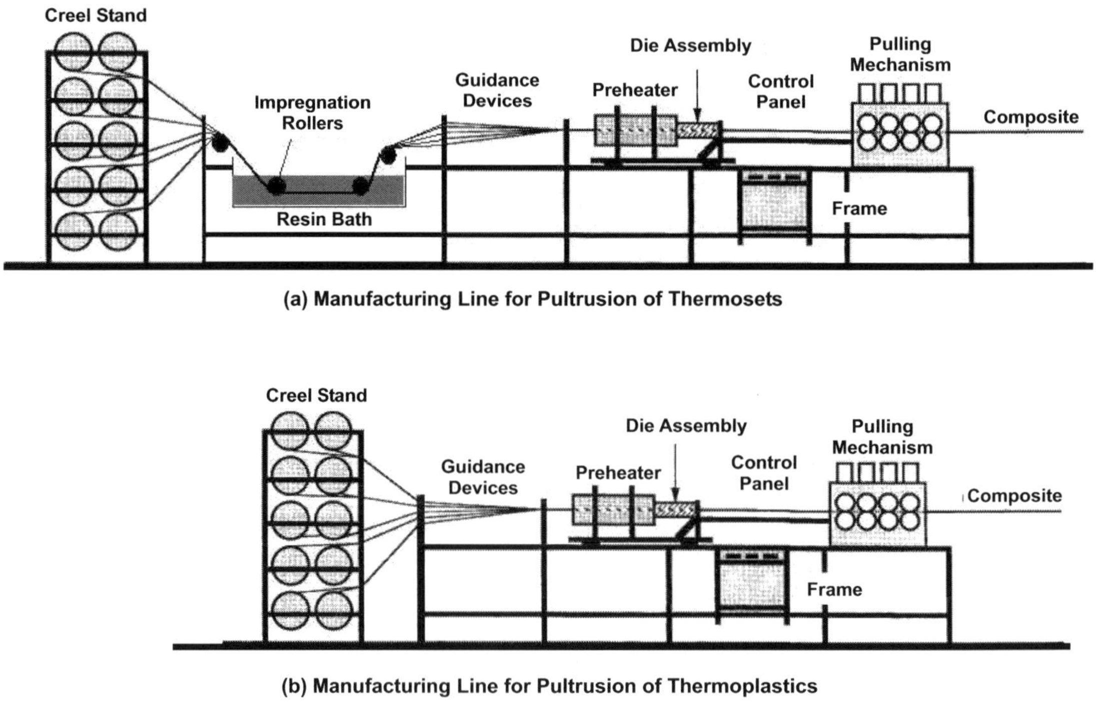

Figure 2.19: Schematic of a pultrusion line [16].

Process

The process involves dragging a combination of fiber and matrix materials from a supply rack through a temperature-controlled tool, which will determine the final part geometry. The design of these tools depends on whether one is going to use a low-viscosity thermoset resin or high-viscosity thermoplastic polymer. For thermosets, one can use bare fiber rovings to go through a liquid thermoset bath before entering the die for curing and cross-linking. On the other hand, thermoplastic resins are preimpregnated with continuous fibers in the form of a tape or the fibers pass through a station of polymer powder bed, where the thermoplastic powder attaches itself to the charged fiber surface. Therefore, thermoplastic pultrusion requires a preheating area, which can be either an infrared or a convection oven preceeding the actual tool. After passing the preheating station, the fibers enter the tapered entry region of the tool where additional heat is introduced to the precursor, assuring complete melting of the thermoplastic matrix and allowing it to spread over the fiber surfaces. Then the fibers covered with resin go through a tapering die, which consolidates the composite into the shape of the die. A schematic of a pultrusion station is shown in Figure 2.19.

The attractive feature of the pultrusion process is that it is a continuous process, and therefore the material efficiency is extremely high. However, after almost 40 years in production, the major application produces only simple profiles as shown in Figure 2.20. Further understanding of the process physics could extend the application of this process to more challenging shapes. Thermoplastics, although difficult to process, offer improved impact strength, and enhanced fracture toughness, and they allow for reshaping and recycling as compared to thermosets [68].

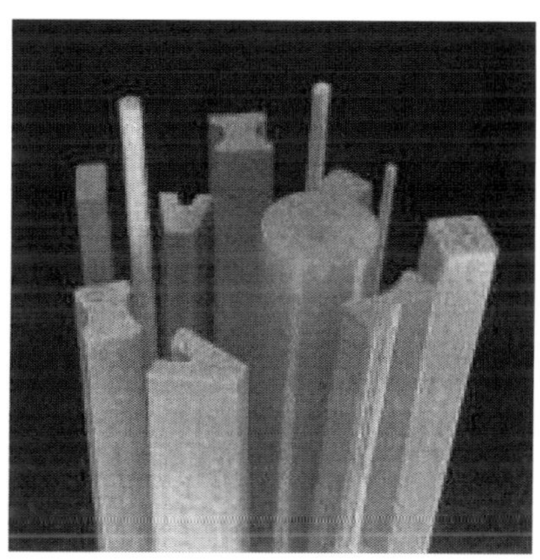

Figure 2.20: Some typical cross-sections made by pultrusion [69].

Transport Issues

The transport phenomenon inside the heated die is of interest as most of the redistribution of the resin and the consolidation of the fibers and the resin takes place there. The precursor material used will influence the transport phenomena modeling in this process. Several different types of thermoplastic preforms can be used with this process. They are either continuous fibers completely impregnated in the shape of thermoplastic tapes (e.g., CF/PEEK) or glass fibers embedded in thermoplastic powder and enclosed by a thermoplastic tube of the same material or commingled polymer and reinforcing fibers. Figure 2.21 depicts available materials for usage in the thermoplastic pultrusion process.

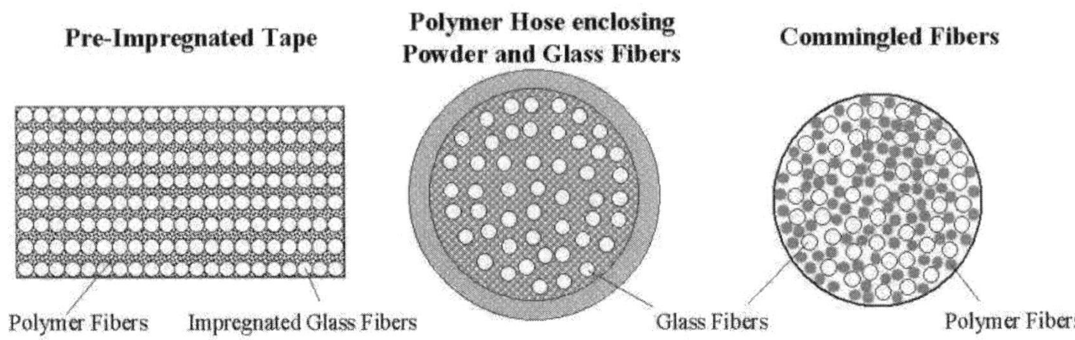

Figure 2.21: Schematic of different available precursor material for the pultrusion process [16].

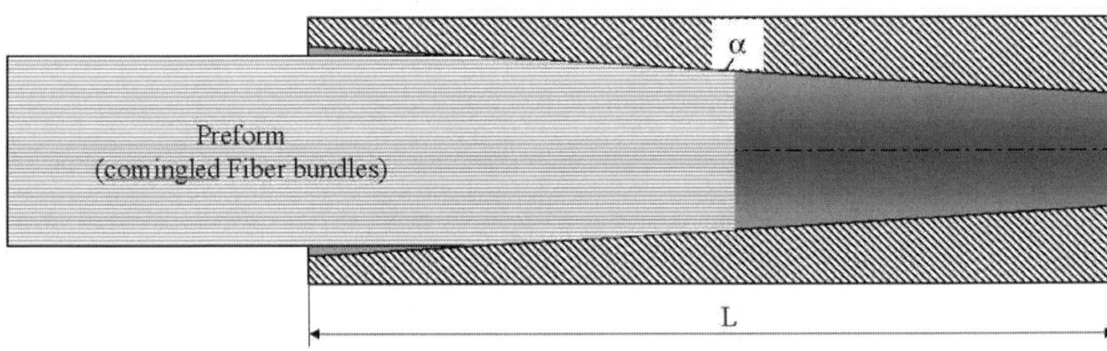

Figure 2.22: Schematic of the tapered section of a standard pultrusion die [16].

The die assembly for this process has two distinct sections. The first section is the heated and the tapered entry region (Figure 2.22), which collects the preheated preform, rearranges the fiber bundles to the desired shape and melts the polymer. The function of the taper is to consolidate the preform, thus encouraging elimination of voids and complete impregnation of the fibers with the polymer matrix. The second section of the die assembly is known as the land region. It is of uniform cross-section without a taper and may have cooling lines attached to it. The role of this region is to solidify the matrix material forming the final shape of the pultruded part. It is usually cooled to approximately 50-80°C (around the glass transition point (T_g) of the thermoplastic matrix). The process velocity is determined by the speed of the pulling system located immediately behind the die assembly. The last step in this production process is to cut the final product into pieces of the desired length.

There are two important parameters the modeling of this process should be able to predict: (1) the pulling force required to run the operation at a reasonable speed to produce parts that are free of voids and contain the desired fiber volume fraction, and (2) the desired level of crystallinity in the matrix with minimal stress concentration in the heating and cooling profile of the die. Hence the viscous flow physics and the heat transfer during the process will play an important role in the determination of these key parameters such as the clamping pressure for the die, preheating temperatures, cooling temperatures, etc. One of the drawbacks of this process has been formation of a fiber nest near the inlet of the die due to insufficient impregnation of fiber bed with resin, which leads to halting of the process to clear the bunching of the fibers and restarting the process. Also it is usually difficult to make a multi-axial composite part with this process as the fiber direction is the direction of the pulling force.

2.4.3 Thermoplastic Tape Lay-Up Process

In this process, the tool, called the tow placement head, is designed to conform to the geometry of the composite structure. Processing of thermoplastic composites is based on melting and solidification of the matrix. The matrix requires energy input for melting and energy extraction for solidification from the system. The method of energy transport can be global where all of the thermoplastic matrix is melted as in compression molding, injection molding, extrusion, pultrusion, etc., or can be localized as in filament winding and tape lay-up where only a portion of the matrix is melted.

Process

In this process, 3-mm to 12-mm wide thermoplastic tape preimpregnated with continuous fibers is placed on the tool surface (if it is the first layer) or on the substrate (previously deposited material on the tool). The incoming tape or tow and the previously deposited material on the head (the substrate) are preheated by laser, gas, or any other methods of concentrated localized energy. Rollers are used to initiate intimate contact and consolidate the incoming tape to the substrate below. The localized nature of heating demands the consolidation process also be localized, and it is commonly referred to as in-situ consolidation. Additionally, local energy may be provided to heat all the layers underneath in the thickness direction to further improve the overall degree of bonding, healing and intimate contact. Void content within the tow and the substrate decreases under the pressure exerted by the consolidation rollers.

In most industrial applications the thermoplastic tape lay-up process is automated and also known as automated tow placement (ATP) process. In this process, a relatively thin tape is consolidated on a substrate under the application of heat and pressure (Figure 2.23).

In most cases, the feed tape, the heater (gas, induction, laser, etc.) and the consolidation rollers/shoes traverse the substrate at a predefined path and velocity

The most commonly used material in aerospace industry is carbon fiber preimpregnated with PEEK or PEKK thermoplastic matrix. Typical applications are the fuselage and wing structures of an aircraft. One of the important objectives and advantages of the ATP process is to eliminate the use of a huge autoclave in order to make the process more cost effective. Also, the ability to make the part out of the mold is attractive, in addition to having the capability to create multi-axial laminates and moderately complex structures. The downside is the investment required in automation and the cost of the tool head.

The demand to be cost effective forces the process to be conducted at the maximum allowable speed. This requires optimization of the process parameters at desired processing speed while maintaining the quality of the product. To achieve such goals, a fundamental understanding of the process is necessary.

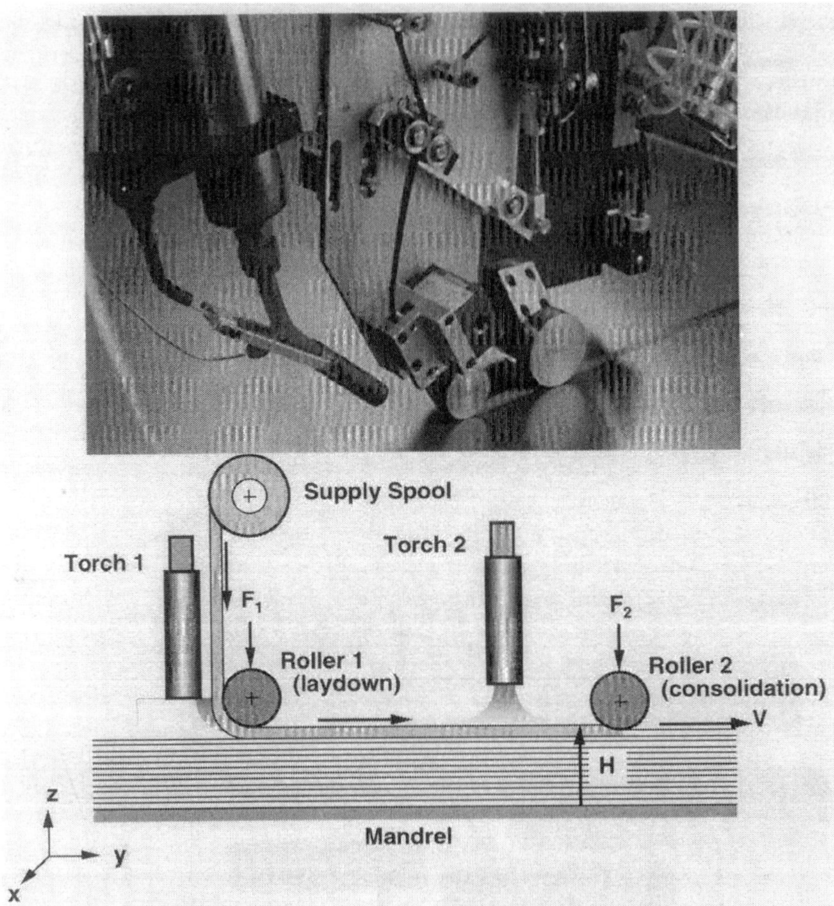

Figure 2.23: In-situ thermoplastic tape lay-up process [70].

Transport Issues

Different aspects of the process that can be modeled are intimate contact, polymer healing and consolidation [71–74]. To model this process, one must quantify the squeeze flow during application of the rollers and the transient heat transfer that governs the temperature of the resin and affects its viscosity. The heat transfer during cooling influences the microstructure, entrapment of voids and the quality of the part. The quality of the part also depends on melting, consolidation, solidification and through the thickness temperature gradient.

Issues related to consolidation are intimate contact, void reduction and migration, gap reduction between adjacent layers, adhesion and diffusion of matrix chains. The key issues for modeling include tow-placement head configuration, consolidation, bonding and the heat transfer between the incoming tape and the substrate interface. A good bond between the substrate and the incoming tape requires the interface temperature to be greater than the melting temperature of the thermoplastic. The temperature gradient through the thickness is responsible for residual stress development in the composite. The critical issues in the thermoplastic tape lay-up (or ATP) process are the heating of the tape above the melt temperature for good bonding with the previous "substrate" layer, without overheating to prevent degradation. So, the rate of heat input is a critical process parameter and will play a role of selection on the type of heater used. The heater width is also a parameter to provide an optimum heating zone for good consolidation.

The consolidation pressure is also a very important parameter. Consolidation pressure is applied for void reduction and adhesion to the previous layer. Low pressure may create pools of resin and poor bonding. Excessive force can squeeze the resin out creating a resin-starved region, and at the same time can deform the fibers, which will reduce the local strength. Residual stress development in the part is an outcome of the localized heating and cooling process described earlier and is an important issue. Intimate contact, diffusion and healing (the movement of polymer chains across the interface of the new tape layer and the previous substrate layer) are other important issues.

As it is common with other composite processes, the critical issues are addressed by trial and error methods on the shop floor. Lack of understanding and lack of process models make it very difficult to quantify the quality of the part. Detailed knowledge of process models, in-situ sensing, controls and feedback can greatly improve the part quality. In the last few years, the need to build large aerospace structures such as wing skins or large space tanks out of the autoclave has accelerated the development of the tape laying process. The process models have been improved substantially and some validation has been carried out [75–77]. However, many critical issues remain such as how to incorporate these models for design and control of the head and how the models can address the variability of the incoming tape such as its width, fiber content and resin-rich areas to not influence the final bonding between the various layers.

2.5 Advanced Thermoset Composite Manufacturing Methods

The major difference between thermoplastic and thermoset advanced composite manufacturing methods from the modeling viewpoint is that one can describe thermoset manufacturing methods using the physics of flow through porous media as the resin viscosity is low enough to move relative to the network of fiber preforms. The three methods we will introduce as examples of this class of transport process are autoclave processing, liquid composite molding and filament winding. In all of the thermoset manufacturing methods, the important phenomenon one should also include during the modeling process is the cure kinetics of the thermoset polymers as it influences the transport mechanisms during processing.

2.5.1 Autoclave Processing

An autoclave is a large pressure vessel with a heating facility, or one can think of an autoclave as a large oven with an integral pressurizing facility. A schematic of an autoclave is shown in Figure 2.24.

Process

Autoclave process is the earliest method used to make advanced composites for aerospace applications. The four stages involved are shown in Figure 2.25. The material and tool preparation stage is initiated by first covering the tool surface with a release film that allows one to detach the composite from the tool surface readily. The next stage involves cutting the prepreg (continuous unidirectional fibers partially impregnated with the uncured thermoset resin) layers and stacking them in a desired sequence on the tool surface to form

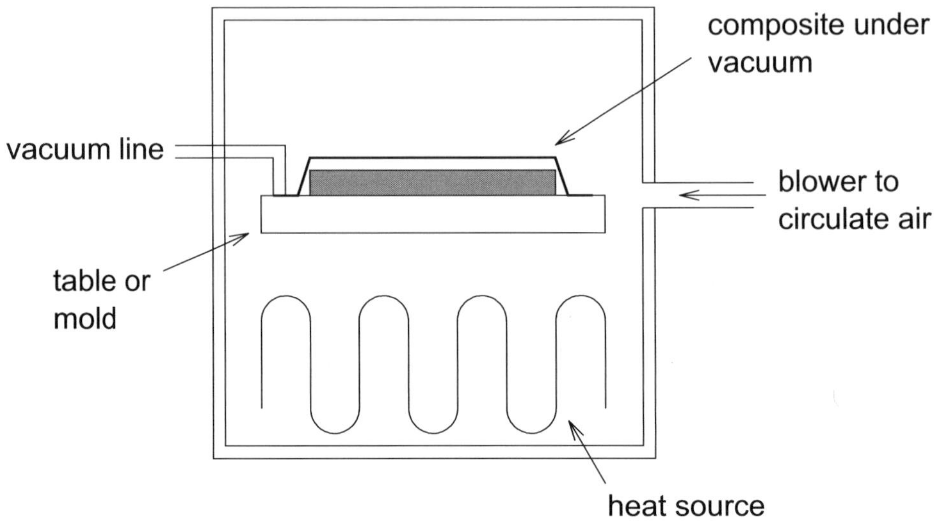

Figure 2.24: Schematic of an autoclave.

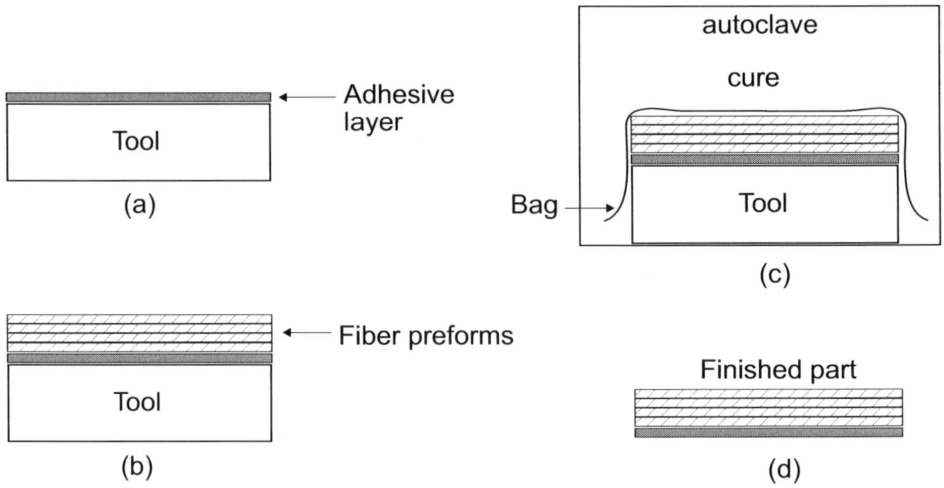

Figure 2.25: Schematic of four main stages of an autoclave: (a) placement of adhesive, (b) placement of fiber preform, (c) autoclave cure, and (d) finished part.

the composite lay-up. This is accomplished usually by hand lay-up, although great advances have been made in the use of automated tow placement and automated tape lay-up for the stacking sequence step. For example, the Boeing 777 aircraft tail assembly used a variety of automated processes [78]. Despite these advances, many parts in the aerospace industry still rely on hand lay-up for this step. The composite lay-up is covered by peel plies, release fabric and bleeder material in that sequence. Peel plies provide surface texture and release fabric allows resin to flow into the bleeders. On top of the bleeder material is the breather material. The breather material distributes the vacuum over the surface area. A vacuum bag envelopes the tool, the part and ancillary materials for vacuuming.

Figure 2.26: A typical autoclave.

The third stage, as shown in Figure 2.25, involves transferring the part into the autoclave and initiating the curing step by exposing the assembly to elevated temperatures and pressures for a predetermined length of time. The goal is to consolidate and solidify. The elevated temperatures provide the heat to initiate the cure reaction, and the applied pressure provides the force needed to drain the excess resin out of the composite, consolidate individual plies or prepreg layers and compress the voids. For thermoset composites this step is irreversible. Hence, it is necessary to subject the composite part to the correct processing window of temperatures and pressures to ensure a quality part.

The temperatures and pressures are of the order of 100-200°C and 500-600 kPa, respectively. Also, if one wants to manufacture a large part, a large autoclave is necessary. As the autoclave is a pressure vessel, it is usually made as a cylindrical or axisymmetric tube with a door at one end. As the autoclave must be strong at high temperatures as well, the autoclave is an expensive piece of equipment usually made out of welded steel. An example is shown in Figure 2.26. Once the part is cured, it is removed from the autoclave and inspected visually and by ultrasound or x-ray for defects, trimmed usually by a router or waterjet.

Curing is the most important autoclave processing step. Hence it has been the main focus of modeling. The magnitude and duration of the temperatures and pressures to which the composite is subjected to during the curing step affect the final quality in terms of thickness variation in the composite, warpage and void content in the composite.

Transport Issues

To address these critical issues, one needs to understand the mass, momentum and heat transfer that the composite undergoes during the curing cycle. The temperature and the pressure of the autoclave influences the temperature of the composite, the degree of cure of the resin, the resin viscosity, the resin flow, fiber volume fraction of the composite, the change in the void sizes, residual stresses and strains in the composite, and the cure time. We will discuss in a later chapter how to develop a model for autoclave processing. However, due to the extensive requirements of material data and batch-to-batch variation in properties, one can apply these models effectively only to simple geometries. For complex geometries one must combine models with sensing and control to produce successful parts.

The advantages of autoclave processing are that it can produce composite structures with very high fiber volume fraction. Also, a lot of empirical data is available on this process which makes it attractive when reliability outweighs the cost. Also once the most time consuming part of the process, the production of the master or the prototype, is completed, duplication can be carried out at relatively low cost. The major disadvantage of autoclave processing is the high cost associated with the initial investment of an autoclave and the limitation on the size of the part due to the size of the autoclave. Also, hand lay-up is an expensive proposition and can introduce many human errors in the manufacturing of the composite and can lead to a variety of defects. Also, it is difficult to calculate the time and cost to design a new master prototype as most of the knowledge base is empirical. Hence, different industries are reluctant to replace other materials with composites and use autoclave processing to make the part.

2.5.2 Liquid Composite Molding

All liquid molding processes involve impregnation of the resin into a fibrous network bed. The goal in these processes is to saturate all the empty space (pores) between the fibers with the resin before the resin gels. This would be difficult to accomplish with highly viscous thermoplastic resins as their viscosity is very high and impregnation would require very high pressures. Thermoplastic resins are not generally used with these processes as one of the attractive features of liquid molding processes is to limit the equipment to low pressures and room temperature impregnation and cure. Though there is recent interest in using cyclic thermoplastics which have low viscosities; however, they still do need to be processed at elevated temperatures.

Liquid composite molding processes encompass

- Resin Transfer Molding (RTM),
- Vacuum Assisted Resin Transfer Molding (VARTM),
- Structural Reaction Injection Molding (S-RIM),
- Co-Injection Resin Transfer Molding (CIRTM),

- Seemann's Composite Resin Infusion Molding Process (SCRIMP),
- RTM light,
- Compression Resin Transfer Molding (CTRM),
- Vacuum Induced Preform Relaxation (VIPR),
- Resin Infusion between Double Flexible Tooling (RIDFT)

and other subsets where the basic approach is to separately inject the liquid resin into a bed of stationary preforms. There are minor differences in the above processes that lead to slight or sometimes formidable modifications in the modeling of the process.

Process

Resin transfer molding consists of a mold cavity that is in the shape of the part to be manufactured. The fiber preform is placed in the cavity. The mold is closed and clamped or held under pressure in a press. The resin is injected into the compressed preform through one or more gates from a pressurized container. Once the mold is full, the injection is discontinued and the resin is allowed to cure. This cure may be initiated by either heating the mold which heats the resin as it flows into it, or by addition of inhibitors that initiate the cure after a time interval allowing the resin to first complete the impregnation of the preform. The mold is opened once the cure is complete or the part is sufficiently hardened to be demolded. These steps are depicted in Figure 2.27.

RTM offers the promise of producing low-cost composite parts with complex structures and large near net shapes. Relatively fast cycle times with good surface definition and appearance are achievable. The ability to consolidate parts allows considerable time saving over conventional lay-up processes. Since RTM is not limited by the size of the autoclave or by pressure, new tooling approaches can be utilized to fabricate large, complicated structures. However, the development of the RTM process has not fulfilled its full potential. For example, the RTM process has yet to be automated in operations such as preforming, reinforcement loading, demolding, and trimming. Therefore, RTM can be considered an intermediate volume molding process [79, 80].

VARTM and SCRIMP are slight modifications of the process where the top half of the mold is replaced by a vacuum bag (as in the autoclave process) and in SCRIMP a permeable layer is introduced at the top or the bottom to facilitate the distribution of the resin throughout the part quickly. The process relies more on pulling the resin by creating a vacuum from a container of resin at atmospheric pressure. Figure 2.28 shows the steps in the VARTM process.

This process has replaced RTM for some applications due to its simplicity, low initial capital investment and the ability to manufacture large structures such as bridge sections, wind blades and rail carriages as shown in Figure 2.29.

Also, the cost is kept low due to low pressures used in the manufacturing process and the reactions being carried out at room temperature. It only needs one tool surface, and the top surface is bagged as in autoclave processing which also cuts down on the tooling costs. The disadvantages of VARTM are poor surface finish on the bagging side, long time involved in material preparation, poor dimensional tolerances and lack of automation. The co-injection process, as shown in Figure 2.30, can use RTM or VARTM process where two different immiscible resins are injected and co-cured to form a composite containing different resins. CRTM is a process which is quite attractive for automobile industry due to its potential for high volume production. Unlike RTM, VARTM or SCRIMP, the resin flow in CRTM compression molding exhibits three distinct stages [86]. In the first stage, the

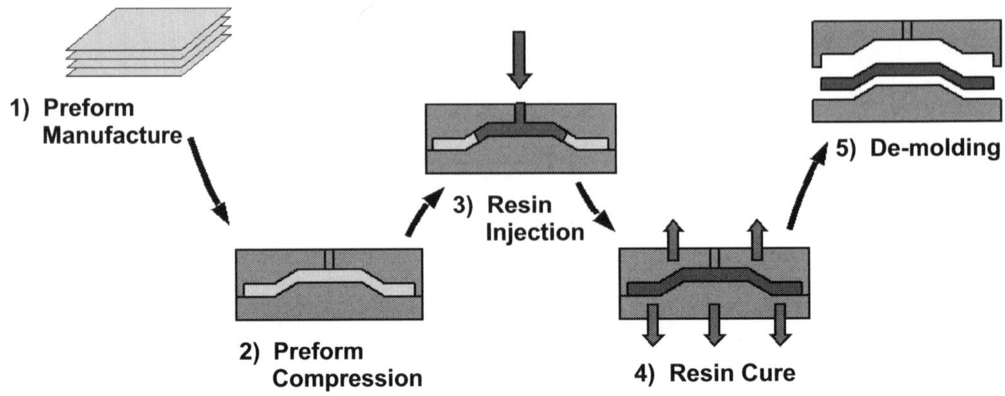

Figure 2.27: Schematic of RTM process steps [81].

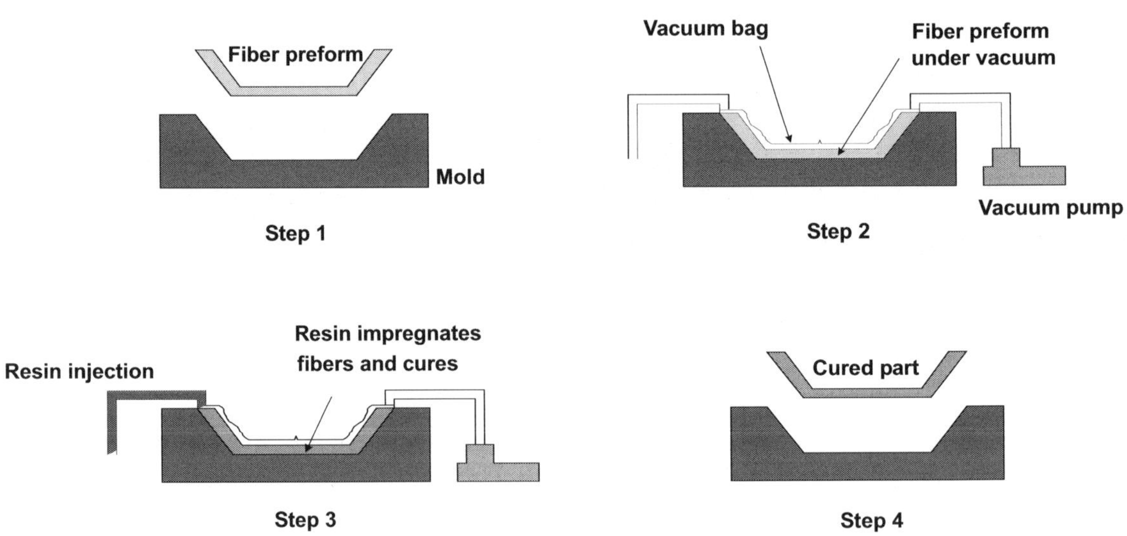

Figure 2.28: Schematic of VARTM process steps [82].

Figure 2.29: Bridge section [83], railway carriage [84], and wind blade by REpower Systems AG [85].

resin is injected into the gap between the mold platen and the fiber preform in the mold. In the second stage, the gap is closed by moving the mold platen closer to the preform until it touches the preform. In the third and final stage, the preform undergoes compaction by the mold platen which displaces resin from filled domain to impregnate the rest of the preform. These are shown in detail in Figures 2.31 and 2.32 [86]. All of the phases can be modeled as a flow through porous media as the flow is still creeping flow due to high viscosity and very small gaps. The first stage resembles injection in SCRIMP, with the flow enhancement layer, also known as the distribution media layer, replaced by the permeable gap where as the final stage resembles traditional compression molding but due to non-uniform distribution of resin at the beginning of this stage the usual compression molding models are not directly applicable.

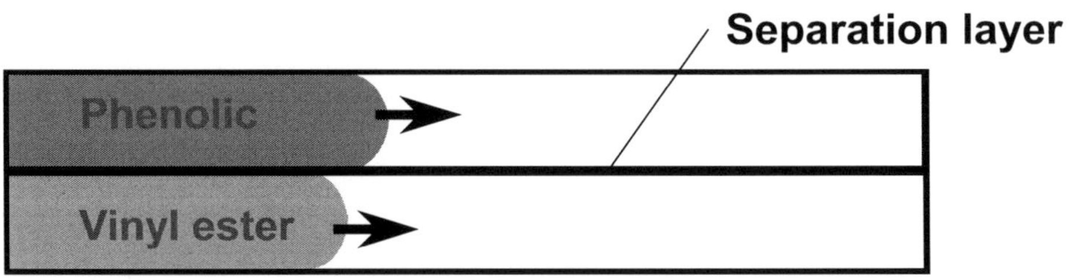

Figure 2.30: Co-injection process which can be RTM or VARTM where two different immiscible resins are injected into a fiber preform and co-cured to form a composite [87].

Transport Issues

Several unresolved issues in LCM encountered by composite engineers are mainly in the areas of process automation, preforming, tooling, mold flow analysis, and resin chemistry. During the last decade, there have been rapid advances in LCM technology that demonstrate the potential of the LCM processes for producing advanced composite parts.

In the last decade, one of the major issues faced by manufacturing engineers using these processes was how and from what locations they should inject the resin into the fiber network. The goal is to produce a void-free part with uniform distribution of the resin in between the spaces of the fibrous network. Of course, this would be a function of the fibrous network, of the part geometry and the fiber volume fraction needed to be achieved, and also of the maximum pressure available to accomplish the task.

To manufacture parts, practice had converged on trial and error methods in which the modus operandi was that as certain areas were seen to be resin starved, the injection location was moved closer to those areas. One of the difficulties was that the resin movement and impregnation could not be seen inside a closed mold, so the only way to check if the filling was successful was to wait until the part had cured. Hence, the part had to be rejected if there were big unfilled areas, and a new location for the gate was chosen based on conjecture about the impregnation process. This trial and error procedure was repeated for every new prototype to be attempted by liquid molding. Many net shape composite parts that were good candidates for this process were too challenging to attempt by trial and error methods as the possibilities to fill the empty spaces between the fibers were numerous. The manufacturing engineer realized the potential of mathematical models and numerical simulations.

Hence the transport issues to be addressed in this process include complete impregnation

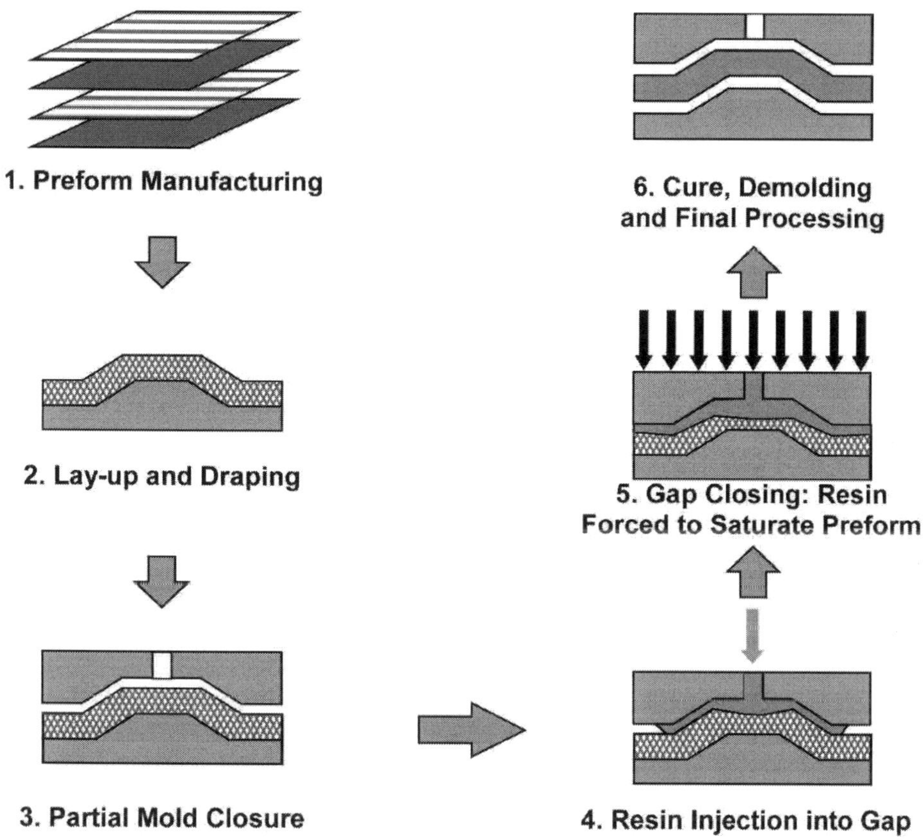

Figure 2.31: Schematic of the CRTM process [86].

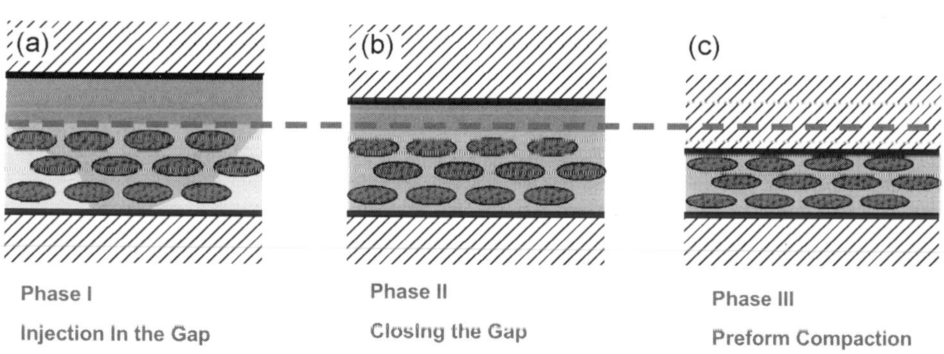

Figure 2.32: Resin injection in the CRTM process [86].

of the resin inside the fiber preform. The fiber preform may be highly anisotropic and heterogeneous and may have more than one scale of pore sizes. Heating of the resin by the mold to invoke resin cure makes the process nonisothermal, and one must account for heat conduction between the mold, fibers and the resin. The flowing resin changes the heat transfer picture as the heat is also convected due to the movement of the resin. The viscosity of the resin will change due to the heating of the resin and initiation of the cure, which will influence the flow dynamics. Hence, the flow equation may be coupled with the heat transfer equation and the cure kinetics. Flow and heat transfer through porous media forms the basis of modeling such processes, where one has to consider empirical relations such as Darcy's law and heat dispersion coefficients to explain the distribution of resin and heat in such processes, discussed in detail in Chapter 8.

2.5.3 Filament Winding

Process

The filament winding process is usually used for the manufacture of cylindrical and axisymmetric hollow composite parts. In this process, either wet filaments or preimpregnated tapes are laid on a rotating mandrel. Schematics of a filament winding machine and the process line are shown in Figures 2.33 and 2.34, respectively. The fibers and parts made from filament winding are shown in Figure 2.35.

The winding may be accomplished either by depositing preimpregnated prepregs on the mandrel (dry winding) or with the fibers being impregnated by passing them through a resin bath (wet winding). The fiber tows or prepregs are placed on the mandrel under a fiber tension with a prescribed speed guided by a crosshead. If the resin is a thermoplastic resin, heat is applied to the tape simultaneously with the winding (usually automated tape placement is used for thermoplastics). In the case of thermosets, the mandrel is wound and may be cured in an autoclave or on-line. During this process, the physical phenomena that occur are the curing of the resin and dissemination of the heat generated to the ambient. There is also fiber slippage due to the presence of the uncured resin between the fibers while the fibers are still under tension. Moreover, voids may form due to air trapped between the bands of fibers. The mandrel and the composite expand due to changes in temperature, which leads to the development of thermal stresses and strains, adding to the stresses developed due to the fiber tension.

Issues

The process variables that can be selected and controlled independently are the winding speed, fiber tension, and external temperature or heating rates. Hence, the process model should provide information regarding the mandrel temperature and the temperatures inside the composite, degree of cure, viscosity, fiber positions, stresses and strains, porosity, and winding and curing times. The transport of the resin through a moving fiber bed and also the cure kinetics reaction that changes the viscosity of the resin requires one to address the flow, energy and reaction kinetics.

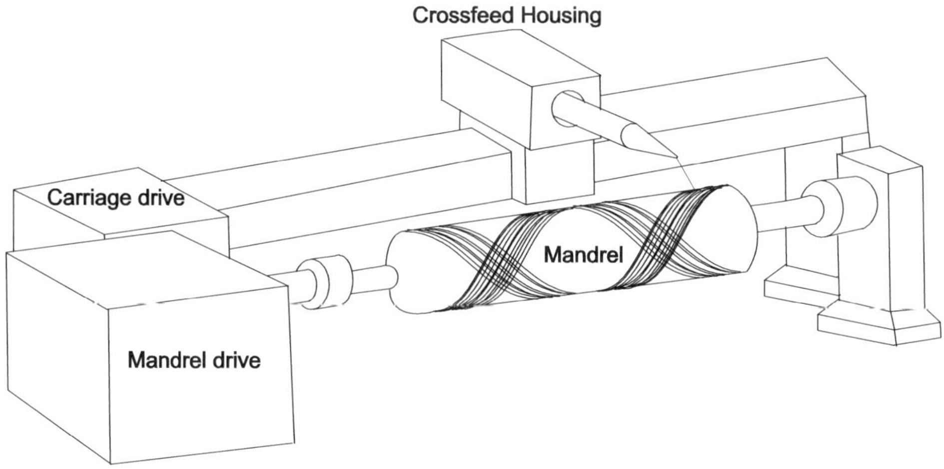

Figure 2.33: Schematic of a filament winding machine [88].

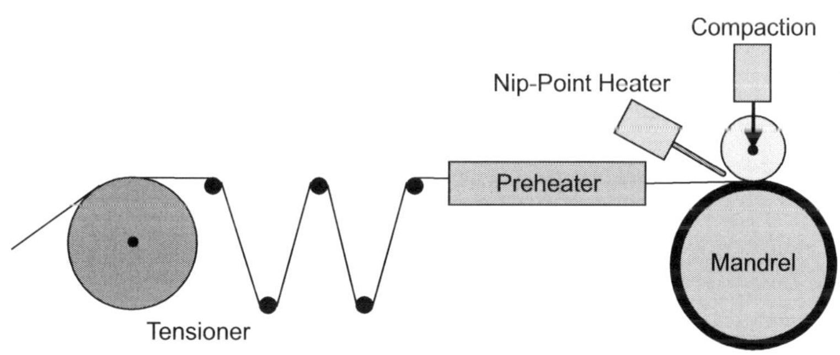

Figure 2.34: Schematic of the filament winding process [88].

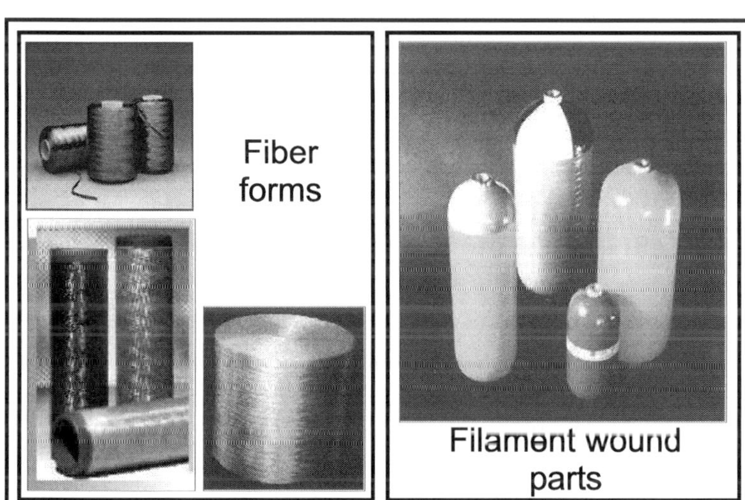

Figure 2.35: Fiber forms and filament wound parts [89].

2.6 Exercises

2.6.1 Questions

1. What is the main difference between metal and composite parts in terms of the properties of finished parts and the raw materials that are used to manufacture them?

2. Why were composite materials used mainly in some selected industries, such as the aerospace industry, during the early stages of composite material development?

3. What are the two major classes of composite manufacturing processes in terms of types of molds being used? Give examples.

4. Describe the injection molding process by using all of these words and terms: high-volume production, thermoplastic resin, solid pellets, short fibers, fillers, feed, hopper, barrel, screw, melt, mix, functions as a piston, sprues, runners, solidification, and eject.

5. What is a pellet? What are the ingredients? What is the size of a typical pellet?

6. Why is the length of fibers that solidify in the final part limited in the traditional injection molding process? What is an alternative approach to overcome this? What is the advantage of using longer fibers in this new approach?

7. While modeling the transport phenomena in the injection molding process, what physical conservation laws are used?

8. What is "fountain flow" in injection molding? What process and material parameters determine its significance? What is the result of this flow?

9. What are "skin" and "core" layers in injection molding?

10. Does the length of the fibers change during the flow of suspension in injection molding? Explain why.

11. What makes injection molding so popular that nearly 20% of the goods are manufactured by this process? What are the advantages and disadvantages of this process?

12. Give some examples of common products that are injection molded.

13. Describe the extrusion process by using all of these words and terms: hopper, polymer pellets, barrel, screw, die, plasticize, compound, cross-sections, puller, sizer, and cutter or coiler.

14. How does the screw of an extrusion machine work?

15. What are the similarities and differences between the extrusion and injection molding processes?

16. Give some examples of common products that are manufactured with the extrusion process.

17. In the last few years, the extrusion process has been modified to allow extrusion of polymers containing reinforcement. What is the reason for this modification?

18. What are the two phenomena that help to soften and then melt the solid polymer pellet in an extrusion machine before it is pushed by the screw? Which one creates more heat?

19. When is a cooling system needed in an extrusion machine? If we turn off the heaters, do we still need a cooling system? Why?

20. What is "plasticating?"

21. What is "die swelling?" What causes it?

22. If you are asked to design an extrusion machine, how would you calculate the power needed to run the screw and the pumping rate? What types of suspension flow would you solve in your model?

23. Describe the compression molding process by using these words and terms: composite material, mold cavity, pressure, and deform.

24. Compare the injection molding and compression molding processes in terms of (i) ease in material handling, (ii) automation, (iii) amount of material deformation, (iv) regions of very high stress in material, (v) need for gates in the mold, (vi) damage of fibers, (vii) using longer fibers, (viii) achieving higher fiber volume fractions, (ix) cycle time, (x) repeatability, (xi) dimensional tolerances, (xii) amount of initial investment, (xiii) storage time and conditions for material, (xiv) difficulty of mold design, (xv) residual stresses in the parts, (xvi) delamination, and (xvii) warpage.

25. What are the most commonly used materials for the compression molding process? What are the different versions of them?

26. Describe the procedure of forming the sheets of SMC material.

27. What is "initial charge" in the compression molding process? Why is it crucial to properly place it inside the mold?

28. In compression molding, what are the typical values for temperature and pressure within the SMC material during heating and compression? What is a typical cycle time?

29. Although it is small, why is the flow in compression molding critical for determining the physical and mechanical properties of the composite?

30. Describe "post-cure" in compression molding. Where does it take place? Why is it needed?

31. Why is the mold design very important, and why is the overall cost of the molds usually high in compression molding?

32. What are the most commonly used resin systems in SMC materials for the automotive and aerospace industries?

33. What is "in-mold coating?" Which industry usually uses it?

34. Why are the temperature distribution and rate of cooling important in the compression molding process?

35. The tensile strength and elastic modulus of compression molded parts might exhibit significant variations from one molding to another. What are the two main reasons?

36. What are the three advanced thermoplastic manufacturing methods? What is the common assumption used to model them?

37. What are the main issues in advanced thermoplastic manufacturing methods?

38. Describe the sheet forming process by using all of these words and terms: thermoplastic matrix, short and long fibers, unidirectional or multi-axial sheet preform, stacked or preconsolidated, forming temperature, mechanical or hydraulic press, form, curvilinear tool surface, and cooling.

39. What is "forming temperature" in the sheet forming process?

40. How is incremental reforming of a composite part performed in the sheet forming process until the final desired geometry of the part is attained?

41. What are the unique advantages of sheet forming over other thermoplastic processing methods?

42. What are the three major composite sheet forming processing methods?

43. In the stretch forming process, the major issues are wrinkles and folds in the final parts. What causes them? What type of mathematical model would help the manufacturer to overcome this hurdle?

44. Describe the pultrusion process by using these words and terms: drag, fibers, thermoset or thermoplastic matrix, supply rack, temperature control, and tool.

45. What are the main differences between thermoset and thermoplastic pultrusion?

46. The die assembly for the pultrusion process has two distinct sections: (i) the heated and tapered entry region, and (ii) the "land" region. What are the functions of these sections?

47. Where and how does a fiber nest occur in the pultrusion process?

48. Describe the thermoplastic tape lay-up process by using these words and terms: thermoplastic tape, continuous fibers, tow placement head, substrate, preheating, rollers, consolidation, conform, and local or global melting.

49. What are the major issues in the thermoplastic tape lay-up process?

50. Can one single mathematical model be applicable for both thermoplastic and thermoset advanced composite manufacturing methods? From the modeling viewpoint, what are the major differences between these two methods?

51. Describe the autoclave process.

52. What are the typical temperatures and pressures in the autoclave process to consolidate and solidify the material? Why is it important to have a proper processing window of temperatures and pressures for thermoset materials?

53. In the autoclave process, what are the functions of peel plies, release fabric, bleeder material, breather material, and vacuum bag?

54. During the curing step in the autoclave process, how is the final quality of the composite part affected by the magnitude and duration of the temperatures and pressures to which the composite is subjected? Explain in terms of the process parameters and variables used in the mathematical models.

55. What are the advantages and disadvantages of autoclave processing compared to other methods?

56. Describe the liquid composite molding process by using these words and terms: bed of fibrous network (preform), resin, impregnation, cure, and demolding.

57. What do RTM, VARTM, S-RIM, and CIRTM stand for?

58. What are the advantages and disadvantages of VARTM over RTM?

59. What are the critical issues in the liquid composite molding process?

60. Describe the filament winding process

61. What are the two types of filament winding processes? What is the difference between them?

62. What are the major issues for filament winding?

63. What is the major difference between VARTM and RTM processes in terms of mold and injection system used?

64. You have been chosen to select a composite manufacturing process due to your familiarity with the processes as a result of the course you took at the University of Delaware. Your company is looking at making the following six components and would like you to recommend which process should be considered with a single sentence explanation as to why you selected that process. 1.) Short fiber reinforced dashboards for the new Acura car. 2.) Telephone poles for the city of Newark. 3.) I-beams for Ford Passenger Vans. 4.) Axi-symmetric casing for the rocket motor. 5.) Recycleable door panels for the Mercedes Benz. 6.) A composite spring for a helicopter.

65. Which of the cross-sections in Figure 2.36 cannot be filament wound?

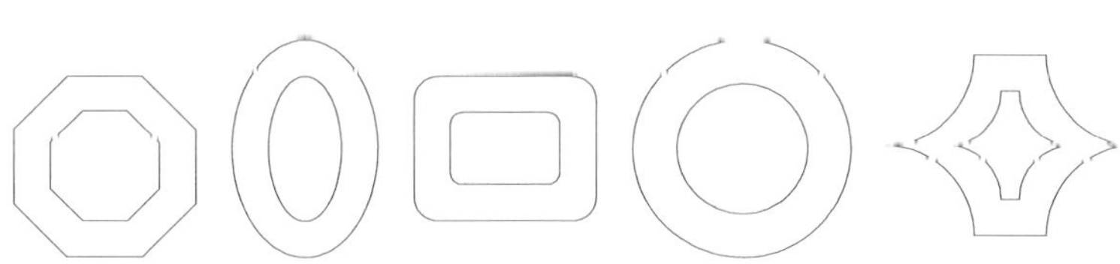

Figure 2.36: Cross-sections for filament winding [89].

2.6.2 Fill in the Blanks

1. In order for composites to be widely used, especially by consumer goods industries such as automotive and sporting goods, two major goals had to be achieved: (i) the ___________ of raw materials had to be reduced, and (ii) manufacturing methods had to be developed to achieve high ___________ production by reducing the manufacturing ___________ time.

2. As composites are heterogeneous materials, there is simultaneous transfer of ___________, ___________ and ___________ at ___________, ___________ and ___________ scales, often along with ___________ reaction, in a multi-phase system with time dependent material properties and boundary conditions.

3. Composites manufacturing processes can be broadly grouped into three categories. (i) short fiber ___________ manufacturing methods: involve the transport of ___________ or ___________ fibers and ___________ as a ___________ into a ___________ or through a ___________. (ii) ___________ flow or advanced thermoplastic composites manufacturing methods: involve deformation of ___________ or long aligned ___________ fibers preimpregnated with ___________ resin under applied stress. (iii) ___________ media or advanced thermoset composite manufacturing methods: involve impregnating ___________ and nearly stationary fiber networks with resin.

4. As the making of metal molds can be very expensive, one can justify the use only for high-volume production parts. Recently, researchers are exploring the use of ___________ molds reinforced with ___________ powder for small-volume production or for prototype development.

5. The molding time in injection molding is usually of the order of a few ___________.

6. In the injection molding process, the important process parameters that can be controlled on the injection units are the ___________ temperature, ___________ and ___________ speeds, ___________ pressure, and in some instances the ___________ temperature.

7. In the injection molding process, the material and geometric parameters that will influence the manufacturing process and the final properties of the part are the ___________ rheology, the ___________ type and content, mold cavity ___________ and ___________, the locations of _______________ and _______________ on the mold.

8. In the injection molding process, the filled thermoplastic pellets usually contain a second, discontinuous, usually more rigid phase blended into the polymer. When the aspect ratio (ratio of largest to smallest dimension) of the second component is around one, it is referred to as a _______________. If the aspect ratio is one to two orders of magnitude larger, then it is called _______________.

9. In the injection molding process, the most commonly used reinforcements are ___________, ___________ and short ___________ usually less than ___________ in length.

10. In the injection molding process, the parts usually have a fiber volume fraction between ___________ and ___________ %. Filled or reinforced materials provide much different properties than the base resin. For example, reinforced polypropylene provides higher ___________ and lower ___________ characteristics than neat polypropylene. In practice, fibrous reinforcements used with glass fibers dominate the market although the ___________ and ___________ fibers provide higher stiffness and strength but are seldom used due to the high ___________ of raw materials.

11. The traditional injection molding process limits the length of fibers that solidify in the final part since the high ____________ rates in the barrel and the passage of fibers through narrow gates and openings in the mold cause significant fiber ____________. Usually, the fiber diameter is of the order of few ____________, and the final length distribution, irrespective of the starting fiber length, is of the order of ____________ to ____________ μm. The starting length of these fibers in the log-like pellets is usually of the order of ____________ to ____________ mm. As a result, new methods to produce pellets containing longer fibers were developed in which the fibers were pultruded and stayed bundled together and were not dispersed in the pellet by the action of compounding. These pellets produced final parts that retained a higher percentage of longer fibers and consequently showed a significant increase in ____________ and ____________ toughness.

12. In the injection molding process, the ________________________ material selected also plays a role in the final physical and optical properties.

13. The issues that relate to transport phenomena in the injection molding process are the ____________ of fiber suspensions as they occupy the closed mold, the ____________ of the fibers during the flow, fiber ____________ distribution, fiber ____________, and the ____________ transfer that changes the microstructure of the resin.

14. One can account for the mass balance of the suspension in the injection molding process which can be treated as a ____________ material, at least for the short fiber materials.

15. The conservation of the fiber orientation field in the injection molding process simply states that if the orientation of the fibers disappears in one direction, it should ____________ in some ____________ direction.

16. One can account for the momentum balance of the suspension in the injection molding process to describe the ____________ and the ____________ fields during the flow process. This requires one to describe the constitutive equation between the ____________ applied and the ____________ rate experienced by the material. As the thermoplastic melts are ____________ thinning, the viscosity is known to ____________ with ____________ rate. Addition of fibers can change the stress strain rate behavior and even make it ____________.

17. The energy conservation model in the injection molding process allows one to describe the ____________ history of the melt in the channel between the screw and the barrel, where it gets its ____________ input from the ____________ on the barrel and due to viscous ____________ caused by the ____________ of the suspension. It also allows one to keep track of the ____________ history in the closed mold as the suspension enters into it, which plays a crucial role in the resulting ____________.

18. In the injection molding process, there are some microscale phenomena occurring simultaneously: molecular ____________ and spherulitic ____________ of polymers during solidification and fiber ____________ due to shearing action in the screw that results in a ____________ distribution. For long fiber suspensions, the constitutive equation may change, and issues such as fiber ____________ may also need to be addressed.

19. The coupling between the transports of ____________ and ____________ creates a ____________ flow mechanism in injection molding. As the walls are ____________ than the core, the suspension viscosity is ____________ near the walls as compared to the core. Hence, under the same pressure, the suspension in the core moves ____________ of the suspension near the walls, spreading from the ____________ ____________ like a ____________.

20. In the injection molding process, the shear flow near the mold walls aligns the fibers in the direction of the ____________ and is called the ____________. Below this layer, the suspension continues to experience shear and fibers orient along the shear lines; this layer is known as the ____________ layer. Finally, the core has fibers that are influenced by the bulk deformation of the flow in the mold which usually has an elongated component, causing the material to ____________ in and out of the paper direction aligning the fibers.

21. Lack of control of fiber ____________ and ____________ causes variation in part properties from one injection to another in injection molding.

22. The screw of an extrusion machine is usually machined from a solid steel rod and fits within the barrel with less than a ____________ clearance. To pump a suspension through a die, the screw is designed to generate over ____________ to ____________ atmospheres of pressure in the suspension.

23. For a simple and crude explanation of an extrusion machine, one may think of a bolt as the ____________, the nut as the ____________, and the wrench as the ____________. If one turns the bolt and holds the wrench in place, the nut will move forward.

24. ________________________ (SMC), the most commonly used material for ________________________ process, may involve either (i) compounding a resin, combining it with ____________ and ____________, or (ii) prepregs (impregnating a fiber ____________ with a resin).

25. SMC material is prepared as follows: ____________ is placed on a nonporous nylon sheet. ____________ are added to it. A cover sheet is applied onto the fibers. This sheet, enclosed between the nylon sheet and the cover sheet, is then passed through several compaction ____________. These mix the resin and fibers together, and also ____________ the sheets. The resin cures slowly, and it takes approximately ____________ for the SMC sheet to be ready to be used in the compression molding process.

26. Several types of SMC are currently used in industry: SMC-R (reinforced with fibers oriented ____________), SMC-C (reinforced with ____________ unidirectional fibers), SMC-C/R (reinforced with both ____________ oriented and ____________ unidirectional fibers), SMC-D (reinforced with ____________ but discontinuous fibers).

27. It is possible to use both thermoplastics and thermosets in SMC, but the majority of SMC is done using ____________.

28. The final properties of composite parts are influenced by the ____________ and ____________ of the initial charge in the mold.

29. In the compression molding process, after SMC material is placed and the mold is closed, the heated top and bottom platens containing the two halves of the mold cavity are brought together. This generates ____________ and ____________ to initiate the ____________ of material.

30. In ____________________, after the part is partially cured inside the mold, the mold is opened slightly and a ____________ is injected in the mold. Subsequently, the mold is ____________ again, causing the resin to coat the outside of the part filling any ____________ on it. This greatly improves the ________________________ on the part and can save several stages in the painting process.

31. Due to their high viscosity, it is extremely difficult to process thermoplastic resins if they have to be forced to occupy the empty spaces between fibers to form an advanced thermoplastic composite. To circumvent this problem, the resin is ____________ into the fiber bundles, or is sprayed as commingled ____________ or resin fibers are ____________ with the reinforcing fibers so that the resin does not have to travel far when ____________ and ____________ are applied to occupy the empty spaces between the fibers.

32. In sheet forming of nonreinforced thermoplastic sheets, the sheets are usually held in place along the edges over a tool surface and brought up to their material softening temperature. The most common forming methods are ____________________ where the sheet is between matching dies, and ____________________ in which a vacuum is pulled through small holes in the tool face, pulling and spreading the sheet down over the surface.

33. In composite sheet stamping or matched-die press forming, the composite blank is heated to the ____________ temperature and then stamped against tool surfaces. A variation on this is ____________ tool stamping wherein one or both sides of the die are made compliant. This helps maintain an even consolidation ____________ across the part in case of any tool ____________.

34. In diaphragm forming, the blank is held between two disposable, plastically deformable diaphragms of either ____________ or ____________. During the forming cycle, the diaphragm edges are clamped, heated along with the blank and deformed through the use of ____________ pressure to the tool surface. The diaphragms serve to hold the blank in tension and prevent fiber ____________ that can occur under compressive stresses. When forming parts containing continuous fiber reinforcement, the diaphragms are clamped but the blank cannot be. This is due to the ____________ of the fiber reinforcement.

35. Hydroforming is a process similar to diaphragm forming. However, ____________________ is used instead of ____________ to provide the pressure behind a permanent rubber diaphragm.

36. Pultrusion is one of the oldest composite manufacturing processes as it has been around since the late 1940s. It was originally designed to manufacture ________________.

37. For thermoset pultrusion, one can use bare fiber rovings to go through a liquid thermoset ____________ before entering the ____________ for curing and cross-linking. On the other hand, thermoplastic resins are preimpregnated with ____________ fibers in the form of a tape or the fibers pass through a station of polymer bed, where the thermoplastic powder attaches itself to the charged fiber surface. Therefore, thermoplastic pultrusion requires a ____________ area.

38. There are two important aspects to modelling the pultrusion process. The first is to determine the ____________________ required to run the operation at a reasonable speed to produce parts that are free of ____________ and contain the desired fiber volume fraction. The second is to achieve the desired level of ____________ in the matrix with minimal ____________ concentration. Hence, the viscous flow physics and the heat transfer during the process will play an important role in the determination of these key parameters such as the ____________ pressure for the die, ____________ temperatures, ____________ temperatures, etc.

39. The key issues for modeling the thermoplastic tape lay-up process are: (i) tow-placement head ____________, (ii) ____________, (iii) ____________, (iv) ____________ during melting of the resin in the incoming tape and the substrate interface without ____________ the matrix. A good bonding between the substrate and the incoming tape requires the interface temperature to be ____________ than the ____________ temperature of the thermoplastic. The temperature gradient through the thickness is responsible for ________________________ development in the composite.

40. The winding process variables that can be selected and controlled independently are the (i) winding ____________, (ii) fiber ____________, and (iii) either external ____________ or ____________ rates. Hence, the process model should provide information regarding the temperatures of ____________ and ____________, degree of ____________, ____________, fiber ____________, ____________ and ____________, ____________, and winding and curing ____________. The transport of the resin here through a moving fiber bed and also the cure kinetics reaction that changes the viscosity of the resin requires one to couple the equations of ____________, ____________ and ____________.

Chapter 3

Transport Equations for Composite Processing

3.1 Introduction to Process Models

A *model* is an idealized mathematical representation of a physical system or process. To describe a model, one needs physical laws, constitutive equations and boundary conditions once the system to be modeled is defined and outlined. In this chapter, we will derive the physical laws of fluid flow and heat transfer that are encountered in composite manufacturing processes. Models are useful in expressing the understanding and codifying the knowledge about the manufacturing operation. Models provide detailed information about a process such as flow front location, resin pressure, temperature and flow rate. A model is like a scientific hypothesis which should be validated with experiments. If experiments conflict with the hypothesis, we continue to revise the model until it agrees with the experiments. The models also help in revising and designing the experiments so we can extract the critical information from well-posed experiments. Models help us to either eliminate or reduce the trial and error approach used during composite processing. A scientist is more interested in understanding the physical world by using models, whereas a process engineer is usually interested in manipulating the model.

A model is formulated using the following six elements:

1. *Model or system boundary:* Region in which one should consider physical and constitutive laws.
2. *Physical laws:* In this chapter, we will derive, and then use the conservation equations for mass, momentum and energy as the physical laws.
3. *Constitutive laws:* Deformation of materials, transfer of heat, resin cure chemistry, etc., are expressed through phenomenological laws that are formed based on certain assumptions and experimental observations in simple geometries.
4. *Boundary conditions:* The external influences that affect the system or process are expressed by formulating boundary conditions for the governing variables.
5. *Assumptions:* In order to simplify the models, assumptions need to be made. This will allow one to neglect some of the terms in the physical laws. The assumptions may be about constitutive equations or the geometry of the system and boundary conditions to simplify them.

6. *Experimental validation:* For a model that is expected to describe the physics, it is important to conduct controlled experiments and measure relevant variables and compare them with the model predictions.

The philosophy behind a good model should be "prepare for the worst and hope for the best." The assumptions made must still maintain intact the behavior and features of the system we are trying to model. Otherwise, the model may not properly represent the physical system.

The general approach to building a model is illustrated in Figure 3.1.

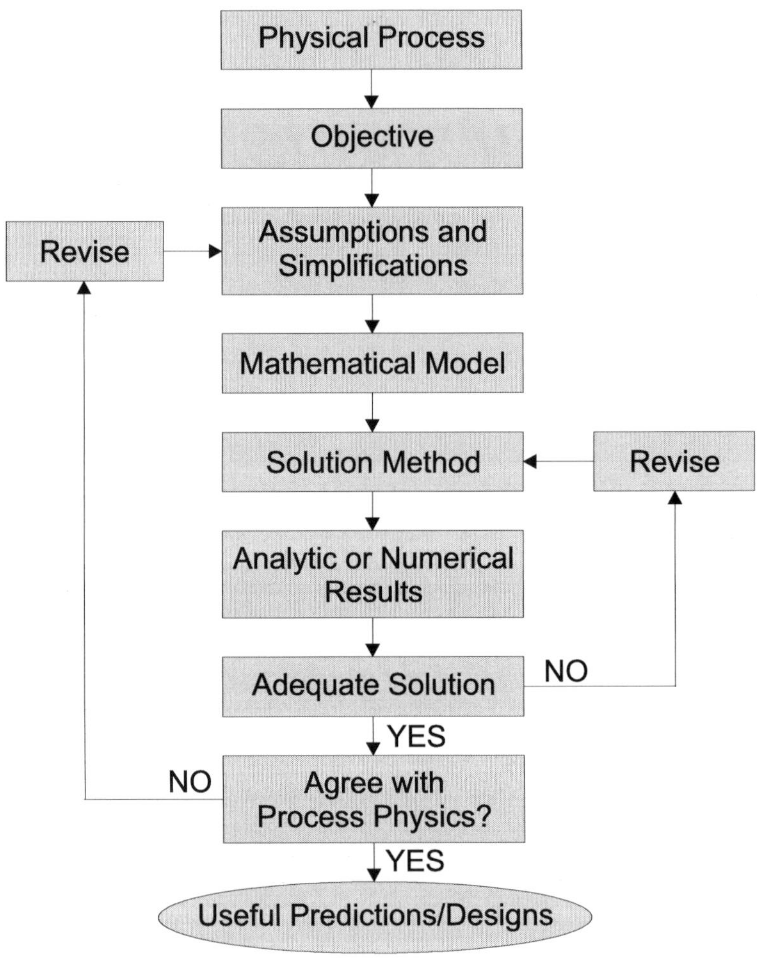

Figure 3.1: Flowchart to build a model.

3.2 Conservation of Mass (Continuity Equation)

In this subsection, we will derive the differential equation for the physical law of *conservation of mass* which is also known as the *continuity equation* for any fluid flow. Then, we will modify this equation to address resin flow in composite processing applications to include the presence of fibers.

3.2.1 Conservation of Mass

We will derive the mass conservation equation in two different ways: (i) using Gauss' Divergence Theorem on any arbitrary shaped control volume, and (ii) a pseudo derivation with a prism control volume.

Derivation 1: Divergence Theorem on Any Arbitrary Shaped Control Volume

Let's consider flow of a fluid within a region with velocity $\mathbf{U}$ and density ρ. Although the derivation is independent of the coordinate system used, let's use the Cartesian coordinate system here, and let $\mathbf{U}$ and ρ be functions of x, y, z and time t hence we can represent them as $\mathbf{U}(x, y, z, t)$ and $\rho(x, y, z, t)$, respectively. The mass of the fluid within any arbitrary fixed control volume $\mathcal{V}$ at any time t is calculated as $M = \int_{\mathcal{V}} \rho \, dV$. The rate of increase of M is calculated as

$$\frac{dM}{dt} = \frac{d}{dt}\int_{\mathcal{V}} \rho(x, y, z, t)\, dV = \int_{\mathcal{V}} \frac{\partial \rho}{\partial t}\, dV. \tag{3.1}$$

Here, the total time derivative term on the left-hand side of Equation (3.1) is carried inside the integral, by using the Leibniz rule,[1] as a partial time derivative. The additional terms from the Leibniz rule dropped out since the control volume is fixed in space.

The rate at which fluid mass enters the control volume $\mathcal{V}$ through its boundary $\mathcal{S}$ is the flux integral and is given by

$$q = -\int_{\mathcal{S}} \rho \, \mathbf{n} \cdot \mathbf{U} \, dA, \tag{3.2}$$

where $\mathbf{n}$ is the unit outward normal to $\mathcal{S}$ as shown in Figure 3.2.

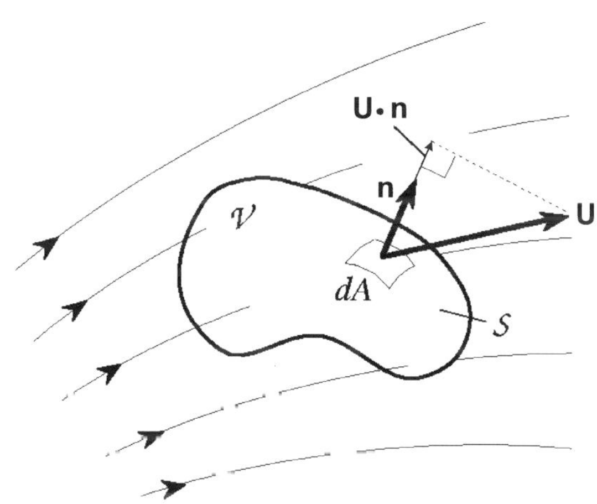

Figure 3.2: Control volume for the derivation of conservation of mass equation.

[1]The *Leibniz rule* states that

$$\frac{d}{dt}\int_{A(t)}^{B(t)} f(x,t)\,dx = \int_{A(t)}^{B(t)} \frac{\partial}{\partial t} f(x,t)\,dx + \frac{dB(t)}{dt} f(B(t),t) - \frac{dA(t)}{dt} f(A(t),t)$$

for any continuous function f and its time derivative $\partial f/\partial t$.

In Chapter 8, we will study the micro mold filling issues which may lead to microvoids. We will model the delayed saturation of the inter-fiber bundles after the flow front passes, as having imaginary sinks within the reinforcing preform. If there is a sink within $\mathcal{V}$ that absorbs fluid mass at a rate of $s(x, y, z, t)$ mass per unit volume per unit time, then

$$\text{Net rate of mass loss in } \mathcal{V} \;=\; \int_{\mathcal{V}} s(x, y, z, t)\, dV \tag{3.3}$$

mass per unit time.

The rate of increase of mass within $\mathcal{V}$ is equal to the rate at which mass enters $\mathcal{V}$ through $\mathcal{S}$, minus the rate at which mass is lost:

$$\int_{\mathcal{V}} \frac{\partial \rho}{\partial t}\, dV = -\int_{\mathcal{S}} \rho\, \mathbf{n} \cdot \mathbf{U}\, dA - \int_{\mathcal{V}} s\, dV. \tag{3.4}$$

In order to combine these integrals, we need to convert the surface integral to a volume integral so that all the integrals will be volume integrals. For this purpose, Gauss' divergence theorem will be used. This theorem states that

$$\int_{\mathcal{S}} \mathbf{n} \cdot (a\mathbf{B})\, dA = \int_{\mathcal{V}} \nabla \cdot (a\mathbf{B})\, dV \tag{3.5}$$

for any scalar a and vector $\mathbf{B}$ within a volume $\mathcal{V}$ bounded by surface $\mathcal{S}$. $\mathbf{n}$ is the unit outward normal vector on $\mathcal{S}$. By using Gauss' divergence theorem, the first term on the right-hand side of Equation (3.4) could be rewritten as

$$\int_{\mathcal{S}} \rho\, \mathbf{n} \cdot \mathbf{U}\, dA = \int_{\mathcal{S}} \mathbf{n} \cdot (\rho \mathbf{U})\, dA = \int_{\mathcal{V}} \nabla \cdot (\rho \mathbf{U})\, dV \tag{3.6}$$

where the differential vector operator ∇ in Cartesian coordinates is given by

$$\nabla = \mathbf{i}\frac{\partial()}{\partial x} + \mathbf{j}\frac{\partial()}{\partial y} + \mathbf{k}\frac{\partial()}{\partial z}.$$

Equation (3.4) takes the following form after Equation (3.6) is substituted into it:

$$\int_{\mathcal{V}} \left[\frac{\partial \rho}{\partial t} + \nabla \cdot (\rho \mathbf{U}) + s \right] dV = 0. \tag{3.7}$$

Since the considered control volume $\mathcal{V}$ is assumed to be of any arbitrary shape, and at any arbitrary location of the flow region, not only the integral, but actually the integrand, $\partial\rho/\partial t + \nabla \cdot (\rho \mathbf{U}) + s$ must be zero, too, then

$$\frac{\partial \rho}{\partial t} + \nabla \cdot (\rho \mathbf{U}) = -s. \tag{3.8}$$

This is the *conservation of mass* (or, also known as continuity equation) of fluid mechanics. In a Cartesian coordinate system, it takes the following form by expanding $\nabla \cdot (\rho \mathbf{U})$

$$\frac{\partial \rho}{\partial t} + \frac{\partial}{\partial x}(\rho u_x) + \frac{\partial}{\partial y}(\rho u_y) + \frac{\partial}{\partial z}(\rho u_z) = -s. \tag{3.9}$$

This equation is valid at every location for any fluid flow domain. Here u_x, u_y and u_z are scalar components of the velocity vector $\mathbf{U}$ in the x, y and z directions, respectively. For most resins, the density is constant and if there is no mass loss, Equation (3.9) reduces to

$$\frac{\partial u_x}{\partial x} + \frac{\partial u_y}{\partial y} + \frac{\partial u_z}{\partial z} = 0. \tag{3.10}$$

Derivation 2: Pseudo derivation with a Control Volume

This alternative derivation requires additional steps, but is simpler to understand and follow than the first one since it doesn't require knowledge of the Leibniz Rule and Gauss' Divergence Theorem. This derivation also makes it easier for the reader to understand the physical meaning of the individual terms in the conservation of mass equation.

Consider a fluid flow within a region with velocity $\mathbf{U}$ and density ρ. Let us again use the Cartesian coordinate system here, and let $\mathbf{U}$ and ρ be functions of x, y, z and time t represented as $\mathbf{U}(x, y, z, t)$ and $\rho(x, y, z, t)$, respectively. However, this time, the control volume, $\mathcal{V}$ has sides of Δx, Δy and Δz as shown in Figure 3.3. The mass of the fluid within this control volume at any time t is equal to $M = \int_{\mathcal{V}} \rho dV = \rho \Delta x \Delta y \Delta z$. You might object to this calculation since we treated the density as a constant. Although we let density be $\rho(x, y, z, t)$, consider that ρ appearing in $M = \rho \Delta x \Delta y \Delta z$ is the volume averaged density which is within the lower and upper bounds of ρ within $\mathcal{V}$. The balance of rate of mass increase (accumulation) inside $\Delta x \Delta y \Delta z$ is given as

$$\left(\begin{array}{c}\text{Rate of}\\ \text{mass increase}\end{array}\right) = \left(\begin{array}{c}\text{Rate of}\\ \text{mass inflow}\end{array}\right) - \left(\begin{array}{c}\text{Rate of}\\ \text{mass outflow}\end{array}\right) - \left(\begin{array}{c}\text{Rate of mass lost}\\ \text{due to a sink}\end{array}\right). \tag{3.11}$$

The rate of increase of M is calculated as

$$\frac{\partial M}{\partial t} = \frac{\partial}{\partial t}(\rho \Delta x \Delta y \Delta z) = \Delta x \Delta y \Delta z \frac{\partial \rho}{\partial t} \tag{3.12}$$

since the control volume and hence its sides Δx, Δy and Δz are fixed, i.e., not changing with time. The rate of fluid mass inflow into the control volume $\mathcal{V}$ through its face at x is $(\rho u_x)|_x \Delta y \Delta z$. The rate of fluid mass outflow from the control volume $\mathcal{V}$ through its face at $x + \Delta x$ is $(\rho u_x)|_{x+\Delta x} \Delta y \Delta z$. Similar expressions can be written through the faces at y and z for inflows, and at $y + \Delta y$ and $z + \Delta z$ for outflows as shown in Figure 3.3. Hence, the net rate of flow is

$$\begin{aligned}\left(\begin{array}{c}\text{Rate of}\\ \text{mass inflow}\end{array}\right) - \left(\begin{array}{c}\text{Rate of}\\ \text{mass outflow}\end{array}\right) &= \Delta y \Delta z \left[(\rho u_x)|_x - (\rho u_x)|_{x+\Delta x}\right] \\ &\quad + \Delta x \Delta z \left[(\rho u_y)|_y - (\rho u_y)|_{y+\Delta y}\right] \\ &\quad + \Delta x \Delta y \left[(\rho u_z)|_z - (\rho u_z)|_{z+\Delta z}\right].\end{aligned} \tag{3.13}$$

If there is a sink within $\mathcal{V}$ which absorbs fluid mass at a rate of $s(x, y, z, t)$ mass per unit volume per unit time, then the net rate of lost mass within $\mathcal{V}$ is equal to

$$\int_{\mathcal{V}} s(x, y, z, t)\, dV = s \Delta x \Delta y \Delta z \tag{3.14}$$

mass per unit time. On the right-hand side of Equation (3.14), s is volume averaged. Substituting Equations (3.12)-(3.14) into (3.11) results in

$$\begin{aligned}\frac{\partial \rho}{\partial t} \Delta x \Delta y \Delta z &= \Delta y \Delta z \left[(\rho u_x)|_x - (\rho u_x)|_{x+\Delta x}\right] \\ &\quad + \Delta x \Delta z \left[(\rho u_y)|_y - (\rho u_y)|_{y+\Delta y}\right] \\ &\quad + \Delta x \Delta y \left[(\rho u_z)|_z - (\rho u_z)|_{z+\Delta z}\right] \\ &\quad - s \Delta x \Delta y \Delta z.\end{aligned} \tag{3.15}$$

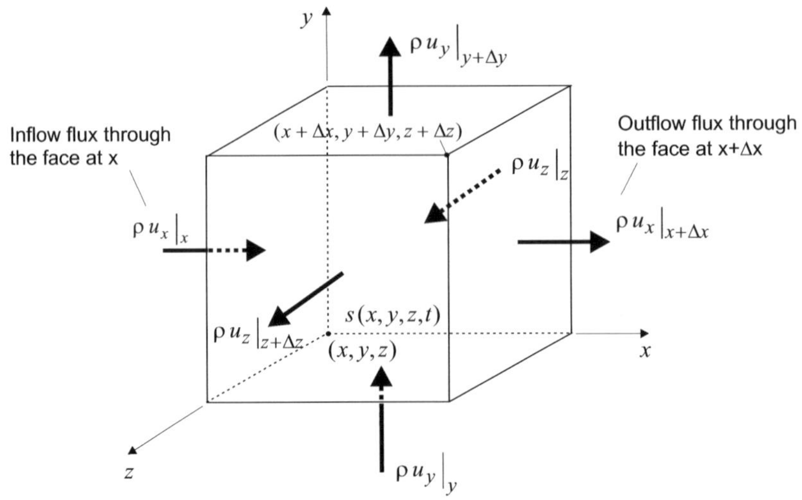

Figure 3.3: Control volume for the pseudo derivation of conservation of mass equation in Cartesian coordinate reference frame.

Dividing both sides of Equation (3.15) by the control volume $\Delta x \Delta y \Delta z$ leads to

$$\frac{\partial \rho}{\partial t} = \frac{(\rho u_x)\,|_x - (\rho u_x)\,|_{x+\Delta x}}{\Delta x} + \frac{(\rho u_y)\,|_y - (\rho u_y)\,|_{y+\Delta y}}{\Delta y} + \frac{(\rho u_z)\,|_z - (\rho u_z)\,|_{z+\Delta z}}{\Delta z} - s. \quad (3.16)$$

As Δx, Δy and $\Delta z \to 0$, Equation (3.16) converges to

$$\frac{\partial \rho}{\partial t} = -\frac{\partial}{\partial x}(\rho u_x) - \frac{\partial}{\partial y}(\rho u_y) - \frac{\partial}{\partial z}(\rho u_z) - s. \quad (3.17)$$

This is identical to Equation (3.10) in Derivation 1. Or, using the differential vector operator, ∇, we can rewrite it as

$$\frac{\partial \rho}{\partial t} + \nabla \cdot (\rho \mathbf{U}) + s = 0 \quad (3.18)$$

which is the same as Equation (3.8). Here, $\partial \rho / \partial t$ is the partial time derivative of ρ at a fixed spatial location. Alternatively, the continuity equation can be written in terms of the substantial time derivative $D\rho/Dt$ following a fluid particle. Since

$$\begin{aligned} \frac{D()}{Dt} &= \frac{\partial()}{\partial t} + \frac{\partial()}{\partial x}\frac{dx}{dt} + \frac{\partial()}{\partial y}\frac{dy}{dt} + \frac{\partial()}{\partial z}\frac{dz}{dt} \\ &= \frac{\partial()}{\partial t} + \frac{\partial()}{\partial x}u_x + \frac{\partial()}{\partial y}u_y + \frac{\partial()}{\partial z}u_z \\ &= \frac{\partial()}{\partial t} + \mathbf{U} \cdot \nabla() \end{aligned} \quad (3.19)$$

then, $\partial \rho / \partial t = D\rho/Dt - \mathbf{U} \cdot \nabla \rho$. Substituting this into Equation (3.18), and rewriting $\nabla \cdot (\rho \mathbf{U})$ as $\nabla \rho \cdot \mathbf{U} + \rho \nabla \cdot \mathbf{U}$, we obtain the following

$$\frac{D\rho}{Dt} + \rho \nabla \cdot \mathbf{U} + s = 0. \quad (3.20)$$

Most fluid flows in composite manufacturing processes will be treated as quasi-steady state ($\partial \rho/\partial t = 0$), and incompressible ($\rho =$ constant). Hence, the continuity equation will simplify to $\rho \nabla \cdot \mathbf{U} = -s$. In many applications, there won't be any significant sink effects, so the continuity equation could be further simplified to

$$\nabla \cdot \mathbf{U} = 0 \tag{3.21}$$

which can be written in open form as

$$\frac{\partial u_x}{\partial x} + \frac{\partial u_y}{\partial y} + \frac{\partial u_z}{\partial z} = 0, \tag{3.22}$$

$$\frac{1}{r}\frac{\partial (r\, u_r)}{\partial r} + \frac{1}{r}\frac{\partial u_\theta}{\partial \theta} + \frac{\partial u_z}{\partial z} = 0, \tag{3.23}$$

$$\frac{1}{r^2}\frac{\partial (r^2\, u_r)}{\partial r} + \frac{1}{r \sin\theta}\frac{\partial (u_\theta \sin\theta)}{\partial \theta} + \frac{1}{r\sin\theta}\frac{\partial u_\phi}{\partial \phi} = 0 \tag{3.24}$$

in xyz Cartesian, $r\theta z$ cylindrical, and $r\theta\phi$ spherical coordinate systems, respectively.

3.2.2 Mass Conservation for Resin with Presence of Fibers

Here, we will derive the equation of mass conservation for resin flow with the presence of fiber phase. We will assume the following:

- Fibers are incompressible.
- The velocity and stress-strain changes are small in the control volume.
- The process is quasi-static.
- Body forces such as self-weight are negligible.

Consider a fluid (resin) flow within a region consisting of fibers as shown in Figure 3.4. The derivation is very similar to the one in the previous subsection, except that we will use $\rho_b =$ (mass of resin)/(control volume containing both resin and fibers) as the resin density, and $\mathbf{U} = u_x\mathbf{i} + u_y\mathbf{j} + u_z\mathbf{k}$ as the interstitial velocity of the resin within the composite. Inside a partially filled fibrous region, $\rho_b = \epsilon S \rho$, where ϵ is the porosity of the control volume or that fraction of the control volume available for resin to occupy ($\epsilon =$ porous volume/control volume), S is the saturation fraction of the porous space filled by resin, and ρ is the actual resin density (resin mass divided by resin volume). In many other textbooks and in some chapters of this book, the porosity will be written as $\epsilon = 1 - V_f$, where V_f is the fiber volume fraction. The porous volume within the control volume is completely empty when $S = 0$, and completely filled when $S = 1$.

The mass balance for a control volume can be expressed as

$$\begin{aligned}
\frac{\partial \rho_b}{\partial t}\Delta x \Delta y \Delta z &= \Delta y \Delta z \left[(\rho_b u_x)|_x - (\rho_b u_x)|_{x+\Delta x}\right] \\
&+ \Delta x \Delta z \left[(\rho_b u_y)|_y - (\rho_b u_y)|_{y+\Delta y}\right] \\
&+ \Delta x \Delta y \left[(\rho_b u_z)|_z - (\rho_b u_z)|_{z+\Delta z}\right] \\
&- s \Delta x \Delta y \Delta z.
\end{aligned} \tag{3.25}$$

Dividing both sides of Equation (3.25) by $\Delta x \Delta y \Delta z$ and then taking the limit as Δx, Δy and $\Delta z \to 0$, Equation (3.25) converges to

$$\frac{\partial \rho_b}{\partial t} = -\frac{\partial}{\partial x}(\rho_b u_x) - \frac{\partial}{\partial y}(\rho_b u_y) - \frac{\partial}{\partial z}(\rho_b u_z) - s. \tag{3.26}$$

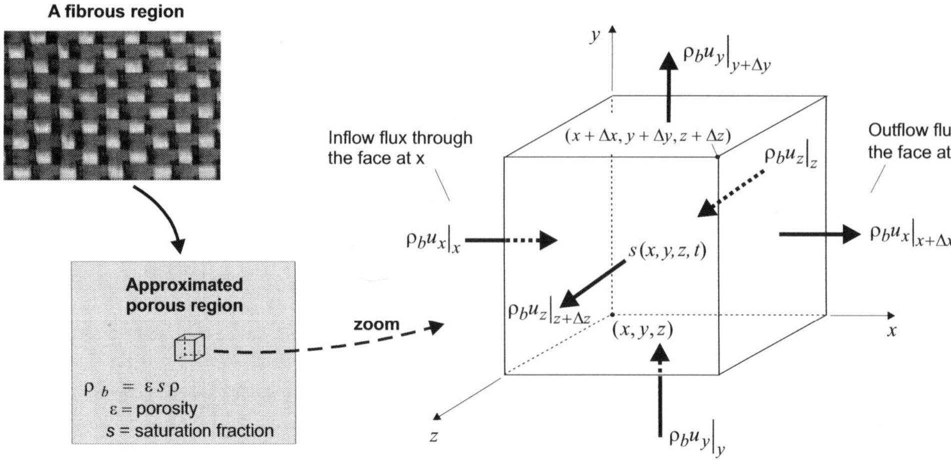

Figure 3.4: Control volume for the pseudo derivation of conservation of mass equation for resin flow through a fibrous region.

Since $\rho_b = \epsilon S \rho$,

$$\frac{\partial(\epsilon S\rho)}{\partial t} = -u_x \frac{\partial(\epsilon S\rho)}{\partial x} - u_y \frac{\partial(\epsilon S\rho)}{\partial y} - u_z \frac{\partial(\epsilon S\rho)}{\partial z} - \epsilon S\rho \frac{\partial u_x}{\partial x} - \epsilon S\rho \frac{\partial u_y}{\partial y} - \epsilon S\rho \frac{\partial u_z}{\partial z} - s \quad (3.27)$$

which can be written in vectorial form as

$$\frac{\partial(\epsilon S\rho)}{\partial t} + \mathbf{U} \cdot \nabla(\epsilon S\rho) + (\epsilon S\rho)\nabla \cdot \mathbf{U} = s. \quad (3.28)$$

Note that even if the resin density ρ is constant, the porosity and/or the saturation of fibrous media may not be constant within the entire region. Hence, $\nabla(\epsilon S\rho)$ may not be zero everywhere. The reader must keep this fact in mind. Further simplification can be obtained depending upon the manufacturing process being considered [90].

3.3 Conservation of Momentum (Equation of Motion)

As we have done in Derivation 2 of the previous section, we will consider a control volume as shown in Figure 3.5. In the figure, only the forces in x direction induced by stresses are shown. Now, let's take a diversion to understand what a stress on a fluid is. Stress is a measure of force per unit area transmitted by physical contact either within a body (of a continuum system, which is a fluid here) or by contact with the body at its external surface. $\underline{\underline{\sigma}}$ is the stress tensor whose components are σ_{ij}. Each component σ_{ij} has two directions associated with it. The second index j gives the direction of the force per unit area, and the first index i gives the normal direction of the surface. The surface might be an external one, or a hypothetical interior one within the fluid domain. The diagonal stress components, when $i = j$ are called *normal stresses*, and the off-diagonal components when $i \neq j$ are called *shear stresses*.

Usually, the stress tensor is expressed as the sum of two components

$$[\text{stress tensor}] \quad = \quad -\ \text{pressure}\,[\text{unit tensor}] + [\text{viscous stress tensor}]$$

$$\begin{aligned}\sigma_{ij} &= -P\delta_{ij} + \tau_{ij} \\ \begin{bmatrix} \sigma_{xx} & \sigma_{yx} & \sigma_{zx} \\ \sigma_{xy} & \sigma_{yy} & \sigma_{zy} \\ \sigma_{xz} & \sigma_{yz} & \sigma_{zz} \end{bmatrix} &= -P \begin{bmatrix} 1 & 0 & 0 \\ 0 & 1 & 0 \\ 0 & 0 & 1 \end{bmatrix} + \begin{bmatrix} \tau_{xx} & \tau_{yx} & \tau_{zx} \\ \tau_{xy} & \tau_{yy} & \tau_{zy} \\ \tau_{xz} & \tau_{yz} & \tau_{zz} \end{bmatrix}. \end{aligned} \tag{3.29}$$

The motion of an incompressible substance only determines its stress state to within an arbitrary isotropic constant (a scalar multiple of δ_{ij}), and that is the negative pressure, $-P$. A positive pressure generates a negative normal stress. The viscous stress arises from the fluid motion, and is related to the deformation of the fluid by constitutive equations as we will study in the next section. If there is no fluid motion (hydrostatic case), then all viscous stresses τ_{ij} are zero since there is no fluid deformation. In this case, all shear stresses σ_{ij} are zero, and normal stresses σ_{ii} are equal to $-P$.

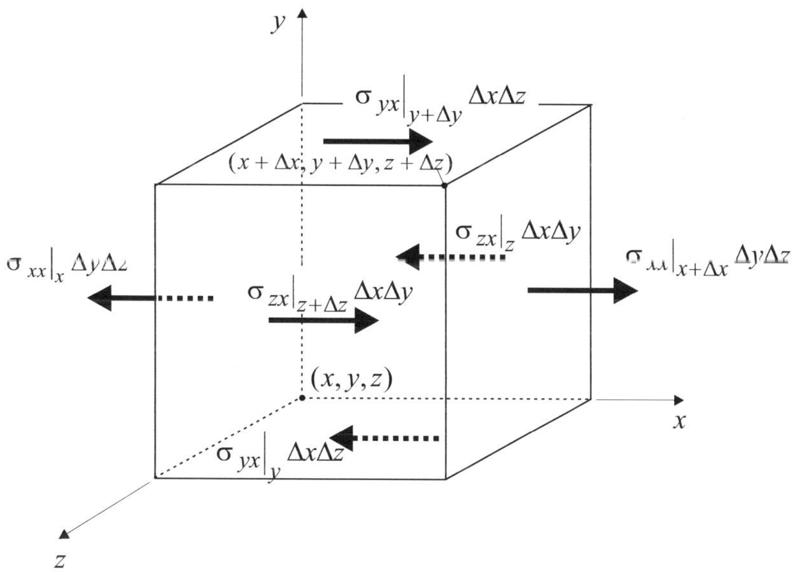

Figure 3.5: Control volume for derivation of conservation of momentum equation for fluid flow.

Let's consider the forces, in the x direction only, acting on the control volume shown in Figure 3.5. Once we have the force balance in the x direction, the procedure can be repeated in the y and z directions as well. The force balance on the control volume yields

$$\begin{aligned}\mathbf{F}_x = m\frac{Du_x}{Dt} &= \begin{pmatrix} x \text{ component of} \\ \text{total stress forces} \\ \text{on all six surfaces} \end{pmatrix} + \begin{pmatrix} x \text{ component of} \\ \text{body forces, } \mathbf{F}_B \text{ due to} \\ \text{gravity, magnetism, etc.} \end{pmatrix} \\ \rho\Delta x\Delta y\Delta z\frac{Du_x}{Dt} &= \sigma_{xx}\Big|_{x+\Delta x}\Delta y\Delta z - \sigma_{xx}\Big|_{x}\Delta y\Delta z \\ &+\sigma_{yx}\Big|_{y+\Delta y}\Delta x\Delta z - \sigma_{yx}\Big|_{y}\Delta x\Delta z \\ &+\sigma_{zx}\Big|_{z+\Delta z}\Delta x\Delta y - \sigma_{zx}\Big|_{z}\Delta x\Delta y \\ &+F_{Bx}\Delta x\Delta y\Delta z \end{aligned} \tag{3.30}$$

Here, Du_x/Dt is the substantial derivative of the velocity component u_x and given by $Du_x/Dt = \partial u_x/\partial t + \mathbf{U}\cdot\nabla u_x$, and $\mathbf{F}_B$ is the body force which could be the total of gravitational, magnetic, etc. forces per unit volume acting on the fluid. For example, if there is only gravitational force acting on, then $\mathbf{F}_B = \rho g\hat{\mathbf{e}}$ where, ρ is the fluid density, g is the gravitational acceleration, and $\hat{\mathbf{e}}$ is the unit vector along the direction of gravitational force which is dependent on the coordinate system (e.g., if z axis is the radial axis drawn from the center of the earth, then $\hat{\mathbf{e}} = -\hat{\mathbf{k}}$, $F_{Bx} = F_{By} = 0$, and $F_{Bz} = -\rho g$).

By dividing both sides of Equation (3.30) by $\Delta x\Delta y\Delta z$, we obtain the following:

$$\rho\frac{Du_x}{Dt} = -\frac{\partial P}{\partial x} + \frac{\partial\tau_{xx}}{\partial x} + \frac{\partial\tau_{yx}}{\partial y} + \frac{\partial\tau_{zx}}{\partial z} + F_{Bx} \tag{3.31}$$

as Δx, Δy and $\Delta z \to 0$ in the limit. Similarly, one can repeat the force balance in the y and z directions as well, yielding

$$\rho\frac{Du_y}{Dt} = -\frac{\partial P}{\partial y} + \frac{\partial\tau_{xy}}{\partial x} + \frac{\partial\tau_{yy}}{\partial y} + \frac{\partial\tau_{zy}}{\partial z} + F_{By} \tag{3.32}$$

$$\rho\frac{Du_z}{Dt} = -\frac{\partial P}{\partial z} + \frac{\partial\tau_{xz}}{\partial x} + \frac{\partial\tau_{yz}}{\partial y} + \frac{\partial\tau_{zz}}{\partial z} + F_{Bz}. \tag{3.33}$$

One can rewrite the momentum equation with substantial derivative terms in expanded form as follows

$$\rho\left(\frac{\partial u_x}{\partial t} + u_x\frac{\partial u_x}{\partial x} + u_y\frac{\partial u_x}{\partial y} + u_z\frac{\partial u_x}{\partial z}\right) = -\frac{\partial P}{\partial x} + \frac{\partial\tau_{xx}}{\partial x} + \frac{\partial\tau_{yx}}{\partial y} + \frac{\partial\tau_{zx}}{\partial z} + F_{Bx} \tag{3.34}$$

$$\rho\left(\frac{\partial u_y}{\partial t} + u_x\frac{\partial u_y}{\partial x} + u_y\frac{\partial u_y}{\partial y} + u_z\frac{\partial u_y}{\partial z}\right) = -\frac{\partial P}{\partial y} + \frac{\partial\tau_{xy}}{\partial x} + \frac{\partial\tau_{yy}}{\partial y} + \frac{\partial\tau_{zy}}{\partial z} + F_{By} \tag{3.35}$$

$$\rho\left(\frac{\partial u_z}{\partial t} + u_x\frac{\partial u_z}{\partial x} + u_y\frac{\partial u_z}{\partial y} + u_z\frac{\partial u_z}{\partial z}\right) = -\frac{\partial P}{\partial z} + \frac{\partial\tau_{xz}}{\partial x} + \frac{\partial\tau_{yz}}{\partial y} + \frac{\partial\tau_{zz}}{\partial z} + F_{Bz}. \tag{3.36}$$

We can combine Equations (3.31)-(3.33) in vectorial form. The physical meaning of each term is explained below the equation

$$\underbrace{\rho\frac{D\mathbf{U}}{Dt}}_{\text{Inertia force}} = \underbrace{-\nabla P}_{\text{Hydrodynamic force}} + \underbrace{\nabla\cdot\tau}_{\text{Force due to stresses}} + \underbrace{\mathbf{F}_B.}_{\text{Body force}} \tag{3.37}$$

This is called the conservation of momentum equation, also known as the equation of motion. These equations along with the continuity equation describe the physical laws for the selected isothermal system. Considering gravitational force as the only body force (i.e., $\mathbf{F} = \rho\mathbf{g}$), the equation of motion can be written in Cartesian, cylindrical and spherical coordinates (see Figure 3.6) as listed in Tables 3.1, 3.2 and 3.3, respectively. However, one does need a relationship between stresses and the deformation of the fluid before one can solve for either pressure or velocities experienced by the fluid. The equations that describe these are discussed in the next section.

Table 3.1: The Equation of Motion in Terms of τ in Rectangular Coordinates in x, y and z Directions

$$(x\text{-dir.:})\quad \rho\left[\frac{\partial v_x}{\partial t}+v_x\frac{\partial v_x}{\partial x}+v_y\frac{\partial v_x}{\partial y}+v_z\frac{\partial v_x}{\partial z}\right]=-\frac{\partial P}{\partial x}+\left[\frac{\partial \tau_{xx}}{\partial x}+\frac{\partial \tau_{yx}}{\partial y}+\frac{\partial \tau_{zx}}{\partial z}\right]+\rho g_x$$

$$(y\text{-dir.:})\quad \rho\left[\frac{\partial v_y}{\partial t}+v_x\frac{\partial v_y}{\partial x}+v_y\frac{\partial v_y}{\partial y}+v_z\frac{\partial v_y}{\partial z}\right]=-\frac{\partial P}{\partial y}+\left[\frac{\partial \tau_{xy}}{\partial x}+\frac{\partial \tau_{yy}}{\partial y}+\frac{\partial \tau_{zy}}{\partial z}\right]+\rho g_y$$

$$(z\text{-dir.:})\quad \rho\left[\frac{\partial v_z}{\partial t}+v_x\frac{\partial v_z}{\partial x}+v_y\frac{\partial v_z}{\partial y}+v_z\frac{\partial v_z}{\partial z}\right]=-\frac{\partial P}{\partial z}+\left[\frac{\partial \tau_{xz}}{\partial x}+\frac{\partial \tau_{yz}}{\partial y}+\frac{\partial \tau_{zz}}{\partial z}\right]+\rho g_z$$

Table 3.2: The Equation of Motion in Terms of τ in Cylindrical Coordinates in r, θ and z Directions

$$(r\text{-dir.:})\quad \rho\left[\frac{\partial v_r}{\partial t}+v_r\frac{\partial v_r}{\partial r}+\frac{v_\theta}{r}\frac{\partial v_r}{\partial \theta}-\frac{v_\theta^2}{r}+v_z\frac{\partial v_r}{\partial z}\right]$$

$$=-\frac{\partial P}{\partial r}+\left[\frac{1}{r}\frac{\partial}{\partial r}(r\tau_{rr})+\frac{1}{r}\frac{\partial \tau_{\theta r}}{\partial \theta}+\frac{\partial \tau_{zr}}{\partial z}-\frac{\tau_{\theta\theta}}{r}\right]+\rho g_r$$

$$(\theta\text{-dir.:})\quad \rho\left[\frac{\partial v_\theta}{\partial t}+v_r\frac{\partial v_\theta}{\partial r}+\frac{v_\theta}{r}\frac{\partial v_\theta}{\partial \theta}+\frac{v_r v_\theta}{r}+v_z\frac{\partial v_\theta}{\partial z}\right]$$

$$=-\frac{1}{r}\frac{\partial P}{\partial \theta}+\left[\frac{1}{r^2}\frac{\partial}{\partial r}(r^2\tau_{r\theta})+\frac{1}{r}\frac{\partial \tau_{\theta\theta}}{\partial \theta}+\frac{\partial \tau_{z\theta}}{\partial z}+\frac{\tau_{\theta r}-\tau_{r\theta}}{r}\right]+\rho g_\theta$$

$$(z\text{-dir.:})\quad \rho\left[\frac{\partial v_z}{\partial t}+v_r\frac{\partial v_z}{\partial r}+\frac{v_\theta}{r}\frac{\partial v_z}{\partial \theta}+v_z\frac{\partial v_z}{\partial z}\right]$$

$$=-\frac{\partial P}{\partial z}+\left[\frac{1}{r}\frac{\partial}{\partial r}(r\tau_{rz})+\frac{1}{r}\frac{\partial \tau_{\theta z}}{\partial \theta}+\frac{\partial \tau_{zz}}{\partial z}\right]+\rho g_z$$

Table 3.3: The Equation of Motion in Terms of τ in Spherical Coordinates in r, θ and ϕ Directions

(r-dir.:)

$$\rho\left[\frac{\partial v_r}{\partial t}+v_r\frac{\partial v_r}{\partial r}+\frac{v_\theta}{r}\frac{\partial v_r}{\partial \theta}+\frac{v_\phi}{r\sin\theta}\frac{\partial v_r}{\partial \phi}-\frac{v_\theta^2+v_\phi^2}{r}\right]$$

$$=-\frac{\partial P}{\partial r}+\left[\frac{1}{r^2}\frac{\partial}{\partial r}(r^2\tau_{rr})+\frac{1}{r\sin\theta}\frac{\partial}{\partial\theta}(\tau_{\theta r}\sin\theta)+\frac{1}{r\sin\theta}\frac{\partial\tau_{\phi r}}{\partial\phi}-\frac{\tau_{\theta\theta}+\tau_{\phi\phi}}{r}\right]+\rho g_r$$

(θ-dir.:)

$$\rho\left[\frac{\partial v_\theta}{\partial t}+v_r\frac{\partial v_\theta}{\partial r}+\frac{v_\theta}{r}\frac{\partial v_\theta}{\partial \theta}+\frac{v_\phi}{r\sin\theta}\frac{\partial v_\theta}{\partial \phi}+\frac{v_r v_\theta}{r}-\frac{v_\phi^2\cot\theta}{r}\right]$$

$$=-\frac{1}{r}\frac{\partial P}{\partial \theta}+\left[\frac{1}{r^3}\frac{\partial}{\partial r}(r^3\tau_{r\theta})+\frac{1}{r\sin\theta}\frac{\partial}{\partial\theta}(\tau_{\theta\theta}\sin\theta)\right.$$

$$\left.+\frac{1}{r\sin\theta}\frac{\partial\tau_{\phi\theta}}{\partial\phi}+\frac{(\tau_{\theta r}-\tau_{r\theta})-\tau_{\phi\phi}\cot\theta}{r}\right]+\rho g_\theta$$

(ϕ-dir.:)

$$\rho\left[\frac{\partial v_\phi}{\partial t}+v_r\frac{\partial v_\phi}{\partial r}+\frac{v_\theta}{r}\frac{\partial v_\phi}{\partial \theta}+\frac{v_\phi}{r\sin\theta}\frac{\partial v_\phi}{\partial \phi}+\frac{v_\phi v_r}{r}+\frac{v_\theta v_\phi}{r}\cot\theta\right]$$

$$=-\frac{1}{r\sin\theta}\frac{\partial P}{\partial \phi}+\left[\frac{1}{r^3}\frac{\partial}{\partial r}(r^3\tau_{r\phi})+\frac{1}{r\sin\theta}\frac{\partial}{\partial\theta}(\tau_{\theta\phi}\sin\theta)\right.$$

$$\left.+\frac{1}{r\sin\theta}\frac{\partial\tau_{\phi\phi}}{\partial\phi}+\frac{(\tau_{\phi r}-\tau_{r\phi})+\tau_{\phi\theta}\cot\theta}{r}\right]+\rho g_\phi$$

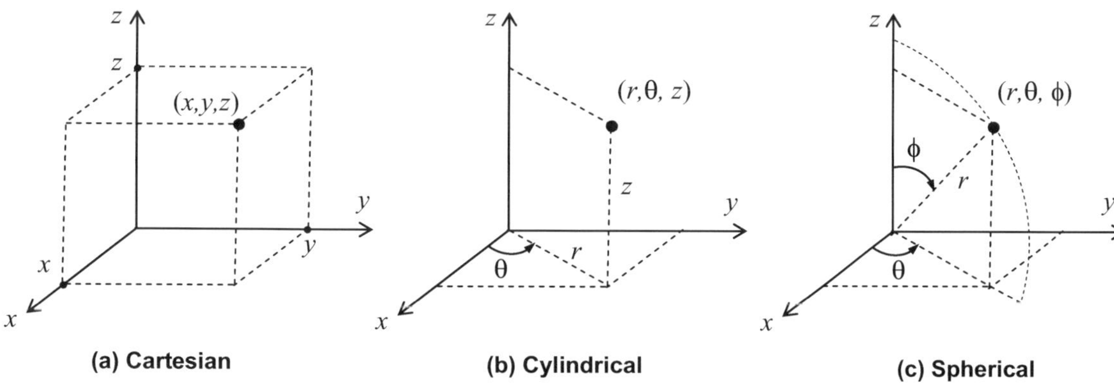

Figure 3.6: Cartesian, cylindrical and spherical coordinate systems.

3.4 Stress-Strain Rate Relationship

In the previous sections, the conservation equations of mass and momentum were derived. The unknown variables appearing in these equations are the velocity vector $\mathbf{U}$, the fluid pressure P, and the viscous stress tensor τ. Pressure P is a scalar, $\mathbf{U}$ has three scalar components, and viscous stress tensor τ has nine scalar components; however, only six of them are independent scalar components if it is a symmetric tensor, as it will be assumed in this book. A symmetric stress tensor is assumed in order to satisfy zero net angular momentum of control volumes. Thus, we have ten unknown scalar variables. In the Cartesian coordinate system, they are $u_x, u_y, u_z, P, \tau_{xx}, \tau_{yy}, \tau_{zz}, \tau_{xy}, \tau_{xz}, \tau_{yz}$. These variables can be functions of spatial independent variables x, y, z and time t. However, we have only four scalar equations, one continuity and three equations of motion. We need six more equations. In this section, we will study the relation between the stress tensor and deformation of fluid. They will provide six additional scalar equations to complete the set of independent equations.

3.4.1 Kinematics of Fluid

Consider an infinitesimal fluid element at an instant of flow (say at t) as shown in Figure 3.7. Note that a 2D element (with dimensions of dx and dy in x and y directions, respectively) is shown in the figure for simplicity, however a 3D formulation will be given later. As the fluid particles flow, the element moves and may take a different shape and size at the next instant (at $t+dt$). To formulate the motion of the element, one may superpose different components of the fluid motion. The motion of a continuum can include "translation," "rotation" and "deformation." Deformation may be due to "angular distortion" and "volume distortion" as discussed in Sabersky et al. [91]. These different components are schematically illustrated in Figure 3.8.

Translation:

The velocity vector $\mathbf{U}$ itself describes the translation of the continuum. Let's write the velocity vector in Cartesian coordinates, xyz as $\mathbf{U} = u_x\mathbf{i} + u_y\mathbf{j} + u_z\mathbf{k}$, or as $\mathbf{U} = u_i\mathbf{i}$ in short tensorial notation where i can take values of x, y, z, and $\mathbf{i}$ will correspond to unit vectors $\mathbf{i}$, $\mathbf{j}$, or $\mathbf{k}$, respectively. As seen in Figure 3.8(a), the coordinate system attached to the fluid element translates by $u_x\,dt$ and $u_y\,dt$ in x and y directions, respectively (and $u_z\,dt$ in z direction for a 3D flow) as time increases from t to $t + dt$.

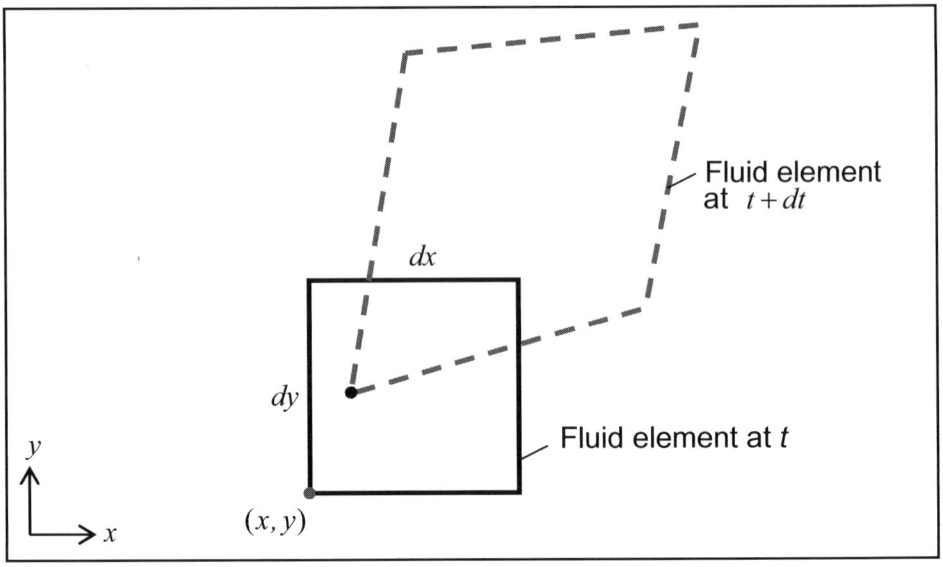

Figure 3.7: Motion of a fluid element.

Rotation:

Unlike the translation, the rotation and deformation of a continuum are dependent on the spatial derivatives of velocity as seen in Figure 3.8(b, c and d).

The bottom edge rotates an angle of $\frac{\partial u_y}{\partial x}dt$ in CCW (counterclockwise) direction in a time interval of dt. Similarly, the left edge rotates an angle of $-\frac{\partial u_x}{\partial y}dt$ in CCW direction in a time interval of dt. The rotation of the diagonal may be taken as the average of these two edges' rotations in xy plane (i.e., about z axis):

$$\text{Angular velocity about } z \text{ axis} = \Omega_z = \frac{1}{2}\left(\frac{\partial u_y}{\partial x} - \frac{\partial u_x}{\partial y}\right). \tag{3.38}$$

Similarly, the rotations in x and y axes are given by

$$\text{Angular velocity about } x \text{ axis} = \Omega_x = \frac{1}{2}\left(\frac{\partial u_z}{\partial y} - \frac{\partial u_y}{\partial z}\right) \tag{3.39}$$

$$\text{Angular velocity about } y \text{ axis} = \Omega_y = \frac{1}{2}\left(\frac{\partial u_x}{\partial z} - \frac{\partial u_z}{\partial x}\right). \tag{3.40}$$

Angular Distortion:

In the rotational motion formulation, we did not look at the details of the derivation; however, we will do so now. To investigate what terms contribute to the angular distortion, the reader is suggested to determine the six unknown distances shown in Figure 3.9. The answers are given in Figure 3.10.

The bottom edge rotates an angle of $d\alpha$ in CCW direction in a time interval of dt, and it is given by

$$d\alpha = \tan^{-1}\left(\frac{\frac{\partial u_y}{\partial x}dx\,dt}{dx}\right) = \tan^{-1}\left(\frac{\partial u_y}{\partial x}dt\right) \cong \frac{\partial u_y}{\partial x}dt \tag{3.41}$$

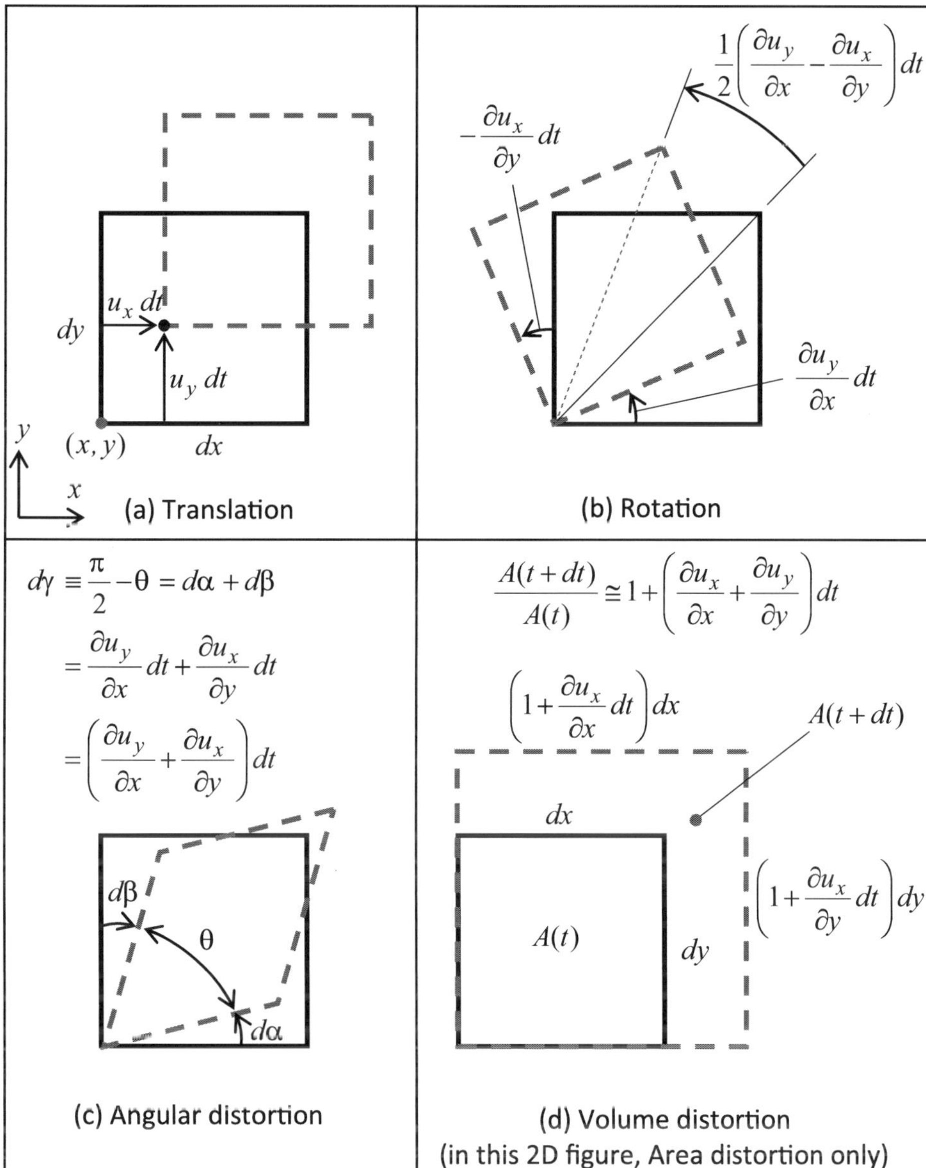

Figure 3.8: Four different components of the motion of a fluid element.

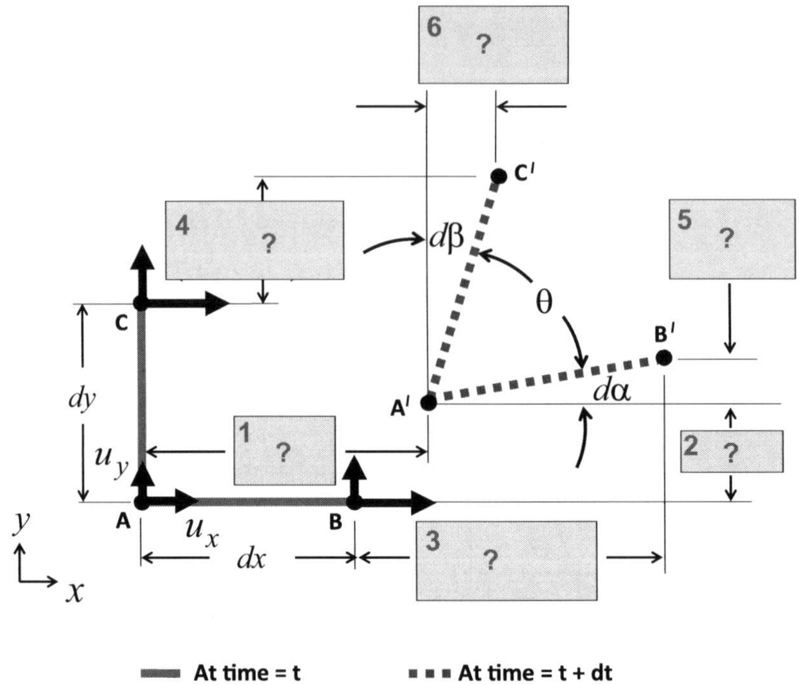

Figure 3.9: Angular distortion of a fluid element (only the bottom and left edges have been shown at times t and $t + dt$). The velocity is $\mathbf{U} = u_x\mathbf{i} + u_y\mathbf{j}$ at point A which has the coordinates of (x, y). As an exercise, determine the six unknown distances in terms of velocity components and their derivatives. Hint: first write the velocity components at points B and C.

as the infinitesimal angle $d\alpha$ is expected to be small, the inverse tangent of the above expression is approximately equal to its argument. If one divides $d\alpha$ by the infinitesimal time interval, dt one obtains the angular speed of the bottom edge:

$$\frac{d\alpha}{dt} = \frac{\dfrac{\partial u_y}{\partial x}\,dt}{dt} = \frac{\partial u_y}{\partial x}. \tag{3.42}$$

The angular speed of the left edge in CW direction is calculated as

$$\frac{d\beta}{dt} = \frac{\dfrac{\partial u_x}{\partial y}\,dt}{dt} = \frac{\partial u_x}{\partial y}. \tag{3.43}$$

The infinitesimal shear strain is defined as the angular distortion of a right angle per dt, thus

$$\begin{aligned} d\gamma_{xy} &\equiv \frac{\pi}{2} - \theta \\ &= d\alpha + d\beta \\ &= \frac{\partial u_y}{\partial x}\,dt + \frac{\partial u_x}{\partial y}\,dt \end{aligned}$$

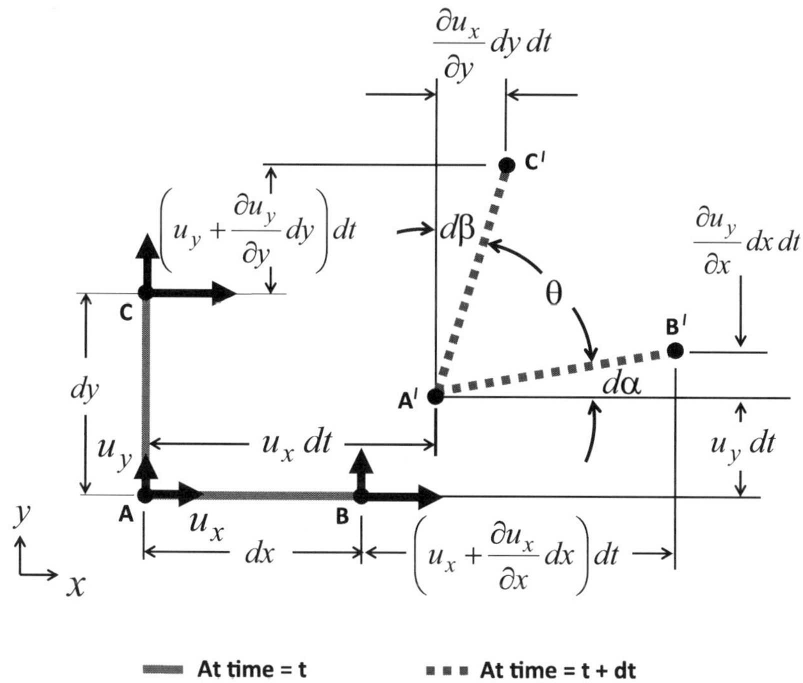

Figure 3.10: Angular distortion of a fluid element.

$$= \left(\frac{\partial u_y}{\partial x} + \frac{\partial u_x}{\partial y}\right) dt. \tag{3.44}$$

Shear strain rate is calculated dividing $d\gamma$ by dt as follows:

$$\frac{d\gamma_{xy}}{dt} = \frac{\partial u_y}{\partial x} + \frac{\partial u_x}{\partial y}. \tag{3.45}$$

This is the rate of angular distortion in xy plane.

Similarly, the rate of angular distortions in xz and yz planes are given by

$$\frac{d\gamma_{xz}}{dt} = \frac{\partial u_z}{\partial x} + \frac{\partial u_x}{\partial z} \tag{3.46}$$

$$\frac{d\gamma_{yz}}{dt} = \frac{\partial u_z}{\partial y} + \frac{\partial u_y}{\partial z}. \tag{3.47}$$

Volume Distortion:

As shown in Figure 3.8 for 2D flow, the area of the fluid element, A may be distorted:

$$\begin{aligned} A(t+dt) &= \left[\left(1 + \frac{\partial u_x}{\partial x} dt\right) dx\right] \left[\left(1 + \frac{\partial u_y}{\partial y} dt\right) dy\right] \\ &= \left(1 + \frac{\partial u_x}{\partial x} dt\right)\left(1 + \frac{\partial u_y}{\partial y} dt\right) dx\, dy \\ &= \left(1 + \frac{\partial u_x}{\partial x} dt\right)\left(1 + \frac{\partial u_y}{\partial y} dt\right) A(t) \end{aligned} \tag{3.48}$$

The rate of element's area distortion is given by

$$\begin{aligned}
\frac{1}{A}\frac{dA}{dt} &= \frac{1}{A(t)}\frac{A(t+dt)-A(t)}{dt} \\
&= \frac{1}{A(t)}\frac{\left[\left(1+\frac{\partial u_x}{\partial x}dt\right)\left(1+\frac{\partial u_y}{\partial y}dt\right)A(t)\right]-A(t)}{dt} \\
&= \frac{1}{A(t)}\frac{\left[1+\left(\frac{\partial u_x}{\partial x}+\frac{\partial u_y}{\partial y}\right)dt+\frac{\partial u_x}{\partial x}\frac{\partial u_y}{\partial y}(dt)^2\right]A(t)-A(t)}{dt} \\
&\cong \frac{\partial u_x}{\partial x}+\frac{\partial u_y}{\partial y}
\end{aligned} \tag{3.49}$$

as $dt \to 0$.

In 3D, the rate of element's volume distortion is given by

$$\begin{aligned}
\frac{1}{V}\frac{dV}{dt} &= \frac{1}{V(t)}\frac{\left[\left(1+\frac{\partial u_x}{\partial x}dt\right)\left(1+\frac{\partial u_y}{\partial y}dt\right)\left(1+\frac{\partial u_z}{\partial z}dt\right)V(t)\right]-V(t)}{dt} \\
&\cong \frac{\partial u_x}{\partial x}+\frac{\partial u_y}{\partial y}+\frac{\partial u_z}{\partial z}
\end{aligned} \tag{3.50}$$

as $dt \to 0$.

One recognizes the term $(\frac{\partial u_x}{\partial x}+\frac{\partial u_y}{\partial y}+\frac{\partial u_z}{\partial z})$ in Equation (3.50) as $\nabla \cdot \mathbf{U}$. Also, one recalls that $\nabla \cdot \mathbf{U} = 0$ if the fluid is incompressible, as studied earlier in this chapter. Thus, Equation (3.50) states that $dV/dt = 0$, or in other words $V(t+dt) = V(t) =$ constant $= V_o$ for an incompressible fluid. This is not a surprising result because an incompressible fluid element may deform, but its mass and thus volume must be conserved during the flow.

We have seen what terms contribute to the deformation and rotation of the fluid element as it moves. Now, we will obtain "angular distortion" and "rotation" tensors by a mathematical operation on the velocity vector.

We described the differential vector operator, ∇, at the beginning of this chapter. The velocity gradient is given as

$$\begin{aligned}
\nabla\mathbf{U} &= \left(\mathbf{i}\frac{\partial()}{\partial x}+\mathbf{j}\frac{\partial()}{\partial y}+\mathbf{k}\frac{\partial()}{\partial z}\right)(u_x\mathbf{i}+u_y\mathbf{j}+u_z\mathbf{k}) \\
&= \mathbf{ii}\frac{\partial u_x}{\partial x}+\mathbf{ij}\frac{\partial u_y}{\partial x}+\mathbf{ik}\frac{\partial u_z}{\partial x} \\
&\quad +\mathbf{ji}\frac{\partial u_x}{\partial y}+\mathbf{jj}\frac{\partial u_y}{\partial y}+\mathbf{jk}\frac{\partial u_z}{\partial y} \\
&\quad +\mathbf{ki}\frac{\partial u_x}{\partial z}+\mathbf{kj}\frac{\partial u_y}{\partial z}+\mathbf{kk}\frac{\partial u_z}{\partial z}
\end{aligned} \tag{3.51}$$

or, using tensorial notation,

$$\nabla\mathbf{U} = \begin{bmatrix} \frac{\partial u_x}{\partial x} & \frac{\partial u_y}{\partial x} & \frac{\partial u_z}{\partial x} \\ \frac{\partial u_x}{\partial y} & \frac{\partial u_y}{\partial y} & \frac{\partial u_z}{\partial y} \\ \frac{\partial u_x}{\partial z} & \frac{\partial u_y}{\partial z} & \frac{\partial u_z}{\partial z} \end{bmatrix} = \begin{bmatrix} \frac{\partial u_1}{\partial x_1} & \frac{\partial u_2}{\partial x_1} & \frac{\partial u_3}{\partial x_1} \\ \frac{\partial u_1}{\partial x_2} & \frac{\partial u_2}{\partial x_2} & \frac{\partial u_3}{\partial x_2} \\ \frac{\partial u_1}{\partial x_3} & \frac{\partial u_2}{\partial x_3} & \frac{\partial u_3}{\partial x_3} \end{bmatrix} \tag{3.52}$$

or, as $(\nabla\mathbf{U})_{ij} = \partial u_j/\partial x_i = u_{j,i}$ in short. By adding and subtracting $\frac{1}{2}(\nabla\mathbf{U})^{\mathrm{T}}$, velocity gradient can be written as

$$\begin{aligned}
\nabla\mathbf{U} &= \left(\frac{1}{2}\nabla\mathbf{U} + \frac{1}{2}\nabla\mathbf{U}\right) + \frac{1}{2}(\nabla\mathbf{U})^{\mathrm{T}} - \frac{1}{2}(\nabla\mathbf{U})^{\mathrm{T}} \\
&= \frac{1}{2}\underbrace{\left[\nabla\mathbf{U} + (\nabla\mathbf{U})^{\mathrm{T}}\right]}_{\text{rate of strain tensor } \underline{\underline{\dot{\gamma}}}} + \frac{1}{2}\underbrace{\left[\nabla\mathbf{U} - (\nabla\mathbf{U})^{\mathrm{T}}\right]}_{\text{vorticity tensor } \underline{\underline{\omega}}} \\
&= \frac{1}{2}(u_{j,i} + u_{i,j}) + \frac{1}{2}(u_{j,i} - u_{i,j}) \\
&= \frac{1}{2}\begin{bmatrix}
2\frac{\partial u_1}{\partial x_1} & \left(\frac{\partial u_2}{\partial x_1} + \frac{\partial u_1}{\partial x_2}\right) & \left(\frac{\partial u_3}{\partial x_1} + \frac{\partial u_1}{\partial x_3}\right) \\
\left(\frac{\partial u_1}{\partial x_2} + \frac{\partial u_2}{\partial x_1}\right) & 2\frac{\partial u_2}{\partial x_2} & \left(\frac{\partial u_3}{\partial x_2} + \frac{\partial u_2}{\partial x_3}\right) \\
\left(\frac{\partial u_1}{\partial x_3} + \frac{\partial u_3}{\partial x_1}\right) & \left(\frac{\partial u_2}{\partial x_3} + \frac{\partial u_3}{\partial x_2}\right) & 2\frac{\partial u_3}{\partial x_3}
\end{bmatrix} \\
&\quad + \frac{1}{2}\begin{bmatrix}
0 & \left(\frac{\partial u_2}{\partial x_1} - \frac{\partial u_1}{\partial x_2}\right) & \left(\frac{\partial u_3}{\partial x_1} - \frac{\partial u_1}{\partial x_3}\right) \\
\left(\frac{\partial u_1}{\partial x_2} - \frac{\partial u_2}{\partial x_1}\right) & 0 & \left(\frac{\partial u_3}{\partial x_2} - \frac{\partial u_2}{\partial x_3}\right) \\
\left(\frac{\partial u_1}{\partial x_3} - \frac{\partial u_3}{\partial x_1}\right) & \left(\frac{\partial u_2}{\partial x_3} - \frac{\partial u_3}{\partial x_2}\right) & 0
\end{bmatrix} .
\end{aligned} \tag{3.53}$$

The manipulation above was done to decompose the velocity gradient $\nabla\mathbf{U}$ into a symmetric part $\frac{1}{2}(\nabla\mathbf{U}+(\nabla\mathbf{U})^{\mathrm{T}})$, and an anti-symmetric part $\frac{1}{2}(\nabla\mathbf{U}-(\nabla\mathbf{U})^{\mathrm{T}})$. Thus, one can express the strain rate tensor as

$$\dot{\gamma}_{ij} = (u_{j,i} + u_{i,j}) . \tag{3.54}$$

The rate of strain tensor, $\underline{\underline{\dot{\gamma}}}$ describes the rate at which the material (fluid in our case) is changing its shape irrespective of the translational or rotational motion. If one considers two close points P and Q within the fluid domain, then the distance between them is $ds = \sqrt{d\mathbf{x}\cdot d\mathbf{x}}$, where $d\mathbf{x}$ is the position vector from one point to the other. One can show that

$$\frac{d(ds)^2}{dt} = d\mathbf{x}\cdot\underline{\underline{\dot{\gamma}}}\cdot d\mathbf{x} = dx_i\dot{\gamma}_{ij}dx_j . \tag{3.55}$$

Hence, the rate of change of the squared distance can be written in terms of $\underline{\underline{\dot{\gamma}}}$ and $d\mathbf{x}$ only. For that reason, $\underline{\underline{\dot{\gamma}}}$ is also called the rate of deformation tensor. It is a symmetric tensor since $\underline{\underline{\dot{\gamma}}} = \nabla\mathbf{U} + (\nabla\mathbf{U})^{\mathrm{T}}$, and $\mathbf{A}+\mathbf{A}^{\mathrm{T}}$ is symmetric for any tensor $\mathbf{A}$. The scalar magnitude of $\underline{\underline{\dot{\gamma}}}$ tensor, which is independent of the coordinate system is given by

$$\dot{\gamma} = \sqrt{\frac{1}{2}\dot{\gamma}_{ij}\dot{\gamma}_{ji}} \tag{3.56}$$

and is called the "strain rate."

Components of the rate of strain tensor are listed in Tables 3.4, 3.5 and 3.6 in Cartesian, cylindrical and spherical coordinates, respectively.

In addition to deformation, the motion of a continuum may include solid body rotation. Vorticity tensor, $\underline{\underline{\omega}}$ is the curl of the velocity vector:

$$\omega_{ij} = u_{j,i} - u_{i,j} . \tag{3.57}$$

Table 3.4: Rate of Strain Tensor in Cartesian Coordinates

$\dot{\gamma}_{xx} = 2\,\dfrac{\partial v_x}{\partial x}$
$\dot{\gamma}_{yy} = 2\,\dfrac{\partial v_y}{\partial y}$
$\dot{\gamma}_{zz} = 2\,\dfrac{\partial v_z}{\partial z}$
$\dot{\gamma}_{xy} = \dot{\gamma}_{yx} = \dfrac{\partial v_y}{\partial x} + \dfrac{\partial v_x}{\partial y}$
$\dot{\gamma}_{yz} = \dot{\gamma}_{zy} = \dfrac{\partial v_z}{\partial y} + \dfrac{\partial v_y}{\partial z}$
$\dot{\gamma}_{zx} = \dot{\gamma}_{xz} = \dfrac{\partial v_x}{\partial z} + \dfrac{\partial v_z}{\partial x}$

Table 3.5: Rate of Strain Tensor in Cylindrical Coordinates

$\dot{\gamma}_{rr} = 2\,\dfrac{\partial v_r}{\partial r}$
$\dot{\gamma}_{\theta\theta} = 2\left(\dfrac{1}{r}\dfrac{\partial v_\theta}{\partial \theta} + \dfrac{v_r}{r}\right)$
$\dot{\gamma}_{zz} = 2\,\dfrac{\partial v_z}{\partial z}$
$\dot{\gamma}_{r\theta} = \dot{\gamma}_{\theta r} = r\,\dfrac{\partial}{\partial r}\left(\dfrac{v_\theta}{r}\right) + \dfrac{1}{r}\dfrac{\partial v_r}{\partial \theta}$
$\dot{\gamma}_{\theta z} = \dot{\gamma}_{z\theta} = \dfrac{1}{r}\dfrac{\partial v_z}{\partial \theta} + \dfrac{\partial v_\theta}{\partial z}$
$\dot{\gamma}_{zr} = \dot{\gamma}_{rz} = \dfrac{\partial v_r}{\partial z} + \dfrac{\partial v_z}{\partial r}$

Table 3.6: Rate of Strain Tensor in Spherical Coordinates

$\dot{\gamma}_{rr} = 2\frac{\partial v_r}{\partial r}$
$\dot{\gamma}_{\theta\theta} = 2\left(\frac{1}{r}\frac{\partial v_\theta}{\partial \theta} + \frac{v_r}{r}\right)$
$\dot{\gamma}_{\phi\phi} = 2\left(\frac{1}{r\sin\theta}\frac{\partial v_\phi}{\partial \phi} + \frac{v_r}{r} + \frac{v_\theta \cot\theta}{r}\right)$
$\dot{\gamma}_{r\theta} = \dot{\gamma}_{\theta r} = r\frac{\partial}{\partial r}\left(\frac{v_\theta}{r}\right) + \frac{1}{r}\frac{\partial v_r}{\partial \theta}$
$\dot{\gamma}_{\theta\phi} = \dot{\gamma}_{\phi\theta} = \frac{\sin\theta}{r}\frac{\partial}{\partial \theta}\left(\frac{v_\phi}{\sin\theta}\right) + \frac{1}{r\sin\theta}\frac{\partial v_\theta}{\partial \phi}$
$\dot{\gamma}_{\phi r} = \dot{\gamma}_{r\phi} = \frac{1}{r\sin\theta}\frac{\partial v_r}{\partial \phi} + r\frac{\partial}{\partial r}\left(\frac{v_\phi}{r}\right)$

$\underline{\underline{\omega}}$ is twice the angular velocity tensor ($\underline{\underline{\omega}} = 2\underline{\underline{\Omega}}$) and it describes this solid body rotation. The fluid flow is called irrotational if the vorticity tensor is a zero tensor, i.e., all the components are zero.

Example 3.1: Simple Shear Flow

Consider flow of resin within a region in which the velocity field is given by $u_x = cy$, $u_y = 0$, and $u_z = 0$ where c is a constant (see Figure 3.11). Find the strain rate and vorticity tensor for this flow. Also calculate the magnitude of the strain rate.

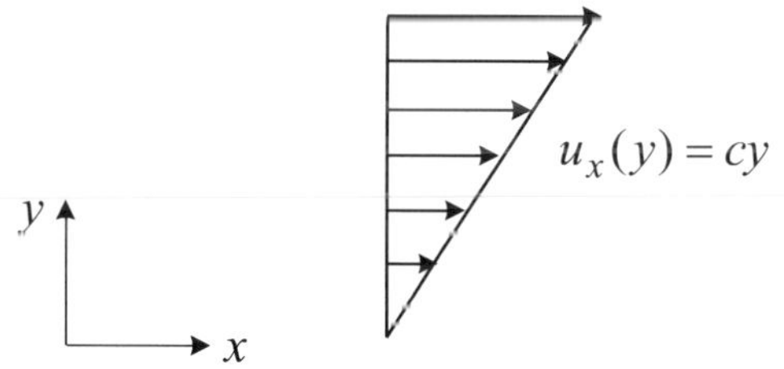

Figure 3.11: Velocity profile in steady simple shear flow.

Solution

The rate of strain tensor $\underline{\underline{\dot{\gamma}}}$ and the vorticity tensor $\underline{\underline{\omega}}$ can be found using Equations (3.54) and (3.57), respectively, and are calculated to be

$$\underline{\underline{\dot{\gamma}}} = \begin{bmatrix} 0 & c & 0 \\ c & 0 & 0 \\ 0 & 0 & 0 \end{bmatrix}, \qquad \underline{\underline{\omega}} = \begin{bmatrix} 0 & -c & 0 \\ c & 0 & 0 \\ 0 & 0 & 0 \end{bmatrix}. \tag{3.58}$$

For this flow, the scalar magnitude of the strain rate tensor is equal to $\dot{\gamma} = \sqrt{\frac{1}{2}(c^2 + c^2)} = c$.
■

Example 3.2: Two-Dimensional Simple Elongational Flow

Consider stretching of a thin sheet of viscous material in the x direction as shown in Figure 3.12. This flow is also called *simple extensional flow.* The velocity field in the sheet is given by $u_x = cx$, $u_y = -cy$, and $u_z = 0$ where c is a positive constant. Find the strain rate and vorticity tensor. Calculate the strain rate. Is the flow irrotational?

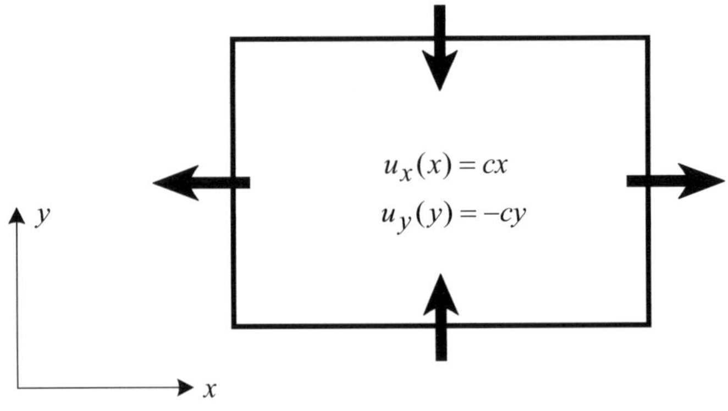

Figure 3.12: Two-dimensional elongational flow.

Solution

The rate of strain tensor $\underline{\underline{\dot{\gamma}}}$ and the vorticity tensor $\underline{\underline{\omega}}$ are calculated to be

$$\underline{\underline{\dot{\gamma}}} = \begin{bmatrix} 2c & 0 & 0 \\ 0 & -2c & 0 \\ 0 & 0 & 0 \end{bmatrix}, \qquad \underline{\underline{\omega}} = \begin{bmatrix} 0 & 0 & 0 \\ 0 & 0 & 0 \\ 0 & 0 & 0 \end{bmatrix}. \tag{3.59}$$

Since $\underline{\underline{\omega}}$ is zero, the flow is irrotational; hence, there is no vorticity and because of this, the simple elongational flow is sometimes called *pure shear flow.* The strain rate is equal to $\dot{\gamma} = \sqrt{\frac{1}{2}(4c^2 + 4c^2)} = 2c$. ■

Example 3.3: Three-Dimensional Simple Elongational Flow

Consider a cubic rod of viscous material being stretched in the x-direction by velocity component in the x direction being $u_x = cx$, as shown in Figure 3.13. The flow within this material can be described by the velocity field given by $u_x = cx$, $u_y = -cy/2$, and $u_z = -cz/2$ where c is a positive constant. Find the strain rate and vorticity tensor. Calculate the magnitude of the strain rate tensor. Is this flow irrotational?

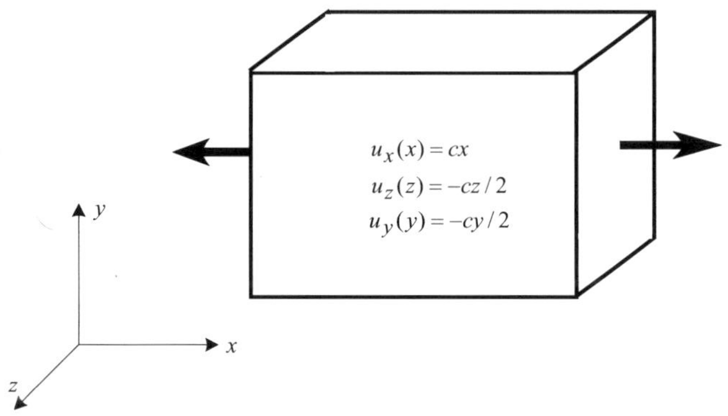

Figure 3.13: Three-dimensional elongational flow.

Solution

The rate of strain tensor $\underline{\underline{\dot{\gamma}}}$ and the vorticity tensor $\underline{\underline{\omega}}$ are calculated to be

$$\underline{\underline{\dot{\gamma}}} = \begin{bmatrix} 2c & 0 & 0 \\ 0 & -c & 0 \\ 0 & 0 & -c \end{bmatrix}, \qquad \underline{\underline{\omega}} = \begin{bmatrix} 0 & 0 & 0 \\ 0 & 0 & 0 \\ 0 & 0 & 0 \end{bmatrix}. \tag{3.60}$$

The strain rate for this case is equal to $\dot{\gamma} = \sqrt{\frac{1}{2}(4c^2 + c^2 + c^2)} = \sqrt{3}\,c$. Again, this is an elongational flow and irrotational as $|\omega| = 0$. ■

3.4.2 Newtonian Fluids

Earlier in this chapter, the momentum equation was derived in terms of the viscous stress tensor $\underline{\underline{\tau}}$. In this section, $\underline{\underline{\tau}}$ will be related to the rate of strain tensor $\underline{\underline{\dot{\gamma}}}$ in order to complete the set of equations for the fluid flow. (Recall that, so far the number of equations is less than the number of unknowns in the equations.) There is no unique relation between $\underline{\underline{\tau}}$ and $\underline{\underline{\dot{\gamma}}}$ that is valid for all fluids. However, the simplest relation between them, which is valid for many fluids, is a linear relation. This linear relation idea was first proposed by Newton; hence, the fluids which obey this relation are called Newtonian fluids. Later, Navier and Stokes derived the equations for the flow of Newtonian fluids independently. These equations are called the Navier Stokes equations. The linear relation is given by

$$\tau_{ij} = \lambda(\nabla \cdot \mathbf{U})\delta_{ij} + \mu\dot{\gamma}_{ij} \tag{3.61}$$

where μ is the viscosity of the fluid. The assumption of mechanical and thermodynamic pressures being equal (known as Stokes' assumption) yields

$$\lambda = -\frac{2}{3}\mu. \tag{3.62}$$

Hence, a normal viscous stress τ_{ii} and a shear viscous stress τ_{ij} are given by

$$\tau_{ii} = \mu\left[-\frac{2}{3}(\nabla\cdot\mathbf{U}) + 2\frac{\partial u_i}{\partial x_i}\right] \tag{3.63}$$

$$\tau_{ij} = \mu\left(\frac{\partial u_i}{\partial x_j} + \frac{\partial u_j}{\partial x_i}\right). \tag{3.64}$$

Although many fluids are characterized very well as Newtonian fluids, there are many other fluids which behave differently. The relation between the viscous stress and the rate of strain tensors is not a linear one as in the Newtonian fluids. Some typical non-Newtonian fluids are "shear-thinning," "shear-thickening," and "Bingham plastic" fluids. These fluids will be discussed in detail in the next chapter.

Let's return to Newtonian fluids. If the fluid density is assumed to be constant (incompressible fluid), then according to the conservation of mass equation, $\partial\rho/\partial t + \nabla\cdot(\rho\mathbf{U}) = 0 + \rho\nabla\cdot\mathbf{U} = 0$, which requires that $\nabla\cdot\mathbf{U} = 0$. Then, the $\nabla\cdot\mathbf{U}$ term in the normal stresses drops out. So, for constant density fluids, Equation (3.61) reduces to

$$\tau_{ij} = \mu\dot{\gamma}_{ij} \tag{3.65}$$

and the normal and shear stresses are written as

$$\tau_{ii} = 2\mu\frac{\partial u_i}{\partial x_i} \tag{3.66}$$

$$\tau_{ij} = \mu\left(\frac{\partial u_i}{\partial x_j} + \frac{\partial u_j}{\partial x_i}\right) \tag{3.67}$$

For constant-viscosity and incompressible fluids, momentum equations (3.34)-(3.36) reduce to the following equations after substituting viscous stress equations (3.66) and (3.67):

$$\rho\left(\frac{\partial u_x}{\partial t} + u_x\frac{\partial u_x}{\partial x} + u_y\frac{\partial u_x}{\partial y} + u_z\frac{\partial u_x}{\partial z}\right) = -\frac{\partial P}{\partial x} + \mu\left[\frac{\partial^2 u_x}{\partial x^2} + \frac{\partial^2 u_x}{\partial y^2} + \frac{\partial^2 u_x}{\partial z^2}\right] + F_{Bx} \tag{3.68}$$

$$\rho\left(\frac{\partial u_y}{\partial t} + u_x\frac{\partial u_y}{\partial x} + u_y\frac{\partial u_y}{\partial y} + u_z\frac{\partial u_y}{\partial z}\right) = -\frac{\partial P}{\partial y} + \mu\left[\frac{\partial^2 u_y}{\partial x^2} + \frac{\partial^2 u_y}{\partial y^2} + \frac{\partial^2 u_y}{\partial z^2}\right] + F_{By} \tag{3.69}$$

$$\rho\left(\frac{\partial u_z}{\partial t} + u_x\frac{\partial u_z}{\partial x} + u_y\frac{\partial u_z}{\partial y} + u_z\frac{\partial u_z}{\partial z}\right) = -\frac{\partial P}{\partial z} + \mu\left[\frac{\partial^2 u_z}{\partial x^2} + \frac{\partial^2 u_z}{\partial y^2} + \frac{\partial^2 u_z}{\partial z^2}\right] + F_{Bz}. \tag{3.70}$$

or, in vectorial form as

$$\rho\frac{D\mathbf{U}}{Dt} = -\nabla P + \mu\nabla^2\mathbf{U} + \mathbf{F}_B \tag{3.71}$$

where $\nabla^2() \equiv \partial^2()/\partial x^2 + \partial^2()/\partial y^2 + \partial^2()/\partial z^2$ in Cartesian coordinates. Equation (3.71) is the conservation of momentum equation for Newtonian fluids with constant density and constant viscosity. Tables 3.7, 3.8 and 3.9 list the N-S equations for incompressible resins that behave in a Newtonian manner in Cartesian, cylindrical and spherical coordinate systems.

Table 3.7: The Navier-Stokes Equations for Incompressible Resins in x, y and z Directions of Cartesian Coordinates

$$\text{(}x\text{-dir.:)}\quad \rho\left[\frac{\partial u_x}{\partial t}+u_x\frac{\partial u_x}{\partial x}+u_y\frac{\partial u_x}{\partial y}+u_z\frac{\partial u_x}{\partial z}\right] = -\frac{\partial P}{\partial x}+\mu\left[\frac{\partial^2 u_x}{\partial x^2}+\frac{\partial^2 u_x}{\partial y^2}+\frac{\partial^2 u_x}{\partial z^2}\right]+F_{Bx}$$

$$\text{(}y\text{-dir.:)}\quad \rho\left[\frac{\partial u_y}{\partial t}+u_x\frac{\partial u_y}{\partial x}+u_y\frac{\partial u_y}{\partial y}+u_z\frac{\partial u_y}{\partial z}\right] = -\frac{\partial P}{\partial y}+\mu\left[\frac{\partial^2 u_y}{\partial x^2}+\frac{\partial^2 u_y}{\partial y^2}+\frac{\partial^2 u_y}{\partial z^2}\right]+F_{By}$$

$$\text{(}z\text{-dir.:)}\quad \rho\left[\frac{\partial u_z}{\partial t}+u_x\frac{\partial u_z}{\partial x}+u_y\frac{\partial u_z}{\partial y}+u_z\frac{\partial u_z}{\partial z}\right] = -\frac{\partial P}{\partial z}+\mu\left[\frac{\partial^2 u_z}{\partial x^2}+\frac{\partial^2 u_z}{\partial y^2}+\frac{\partial^2 u_z}{\partial z^2}\right]+F_{Bz}$$

Table 3.8: The Navier-Stokes Equations for Incompressible Resins in r, θ and z Directions of Cylindrical Coordinates

$$\text{(}r\text{-dir.:)}\quad \rho\left[\frac{\partial u_r}{\partial t}+u_r\frac{\partial u_r}{\partial r}+\frac{u_\theta}{r}\frac{\partial u_r}{\partial \theta}-\frac{u_\theta^2}{r}+u_z\frac{\partial u_r}{\partial z}\right] = -\frac{\partial P}{\partial r}+\mu\left[\frac{\partial}{\partial r}\left(\frac{1}{r}\frac{\partial}{\partial r}(ru_r)\right)+\frac{1}{r^2}\frac{\partial^2 u_r}{\partial \theta^2}+\frac{\partial^2 u_r}{\partial z^2}-\frac{2}{r^2}\frac{\partial u_\theta}{\partial \theta}\right]+F_{Br}$$

$$\text{(}\theta\text{-dir.:)}\quad \rho\left[\frac{\partial u_\theta}{\partial t}+u_r\frac{\partial u_\theta}{\partial r}+\frac{u_\theta}{r}\frac{\partial u_\theta}{\partial \theta}+\frac{u_r u_\theta}{r}+u_z\frac{\partial u_\theta}{\partial z}\right] = -\frac{1}{r}\frac{\partial P}{\partial \theta}+\mu\left[\frac{\partial}{\partial r}\left(\frac{1}{r}\frac{\partial}{\partial r}(ru_\theta)\right)+\frac{1}{r^2}\frac{\partial^2 u_\theta}{\partial \theta^2}+\frac{\partial^2 u_\theta}{\partial z^2}+\frac{2}{r^2}\frac{\partial u_r}{\partial \theta}\right]+F_{B\theta}$$

$$\text{(}z\text{-dir.:)}\quad \rho\left[\frac{\partial u_z}{\partial t}+u_r\frac{\partial u_z}{\partial r}+\frac{u_\theta}{r}\frac{\partial u_z}{\partial \theta}+u_z\frac{\partial u_z}{\partial z}\right] = -\frac{\partial P}{\partial z}+\mu\left[\frac{1}{r}\frac{\partial}{\partial r}\left(r\frac{\partial u_z}{\partial r}\right)+\frac{1}{r^2}\frac{\partial^2 u_z}{\partial \theta^2}+\frac{\partial^2 u_z}{\partial z^2}\right]+F_{Bz}$$

Table 3.9: The Navier-Stokes Equations for Incompressible Resins in r, θ and ϕ Directions of Spherical Coordinates

(r-dir.:)

$$\rho\left[\frac{\partial u_r}{\partial t}+u_r\frac{\partial u_r}{\partial r}+\frac{u_\theta}{r}\frac{\partial u_r}{\partial \theta}+\frac{u_\phi}{r\sin\theta}\frac{\partial u_r}{\partial \phi}-\frac{u_\theta^2+u_\phi^2}{r}\right]$$

$$=-\frac{\partial P}{\partial r}+\mu\left[\frac{\partial}{\partial r}\left(\frac{1}{r^2}\frac{\partial}{\partial r}(r^2u_r)\right)+\frac{1}{r^2\sin\theta}\frac{\partial}{\partial\theta}\left(\sin\theta\frac{\partial u_r}{\partial\theta}\right)\right.$$

$$\left.+\frac{1}{r^2\sin^2\theta}\frac{\partial^2 u_r}{\partial\phi^2}-\frac{2}{r^2\sin\theta}\frac{\partial}{\partial\theta}(u_\theta\sin\theta)-\frac{2}{r^2\sin\theta}\frac{\partial u_\phi}{\partial\phi}\right]+F_{Br}$$

(θ-dir.:)

$$\rho\left[\frac{\partial u_\theta}{\partial t}+u_r\frac{\partial u_\theta}{\partial r}+\frac{u_\theta}{r}\frac{\partial u_\theta}{\partial \theta}+\frac{u_\phi}{r\sin\theta}\frac{\partial u_\theta}{\partial \phi}+\frac{u_ru_\theta}{r}-\frac{u_\phi^2\cot\theta}{r}\right]$$

$$=-\frac{1}{r}\frac{\partial P}{\partial \theta}+\mu\left[\frac{1}{r^2}\frac{\partial}{\partial r}\left(r^2\frac{\partial u_\theta}{\partial r}\right)+\frac{1}{r^2}\frac{\partial}{\partial\theta}\left(\frac{1}{\sin\theta}\frac{\partial}{\partial\theta}(u_\theta\sin\theta)\right)\right.$$

$$\left.+\frac{1}{r^2\sin^2\theta}\frac{\partial^2 u_\theta}{\partial\phi^2}+\frac{2}{r^2}\frac{\partial u_r}{\partial\theta}-\frac{2\cot\theta}{r^2\sin\theta}\frac{\partial u_\phi}{\partial\phi}\right]+F_{B\theta}$$

(ϕ-dir.:)

$$\rho\left[\frac{\partial u_\phi}{\partial t}+u_r\frac{\partial u_\phi}{\partial r}+\frac{u_\theta}{r}\frac{\partial u_\phi}{\partial \theta}+\frac{u_\phi}{r\sin\theta}\frac{\partial u_\phi}{\partial \phi}+\frac{u_\phi u_r}{r}+\frac{u_\theta u_\phi}{r}\cot\theta\right]$$

$$=-\frac{1}{r\sin\theta}\frac{\partial P}{\partial \phi}+\mu\left[\frac{1}{r^2}\frac{\partial}{\partial r}\left(r^2\frac{\partial u_\phi}{\partial r}\right)+\frac{1}{r^2}\frac{\partial}{\partial\theta}\left(\frac{1}{\sin\theta}\frac{\partial}{\partial\theta}(u_\phi\sin\theta)\right)\right.$$

$$\left.+\frac{1}{r^2\sin^2\theta}\frac{\partial^2 u_\phi}{\partial\phi^2}+\frac{2}{r^2\sin\theta}\frac{\partial u_r}{\partial\phi}+\frac{2\cot\theta}{r^2\sin\theta}\frac{\partial u_\theta}{\partial\phi}\right]+F_{B\phi}$$

Example 3.4: Derivation of Viscous Term in Equation (3.71)

Show how the viscous term in Equation (3.71) reduces to $\mu\nabla^2\mathbf{U}$ (i.e., not $2\mu\nabla^2\mathbf{U}$) starting from Equation (3.34).

Solution

As seen in Equation (3.34), the x-component of force due to viscous stresses is $\partial\tau_{xx}/\partial x + \partial\tau_{yx}/\partial y + \partial\tau_{zx}/\partial z$, and for constant density fluids, $\tau_{xx} = 2\mu\,\partial u_x/\partial x$ as derived from Equation (3.63). The x derivative on τ_{xx} results in $\partial\tau_{xx}/\partial x = \frac{\partial}{\partial x}(2\mu\,\partial u_x/\partial x) = 2\mu\,\partial^2 u_x/\partial x^2$. However, we have the resultant force as $\mu[\partial^2u_x/\partial x^2 + (\text{etc.})]$ as seen in Equation (3.68). To eliminate the coefficient 2 in the momentum equation, one needs to look at the entire terms as

$$\frac{\partial\tau_{xx}}{\partial x}+\frac{\partial\tau_{yx}}{\partial y}+\frac{\partial\tau_{zx}}{\partial z} = \mu\left\{\frac{\partial}{\partial x}\left[2\frac{\partial u_x}{\partial x}\right]+\frac{\partial}{\partial y}\left[\frac{\partial u_y}{\partial x}+\frac{\partial u_x}{\partial y}\right]+\frac{\partial}{\partial z}\left[\frac{\partial u_z}{\partial x}+\frac{\partial u_x}{\partial z}\right]\right\}$$

$$
\begin{aligned}
&= \mu\left\{2\frac{\partial^2 u_x}{\partial x^2}+\frac{\partial^2 u_x}{\partial y^2}+\frac{\partial^2 u_y}{\partial x\partial y}+\frac{\partial^2 u_z}{\partial x\partial z}+\frac{\partial^2 u_x}{\partial z^2}\right\}\\
&= \mu\left\{\frac{\partial^2 u_x}{\partial x^2}+\frac{\partial^2 u_x}{\partial y^2}+\frac{\partial^2 u_x}{\partial z^2}+\left[\frac{\partial^2 u_x}{\partial x^2}+\frac{\partial^2 u_y}{\partial x\partial y}+\frac{\partial^2 u_z}{\partial x\partial z}\right]\right\}\\
&= \mu\left\{\frac{\partial^2 u_x}{\partial x^2}+\frac{\partial^2 u_x}{\partial y^2}+\frac{\partial^2 u_x}{\partial z^2}+\frac{\partial}{\partial x}\left[\frac{\partial u_x}{\partial x}+\frac{\partial u_y}{\partial y}+\frac{\partial u_z}{\partial z}\right]\right\}\\
&= \mu\left\{\frac{\partial^2 u_x}{\partial x^2}+\frac{\partial^2 u_x}{\partial y^2}+\frac{\partial^2 u_x}{\partial z^2}+\frac{\partial}{\partial x}[0]\right\}\\
&= \mu\left\{\frac{\partial^2 u_x}{\partial x^2}+\frac{\partial^2 u_x}{\partial y^2}+\frac{\partial^2 u_x}{\partial z^2}\right\}
\end{aligned}
\tag{3.72}
$$

The three terms drop out since their summation is zero because $\frac{\partial u_x}{\partial x}+\frac{\partial u_y}{\partial y}+\frac{\partial u_z}{\partial z}=0$ for incompressible (constant density) fluids. Similar derivations can be repeated in the y and z directions to show that $\nabla\cdot\tau=\mu\nabla^2\mathbf{U}$ for incompressible Newtonian fluids. ■

3.5 Examples on Use of Conservation Equations to Solve Viscous Flow Problems

The conservation of mass equation (a partial differential equation on scalar variables) and the conservation of momentum equations (three partial differential equations on scalar variables) were derived earlier in this chapter. Then, constitutive equations were studied to relate the viscous stress tensor to the rate of strain tensor. We substituted these constitutive relations into the momentum equations for Newtonian fluids with constant density and viscosity and formulated four equations containing the four unknown dependent variables: P, u_x, u_y, u_z and four independent variables x, y, z and t. For non-Newtonian fluids, or, for Newtonian fluids with non-constant density and/or viscosity, the corresponding constitutive relations must be used together with the general momentum equations, Equations (3.34)-(3.36), instead of using Equations (3.68)-(3.70). The constitutive relations for some of the non-Newtonian fluids will be studied in the next chapter. However, before we can solve a fluid flow problem, we need the same number of variables and equations and a sufficient number of boundary conditions to well pose a problem to obtain a unique solution.

3.5.1 Boundary Conditions

In general, there are two types of boundary conditions: kinematic and dynamic. Kinematic and dynamic boundary conditions deal with the velocity and stress fields, respectively.

Considering the physical contact of the fluid at the boundary, boundary conditions can be further divided into five groups [5]:

1. Liquid-solid interface (contact at solid surface),
2. Liquid-liquid interface,
3. Liquid-vapor interface,
4. Free surface, and

5. Specified inlet and exit boundary conditions.

At the liquid-solid interfaces, the relative velocity of a viscous fluid with respect to the solid boundary is assumed to be zero. That is, $\mathbf{U} = \mathbf{U}_{\text{solid}}$. This is known as the *no-slip* condition. Although this assumption is not always correct, i.e., fluid may slip on the solid surface, the idea of a no-slip approximation is supported by many experiments in the literature [5].

Some examples of fluid flow problems with "liquid-solid interface" type boundary conditions are shown in Figure 3.14. In all four problems, it was assumed that the fluid domain in the z direction is so long that there is no variation in the z direction. In (a) and (b), a fluid is bounded by two infinitely long parallel plates. The lower plate is stationary, but the upper one moves horizontally in the x direction. Due to the disturbance from the moving upper plate, there will be fluid flow. The upper plate is pulled with a constant speed V in (a). In (b), the pulling force is specified instead of the speed.

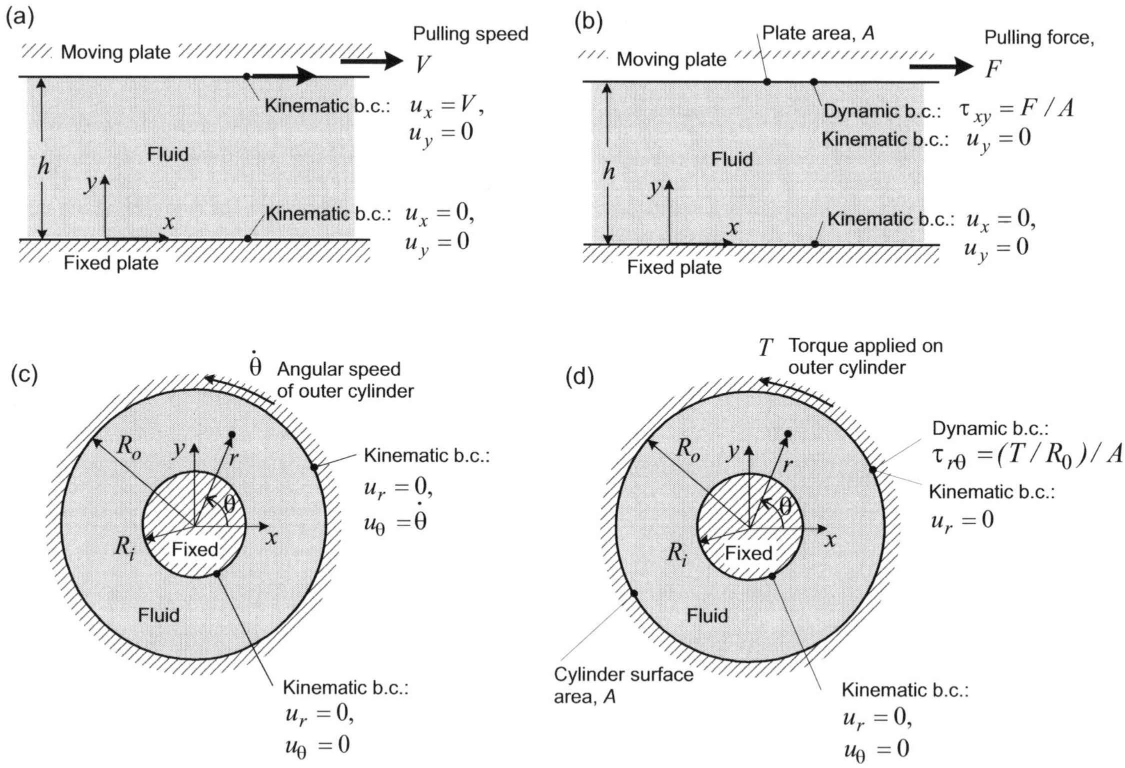

Figure 3.14: Four example situations for different liquid-solid interface boundary conditions. (a) Kinematic boundary conditions at the upper and lower boundaries for fluid flow within a rectangular channel. (b) The same as in (a), except that the pulling force F on the plate is specified instead of the speed. So, this boundary condition on the upper boundary of (b) is dynamic. (c) Kinematic boundary conditions on the outer and inner cylinder boundaries of fluid flow within a concentric domain. (d) The same as in (c), except that the torque T applied to the outer cylinder is specified instead of the angular speed. So, this boundary condition on the outer boundary of (d) is dynamic.

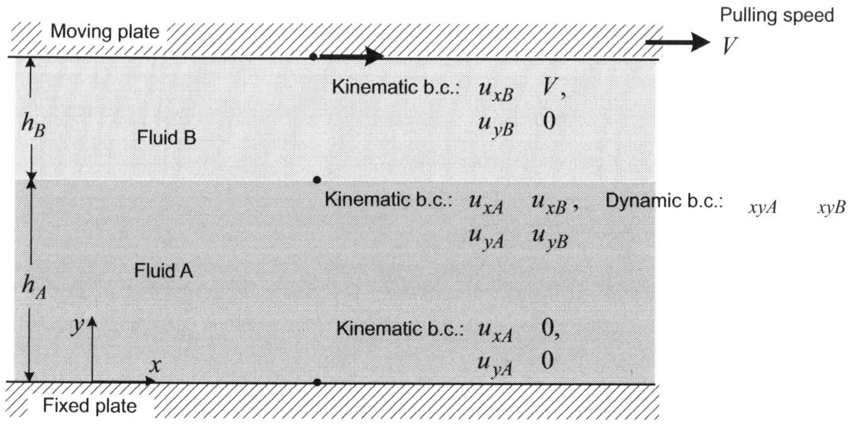

Figure 3.15: Boundary condition at a liquid-liquid interface.

The no-slip condition can be directly applied in (a). If the plates are impermeable (non-porous), then there is no fluid flow in the normal directions on the plates, which can be written as $\mathbf{U} \cdot \mathbf{n} = u_y = 0$ at $y = 0$ and $y = h$, where $\mathbf{U}$ is the fluid velocity vector, and $\mathbf{n}$ is the normal vector on the plate. The tangential velocity component, $u_x = V$ at $y = h$, and $u_x = 0$ at $y = 0$, considering the no-slip condition. These are kinematic boundary conditions since they only involve the velocity field.

The boundary condition on the upper plate in (b) is a dynamic condition. The pulling force applied on the plate is specified. This force can be related to the shear stress, which is usually needed in the dynamic boundary condition. In fact, this problem is a one-dimensional problem, i.e., velocity vector and stress tensor are functions of only y, not x. Hence, τ_{xy} is constant at $y = h$. The dynamic boundary condition is then $\tau_{xy} = F/A$, where

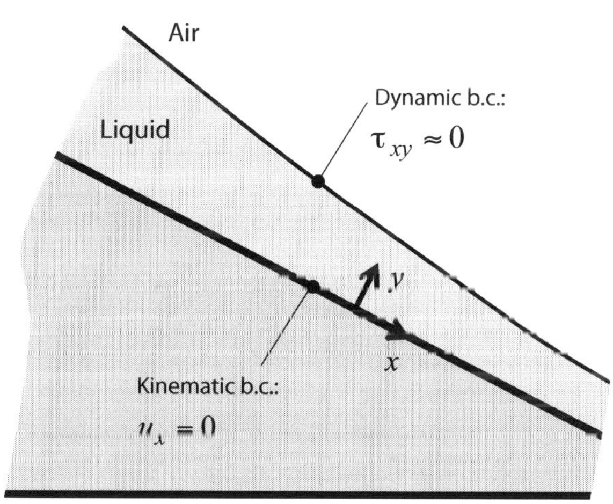

Figure 3.16: Boundary condition at a liquid-gas interface. A thin film of liquid is flowing down an incline, the boundary condition at the liquid-gas interface is that the shear stresses on both sides of the interface are equal and as the shear stress of a gas at low velocities is very small, the shear stress for the liquid (resin) at the interface is negligible.

A is the area of the plate.

In (c) and (d), the fluid domain is a concentric gap between a fixed inner cylinder and a rotating outer cylinder. Since the inner cylinder is fixed, (i.e., has a zero velocity vector), the fluid will also have the same velocity vector due to the no-slip condition. In (c), the angular velocity of the outer cylinder is specified as $\dot{\theta}$. Hence, the fluid will also have the same velocity, $u_\theta = \dot{\theta} R_0$.

In (d), the external torque applied to the outer cylinder is specified as T. To relate T to the shear stress $\tau_{r\theta}$, one can show that $\tau_{r\theta}$ is constant on the surface area $2\pi R_0 L$ as this problem is also one-dimensional; the velocity and stress fields are functions of r only, and

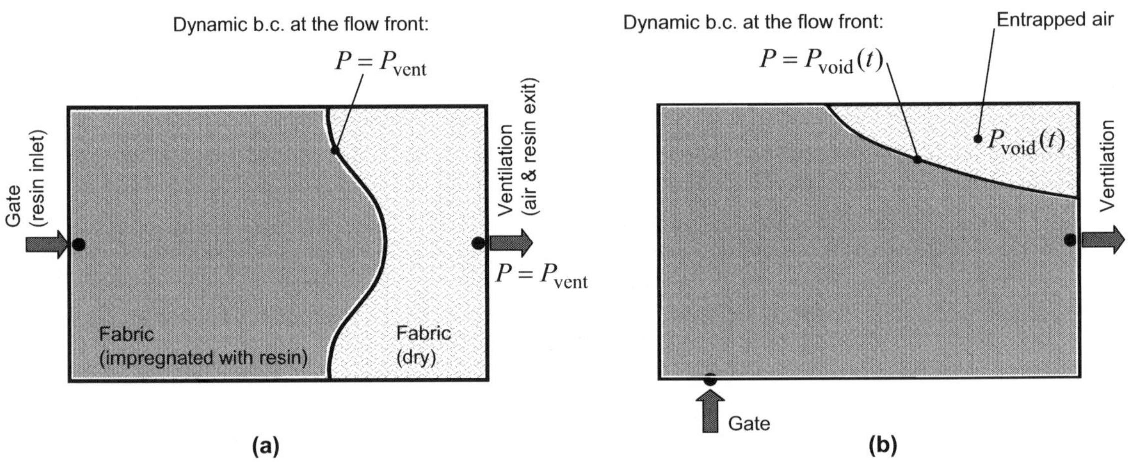

Figure 3.17: Free surface boundary condition at the flow front of the propagating resin in liquid molding process. In (a), the pressure of the free surface is equal to the ventilation pressure which may be either the atmospheric pressure (if the exit tube is open directly to atmosphere) or vacuum pressure (if a vacuum pump is being used). The air in the mold cavity is entrapped in (b) due to incomplete mold filling when the flow front reaches the ventilation. The pressure in this dry region increases with time as its volume decreases.

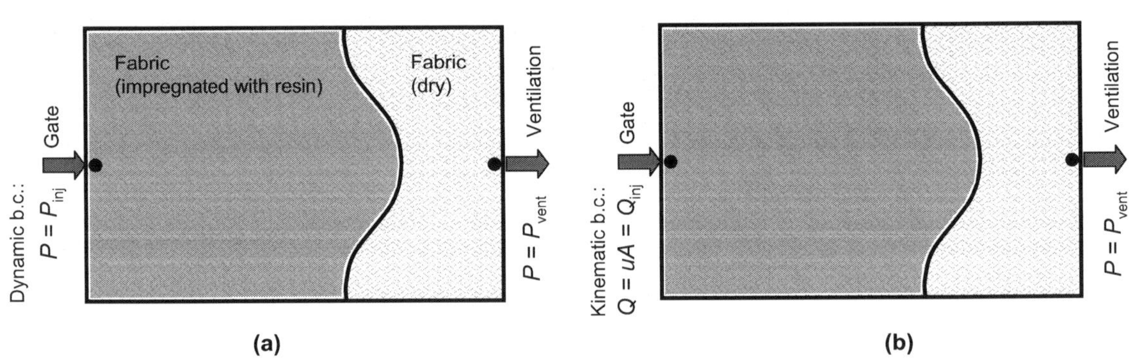

Figure 3.18: Specified inlet and exit boundary conditions at a gate and ventilation, respectively. The resin injection is done at (a) a constant pressure, P_{inj}, and (b) a constant flow rate, Q_{inj}.

hence $\tau_{r\theta}$ is constant at $r = R_0$. The corresponding dynamic boundary condition is then $\tau_{r\theta} = T/2\pi R_0^2 L$. Here, L is the length of the cylinders in the z direction.

Consider the two immiscible fluids A and B within a rectangular channel as shown in Figure 3.15. The kinematic boundary conditions on the upper and lower plates are $u_{xB} = V$ at $y = h_A + h_B$, and $u_{xA} = 0$ at $y = 0$, respectively. We need two more boundary conditions at the liquid-liquid interface in order to uniquely solve this problem. For viscous fluids, the velocities are assumed to be continuous at their interface. So, $\mathbf{U}_A = \mathbf{U}_B$ in vectorial form, or $u_{xA} = u_{xB}$ and $u_{yA} = u_{yB}$ in scalar form at $y = h_A$. Considering the dynamical interactions of adjacent control volumes of fluids A and B, the viscous shear stress τ_{xy} must be continuous across the interface. Hence, $\tau_{xyA} = \tau_{xyB}$ which can also be written as $\mu_A\left(\frac{\partial u_{xA}}{\partial y} + \frac{\partial u_{yA}}{\partial x}\right) = \mu_B\left(\frac{\partial u_{xB}}{\partial y} + \frac{\partial u_{yB}}{\partial x}\right)$ at $y = h_A$.

Some examples of fluid flow problems with "liquid-gas interface," "free surface," and "specified inlet and exit" type boundary conditions are shown in Figures 3.16, 3.17 and 3.18.

3.5.2 Solution Procedure

In this chapter, so far we have studied the necessary tools to solve isothermal flow problems. These tools are conservation equations for mass and momentum which are mathematical models of physical laws, and constitutive equations to relate the stress tensor to the fluid deformation in terms of rate of strain tensor. These equations are valid everywhere in the fluid domain. Different boundary conditions and assumptions made in different fluid flow problems are the reasons why different solutions are obtained in different situations. We can generalize the solution procedure as follows:

1. Choose coordinate system and verify parameters,
2. State any physical conditions (such as symmetry about a boundary, one-dimensionality, two-dimensionality, steady state) in mathematical terms,
3. Write governing equations (conservation of mass, momentum and constitutive equations) and neglect terms that are zero or negligible compared to other terms,
4. Write the boundary conditions,
5. Solve for pressure, velocities and stresses, and
6. Calculate other quantities of interest such as flow rate, forces, etc., by integrating velocities, pressure or stresses.

Below we will solve several fluid flow problems by following this solution procedure assuming Newtonian constitutive law, and then in the following chapters, the procedure will be applied to the solution of different composite manufacturing processes.

Example 3.5: Plane Couette Flow

Couette flow, named after M. F. A. Couette, is flow in a rectangular channel induced by a moving upper boundary at a constant speed V. The geometry and the corresponding boundary conditions (Figure 3.14(a)) were stated in the previous section. In reality, the length of the channel in the x direction is never infinite, despite the assumption made. However, we will assume that the channel is long enough that the end effects are negligible so a fully developed steady-state velocity profile is created. Now, let's demonstrate the solution procedure for this problem.

Solution

Step 1: Cartesian coordinate system is chosen and attached at a convenient location as shown in Figure 3.14(a). The parameters are the speed of plate V, height of the channel h, and viscosity of the fluid μ.

Step 2: As we stated in the previous section, we assume that the problem is two-dimensional in the xy plane, i.e., the z component of velocity u_z is zero, and the variables do not vary in the z direction ($\partial()/\partial z = 0$). We also assume that the fluid is Newtonian.

Step 3: The governing differential equations are the conservation of mass and the conservation of momentum equations, Equations (3.10) and (3.68). We will consider steady-state flow ($\partial()/\partial t = 0$ for any variable). Note that steady state doesn't mean $D()/Dt = 0$ for a variable since it is the substantial derivative following a fluid particle. Also, a fully developed velocity field in the x direction will be considered in the channel, hence $\partial()/\partial x = 0$.

Step 4: The boundary conditions have already been shown in Figure 3.14(a) as $u_x = V$, $u_y = 0$ at $y = h$, and $u_x = 0$, $u_y = 0$ at $y = 0$.

Step 5: For a steady state with no-sink effects, the conservation equation can be simplified to

$$\frac{\partial u_x}{\partial x} + \frac{\partial u_y}{\partial y} + \frac{\partial u_z}{\partial z} = 0 \tag{3.73}$$

where $u_z = 0$ due to 2D assumption, and $\partial u_x/\partial x = 0$ due to the fully developed velocity assumption. Then, Equation (3.73) reduces to $\partial u_y/\partial y = 0$. Integration of $\partial u_y/\partial y = 0$ yields that $u_y = u_y(x)$, i.e., u_y is not a function of y, but can be a function of x. The boundary conditions show that $u_y = 0$ for all x values, and hence u_y must be zero within the entire domain as well since it is not a function of y either. The only non-zero velocity component is then u_x which is a function of y only. How do we find $u_x(y)$? We use the momentum equation in the x direction:

$$\rho\left(\frac{\partial u_x}{\partial t} + u_x\frac{\partial u_x}{\partial x} + u_y\frac{\partial u_x}{\partial y} + u_z\frac{\partial u_x}{\partial z}\right) = -\frac{\partial P}{\partial x} + \left[\frac{\partial \tau_{xx}}{\partial x} + \frac{\partial \tau_{yx}}{\partial y} + \frac{\partial \tau_{zx}}{\partial z}\right] + F_{Bx} \tag{3.74}$$

If the fluid is a Newtonian fluid, needed components of the stress tensor are equal to

$$\tau_{xx} = 2\mu\frac{\partial u_x}{\partial x} = 0 \tag{3.75a}$$

$$\tau_{yx} = \mu\left(\frac{\partial u_x}{\partial y} + \frac{\partial u_y}{\partial x}\right) = \frac{\partial u_x}{\partial y} \tag{3.75b}$$

$$\tau_{zx} = \mu\left(\frac{\partial u_x}{\partial z} + \frac{\partial u_z}{\partial x}\right) = 0. \tag{3.75c}$$

Hence,

$$\begin{aligned}\left[\frac{\partial \tau_{xx}}{\partial x} + \frac{\partial \tau_{yx}}{\partial y} + \frac{\partial \tau_{zx}}{\partial z}\right] &= \mu\left[\frac{\partial}{\partial x}(0) + \frac{\partial}{\partial y}\left(\frac{\partial u_x}{\partial y}\right) + \frac{\partial}{\partial z}(0)\right] \\ &= \mu\frac{\partial^2 u_x}{\partial y^2}\end{aligned} \tag{3.76}$$

After eliminating the terms which are zero and considering that $\mathbf{F}_B = -\rho g\mathbf{k}$, hence $F_{Bx} = 0$, Equation (3.74) reduces to

$$0 = -\frac{\partial P}{\partial x} + \mu \frac{\partial^2 u_x}{\partial y^2}. \tag{3.77}$$

The momentum equations in the y and z directions reveal that $\frac{\partial P}{\partial y}$ and $\frac{\partial P}{\partial z}$ are equal to zero if one ignores gravitational effects.

As the fluid pressure is the same at the inlet and exit and there is no pressure build-up inside the geometry, $\partial P/\partial x = 0$ as well, which reduces Equation (3.77) to

$$\frac{\partial^2 u_x}{\partial y^2} = 0. \tag{3.78}$$

Integrating Equation (3.78) with respect to y twice yields $u_x(y) = c_1 + c_2 y$ where c_1 and c_2 are constants to be determined by using the boundary conditions. $u_x(0) = 0$ requires that $c_1 = 0$, and $u_x(h) = V$ requires that $c_2 = V/h$, so

$$u_x(y) = \frac{V}{h} y \tag{3.79}$$

which is a linear velocity profile as shown in Figure 3.19.

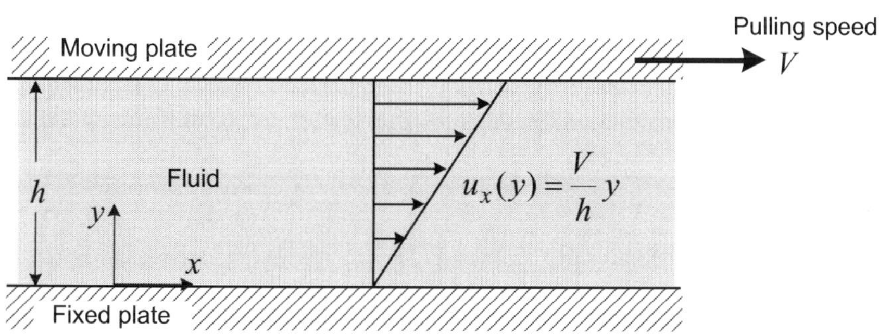

Figure 3.19: Velocity profile in a rectangular channel induced by a moving plate at constant velocity V.

Step 6: The viscous shear stress τ_{xy} on fluid at any location is $\tau_{xy} = \mu[(\partial u_x/\partial y) + (\partial u_y/\partial x)] = \mu(V/h + 0) = \mu V/h$ which is a constant. The flow rate across a vertical cross-section of the channel is $Q = \int_0^h a u_x(y) dy = \int_0^h a(Vy/h) dy = aVh/2$ where a is the width of the channel in the z direction. The flow rate is linearly proportional to V and h. The shear stress τ_{xy}, which is equal to the required pulling force per unit plate area, is linearly proportional to V, but inversely proportional to h. ■

Example 3.6: Combined Couette-Poiseuille Flow

Couette flow was studied in the previous example. Another flow problem that was studied by and named after Poiseuille has the same geometry as in the Couette problem, except that both plates are fixed, but there is a pressure drop applied across the two ends of the channel. We will introduce "Combined Couette-Poiseuille Flow" in this example. The problem geometry and the boundary conditions are shown in Figure 3.20. The fluid pressure is $P = P_i$ at $x = 0$ and $P = P_e$ at $x = L$.

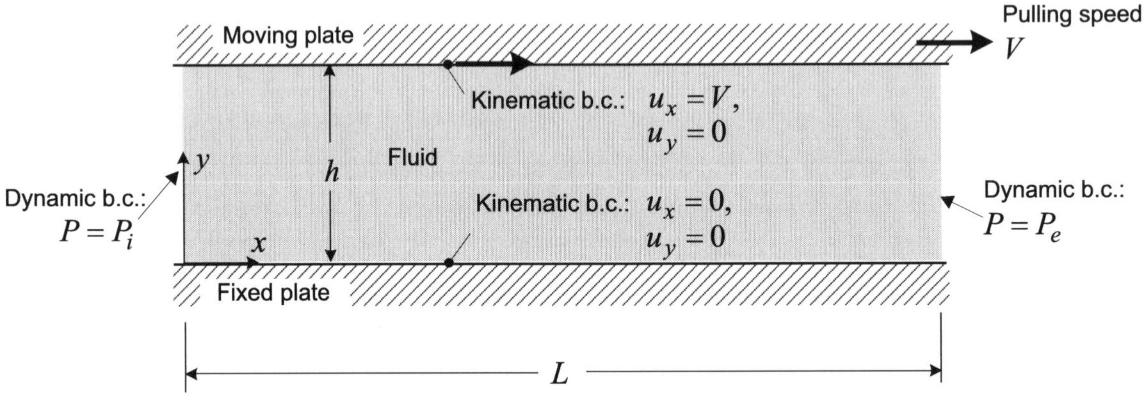

Figure 3.20: Combined Couette-Poiseuille flow.

Solution

Step 1: Cartesian coordinate system is chosen as shown in Figure 3.20. The parameters are V, h, L, μ, P_i, P_e.

Step 2: Two-dimensional steady-state flow in xy plane is considered with a Newtonian fluid. The end effects are neglected.

Step 3: The governing differential equations are Equations (3.10), (3.68), and (3.69) with the assumptions of a fully developed velocity field in the channel.

Step 4: The boundary conditions are shown in Figure 3.20 as $u_x = V$, $u_y = 0$ at $y = h$, $u_x = 0$, $u_y = 0$ at $y = 0$, $P = P_i$ along $x = 0$, and $P = P_e$ at $x = L$.

Step 5: The solution is exactly the same as in Step 5 of Example 3.5 up to the momentum equation in the x direction:

$$\frac{\partial P}{\partial x} = \mu \frac{\partial^2 u_x}{\partial y^2}. \tag{3.80}$$

In this case, $\partial P/\partial x \neq 0$ as specified by the boundary conditions. The momentum equation in the y direction, Equation (3.69) yields

$$0 = -\frac{\partial P}{\partial y} \tag{3.81}$$

neglecting the body force. Thus P is a function of x only. u_x was found to be a function of y only from the continuity equation (see Example 3.5). The left-hand side of Equation (3.80) is a function of x only, and the right-hand side is a function of y only. In order to hold this equality for all x and y within the fluid domain, both sides must be constants. Hence,

$\partial P/\partial x = c$, which means that the pressure changes linearly in the x direction. Using the dynamic boundary conditions at both ends of the channel, $\partial P/\partial x = (P_e - P_i)/L$. Equation (3.80) reduces to

$$\frac{\partial^2 u_x}{\partial y^2} = \frac{P_e - P_i}{\mu L}. \tag{3.82}$$

Integrating Equation (3.82) twice with respect to y yields $u_x(y) = [(P_e - P_i)/(2\mu L)]y^2 + c_1 y + c_2$. Boundary condition $u_x(0) = 0$ yields $c_2 = 0$. c_1 is found by using the boundary condition, $u_x(h) = V$. Hence,

$$u_x(y) = \underbrace{\frac{P_e - P_i}{2\mu L}(y^2 - hy)}_{\text{pressure driven flow}} + \underbrace{\frac{V}{h}y}_{\text{drag flow}}. \tag{3.83}$$

It is very interesting that the velocity profile given by Equation (3.83) is the superposition of the two $u_x(y)$'s: the first one is induced by a pressure gradient without a moving boundary

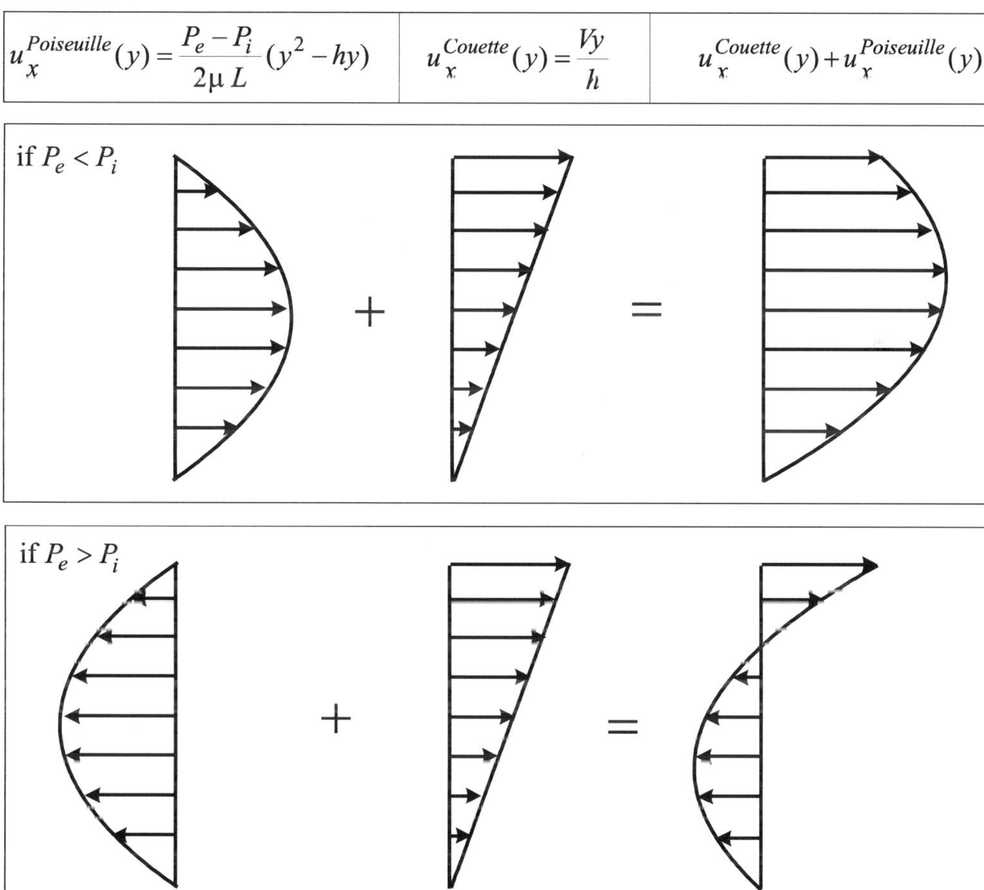

Figure 3.21: Velocity profiles for combined Couette-Poiseuille flow.

plate (Poiseuille flow), and the second one is induced by drag flow induced by a moving plate (Couette flow) as we studied in Example 3.5. The individual and superposed velocity profiles for two typical cases, where $P_e < P_o$ and $P_e > P_o$, are shown in Figure 3.21.

Step 6: What is the flow rate in this flow? Flow rate can be calculated as $Q = \int_0^h u_x(y)\, a\, dy$, where a is the width of the channel in z direction. For this flow, it will be

$$Q = \int_0^h \left[\frac{P_e - P_i}{2\mu L}(y^2 - hy) + \frac{V}{h} y \right] a\, dy = \frac{(P_i - P_e) a h^3}{12\mu L} + \frac{ahV}{2}. \tag{3.84}$$

■

Example 3.7: Flow in Concentric Cylinders due to Longitudinally Moving Inner Cylinder

Consider an incompressible fluid between inner and outer cylinders with radii R_i and R_o, respectively, as shown in Figure 3.22. The inner cylinder is pulled along the z direction with a constant velocity V. The fluid is assumed to be Newtonian with viscosity μ. Assume that the cylinders are very long in the z direction, and consider steady-state, fully developed flow. Find the velocity field, viscous stress tensor, flow rate and required pulling force per unit length.

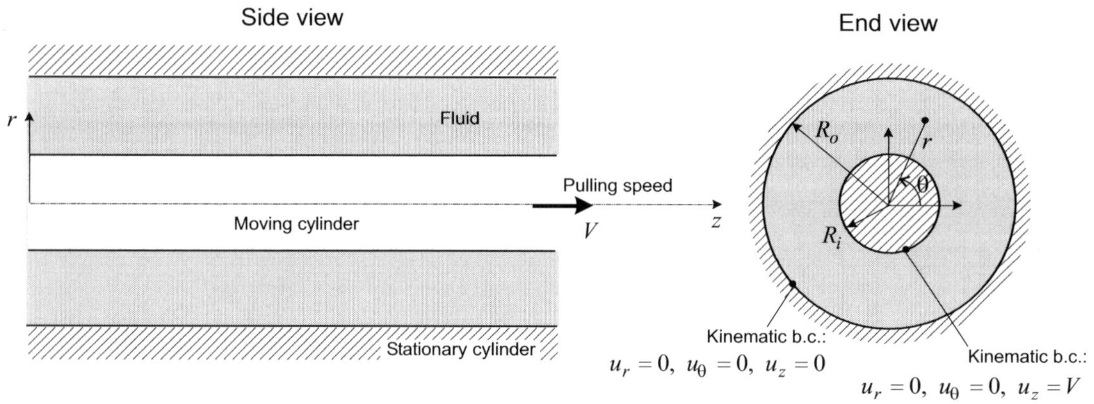

Figure 3.22: Incompressible fluid flow between two concentric cylinders induced by the motion of inner cylinder in the z direction [92].

Solution

Step 1: Cylindrical coordinate system is chosen as shown in Figure 3.22. The relevant parameters are R_i, R_o, V and μ.

Step 2: Steady state ($\partial()/\partial t = 0$), fully developed (no end effects, $\partial \mathbf{U}/\partial z = \mathbf{0}$), and incompressible flow ($D\rho/Dt = 0$, hence $\nabla \cdot \mathbf{U} = 0$) is considered in this problem. Due to symmetry, $u_\theta = 0$ and $\partial()/\partial\theta = 0$. Also, no body force (including gravity) will be considered here.

Step 3: Conservation of mass equation in the cylindrical coordinates is

$$\frac{\partial \rho}{\partial t} + \frac{1}{r}\frac{\partial}{\partial r}(\rho r u_r) + \frac{1}{r}\frac{\partial}{\partial \theta}(\rho u_\theta) + \frac{\partial}{\partial z}(\rho u_z) = 0. \tag{3.85}$$

The conservation of momentum equation in the r direction:

$$\rho\frac{\partial u_r}{\partial t} + \rho\left(u_r\frac{\partial u_r}{\partial r} + \frac{u_\theta}{r}\frac{\partial u_r}{\partial \theta} - \frac{u_\theta^2}{r} + u_z\frac{\partial u_r}{\partial z}\right)$$
$$= -\frac{\partial P}{\partial r} + \left[\frac{1}{r}\frac{\partial}{\partial r}(r\tau_{rr}) + \frac{1}{r}\frac{\partial}{\partial \theta}(r\tau_{\theta r}) + \frac{\partial}{\partial z}(\tau_{zr}) - \frac{\tau_{\theta\theta}}{r}\right] + F_{Br} \tag{3.86}$$

and the conservation of momentum equation in the z direction:

$$\rho\frac{\partial u_z}{\partial t} + \rho\left(u_r\frac{\partial u_z}{\partial r} + \frac{u_\theta}{r}\frac{\partial u_z}{\partial \theta} + u_z\frac{\partial u_z}{\partial z}\right)$$
$$= -\frac{\partial P}{\partial z} + \left[\frac{1}{r}\frac{\partial}{\partial r}(r\tau_{rz}) + \frac{1}{r}\frac{\partial}{\partial \theta}(r\tau_{\theta z}) + \frac{\partial}{\partial z}(\tau_{zz})\right] + F_{Bz}. \tag{3.87}$$

Step 4: No-slip boundary conditions are

$$u_r = 0, \quad u_z = V \quad \text{at } r = R_i \tag{3.88}$$
$$u_r = 0, \quad u_z = 0 \quad \text{at } r = R_o. \tag{3.89}$$

Step 5: After eliminating the terms that are equal to zero, Equation (3.85) reduces to

$$\frac{\partial}{\partial r}(r\, u_r) = 0. \tag{3.90}$$

Integrating Equation (3.90) with respect to r yields

$$r\, u_r = f(z) \tag{3.91}$$

as ru_r is not a function of r, but could be a function of z. But, as the flow is fully developed (so $\partial u_r/\partial z = 0$), $f(z)$ has to be equal to a constant

$$r\, u_r = c_1. \tag{3.92}$$

Applying boundary condition either (3.88) or (3.89), $c_1 = 0$, hence $u_r = 0$. The only non-zero velocity component is then u_z. For an incompressible Newtonian fluid, one can now find the components of the stress tensor as follows:

$$\tau_{rr} = 2\mu\frac{\partial u_r}{\partial r} = 0 \tag{3.93a}$$

$$\tau_{\theta\theta} = 2\mu\left(\frac{1}{r}\frac{\partial u_\theta}{\partial \theta} + \frac{u_r}{r}\right) = 0 \tag{3.93b}$$

$$\tau_{zz} = 2\mu \frac{\partial u_z}{\partial z} = 0 \tag{3.93c}$$

$$\tau_{r\theta} = \mu \left[r \frac{\partial}{\partial r} \left(\frac{u_\theta}{r} \right) + \frac{1}{r} \frac{\partial u_r}{\partial \theta} \right] = 0 \tag{3.93d}$$

$$\tau_{\theta z} = \mu \left[\frac{\partial u_\theta}{\partial z} + \frac{1}{r} \frac{\partial u_z}{\partial \theta} \right] = 0 \tag{3.93e}$$

$$\tau_{rz} = \mu \left[\frac{\partial u_z}{\partial r} + \frac{\partial u_r}{\partial z} \right] = \mu \frac{\partial u_z}{\partial r}. \tag{3.93f}$$

The only non-zero stress component is τ_{rz}, and its derivative with respect to z is also zero. Hence, one can simplify the momentum Equations (3.86) and (3.87) to

$$0 = -\frac{\partial P}{\partial r} \tag{3.94}$$

$$0 = -\frac{\partial P}{\partial z} + \mu \frac{1}{r} \frac{\partial}{\partial r} \left(r \frac{\partial u_z}{\partial r} \right). \tag{3.95}$$

Equation (3.94) implies that $P \neq P(r)$. There is no external pressure drop applied in the z direction either; hence $P \neq P(z)$. Equation (3.95) reduces to

$$\frac{\partial}{\partial r} \left(r \frac{\partial u_z}{\partial r} \right) = 0. \tag{3.96}$$

Integrating Equation (3.96) with respect to r yields

$$r \frac{\partial u_z}{\partial r} = c_1, \tag{3.97}$$

and further integration results in

$$u_z = c_1 \ln r + c_2. \tag{3.98}$$

Boundary conditions (3.88) and (3.89) can be applied which allows us to find c_1 and c_2:

$$u_z(R_i) = c_1 \ln R_i + c_2 = V, \tag{3.99}$$

$$u_z(R_o) = c_1 \ln R_o + c_2 = 0. \tag{3.100}$$

Subtraction of Equation (3.100) from Equation (3.99) yields $c_1 = V/\ln(R_i/R_o)$ and $c_2 = -c_1 \ln R_o$. Substitution of c_1 and c_2 in Equation (3.98) gives the velocity profile

$$u_z(r) = V \frac{\ln(r/R_o)}{\ln(R_i/R_o)}. \tag{3.101}$$

This velocity profile is sketched in Figure 3.23.

Step 6: The flow rate is

$$\begin{aligned} Q &= \int_{R_i}^{R_o} u_z(r)\, 2\pi r\, dr \\ &= -\frac{2\pi V}{\ln(R_i/R_o)} \left[\frac{1}{4}(R_o^2 - R_i^2) - \frac{1}{2} R_i^2 \ln(R_i/R_o) \right]. \end{aligned} \tag{3.102}$$

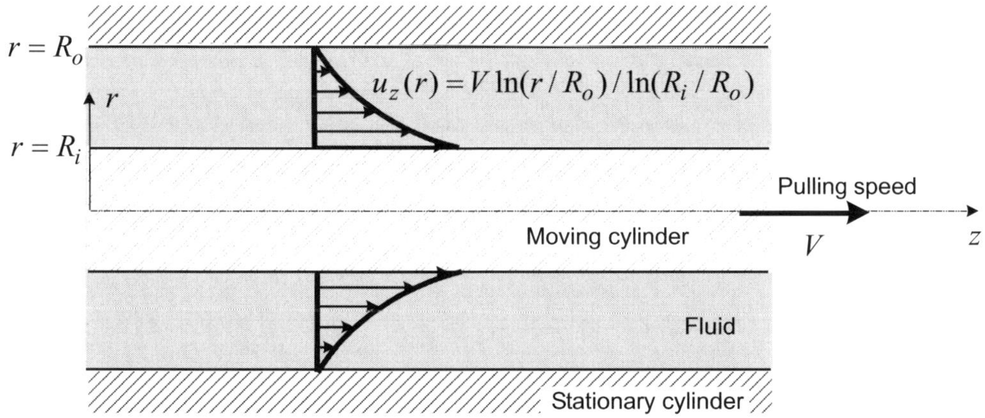

Figure 3.23: Velocity profile due to inner cylinder moving with speed V in the z direction.

The only non-zero stress component τ_{zr} is calculated as follows:

$$\begin{aligned}\tau_{zr} &= \mu\left(\frac{\partial u_r}{\partial z}+\frac{\partial u_z}{\partial r}\right)\\ &= \mu\left\{0+\frac{\partial}{\partial r}\left[V\frac{\ln(r/R_o)}{\ln(R_i/R_o)}\right]\right\}\\ &= \frac{\mu V}{r\,\ln(R_i/R_o)}.\end{aligned} \tag{3.103}$$

The external force per unit z length, F_i, required to pull the cylinder to maintain its speed at velocity V is calculated as

$$\begin{aligned}F_i &= A\,\tau_{zr}\Big|_{r=R_i} = (2\pi R_i)\left.\frac{\mu V}{r\,\ln(R_i/R_o)}\right|_{r=R_i}\\ &= \frac{2\pi\mu V}{\ln(R_i/R_o)}.\end{aligned} \tag{3.104}$$

Note that $F_o = (2\pi R_o)\tau_{zr}|_{r=R_o} = 2\pi\mu V/\ln(R_i/R_o)$ acts on the stationary outer cylinder, which is the same as F_i. ■

Example 3.8: Flow in a Tapered Unit Cell due to Longitudinally Moving Inner Cylinder

Consider an incompressible fluid between an inner cylinder and an outer tapered cylinder as shown in Figure 3.24. This problem models the "unit cell" flow in the pultrusion process in which a resin matrix and many parallel fiber rovings are pulled together with a speed of U through a tapered and heated die for the consolidation of a composite part. Assuming that the flow is dominantly in the pulling direction, one may solve the flow problem in a periodic tapered annulus around a single fiber roving as shown in Figure 3.24. Here in this example, we will solve for the velocity field only. In Chapter 7, this problem will be revisited for the calculation of pressure field and flow rate for a typical window of process parameters.

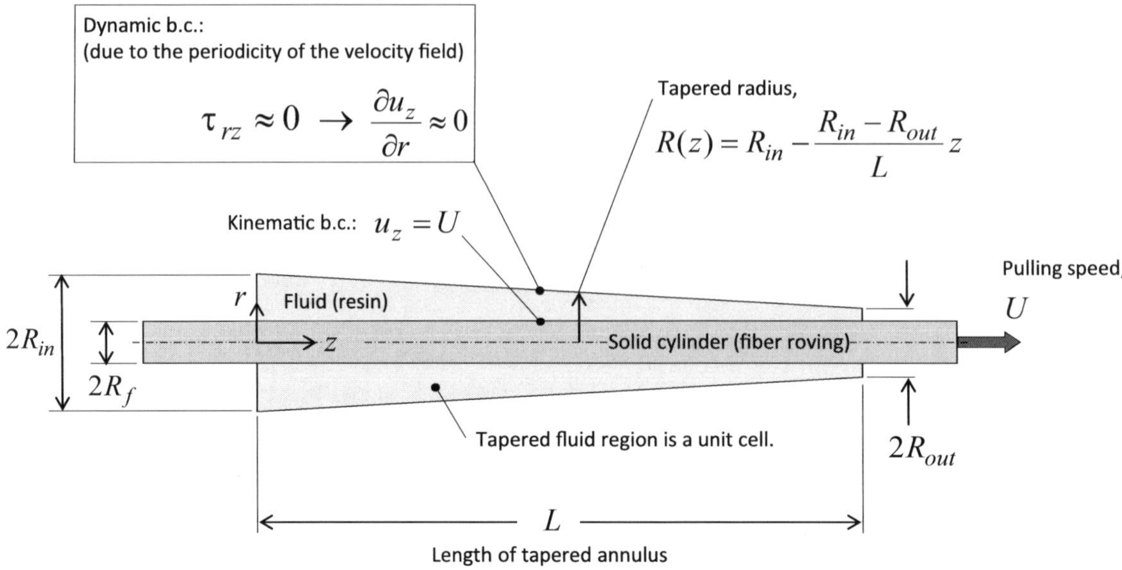

Figure 3.24: Incompressible fluid flow in a tapered unit cell.

The fiber roving has a radius of R_f and the tapered annulus has an inlet and outlet radii of R_{in} and R_{out}, respectively. The inner cylinder (fiber roving) is pulled along the z direction with a constant velocity U. The fluid is assumed to be Newtonian with a viscosity μ. Assume that the tapered annulus is very long in the z direction compared to the transverse dimensions (that means the tapered die's length, L is much bigger than its diameter), and consider steady-state flow. Notice that we will not look for a fully developed flow due to the tapered domain. Indeed, we expect to see that the velocity field changes with z as well as r.

Solution

Step 1: Cylindrical coordinate system is chosen as shown in Figure 3.24. The relevant parameters are R_f, R_{in}, R_{out}, U and μ.

Step 2: Steady state ($\partial(\,)/\partial t = 0$) and incompressible flow ($D\rho/Dt = 0$, hence $\nabla \cdot \mathbf{U} = 0$) is considered. No body force (including gravity) is considered. Due to symmetry, $u_\theta = 0$ and $\partial(\,)/\partial\theta = 0$.

Step 3: Conservation of mass equation in the cylindrical coordinates is

$$\frac{\partial \rho}{\partial t} + \frac{1}{r}\frac{\partial}{\partial r}(\rho r u_r) + \frac{1}{r}\frac{\partial}{\partial \theta}(\rho u_\theta) + \frac{\partial}{\partial z}(\rho u_z) = 0. \tag{3.105}$$

The conservation of momentum equation in the r direction:

$$\rho\left[\frac{\partial u_r}{\partial t} + u_r\frac{\partial u_r}{\partial r} + \frac{u_\theta}{r}\frac{\partial u_r}{\partial \theta} - \frac{u_\theta^2}{r} + u_z\frac{\partial u_r}{\partial z}\right]$$
$$= -\frac{\partial P}{\partial r} + \mu\left[\frac{\partial}{\partial r}\left(\frac{1}{r}\frac{\partial}{\partial r}(r u_r)\right) + \frac{1}{r^2}\frac{\partial^2 u_r}{\partial \theta^2} + \frac{\partial^2 u_r}{\partial z^2} - \frac{2}{r^2}\frac{\partial u_\theta}{\partial \theta}\right] + F_{Br} \tag{3.106}$$

and the conservation of momentum equation in the z direction:

$$\rho\left[\frac{\partial u_z}{\partial t}+u_r\frac{\partial u_z}{\partial r}+\frac{u_\theta}{r}\frac{\partial u_z}{\partial \theta}+u_z\frac{\partial u_z}{\partial z}\right]$$
$$=-\frac{\partial P}{\partial z}+\mu\left[\frac{1}{r}\frac{\partial}{\partial r}\left(r\frac{\partial u_z}{\partial r}\right)+\frac{1}{r^2}\frac{\partial^2 u_z}{\partial \theta^2}+\frac{\partial^2 u_z}{\partial z^2}\right]+F_{Bz} \quad (3.107)$$

Step 4: The kinematic boundary condition (no-slip) on the fiber roving surface is considered:

$$u_z = U \quad \text{at } r = R_f. \quad (3.108)$$

In adjacent two unit cells, the velocity fields are assumed to be symmetric about the annulus outer boundary. This suggests that

$$\frac{\partial u_z}{\partial r} = 0 \quad \text{at } r = R(z). \quad (3.109)$$

Step 5: For this problem, we will neither use conservation of mass, nor the conservation of momentum in r direction. All we need is the conservation of momentum in z direction. After eliminating the terms that are equal to zero and also assuming that the term $\rho(u_r\frac{\partial u_z}{\partial r}+u_z\frac{\partial u_z}{\partial z})$ is relatively small compared to $\mu\frac{1}{r}\frac{\partial}{\partial r}(r\frac{\partial u_z}{\partial r})$, Equation (3.107) reduces to

$$\frac{\partial}{\partial r}\left(r\frac{\partial u_z}{\partial r}\right)=\frac{1}{\mu}\frac{\partial P}{\partial z}r. \quad (3.110)$$

We assume that the pressure is a function of z (i.e., it does not vary in r direction as much as in z direction). Thus, the partial derivative on the right-hand side of Equation (3.110) becomes an ordinary derivative:

$$\frac{\partial}{\partial r}\left(r\frac{\partial u_z}{\partial r}\right)=\frac{1}{\mu}\frac{dP}{dz}r. \quad (3.111)$$

Integrating Equation (3.111) with respect to r yields

$$r\frac{\partial u_z}{\partial r}=\frac{1}{\mu}\frac{dP}{dz}\frac{r^2}{2}+A \quad (3.112)$$

or,

$$\frac{\partial u_z}{\partial r}=\frac{1}{2\mu}\frac{dP}{dz}r+\frac{A}{r}. \quad (3.113)$$

Integrating Equation (3.113) with respect to r yields

$$u_z=\frac{1}{4\mu}\frac{dP}{dz}r^2+A\ln r+B. \quad (3.114)$$

To calculate the integration constants A and B, one needs to apply the boundary conditions. Let's apply the boundary condition, Equation (3.109) first:

$$\left.\frac{\partial u_z}{\partial r}\right|_{r=R(z)} = 0 = \frac{1}{2\mu}\frac{dP}{dz}R(z) + \frac{A}{R(z)} \tag{3.115}$$

which results in $A = A(z) = -\frac{1}{2\mu}\frac{dP}{dz}R^2(z)$. Notice that A is a function of z. This is OK since the integration had been carried out in r in Equation (3.111) when A appeared in Equation (3.112). The velocity field becomes

$$u_z = \frac{1}{4\mu}\frac{dP}{dz}\left(r^2 - 2R^2(z)\ln r\right) + B. \tag{3.116}$$

Let's apply the other boundary condition, Equation (3.108) to calculate B:

$$u_z(R_f) = U = \frac{1}{4\mu}\frac{dP}{dz}\left(R_f^2 - 2R^2(z)\ln R_f\right) + B. \tag{3.117}$$

By substituting B in Equation (3.116),

$$u_z(r,z) = \frac{1}{4\mu}\frac{dP}{dz}\left[r^2 - R_f^2 + 2R^2(z)\ln\left(\frac{R_f}{r}\right)\right] + U. \tag{3.118}$$

Step 6: The velocity profile, Equation (3.118) will be used in Chapter 7 while modeling pultrusion process. One can design the process parameters (within the constraints and allowed process parameter window) to achieve the desired manufacturing goals.

3.6 Conservation of Energy

In the previous sections, the mathematical modeling was based on the assumption of isothermal flow. For isothermal flows, all we need are conservation equations of mass and momentum, constitutive equations to relate the stress tensor to fluid deformation, and a set of boundary conditions. In manufacturing, as one has to heat and cool the resin, the process is hardly isothermal although some of the manufacturing operations may be. The temperature will play a very significant role due to the sensitivity of the resin viscosity to the temperature and exothermic reactions in case of thermoset resins.

Hence, for non-isothermal cases, there is a need for an additional equation in order to solve for the flow parameters. Considering the physical law of the conservation of energy within any control volume, this equation will be derived here. Although a relatively detailed derivation will be presented here, readers are referred to Richardson [93] since the derivation requires some thermodynamics knowledge.

The momentum equation was given in Equation (3.37) for any type of fluid flow. Although the next step may not be obvious to some readers at the moment, the dot product of Equation (3.37) with velocity $\mathbf{U}$ results in:

$$\mathbf{U}\cdot\left(\rho\frac{D\mathbf{U}}{Dt}\right) = \mathbf{U}\cdot(-\nabla P) + \mathbf{U}\cdot(\nabla\cdot\tau) + \mathbf{U}\cdot(\mathbf{F}_B). \tag{3.119}$$

Since $(D/Dt)(\mathbf{A}\cdot\mathbf{B}) = \mathbf{B}\cdot(D\mathbf{A}/Dt) + \mathbf{A}\cdot(D\mathbf{B}/Dt)$ for any two vectors $\mathbf{A}$ and $\mathbf{B}$, one can write $(D/Dt)(\mathbf{U}\cdot\mathbf{U}) = 2\mathbf{U}\cdot(D\mathbf{U}/Dt)$. Hence, the first term in Equation (3.119) can be rewritten as $\mathbf{U}\cdot[\rho(D\mathbf{U}/Dt)] = \rho\, D(\frac{1}{2}\mathbf{U}\cdot\mathbf{U})/Dt = \rho\, D(\frac{1}{2}|\mathbf{U}|^2)/Dt$ where $\frac{1}{2}\mathbf{U}\cdot\mathbf{U} = \frac{1}{2}|\mathbf{U}|^2$ is the kinetic energy of fluid particles per unit mass. Hence, Equation (3.119) can be recast as

$$\rho\frac{D}{Dt}\left(\frac{1}{2}|\mathbf{U}|^2\right) = -\mathbf{U}\cdot(\nabla P) + \mathbf{U}\cdot(\nabla\cdot\tau) + \mathbf{U}\cdot\mathbf{F}_B \qquad (3.120)$$

which is called the conservation of mechanical energy equation. Recall that $-\nabla P + \nabla\cdot\tau = \nabla\cdot\sigma$ is the surface force per unit volume due to the total stresses, and $\mathbf{F}_B$ is the body force per unit volume. Also recall from the introduction to physics and dynamics courses that $\mathbf{U}\cdot\mathbf{F}$ is the rate of work done by any force $\mathbf{F}$ on the material (fluid particles here) moving with the velocity $\mathbf{U}$. Hence, one can physically describe the conservation of mechanical energy equation (Equation (3.120)) as follows: the rate of increase (accumulation) of kinetic energy per unit volume at any point in the fluid domain is equal to the summation of the rate of work done per unit volume by the forces due to the fluid stresses and body forces.

Consider the control volume shown in Figure 3.2 which is fixed in space. One can write the energy balance for this control volume as follows:

$$\begin{pmatrix}\text{Rate of increase}\\ \text{of internal and}\\ \text{kinetic energy}\end{pmatrix} = \begin{pmatrix}\text{Inflow flux}\\ \text{of internal and}\\ \text{kinetic energy}\end{pmatrix} + \begin{pmatrix}\text{Inflow flux}\\ \text{of heat}\end{pmatrix} + \begin{pmatrix}\text{Rate of energy}\\ \text{increase due to}\\ \text{total stress } \sigma\end{pmatrix}$$
$$+ \begin{pmatrix}\text{Rate of energy}\\ \text{increase due to}\\ \text{body forces } \mathbf{F}_B\end{pmatrix} + \begin{pmatrix}\text{Rate of energy}\\ \text{generation}\end{pmatrix} \qquad (3.121)$$

The individual terms can be expressed mathematically as follows:

$$\begin{pmatrix}\text{Rate of increase}\\ \text{of internal and}\\ \text{kinetic energy}\end{pmatrix} = \frac{d}{dt}\int_{\mathcal{V}}\left(\rho E + \rho\frac{1}{2}|\mathbf{U}|^2\right)dV = \int_{\mathcal{V}}\frac{\partial}{\partial t}\left(\rho E + \rho\frac{1}{2}|\mathbf{U}|^2\right)dV \qquad (3.122)$$

where E is the internal energy of fluid particles per unit mass. The total time derivative was moved inside the integral as a partial time derivative, and the additional terms were eliminated using the Leibniz rule since $\mathcal{V}$ is fixed. The first term on the right-hand side of Equation (3.121) can be mathematically represented as

$$\begin{pmatrix}\text{Inflow flux}\\ \text{of internal and}\\ \text{kinetic energy}\end{pmatrix} = \int_{\mathcal{S}} -\left(\rho E + \rho\frac{1}{2}|\mathbf{U}|^2\right)\mathbf{n}\cdot\mathbf{U}\,dA. \qquad (3.123)$$

The minus sign is due to the fact that $\mathbf{n}$ is the outward normal vector, and $-\mathbf{n}\cdot\mathbf{U}\,dA$ is the infinitesimal inflow velocity flux of fluid particles through the infinitesimal surface area dA. The second term on the right-hand side of Equation (3.121) can be expressed as

$$\begin{pmatrix}\text{Inflow flux}\\ \text{of heat}\end{pmatrix} = \int_{\mathcal{S}} -\mathbf{n}\cdot\mathbf{q}\,dA \qquad (3.124)$$

where $\mathbf{q}$ is the heat flux vector. The third term on the RHS of Equation (3.121) will be

$$\begin{pmatrix} \text{Rate of energy} \\ \text{increase due to} \\ \text{total stress } \sigma \end{pmatrix} = \int_{\mathcal{S}} \mathbf{n} \cdot (\sigma \cdot \mathbf{U}) \, dA \tag{3.125}$$

where σ is the total stress tensor. The energy contribution due to the body force term will be

$$\begin{pmatrix} \text{Rate of energy} \\ \text{increase due to} \\ \text{body forces } \mathbf{F}_B \end{pmatrix} = \int_{\mathcal{V}} \mathbf{U} \cdot \mathbf{F}_B \, dV \tag{3.126}$$

and finally, the last contribution will be

$$\begin{pmatrix} \text{Rate of energy} \\ \text{generation} \end{pmatrix} = \int_{\mathcal{V}} \dot{R} \, dV \tag{3.127}$$

where $\dot{R}$ is the rate of energy generation per unit volume due to internal energy sources (such as chemical and nuclear reactions). Equation (3.121) can be assembled in mathematical form as follows:

$$\begin{aligned} \int_{\mathcal{V}} \frac{\partial}{\partial t}\left(\rho E + \rho \frac{1}{2}|\mathbf{U}|^2\right) dV &= -\int_{\mathcal{S}} \left(\rho E + \rho \frac{1}{2}|\mathbf{U}|^2\right) \mathbf{n} \cdot \mathbf{U} \, dA - \int_{\mathcal{S}} \mathbf{n} \cdot \mathbf{q} \, dA \\ &\quad + \int_{\mathcal{S}} \mathbf{n} \cdot (\sigma \cdot \mathbf{U}) \, dA + \int_{\mathcal{V}} \mathbf{U} \cdot \mathbf{F}_B \, dV + \int_{\mathcal{V}} \dot{R} \, dV. \end{aligned} \tag{3.128}$$

Applying Gauss' divergence theorem (Equation (3.5)), the first three surface integrals on the right-hand side of Equation (3.128) are converted to volume integrals, and Equation (3.128) becomes

$$\begin{aligned} &\int_{\mathcal{V}} \left\{ \frac{\partial}{\partial t}\left(\rho E + \rho \frac{1}{2}|\mathbf{U}|^2\right) + \nabla \cdot \left[\left(\rho E + \rho \frac{1}{2}|\mathbf{U}|^2\right) \mathbf{U}\right] \right. \\ &\qquad \left. + \nabla \cdot \mathbf{q} + \nabla \cdot (\sigma \cdot \mathbf{U}) + \mathbf{U} \cdot \mathbf{F}_B + \dot{R} \right\} dV \\ &= 0. \end{aligned} \tag{3.129}$$

Since the control volume $\mathcal{V}$ is arbitrary, not only the integral but actually the integrand must be zero:

$$\frac{\partial}{\partial t}\left(\rho E + \rho \frac{1}{2}|\mathbf{U}|^2\right) + \nabla \cdot \left[\left(\rho E + \rho \frac{1}{2}|\mathbf{U}|^2\right) \mathbf{U}\right] + \nabla \cdot \mathbf{q} + \nabla \cdot (\sigma \cdot \mathbf{U}) + \mathbf{U} \cdot \mathbf{F}_B + \dot{R} = 0. \tag{3.130}$$

The first two terms in Equation (3.130) can be rewritten as

$$\begin{aligned} &\frac{\partial}{\partial t}\left[\rho\left(E + \frac{1}{2}|\mathbf{U}|^2\right)\right] + \nabla \cdot \left[\rho\left(E + \frac{1}{2}|\mathbf{U}|^2\right) \mathbf{U}\right] \\ &= \left(E + \frac{1}{2}|\mathbf{U}|^2\right) \frac{\partial \rho}{\partial t} + \rho \frac{\partial}{\partial t}\left(E + \frac{1}{2}|\mathbf{U}|^2\right) \\ &\quad + \mathbf{U} \cdot \nabla \left[\rho\left(E + \frac{1}{2}|\mathbf{U}|^2\right)\right] + \rho\left(E + \frac{1}{2}|\mathbf{U}|^2\right) \nabla \cdot \mathbf{U} \end{aligned}$$

$$\begin{aligned}
&= \left(E+\frac{1}{2}|\mathbf{U}|^2\right)\left[\frac{\partial \rho}{\partial t}+\rho\nabla\cdot\mathbf{U}\right]+\rho\frac{\partial}{\partial t}\left(E+\frac{1}{2}|\mathbf{U}|^2\right)+\mathbf{U}\cdot\nabla\left[\rho\left(E+\frac{1}{2}|\mathbf{U}|^2\right)\right]\\
&= \left(E+\frac{1}{2}|\mathbf{U}|^2\right)(-\mathbf{U}\cdot\nabla\rho-s)+\rho\frac{\partial}{\partial t}\left(E+\frac{1}{2}|\mathbf{U}|^2\right)\\
&\quad +\mathbf{U}\cdot\left[\left(E+\frac{1}{2}|\mathbf{U}|^2\right)\nabla\rho+\rho\nabla\left(E+\frac{1}{2}|\mathbf{U}|^2\right)\right]\\
&= -\left(E+\frac{1}{2}|\mathbf{U}|^2\right)(\mathbf{U}\cdot\nabla\rho)-s\left(E+\frac{1}{2}|\mathbf{U}|^2\right)+\rho\frac{\partial}{\partial t}\left(E+\frac{1}{2}|\mathbf{U}|^2\right)\\
&\quad +\left(E+\frac{1}{2}|\mathbf{U}|^2\right)(\mathbf{U}\cdot\nabla\rho)+\rho\left[\mathbf{U}\cdot\nabla\left(E+\frac{1}{2}|\mathbf{U}|^2\right)\right]\\
&= -s\left(E+\frac{1}{2}|\mathbf{U}|^2\right)+\rho\left[\frac{\partial}{\partial t}\left(E+\frac{1}{2}|\mathbf{U}|^2\right)+\mathbf{U}\cdot\nabla\left(E+\frac{1}{2}|\mathbf{U}|^2\right)\right]\\
&= -s\left(E+\frac{1}{2}|\mathbf{U}|^2\right)+\rho\frac{D}{Dt}\left(E+\frac{1}{2}|\mathbf{U}|^2\right).
\end{aligned} \tag{3.131}$$

In the third equality line of Equation (3.131), $\frac{\partial\rho}{\partial t}+\rho\nabla\cdot\mathbf{U}$ was replaced by $-\mathbf{U}\cdot\nabla\rho-s$ by using the conservation equation (Equation (3.8)). After substituting Equation (3.131) into Equation (3.130), it reduces to

$$\rho\frac{D}{Dt}\left(E+\frac{1}{2}|\mathbf{U}|^2\right)-s\left(E+\frac{1}{2}|\mathbf{U}|^2\right)+\nabla\cdot\mathbf{q}-\nabla\cdot(\sigma\cdot\mathbf{U})-\mathbf{U}\cdot\mathbf{F}_B-\dot{R}=0. \tag{3.132}$$

Equation (3.132) is expanded by writing the total stress tensor as $\sigma=-P\mathbf{I}+\tau$

$$\rho\frac{DE}{Dt}+\rho\frac{D}{Dt}\left(\frac{1}{2}|\mathbf{U}|^2\right)-s\left(E+\frac{1}{2}|\mathbf{U}|^2\right)+\nabla\cdot\mathbf{q}+\nabla\cdot(P\mathbf{U})-\nabla\cdot(\tau\cdot\mathbf{U})-\mathbf{U}\cdot\mathbf{F}_B-\dot{R}=0. \tag{3.133}$$

$\nabla\cdot(\tau\cdot\mathbf{U})$ that appears in Equation (3.133) can be expanded as $\tau:(\nabla\mathbf{U})+\mathbf{U}\cdot(\nabla\cdot\tau)$. The double dot product between two tensors, say $\mathbf{A}$ and $\mathbf{B}$, is defined as $\mathbf{A}:\mathbf{B}=A_{ij}B_{ji}=A_{11}B_{11}+A_{12}B_{21}+A_{13}B_{31}+A_{21}B_{12}+A_{22}B_{22}+A_{23}B_{32}+A_{31}B_{13}+A_{32}B_{23}+A_{33}B_{33}$, which is a scalar. Equation (3.133) then reduces to

$$\begin{aligned}
\rho\frac{DE}{Dt}+\rho\frac{D}{Dt}\left(\frac{1}{2}|\mathbf{U}|^2\right)-s\left(E+\frac{1}{2}|\mathbf{U}|^2\right)+\nabla\cdot\mathbf{q}+\nabla\cdot(P\mathbf{U})&\\
-\tau:(\nabla\mathbf{U})-\mathbf{U}\cdot(\nabla\cdot\tau)-\mathbf{U}\cdot\mathbf{F}_B-\dot{R}=0.&
\end{aligned} \tag{3.134}$$

The second term on the left-hand side of Equation (3.134) is rewritten by substituting the expansion from Equation (3.120). Thus, Equation (3.134) expands to the form

$$\begin{aligned}
&\rho\frac{DE}{Dt}+[-\mathbf{U}\cdot(\nabla P)+\mathbf{U}\cdot(\nabla\cdot\tau)+\mathbf{U}\cdot\mathbf{F}_B]\\
&-s\left(E+\frac{1}{2}|\mathbf{U}|^2\right)+\nabla\cdot\mathbf{q}+\nabla\cdot(P\mathbf{U})-\tau:(\nabla\mathbf{U})-\mathbf{U}\cdot(\nabla\cdot\tau)\\
&-\mathbf{U}\cdot\mathbf{F}_B-\dot{R}=0.
\end{aligned} \tag{3.135}$$

Using the fact that $\mathbf{U}\cdot(\nabla P)-\nabla\cdot(P\mathbf{U})=-P\nabla\cdot\mathbf{U}$, and simplifying Equation (3.135) results in

$$\rho\frac{DE}{Dt} = -\nabla\cdot\mathbf{q}-P\nabla\cdot\mathbf{U}+\tau:(\nabla\mathbf{U})+s\left(E+\frac{1}{2}|\mathbf{U}|^2\right)+\dot{R}. \tag{3.136}$$

This is the conservation of energy equation, but it is not very useful, since we cannot evaluate the first term, DE/Dt directly.

The internal energy E will be written by using thermodynamic relations, and then DE/Dt will be substituted into Equation (3.136) to obtain the final form of the conservation of energy equation.

One can use the thermodynamics relation $H = E + P/\rho$ in order to write the internal energy E in terms of enthalpy H. Then DE/Dt is equal to

$$\begin{aligned}\frac{DE}{Dt} &= \frac{D}{Dt}\left(H - \frac{P}{\rho}\right) = \frac{DH}{Dt} - \frac{1}{\rho}\frac{DP}{Dt} + \frac{P}{\rho^2}\frac{D\rho}{Dt} \\ &= \frac{DH}{Dt} - \frac{1}{\rho}\frac{DP}{Dt} + \frac{P}{\rho^2}(-\rho\nabla\cdot\mathbf{U}) \\ &= \frac{DH}{Dt} - \frac{1}{\rho}\frac{DP}{Dt} - \frac{P}{\rho}\nabla\cdot\mathbf{U}. \end{aligned} \tag{3.137}$$

The conservation of mass equation was used on the second line of Equation (3.137) to replace $D\rho/Dt$ by $-\rho\nabla\cdot\mathbf{U}$ in the absence of sink term s. Hence, Equation (3.136) reduces to

$$\rho\left(\frac{DH}{Dt} - \frac{1}{\rho}\frac{DP}{Dt} - \frac{P}{\rho}\nabla\cdot\mathbf{U}\right) = -\nabla\cdot\mathbf{q} - P\nabla\cdot\mathbf{U} + \tau : (\nabla\mathbf{U}) + \dot{R}. \tag{3.138}$$

Canceling the two $-P\nabla\cdot\mathbf{U}$ terms, the equation simplifies to

$$\rho\frac{DH}{Dt} - \frac{DP}{Dt} = -\nabla\cdot\mathbf{q} + \tau : (\nabla\mathbf{U}) + \dot{R}. \tag{3.139}$$

Considering that H is a function of P and T,

$$dH = \left.\frac{\partial H}{\partial P}\right|_T dP + \left.\frac{\partial H}{\partial T}\right|_P dT \tag{3.140}$$

and

$$\begin{aligned}\frac{DH}{Dt} &= \left.\frac{\partial H}{\partial P}\right|_T \frac{DP}{Dt} + \left.\frac{\partial H}{\partial T}\right|_P \frac{DT}{Dt} \\ &= \left.\frac{\partial H}{\partial P}\right|_T \frac{DP}{Dt} + c_p\frac{DT}{Dt}\end{aligned} \tag{3.141}$$

where c_p $(= \partial H/\partial T|_P)$ is the specific heat evaluated at constant pressure. After substituting Equation (3.141), Equation (3.139) takes the following form:

$$\rho\left.\frac{\partial H}{\partial P}\right|_T \frac{DP}{Dt} + \rho c_p\frac{DT}{Dt} - \frac{DP}{Dt} = -\nabla\cdot\mathbf{q} + \tau : (\nabla\mathbf{U}) + \dot{R}. \tag{3.142}$$

Also, expanding the substantial derivative DT/Dt as $\partial T/\partial t + \mathbf{U}\cdot\nabla T$, Equation (3.142) can be rewritten as

$$\rho c_p\frac{\partial T}{\partial t} + \rho c_p\mathbf{U}\cdot\nabla T = -\nabla\cdot\mathbf{q} + \tau : (\nabla\mathbf{U}) + \dot{R} + \left(1 - \rho\left.\frac{\partial H}{\partial P}\right|_T\right)\frac{DP}{Dt}. \tag{3.143}$$

This is called the *conservation of energy equation.* In many processes, the last term $\left(1 - \rho\frac{\partial H}{\partial P}|_T\right)\frac{DP}{Dt}$ is very small, so it is neglected, which then further simplifies the conservation of energy equation to

$$\underbrace{\rho c_p \frac{\partial T}{\partial t}}_{\text{transient}} + \underbrace{\rho c_p \mathbf{U}\cdot\nabla T}_{\text{convection}} = \underbrace{-\nabla\cdot\mathbf{q}}_{\text{conduction}} + \underbrace{\tau : (\nabla\mathbf{U})}_{\text{viscous dissipation}} + \underbrace{\dot{R}}_{\text{energy generation}} \tag{3.144}$$

The physical meaning of individual terms in Equation (3.144) is written underneath them. Note that the summation of the two terms on the left-hand side of Equation (3.144) is simply $\rho c_p DT/Dt$. $\dot{R}$ is usually due to chemical reactions of the resin and/or microwave induction heating of the polymer.

Example 3.9: An Alternative Derivation of Energy Equation

In this section, the derivation of the conservation of energy equation was carried out by expressing the internal energy E in terms of enthalpy H as $E = H - P/\rho$. Alternatively, one could express the internal energy E by using another thermodynamic relation and derive the energy equation in terms of different variables. In this example, we ask to you to use the infinitesimal internal energy given by

$$dE = -P\,dv + T\,dS \tag{3.145}$$

in the derivation of energy equation. Here $v = 1/\rho$ is the volume of fluid per unit mass, and S is the entropy of fluid per unit mass.

Solution

As E is a function of v and T, dE can also be written as

$$dE = \left.\frac{\partial E}{\partial v}\right|_T dv + \left.\frac{\partial E}{\partial T}\right|_v dT. \tag{3.146}$$

Hence, the total time derivative of internal energy is equal to

$$\begin{aligned}\frac{DE}{Dt} &= \left.\frac{\partial E}{\partial v}\right|_T \frac{Dv}{Dt} + \left.\frac{\partial E}{\partial T}\right|_v \frac{DT}{Dt} \\ &= \left(-P + T\left.\frac{\partial S}{\partial v}\right|_T\right)\frac{Dv}{Dt} + \left.\frac{\partial E}{\partial T}\right|_v \frac{DT}{Dt}.\end{aligned} \tag{3.147}$$

Now Dv/Dt can be written explicitly as

$$\begin{aligned}\frac{Dv}{Dt} &= \frac{D(\rho^{-1})}{Dt} \\ &= -\frac{1}{\rho^2}\frac{D\rho}{Dt} \\ &= -\frac{1}{\rho^2}\left(-\rho\nabla\cdot\mathbf{U} - s\right) \\ &= \frac{1}{\rho}\nabla\cdot\mathbf{U} + \frac{s}{\rho^2}\end{aligned} \tag{3.148}$$

where $D\rho/Dt$ was replaced by $-\rho\nabla\cdot\mathbf{U} - s$ using the conservation of mass equation. For brevity, no sink case ($s = 0$) will be considered from now on, and the substitution of Equation (3.148) into Equation (3.147) yields

$$\begin{aligned}\frac{DE}{Dt} &= -\frac{P}{\rho}\nabla\cdot\mathbf{U} + \frac{T}{\rho}\left.\frac{\partial S}{\partial v}\right|_T \nabla\cdot\mathbf{U} + \left.\frac{\partial E}{\partial T}\right|_v \frac{DT}{Dt} \\ &= -\frac{P}{\rho}\nabla\cdot\mathbf{U} + \frac{T}{\rho}\left.\frac{\partial P}{\partial T}\right|_\rho \nabla\cdot\mathbf{U} + c_v\frac{DT}{Dt} \end{aligned} \quad (3.149)$$

where c_v $(= \partial E/\partial T|_v)$ is the specific heat evaluated at constant volume, and $\frac{\partial S}{\partial v}|_T = \frac{\partial P}{\partial T}|_v = \frac{\partial P}{\partial T}|_\rho$ relation (Maxwell equation) was used to replace $\frac{\partial S}{\partial v}|_T$. After substituting Equation (3.149), Equation (3.136) reduces to

$$\rho\left(-\frac{P}{\rho}\nabla\cdot\mathbf{U} + \frac{T}{\rho}\left.\frac{\partial P}{\partial T}\right|_\rho \nabla\cdot\mathbf{U} + c_v\frac{DT}{Dt}\right) = -\nabla\cdot\mathbf{q} - P\nabla\cdot\mathbf{U} + \tau : (\nabla\mathbf{U}) + \dot{R}. \quad (3.150)$$

$-P\nabla\cdot\mathbf{U}$ on the left-hand side cancels $-P\nabla\cdot\mathbf{U}$ on the right-hand side, and expanding the substantial derivative as $DT/Dt = \partial T/\partial t + \mathbf{U}\cdot\nabla T$, Equation (3.150) reduces to

$$\rho c_v\frac{\partial T}{\partial t} + \rho c_v\mathbf{U}\cdot\nabla T = -\nabla\cdot\mathbf{q} + \tau : (\nabla\mathbf{U}) \underbrace{- T\left.\frac{\partial P}{\partial T}\right|_\rho \nabla\cdot\mathbf{U}} + \dot{R}. \quad (3.151)$$

This is the conservation of energy equation, but it is seldom used due to the need for calculation of the underlined term. For incompressible fluids, we do recover Equation (3.144). ■

3.6.1 Heat Flux-Temperature Gradient Relationship

Previously, a constitutive equation was studied to relate the stress tensor to the strain rate tensor, since the stress tensor appears within the momentum equations. In the energy equation, the heat flux $\mathbf{q}$ appears. In this section, we will study another constitutive equation to relate the heat flux to the temperature field. $\mathbf{q}$ is the conductive flux through the polymeric liquid and the reinforcing fibers. In general, heat can be transmitted in terms of conduction, convection and radiation. However, here $\mathbf{q}$ doesn't include the convection part, and in many composite manufacturing processes, the radiation term is negligible. For most materials (including many fluids and solids), the heat flux is linearly proportional to the temperature gradient and was first proposed by Fourier [94]

$$\mathbf{q} = -\mathbf{k}\cdot\nabla T \quad (3.152)$$

where $\mathbf{k}$ is the thermal conductivity tensor of the composite material. The minus sign is because heat is transmitted from hotter to colder locations. Hence, the energy equation can be rewritten as

$$\rho c_p\frac{\partial T}{\partial t} + \rho c_p\mathbf{U}\cdot\nabla T = \nabla\cdot(\mathbf{k}\cdot\nabla T) + \tau : (\nabla\mathbf{U}) + \dot{R}. \quad (3.153)$$

For an isotropic thermal conductivity which is constant within the entire domain, $\mathbf{k} \cdot \nabla T$ reduces to $k\nabla T$ where k is the scalar thermal conductivity. Hence, the energy equation further simplifies to

$$\rho c_p \frac{\partial T}{\partial t} + \rho c_p \mathbf{U} \cdot \nabla T = k\nabla^2 T + \tau : (\nabla \mathbf{U}) + \dot{R} \tag{3.154}$$

for materials with constant (uniform) isotropic thermal conductivity. However, composites may have nonisotropic thermal conductivity especially when carbon fabric is used, in which case the conduction term is expanded into

$$\begin{aligned} \nabla \cdot (\mathbf{k} \cdot \nabla T) = {} & \frac{\partial}{\partial x_1}\left(k_{11}\frac{\partial T}{\partial x_1} + k_{12}\frac{\partial T}{\partial x_2} + k_{13}\frac{\partial T}{\partial x_3}\right) \\ & + \frac{\partial}{\partial x_2}\left(k_{21}\frac{\partial T}{\partial x_1} + k_{22}\frac{\partial T}{\partial x_2} + k_{23}\frac{\partial T}{\partial x_3}\right) \\ & + \frac{\partial}{\partial x_3}\left(k_{31}\frac{\partial T}{\partial x_1} + k_{32}\frac{\partial T}{\partial x_2} + k_{33}\frac{\partial T}{\partial x_3}\right). \end{aligned} \tag{3.155}$$

■

Example 3.10: Heat Transfer in a Material with Nonisotropic Thermal Conductivity

Consider a composite of thickness h having thermal conductivities of k'_1, k'_2 and k'_3 along the principal directions of the fibers and perpendicular to the fiber direction. This composite has fibers at an angle θ with respect to the large faces of the composite as shown in Figure 3.25. The composite is held between two highly conductive materials maintained at temperatures T_U and T_L as shown in Figure 3.25.

How will you calculate the heat flux flowing normal to the large faces of the plate? Will the heat flux vector be parallel to the temperature gradient ∇T as is the case for isotropic materials?

Solution

Note that as $h \ll L, W$, one can assume isothermal surfaces parallel to the large faces of the composite. This is usually true everywhere except near the edges. If the plate thickness h is small compared to L and W, edge effects are negligible. This ∇T will be perpendicular to the lateral faces. Hence $\frac{\partial T}{\partial x_2} = 0$ and $\frac{\partial T}{\partial x_3} = 0$.

For an anisotropic composite, one could write the relationship between the heat flux and temperature gradients as follows:

$$-q_1 = k_{11}\frac{\partial T}{\partial x_1} + k_{12}\frac{\partial T}{\partial x_2} + k_{13}\frac{\partial T}{\partial x_3} \tag{3.156}$$

$$-q_2 = k_{21}\frac{\partial T}{\partial x_1} + k_{22}\frac{\partial T}{\partial x_2} + k_{23}\frac{\partial T}{\partial x_3} \tag{3.157}$$

$$-q_3 = k_{31}\frac{\partial T}{\partial x_1} + k_{32}\frac{\partial T}{\partial x_2} + k_{33}\frac{\partial T}{\partial x_3}. \tag{3.158}$$

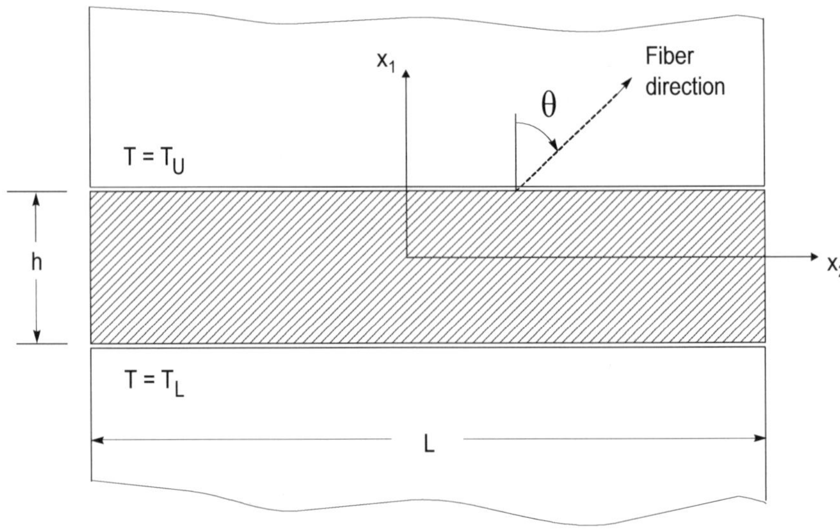

Figure 3.25: A composite section being heated by two isothermal surfaces at uniform temperatures T_U and T_L. The fiber direction in the composite is at angle θ with the vertical direction x_1 as shown.

Now as $\frac{\partial T}{\partial x_2} = 0$ and $\frac{\partial T}{\partial x_3} = 0$, Equations (3.156)-(3.158) reduce to

$$-q_1 = k_{11}\frac{\partial T}{\partial x_1} \tag{3.159}$$

$$-q_2 = k_{21}\frac{\partial T}{\partial x_1} \tag{3.160}$$

$$-q_3 = k_{31}\frac{\partial T}{\partial x_1} \tag{3.161}$$

where q_1, q_2, and q_3 are the components of the heat flux vector $\mathbf{q} = q_1\mathbf{i} + q_2\mathbf{j} + q_3\mathbf{k}$. Thus q_1 will be heat flux in the $\mathbf{i}$ direction. Also,

$$\frac{\partial T}{\partial x_1} \approx \frac{T_U - T_L}{h} \tag{3.162}$$

and k_{11} will be related to k_1', k_2' and k_3' as follows:

$$k_{11} = l_1^2 k_1' + l_2^2 k_2' + l_3^2 k_3' \tag{3.163}$$

where k_1', k_2' and k_3' are the principal conductivities and l_1, l_2 and l_3 are direction cosines of the 1-axis relative to the principal axes $1'$, $2'$ and $3'$. For this case, it simplifies to $k_{11} = \cos\theta^2 k_1' + \sin\theta^2 k_2'$. Note that as $\mathbf{q}$ has non-zero q_2 and q_3, it will not be along the same direction as ∇T. ■

3.6.2 Thermal Boundary Conditions

In general, three types of boundary conditions exist when one is interested in calculating the temperature from the conservation of energy equation. They are:

1. Boundary temperature T is prescribed.

2. The external heat flux, $q = -k\partial T/\partial n$ is prescribed, where n is the outward normal direction on the boundary.

3. Convection heat transfer coefficient h is prescribed. The related boundary condition is $q = -k\partial T/\partial n = h(T_\infty - T)$, where T_∞ is the ambient temperature adjacent to the boundary, since the convected heat flux is equal to the conducted heat flux.

Some examples of thermal boundary conditions are shown in Figure 3.26.

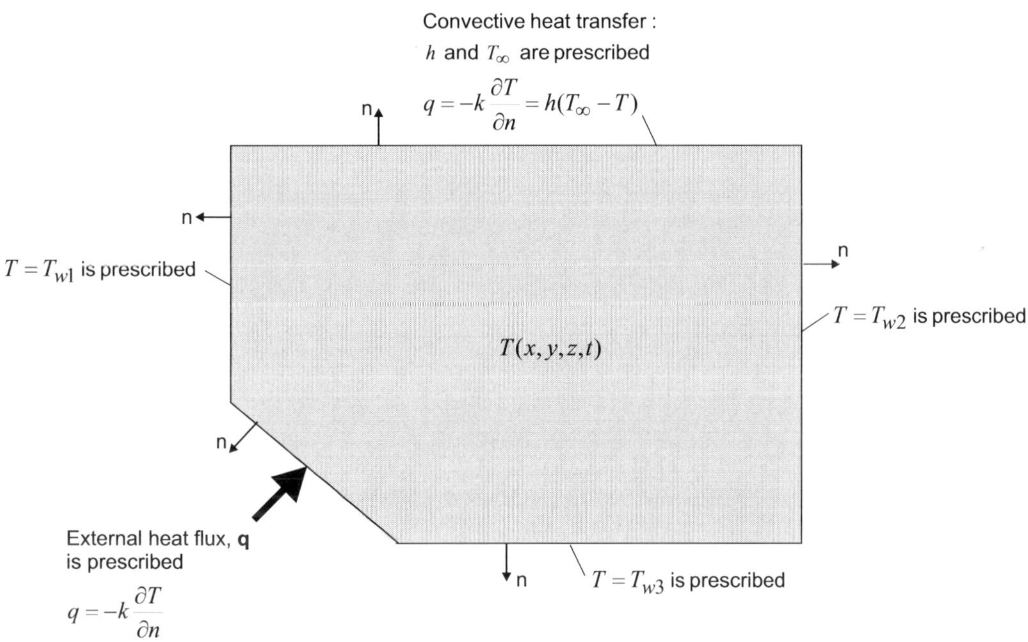

Figure 3.26: Examples of three types of thermal boundary conditions.

Example 3.11: Temperature Distribution Within Concentric Cylinders Due to Different Wall Temperatures

Consider an incompressible resin between two concentric cylinders with radii R_i and R_o, as shown in Figure 3.27(a). Assume that the gap between the two cylinders is much smaller than the radii, $(R_o - R_i) \ll R_i$. The outer cylinder is rotated around the z axis with a constant angular speed of ω. The fluid is assumed to be Newtonian, and the viscosity μ doesn't change significantly with temperature, $\mu \neq \mu(T)$. Assume that the cylinders are very long in the z direction, and consider steady-state flow. Find the resin temperature distribution if the inner and outer cylinder walls are kept at constant temperatures T_i and T_o, respectively. Also find the resultant heat flux through the system.

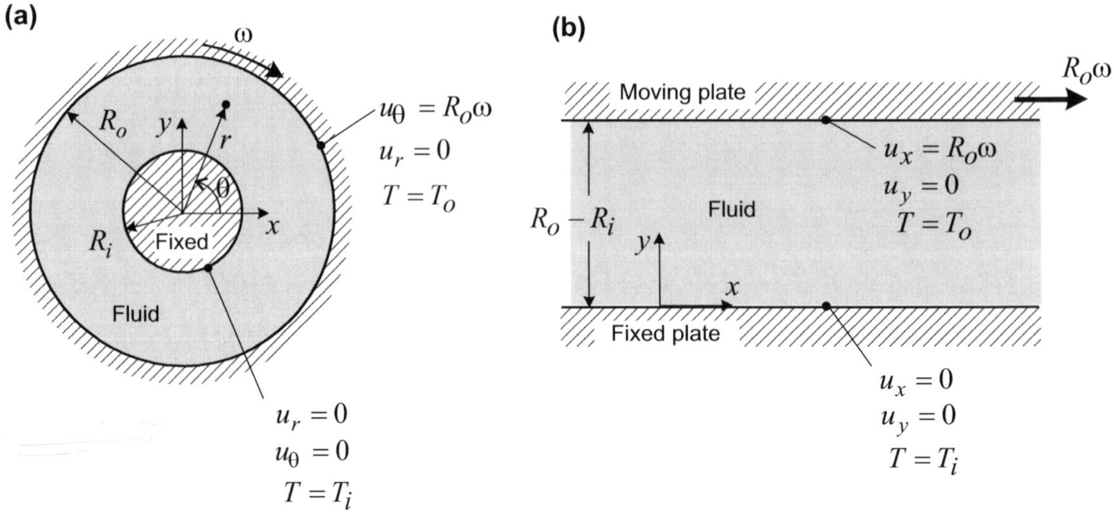

Figure 3.27: (a) Geometry for two concentric cylinders and boundary conditions. The outer cylinder rotates while the inner one is stationary. (b) The geometry can be converted into one-dimensional planar Couette flow if $R_0 - R_i \ll R_i$.

Solution

Steps 1-5: Although the cylindrical coordinate system seems to be the logical choice here, the Cartesian coordinate system is preferred because of its simplicity. The justification for this choice is that since $(R_o - R_i) \ll R_i$, the velocity component in the radial direction u_r is expected to be much smaller than the one in the tangential θ direction, u_θ. Hence the flow is assumed to be only in the θ direction. With a coordinate transformation such as $x = R_i\theta$ and $y = r - R_i$, the problem is simplified to one-dimensional in the x direction. The simplified problem geometry and the corresponding boundary conditions are shown in Figure 3.27(b). This problem is exactly the same as the one studied in Example 3.5, except that this one is nonisothermal due to the different thermal boundary conditions applied at $y = 0$ and $y = R_o - R_i$. However, since $\mu \neq \mu(T)$, the velocity distribution solutions will not change, so the result can be taken directly from Example 3.5:

$$u_x(y) = \frac{V}{h}y = \frac{R_o\omega}{R_o - R_i}y \tag{3.164}$$

since $V = R_o\omega$ and $h = R_o - R_i$.

Assuming an isotropic thermal conductivity k for the resin, the energy equation (Equation (3.154)) is written as

$$\rho c_p \frac{\partial T}{\partial t} + \rho c_p \mathbf{U} \cdot \nabla T = k\nabla^2 T + \tau : (\nabla \mathbf{U}) + \dot{R}. \tag{3.165}$$

Here $\frac{\partial T}{\partial t} = 0$ considering steady state, and $\dot{R} = 0$ since there is no heat generation within the resin. The only non-zero velocity component is u_x as found above. Considering that the resin should have the same properties at every θ (hence x) value, the temperature T is expected to be a function of y only. Hence, after dropping the zero terms, Equation (3.165) reduces to

$$\begin{aligned}
\rho c_p \left[u_x \frac{\partial T}{\partial x} + u_y \frac{\partial T}{\partial y} + u_z \frac{\partial T}{\partial z}\right] &= k\left(\frac{\partial^2 T}{\partial x^2} + \frac{\partial^2 T}{\partial y^2} + \frac{\partial^2 T}{\partial z^2}\right) + \tau : (\nabla \mathbf{U}) \\
\rho c_p \left[u_x(0) + (0)\frac{\partial T}{\partial y} + (0)(0)\right] &= k\left[(0) + \frac{\partial^2 T}{\partial y^2} + (0)\right] + \tau : (\nabla \mathbf{U}) \\
0 &= k\frac{\partial^2 T}{\partial y^2} + \tau : (\nabla \mathbf{U}) \qquad (3.166)
\end{aligned}$$

The only non-zero terms in the viscous stress tensor are τ_{xy} and τ_{yx}:

$$\tau_{xy} = \tau_{yx} = \mu\left(\frac{\partial u_x}{\partial y} + \frac{\partial u_y}{\partial x}\right) = \mu\left(\frac{R_o\omega}{R_o - R_i} + \frac{\partial(0)}{\partial x}\right) = \frac{\mu R_o \omega}{R_o - R_i}. \qquad (3.167)$$

Hence the viscous dissipation term can be calculated as

$$\begin{aligned}
\tau : (\nabla \mathbf{U}) &= \tau_{xy}\frac{\partial u_x}{\partial y} + \tau_{yx}\frac{\partial u_y}{\partial x} \\
&= \left(\mu\frac{R_o\omega}{R_o - R_i}\right)\left(\frac{R_o\omega}{R_o - R_i}\right) + \left(\mu\frac{R_o\omega}{R_o - R_i}\right)(0) \\
&= \mu\left(\frac{R_o\omega}{R_o - R_i}\right)^2 = A. \qquad (3.168)
\end{aligned}$$

The constant A is the energy generated within the fluid due to the viscous dissipation.

By substituting $\tau : (\nabla \mathbf{U})$ from Equation (3.168) and replacing $\partial^2 T/\partial y^2$ with d^2T/dy^2 as T is only a function of y, Equation (3.166) reduces to

$$\frac{d^2 T}{dy^2} = -\frac{A}{k} = B. \qquad (3.169)$$

By using the non-dimensionalized variables, $\bar{T} = T/(T_o - T_i) = T/\Delta T$ and $\bar{y} = y/(R_o - R_i) = y/h$, Equation (3.169) takes the following form:

$$\frac{d^2\bar{T}}{d\bar{y}^2} = B\left(\frac{h^2}{\Delta T}\right) = D \qquad (3.170)$$

where $D = -\frac{\mu}{k}(\frac{R_o\omega}{\Delta T})^2 = \frac{\mu}{k}(\frac{V}{\Delta T})^2$. Integrating Equation (3.170) with respect to $\bar{y}$ twice yields

$$T(\bar{y}) = \frac{1}{2}D\bar{y}^2 + c_1\bar{y} + c_2 \qquad (3.171)$$

where the constants c_1 and c_2 are to be determined by using the thermal boundary conditions. Boundary condition $\bar{T}(0) = T_i/\Delta T$ results in $c_2 = T_i/\Delta T$, and the other boundary condition $\bar{T}(1) = T_o/\Delta T$ results in $c_1 = 1 - D/2$. Hence, the resin temperature distribution is given by

$$\bar{T}(\bar{y}) = \frac{1}{2}D\bar{y}^2 + (1 - D/2)\bar{y} + \frac{T_i}{\Delta T} \qquad (3.172)$$

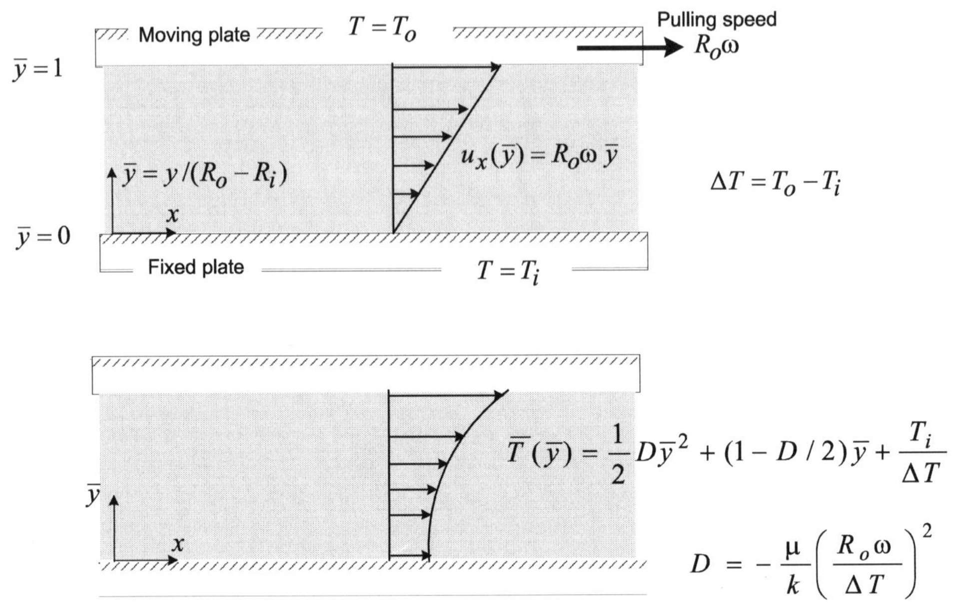

Figure 3.28: Velocity and temperature profiles.

For $T_o > T_i$ case, a typical non-dimensional temperature profile is shown in Figure 3.28 underneath the velocity profile. The non-linearity is due to heat generation induced by viscous dissipation.

Step 6: Conductive heat flux in the x and z directions is zero since the temperature doesn't change in those directions. The flux q (which is dimensional here) through the top and bottom plates (where $n = y$ and $n = -y$, respectively) are

$$
\begin{aligned}
q\Big|_{\bar{y}=1} &= -k\frac{\partial T}{\partial n} = -k\frac{\partial T}{\partial y} = -\frac{k\Delta T}{h}\frac{\partial \bar{T}}{\partial \bar{y}} \\
&= -\frac{k\Delta T}{h}\,[D(1) + 1 - D/2] = -\frac{k\Delta T(1 + D/2)}{h} \qquad (3.173) \\
q\Big|_{\bar{y}=0} &= -k\frac{\partial T}{\partial n} = k\frac{\partial T}{\partial y} = \frac{k\Delta T}{h}\frac{\partial \bar{T}}{\partial \bar{y}} \\
&= \frac{k\Delta T}{h}\,[D(0) + 1 - D/2] = \frac{k\Delta T(1 - D/2)}{h} \qquad (3.174)
\end{aligned}
$$

Note that, $q\Big|_{\bar{y}=0} \neq q\Big|_{\bar{y}=1}$.

Although there is no energy source inside the fluid, it was assumed that the velocity and temperature profiles do not change along the x direction, and the steady state was assumed, the conductive heat flux at the two walls is not equal. This is because mechanical energy is converted into heat due to viscous dissipation within the fluid. ■

3.7 Exercises

3.7.1 Questions

1. Why does one model a manufacturing process?
2. What are the main ingredients needed to create a mathematical model?
3. The conservation of momentum equation can be succinctly written as

$$\rho \frac{D\mathbf{U}}{Dt} = -\nabla P + \nabla \cdot \tau + \mathbf{F}_B. \tag{3.175}$$

   Describe the physical significance of each term.

4. How is the rate of strain tensor calculated from the fluid velocity vector? What is the scalar magnitude of this tensor, called *the strain rate.*
5. How is the vorticity tensor calculated from the fluid velocity vector? What type of flow results if all the components of this tensor are zero?
6. Why are the conservation equations of mass and momentum not sufficient to solve flow problems? What additional equation is needed?
7. Explain the kinetic energy of fluid particles.
8. What is the substantial derivative of internal energy of a fluid, DE/Dt in terms of fluid material parameters, temperature T, pressure P and velocity $\mathbf{U}$?
9. What are the three types of thermal boundary conditions?

3.7.2 Problems

1. The conservation of mass equation is

$$\frac{\partial \rho}{\partial t} + \nabla \cdot (\rho \mathbf{U}) + s = 0 \tag{3.176}$$

   or, alternatively

$$\frac{D\rho}{Dt} + \rho \nabla \cdot \mathbf{U} + s = 0. \tag{3.177}$$

   What is the difference between $\partial \rho/\partial t$ and $D\rho/Dt$? The second term in the first equation is $\nabla \cdot (\rho \mathbf{U})$ whereas it is $\rho \nabla \cdot \mathbf{U}$ in the second equation. Are they equivalent? Explain.

2. For a two-dimensional, constant density and steady fluid flow, the velocity components are given as $u_x = ax$ and $u_y = by$. What is the relation between a and b? Can $a = 2b$? Explain.
3. Construct numerical examples of $\mathbf{U}$ in which $\dot{\gamma}$ and ω (i) are both zero, (ii) are both non-zero, and (iii) one of them is zero and the other one is non-zero. Make sure that $\mathbf{U}$ satisfies the conservation of mass at steady flow: $\nabla \cdot \mathbf{U} = 0$.

4. In a plane Couette flow configuration, the fluid is bounded by two parallel plates with a depth of h. For a Newtonian fluid with viscosity μ, what are the velocity, viscous stress tensor, and vortex tensor if the upper plate is pulled with a constant speed of V in the positive x direction, and the lower plate is pulled with velocity V in the negative x direction? If both plates are pulled at the same speed in the same direction, what is the fluid velocity profile? What is the viscous stress tensor, τ_{ij} in this case?

5. What is the required force per unit plate area in order to pull a plate in plane Couette flow? How does the force change with the viscosity and the density of the fluid?

6. In a combined Couette-Poiseuille flow (Example 3.6), if the flow rate is $Q = 0$, at which y location does $u_x(y) = 0$? Explain the importance of your result. In this case ($Q = 0$), does it mean that there is no fluid flow? Explain.

7. Repeat Example 3.11, if viscosity changes with temperature as $\mu = \mu_0 \exp(-bT)$.

8. The three types of thermal boundary conditions for an isotropic material are (i) $T = T_W$, (ii) $q = -k\,\partial T/\partial n$, and (iii) $q = -k\,\partial T/\partial n = h(T_\infty - T)$. How would one write these boundary conditions if the material was anisotropic?

Chapter 4

Constitutive Laws and Their Characterization

4.1 Introduction

In the previous chapter, we familiarized ourselves with the physical laws needed to create a mathematical model for any manufacturing process and tailored it for composites processing. The important pieces needed to construct the model are

- identification of the system along with its boundaries,
- the governing equations for conservation of mass, momentum and energy,
- constitutive equations to describe the materials and their phenomenological behavior,
- boundary conditions to tailor the model to a specific composite manufacturing process.

We need constitutive laws because we cannot completely describe, from first principles, some of the transport phenomena such as the nonlinear material behavior of resin, the resin and fiber interactions and resin cure kinetics at the macroscale.

Constitutive equations are empirical relations between parameters of interest. They endeavor to incorporate the physics observed from the experiments or studied and analyzed at the micron and molecular level into the equations at the macroscale. They have to be objective such that the relationship and the results do not change with the coordinate frame. Almost all constitutive equations require the researchers to characterize constants needed in the equation that are specific to the material and its state. For example, when a fluid is subjected to a stress, if the stress is always directly proportional to the strain rate the fluid undergoes, the constant of proportionality (which can be characterized by experiments) is called viscosity. This constitutive relation is known as Newtonian law and the fluids that exhibit this behavior are called Newtonian fluids. Constitutive equations should not exhibit any singularity or instability that is an artifact in the processing regime to be consistent with the physics.

Constitutive equations are necessary to describe the processing of composite materials due to their heterogeneous nature, complex chemistry of the resin, its interaction with fibers and fillers and simultaneous transport of mass, momentum and energy at the micro-, meso- and macrolevels. Also, issues that are analyzed at the microscale, such as growth rate of a micron-size spherulite, can be represented at the macroscale level in terms of crystallization kinetics, by use of appropriate constitutive equations. Constitutive equations thus endeavor to describe the global picture of the process and the relationship between the process parameters instead of the details at the micron level, which is of little interest to the process engineer. However, constitutive equations invariably require the researchers to determine constants that may be functions of the material or even process conditions. Thus, independent characterization and cataloguing of material constants for constitutive equations can become a daunting task and slow down and undermine the modeling and simulation of such processes. Two important guidelines to keep in mind when developing or using a constitutive equation in a process model are that

1. As one increases the complexity of the constitutive equation, the number of parameters to be determined will also increase.
2. The model must always weigh the trade-offs between using a simpler constitutive model with less uncertainties in the measurement of the constants needed or adopting a more complex but more accurate constitutive model but with higher uncertainty in the measurement of the constants needed.

It is entirely possible to use a more complex constitutive model but because of higher uncertainty in the measurements of the constants, the final answer could be less accurate as compared to use of a simple constitutive equation with more confidence in the measured values of the constants.

In this chapter, our goal will be to introduce to the reader various constitutive equations that are used to describe the material behavior, or the manufacturing process behavior. We will also comment as to the characterization, development and availability of the empirical constants for such constitutive equations.

4.2 Resin Viscosity

Every fluid substance offers resistance if one tries to make it deform or flow. This resistance arises because one is trying to change the arrangement of molecules of that material in its fluid state. This property has been termed viscosity. The dimensions of viscosity are $ML^{-1}T^{-1}$. For units, one can use Poise (P) or Pascal-second (Pa.s). One Poise is equal to 0.1 Pa.s. The viscosity of water is one centiPoise (cP) which is one hundredth of a Poise. Water, corn syrup, liquid metals and other short chain molecules are called Newtonian fluids because the viscosity of such fluids can be characterized by a single value which does not change with its shear rate. As illustrated in Figure 4.1, in general, the viscosity is a function of pressure, temperature, degree of cure, and shear rate. If there are particles or fibers suspended in the resin, then the fiber volume fraction and orientation of the particles can influence the viscosity if their aspect ratio is greater than one.

Resin velocity and pressure are calculated by solving the equation of motion which contains the viscosity. Hence the velocity and pressure are functions of viscosity.

Most thermoset resins can be treated as Newtonian, although this may not be strictly true under high shear rates. However, their viscosity is affected by temperature and cure kinetics. Thermoset molecules will cross-link and undergo exothermic reactions when initiated by a catalyst that can be invoked by heat or chemicals. This phenomenon is known as curing. As the thermosets cure, their viscosity increases since one would need more force to move a molecule that is cross-linked. Thus, constitutive equations to describe the cure kinetics and dependence of viscosity on degree of cure are necessary for thermosets. Before the initiation of the cure, viscosities of thermosets are usually between 100 cP and 500 cP.

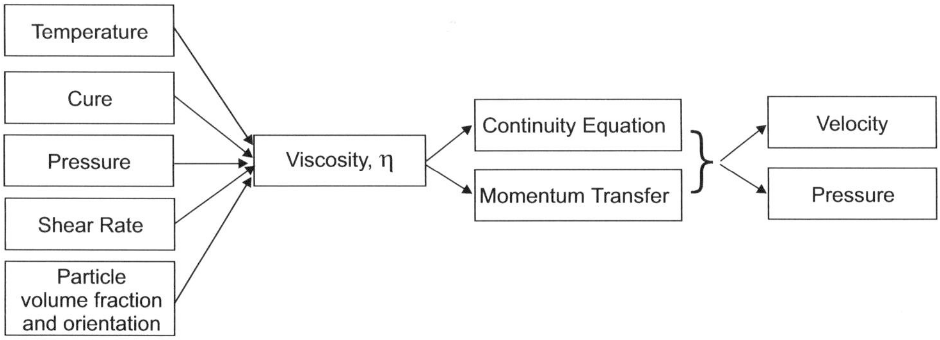

Figure 4.1: Flowchart illustrating the dependence of viscosity on temperature, cure, pressure, shear rate, particle volume fraction and orientation. Velocity and pressure are dependent on viscosity through the mass and momentum conservation.

Hence, most of the polymer composite processing and manufacturing methods are modeled using the low Reynolds number flow assumption or creeping flow assumption, in which one neglects the inertial force in favor of the viscous force. This argument is even stronger when one has to process thermoplastic resins as their viscosities under quiescent state when they are in a melt form can be as high as 10,000 P. This is primarily because at the molecular level thermoplastic resin forms a mosaic of long-chained molecules all entangled together in the quiescent liquid state [92]. As soon as one subjects them to shear, these molecules start to align in the direction of the shear; thus they can slip over each other much more easily as they are no longer entangled, as shown below in Figure 4.2.

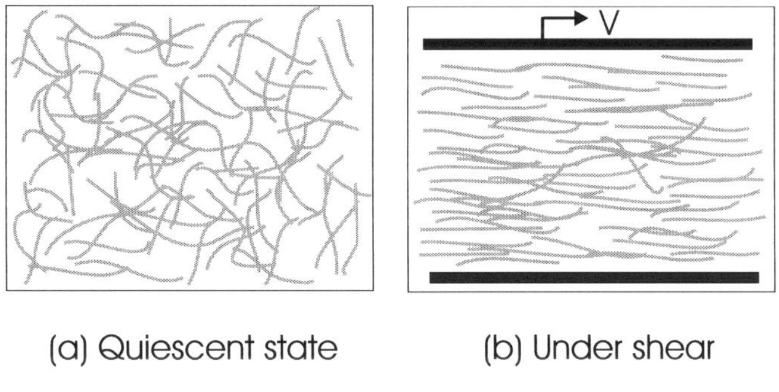

Figure 4.2: Applied shear causing the molecules to align along the flow direction.

Thus, the resistance to flow can dramatically reduce due to the change in the structure under applied shear load, reducing the viscosity by orders of magnitude. Such liquids are known as shear thinning fluids, where the viscosity decreases with shear rate. Thus thermoplastics belong to a class of non-Newtonian fluids. Thermoplastics may also display viscoelastic behavior due to the long chained spring like molecules; however, in composites processing, due to the presence of the fibers and low flow rates, one can ignore the viscoelastic effects [95, 96]. Usually, one checks the Deborah number (De) to gauge the importance of viscoelasticity. De is the ratio of elastic effects to viscous effects and if De $\ll 1$, one can safely ignore the elastic effects of the thermoplastic polymers. This is usually true for almost all composites manufacturing processes. As there is no cross-linking or reactions that the thermoplastics undergo in their melt state, cure kinetics characterization for thermoplastics is not necessary.

The viscosity of the resin also changes with temperature for both thermoplastics and thermosets. This effect is more significant for thermoplastics. The viscosity will usually reduce by one to two orders of magnitude as the temperature of the material increases from just below the melt temperature to a few degrees higher than the melt temperature. This is primarily due to reduction in the resistance to flow as the molecules at high temperature can move much more freely. Molecular dynamic simulations can be performed to monitor the motion of each and every molecule under applied shear and temperature [97]. These simulations confirm the macroscopically observed behavior. This information is not very useful from the processing viewpoint, but can help in development of constitutive equations for temperature dependence of viscosity.

4.2.1 Shear Rate Dependence

For process modeling, one needs a relationship between the stress the material is subjected to and its deformation rate. Usually, one would express this relationship as

$$\underline{\underline{\tau}} = \eta \underline{\underline{\dot{\gamma}}} \tag{4.1}$$

where $\underline{\underline{\tau}}$ is the viscous stress tensor, $\underline{\underline{\dot{\gamma}}}$ is the strain rate tensor, and η is the viscosity of the material. For a Newtonian fluid, the viscosity is a constant and can be expressed as

$$\eta = \mu. \tag{4.2}$$

However, for shear thinning materials, such as thermoplastics, one must be able to express the change in viscosity with shear rate. There are three phenomenological models that are commonly used to describe the shear thinning nature of the material. They are known as the power-law model [98,99], the Spriggs truncated model [100] and the Carreau

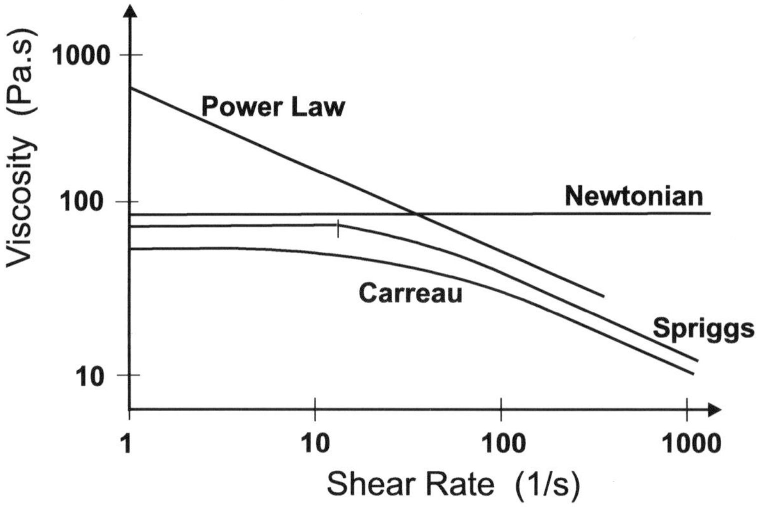

Figure 4.3: Graphical depiction of viscosity changes with the shear rate on a log-log scale as described by various viscosity models.

model [101, 102]. Their form for viscosity can be expressed as shown below:

$$\eta(\dot{\gamma}) = m\dot{\gamma}^{n-1} \qquad \text{(power-law model)} \tag{4.3}$$

$$\eta(\dot{\gamma}) = \begin{cases} \eta_0 & \text{if} \dot{\gamma} \leq \dot{\gamma}_0 \\ \eta_0 \left(\dfrac{\dot{\gamma}}{\dot{\gamma}_0}\right)^{n-1} & \text{if} \dot{\gamma} \geq \dot{\gamma}_0 \end{cases} \qquad \text{(Spriggs model)} \tag{4.4}$$

$$\eta(\dot{\gamma}) = \eta_\infty + (\eta_0 - \eta_\infty)\left[1 + (\lambda\dot{\gamma})^2\right]^{\frac{n-1}{2}} \qquad \text{(Carreau model)} \tag{4.5}$$

Figure 4.3 graphically depicts how the viscosity changes with the shear rate ($\dot{\gamma}$). The unit of shear rate is 1/s. Here, $\dot{\gamma}$ must be independent of the coordinate system to maintain objectivity, chosen as the magnitude of the strain rate tensor, and given by

$$\dot{\gamma} = \sqrt{\frac{1}{2}\sum_{i,j} \dot{\gamma}_{ij}\dot{\gamma}_{ji}}. \tag{4.6}$$

Experiments are conducted in a "cone and plate" device or a capillary rheometer to measure the viscosity under a prescribed shear rate. Cone and plate, as shown in Figure 4.4, is usually a suitable device for viscosity measurement at low shear rates. Capillary rheometer as shown in Figure 4.5 is suitable to measure viscosity under high shear rates.

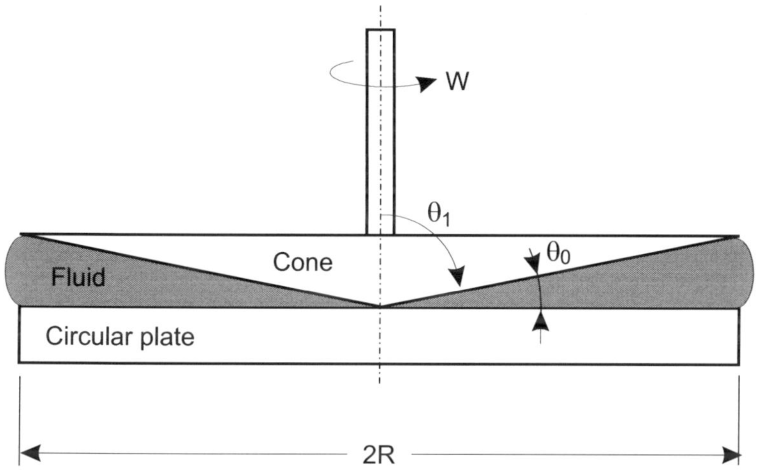

Figure 4.4: Schematic of a cone and plate viscometer

For Newtonian fluids, the value of the viscosity should be the same at different shear rates. However, for thermoplastic resins, this is not true. At different shear rates, one measures different viscosity values. To measure the change in the viscosity as a function of the shear rate, one has to subject the material to different shear rates. Almost all of the thermoplastics display a Newtonian plateau at low shear rates, then a logarithmic decrease with increase in shear rate and again a plateau at high shear rates as shown in Figure 4.6. Figure 4.3 shows how the constitutive Equations (4.2)-(4.5) try to capture the observed behavior. The Carreau model correctly captures the behavior of shear thinning materials. However, one needs to determine four parameters by fitting the form of Equation (4.5) to the experimental data. One can assume that when the shear rate is infinity, the viscosity

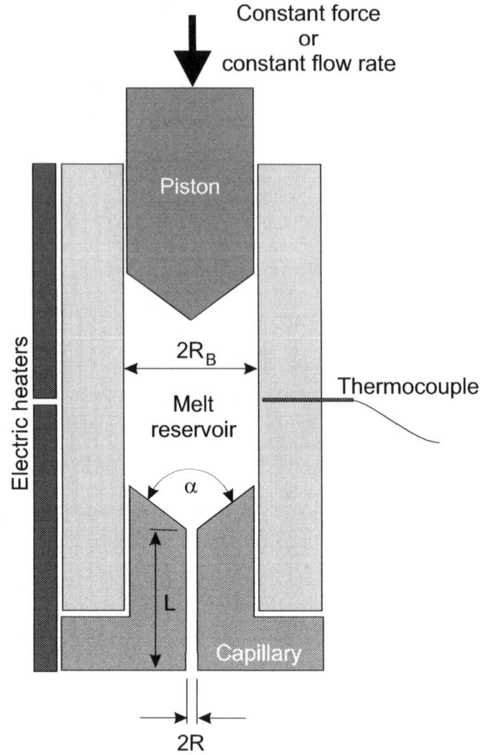

Figure 4.5: Schematic of a capillary rheometer.

goes to zero, i.e., $\eta(\infty) = 0$, and thus reduces the Carreau model to three parameters as follows: $\eta = \eta_0 + (1 + (\lambda\dot{\gamma})^2)^{\frac{n-1}{2}}$. The Spriggs model needs only three parameters too, but does not have a smooth transition from the Newtonian plateau to the power-law unlike the Carreau model. The power-law model requires only two parameters and does not predict any Newtonian plateau. Thus, at low and high shear rates, the power-law model will give physically incorrect results, whereas the truncated Spriggs model does not have a smooth transition from the Newtonian plateau to shear thinning behavior. However, the power-law model is popular for its simplicity and captures the essential behavior of shear thinning in the intermediate shear rate region, which is usually the case when processing such materials.

Example 4.1: Axial Annular Flow of a Power-Law Polymeric Fluid [92]

A fiber of radius R_i is dragged through a cylinder of radius R_o to transfer resin from container A to container B as shown in Figure 4.7. The resin viscosity can be described by a power-law fluid.

1. Find the velocity profile of the resin between the cylinder and the fiber, assuming entrance and exit effects are negligible.
2. Find the volumetric flow rate Q.
3. What is the force necessary to drag the fiber with velocity V_o?

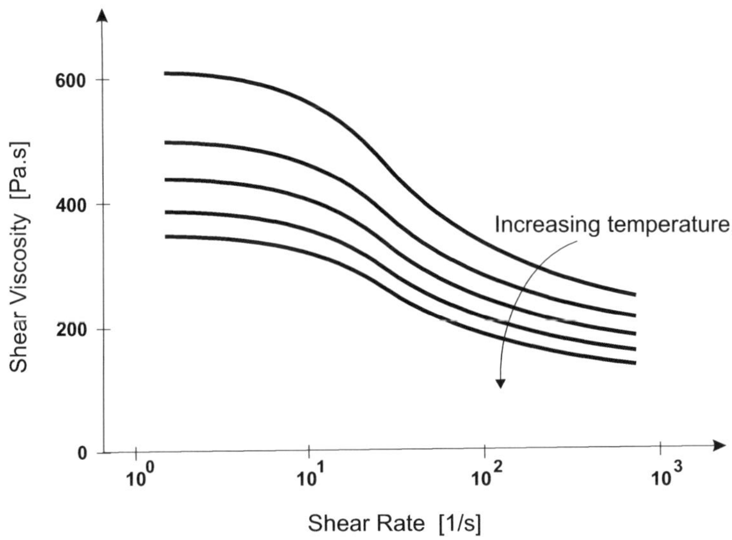

Figure 4.6: Schematic of viscosity versus shear rate at various temperatures for thermoplastic resins.

Solution

1. Attach the cylindrical coordinate frame such that the z direction is along the center of the rod as shown in Figure 4.7. Postulate that $v_r = v_\theta = 0$ and $v_z = v_z(r)$ (satisfying the continuity equation), which implies that

$$\tau = \tau(r). \tag{4.7}$$

To find the velocity profile, one can deduce from the z direction equation of motion

$$\frac{1}{r}\frac{d}{dr}(r\tau_{rz}) = 0 \tag{4.8}$$

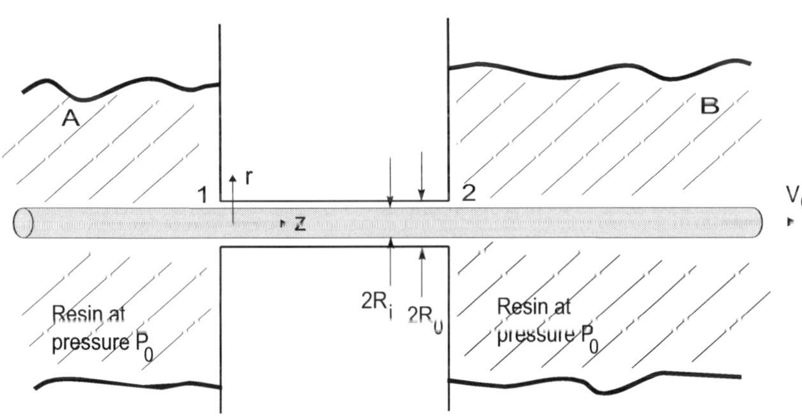

Figure 4.7: Axial annular flow of a power-law polymeric fluid.

as all other terms are zero, if we ignore inertia and unsteady effects due to low Reynolds number flow.

Note that now we need to substitute the power-law constitutive model to relate stresses to velocity gradients

$$\tau_{rz} = \eta \dot{\gamma}_{rz} \tag{4.9}$$

where

$$\eta = m \dot{\gamma}^{n-1} \tag{4.10}$$

and

$$\dot{\gamma} = \sqrt{\frac{1}{2}\left(\dot{\gamma}_{rz}\,\dot{\gamma}_{zr} + \dot{\gamma}_{rz}\,\dot{\gamma}_{zr}\right)}. \tag{4.11}$$

Since

$$\dot{\gamma}_{rz} = \dot{\gamma}_{zr} = \frac{\partial v_z}{\partial r}, \tag{4.12}$$

Equation (4.11) reduces to

$$\dot{\gamma} = \left|\frac{\partial v_z}{\partial r}\right| = -\frac{\partial v_z}{\partial r} \tag{4.13}$$

as $\frac{\partial v_z}{\partial r} < 0$ and $\dot{\gamma}$ should be always positive.

Substitution of Equation (4.13) into Equation (4.10) results in

$$\eta = m\left(-\frac{\partial v_z}{\partial r}\right)^{n-1}. \tag{4.14}$$

Substitution of Equation (4.14) into Equation (4.9) gives

$$\tau_{rz} = m\left(-\frac{dv_z}{dr}\right)^{n}. \tag{4.15}$$

Now if one substitutes Equation (4.15) into Equation (4.8), the following differential equation for v_z is obtained:

$$\frac{d}{dr}\left[mr\left(-\frac{dv_z}{dr}\right)^{n}\right] = 0. \tag{4.16}$$

Integrating twice and using the boundary conditions

$$v_z = V_o \quad \text{at } r = R_i \tag{4.17}$$

$$v_z = 0 \quad \text{at } r = R_o \tag{4.18}$$

results in

$$\frac{v_z}{V_o} = \frac{\left(\frac{r}{R_o}\right)^{1-\frac{1}{n}} - 1}{\kappa^{1-\frac{1}{n}} - 1} \tag{4.19}$$

where $\kappa = R_i/R_o$.

2. The volumetric flow rate can be found as follows:

$$Q = \int_{R_i}^{R_o} 2\pi r\, v_z\, dr = \frac{2\pi R_o^2 V_o}{\kappa^{1-\frac{1}{n}} - 1}\left[\frac{1-\kappa^{3-\frac{1}{n}}}{3-\frac{1}{n}} - \frac{1-\kappa^2}{2}\right]. \tag{4.20}$$

3. The force is calculated as

$$\begin{aligned} F_z &= 2\pi\kappa R_o L \tau_{rz}\Big|_{r=R_i} \\ &= 2\pi\kappa R_o L m \left[-\frac{V_o\left(1-\frac{1}{n}\right)}{(\kappa-\kappa^{1/n})R_o} \right]^n . \end{aligned} \tag{4.21}$$

■

4.2.2 Temperature and Cure Dependence

The viscosity of all resins decreases with temperature. For thermoplastics, the constitutive equation usually follows an Arrhenius law which describes the drop in viscosity with an exponential function. The generic form usually is represented as

$$\eta(T) - \eta_0 \exp \frac{c_1}{c_2 T} \tag{4.22}$$

where η_0 is the viscosity at temperature T_0 and c_1 and c_2 are constants that depend on the material. Temperature dependence of some of the common thermoplastics is sketched in Figures 4.6 and 4.8.

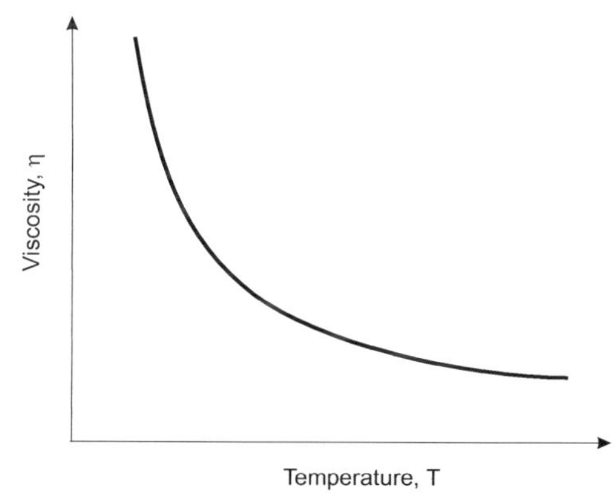

Figure 4.8: Sketch of viscosity dependence on temperature for thermoplastics.

Thermosets also exhibit similar temperature dependence, but their temperature behavior is coupled with the cure kinetics. Usually, the viscosity of thermosets reduces with temperature, T and increases with degree of cure, α. Both phenomena can be described well with exponential curves. Hence, the viscosity of a thermoset can be expressed in a generic constitutive equation as

$$\eta(T,\alpha) = \eta_0 \exp\left(\frac{c_1}{c_2 T} + c_3\alpha\right) \tag{4.23}$$

where c_1, c_2 and c_3 depend on the thermoset material under consideration. There are many other ways to characterize the viscosity dependence of reacting systems.

Two concurrent phenomena govern the rheological behavior of a reacting system: one associated with the intensification of the mobility because of the increase in temperature, responsible for decreasing the viscosity, and another one related to the growing size of the molecules during cure, responsible for increasing the viscosity of the resin [103]. The empirical Williams-Landel-Ferry (WLF) equation [104], valid near the glass transition and based on the free-volume theory, is frequently used to represent the first phenomenon:

$$\ln \frac{\eta}{\eta_{T_o}} = \frac{-C_1(T - T_o)}{C_2 + T - T_o}, \tag{4.24}$$

where C_1, C_2 are adjustable parameters and T_o is the reference temperature.

The underlying concept of free-volume theory is that the movement of the molecules is intrinsically conditioned to the amount of free volume in a molecular ensemble; the less the unoccupied space, the more the collisions among the molecules, resulting in a slow response to a perturbation in an equilibrium state [105].

With the objective of accounting for the second phenomenon mentioned earlier, Enns and Gillham [106] proposed the following equation:

$$\ln \eta = \ln \eta_\infty + \ln \overline{M}_w + \frac{E_\eta}{RT_o} - \frac{C_1(TT_o)}{C_2 + T - T_o}, \tag{4.25}$$

with η_∞ as the extrapolated viscosity as $T \to \infty$, E_η as the activation energy and $\overline{M}_w$ as the average molecular weight. The reference temperature, T_o, is arbitrarily chosen to be the temperature above which the Enns equation is valid to describe the viscosity-temperature dependence. There are many relations in the literature that describe the average molecular weight as a function of degree of cure and the molecular weight of monomers [106–109].

Many authors [110–116] have followed an empirical equation, which was proposed by Stolin et al. [117] and is similar to Equation (4.23), in order to calculate the resin viscosity:

$$\eta(T, \alpha) = \eta_o \exp\left(\frac{U}{RT} + K\alpha\right), \tag{4.26}$$

where U is the activation energy of the viscous fluid, R is the gas constant and K is a constant which accounts for the effect of the chemical reaction on the change in the reacting mass viscosity and, consequently, on the dissipation intensity. It is assumed that U is independent of the degree of cure. The values for these parameters for two resin systems are listed in Table 4.1.

Table 4.1: Parameters for Equation (4.26)

Author	System	η_o (Pa.s)	U (J/mol)	K	Range of Validity
Lee et al. [114]	Hercules 3501-6	$7.93\text{x}10^{-14}$	$9.08\text{x}10^4$	14.1±1.2	$\alpha < 0.5$
Dusi et al. [116]	Fiberite 976	$1.06\text{x}10^{-6}$	$3.76\text{x}10^4$	18.8±1.2	$\alpha < 0.2$

One commonly used empirical relation is [118–120]:

$$\eta = A_\eta \exp\left(\frac{E_\eta}{RT}\right)\left(\frac{\alpha_g}{\alpha_g - \alpha}\right)^{a+b\alpha}. \tag{4.27}$$

Table 4.2: Parameters for Equation (4.27)

Authors	System	A_η (Pa.s)	E_η (J/mol)	α_g	a	b	Range of Validity
Castro and Macosko [118] (1980)	RIM2200	10.3x10^{-8}	41.3	0.65	1.5	1.0	-
Kenny et al. [119] (1990)	Commercial grade thermoset DSM resin for RTM	3.416x10^{-11}	64900	0.088	3.706	-34.62	30-60°C
Kim and Kim [120] (1994)	DGEBA and TETA (10phr)	3.60x10^{-10}	58600	0.765	1.0	5.2	70-90°C

The parameters for a few common resin systems are listed in Table 4.2.

Another related form, proposed by Lee and Han [121–123] is:

$$\eta = A_\eta \exp\left(\frac{E_\eta}{RT}\right) \tag{4.28}$$

where the fluid activation energy, E_η, and the frequency factor, A_η, are given by:

$$E_\eta = a + b\alpha \qquad A_\eta = a_o \exp(-b_o\alpha), \tag{4.29}$$

with a, b, a_o and b_o as specific constants for each resin. For example, consider a system composed of a partially cured unsaturated polyester resin, OC-E701 [121], 43 wt % of styrene and t-butyl perbenzoate as an initiator. The constants for this system are: a=7.8 kcal/mole; b=19.7 kcal/mole; a_o=6.41x10^{-5} N.s/m^2; b_o=23.1. However, Lee and Han [121] pointed out that for these values, at temperatures higher than 156.2°C, the viscosity calculated by Equation (4.28) decreases with an increase in α, which is not physically correct. Hence, they proposed two other equations, valid for the system used: one based on the Tajima and Crozier [124] model:

$$\log \eta(T, \alpha) = (23.64 + 3.82\alpha) = \frac{26.8(41.63 + T - 90.19\alpha)}{55.03 + T - 90.19\alpha} \tag{4.30}$$

and another one based on the Hou [125] model:

$$\log \eta(T, \alpha) = -(6.06 + 4.78\alpha) + \frac{51.6(22.11 + 29.92\alpha)}{T - 143.3 - 33.06\alpha - 281.4\alpha^2}. \tag{4.31}$$

Hence, it should be clear to the reader that development of a universal formula for viscosity as a function of temperature and cure does not exist. However, forms suggested in Equations (4.26)-(4.28) can accommodate many thermoset resin systems.

Example 4.2: Resin Viscosity as a Function of T and α

Using Equation (4.26) by Stolin et al. [117], plot the viscosity of Hercules 3501-6 [114] and Fiberite 976 [116] resin systems for $0 \leq \alpha < 0.2$ at $T = 350°$K, 375°K and 400°K. Use the listed parameters in Table 4.1, and take the mean value for the parameter K.

Solution

Equation (4.26) has been plotted using MATLAB, and shown in Figure 4.9 for the three temperatures asked.

Especially at low temperatures, the viscosity increases very fast as the degree of cure increases. For example, the viscosity of Hercules 3501-6 [114] at $T = 350°$K increases from 3 Pa.s to 47 Pa.s as α increases from zero to 0.2, and it reaches to 3250 when α is 0.5 (not shown in Figure 4.9). The user of an empirical formula should be aware of the validity range of the formula. For example, Equation (4.26) is not validated for $\alpha > 0.5$ for this particular resin system as mentioned in Table 4.1. The validity range for Fiberite 976 [116] is even shorter. Also, keep in mind that the temperature of the resin was held constant during the characterization experiments; however, during the actual curing stage of the manufacturing, the resin and mold may not be held at a constant temperature due to the heat transfer between resin, mold and ambient. ■

4.3 Viscosity of Aligned Fiber Thermoplastic Laminates

The deformation of a bundle of aligned fibers lubricated with viscous resin is primarily viscous. The deformation response of such materials will be a strong function of the nature of the resin and the fiber arrangement. For transversely isotropic bundles or laminates, there are two dominant shear modes: longitudinal shearing and transverse shearing. The schematics of the two modes are shown in Figure 4.10 [8]. Both shear modes may be important as aligned composite laminates are deformed to form complex shapes. For thermoplastics, these material constants display non-Newtonian behavior and need to be characterized. For thermosets, one must separately consider the elastic deformation of the fibers by characterizing the elastic constants. In this section, we will only consider the transverse deformation of a material, which consists of fibers and resin on the microscale but is treated as a homogeneous material on the macroscale.

Thermoplastic composite sheet forming involves the deformation of a laminate within and out of the plane and is characterized by the flow of resin and fibers. It is, therefore, necessary to have effective constitutive relations describing the highly anisotropic flow behavior of these laminates. Unfortunately, the shear and extensional properties of long discontinuous fiber reinforced composites in their melt state are difficult to obtain using traditional rheological techniques, such as cone and plate and capillary rheometers, due to the presence of continuous or long fibers.

One way to describe their shear behavior is to use squeeze flow to characterize the bulk transverse shear viscosity of these highly filled viscous resin systems. Squeeze flow is invoked by placing the material between two impermeable platens and applying a normal

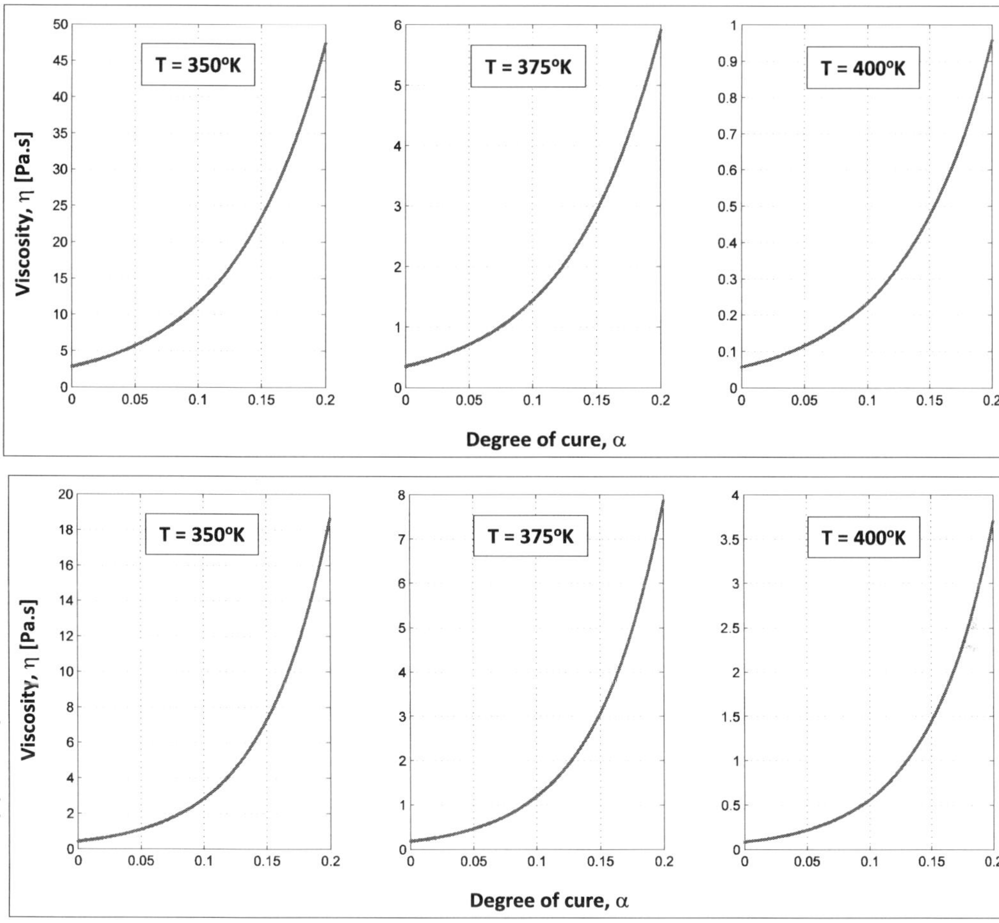

Figure 4.9: Resin viscosity, $\eta(T, \alpha)$ for two resin systems [114, 116] and using the empirical Equation (4.26).

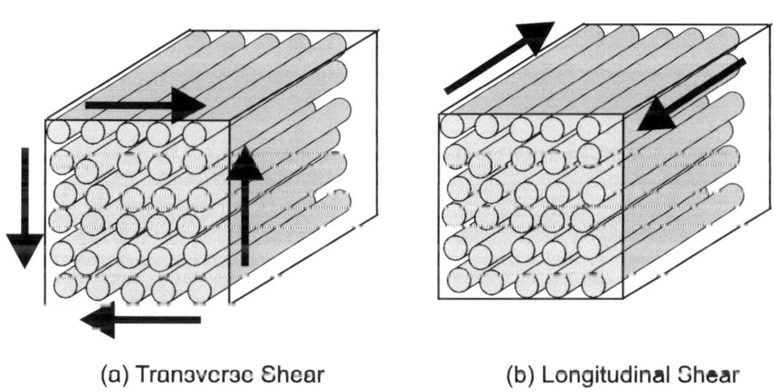

Figure 4.10: The deformation response of a bundle of aligned fibers lubricated with viscous resin. For transversely isotropic bundle or laminates, the two dominant shear modes are (a) transverse shearing, and (b) longitudinal shearing [8].

force. One may characterize the material under constant applied load or under a constant closure rate of the platens. Such industrial testing devices are referred to as the *parallel plate plastometer* or *squeeze plate viscometer* [126, 127]. These devices measure radial flow between two parallel disks and have been used to study the viscosity of polymer melts and other viscous materials.

The advantages of this type of test include mechanical simplicity, inclusion of fibers without the danger of attrition, achievement of very high shear rates, usage at high temperatures and the ease with which high-viscosity materials can be tested [128]. Additionally, this type of flow is of interest since it is encountered within lubrication systems and the stamping of plastic sheets. Finally, it provides a technique for investigators to evaluate rheological equations under transient conditions.

In the parallel plate plastometer, the test material, in the form of a cylinder, is placed between parallel, circular flat plates. With the lower plate fixed, either a constant force or a constant closure rate is applied to the upper plate. If a constant force is applied, measurement of the resulting displacement of the upper plate is used to determine the rheological properties of the specimen. If a constant closure rate is applied, measurement of the resulting force on the upper plate is used to determine the viscosity dependence on shear rate. Although these experiments involve unsteady shear flows, the flow rates are usually modest so that the analyses are based on quasi-steady state solutions. Considering the very high viscosities, the Reynold's number is usually much lower than unity, justifying the quasi-steady state assumption. Two different sample arrangements may be used in the plastometer. Either the specimen sample cross-sectional area can be equal to the cross-sectional area of the plates, in which case the area under compression is constant, or the cross-sectional area of plates may be larger than the specimen sample, in which case the volume of the specimen is constant.

Squeeze flow rheometers have been used to study both unfilled viscous and viscoelastic liquids and have been modified to study fiber filled thermoplastic materials. Experimental squeeze flow studies on polymeric liquids indicate that under slow loading rates, their behavior can be approximated through the shear dependent viscosity [129–132]. However, under rapid loading, viscoelastic effects may become evident [133–138]. The magnitude of the material's Deborah number is often used as a determinant of its response. The Deborah number, De, is defined as the ratio of the fluid's relaxation time to a characteristic time of the squeezing process. Low values of the Deborah number correspond to viscous fluids while high Deborah numbers correspond to fluids with an elastic response.

The flow of sheet molding compounds was investigated in Barone and Caulk [2, 139]. The squeeze flow behavior of aligned fiber reinforced thermoplastic materials has been discussed in references [140–142]. They treated these composites as transversely isotropic materials. Experimental observations of aligned fiber reinforced APC-2 material by these investigators showed that the resulting squeeze flow deformation is strictly perpendicular to the fiber direction. This is due to the high ratio of extensional viscosity in the fiber direction, which is infinite in the case of continuous reinforcement, to the shear viscosity of the composite transverse to the fiber direction. Therefore, with no resin percolation out of the fiber bed, these composites have been viewed as incompressible anisotropic fluids with an effective shear viscosity transverse to the fiber direction. O'Bradaigh [143] has experimentally measured the effects of transverse squeeze flow on the cross-sectional thickness variation of diaphragm formed from APC-2 components. Hull, Rogers and Spencer [144] have examined the evolution of fiber wrinkling during these squeezing flows.

Shuler and Advani [142] used a servohydraulic test frame with specially designed hot platens. These test platens, as depicted in Figure 4.11, consisted of two square steel platens fastened to two specially constructed round steel fixtures attached to the test frame's hydraulic rams. Cartridge heaters were mounted within these fixtures to provide the platen heat and were controlled through feedback controllers that can receive temperature information from thermocouple probes mounted within the steel platens. Shuler and Advani [142] mounted a video camera perpendicular to the sample, recording the material flow during the test. For steady squeeze flow testing, a constant force may be applied and the change in height is measured over time, or a constant closure rate may be applied with the change in force is measured over time through appropriate load cells. If desired, special oscillatory, dynamic or exponential loading rates may also be applied through the servohydraulic system.

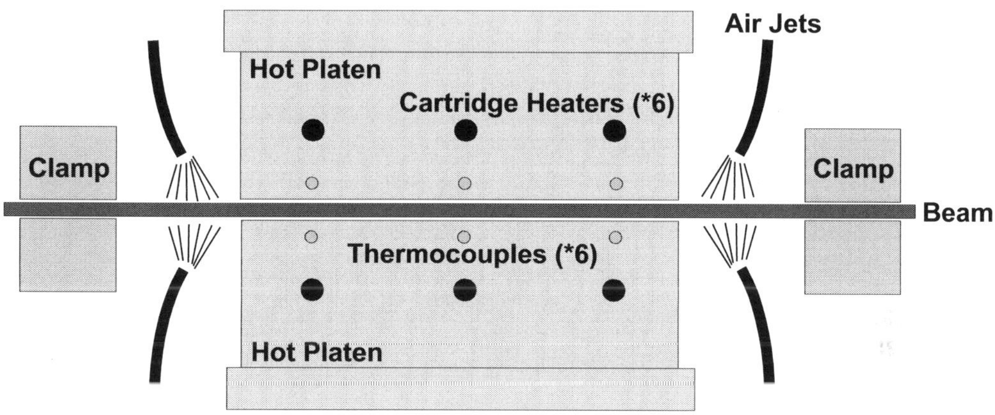

Figure 4.11: High temperature squeeze flow set up for thermoplastic composite laminates in a servohydraulic test frame [65].

With this system, they tested three different combinations of aligned fiber materials: clay matrix with nylon fibers, clay matrix with glass fibers, and PEEK matrix with aligned carbon fibers (APC-2). The model matrix of clay was prepared with various volume fractions of nylon fibers and 60% glass fibers. The thermoplastic PEEK material contained 64% aligned carbon fibers. The sample size was 1 cm in thickness and occupied the complete cross-sectional area of the platen that would squeeze the sample with the hydraulic cylinder. The force and displacement data were recorded. As the clay matrix displayed shear thinning behavior well described by a power-law model, the same constitutive form

$$\eta(\dot{\gamma}) = m\dot{\gamma}^{n-1} \tag{4.32}$$

for the composite was assumed. The analysis showed that one could match the experimental data by modifying just the bulk viscosity constant m, as shown in Table 4.3.

They found that the unfilled clay and the filled samples behaved as shear thinning power-law fluids. Table 4.3 summarizes the results that were plotted over a broad range of shear rates. It reveals the decidedly nonlinear relationship between the composite's transverse shear viscosity and the percentages of fibers. The clay filled with 60% nylon fibers resulted in a transverse shear viscosity 27 times greater than the viscosity of the unfilled clay, while 60 volume percentage of the glass fibers resulted in a transverse shear viscosity 55 times greater. From this information, it seems that adding the same volume percentage of smaller diameter fibers has the effect of increasing the transverse shear viscosity even more than the increase found using larger diameter fibers.

The reinforcing PEEK matrix polymer in the APC-2 laminates is known to behave as a shear thinning Carreau-type fluid. The four parameter Carreau fluid can be reduced to

Table 4.3: Power-Law Viscosity Parameters for Unfilled and Fiber-Filled Clay

Fiber Volume Percentage	Platen Closure Rate [cm/min]	m [Pa s^n]	n
0	0.0254 to 0.762	30,000	0.1
20% Nylon Fibers	0.0254	82,000	0.1
40% Nylon Fibers	0.0254	110,000	0.1
50% Nylon Fibers	0.0254	300,000	0.1
60% Nylon Fibers	0.0254	810,000	0.1
60% Glass Fibers	0.0254	1,650,000	0.1

three parameters by assuming that the fluid viscosity goes to zero at infinite shear rate. Thus, the three-parameter Carreau fluid model [101, 102] describes the viscosity exhibiting Newtonian behavior at low shear rates followed by shear thinning power-law behavior at higher shear rates, and is commonly expressed as:

$$\eta = \eta_0 \left[1 + (\lambda\dot{\gamma})^2\right]^{(n-1)/2} \tag{4.33}$$

where η_0 is the zero shear rate viscosity — the viscosity of the material when deformed infinitely slowly, n is the shear thinning exponent that adjusts the slope of the shear thinning region and λ is a time constant that adjusts the position of the "knee" in the $\eta\,\mathrm{vs}\,\dot{\gamma}$ curve where there is an onset of shear thinning behavior. The Newtonian model is recovered for $n = 1$, and the power-law expression is approached for large values of λ.

For the PEEK matrix, these viscous parameters were measured by Pipes et al. [145] and are listed in Table 4.4. Viewing the transverse shear flow of the fiber-filled matrix as viscous flow with a lubricating fluid between the fibers, the Carreau model was investigated as perhaps an appropriate model to also describe the shear thinning viscous behavior of the APC-2 composites.

Table 4.4: Carreau Viscosity Parameters for PEEK at 370°C

PEEK Carreau Viscosity Parameters at 370°C:
$\eta_0 = 687\,\mathrm{Pa.s}$
$\lambda = 0.0932\,\mathrm{sec}$
$n = 0.787$

Shuler and Advani [142] collected the platen force data for the four closure rates and used least-squares fit to the Carreau model to find the parameters for the bulk transverse shear viscosity of the APC-2 laminates. The three Carreau parameters of Table 4.5 were found to produce loading curves that closely matched the experimental platen force data collected.

The Carreau model curve for the transverse shear viscosity of the APC-2 material at 370°C based on the parameters listed in Table 4.5 are graphed in Figure 4.12. The Carreau model viscosity of the neat PEEK matrix polymer at 370°C is graphed in Figure 4.12 as well. It is noteworthy that the transverse shear viscosity of the composite is approximately 3600 times greater than viscosity of the PEEK matrix resin. Also, the onset of shear thinning behavior occurs at a lower shear rate than for the neat resin. Since the Carreau model

Table 4.5: Carreau Model Parameters for the Transverse Shear Viscosity of APC-2 Material at 370°C

APC-2 Transverse Shear Viscosity Carreau Parameters at 370°C.
$\eta_0 = 2.5 * 10^6$ Pa.s
$\lambda = 50.0$ sec
$n = 0.65$

captures both the Newtonian behavior at low shear rates and the shear thinning behavior at higher shear rates, this three parameter model is able to describe the viscous behavior of the unidirectional APC-2 laminates over a range of shear rates.

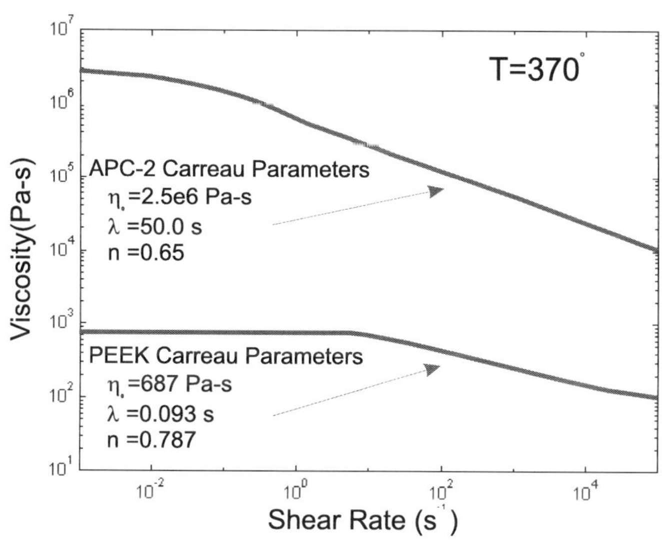

Figure 4.12: The Carreau model curve using the parameters listed that describes the transverse shear viscosity of the APC-2 material and the resin PEEK at 370°C [65].

It is instructional to note that during squeeze flow the material experiences a range of shear rates that vary both over time and throughout the volume of the material. The transverse shear rates encountered in squeeze flow range from very low at the start of each test and at the symmetry planes to a maximum which occurs at the largest closure rates adjacent to the platen contact surfaces near the outer edges of the samples. Because squeeze flow encompasses such a broad range of shear rates throughout the volume of the material, a simple Newtonian or power-law fluid model cannot adequately describe the squeeze flow behavior of the laminate that, depending on the shear rate, follows both Newtonian and shear thinning behavior. Hence the use of the Carreau model is more appropriate for composite laminates.

Example 4.3: Squeeze Flow of GMT Material

A thermoplastic material reinforced with fibers (GMT) [146] is expected to behave as a power-law fluid. To find the power-law constant, m, and the power-law index, n, squeeze flow of GMT between two disks is proposed as shown in Figure 4.13, in which one can measure the force and the displacement. Conduct the analysis to delineate the approach under constant closure rate.

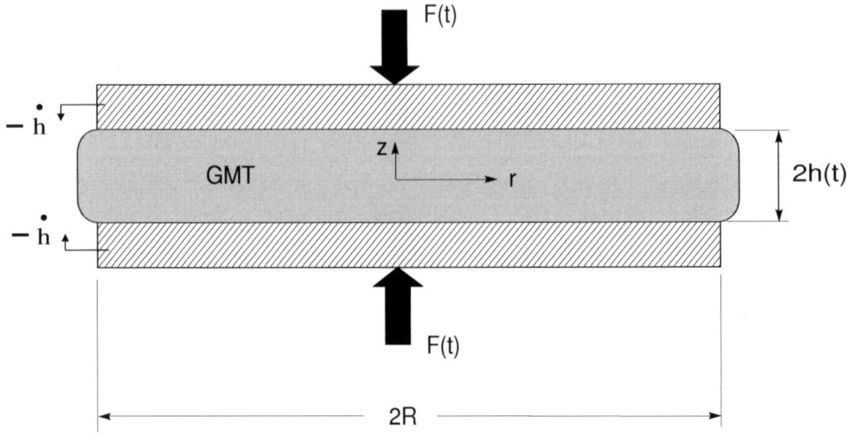

Figure 4.13: Schematic of GMT material between two parallel disks of radius R with cylindrical coordinate system placed midway between the disks.

Solution

As $h \ll R$, one can assume that shear forces will dominate. Also for GMT materials, the viscosity is very high such that the Reynolds number is much less than unity. Hence, one can ignore the inertia forces and assume a quasi-steady state process. A "quasi-steady state" as we will see in Chapter 5, implies a "steady state" at a given instant. This allows us to solve the steady-state problem at any height h, which will then change at the next instant and the velocity profile will change accordingly and achieve steady state instantly.

Using mass conservation for a disk of any radius $r < R$, one can show that

$$-\dot{h}\pi r^2 = 2\pi r \int_0^h v_r \, dz \tag{4.34}$$

where v_r is the radial velocity and $\dot{h} = dh/dt$ is the instantaneous disk velocity. See Figure (4.14) for the derivation of Equation (4.34).

The mass that disappears in the z-direction must appear in the radial direction as shown

$$-\rho\pi r^2 \frac{dh}{dt} = \rho \int_0^h 2\pi r v_r(z) \, dz. \tag{4.35}$$

In the equation of motion, as only the shear stress component τ_{rz} is important, the r component of the equation, ignoring inertia and gravity as well, reduces to

$$\frac{\partial \tau_{rz}}{\partial z} = \frac{\partial P}{\partial r}. \tag{4.36}$$

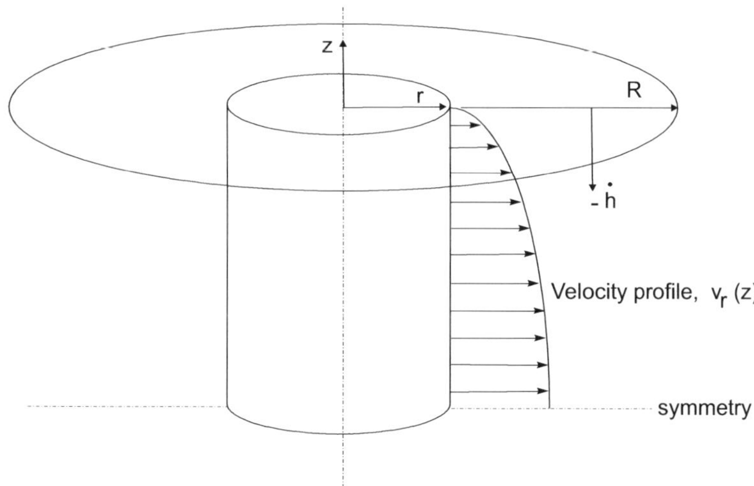

Figure 4.14: Sketch of velocity profile of the material between the midplane and the upper platen in compression molding of a radial charge.

The changing radial velocity $v_r(z)$ is introduced through the boundary condition at $z = h(t)$

$$v_r(z = h) = 0. \tag{4.37}$$

Further, because of symmetry

$$\frac{\partial v_r}{\partial z} = 0 \quad \text{at } z = 0. \tag{4.38}$$

The other two components of the equation of motion reduce to $\frac{\partial P}{\partial \theta} = 0$ and $\frac{\partial P}{\partial z} = 0$; hence P is a function of r only as shown in Figure 4.14. Thus one can integrate Equation (4.36)

$$\tau_{rz} = \left(\frac{dP}{dr}\right) z. \tag{4.39}$$

For power-law fluids,

$$\tau_{rz} = -m \left(\frac{dv_r}{dz}\right)^n. \tag{4.40}$$

Substitution of Equation (4.39) into Equation (4.40) and integration of v_r with respect to z and using boundary conditions (4.37) and (4.38) results in

$$v_r = \frac{h^{1+s}}{1+s}\left(-\frac{1}{m}\frac{dP}{dr}\right)^s \left[1 - \left(\frac{z}{h}\right)^{1+s}\right] \tag{4.41}$$

where $s = 1/n$. Note that the velocity profile is identical to fully developed flow between parallel plates separated by distance $2h$. Now, one substitutes Equation (4.41) into Equation (4.34), and integrates with respect to z, to obtain the following differential equation for $P(r)$:

$$\frac{dP}{dr} = -m\left(\frac{2+s}{2h^{s+2}}\right)^n (-\dot{h})^n r^n. \tag{4.42}$$

One can integrate Equation (4.42) to obtain the pressure profile

$$P - P_a = \frac{m(2+s)^n(-\dot{h})^n R^{1+n}}{2^n(n+1)h^{1+2n}}\left[1 - \left(\frac{r}{R}\right)^{1+n}\right] \tag{4.43}$$

where P_a is atmospheric pressure at $r = R$. The maximum pressure as expected is at the center ($r = 0$).

The total instantaneous force that must be applied to the disk to maintain the closing speed of $-\dot{h}$ is obtained by integrating P over the disk surface

$$F = m\pi\,\frac{(2+s)^n(-\dot{h})^n R^{3+n}}{2^n(3+n)h^{1+2n}}. \tag{4.44}$$

Equation (4.44) is called Scott's equation [132].

If $\dot{h}$ is constant, one can rewrite Equation (4.44) as

$$F = A\left(\frac{1}{h}\right)^b \tag{4.45}$$

where

$$A = m\pi\,\frac{(2+s)^n(-\dot{h})^n R^{3+n}}{2^n(3+n)}, \qquad b = 1 + 2n. \tag{4.46}$$

By taking the natural logarithm of Equation (4.45), one can obtain

$$\ln F = \ln A + b\,\ln\left(\frac{1}{h}\right). \tag{4.47}$$

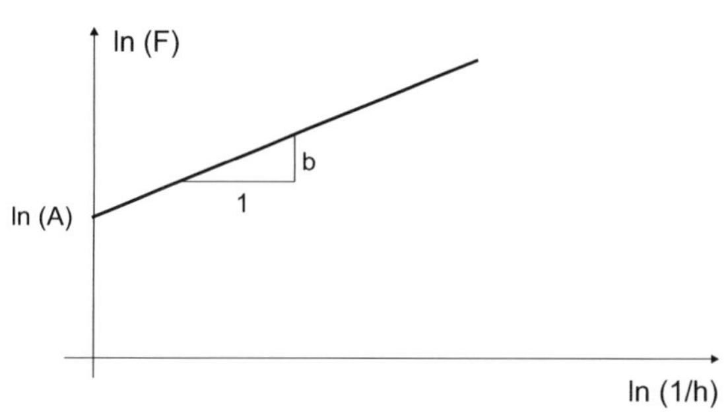

Figure 4.15: Approach to find the power law parameters from squeeze flow compression experiment in which the material is squeezed at the constant rate of $-\dot{h}$ and the required force, F, is measured.

One can find A and b from the plot of $\ln F$ versus $\ln\left(\frac{1}{h}\right)$ as shown in Figure 4.15. From b, one can find n. Knowing n, R, $s = 1/n$ and A, one can find m. ■

The above method can be used to find the shear viscosity in the transverse direction of a composite with aligned fibers or for a thermoplastic material with randomly distributed chopped fibers in the plane to provide an effective bulk viscosity.

4.4 Suspension Viscosity

In processes such as injection molding and extrusion, a short fiber suspension is created when the polymer in the pellets, containing the short fibers, melts. In compression molding of short fiber composites, materials such as sheet molding compound (SMC) that contain chopped glass fibers are heated to soften the resin to a paste to form a suspension which is compressed in the net shape of the mold. In injection molding, fiber suspension is pumped into a mold, whereas in extrusion the suspension flows through a die. In all such flows the fibers are convected with the polymer and solidify to create the short fiber composite part. The important issues here are the microstructure created in the part by the fibers due to their orientation and concentration distribution and the process parameters such as injection pressure and flow rate needed to inject the suspension into the mold or the die. Suspension viscosity plays a crucial role in modeling of these issues.

4.4.1 Regimes of Fiber Suspension

A suspension of uniform, cylindrical rods (approximation for fibers) is characterized by the particle volume fraction c and the fiber aspect ratio $r = L/D$, where L is the fiber length and D is the fiber diameter. A suspension is *dilute* only if

$$c \ll \frac{1}{r^2}. \tag{4.48}$$

Often this requirement is phrased in terms of the number density n (particles per unit volume), and the requirement for a dilute suspension is

$$n \ll \frac{1}{L^3}. \tag{4.49}$$

When the fibers are long and slender ($r \ll 1$) this requires a very small volume fraction. For example, a suspension with fibers of aspect ratio equal to 100 must have a volume fraction less than 0.01% to be dilute. If the fiber volume fraction is increased such that

$$\frac{1}{r^2} \ll c \ll \frac{1}{r} \quad \text{or} \quad \frac{1}{L^3} \ll n \ll \frac{1}{L^2 D} \tag{4.50}$$

then the suspension is said to be *semi-dilute*. In this regime the fibers seldom touch, but fibers experience frequent fluid-mechanical interactions from the perturbation of fluid velocity caused by the nearby fibers. Some researchers call this the semi-concentrated regime.

Increasing the volume fraction still more, to

$$\frac{1}{r} < c \quad \text{or} \quad n > \frac{1}{L^2 D} \tag{4.51}$$

reaches the *concentrated* regime. Here the average interfiber spacing is of the order of a fiber diameter, so that each fiber is always touching, or nearly touching, many other fibers.

Figure 4.16 maps out these regimes in terms of volume fraction and aspect ratio. As shown in the figure, all commercial composites fall into the concentrated regime. No commercial materials are truly dilute. Hence we will discuss here the phenomenological equation to describe the viscosity of concentrated suspensions.

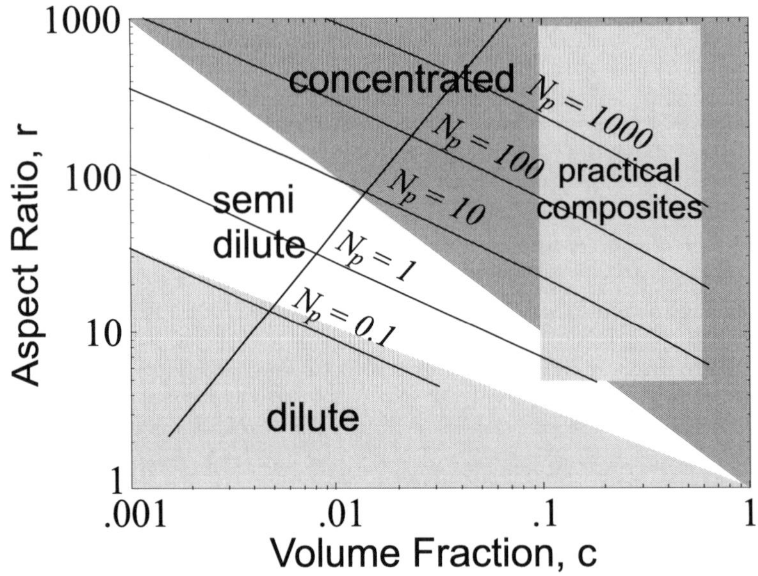

Figure 4.16: Different regimes of concentrations for fiber suspensions [147]. Here N_p is the "particle number" which will be discussed in Section 4.4.2.

The most important action of a fiber in a suspension is to resist stretching of the fluid along the fiber axis. All fiber suspensions have high viscosities in the fiber direction. This effect is greatest in steady uniaxial elongation in which the flow field lines are along the fiber length direction, where the fibers exert maximum resistance to the stretching motion.

In contrast, fibers add very little resistance to a shearing motion either parallel or perpendicular to the fiber axis. If one models the fiber using slender-body theory, the fiber adds no resistance to these motions at all. Shearing flow tends to make the fibers align parallel to the planes of shear, where they exert minimum resistance to the deformation. As a result, fiber suspensions have much lower viscosities in steady simple shear than in steady uniaxial elongation. That is, their Trouton ratio η_e/η_s (the ratio of elongational viscosity over shear viscosity) is much greater than three.

The low shear viscosity relies on fibers being aligned in the flow direction. This is usually not true in a start-up flow. If the fibers are initially random, then the shear viscosity is high. After a small amount of strain the fibers start to align, and at some point many fibers orient around 45° to the flow direction. This is the direction of greatest stretching in a simple shear flow, so at this point the viscosity reaches a maximum. With continued shearing the fibers rotate towards the flow direction, and viscosity drops steadily until it reaches the low steady-state value. There will also be normal stresses during the start-up of steady shear, as each fiber contributes to the stress in the direction of its axis. If the final fiber orientation is not symmetric about the plane of shear, the normal stresses will be present at steady state.

Suspension viscosity, in addition to the volume fraction, c and the aspect ratio, L/D of the fibers, will also depend on the orientation state of the fibers. Hence it is important to characterize the orientation state. It is relatively simple and straightforward to characterize c as it is the ratio of the volume of fibers to that of the total suspension volume. Aspect ratio is also easily characterized using the geometry of the fiber. However, to characterize the orientation state, one needs to develop a description.

Description of the Orientation of a Single Fiber

The orientation of a single fiber relative to a reference coordinate system can be specified by the angles (ϕ, θ) as shown in Figure 4.17.

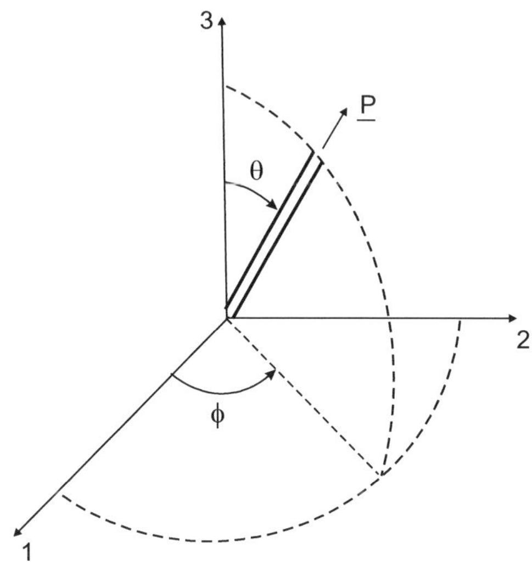

Figure 4.17: Definition of the orientation of a single fiber in a Cartesian coordinate frame [148].

Alternatively, one can describe the orientation of the fiber using a unit vector $\mathbf{p}$ directed along the fiber axis. The two descriptions are related

$$p_1 = \sin\theta\cos\phi \tag{4.52a}$$

$$p_2 = \sin\theta\sin\phi \tag{4.52b}$$

$$p_3 = cos\theta \tag{4.52c}$$

where p_i (i=1, 2, and 3) are Cartesian components of $\mathbf{p}$. Orientation space is the set of all possible directions of $\mathbf{p}$ and corresponds to the surface of a unit sphere. Sometimes, the fibers are long and the parts are thin which constrains the fibers to lie in the plane. Thus the orientation is called two-dimensional or planar orientation. In Figure 4.17, if one selects the 1-2 plane as the planar orientation plane, then θ will be equal to $\pi/2$ or p_3 equal to zero for every fiber. The planar orientation can be described by either ϕ or a two-dimensional vector $\mathbf{p}$. The orientation space for two-dimensional orientation will be a unit circle, whereas for 3 dimensions it will be a unit sphere.

Probability Distribution Function

In a real composite there are fibers oriented in many different directions. Figure 4.18 is a redrawn image of some of the fibers in a composite and denotes the complexity of the situation. It should be clear that, in practice, it will be a formidable task to keep track of the orientation of each individual fiber. It will be useful to use a description that can embody many different directions.

The most basic description is the probability distribution function. If we consider a small region from a composite or a suspension, we would expect to see many fibers with different

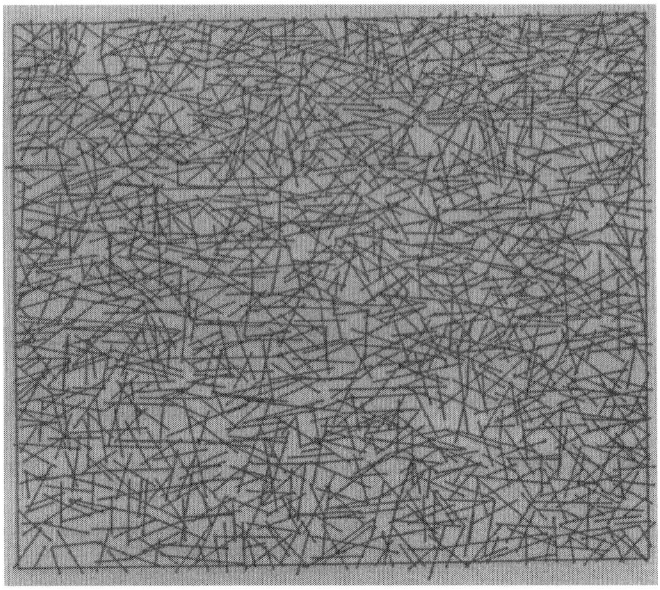

Figure 4.18: Sample of fibers digitized from a radiograph of a plaque of sheet molding compound [149].

orientations of fibers. The probability distribution will give us a measure of the state of fiber orientation at that location. Thus one can define $\psi(\theta, \phi)$ such that the probability of the fiber lying between the angles θ and $(\theta + d\theta)$ and angles ϕ and $(\phi + d\phi)$ is given by $\psi(\theta, \phi) \sin\theta d\theta d\phi$. Here $\sin\theta d\theta d\phi$ is the increment in area on the surface of the unit sphere. This distribution function can also be represented in vector form as $\psi(\mathbf{p})$. For planar cases, as $\sin\theta$ is unity, the distribution function can be represented as $\psi_\phi(\phi)$.

This distribution function must satisfy two obvious conditions. First, one end of the fiber is the same as the other end; hence, ψ must have the property of periodicity,

$$\psi(\theta, \phi) = \psi(\pi - \theta, \phi + \pi). \tag{4.53}$$

For the planar case,

$$\psi_\phi(\phi) = \psi_\phi(\phi + \pi). \tag{4.54}$$

Secondly, every fiber must have some direction, hence the integral of the distribution function over the surface of the orientation space must be unity

$$\int_0^{2\pi} \int_0^{\pi} \psi(\theta, \phi) \sin\theta d\theta d\phi = 1. \tag{4.55}$$

For the planar case,

$$\int_0^{2\pi} \psi_\phi(\phi) d\phi = 1. \tag{4.56}$$

This is known as the orientation normalization condition.

The distribution function ψ is a complete description of the orientation of the fibers at any location. Thus if the orientation state changes with position in a flowing suspension, ψ will be a function of position in addition to being a function of the angles θ and ϕ.

To estimate the distribution function at a given location, one should first capture an image of the fibers at that location, for example, the image shown in Figure 4.18. Let us suppose that the image is a planar one; thus to estimate ψ_ϕ, one would create a histogram as shown in Figure 4.19. This is done by dividing the histogram in n bins, each of width (π/n). Hence to find the orientation average of any function, one can use ψ as the weighting function.

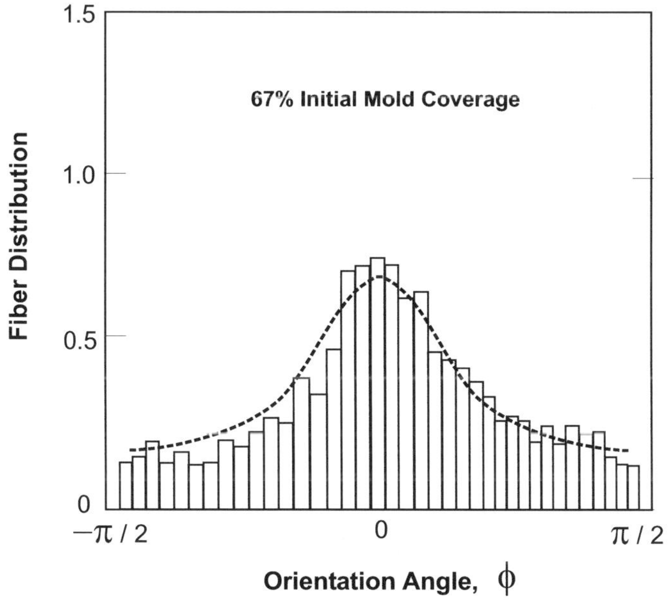

Figure 4.19: Histogram and probability distribution function for the fiber orientation in Figure 4.18 [149].

Therefore, the orientation average of any function $g(\theta, \phi)$ is given as

$$< g > \equiv \int_0^{2\pi} \int_0^{\pi} \psi(\theta, \phi) g(\theta, \phi) \sin\theta d\theta d\phi. \tag{4.57}$$

That is, $< g >$ is the average of g over all directions, weighted by the probability distribution function.

Orientation Parameters

The probability distribution function is a complete description of orientation state, but it is also cumbersome. A number of *orientation parameters* have been defined to provide concise and more easily interpreted measures of orientation. The best known is the Hermans orientation parameter. It concerns the special case of axisymmetric orientation, when the fibers are uniformly distributed in ϕ so that ψ is a function of θ only. The extent of alignment around the $\theta = 0$ axis is then measured by the parameter f, defined as

$$f = \frac{3 < \cos^2\theta > -1}{2}. \tag{4.58}$$

When the orientation is random $f = 0$, while completely aligned fibers give $f = 1$.

Many other orientation parameters have been defined. It can be shown that, if the probability distribution function has certain symmetries about the coordinate axes, then a particular set of orientation parameters provides all the information needed to predict the influence of orientation on the elastic properties of a composite. However, as there is no way to model the influence of processing conditions on the orientation parameters, they cannot be used to predict flow-induced orientation.

Orientation Tensors

A description which combines the generality of the probability distribution function and the concise nature of the orientation parameters is a tensor description of orientation. A second-rank orientation tensor can be defined as

$$a_{ij} \equiv < p_i p_j > \tag{4.59}$$

and a fourth-rank tensor as

$$a_{ijkl} = < p_i p_j p_k p_l > \tag{4.60}$$

That is, the components of the orientation tensors are formed by taking the orientation average of the products of components of the vector $\mathbf{p}$. These tensors are a generalization of the orientation parameters. It is evident from their definition that a_{ij} and a_{ijkl} will transform according to the rules of tensor transformation, so they are tensors. They are also symmetric, i.e.,

$$a_{ij} = a_{ji} \tag{4.61}$$

The normalization condition, Equation (4.55), implies that the trace of a_{ij} is unity,

$$a_{ij} = 1 \tag{4.62}$$

(In this and all subsequent indicial equations, a repeated index implies summation over all possible values.) Equations (4.61) and (4.62) show that only five of the nine components of a_{ij} are independent. For planar orientation, the tensor a_{ij} has four components, but only two are independent. The tensor description of orientation replaces the distribution function with a small number of discrete values.

Figure 4.20 shows some example planar orientation states and the corresponding orientation tensors. When a_{11} equals unity then all the fibers are aligned along the 1-axis. When a_{11} and a_{22} both equal one half, the fibers are randomly oriented. If the direction of principal orientation is other than the 1-axis or the 2-axis then a_{12} will be nonzero.

There are a number of physical interpretations of the orientation tensors. They can be thought of as a generalization of the orientation parameters, as the moments of the probability distribution function or as the coefficients of a series expansion of the distribution function. For a complete review see Advani and Tucker [150]. For the present we note that the orientation tensors are free from assumptions about the shape or symmetry of the distribution function, and can be readily expressed in any convenient coordinate system.

Example 4.4: Orientation Tensor for Uniformly Distributed Fibers in 3D

If the fibers are uniformly distributed in three-dimensional space, what will the components of a be?

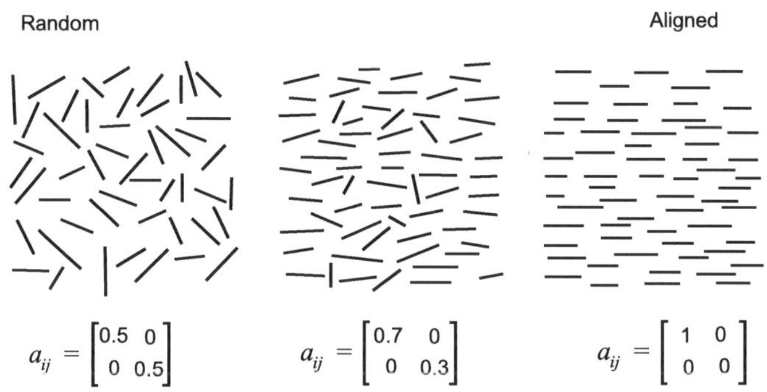

Figure 4.20: Planar fiber orientation states and the corresponding tensor representations.

Solution

As the fibers are uniformly distributed, $a_{11} = a_{22} = a_{33}$ and $a_{12} = a_{23} = a_{31} = 0$. From normalization, we know that $a_{11} + a_{22} + a_{33} = 1$. Thus,

$$a_{ij} = \begin{bmatrix} \frac{1}{3} & 0 & 0 \\ 0 & \frac{1}{3} & 0 \\ 0 & 0 & \frac{1}{3} \end{bmatrix}. \tag{4.63}$$

■

It can also be proved that the nth-rank orientation tensor provides sufficient information to predict the effect of orientation on any nth-rank property of the suspension or composite. Viscosity and elastic stiffness are fourth-rank properties, and one would normally have to know a_{ijkl} to predict them. However, we will later see ways to approximate a_{ijkl} in terms of a_{ij}, so that the second-rank tensor a_{ij} can be used as a description of orientation state.

The orientation tensors can easily be determined from experimental data. If the orientation of N fibers is measured, and if the vector p_i^k describes the orientation of the kth fiber, then the second-rank tensor is approximated by

$$a_{ij} = \frac{\sum_{k=1}^{N} p_i^k p_j^k w^k}{\sum_{k=1}^{N} w^k} \tag{4.64}$$

where w^k is the weighting factor for the kth fiber. If the fibers are almost circular in cross-section, then w^k can be taken as 1, which reduces Equation (4.64) to

$$a_{ij} = \frac{1}{N} \sum_{k=1}^{N} p_i^k p_j^k. \tag{4.65}$$

This equation presumes that the sample is unbiased [151]

4.4.2 Constitutive Equations

For process modeling purposes, one is interested in a general constitutive equation — one that represents the behavior of the material in any type of flow: shear, elongation or mixed;

steady or unsteady. It is also important to account for the effect of fiber orientation. A suspension in a given flow field will exhibit different stresses if its fiber orientation state is different.

Typically the compressibilities of the suspending fluid and the fibers are negligible, so the total stress σ_{ij} is separated into an isotropic pressure P and an extra stress τ_{ij}:

$$\sigma_{ij} = -P\delta_{ij} + \tau_{ij} \tag{4.66}$$

Subsequent attention is focused on relating the extra stress to the fiber orientation state, the rate of deformation, and suspension parameters such as the fiber volume fraction, the fiber aspect ratio and the viscosity of the suspending fluid.

A variety of theories exist to predict the viscosity of suspensions of fibers in a Newtonian fluid. All of them can be written in the general form for τ_{ij} [152]:

$$\tau_{ij} = \eta_s\dot{\gamma}_{ij} + \eta_s c\dot{\gamma}_{kl}a_{ijkl} + B\left[\dot{\gamma}_{ik}a_{kj} + a_{ij}\dot{\gamma}_{kj}\right] + C\dot{\gamma}_{ij} + 2Fa_{ij}D_r \tag{4.67}$$

Here $\dot{\gamma}_{ij}$ is the rate of deformation tensor, η_s is the solvent viscosity, c is the particle volume fraction, A, B, C and F are material constants, and D_r is the rotary diffusivity due to Brownian motion. Note that the extra stress τ_{ij} is a linear function of the instantaneous rate of deformation, except for the rotary diffusion term. This term is usually neglected for suspensions of fibers with large aspect ratios. The a_{ij} and a_{ijkl} are orientation tensors that describe the state of fiber orientation at that location.

Exact expressions for the coefficients A, B, C and F are given in references [153,154] for dilute suspensions. For a semi-dilute suspension, slender-body theory was used following the work of Burgers [155]. Batchelor [156] used this type of theory to model the stress in a fiber suspension undergoing uniaxial elongational flow, assuming that the fibers were all aligned in the direction of stretching. Batchelor modeled the fluid-mechanical interaction between the particles, treating each fiber as lying within a cell created by the surrounding fibers, and thus extended his theory beyond the dilute range. A key physical concept in Batchelor's theory is the idea of *hydrodynamic screening*, which assumes that the velocity disturbance caused by each particle is felt only a distance L away. The particle's nearby neighbors limit the long-range disturbance, and the major effect of the particle is to alter the fluid velocity only in the immediate vicinity. Evans [157, 158] derived the constants by substituting the result in Batchelor [156] into a general constitutive equation. Shaqfeh and Fredrickson [159] have developed a theory that gives this as the correct result for isotropic (3D random) orientation in a dilute suspension. They argue that Batchelor's result for A is correct for any orientation state, provided the suspension is dilute. Dinh and Armstrong [160] used an approach similar to Batchelor [156] and adopted ideas from liquid crystal theory about the distance between a fiber and its nearest neighbors to determine the cell size to predict the constants.

The slender-body approach is quite useful because it allows a way to incorporate hydrodynamic interactions and model nondilute suspensions, but it has some limitations. Slender-body theories are strictly applicable only for very high aspect-ratio fibers, so they might not be very accurate when the fiber aspect ratio is low (say, $r = 10$). Also, slender-body theory cannot represent the interaction between two fibers that pass within a distance on the order of D of each other, so truly concentrated suspensions cannot be modeled. Still, these theories do extend into the semi-dilute range, and give a good idea of the type of constitutive equation one might use, even for a concentrated suspension.

Equation (4.67) can easily be recast into the form

$$\tau_{ij} = \eta_I\left[\dot{\gamma}_{ij} + N_p\dot{\gamma}_{kl}a_{ijkl} + N_s\left\{\dot{\gamma}_{ik}a_{kj} + a_{ij}\dot{\gamma}_{kj}\right\}\right] \tag{4.68}$$

where the coefficients η_I , N_p and N_s depend on η_s, c, A, B and C, and a_{ij} and a_{ijkl} are the second and fourth ranked orientation tensors. Now η_I contains all the isotropic contributions to viscosity (from both the solvent and the particles), while anisotropic contributions by the particles are represented by two dimensionless parameters, N_s and N_p. N_p is called the *particle number* and N_s is the *shear number* [152]. These parameters depend on the particle aspect ratio and volume fraction, but (at least for dilute suspensions) not on the orientation state of the fibers. When both of these parameters equal zero, Equation (4.68) reduces to an isotropic Newtonian fluid.

The particle number N_p is the key rheological parameter to describe the behavior of a fiber suspension. Physically it represents the factor by which the suspension resists elongation parallel to the fiber direction, as compared to other deformations. Evans [157] was the first to identify this parameter and its key role in suspension rheology. For a suspension of slender particles N_p can be quite large. Figure 4.16 shows how N_p varies with volume fraction and aspect ratio according to the theory of Dinh and Armstrong, using the aligned assumption for inter-particle spacing. The other theories give similar predictions.

4.5 Reaction Kinetics

An important processing step in reactive polymer composites, which usually involve thermoset resins, is the reaction that starts from monomers or oligomers and through an addition or condensation mechanism forms a three-dimensional cross-linked network. The reaction can be activated either thermally by heating up the monomer or chemically by the addition of initiators. The phenomenon is generally referred to as curing of the resin. The degree of cure is usually proportional to the fraction of the network formed. The reaction is an exothermic one, so heat evolved during the formation of the network needs to be convected out of the composite.

Some of the important resins that react as they are processed are phenolics, unsaturated polyesters, epoxies and thermoset urethanes. Often, these materials are processed by compression, autoclave or liquid composite molding. The advantages of reactive processing are that the material is shaped into its final form before it reaches high molecular weight (hence viscosities are low and consequently the pressures required for processing are low). One of the disadvantages is that one has to deal with removal of the heat generated during the reaction. In order to understand reactive processing, one must have an appreciation of the reaction kinetics (how the reaction proceeds with time and temperature). Once this is understood and characterized, the familiar concepts of momentum and energy balance as detailed in Chapter 3 can be applied to predict the interaction of cure with flow and heat transfer.

The reaction may be easy to characterize, as in the case of epoxy and polyester resins, or may be very difficult to describe, as in the case of polyamides and polyurethanes that evolve volatile by-products. However, it is important to characterize the cure kinetics of the resin because it influences the viscosity of the resin and may govern the flow of the resin. Secondly, as significant heat is evolved during the curing of the resin, it cannot be ignored when modeling the heat transfer process during manufacturing. Modeling of the heat transfer phenomenon during processing allows the engineer to predict the temperature history of the composite during manufacturing which is important to address the residual stresses, warpage and shrinkage issues and to design appropriate heating and cooling cycles for the material. If an appropriate cooling system is not designed, the polymer can degrade due to the excess heat generated by the reaction. Thirdly, reaction kinetics can forecast the time of resin gelation. The gel point is defined by the degree of cure at which the resin has networked sufficiently that it no longer needs the mold or the die to maintain its final shape and can be demolded. This can help save processing time for the component. The resin is allowed to reach complete cure outside the mold in an oven or on the factory floor.

Hence, there is the need for a constitutive equation that can describe the reaction on a macroscale and can be coupled with the momentum and energy equations to determine the pressure, temperature and degree of cure simultaneously. On the microscale, if there is evolution of volatiles or water vapor during the curing process, it can lead to microvoid formation. To model void formation and growth, one may need many constitutive parameters to describe the mass diffusion process such as mass transfer coefficient, expressions for vapor pressure, partial pressure and solubility which are usually not readily available.

The general approach to the constitutive equation development for resin kinetics is traditionally derived from the type of the reaction and the chemistry of the reaction. Thermosets usually cure by either addition polymerization (for example, unsaturated polyesters) or condensation polymerization (for example, urethanes). Consider, as a typical example, the reaction between two chemical groups denoted by A and B which link together two segments of polymer chain P_1 and P_2:

$$\mathrm{P_1A + P_2B = P_1ABP_2} \tag{4.69}$$

For example, if it was the polyurethane reaction, A would represent an OH group and B an NCO group, resulting in

$$\mathrm{P_1OH + P_2NCO = P_1 - \underset{\displaystyle H}{\underset{|}{N}} - \underset{\displaystyle O}{\underset{|}{C}} - O - P_2} \tag{4.70}$$

The reaction can be tracked by monitoring the instantaneous concentrations of A and B, denoted by C_{A} and C_{B}. They can be measured in moles per gram and will decrease with time as more A's and B's react with one another to polymerize.

The rate at which the constituents A and B react will in general depend on the temperature and on the concentration of the reactants (sometimes catalysts are added to speed up the reaction). Higher the number of A and B molecules in a given space, the more likely it is that some will react. An equation describing the reaction rate as a function of concentrations of the reactants and temperature is called a kinetic equation because it describes the motion of the reaction. Many reactions of this type follow a kinetic equation such as

$$R_a(\text{moles per gram per second}) = -\frac{dC_{\mathrm{A}}}{dt} = k_0 \exp\left(-E/RT\right) C_{\mathrm{A}} C_{\mathrm{B}} \tag{4.71}$$

While in some cases there is a good physical reason for choosing one kinetic type of equation over another, mostly one can think of a kinetic equation as an empirical model which mimics the behavior of the resin. Equation (4.71) is referred to as a second-order kinetic equation, because two concentrations are multiplied together on the right-hand side. A first-order kinetic equation will depend on the concentration of only one of the groups. For example, if it was dependent on group A only, then

$$R_a(\text{moles per gram per second}) = -\frac{dC_{\mathrm{A}}}{dt} = k_1 \exp\left(-E/RT\right) C_{\mathrm{A}} \tag{4.72}$$

This obviously is not appropriate for this reaction since B groups as well as A groups must be present for a reaction to occur. In Equation (4.71), k_0 is the rate constant and is the empirical parameter to be determined from experiments, E is the activation energy for the reaction, R is the universal gas constant and T is the absolute temperature in K.

Once the kinetics are defined, one can analyze how the reaction proceeds with time. First, note that R_a is the rate at which the A groups disappear. Also as one B group reacts with one A group, so

$$\frac{dC_{\mathrm{A}}}{dt} = \frac{dC_{\mathrm{B}}}{dt} \tag{4.73}$$

The initial conditions for Equation (4.71) are the known initial concentrations of A and B, denoted by C_{A0} and C_{B0}. Instead of tracking C_{A} versus time, it is more useful to find the extent of reaction C^*, which can be defined as

$$C^* = \frac{C_{\mathrm{A0}} - C_{\mathrm{A}}}{C_{\mathrm{A0}}} \tag{4.74}$$

C^* is zero when there is no reaction and equals unity when all of the A's have been consumed and the reaction is complete. C^* can also be called the degree of cure or conversion (also sometimes the symbol α is used to denote degree of cure). Substitution of this in Equation (4.71) with some algebraic manipulation gives

$$\frac{dC^*}{dt} = k_0 \exp\left(-E/RT\right) C_{\mathrm{A0}} (1 - C^*)(\Lambda - C^*) \tag{4.75}$$

with an initial condition of $C^* = 0$. Here Λ is equal to $C_{\mathrm{B0}}/C_{\mathrm{A0}}$, and Λ equals one if A and B are present in equal number of moles.

Example 4.5: Degree of Cure with Time

If the temperature of the polymer is held constant at $T = 400°\mathrm{K}$ and $\Lambda = 1$, then using Equation (4.75) plot degree of cure with time for initial concentration of $C_{\mathrm{A0}} = 0.5\,\mathrm{mole/g}$. Take the parameters as $k_0 = 2\,(\mathrm{g/mole})/\mathrm{s}$ and $E/R = 1000°\mathrm{K}$.

- When does C^* reach a value of 0.8?
- At $t = 15\,\mathrm{s}$, calculate C^*. How does the result change, if the temperature is lowered to 350°K?

Solution

One can integrate the ordinary differential Equation (4.75) to find cure as a function of time:

$$C^* = \frac{k_0 \exp\left(-E/RT\right) C_{\mathrm{A0}} t}{1 + k_0 \exp\left(-E/RT\right) C_{\mathrm{A0}} t} \tag{4.76}$$

One can easily plot this equation for C^* with respect to time. Selection of $k_0 = 2\,(\mathrm{g/mole})/\mathrm{s}$ and $E/R = 1000°\mathrm{K}$ results in the plots shown in Figure 4.21. As seen from that figure, $C^* = 0.8$ at $t = 48.7\,\mathrm{s}$. At $t = 15\,\mathrm{s}$, $C^* = 0.552$ if the temperature is held at $T = 350°\mathrm{K}$, and it is 0.463 if $T = 400°\mathrm{K}$. ■

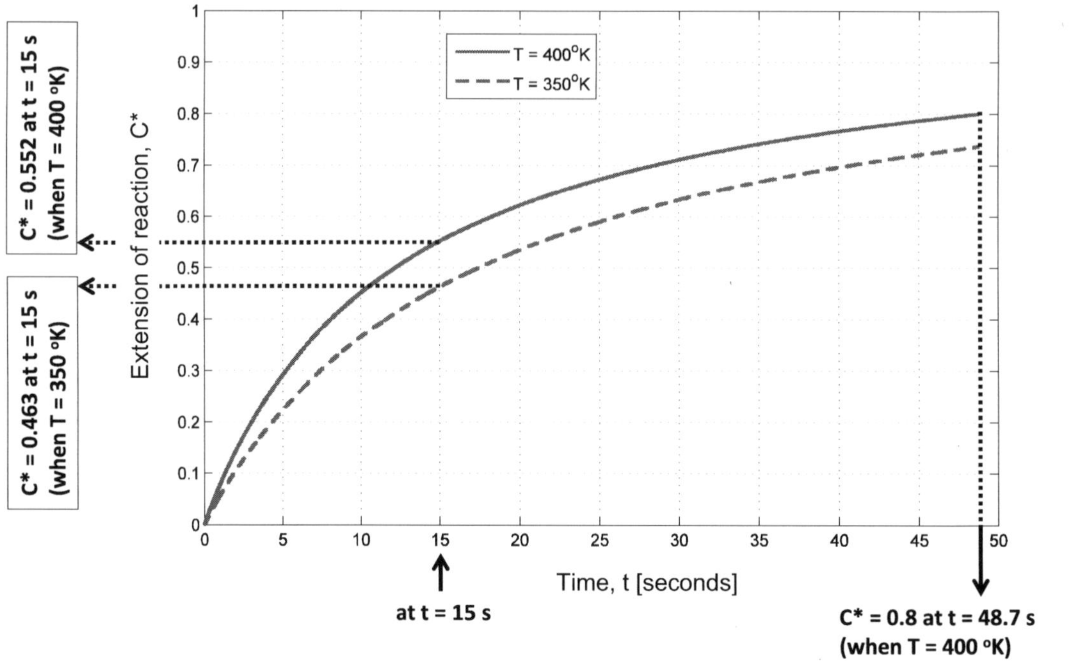

Figure 4.21: The degree of cure as a function of time for the parameters chosen in Example 4.5.

Example 4.6: Re-solve Example 4.5 Using Numerical Method

In the previous example, Equation (4.76) has been obtained by integrating the ordinary differential Equation (4.75) in time, t. This way, a closed form solution has been obtained for the evolution of $C^*(t)$. In the next chapters, we will deal with complicated differential equations, which do not have a closed form solution. Sometimes, the differential equations will be not so complicated, but either the domain or the boundary conditions will make the use of analytical solution methods either fruitless or tedious. To get familiar with numerical methods, here in this example, we will re-solve Example 4.5 numerically instead of analytically.

Solution

The reader is first suggested to review the MATLAB programming language and some basic numerical methods given in Appendix of this book. In this example, we will need to apply a numerical time-integration method. The simplest method is Eulerian method in which, the time domain is discretized by having a time increment of Δt

$$t(i) \equiv t_i = (i-1)\,\Delta t \qquad \text{for } i = 1, 2, 3, \ldots, N \tag{4.77}$$

where t is an array with N elements. In other words, the subscript (or the argument) of the independent variable t indicates the index of its discrete array.

The Eulerian method allows the user to calculate the dependent variable (here $C^*(t)$) at t_i values using the following formula (which is the Taylor series expansion at each discrete point up to the first derivative term only in the series):

$$C^*(i+1) \equiv C^*(t_{i+1}) \equiv C^*_{i+1} = C^*_i + \left(\frac{dC^*}{dt}\right)_i \Delta t + \ldots$$
$$\approx C^*_i + \left(\frac{dC^*}{dt}\right)_i \Delta t \qquad (4.78)$$

where the order of error is $(\Delta t)^2$.

dC^*/dt is given in Equation (4.75), thus Equation (4.78) becomes

$$C^*_{i+1} = C^*_i + (k_0 \exp(-E/RT) C_{A0}(1 - C^*_i)(\Lambda - C^*_i))\, \Delta t. \qquad (4.79)$$

A MATLAB code has been written below to numerically calculate $C^*(t)$ until it reaches 0.8. The corresponding $C^*(t)$ versus time plot is given in Figure 4.22 which agrees with the analytical solution plotted in Figure 4.21.

```
%%%%%%%%%%%%%%%  Matlab code for Example 4.6  %%%%%%%%%%%%%%%
clear all, close all, clc
R      = 8.3145;
CA0    = 0.5;
k0     = 2;
E      = 1000 * R;
LAMBDA = 1;
T      = 400;
C(1)   = 0;
t(1)   = 0;
dt     = 0.01;
i      = 1;
while  C(end) < 0.8
dCdt   = k0 * exp(-E/(R*T)) * CA0 * (1-C(i)) * (LAMBDA-C(i));
C(i+1) = C(i) + dCdt*dt;
t(i+1) = t(i) + dt;
i      =    i + 1;
end
figure(1), plot(t,C,'b-','linewidth',2)
xlabel('Time, t [seconds]'), ylabel('Extent of reaction, C^*')
grid, legend('T = 400^oK',0)
%%%%%%%%%%%%%%%%%%%%%%%%%%%%%%%%%%%%%%%%%%%%%%%%%%%%%%%%%%%%%%%%%%%%%
```

■

For a reaction in which temperature varies with time, which is the case for all composite processing methods, one needs to solve Equation (4.75) once the temperature history $T(t)$ is known.

If the polymerization mechanisms and the order of the reaction are known then it may be straightforward to determine the molecular weight and the gel point explicitly as a function of the extent of reaction at a single temperature. However, the chemistry is usually complex and the reactions are taking place at different temperatures as the curing process by nature is nonisothermal. Hence the order of the reaction is not known and one cannot usually have a single type of constitutive equation to accommodate the chemical complexity of the reaction.

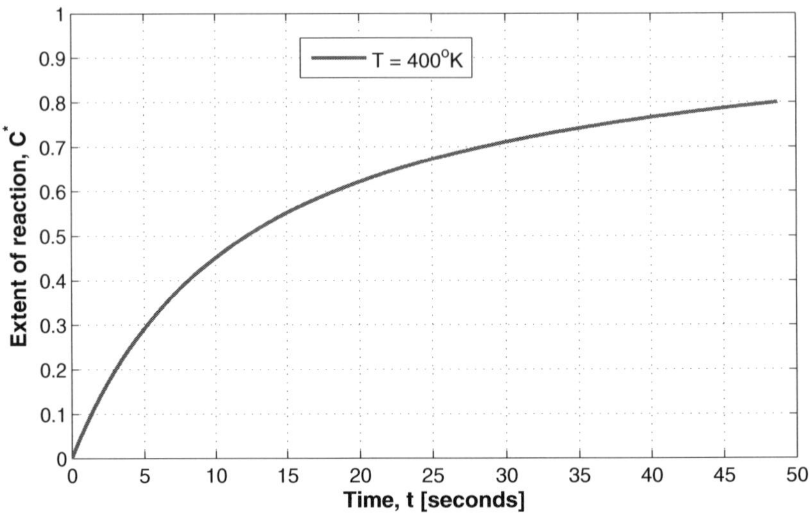

Figure 4.22: Numerically evaluated $C^*(t)$ using Euler method.

Hence, kinetic equations due to complex reactions are strictly empirical. For example, cure of unsaturated polyesters is fairly well represented by a kinetic equation such as

$$\frac{dC^*}{dt} = k_1 \exp\left(-E/RT\right)(C^*)^m(1 - C^*)^n. \tag{4.80}$$

Since $dC^*/dt = 0$ for $C^* = 0$, one assumes that the resin starts with a small degree of cure, say $C^* = 0.0001$. One now needs to measure k_0 and m and n from experiments, where $(m + n)$ represents the reaction order. There are many other deviations from this type of reaction. As outlined in Calado and Advani [161], one can in general decouple temperature and reaction effects and express

$$\frac{dC^*}{dt} = K(T)f(C^*). \tag{4.81}$$

Here $f(C^*)$ is the function that describes the rate of reaction in terms of primary constituents and empirical parameters, K is the rate constant defined by the Arrhenius type of relationship and expressed as

$$K(T) = A\exp(-E/RT) \tag{4.82}$$

where A is the frequency factor. Different forms are proposed for $f(C^*)$ in the literature. We will list two of the most common ones used. The first one corresponds to an nth degree reaction which takes the form

$$f(C^*) = (1 - C^*)^n. \tag{4.83}$$

The second one is related to the autocatalytic reaction as follows:

$$f(C^*) = (1 - k_2C^*)^m(1 - C^*)^n \tag{4.84}$$

in which $m + n$ represents the order of the reaction and k_2 is described by the Arrhenius equation.

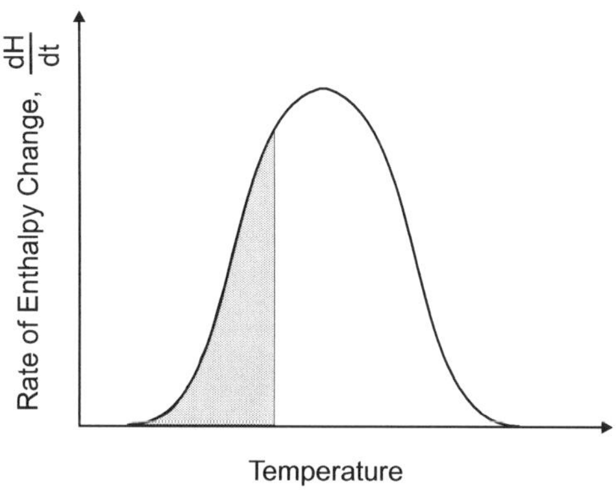

Figure 4.23: Example of a thermogram — rate of enthalpy change with temperature [161].

4.5.1 Techniques to Monitor Cure: Macroscopic Characterization

The widely used method to characterize and validate kinetic models is to experimentally measure degree of cure as a function of time and temperature. One can monitor this at the macroscopic and microscopic levels by recording chemical, physical (refractive index [162], density [163] and viscosity [164]), electrical (electrical resistivity [108,165]), mechanical and thermal property changes with time [166,167]).

Usually a differential scanning calorimeter (DSC) or differential thermal analyzer (DTA) is employed in both isothermal and dynamic modes. The isothermal scans provide the heat of reaction, and the dynamic scans yield degree of cure or the extent of reaction as a function of time at a specific heating rate [168]. A DSC is capable of controlling the temperature of a small sample of material and of simultaneously measuring the rate of heat flow into or out of the sample. Curing reactions can be tracked this way because they are exothermic. Whenever a linkage is formed, a certain amount of chemical energy is converted into heat. If H_R is the total heat of reaction (i.e., calories per gram), then the rate of heat liberated is

$$\dot{Q} = H_R M \frac{dC^*}{dt} \tag{4.85}$$

where M is the mass of the sample. Kinetics are best characterized by reacting the sample at a constant temperature and measuring $\dot{Q}$ with time. Several runs at different temperatures are needed to find the temperature dependence.

DSC is by far the most widely utilized technique to obtain the degree and reaction rate of cure as well as the specific heat of thermosetting resins. It is based on the measurement of the differential voltage (converted into heat flow) necessary to obtain the thermal equilibrium between a sample (resin) and an inert reference, both placed into a calorimeter [169,170]. As a result, a thermogram, as shown in Figure 4.23 is obtained [171]. In this curve, the area under the whole curve represents the total heat of reaction, H_R, and the shadowed area represents the enthalpy at a specific time.

The DSC can operate under isothermal or nonisothermal conditions [172]. In the former mode, two different methods can be used [173]:

i) A sample is placed into a previously heated calorimeter (faster equilibrium [174]) or into an unheated calorimeter, whose temperature is raised as quickly as possible up to the curing temperature.

ii) A sample is cured for various times until no additional curing can be detected; then, the samples are scanned (heating rate ranging from 2 to 20°C/min) in order to measure the residual enthalpy, H_{res}. The degree of cure is calculated directly,

$$\alpha = C^* = \frac{H_R - H_{res}}{H_R}, \tag{4.86}$$

but not the reaction rate, which is obtained by tangents to the curve of α versus time. For reactions with very small exothermal heat, this method should be utilized instead of the first method.

In the nonisothermal or dynamic mode, a sample is set into a calorimeter and the temperature is raised at a certain constant heating rate up to the operational temperature. The total heat of reaction is independent of the heating rate (recommended range is 2-20°C/min [166, 173]). Three methods are possible:

i) All kinetic information A and E in Equation (4.82) can be obtained from only one experiment. Although this method has been successfully applied for some first order reactions, it is not accurate for others; the heat of reaction is higher (around 10-30%) when compared with the isothermal mode [173].

ii) A and E in Equation (4.82) can be accurately measured from the values of the peak exotherm temperature for various heating rates for all reactions [173]. This is the recommended method.

iii) More general than the second method, it measures A and E from the values of the temperature necessary to reach a constant conversion for various heating rates. This technique needs considerable effort to be moderately successful.

Although many researchers use DSC for measuring enthalpies over the entire curing process, it is not accurate after the gel point [175–177].

Several authors have pointed out that the parameters of the reaction rate expression for isothermal and dynamic conditions may be different [175, 178–183].

One possible explanation for different results between isothermal and dynamic analysis was first given by Prime [184], and it is related to the fact that the degree of conversion is a function of the curing time and temperature. Therefore, we have:

$$\frac{d\alpha}{dt} = \left(\frac{\partial \alpha}{\partial t}\right)_T + \left(\frac{\partial \alpha}{\partial T}\right)_t \frac{dT}{dt}, \tag{4.87}$$

where $d\alpha/dt$ represents the dynamic reaction rate, $(\partial\alpha/\partial t)_T$ the isothermal reaction rate and dT/dt, the heating rate.

According to some authors [166, 179, 181, 184–187], the isothermal rate in reality is expressed by Equation (4.81). The physical meaning of the term $(\partial\alpha/\partial T)_t$ was questioned by other authors [188–190]. They affirmed that $(\partial\alpha/\partial T)_t$ has to be zero since if one fixes the time, the positions of all the particles in the system will be fixed, making α constant. On the other hand, by using thermodynamics, Kratochvil and Sesta [186] showed that $(\partial\alpha/\partial T)_t$ exists and is non-zero.

Equation (4.87) can also be recast in a different form:

$$\frac{d\alpha}{dt} = \left(\frac{\partial \alpha}{\partial t}\right)_T \left[1 + a\frac{(\partial\alpha/\partial T)_t}{(\partial\alpha/\partial t)_T}\right] \tag{4.88}$$

where $a = dT/dt$. For isothermal experiments $a = 0$, and there is no difference between Equations (4.81) and (4.88). But as the dynamic measurements are simpler and faster than isothermal ones (for the first case, only one experiment is needed to obtain all kinetic parameters), it is important to calculate the term $(\partial\alpha/\partial T)_t$ in order to acquire a relationship between dynamic measurements and isothermal parameters.

Other thermal techniques to characterize cure are thermogravimetric analysis (TGA) [106, 191], high pressure calorimeter (HPC) [173], thermomechanical analysis (TMA) [173, 192] and differential (or dynamic) thermal analysis (DTA) [193]. These are rarely used and will not be discussed here.

However, we can briefly mention another technique used which is called torsional braid analysis (TBA). This technique is a variation of torsion pendulum (TP) and was developed by Gillham in 1958 as discussed in Lewis [194]. It consists of measuring the frequency and decay constants which characterize each wave resultant of free torsional oscillations (at approximately 1 Hz) subjected to a sample (a glass braid impregnated with a resin). These measurements are converted into elastic (stored energy) and loss (loss energy) moduli, which are related to transition (gelation and vitrification) times and temperatures.

TBA is a sensitive technique for determination of physical changes in the resin [195–200], occurring at and after the gel point; it has a lower limit of detectability [173]. TBA is not capable of measuring the degree of cure.

Other mechanical techniques are dynamic mechanical analysis (DMA) [201, 202] and rheological dynamic spectrometer (RDS) [106, 203–205].

Due to the fact that there is no accurate technique valid for the entire curing reaction, two techniques should be used to follow cure: one applied before gelation (DSC) and another one applied after gelation (TBA, DMA, TMA, d.c. conductivity [206–209]) .

4.5.2 Technique to Monitor Cure: Microscopic Characterization

During curing there are many reactions occurring simultaneously. DSC tries to capture the overall reaction rate. However, if one is interested in understanding details of individual reactions, one must resort to microscopic characterization.

Sophisticated techniques are required to measure the concentration of all components in order to trace individual reactions occurring during cure. Some examples are Fourier transform infrared spectroscopy (FT i.r.) [106, 210–212], solid state nuclear magnetic resonance (n.m.r.) [213–218], near infrared spectroscopy (n.i.r.) [219–222] and chromatography [215, 223].

The spectroscopy methods consist of dispersing radiation from a source and passing it over a slit system which isolates a narrow frequency range falling on the detector [216]. By using a scanning mechanism, the energy transmitted through a sample as a function of frequency, known as the spectrum, is obtained and compared with the spectrum characteristic for each functional group of thermosetting resins. The mid-(m.i.r.) and near-infrared (n.i.r.) regions of the spectrum can be utilized. Although all functional groups involved in cure reactions of epoxy resins have strong characteristic absorptions in the m.i.r. region, the m.i.r. spectra of epoxy resins and hardeners are very complex [222]. The absorption bands of the main functional groups in cure reactions of epoxies are well-isolated in the near infrared region of the spectrum (10,000 and 4,000 cm^{-1}). A graph of absorbance versus wavenumber, as shown in Figure 4.24, is generated, presenting peaks correspondent to functional groups, such as epoxide, primary and secondary amines, and hydroxyl. The values of the absorption bands for these groups can be found in such works as [219–222].

The basic principle of all these methods is the comparison between the spectrum of reference substances and spectra of the reactants and products of a curing reaction subjected to radiation. A qualitative and quantitative identification of the components is then possible. However, microscopic characterization has not yet been linked to the phenomenological parameters used in cure-kinetic equations, but helps in understanding how the components react.

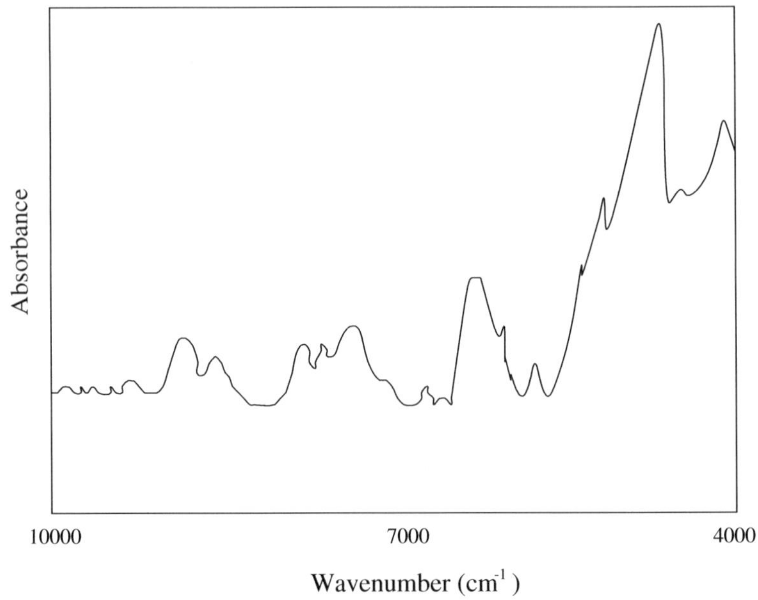

Figure 4.24: Absorbance as a function of wavenumber. Every peak is characteristic of a particular group [161].

4.5.3 Effect of Reinforcements on Cure Kinetics

Usually the resin flows through the fiber reinforcements, and these fibers may be sized and treated with chemical compounds, such as silane, in order to enhance coupling between the fiber and matrix phases [224, 225], which may change the kinetics. It is essential to know to what extent this coupling alters the cure kinetics and thus if it is possible to use the kinetic parameters from neat resins to describe the kinetics for reinforced resin materials. However, there is no consensus with respect to the influence of fillers or reinforcing materials on the cure kinetics. Inorganic fillers are used in thermosetting resins in order to reduce the cost and the shrinkage and as a heat sink to achieve a better temperature control across a molded part during cure [226]. Dutta and Ryan [227] analyzed the influence of carbon black and silica on the cure kinetics of DGEBA/*m*-phenylenediamine, concluding that the heat of reaction is independent of the filler content but is dependent on the type of filler, the overall reaction order of the system is not significantly affected, and the reaction rate changes because of the variations in the rate constants of the autocatalytic model. This effect is more sensitive at higher temperatures and for carbon black, probably because of chemical complexities on the carbon black surface [227].

Whereas McGee [228] did not verify any effect of particulate fillers (glass, calcium carbonate and aluminum) on the reaction kinetics, [226], working with calcium carbonate and clay in an unsaturated polyester resin, he concluded that the reaction rate increases by increasing the filler content.

When working with vinyl ester (specifically DERAKANE 411-C50), Michaud [122] observed a drastic influence of the fibers (E-glass) on the induction time (time required for the reaction to become observable at a specific temperature). An induction time of 90 minutes and 5.5 hours was necessary for RTM runs at 53°C and DSC runs at 50°C, respectively. Lee and Lee [229] affirmed that the age of the resin can also alter the induction time; at 70°C, a reduction time of 50% was found after 10 months of storage. All these effects are more evident at low temperatures.

Han et al. [230] found that the rate of cure of a resin is greatly influenced by the presence of fibers and the type of fibers employed; the rate of reaction for resin/fiber system can be 60% different from that of neat resin, after a ten-minute cure. A similar conclusion was presented by [231] for graphite/epoxy composites based on TGDDM/DDS; they verified large differences — see Table 4.6 — in the kinetic parameters when considering an autocatalytic model.

Table 4.6: Kinetic Parameter Percentage Errors Between Composite and Neat Resin

Graphite Epoxy Composite	A_1 (min^{-1})	A_2 (min^{-1})	E_{a_1} (J/mol)	E_{a_2} (J/mol)	m –	n –
$\left\lvert \frac{\text{Composite–Neat Resin}}{\text{Composite}} \right\rvert$	83.2%	50.3%	5.70%	0.344%	11.1%	36.0%

It is clear that the surface of reinforcement can affect the kinetics. The disagreement is about the extent of the change in kinetics when the resin is introduced into the resin system. The type of resin and the temperature range are important considerations. Generally at low temperatures, one must be cautious about applying the neat resin phenomenological parameters to a system in which the resin will impregnate a network of fibers.

4.6 Thermoplastic Reactive Processing

For the past decade there has been interest to explore reactive processing of thermoplastics instead of melt processing. The main advantage of reactive processing is that one can introduce the resin into the fibers with a low viscosity mono- or oligomeric molecules and then polymerize them with heat or UV radiation thus enabling processing of thermoplastic composites with low pressure processes such as RTM and VARTM.

In reactive processing of thermoplastics, addition polymerizations of mono- and di-functional species, of which vinyl polymerization and ring-opening polymerization are most common routes are pursued to form a high molecular weight linear polymer without the generation of unwanted by-products. A review article by K. van Rijswijk and H.E.N. Bersee [18] describe in detail the chemistry of reactive thermoplastic processing and the current research in this field. According to them, in vinyl polymerization the carbon-carbon double bonds in the vinyl monomers are broken into single bonds during polymerization resulting in two free electrons which join monomers to form a long chain of many carbon atoms containing only single bonds between them. In ring-opening polymerization, cyclic molecules are opened into linear monomers or oligomers and subsequently connected into high molecular weight polymers without generating by-products.

Although reactive thermoplastics have a lot in common with thermoset processing, there are some differences. The obvious difference being that most thermosets are liquid at room temperatures whereas the thermoplastics are solid and need to be heated to get them into the melt state. Although the reaction rate will increase with temperature, the crystallization rate for the thermoplastics decreases with temperature hence they have to be heated to an optimal temperature to balance the crystallization rate and the reaction rate. Probably the most important difference as to why reactive thermoplastic processing is not being used in composite industry is the poor fiber resin interface properties. For thermosets, the fiber sizing with the compatible coupling agent have been used for many years to form a good bond between the fiber surface and the resin. However for reactive thermoplastics, not many compatible coupling agents have been developed yet to apply on the fiber surface thus resulting in poor bond between fibers and resin and poor interphase properties. However, studies are underway to address this issue.

4.7 Crystallization Kinetics

While one needs to model cure kinetics when processing thermosets, one has to account for crystallization for thermoplastics as they solidify. Composites that contain thermoplastic polymers as the binding matrices can be formed or reformed by application of heat to melt or soften the polymer and stress to shape it. It is then subjected to cooling conditions that will solidify the polymer. The solidification of thermoplastics, in general semicrystalline polymers, and its effect on the resulting microstructure does play an important role in composites processing and hence, it is useful to study and model their behavior.

4.7.1 Introduction

The term crystallization has its roots in the word *crystal*, which refers to an ordered arrangement of molecules/units, which can be categorized in a specific way. Thus *crystallization* refers to the phenomenon of turning into a crystal. First, it is important to distinguish between the phenomenon of solidification and crystallization although one uses them as synonyms. Solidification refers to the transition of a substance from the liquid state to the solid state; it doesn't require it to solidify into an ordered structure. On the other hand, crystallization specifically refers to the transition of a body from the liquid state to an ordered solid state or a crystalline state. While solidification refers to all substances in general, crystallization usually refers to long chain structures, such as thermoplastics and proteins for example. "Crystal" refers to solidified metal structures too, due to their regularity. An important point to note is that although the term *crystalline polymers* can be theoretically defined, it is not physically realizable, since there is always a separate noncrystalline phase. Hence thermoplastics can be in either an amorphous state (no crystals) or a semicrystalline state (partial crystallinity). Crystallization evolves either from a solution or from the melt. These types of crystallization entail different concepts. We restrict ourselves here to only crystallization from the melt as thermoplastic composites involve processing of melts.

When we consider the crystallization of thermoplastics, the complex nature of their structure is influenced by the processing conditions. Depending on how the thermoplastic is cooled, it will either crystallize (slow cooling) or solidify into a glassy state (fast cooling); unlike a pure metal melt, which will always crystallize. This occurs due to the structure of polymers, i.e., their chain-like molecules. The ordering of these chains does not occur as easily as does the ordering of the atoms of a pure metal melt, where they organize themselves into a regular lattice (ordered structure/unit) when the thermodynamic conditions are conducive. Instead, various effects come into play at the micro- and the macrolevel of the crystallizing molecules, which makes crystallization in the case of polymers not a truly thermodynamic transition.

4.7.2 Solidification and Crystallization

One may quantify the differences between solidification and crystallization using the concept of phase transition as well. Phase transition refers to a change occurring between the solid, liquid and gaseous states, the three fundamental states of matter. When the change from one state to another occurs at a specific temperature (at a certain pressure), then the transition is termed a *first order transition*. More precisely, the transition occurs at the point when the first derivative of the state functions (volume and enthalpy) exhibits a sudden jump, i.e., when the first partial derivative of the Gibbs free energy (G) shows a discontinuity. In explicit terms: $dG = Sdt + VdP$, hence the entropy, S, and the volume, V, also exhibit a discontinuity [232].

With polymers there is no transition from the liquid to the gaseous state, as the temperature of decomposition of the polymer is much lower than the temperature required to separate the molecules; hence, the polymer disintegrates before it can turn into a gas. Also, for semicrystalline polymers, the solid/liquid transition does not take place abruptly; instead, it occurs over a range of temperatures, i.e., the primary state functions (V, H) change gradually between the values corresponding to solid and the liquid states [233].

When amorphous (irregular or having no order) solid materials or glasses solidify, the transition temperature is identified by a change in the second derivative of the primary state variables such as the Gibbs free energy. Hence the "glass transition" is known as a second order transition. The phase transition concept is illustrated in Figure 4.25.

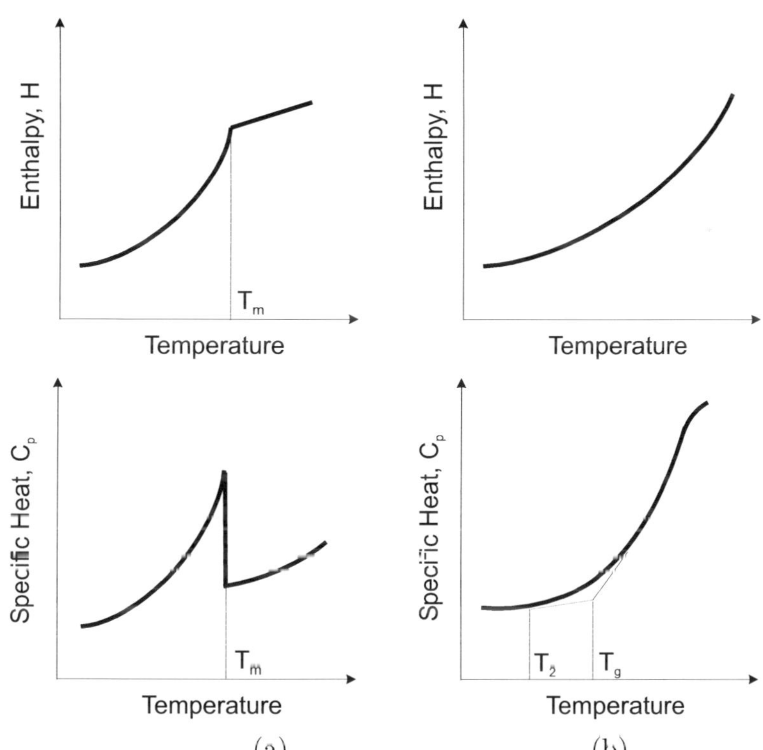

Figure 4.25: Variation of thermodynamic variables with temperature for (a) a semicrystalline polymer, and (b) an amorphous polymer [232].

4.7.3 Background

The knowledge that a thermoplastic could crystallize was not realized until 1912, when Von Laue showed that, because of the repetitive character of their lattices, crystals would scatter a beam of x-rays into regular patterns which were analogous to optical diffraction patterns [234]. Thus, subsequent work on polymeric materials demonstrated that they exhibit crystallinity at the level of molecular packing. Thus the next step was to determine the development and nature of this crystallinity, which to this date is not completely understood.

The chemical properties of a thermoplastic composite product are dictated by its matrix material microstructure, hence the need to model and accurately predict the microstructure evolution of the thermoplastic during its solidification. In semicrystalline polymers, the solidification conditions have a strong influence on the evolution and the resulting microstructure [235]. It has been shown that the kinetics of crystallization is the limiting factor during the growth of crystals from quiescent melts, instead of the growth following a truly thermodynamic progression. Thus, methods that involve the kinetics are often used to model and describe the microstructure instead of classical phase change methods (used to describe the solidification of water, as an example). Several models have been developed to describe the crystallinity variation with temperature in semicrystalline polymers [236]. But these models require the temperature profile in order to calculate the crystallinity, and various schemes have been developed to couple these models with conservation of energy equation for the composite.

4.7.4 Crystalline Structure

It has been shown that crystallization of thermoplastics occurs by the stacking up of individual polymer chains side by side to form a crystal lattice or by the physical phenomenon of the folding of polymer chains. This is shown schematically in Figure 4.26. Polymer chains fold back upon themselves to form a crystal called a *lamella.* Stacks of lamellae connected by regions of amorphous material or interlamellar links make up a semicrystalline polymer.

The crystallization from the melt is also referred to as *bulk crystallization*, as the movement of chains is hindered by other chains in their path and also by crystallized areas in the bulk. Crystallization in the melt or bulk occurs by the formation of crystals in the radial direction from a central nucleus, also known as *spherulites.* A schematic of a spherulite is shown in Figure 4.27.

The formation of spherulites is described by the phenomenological theory first proposed by [237]. They theorized that in order for growth of a spherulite to take place, two conditions must be met: the medium must have high viscosity, and noncrystallizable material (or solute) must be present. Similar to solidification of metals, the solute is rejected from the crystallizing solid and pushed ahead of the solidifying interface, and is assumed to promote the instability of the solid/liquid interface. They suggested that branches of the spherulite penetrate into the melt and grow preferentially in regions rich in crystallizable material.

Initially, the nucleation process must occur in order for the polymer chains to attach themselves to the nuclei and facilitate growth of the crystal. Then nucleation and growth occur together for a certain amount of time, which is followed by a free growth phase of the spherulites. Free growth continues until the spherulites impinge on each other, and the process of secondary crystallization takes place. Secondary crystallization refers to the attachment of chains on a smooth surface of the crystal, where previously no growth occurred. It is much slower compared to primary crystallization because the polymer chains are not able to find nuclei to attach themselves to, and they are also hindered from diffusing

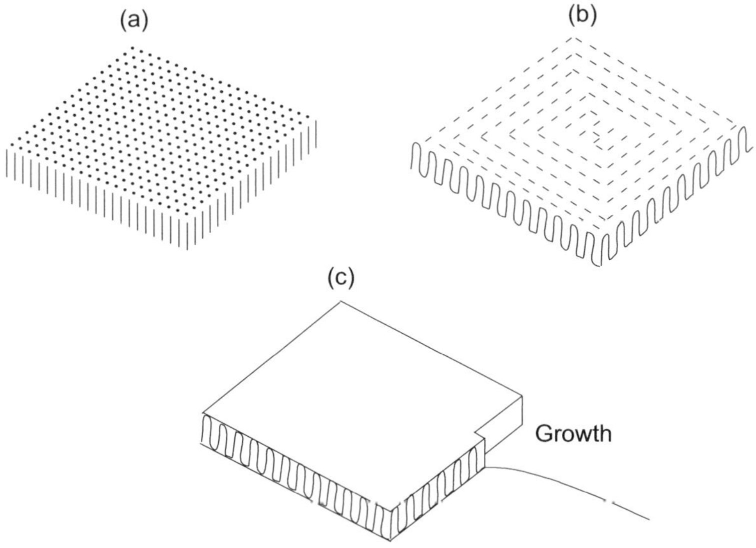

Figure 4.26: Polymeric chains forming a crystal [234].

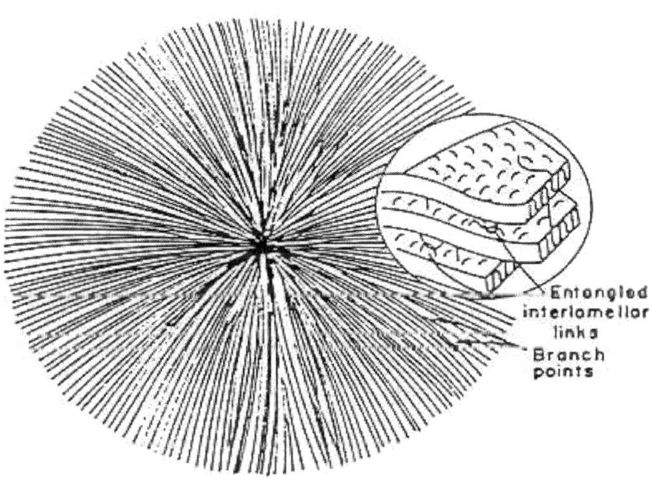

Figure 4.27: Schematic of a spherulite showing lamellae [233].

to other favorable growth sites. Finally, after crystallization is complete, the polymer cools down to the ambient temperature by convection [238].

One can address the crystallization kinetics on two scales that are coupled. At the microscopic scale, one needs to model the nucleation and growth of a spherulite and on a macroscopic scale one needs to determine the overall crystallinity behavior of the polymer. It is obvious that changes at the microscopic level will influence the macroscopic constitutive equation.

4.7.5 Spherulitic Growth

Spherulitic growth is still not fully understood, but it has been shown that their radii often grow linearly with time under isothermal conditions, which supports a kinetics controlled growth theory [238–240].

The growth of the tips of the branches of a spherulite follows the kinetics of attachment of polymer chains given by the Lauritzen and Hoffman theory [241, 242]. Thus, the growth rate of the crystal can be represented by [238]:

$$\frac{\partial R_g}{\partial t} = G_0 \exp\left[\frac{-U^*}{R(T - T_\infty)}\right] \exp\left[\frac{-K_g}{T \Delta T f}\right] \tag{4.89}$$

Here, the first exponential term represents the temperature dependence of the segmental jump rate in the polymer; the second term is the contribution from the net rate of secondary nuclei formation on the surface of a lamella. R_g is the grain radius, U^* is the activation energy, R is the universal gas constant, and T_∞ is the temperature at which all motion associated with viscous flow cease. K_g is a constant related to the growth regime and is specific to a particular regime, f is a correction factor that accounts for the change in value of the heat of fusion at low temperatures (close to T_g), and ΔT is the undercooling. Undercooling is defined here as the difference between the melting temperature and the crystallization temperature. G_0 is a pre-exponential constant that gathers terms not strongly dependent on the temperature and has different values corresponding to the different regimes of crystallization. Table 4.7 lists the kinetic, thermodynamic and material properties for polyethylene.

Table 4.7: Properties for Polyethylene [238].

Kinetic Data
K_g (^{o}K)^2 = 2 × 10^5 G_0 (cm/s) = 4.4 × 10^9 U^* (cal/mole) = 1500
Thermodynamic and Thermal Data
T_∞ = 201 (^{o}K) $\Delta T = T_m - T_g$ = 417.8 - 231 (^{o}K)

4.7.6 Macroscopic Crystallization

On the macroscopic level, one can develop useful constitutive equations to describe isothermal and nonisothermal crystallization. Isothermal crystallization refers to the phenomenon when the spherulite growth process is steady; i.e., the degree of crystallinity (χ) is independent of the temperature [243, 244]. Explicitly stated:

$$\chi = 1 - \exp\left(-Bt^n\right) \tag{4.90}$$

where B is a time-dependent scaling parameter related to the half-time of crystallization, $t_{1/2}$ and depends on the temperature, and n is the Avrami exponent. This equation is commonly known as the Avrami equation of crystallization. Non-isothermal crystallization on the other hand refers to temperature-dependent crystallization kinetics. Modeling of this process has led to a temperature-dependent expression for χ, which is more complex than that given by Equation (4.90) and follows from a modified model proposed for this purpose, the Nakamura model [238]:

$$\frac{\partial \chi}{\partial t} = nK(T)(1-\chi)\ln\left(\frac{1}{1-\chi}\right)^{(n-1)/n} \tag{4.91}$$

where n is the kinetic parameter and $K(T)$ is related to the crystallization rate constant of isothermal crystallization, $k(T)$, by

$$K(T) = k(T)^{1/n}. \tag{4.92}$$

The models for nonisothermal crystallization more accurately describe experimental data collected on the crystallization of various thermoplastics.

The constants don't have a physical meaning and cannot be easily connected to the microscopic analysis. However, the constants can be easily determined through differential scanning calorimetric (DSC) experiments, and no induction time (the time required for the organization of molecules around a nucleus) factor is necessary. DSC is a technique used to determine thermal transition points of various substances. The evolution of the absolute crystallinity profile within a part during cooling can be predicted by coupling the model with a heat transfer analysis [235]. This is done by representing the source term in the energy balance equation by $\rho c_p H_r \partial\chi/\partial t$ which tracks the heat evolved during solidification.

4.8 Permeability

Permeability characterizes the ease with which a fluid can flow through a porous medium. In almost all composite processes that use a thermoset resin as the matrix material, the resin has to be either infused into a fiber preform in processes such as liquid molding or extracted out during consolidation as in autoclave processing. This motion of the resin is usually modeled as flow through a porous medium in which the fiber network constitutes the porous medium. The empirical Darcy's law [245] governs the flow of the resin as follows:

$$Q = -\frac{KA}{\eta}\frac{dP}{dx} \tag{4.93}$$

Here Q is the flow rate across the cross-section A, η is the viscosity of the liquid, dP/dx is the driving pressure gradient and K is the permeability of the porous medium in the x direction.

This law allows the modelers to describe a macroscopic relationship between the flow rate and the pressure gradient needed to drive the flow instead of using the momentum conservation equations. If one wanted to use the momentum equations, one would have to describe the geometry of every channel present in the fibrous network and solve the momentum equations inside them to find the flow rate and pressure inside these channels. There can be a network of over a million channels created by the arrangement of the fibers. This task would be impossible to approach even with a sophisticated computer, and more importantly we only need to know the overall relationship between the pressure drop and the flow rate and not the details of the pressure at each and every location within every channel while processing such materials. Thus, Darcy's law lumps the ease of flow within these channels into a parameter called permeability which characterizes the mobility of the resin through the fibrous porous media.

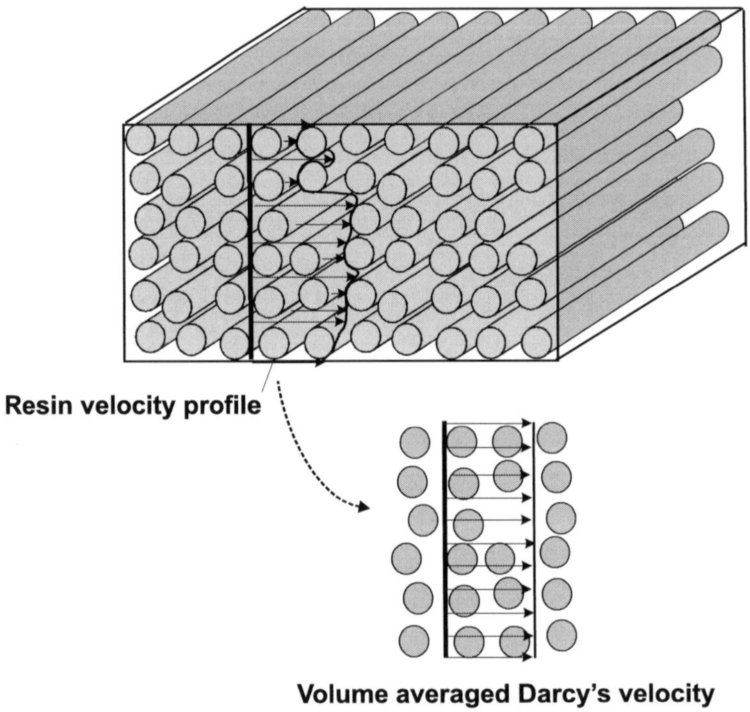

Figure 4.28: Volume averaged Darcy velocity within a porous medium.

Permeability is a function of fiber network, fiber volume fraction and also the type of sizing applied to the fibers. Also, as the porous medium formed by the fibrous network will be anisotropic (it will offer different resistances in different directions), the practice has converged on generalization of Darcy's law to account for it:

$$\bar{\mathbf{u}} = -\frac{\mathbf{K}}{\eta} \cdot \nabla P \tag{4.94}$$

where $\bar{\mathbf{u}}$ is the volume averaged "Darcy velocity" (see Figure 4.28), η is the viscosity of the fluid, ∇P is the pressure gradient, and $\mathbf{K}$ is the permeability tensor of the preform. Here, $\mathbf{K}$ is now a tensor instead of a scalar and can be represented as

$$\mathbf{K} = \begin{pmatrix} K_{xx} & K_{xy} & K_{xz} \\ K_{yx} & K_{yy} & K_{yz} \\ K_{zx} & K_{zy} & K_{zz} \end{pmatrix} \tag{4.95}$$

and the three-dimensional form of Darcy's law is expressed in matrix form as follows:

$$\begin{pmatrix} \bar{u}_x \\ \bar{u}_y \\ \bar{u}_z \end{pmatrix} = -\frac{1}{\eta} \begin{pmatrix} K_{xx} & K_{xy} & K_{xz} \\ K_{yx} & K_{yy} & K_{yz} \\ K_{zx} & K_{zy} & K_{zz} \end{pmatrix} \begin{pmatrix} \partial P/\partial x \\ \partial P/\partial y \\ \partial P/\partial z \end{pmatrix}. \tag{4.96}$$

The Darcy's law that now substitutes the momentum equations can be written down in three directions as follows (by using Cartesian coordinates):

$$x: \quad \bar{u}_x = -\frac{K_{xx}}{\eta}\frac{\partial P}{\partial x} - \frac{K_{xy}}{\eta}\frac{\partial P}{\partial y} - \frac{K_{xz}}{\eta}\frac{\partial P}{\partial z}, \tag{4.97}$$

$$y: \quad \bar{u}_y = -\frac{K_{yx}}{\eta}\frac{\partial P}{\partial x} - \frac{K_{yy}}{\eta}\frac{\partial P}{\partial y} - \frac{K_{yz}}{\eta}\frac{\partial P}{\partial z}, \tag{4.98}$$

$$z: \quad \bar{u}_z = -\frac{K_{zx}}{\eta}\frac{\partial P}{\partial x} - \frac{K_{zy}}{\eta}\frac{\partial P}{\partial y} - \frac{K_{zz}}{\eta}\frac{\partial P}{\partial z}. \tag{4.99}$$

Thus, for a given stationary porous medium of a fiber preform, one needs to know six scalar values (assuming that $K_{ij} = K_{ji}$) to determine the full permeability tensor.

However, if the selected coordinate direction is along the principal directions of the preform as illustrated for a two-dimensional case (where the direction 3 is normal to 1,2

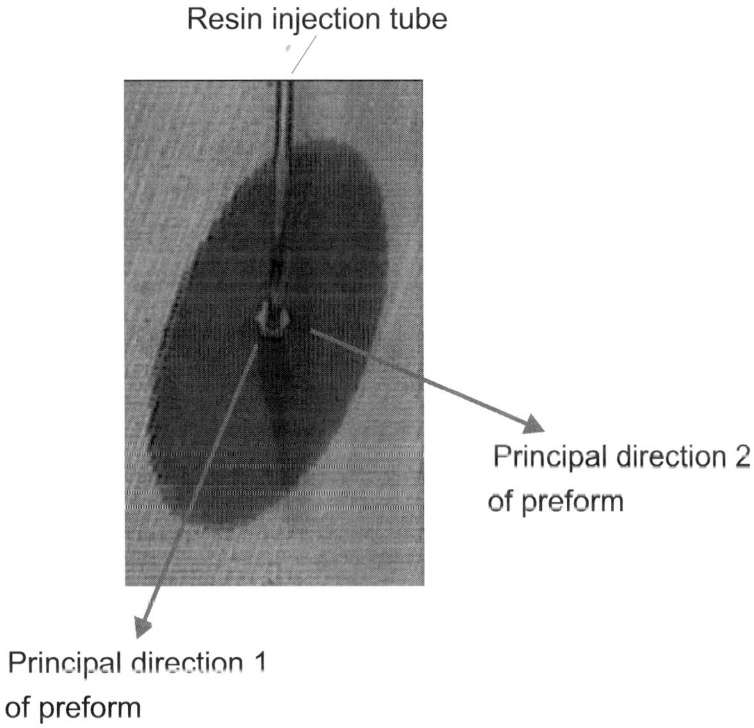

Figure 4.29: Local in-plane principal axes in the plane of a preform along the major and minor axes of the ellipsoid formed by the injected resin volume.

plane) in Figure 4.29, then Equation (4.96) can be simplified to

$$\begin{pmatrix} \bar{u}_1 \\ \bar{u}_2 \\ \bar{u}_3 \end{pmatrix} = -\frac{1}{\eta} \begin{pmatrix} K_{11} & 0 & 0 \\ 0 & K_{22} & 0 \\ 0 & 0 & K_{33} \end{pmatrix} \begin{pmatrix} \partial P/\partial x_1 \\ \partial P/\partial x_2 \\ \partial P/\partial x_3 \end{pmatrix}. \tag{4.100}$$

Here, the local principal coordinate system x_1, x_2, x_3 is along the principal axes of the preform which coincides with the major and minor axes of the ellipsoid formed by the resin volume injected as shown in Figure 4.29. Thus, by choosing the coordinate directions along the principal axes of the fiber preform, one can measure the principal permeability values of the preform. Then by coordinate transformation, one can calculate the non-diagonal values if one knows the flow direction with respect to the principal permeability direction.

Example 4.7: Coordinate Transformation [246]

One-dimensional injection of resin through a preform is shown in Figure 4.30. The resin moves from left to right, and the flow front is a straight line. This can be achieved by having a point injection gate inside an air channel between the preform and the mold cavity wall. Resin fills this section first due to negligible resistance to the flow. The significant flow through the preform starts after this air channel is filled. Note that no significant racetracking channel should exist between the preform and horizontal mold walls (see Figures 8.30, 8.32, 8.33 and 8.34 for definition of racetracking and its effect on the resin flow). Consider the two coordinate systems: (i) the principal axes $x_1 x_2$, and (ii) mold axes

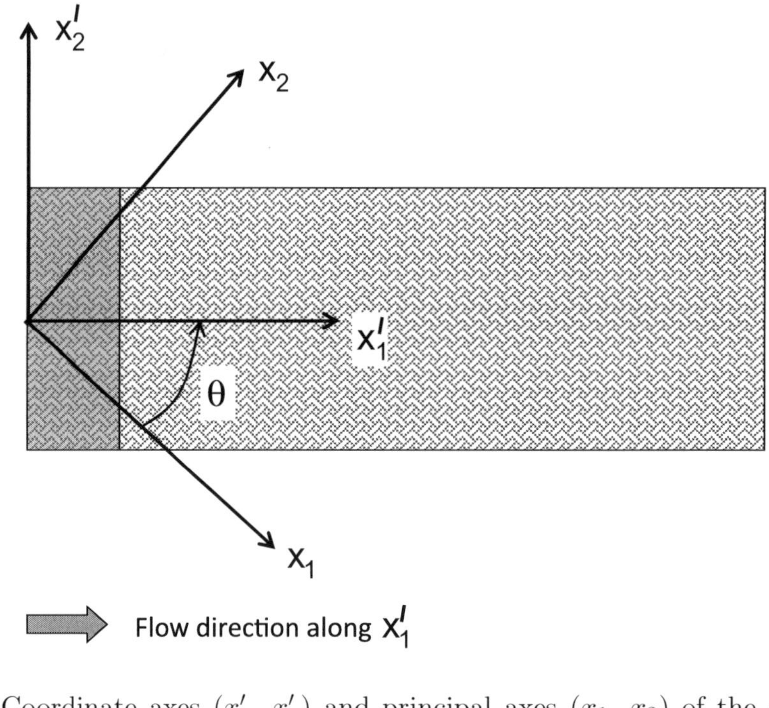

Figure 4.30: Coordinate axes (x_1', x_2') and principal axes (x_1, x_2) of the preform. One-dimensional resin flow from left to right.

$x_1'x_2'$. The two-dimensional (in-plane) permeability tensor is given by

$$\mathbf{K} = \begin{bmatrix} K_{11} & 0 \\ 0 & K_{22} \end{bmatrix} \quad (4.101)$$

for the principal axes. However, if x_1' is at an angle θ to the principal x_1 axis, then the components of the permeability tensor in the $x'y'$ frame can be calculated by tensor transformation

$$\mathbf{K}' = \begin{bmatrix} K_{11}' & K_{12}' \\ K_{21}' & K_{22}' \end{bmatrix} \quad (4.102)$$

where

$$K_{11}' = K_{11}\cos^2\theta + K_{22}\sin^2\theta \quad (4.103)$$

$$K_{22}' = K_{11}\sin^2\theta + K_{22}\cos^2\theta \quad (4.104)$$

$$K_{12}' = K_{21}' = (-K_{11} + K_{22})\sin\theta\cos\theta. \quad (4.105)$$

Thus, if one knows K_{11}, K_{22} and θ, one can find the permeability tensor in any new coordinate frame.

How can you find K_{11} and K_{22} if θ was not known a priori?

Solution

The one-dimensional experiment shown in Figure 4.30 will allow one to relate flow rate to pressure drop as follows:

$$u_1' = -\frac{1}{\eta}\left(K_{11}'\frac{\partial P}{\partial x_1'} + K_{12}'\frac{\partial P}{\partial x_2'}\right) \quad (4.106)$$

$$u_2' = -\frac{1}{\eta}\left(K_{12}'\frac{\partial P}{\partial x_1'} + K_{22}'\frac{\partial P}{\partial x_2'}\right) = 0 \quad \text{(no flow in } 2' \text{ direction).} \quad (4.107)$$

Equation (4.107) can be rearranged to yield

$$\frac{K_{12}'}{K_{22}'} = -\frac{\left(\dfrac{\partial P}{\partial x_2'}\right)}{\left(\dfrac{\partial P}{\partial x_1'}\right)}. \quad (4.108)$$

Equation (4.106) is divided by $\partial P/\partial x_1'$, which results in the following expression:

$$\frac{\eta u_1'}{\left(\dfrac{\partial P}{\partial x_1'}\right)} = -\left(K_{11}' + K_{12}'\frac{\left(\dfrac{\partial P}{\partial x_2'}\right)}{\left(\dfrac{\partial P}{\partial x_1'}\right)}\right). \quad (4.109)$$

Substitute Equation (4.108) in Equation (4.109) to get

$$\frac{\eta u_1'}{\left(\dfrac{\partial P}{\partial x_1'}\right)} = -\left(K_{11}' - K_{12}'\frac{K_{12}'}{K_{22}'}\right). \quad (4.110)$$

Now using Equations (4.103)-(4.105), we can write Equation (4.110) as follows:

$$-\frac{\eta u_1'}{\left(\dfrac{\partial P}{\partial x_1'}\right)} = K_{11}\cos^2\theta\left[\left(1+\beta\tan^2\theta\right) - \frac{(\beta-1)^2}{\dfrac{\beta}{\tan^2\theta}+1}\right] \tag{4.111}$$

where $\beta = K_{22}/K_{11}$. u_1' and $\partial P/\partial x_1'$ can be measured accurately only if the ratio of K_{11}/K_{22} is small or if K_{12} is small. Three experiments should be carried out at angles θ, $\theta + \Delta\theta_1$ and $\theta + \Delta\theta_2$ where $\Delta\theta_1$ and $\Delta\theta_2$ are two distinct and known angles such as 15 and 30 degrees, respectively. This will change the right-hand side and now we will have three equations and three unknowns to solve for θ, K_{11} and K_{22} for a given preform [246]. ■

4.8.1 Permeability and Preform Parameters

In all liquid molding processes we need to know the value of permeability in order to predict the flow behavior of the resin as it impregnates the preform. In processes such as autoclave molding, thermoset pultrusion and filament winding, an estimate of permeability will allow one to predict the resin pressure. If the resin pressure is low, insufficient consolidation could occur and lead to growth of microvoids. Permeability is a function of fiber architecture that characterizes the channel network. As the fibers usually govern the mechanical properties, it is advantageous to increase the fiber volume fraction in the composite. One approach to increase fiber volume fraction is to apply a force on the network of fiber preforms and compress them. As this network is compressed it blocks or constricts many channels of the porous medium and thus reduces the *mobility of resin flow* or the permeability. This reduction of permeability with fiber volume fraction needs to be characterized as it plays a very important role in processing. Many constitutive models that use the physics of lubrication flow or flow through capillary tubes have been developed to describe this relationship. The earliest one was known as the Kozeny-Carman equation and was developed by adopting the capillary model from the soil mechanics literature. It may be expressed as follows:

$$K = A\,\frac{(1-V_f)^3}{V_f^2} \tag{4.112}$$

where V_f is the fiber volume fraction and K is the principal permeability component in the flow direction. The constant A is supposed to be a function of the fiber network. However, it was found that one needed a different value of A for the same network at different fiber volume fractions. Also for flow across an aligned bed of fibers, once all the fibers touch each other, one would expect the permeability to be zero in that direction. However, the model was not able to capture that physics either. More constitutive equations were developed by [247, 248] to explain some of these shortcomings.

Bruschke and Advani [248] found that one could use lubrication theory and the cell model concept to describe permeability across an array of fibers more accurately as a function of fiber volume fraction

$$\frac{K}{r^2} = \frac{1}{3}\,\frac{(1-L^2)^2}{L^3}\left(\frac{3L\tan^{-1}\sqrt{\frac{1+L}{1-L}}}{\sqrt{1-L^2}} + \frac{L^2}{2} + 1\right)^{-1} \tag{4.113}$$

where $L^2 = 4V_f/\pi$, and r is the radius of a fiber. Note that no empirical parameters were needed.

Gebart [247] used similar logic to develop principal permeability values as a function of fiber volume fraction in both along the fiber direction and perpendicular to the fiber direction:

$$K_{\parallel} = \frac{8r^2}{C_{\parallel}} \frac{(1-V_f)^3}{V_f^2} \tag{4.114}$$

$$K_{\perp} = C_{\perp} r^2 \left(\sqrt{\frac{V_a}{V_f}} - 1 \right)^{5/2} . \tag{4.115}$$

In all cases, one finds that permeability of the preform decreases as fiber volume fraction is increased. This results in reduced empty space between the fibers; hence, the resin will face more resistance to impregnating this preform. However, there is no universal formula for the relation between the preform permeability and the fiber volume fraction. This relation is mainly a function of fabric type (i.e., whether it is a glass or carbon) and fabric structure (i.e., whether it is stitched, woven, random, etc.). A typical graph showing the change in permeability with fiber volume fraction is shown in Figure 4.31 for a plain weave glass fabric. Furthermore, the detailed structure of a preform is dependent on whether it is draped over a tool surface with double curvature and/or whether it has some nesting and racetracking channels. All these changes will affect the permeability.

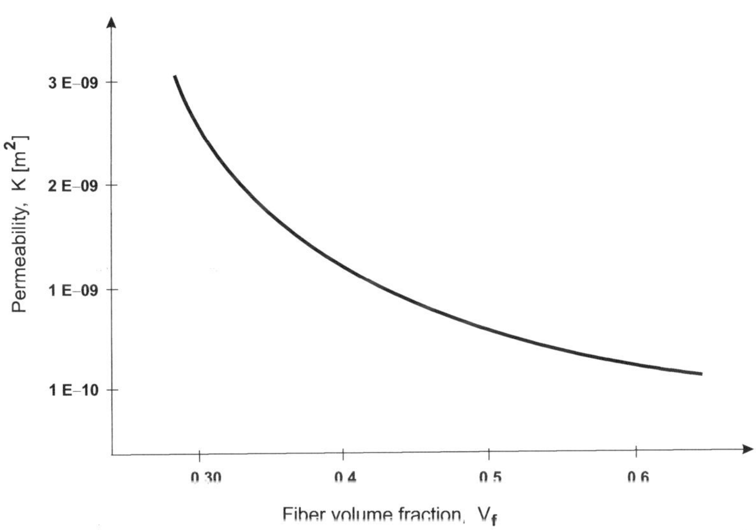

Figure 4.31: A typical graph showing the change in permeability with fiber volume fraction for a plain weave.

4.8.2 Analytic and Numerical Characterization of Permeability

Closed form expressions for the permeability as a function of the fiber diameter, fiber arrangement, and fiber volume fraction have been developed by adopting the capillary model from the soil mechanics literature, and by using an analogy of flow along and across an array of cylindrical tubes. However, as none of these idealized geometric arrangements resemble the preform geometries realistically, computational approaches to solve for the flow/pressure drop relationship in a unit cell with periodic boundary conditions offer the possibility of extending the calculations to find the permeability of complicated fiber arrangements as shown in Figure 4.32 [249–253]. However, these approaches do not comment on the unresolved issues of the influence of surface tension [254], fiber wetting, and void entrapment on permeability predictions. Also nesting of fabrics can influence the permeability, shown in Figure 4.33, as can the addition of binders and stitches that hold the fabrics together, hence the need to express the permeability and its relationship to fiber volume fraction and the preform type empirically.

Since it is impractical to solve for the detailed flow field in the interstices between the medium particles, the basic assumption to formulate a predictive analytic model for permeability is that the porous medium is homogeneous, and one expects to find a repetitive

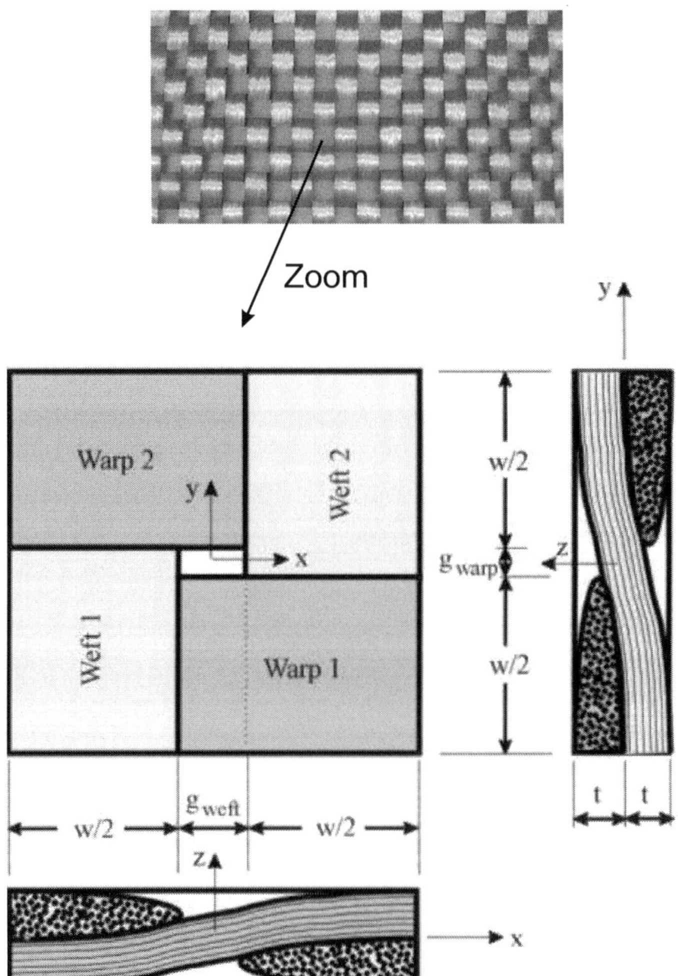

Figure 4.32: Magnified unit cell of a 4 x 5 weave fabric (i.e., 4 and 5 tows per an inch along the warp and weft directions, respectively) for permeability predictions [251].

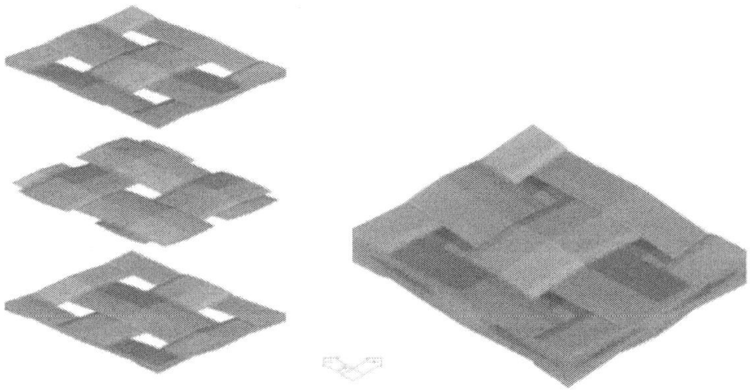

Figure 4.33: Modeling permeability of more than one fabric layer [255].

arrangement of the idealized geometry of the pore structure. This repetitive arrangement is also called a unit cell. All approaches assume that Darcy's law holds on a macroscopic level, and by choosing a geometry that closely resembles the repetitive fiber arrangement a relationship between the flow rate and the pressure drop in the selected geometry of the unit cell may be calculated. By comparing it to Darcy's law, one is able to formulate an expression for the permeability in terms of the parameters of the selected geometry. This approach can improve one's understanding of the flow through small gaps and help develop a more science-based constitutive relationship between the preform permeability and compaction of the preform.

4.8.3 Experimental Characterization of Permeability

Permeability has the units of either cm^2 or m^2. In 1856, Darcy had a column of sand as shown in the Figure 4.34 that was subjected to a constant pressure gradient of water, and he measured the flow rate of the water [245]. Knowing the viscosity of water, he calculated the permeability of the sand bed by using the following simple equation:

$$K = -\frac{Q}{\Delta P/L}\eta \tag{4.116}$$

The porous medium was considered isotropic so one needed just one value of permeability for a given architecture of the porous media.

Fiber preforms offer different resistances in different directions due to their anisotropic architecture. Hence, one needs to find the complete permeability tensor for such preforms. Three different methods have been developed to measure permeability of fiber preforms. They are one-dimensional linear flow, two-dimensional radial flow and three-dimensional hemispherical flow. In all the experiments, one needs a flow rate measurement method and a pressure drop measurement device. The one-dimensional experiment is the easiest to comprehend. By guiding the flow along one of the principal axes of the preform, one can determine the permeability by monitoring the flow rate and the pressure drop, as shown in Figure 4.35(a). The pitfalls to watch out are racetracking along the edges and bending of the mold plates (which makes the fiber volume fraction of the preform uneven).

While one-dimensional flow is from a line source as seen in Figure 4.35(a), two-dimensional flow is usually a radial flow from a point source. The shape of the ellipse and the direction

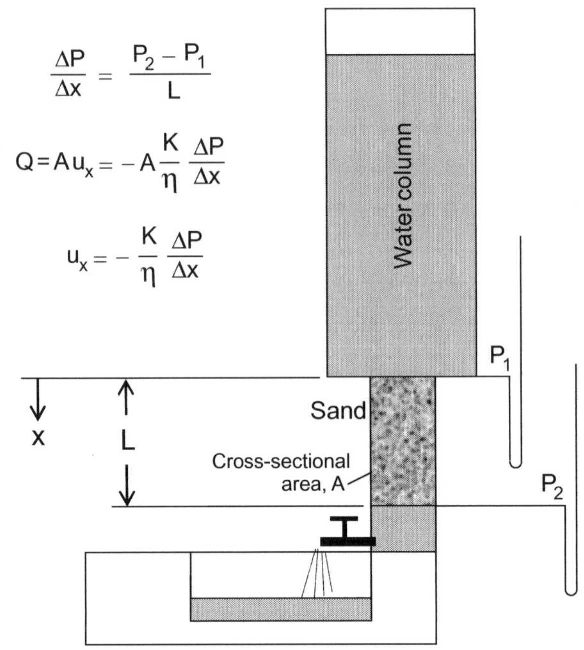

Figure 4.34: A column of sand that was used by Darcy in 1856 in order to measure the permeability of sand [245].

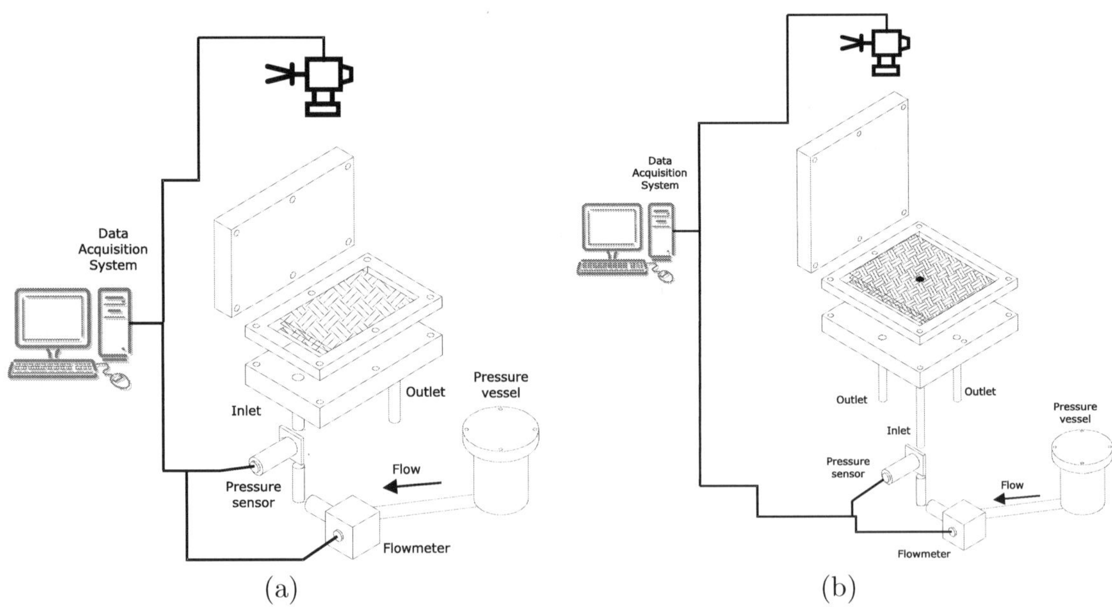

Figure 4.35: Experimental set-ups for preform permeability measurements under (a) one-dimensional, and (b) two-dimensional radial injection [256].

of its major and minor axes allow one to determine the in-plane permeabilities as shown in Figure 4.35(b). However, one needs to conduct some analysis to calculate these values. Details are given in the book chapter by Advani et al. [246]. Three-dimensional permeabilities can be found in a similar way by using a point source as shown in Figure 4.36. The details are given in references [255, 257]. Two- and three-dimensional measurements usually require recording of how the flow front moves with time. This could be accomplished by video-taping the experiment in plane using a transparent mold or digitizing the flow front or using a SMARTWeave sensor system [258] to extract information about the flow front in three dimensions.

Permeability measurements are usually not very accurate or repeatable. Ten to twenty percent error in measurements is common for the same preform in the same laboratory. The results could differ by as much as 50% from one controlled laboratory to another. More analytic and numerical predictions need to be coupled with experimental measurements to get a better handle on values of permeabilities of fiber preforms. Coupled with this issue is that the woven and stitched preforms exhibit two different length scales in the preforms, one of the order of fiber diameter in microns and the other of the order of fiber tows in millimeters. Thus, one is trying to assign a single permeability value for a dual scale porous medium, and this may not be the correct approach. The solution may be to have two different permeability values for the two different scales [259–262].

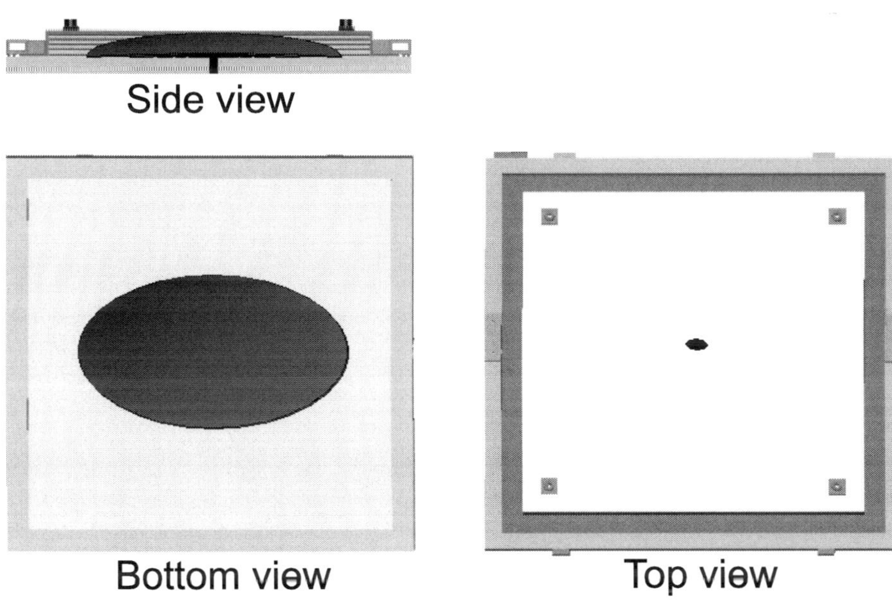

Figure 4.36: Side, bottom and top view of three dimensional permeability measurement set-up [263].

4.9 Fiber Stress

Fiber stress influences processing. However, how one models this contribution is a function of the type of resin and whether it contains short or continuous fibers. If the fibers are short, the contribution of fiber stress appears in enhancement of the viscosity of the resin as expressed by Equation (4.67) or (4.68). For thermoplastic materials containing long continuous fibers, the approach is similar where the fiber and the resin are treated as one material with their combined rheology treated as a continuum and defined by a constitutive equation, as discussed in Section 4.3. However, for thermoset materials in which the fiber beds are subjected to loadings and resin then impregnated, one needs to be able to define the relationship between stress and compression in such fiber networks.

Advanced composites use very high fiber volume fractions. Consequently, the interfiber spacing is very small (of the order of microns) which leads to multiple fiber-fiber contacts when consolidation forces are applied during processing.

Some examples of processing geometries are shown in Figure 4.37. They clearly show that the fiber bundles and the preforms may be subjected to complex three-dimensional stress during processing. For example, when fiber bundles are under tension in processes such as filament winding or pultrusion, their fiber volume fraction increases, reducing their permeability. In molding, the transverse compression will play a key role.

Almost all advanced composites contain fibers in bundles that are woven, stitched or arranged in the desired form by some other means. As the interest is in the average behavior of the fiber bundle, it is treated as a continuum because a section of a fiber bundle is much smaller than the overall dimensions of the composite. As shown in Gutowski [8], one can denote the stresses on a fiber bundle element in a material coordinate system as shown in Figure 4.38. The stresses are carried either by the fibers or by the pressure in the resin follows:

$$T_{ij} = \tau_{ij} - p_r\delta_{ij} + \sigma_{ij} \tag{4.117}$$

Here δ_{ij} is the Kronecker delta, where $\delta_{ij} = 1$ when $i = j$ and zero otherwise. The stresses carried by the fibers are represented by σ_{ij} and are due to elastic deformation stresses. The viscous stresses arise due to the viscous deformation of the material and can be represented by the stress tensor in the fluid as shown in Chapter 3 ($\tau_{ij} = \eta\dot{\gamma}_{ij}$). The elastic deformation stresses can be further subdivided into an axial stress and a bulk stress. The axial stress is along the fiber direction, and the bulk stress is usually the compressive transfer stress in

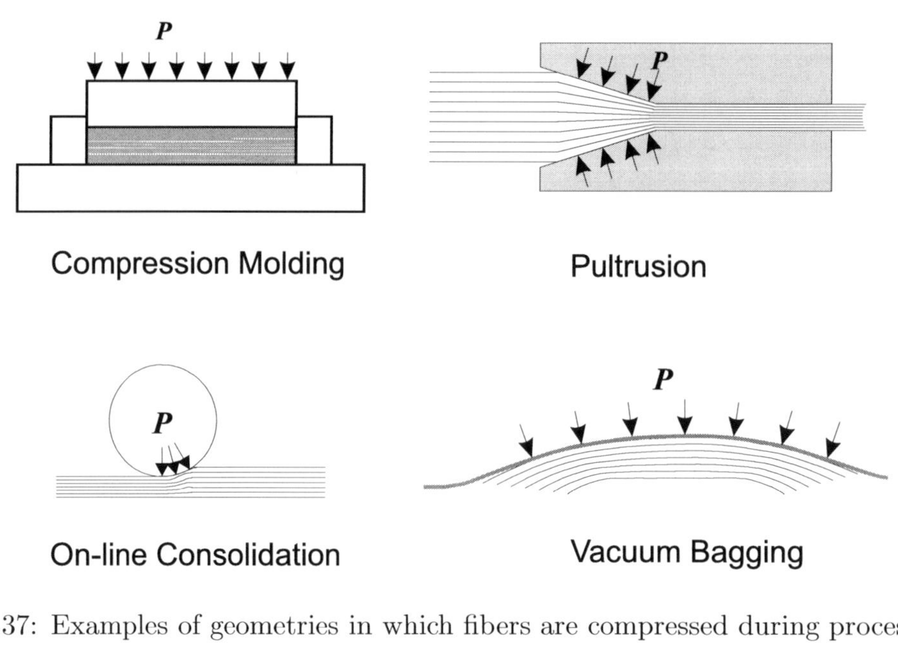

Figure 4.37: Examples of geometries in which fibers are compressed during processing [8].

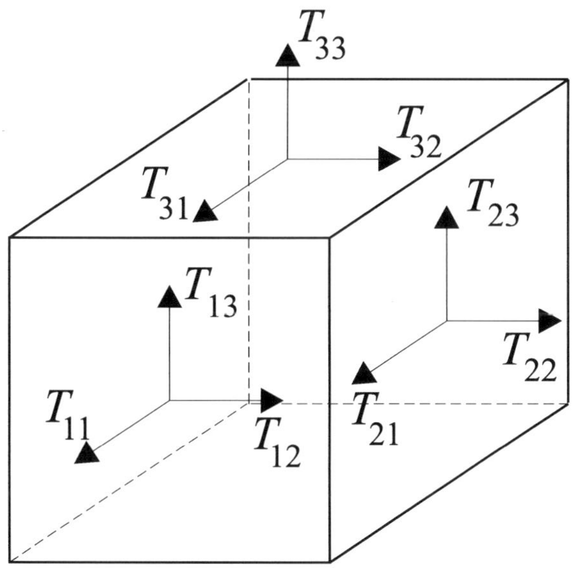

Figure 4.38: Stresses on a fiber bundle element.

the 2-3 plane. The axial stress can be positive or negative, but the bulk stress is always compressive (negative).

The axial behavior of a fiber bundle is important during processes such as making of a prepreg, filament winding and pultrusion. When one applies an axial force or tension to a fiber bundle, Gutowski proposed a nonlinear relationship between the force and the fiber volume fraction [264] as shown in Equation (4.118):

$$\sigma_{11} = \frac{A_0 V_0}{V_f} \frac{1 - \sqrt{\dfrac{V_f}{V_0}}}{\dfrac{16}{\pi^3} \dfrac{\beta^2}{E} \sqrt{\dfrac{V_a}{V_f}} \left(\sqrt{\dfrac{V_a}{V_f}} - 1 \right)^3} \tag{4.118}$$

where modulus E represents the bending stiffness of the fibers, V_0 is the volume fraction under zero stress, V_a is the maximum volume fraction that can be achieved, and β is a constant that defines the state of the bundles. A_0 is the initial cross-sectional area across which the load is applied. This relationship is plotted in Figure 4.39. It shows how tensioning of the fiber bundle increases the fiber volume fraction.

In many other composites manufacturing processes, bulk compressive stresses are applied to the fiber bundle. For example in autoclave processing, one can derive the relationship between the bulk stress σ_{zz} and the fiber volume fraction as

$$\sigma_{zz} = \frac{3\pi E}{\beta^4} \frac{1 - \sqrt{\dfrac{V_f}{V_0}}}{\left(\sqrt{\dfrac{V_a}{V_f}} - 1 \right)^4}. \tag{4.119}$$

The plot reveals the rapid load transfer from the resin to the fibers during processing. Lower resin pressure can promote the growth of the initiated voids, which is detrimental to the composite.

Note that the above two equations are derived but are based on the assumption that the fibers are aligned inside a tow and make multiple contacts and have a slight waviness of a sinusoidal character. This is then translated to a deformation of a slightly curved fiber in a cell that has a length of 100 times that of the fiber arch as shown in Figure 4.40. The slopes at the end are assumed to be zero. Thus, although based on some of the physics of

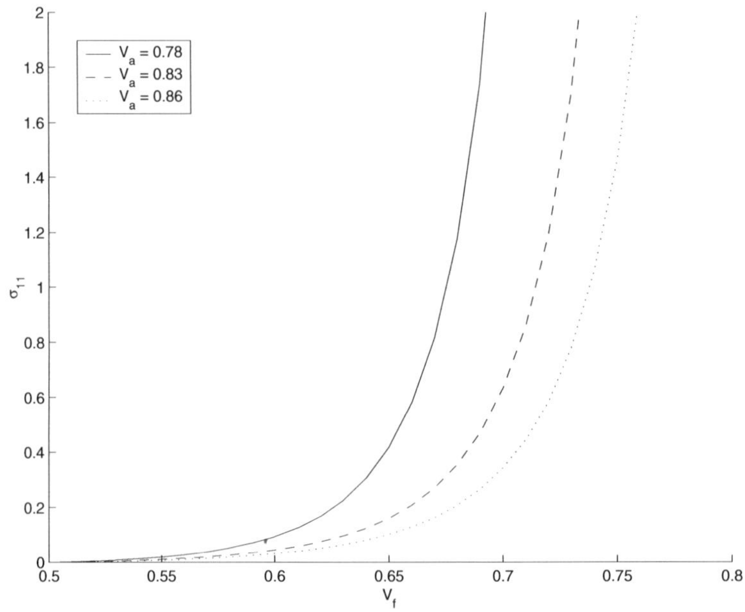

Figure 4.39: σ_{11} versus V_f for various V_a values [8].

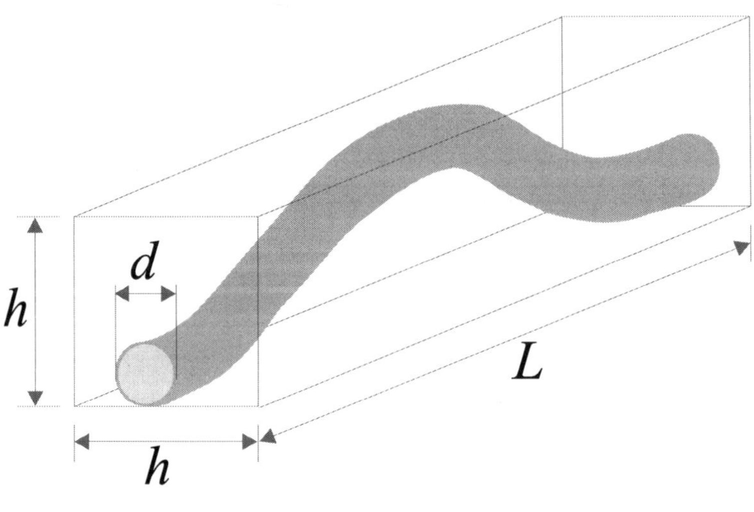

Figure 4.40: Deformation of a slightly curved fiber in a cell (redrawn from [8]).

continuum mechanics, one still needs to find the constant β in Equations (4.118) and (4.119) by conducting experiments at different fiber volume fractions and matching the results with the model results.

Chou and Chen have addressed the compaction of woven and stitched fabrics. They adopted a unit cell approach to develop the stress and compaction relationship and also allow for nesting of the fabric [265–267]. This compaction of fabrics and nesting effects also influence the flow of resin as the permeability of the preforms change. See Sozer et al. [268] for more details.

4.10 Exercises

4.10.1 Questions

1. Why do we need constitutive laws?

2. What is the viscosity of a fluid?

3. What are the dimensions of viscosity in terms of mass (M), length (L) and time (T)?

4. What are the three common constitutive models for viscosity of shear thinning materials, such as thermoplastics? Sketch typical viscosity versus shear rate profiles for these three models as well as for a Newtonian material.

5. What is the mathematical expression for the magnitude of the strain rate tensor?

6. What are the two common devices used to measure the viscosity of fluids?

7. For shear thinning materials at low and high shear rates, the power-law model gives physically incorrect values of viscosity, whereas the Carreau model correctly captures the behavior of the materials. However, the power-law is commonly used. Explain the reason.

8. What device is used to measure the viscosity of long discontinuous fiber reinforced composites in their melt state?

9. What process variables are dependent on the viscosity of a short fiber suspension in processes such as injection molding, extrusion and compression molding?

10. Does the viscosity of a fiber suspension have the same or different values in (i) steady uniaxial elongation flow, and (ii) steady simple shear flow? Explain.

11. If the fibers are initially random, does the viscosity of a fiber suspension have the same or different values in (i) unsteady simple shear flow, and (ii) steady simple shear flow? Explain.

12. Why is a cure model of a resin system needed? Which process parameters and variables are designed using this model?

13. What is the "gel point" of a resin system? What happens if the composite part is de-molded before the gel point? What is the disadvantage of demolding the part much later than the gel point?

14. What is the differential scanning calorimeter (DSC) technique?

15. Describe the two commonly used methods of DSC under isothermal conditions. Which one of these methods should be utilized for reactions with very small exotherm heat?

16. Describe the three commonly used methods of DSC under nonisothermal conditions (dynamic mode) by discussing whether a single or multiple experiments are needed, and what different values are measured in those methods. Which one of these methods should be utilized in order to get reliable results?

17. Suppose that there are two experimental setups available in your laboratory for the measurement of degree of cure during resin systems for the entire cure cycle: DSC (differential scanning calorimeter) and TBA (torsional braid analysis). Which one would you use? Or would you use both of them for different stages of the cure reaction? Explain.

18. What are the reasons for using inorganic fillers in thermosetting resins?

19. Suppose that you run two sets of experiments to measure the rate of cure of (i) a resin alone and (ii) a resin with the presence of fibers. Would you expect a significant variation in the results, or not?

20. What is the difference between solidification and crystallization?

21. Sketch the graphs of (i) enthalpy versus temperature and (ii) specific heat versus temperature, during the phase change for both a semicrystalline polymer and an amorphous polymer.

22. What are lamellae?

23. Describe the formation of spherulites.

24. Describe the primary and secondary crystallizations.

25. What does the permeability of a porous medium mean? If a fabric preform is assumed to be a porous medium, then what affects the permeability of the preform?

26. Write down one-dimensional Darcy's law. Explain each term. What are the dimensions of each term?

27. Write down three-dimensional Darcy's law. What do principal axes physically denote? Rewrite three-dimensional Darcy's law using the directions of principal axes.

28. What is the difference between isotropic and anisotropic porous madia in terms of (i) the permeability tensor and (ii) the flow of a resin through these media?

29. Visually how do you find the principal axes of permeability tensor of a reinforcing preform during its impregnation with a resin?

30. Is there a unique set of principal axes of permeability tensor for a reinforcing preform? Explain.

31. One student writes the following on his/her lab report: "My supervisor asked me to measure the permeability of a preform with 10 layers of fabric 'X' which has 4 and 5 tows per inch in the warp and weft directions, respectively. I measured the permeability of a single layer fabric as $4.0 * 10^{-9}$ m^2 by using a single one-dimensional

experiment. Then, I stacked 10 layers of this fabric and placed them inside a mold with varying thickness. I injected resin from a gate. My experimental results (flow front shape and injection flow rate under constant injection pressure) deviated very significantly from the mold filling simulation results in which I used $K_{xx} = K_{yy} = 4.0 * 10^{-9}$ m^2. I wonder what I did wrong." Can you help him/her? As a supervisor, what additional work would you ask the student to do?

32. Describe the commonly used constitutive equations for the permeability K of a porous layer when a viscous fluid flows through it. Do they accurately represent typical fabric preforms?

33. The advantage of constitutive equations for the permeability of porous layers is that they have simple closed-form solutions. Are there any shortcomings?

34. What is the "unit cell" method in calculation of permeability of a fabric preform? What are the assumptions? What are the applied boundary conditions?

35. Explain how Darcy measured the permeability of a sand bed for the first time in 1856.

36. What are the different methods to measure the permeability of preforms? What are their advantages and disadvantages compared to other methods?

37. Suppose that you are given an isotropic fabric preform and you are asked to measure the in-plane components of the permeability tensor at a specified and uniform fiber volume fraction. Explain your procedure if your laboratory has (i) only a one-dimensional experimental setup, or (ii) a two-dimensional experimental setup?

38. Explain the reasons why the values of permeability measurements have big variations from one laboratory to another, and from one experiment to another in the same laboratory.

39. Describe the three-dimensional permeability measurement method. For what applications is the three-dimensional permeability tensor needed instead of the usual two-dimensional?

40. What causes the fiber stresses in composite manufacturing processes? Are these effects the same or different in processes such as compression molding, RTM, filament winding and pultrusion?

41. Describe how the fiber stress is shared between the resin and the fiber preform during autoclave processing.

4.10.2 Fill in the Blanks

1. Commonly used units of viscosity are Poise (P), centiPoise (cP), and Pascal-times-second (Pa.s). The relations between them are given as follows: 1 P = ________ cP, 1 P = ________ Pa.s.

2. Water, corn syrup, liquid metals and other short chain molecules are called ______________ fluids because the viscosity of such fluids can be characterized by a single value which does not change with its _____________ rate.

3. In general the viscosity is a function of ______________, ______________, ______________, ______________ rate, and if there are particles or fibers suspended in the resin then the ______________ fraction and ______________ of the particles if their aspect ratio is greater than one.

4. As thermosets cure, their viscosity ______________ as one would need more force to move a molecule that is ______________. Thus, ______________ equations to describe the cure kinetics and dependence of viscosity on degree of cure are necessary for thermosets. Before the initiation of the cure, viscosities of thermosets are usually between __________ cP and __________ cP.

5. Most of the polymer composite processing and manufacturing methods are modeled using the ______________ Reynolds number flow assumption or ______________ flow assumption, in which one neglects the ______________ force in favor of the ______________ force.

6. The resistance to flow can dramatically decrease due to the change in the structure under applied shear load reducing the viscosity by orders of magnitude. Such liquids are known as shear ______________ fluids, where the viscosity ______________ with shear rate.

7. Thermoplastics belong to a class of ______________ fluids.

8. One checks the Deborah number (De) to gauge the importance of viscoelasticity. De is the ratio of ______________ effects to ______________ effects. If De $\ll 1$, one can safely ignore the ______________ effects of the thermoplastic polymer.

9. The viscosity of thermoplastics will usually decrease by one to two orders of magnitude as the temperature of the material increases from the ______________ temperature to a few degrees higher than the ______________ temperature.

10. The relationship between the stress tensor of a material $\underline{\underline{\tau}}$ and the strain rate tensor of that material $\underline{\underline{\dot{\gamma}}}$ is given as $\underline{\underline{\tau}} = \eta \underline{\underline{\dot{\gamma}}}$ where η is the viscosity of the material. For a Newtonian fluid, the viscosity is a ______________ and can be expressed as $\eta =$ ________.

11. For shear thinning materials, such as thermoplastics, the change in viscosity with shear rate is expressed with one of the three commonly used phenomenological models. These models are called the power-law model, ______________ truncated model, and the ______________ model. In the power-law model, $\eta(\dot{\gamma}) =$ ______________ where m and n are material constants of the resin system and determined experimentally.

12. For Newtonian fluids, the value of the viscosity should be ______________ at different shear rates.

13. ____________________ is usually a suitable device for viscosity measurement at low shear rate. ____________________ is suitable to measure viscosity at high shear rates.

14. Almost all of the thermoplastics display a ____________________ at low shear rates, and then a ____________________ with the increase in shear rate and again a ____________________ at the high shear rates.

15. The shear and extensional properties of long discontinuous fiber reinforced composites in their melt state are difficult to obtain using traditional rheological techniques such

as cone and plate and capillary rheometers due to the presence of continuous or long fibers. One way to characterize their shear behavior is to use ________ flow to characterize the bulk transverse shear viscosity of these highly filled viscous resin systems. This flow is invoked by placing the material between two impermeable platens and applying a normal force. One may characterize the material under constant applied ________ or under a constant ________ rate of the platens. Such industrial testing devices are referred to as the "________" or "________."

16. For transversely isotropic bundle or laminates, the two dominant shear modes are ________ shearing and ________ shearing.

17. Experimental squeeze flow studies on polymeric liquids indicate that under slow loading rates, their behavior can be well approximated through the shear dependent viscosity. However, under rapid loading, viscoelastic effects may become evident. The magnitude of the material's Deborah number (De) is often used as a determinant of its response. De is defined as the ratio of the fluid's ________ time to a characteristic time of the squeezing process. Low values of De correspond to ________ fluids, while high De corresponds to fluids with an ________ response.

18. Fibers in a suspension can be approximated as uniform cylindrical rods with a length L and diameter D. The number density of the suspension, n is the number of fibers per unit volume. There are three regimes: dilute, semi-dilute and concentrated. The suspension is dilute if n is much less than ________. The suspension is concentrated if n is greater than ________. If n is between these two values, then the suspension is said to be in the semi-dilute regime.

19. All commercial composites fall into the ________ regime.

20. The most important action of a fiber in a suspension is to resist stretching of the fluid along the ________ axis. All fiber suspensions have high viscosities for stretching in the ________ direction.

21. Fibers add very ________ resistance to a shearing motion either parallel or perpendicular to the fiber axis. Shearing flow tends to make the fibers align ________ to the planes of shear, where they exert ________ resistance to the deformation. As a result, fiber suspensions have much ________ viscosities in steady simple shear than in steady uniaxial elongation.

22. The curing reaction of thermoset resins forms a three dimensional cross-linked network. The reaction can be activated either (i) thermally by heating up the or (ii) chemically by the addition of ________.

23. The degree of cure is usually proportional to the fraction of the ________. The reaction is an ________ one, so the evolved ________ needs to be convected out of the composite.

24. Modeling of resin cure and heat transfer between the mold and composite allows the engineer to predict the ________ history of the composite during manufacturing and to design appropriate ________ and ________ cycles for the material. If an appropriate ________ system is not designed, the polymer can ________ due to the excess heat generated by the reaction.

25. When a pure metal melt is cooled, it always crystallizes. When a thermoplastic is cooled it either (i) crystallizes, or (ii) solidifies into a glassy state, depending on whether the cooling is ________________ or ________________, respectively.

26. The two-dimensional flow of resin that is injected from a point source to impregnate a fabric preform, will have ________________ or ________________ flow front shapes if the preform permeability tensor is isotropic or anisotropic, respectively.

4.10.3 Problems

1. The following empirical equation was proposed by [118–120] for the viscosity of a thermoset resin system:

$$\eta = A_\eta \exp\left(\frac{E_\eta}{RT}\right)\left(\frac{\alpha_g}{\alpha_g - \alpha}\right)^{a+b\alpha}$$

where the constants of the equation are listed in Table 4.2. (i) For RIM2200 resin, plot η as a function of T, $30 \le T \le 90$ for three different values of degree of cure, $\alpha = 0$, 0.05 and 0.10. (ii) Repeat (i) except that the resin is commercial DSM resin, and temperature domain is $30 \le T \le 60$. (iii) Repeat (i) except that the resin is DGEBA and TETA resin, and the temperature domain is in $70 \le T \le 90$.

2. Consider the reaction between two chemical groups denoted by A and B which link together two segments of polymer chain P_1 and P_2 as follows: $P_1A + P_2B = P_1ABP_2$. The kinetic equation is $\frac{dC^*}{dt} = k_0 \exp(-E/RT) C_{A0}(1 - C^*)(\Lambda - C^*)$ where $C^* = \frac{C_{A0} - C_A}{C_{A0}}$ and $\Lambda = C_{B0}/C_{A0}$. Plot $C^*(t)$ if the rate constant $k_0 = 0.03$, the ratio of activation energy for the reaction E to the universal gas constant R is $E/R = 500$, the temperature of the polymer is held constant at $T = 450$K, $\Lambda = 1$, and the initial concentration is $C_{A0} = 0.4$, How long does it take for C^* to reach values of 0.50 and 0.80? How do the results change if the temperature is increased to 500K?

3. Write the Kozeny-Carman equation for the permeability of a capillary tube arrangement. Is this equation valid for the flow along or across a set of parallel capillary tube arrangement? Plot K/A versus V_f when $0.50 \le V_f \le 0.80$ by assuming that the coefficient A in this equation doesn't change with the fiber volume fraction. In this model, what is the upper limit of V_f when the capillary tubes are arranged so that they touch each other? At this critical value of V_f, what does the Kozeny-Carman equation predict for K? What would you expect K to be at this critical V_f? What are the reasons for this deviation?

4. One decides to use the Kozeny-Carman equation for a nominal fiber volume fraction of $V_f = 0.5$. If V_f has a tolerance of ± 0.01, what is the corresponding tolerance on K?

5. What is Darcy velocity? Write an equation to calculate the Darcy velocity by using the local velocity profile $u(y)$ of a one-dimensional flow along x direction within a channel of height h.

6. A resin is injected into a preform. The permeability tensor of the preform is given as

$$\mathbf{K} = \begin{pmatrix} K_{xx} & K_{xy} \\ K_{yx} & K_{yy} \end{pmatrix} = \begin{pmatrix} 5 & 1 \\ 1 & 3 \end{pmatrix} \times 10^{-10}\text{m}^2.$$

(i) Calculate the principal permeability components K_{11} and K_{22}. (ii) Draw the x-y and 1-2 coordinate systems, and indicate the angle between the two coordinate systems. (iii) Draw a typical flow front indicating the aspect ratio of the elliptical shape.

7. You are asked to conduct two experiments of mold filling in a rectangular mold. The flow will be constrained to be one-dimensional as the injection will be through a line injection gate as shown in Figure 4.41. The length of the mold is one meter and the viscosity of the fluid is 0.1 Pa.s. The injection pressure is held constant at 2 bars.

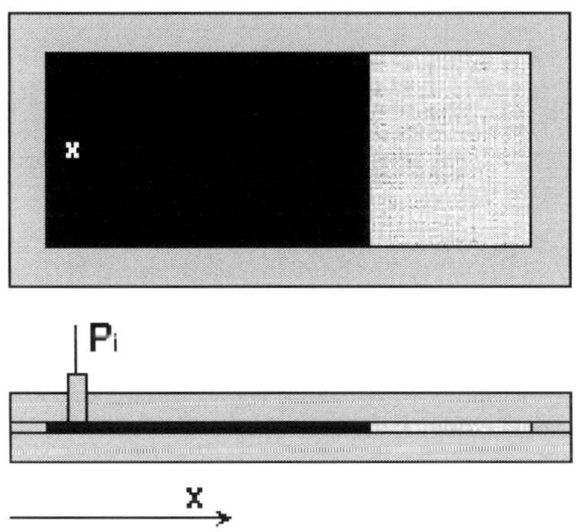

Figure 4.41: One-dimensional flow in a rectangular mold.

In the first experiment, you place in the mold a preform that occupies 25% of the mold, and you are given that its permeability for a fiber volume fraction of 25% is $1 \times 10^{-9}\,\mathrm{m}^2$ in the direction of the flow. Find the time it takes to fill the mold. In the second experiment, you use a different preform that occupies 50% of the same mold, and you are told that its permeability is also $1 \times 10^{-9}\,\mathrm{m}^2$ when it occupies 50% of the mold. Do you expect the filling time to be the same as in the first experiment, as they have the same permeability? Explain your answer.

Chapter 5

Model Simplifications and Solutions

5.1 Introduction

A model is an idealized mathematical representation of a physical process or a system. The intent of this book is to teach the fundamentals involved in building process models for polymer composite manufacturing methods such as injection molding, pultrusion and resin transfer molding.

Manufacturing of composites has relied on experience and trial-and-error methods to design, develop and fabricate a product. However, this approach has proven to be expensive in time and money to develop new prototype geometries, which usually translates into high risk, and therefore has hindered the use of composite materials in many potential industrial applications. Use of process models can accelerate the path from conception to prototype development and make these materials and their processing operations competitive with metals and other materials in terms of their cost.

Composites processing models are built on the foundation of physical laws and appropriate assumptions and boundary conditions based on the understanding of the physical phenomena and constitutive laws derived from experimental data. Once the model is well posed with mathematical equations, the next step is to examine the behavior of the model in response to changes in the process and material variables. This information can prove to be very useful to design a mold or a die for the manufacturing method or to alter the manufacturing process to create a successful part. To investigate the behavior of a composite manufacturing process in a routine manner and with minimum effort, the process model can be incorporated into a computer simulation. A computer simulation or a virtual processing scenario is a combination of an idealized process model expressed in mathematical equations, a numerical method to solve the equations, and the computer software to carry out the solution and display the results graphically, mimicking the physical behavior of the process. Thus, such virtual composite process scenarios can provide valuable and detailed information about the process and improve the understanding.

Example 5.1: Filling a Simple Mold

An L-shaped mold needs to be filled without any "dry spots" (regions with no resin) and in minimum time. You have three choices for injections as shown in Figure 5.1. Which one will you choose?

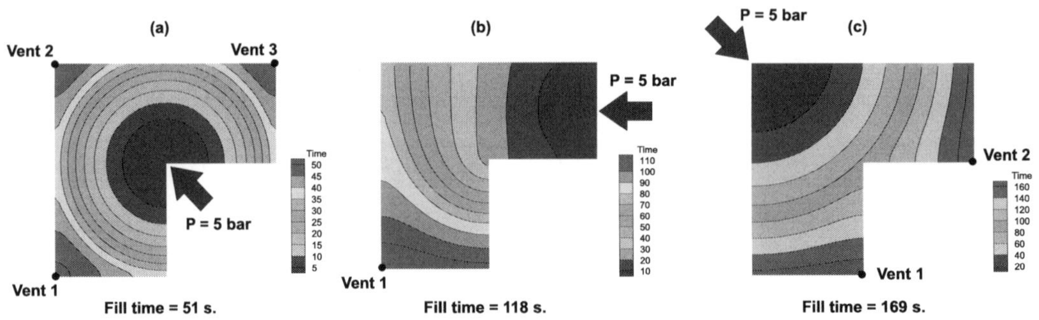

Figure 5.1: Different mold filling scenarios with resin injection from three different locations in an L-shaped mold. The mold filling simulation depicts the location of the flow fronts and can indicate the filling time to completely saturate the mold. The filling times are 51, 118 and 169 seconds for the three selected injection gates.

Solution

Different gate locations/conditions can be investigated to achieve the mold filling goals. In the L-shaped mold shown in Figure 5.1, three different gate locations under the same injection pressure are filled to find which location results in the minimum filling time. A mold filling simulation based on flow through porous media may be used to simulate all three cases. The results show that the filling times vary considerably: 51, 118 and 169 seconds, in (a), (b), and (c), respectively. Instead of conducting trial and error in actual mold filling, investigating different scenarios with simulations will save considerable design time and hence filling time. It can also save the part from "dry regions without resin" by choosing the gate location appropriately, and placing the vent(s) at the last point(s) to be filled. In this example, 3, 1 and 2 vents are needed in (a), (b) and (c), respectively, as shown in the same figure. ■

5.1.1 Usefulness of Models

Process models are very useful as they can quantify understanding and the knowledge about a process. C. L. Tucker in his chapter "Introduction to Process Modeling" provides a brief outline on the usefulness of models and how to build them for polymer processing [269]. Here we will expand on that outline with examples in composites processing. A reliable mathematical model will reduce the scope of experiments by combining the effect of different variables and testing unproven design ideas. Finally, models provide more detailed information than the experiments and thus can be used to investigate which experiments will be interesting to conduct. Models can provide information about important variables that cannot be directly measured. Models do not replace experiments, but together with experiments, mathematical theory and physical laws, models assist engineers and scientists to understand the process and manipulate and tailor it to meet their requirements.

Resin Transfer Molding (RTM) provides an excellent example of the uses of process models. Even if we consider just the filling and curing of an RTM part, many physical phenomena transpire. For example, a viscous polymer flows through a network of channels created between the fiber preforms to fill the mold cavity. Heat may enter the cavity by conduction from the mold walls. The resin molecules may start to react and form a cross-linked network, increasing the viscosity of the resin and also releasing heat, which needs to be convected away. All these factors will determine if the cavity will be filled completely, how long the part should be left in the mold, whether the part will warp due to residual thermal stresses, and what the final properties of the manufactured part will be.

As one cannot "see" inside a closed mold, it is difficult to know if the resin occupied all the empty spaces between the fibers and if the part had cured sufficiently to retain its shape while de-molding the part. A process model can let the engineer or the scientist "see" the filling pattern and the extent of curing, thus quantifying the knowledge of the filling process. As the engineer learns more about the filling from the process model, he or she does not have to conduct many experiments to find the best location for injection by trial and error using prior experience. The model can provide a few possible locations which reduces the scope of experiments. Thus, a reliable model can reveal the performance and outcome of different mold designs, just as an experiment would, without the added expense and time of fabricating and testing every design. So, process models can help save cost and time in developing new designs.

RTM simulations provide further benefits. From experiments, one can measure the inlet pressure and maybe the flow rate. If one had a few sensors inside the mold, one could monitor pressure or temperature at those locations. A process model provides detailed information about the pressure field, flow front shapes, temperature, cure profiles and velocity fields. These can be graphically displayed and easily interpreted to improve the understanding of the process and identify difficult issues. However, computer simulations must be validated and verified if they are to represent the physical world. Also, one needs to conduct experiments to characterize permeability and cure kinetics that are used as inputs into the model. Thus, experiments are a necessary component in the construction of a process model.

In this section, we have explained why models are useful, and next we will provide the approach to build them, introduce simplifications and represent material and phenomenological behavior. We will also present the tools to simplify and analyze the behavior of a process model. It is important to validate the model to justify the assumptions and the constitutive laws used to construct the model. Various approaches to validation and revisions will also be discussed.

5.2 Formulation of Models

Models may vary from each other, but one can build them by following the plan outlined in Figure 5.2. As a model is a mathematical representation of the physical process, one must be able to define it, develop the equations and boundary conditions that describe the process, and solve the equations to calculate the important variables of interest so that one can assess the accuracy and usefulness of the model. Finally, model revision is an important step as it allows one to refine the model and move it a step closer towards the physical process.

5.2.1 Problem Definition

The purpose of a model is to predict the behavior of a process for a range of inputs and conditions [270]. Hence, one must define the physical process and its inputs. The physical

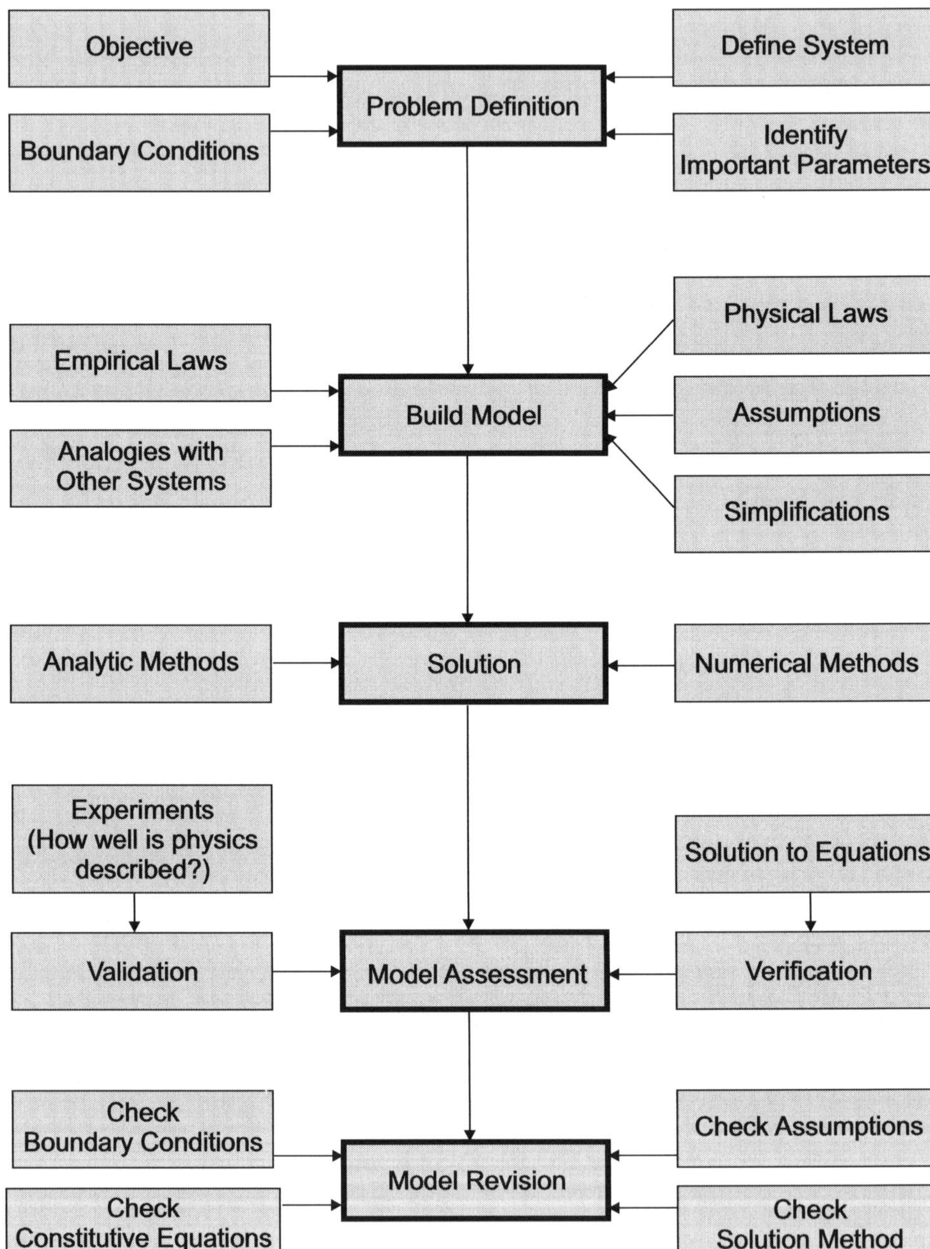

Figure 5.2: Modeling flowchart.

process may not represent the complete composite manufacturing process but may focus on a small aspect of the process, or on a processing operation during manufacturing. For example, in injection molding, mold filling is only one aspect of the manufacturing process. The model here may treat the screw and the runner and the spruce as only a source of fiber suspensions at a given flow rate. Thus, an important step is to define the system that one is modeling and draw a boundary around it. The other parts of the process, which can be modeled as other sub-systems, can interact with the system only by what they supply across the boundary [132].

The aim of the model will influence what aspect of the model is emphasized and hence should also be defined. It is useful to identify the important responses of the process and what conditions and inputs the model requires. Depending on the aim, different models for the same physical process give different answers. For example, in processes such as resin transfer molding, some models only relate the flow rate to the pressure, while others may provide detailed information about the flow front locations in a complex geometry. Other models may be limited to a few cavity shapes. Some models may even provide temperature histories and predict residual stresses. Other models may only focus on the curing phase and ignore filling. Thus, there can be a multiplicity of models for the same process. Which model one wants to choose or build will depend on what issues are important and how much time and effort one can devote to finding the answers. Thus, it is crucial to choose or build a model for the situation of interest if it is to prove useful in enhancing the understanding of that aspect of the process.

5.2.2 Building the Mathematical Model

Once the system is defined and the important quantities are identified, one can build the model from analogies with other systems or from one or more of the following: physical laws or empirical and phenomenological observations. A mathematical model, which represents the physical process, describes the system behavior by a set of mathematical equations with the quantities of interest as primary or secondary variables [269]. Interactions between the selected system and the surroundings are introduced as boundary conditions for the mathematical equations.

Most composites manufacturing process models combine the fundamental physical laws of mass, momentum and energy conservation with empirical observations such as viscosity and permeability laws. Sometimes, the models can become complicated if one wants to include every known effect. The skilled modeler will try to simplify the model by making assumptions that will retain the important effects. This will make it easier to solve the mathematical equations and also easier to understand the system. Typically, simplifying the model will also require fewer parameters to be determined. For example, the flow in resin transfer molding is important only in the in-plane direction when filling thin cavities; thus one can simplify the governing equations from three dimensions to two dimensions. To solve the equations, one needs to provide the permeability tensor. For a two-dimensional case, one needs only two permeability values, whereas for a three dimensional case, one will need the permeability value in the thickness direction in addition. The case for use of a resin viscosity model is similar. If one simplifies the material behavior to a Newtonian fluid, one needs only one material parameter for the resin. However, if one chooses to use the Carreau fluid model, one needs to determine four material parameters. In addition, the equations become nonlinear and difficult to solve. Thus, simplifications are necessary in a model, but one must make sure that simplification does not lose the important effects to be captured.

5.2.3 Solution of the Equations

The formulated mathematical model may be a simple ordinary differential equation with one independent variable and one dependent variable, or it may be a system of complicated and coupled nonlinear partial differential equations with two or more dependent variables (such as resin pressure, temperature and degree of cure, and fiber orientation) that rely on more than one independent variable (such as time and spatial coordinates). The simple equations could be solved analytically if appropriate boundary or initial conditions are specified. For models with partial differential equations to be solved in a complex geometry, one can select standard numerical methods such as finite difference, boundary element or finite element. Special techniques may be required in moving boundary problems of processes involving mold filling. The objective of the model also determines the method to be selected for solution, since the numerical method determines the time and cost of setting up and running a simulation. When implementing a numerical solution, it is important to pay attention to pragmatic issues such as reliability, portability, efficiency and user interface of the computer software.

5.2.4 Model Assessment

Model assessment must consist of two parts: verification and validation. Verification alludes to whether or not the selected method provides the correct solution to the formulated set of equations with the prescribed boundary conditions. Hence, one must always verify that the numerical technique does not have any programming errors or inconsistencies that can lead to inaccurate solution of the governing equations. There is a cliché about "garbage in–garbage out." Verification studies can definitely address this type of criticism. There are many approaches to verify the method. If one has implemented a numerical solution, it can be compared to analytic or classical solutions for simplified geometries or for steady-state situations. If no analytic solution is available, one can refine the time step or the grid size over the domain and check for convergence of the solution. One can also perform global mass and energy balances over the domain to check if they are being conserved on a global scale. This will help the modeler to verify if his/her numerical method is implemented properly.

Validation refers to how well the process model describes the physics of the selected phenomenon in the process mathematically. In formulation of the process model and its simplification, one makes many assumptions and introduces empirical relationships and approximate boundary conditions; hence, it is imperative to check that the important physics we want to model is still retained in the model. Validation allows us to also check if we have missed any important phenomenon in the process. At the same time, validation will uncover insignificant influences that we have included in the process model that may not be necessary.

How does one validate a process model? The most common approach is with experiments. One can approach this at two levels.

First, to ensure that the process model is capturing the correct physics, it is important to carry out a controlled laboratory experiment or a model experiment. For example, if the goal is to validate the fiber orientation model that predicts fiber orientation due to flow in a Newtonian fluid, one could use a model fluid such as silicone oil and nylon fibers with tracer fibers to understand and record the orientation behavior. If one compared the model results directly with fiber orientation measurements taken from injection molded plaques, other effects such as viscoelasticity, fiber breakage, cooling of the polymer and fiber clustering may play a role in influencing fiber orientation in addition to the flow. Hence, it would be difficult to separate the influence due to flow on fiber orientation from these other effects which were not included in the process model.

Thus, model experiments can usually reveal the underlying physics in a processing situation and block out the noise or deviations due to other effects. However, it is important to make sure that other factors are only noise and not important effects in the process that our model is ignoring or overlooking completely. It is necessary to perform controlled experiments of the process and identify and monitor the important variables to make sure that the process model is capturing the relevant physics because a successful process model will always predict the outcome of experiments.

Although model experiments will provide useful information and also help in validation of the "idealized model," one would still like to know how well the idealized model represents the reality of the factory floor. Hence, the second level of experiments should mimic the manufacturing situation to see how good a job the idealized model does. For example, one can use random preforms in model experiments to demonstrate how well Darcy's law can capture the physics of the flow in closed molds in thermoset processing. However, in the manufacturing of composites, stitched or woven preforms are often used as the reinforcement choice as one can increase the fiber volume fraction in the composite part. Hence, it is important to validate if these materials also exhibit Darcy's law, and if not, how far do they deviate from it? Questions such as "Is it necessary to develop a new model for such materials?" or "Can one estimate the error introduced due to the dual scale nature of the preform?" have to be addressed.

5.2.5 Revisions of the Model

If the model agrees with the experimental results, one has a successful model that can make useful predictions. However, in most cases there will be disagreement between the model and the experiments. It is important to methodically explore the cause of the discrepancy, as there can be many. First, one must check that the solution method used is not imperfect. This requires looking for errors in the implementation of the solution method for the governing equations. Next, one must check if there are deficiencies in the model. The model may have used an erroneous assumption, an inappropriate empirical law, a false hypothesis, an oversimplification, or an inaccurate boundary condition. For example, if one assumed that the fiber-fiber interactions during the flow are not important and ignored them in the fiber orientation model and the results did not agree with the experimental outcome, one might have to reexamine this assumption in the model. Thus, it is important to separate the deficiencies in the model from those in the numerical methods as both of these are packaged into a computer program and it is very easy to regard them as one entity [269].

Other simple causes of disagreement could be inaccurate characterization of material parameters. For example, in RTM, the model may use a constant value for the permeability of the preform in the mold. However, in the experiment one may have slight variations in the permeability, or in some cases large variations. The result is that the experimental fill times or inlet pressures will be very different from the predicted ones due to these variations. Thus, the process model may be correct, but incorrect specification of material parameters can cause disagreements as well. It is also useful to make certain that the physics embedded in the process model is the physics being described by the experiment. There is a common saying that when a modeler is presenting results from a model, nobody believes them except for the modeler. However, when an experimentalist presents the results, everyone believes them except for the experimentalist. Thus, it is important to make sure that the experiment was conducted under the same conditions assumed in the model and that it was free from issues such as inaccurate calibration of the measurement probes, other experimental issues such as leakage of the resin from the mold sealing, fluctuations in the flow rate (although it is assumed to be constant in the model), deviation of the actual resin flow from what is assumed in the model (such as 1D or radial flow) etc., as these can make the experimental boundary conditions different from the ones assumed in the model and could be the root cause of disagreement.

The moral is that models routinely disagree with experiments and revisions should be considered as a natural and necessary component of the modeling approach. The revisions sometimes can also focus on inclusion of additional physics in the model to the observations made in the experiments, and thus, continue to refine the model and bring it closer to mirroring the physical process.

5.3 Model and Geometry Simplifications

Once the model has been posed correctly in mathematical terms with appropriate boundary conditions and before one plunges into solving the equations, it is always beneficial to explore simplifications. However, the simplifications depend not only on the process, but also on the material being processed, the phenomena of interest and the objective. There are many approaches to simplifications and one can realize them by reducing one or more of the following [271]:

1. The number of equations to be solved
2. The number of terms in the governing equations and/or in boundary conditions
3. The degree of non-linearity in the equations and/or boundary conditions
4. The degree of coupling between the equations
5. Geometric complexities in the solution domain and boundary conditions
6. Complexities in the material behaviour constitutive law
7. Complexities in the mathematical and solution methods.

However, one must guard against oversimplification as it could result in omission of the important effects one wants to capture in the process model [271]. Several tools can be used to simplify the model. For example, consider the simple shear flow of a composite material between two parallel plates as shown in Figure 5.3.

Resin temperature is T_o everywhere initially. The lower plate is at temperature T_L and the upper plate is moving at a velocity U_0 in x direction, and is at temperature T_H. As

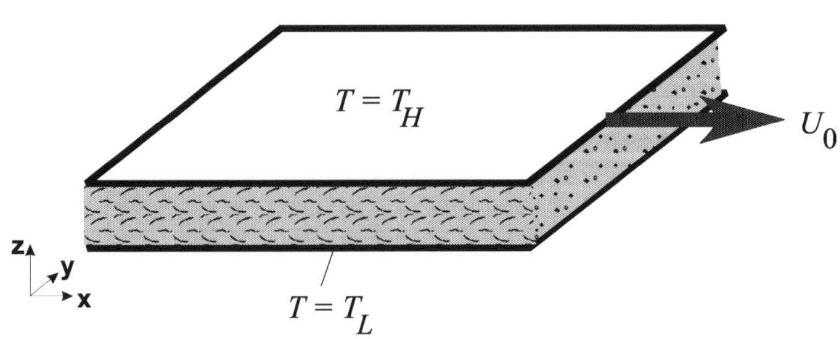

Figure 5.3: Non-isothermal simple shear flow of a fiber suspension between two parallel plates.

the resin viscosity is a strong function of temperature and there may be appreciable viscous dissipation, the momentum and energy equations are coupled. To find the velocity and temperature profiles of the resin, one has many options. One can solve the coupled systems of PDE (partial differential equations)

$$\frac{\partial \rho}{\partial t} + \nabla \cdot (\rho \mathbf{U}) = -s \tag{5.1}$$

$$\rho \frac{D\mathbf{U}}{Dt} = -\nabla P + \nabla \cdot \tau + \mathbf{F}_B \tag{5.2}$$

$$\rho c_p \frac{\partial T}{\partial t} + \rho c_p \mathbf{U} \cdot \nabla T = k \nabla^2 T + \tau : (\nabla \mathbf{U}) + \dot{R} \tag{5.3}$$

which are the conservation of mass, momentum and energy equations, respectively with the following boundary and initial conditions:

$$u_x = U_o, \quad u_y = u_z = 0 \qquad \text{at } y = h \tag{5.4}$$

$$u_x = u_y = u_z = 0 \qquad \text{at } y = 0 \tag{5.5}$$

$$T = T_H \qquad \text{at } y = h \tag{5.6}$$

$$T = T_L \qquad \text{at } y = 0 \tag{5.7}$$

$$T = T_o \qquad \text{at } t = 0. \tag{5.8}$$

This full system will require investment of many hours and computer resources. One could simplify the system by assuming that the flow and/or temperature profile is fully developed. One could simplify it further by assuming that viscous dissipation is negligible and solve for the temperature profile before solving for the momentum equation, or one could assume that the dependence of viscosity on temperature has no effect on the solution. If the aim of the model was to capture the effect of temperature boundary conditions on the flow rate and if one chooses the last option, one has made an assumption that will not result in capturing this effect.

Example 5.2: Temperature-Dependent Viscosity

Consider flow of a polymer suspension between the extruder screw and the barrel. As the gap between the barrel and the screw is usually one to two orders of magnitude smaller than the radius of the barrel, one can model this as flow between two parallel plates separated by a distance h as shown in Figure 5.4. The velocity of the barrel is U_o. The viscosity of the suspension is temperature dependent. The objective is to find the influence of temperature-dependent viscosity on the velocity profile.

Solution

If we assume steady-state and fully developed velocity and temperature profiles, it will allow us to represent velocity and temperature profiles as

$$u = u(y), \qquad v = w = 0 \tag{5.9}$$

$$T = T(y) \tag{5.10}$$

where u is the component of velocity in the x-direction. If the inlet and exit pressures are equal, and thus pressure gradient is zero, the equation of motion (Equation (5.2)) simplifies

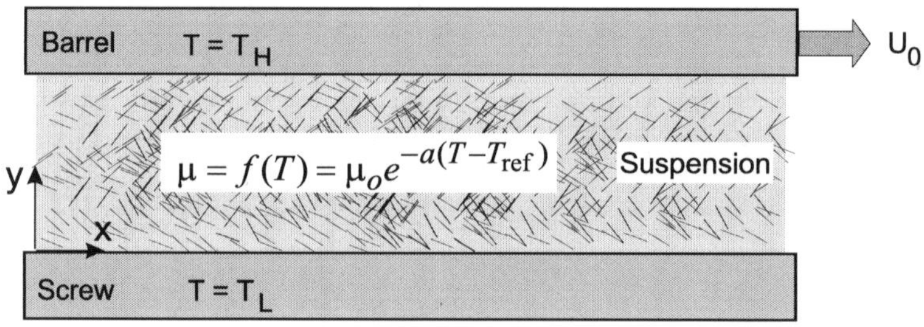

Figure 5.4: Flow of a polymer suspension between the extruder screw and the barrel.

to

$$\frac{d\tau_{xy}}{dy} = 0 \tag{5.11}$$

which results in the following:

$$\tau_{xy} = c_1. \tag{5.12}$$

One can assume a linear stress-strain rate constitutive law

$$\tau_{xy} = \mu \frac{du}{dy} \qquad \text{(Newtonian Law)}. \tag{5.13}$$

As viscosity is temperature dependent, one can represent it as a constitutive law

$$\mu = \mu_o e^{-a(T-T_0)} \tag{5.14}$$

where μ_o is the viscosity at a reference temperature T_0, and a is a material constant. Substitution of Equation (5.14) into Equation (5.13) leads to

$$\tau_{xy} = \mu_o e^{-a(T-T_0)} \frac{du}{dy} = c_1 \qquad \text{(i.e., not a function of } y\text{)}. \tag{5.15}$$

The energy equation (see Chapter 3) in which we use steady-state and fully developed assumptions and retain the influence on temperature due to viscous dissipation allow us to simplify the equation as follows:

$$k \frac{d^2T}{dy^2} = -\mu \left(\frac{du}{dy}\right)^2. \tag{5.16}$$

Equations (5.15) and (5.16) are coupled as u depends on viscosity which is a function of temperature and temperature depends on u. If one further assumes that viscous dissipation was not important (a good assumption for low viscosity materials but questionable for high viscosity materials; usually one could estimate the importance of viscous dissipation with Brinkman's number as we will discuss in a later section) then

$$k \frac{d^2T}{dy^2} = 0 \tag{5.17}$$

which results in

$$\frac{T - T_L}{T_H - T_L} = \frac{y}{h} \tag{5.18}$$

with the use of boundary conditions shown in Figure 5.4. One can now substitute the temperature dependence in terms of y dependence in Equation (5.15) and integrate

$$\int \frac{c_1}{\mu_o} e^{a(\theta_1 \frac{y}{h} + \theta_2)} \, dy = \int du \tag{5.19}$$

where $\theta_1 = T_H - T_L$ and $\theta_2 = T_L - T_0$. Integrating Equation (5.19) results in

$$u = \frac{c_1}{\mu_o} \frac{h}{a\theta_1} e^{(a\theta_1 \frac{y}{h} + \theta_2)} + c_2 = c_3 e^{(a\theta_1 \frac{y}{h} + \theta_2)} + c_2. \tag{5.20}$$

Using the boundary conditions,

$$u = 0 \quad \text{at } y = 0 \tag{5.21}$$

$$u = U_o \quad \text{at } y = h \tag{5.22}$$

allows one to find the velocity profile

$$u = U_o \left[\frac{1 - e^{a\theta_1 \frac{y}{h}}}{1 - e^{a\theta_1}} \right]. \tag{5.23}$$

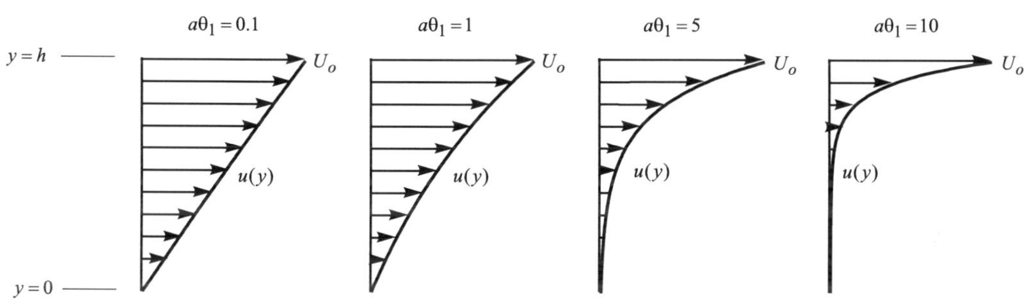

Figure 5.5: Velocity profile for positive θ_1 ($T_H > T_L$).

For $T_H > T_L$, (i.e., θ_1 is positive), the velocity profile $u(y)$ is sketched in Figure 5.5 for different $a\theta_1$ values. Notice that, the velocity profile becomes linear as $a\theta_1 \to 0$. This is not surprising, because as $a\theta_1 \to 0$, $\mu = \mu_o e^{-a(T-T_0)} \to \mu_o$; hence, the dependence of viscosity on temperature is very weak, and the thermal boundary conditions will not cause deviation on the linear velocity profile. On the other hand, as $a\theta_1$ increases, the velocity profile becomes very nonlinear. For a better approximation of the temperature profile with viscosity dissipation included, substitute the derivative of u from Equation (5.23) into Equation (5.16) and integrate to find the nonlinear temperature profile due to viscous dissipation. ■

Important aspects in process modeling are

- What simplifications to introduce in the model?
- How to correctly pose the problem?

- What approach to use to find the solution?

We will focus on some important tools that will be helpful in doing so. They are

- Dimensional analysis,
- Common assumptions in polymer composites processing,
- Various simplifications to represent boundary conditions,
- Simplified geometric models,
- Mathematical manipulations to simplify the solutions.

5.4 Dimensionless Analysis and Dimensionless Numbers

Dimensionless analysis is a very important tool for reducing the number of equations and the number of terms in the governing equations and also for estimating the degree of coupling and nonlinearity. In addition, it can allow us to develop scaling laws for the phenomena of interest in the process to scale the geometry from the laboratory scale to the production scale. Every undergraduate fluid mechanics textbook devotes a section or a chapter to dimensionless analysis and important dimensionless numbers. In this section, we will discuss the approach to the analysis and the important dimensionless numbers in polymers and polymer composites processing [271].

First, one must identify all the independent and dependent variables in the set of governing equations and the boundary conditions. This should be a straightforward task to accomplish. The second task, which is not so obvious, is to find a characteristic value with the same dimensional units for each identified variable in the equation. The candidates for the characteristic value can be an aggregation of known parameters usually related to the material and the process conditions. For example, if one wanted a characteristic value for the velocity in a flow through a die, the average flow rate divided by the cross-sectional area would be a good candidate if the flow rate were constant for the process. Another qualification the characteristic value should try to satisfy is that the variable divided by the characteristic value should be of the order unity. If one adheres to this criterion, every nondimensional term in the equation will be of the order unity. One can then compare the coefficients in front of each term in the equation to decide which terms play a dominant role and eliminate the ones that are insignificant.

Sometimes, it may be difficult to identify an obvious choice for a characteristic value, and much thought should be given to this task. In some cases, it is known that a term is important but the physical situation does not provide a value for its magnitude. In such situations, one can use scaling analysis to find its characteristic magnitude. The general approach is to assign an unknown characteristic value that can later be determined from the nondimensional form of the governing equation by forcing the coefficient containing the unknown characteristic value to be unity. For example, consider a pressure driven flow through a pultrusion die. Usually, the pulling speed is the independent variable; hence the flow rate is known, so one has a choice for a characteristic velocity but not for the pressure. If one nondimensionalizes the equation of motion and allows the coefficients in front of the nondimensional viscous stress terms and the pressure term both to equal unity, it will give the correct characteristic value for the pressure as will be shown in an example.

However, before we present further examples of dimensionless analysis in composite processing, we present here general guidelines for casting a dimensional equation into a nondimensional form [271]. Consider an equation

$$k_1 \frac{\partial T}{\partial t} + k_2 \frac{\partial T}{\partial x} + k_3 \frac{\partial^2 T}{\partial x^2} + k_4 = 0, \tag{5.24}$$

where T is the dependent variable and t and x are independent variables. k_1 to k_4 are coefficients in front of the governing equation. As explained above, to cast this equation in nondimensional form, one must find characteristic values for each of the variables in the above equation. The variables are T, t and x. Thus, one nondimensionalizes them using characteristic values T_c, t_c and x_c

$$T^* = \frac{T}{T_c}, \qquad t^* = \frac{t}{t_c} \qquad \text{and} \qquad x^* = \frac{x}{x_c}. \tag{5.25}$$

Here T_c, t_c, and x_c will be related to some physical situation of the problem. For example, if x was the variable along the length of a mold geometry, L, then $x_c = L$. If T was the variable temperature and represented the temperature of the resin, then T_c could be the temperature difference between the initial temperature of the resin T_i and the mold wall temperature T_w. Thus $T_c = T_i - T_w$; this will ensure that the T^* is also of the order unity. When possible, have all dimensionless variables range from zero to unity.

Thus, the nondimensional form of Equation (5.24) becomes

$$k_1 \frac{T_c}{t_c} \frac{\partial T^*}{\partial t^*} + k_2 \frac{T_c}{x_c} \frac{\partial T^*}{\partial x^*} + k_3 \frac{T_c}{x_c^2} \frac{\partial^2 T^*}{\partial (x^*)^2} + k_4 = 0. \tag{5.26}$$

Here $k_1 T_c/t_c$, $k_2 T_c/x_c$, $k_3 T_c/x_c^2$ and k_4 are coefficients of the four dimensionless terms in the governing equation. At least one of the dimensionless terms should have a coefficient of unity and it should be the dominant of all the terms in the equation. Divide the non-dimensional equation by the coefficient of one of the dominant terms, which hopefully should lead to all the other coefficients in the equation being smaller than or of the order unity. For example, if $k_1 T_c/t_c$ was the dominant coefficient (transient process was important in the model), then one would divide every coefficient in Equation (5.26) by $k_1 T_c/t_c$ which will result in

$$\frac{\partial T^*}{\partial t^*} + \frac{k_2}{k_1} \frac{t_c}{x_c} \frac{\partial T^*}{\partial x^*} + \frac{k_3}{k_1} \frac{t_c}{x_c^2} \frac{\partial^2 T^*}{\partial (x^*)^2} + \frac{k_4}{k_1} \frac{t_c}{T_c} = 0. \tag{5.27}$$

Note that, as the equation is dimensionless and as the coefficient in front of the first term is unity and dimensionless, all other coefficients ($\frac{k_2}{k_1} \frac{t_c}{x_c}$, $\frac{k_3}{k_1} \frac{t_c}{x_c^2}$, and $\frac{k_4}{k_1} \frac{t_c}{T_c}$) will also be dimensionless. Now, one can neglect the terms that have coefficients much less than one. For example if $\frac{k_2}{k_1} t_c \ll x_c$, and $\frac{k_3}{k_1} t_c \ll x_c^2$ one could neglect the terms with $\partial T^*/\partial x^*$ and $\partial^2 T^*/\partial (x^*)^2$ as the coefficients in front of them $\frac{k_2}{k_1} \frac{t_c}{x_c^2}$ and $\frac{k_3}{k_1} \frac{t_c}{x_c^2}$, respectively, will be small. Thus, the partial differential equation (5.27) reduces to the ordinary differential equation given below

$$\frac{dT^*}{dt^*} + \frac{k_4}{k_1} \frac{t_c}{T_c} = 0. \tag{5.28}$$

Also, if one did not have a characteristic value for t_c, one could choose it to be $t_c = k_1 T_c / k_4$ by making the coefficient equal to 1. This reduces the nondimensional equation to

$$\frac{dT^*}{dt^*} + 1 = 0 \tag{5.29}$$

if $k_2 T_c/k_4 \ll x_c$ and $k_3 T_c/k_4 \ll x_c^2$.

Thus, by scaling analysis, one can simplify equations and also find characteristic values for some of the variables. However, beware that one may not be able to satisfy all the boundary conditions. In the above example, we started with a second order PDE which should have two boundary conditions in x and one in t, but we simplified the equation to first order ODE so all we need is one condition in t. The solution will not be far from the solution to the complete PDE in most of the domain but will be poor near the boundaries where the boundary conditions were ignored. Near the boundaries, the terms neglected will be of the same order as the terms retained and should be considered if one wants to know the exact solution in this subregion. So, we can have two solutions, one in the interior of the domain, where the other terms are not so important, and the complete solution near the boundaries of the domain. Thus as one goes from the interior to the boundary, one can match the two solutions using asymptotic expansions in order to get the complete solution.

Example 5.3: Curing of a Composite Plate

Consider the process in which epoxy-carbon fiber prepregs are stacked and placed inside a metal tool. The metal tool is then placed in a press to heat the composite to initiate cure. Formulate the simplified equation that you would have to solve to find the temperature. Find the temperature profile if you assumed that the cure rate was constant throughout the domain and that the composite had reached a steady state. The dimensions, coordinate system, thermal coefficients and thermal boundary conditions are shown in Figure 5.6.

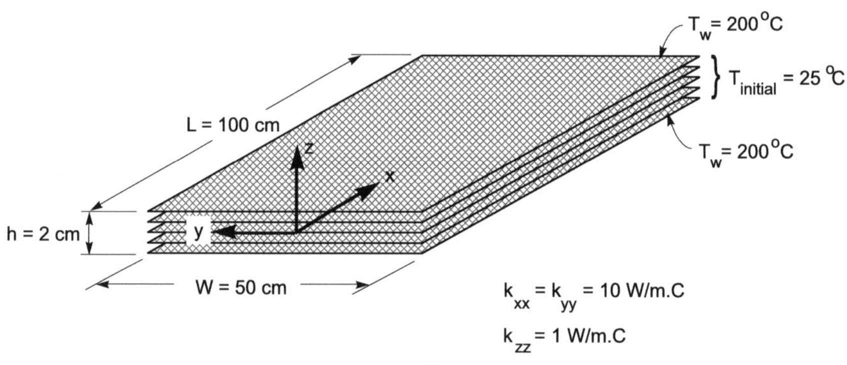

Figure 5.6: Epoxy-carbon fiber prepregs that are stacked and placed inside a metal tool with mold walls at temperature $T_w = 200°$C.

Solution

The energy equation reduces to

$$k_{xx}\frac{\partial^2 T}{\partial x^2} + k_{yy}\frac{\partial^2 T}{\partial y^2} + k_{zz}\frac{\partial^2 T}{\partial z^2} + \dot{R} = 0 \tag{5.30}$$

as there is no flow and no viscous dissipation. Here, k_{xx}, k_{yy} and k_{zz} are thermal conductivities of the composite in the x, y and z directions, respectively, and $\dot{R}$ is the volumetric rate of heat generated inside the composite due to the curing reaction. Also, steady state is assumed (usually not a valid assumption).

One can further simplify Equation (5.30) and find its solution as detailed using the following procedure. First identify independent (x, y and z) and dependent (T) parameters. Then, form the dimensionless variables:

$$x^* = \frac{x}{x_c}, \quad y^* = \frac{y}{y_c}, \quad z^* = \frac{z}{z_c}, \quad \text{and} \quad T^* = \frac{T - T_i}{T_c} \tag{5.31}$$

here $T_c = T_w - T_i$ (characteristic temperature is always expressed as a difference), $x_c = L$, $y_c = W$, and $z_c = h$. Rewriting Equation (5.30) using the nondimensional variables in Equation (5.31) leads to

$$\frac{k_{xx}T_c}{x_c^2}\frac{\partial^2 T^*}{\partial x^{*2}} + \frac{k_{yy}T_c}{y_c^2}\frac{\partial^2 T^*}{\partial y^{*2}} + \frac{k_{zz}T_c}{z_c^2}\frac{\partial^2 T^*}{\partial z^{*2}} + \dot{R} = 0. \tag{5.32}$$

One can drop the first and second term in Equation (5.32) since

$$\frac{k_{xx}(T_w - T_i)}{L^2} \quad \text{and} \quad \frac{k_{yy}(T_w - T_i)}{W^2} \ll \frac{k_{zz}(T_w - T_i)}{h^2}. \tag{5.33}$$

Physically, this implies that most of the heat is conducted through the thickness direction as $h \ll L, W$ even if k_{zz} is an order of magnitude less than k_{xx} and k_{yy}. Thus, Equation (5.32) simplifies to

$$\frac{d^2 T^*}{dz^{*2}} = -\frac{\dot{R}}{k_{zz}}\frac{h^2}{T_w - T_i}. \tag{5.34}$$

As the left-hand side is dimensionless, the right-hand side also should be dimensionless:

$$\frac{d^2 T^*}{dz^{*2}} = -\dot{R}^* \tag{5.35}$$

where

$$\dot{R}^* = \frac{\dot{R}}{k_{zz}\dfrac{T_w - T_i}{h^2}}. \tag{5.36}$$

Thus, R^* is a dimensionless number that compares the volumetric heat generation rate with the heat being conducted through the composite. The solution to Equation (5.35) is straightforward if $\dot{R}$ is assumed to be constant:

$$\frac{dT^*}{dz^*} = -\dot{R}^* z^* + c_1 \tag{5.37}$$

$$T^* = -\frac{\dot{R}^*}{2} z^{*2} + c_1 z^* + c_2. \tag{5.38}$$

The boundary conditions are

$$T^* = 1 \quad \text{at } z^* = 1 \tag{5.39}$$

$$T^* = 1 \quad \text{at } z^* = -1 \qquad \text{or} \qquad \frac{dT^*}{dz^*} = 0 \quad \text{at } z^* = 0. \tag{5.40}$$

One can solve c_1 and c_2 by using these boundary conditions. Hence, Equation (5.38) is rewritten as

$$\begin{aligned} T^* &= -\frac{\dot{R}^*}{2}z^{*2} + \frac{\dot{R}^*}{2} + 1 \\ &= -\frac{\dot{R}^*}{2}(z^{*2} - 1) + 1. \end{aligned} \tag{5.41}$$

The temperature gradient is determined by taking the z^*-derivative of Equation (5.41)

$$\frac{dT^*}{dz^*} = -\dot{R}^* z^*. \tag{5.42}$$

Thus, the absolute temperature gradient is maximum near the edges, and zero at the center. The larger the value of the gradient, the greater the chances of developing of thermal stresses. Hence, it is important to have a low value of $\dot{R}^*$. ■

Example 5.4: Flow in Thin Gaps

Formulate nondimensional conservation of mass and momentum equations for flow of viscous Newtonian fiber suspensions in long and narrow channels.

Solution

Injection molding of thermoplastics and fiber suspensions involves flow of viscous fluids through narrow gaps ($H \ll L$) as shown in Figure 5.7. Before conducting the dimensionless analysis, one must first list the parameters and characteristic values for all dependent and independent variables. For the parameters for which a characteristic value is not intuitively obvious, we use the subscript c to denote the characteristic value that will be selected later. Assume the flow is uniform in the y (the width) direction. Hence, we will consider only 2D analysis (x and z directions).

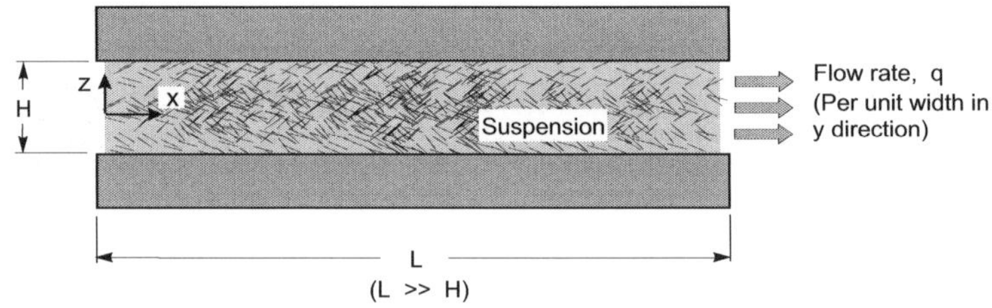

Figure 5.7: Steady-state flow of viscous fiber suspensions in narrow gaps.

Parameters:

$$\begin{aligned} \text{geometry}: &\quad H, L \\ \text{material parameters (viscosity, density)}: &\quad \mu, \rho \\ \text{flow rate}: &\quad q \\ \text{stresses}: &\quad \tau_{xx}, \tau_{xz}, \tau_{zz}. \end{aligned}$$

Dimensionless variables (shown with hats) and characteristic values:

$$\begin{aligned}
\hat{x} &= \frac{x}{L} \\
\hat{z} &= \frac{z}{H} \\
\hat{u} &= \frac{u}{q/H} \qquad \text{(note that } q \text{ is per unit width)} \\
\hat{w} &= \frac{w}{w_c} \\
\hat{P} &= \frac{P}{P_c} \\
\hat{\tau}_{xx} &= \frac{\tau_{xx}}{(\tau_{xx})_c} \\
\hat{\tau}_{zz} &= \frac{\tau_{zz}}{(\tau_{zz})_c} \\
\hat{\tau}_{xz} &= \frac{\tau_{xz}}{(\tau_{xz})_c}.
\end{aligned} \tag{5.43}$$

The continuity equation in 2D incompressible flow,

$$\frac{\partial u}{\partial x} + \frac{\partial w}{\partial z} = 0 \tag{5.44}$$

can be written as follows in terms of dimensionless variables

$$\frac{q}{HL}\frac{\partial \hat{u}}{\partial \hat{x}} + \frac{w_c}{H}\frac{\partial \hat{w}}{\partial \hat{z}} = 0. \tag{5.45}$$

Thus, we can make the order of both terms of the order unity as follows:

$$\frac{\partial \hat{u}}{\partial \hat{x}} + \frac{\partial \hat{w}}{\partial \hat{z}} = 0 \tag{5.46}$$

if we select $w_c = q/L$. Thus, nondimensionalization of the continuity equation can provide an estimate of flow in the z direction. Note that if $H \ll L$, then characteristic flow in the x direction $u_c \gg w_c$ (the characteristic flow in the z direction).

If we use Newtonian law, constitutive equations provide an estimate for characteristic stresses:

$$\hat{\tau}_{xx} = 2\frac{\partial \hat{u}}{\partial \hat{x}} \tag{5.47}$$

if $(\tau_{xx})_c$ is chosen as $\frac{\mu q}{HL}$,

$$\hat{\tau}_{zz} = 2\frac{\partial \hat{w}}{\partial \hat{z}} \tag{5.48}$$

if $(\tau_{zz})_c$ is chosen as $\frac{\mu q}{HL}$,

$$\hat{\tau}_{xz} = \hat{\tau}_{zx} = \left[\frac{\partial \hat{u}}{\partial \hat{z}} + \left(\frac{H}{L}\right)^2 \frac{\partial \hat{w}}{\partial \hat{x}}\right] \tag{5.49}$$

if $(\tau_{xz})_c$ is chosen as $\frac{\mu q}{H^2}$. As $H \ll L$, then $(\tau_{xz})_c \gg (\tau_{zz})_c$ and $(\tau_{xx})_c$. Thus, nondimensionalization suggests that shear stress (τ_{xz}) is dominant as compared to normal stresses (τ_{xx}, τ_{zz}).

As viscous materials achieve a steady-state and fully developed profile very quickly, steady-state flow is not an unreasonable assumption when considering conservation of momentum. Also, one can ignore body (gravity) forces in the equation as viscous forces dominate. Thus, the 2D equation of motion simplifies to

$$\rho\left(u\frac{\partial u}{\partial x}+w\frac{\partial u}{\partial z}\right) = -\frac{\partial P}{\partial x}+\frac{\partial \tau_{xx}}{\partial x}+\frac{\partial \tau_{xz}}{\partial z} \tag{5.50}$$

$$\rho\left(u\frac{\partial w}{\partial x}+w\frac{\partial w}{\partial z}\right) = -\frac{\partial P}{\partial z}+\frac{\partial \tau_{zx}}{\partial x}+\frac{\partial \tau_{zz}}{\partial z} \tag{5.51}$$

Equations (5.50) and (5.51) can be rewritten in terms of dimensionless variables as follows:

$$\frac{\rho q^2}{H^2 L}\left(\hat{u}\frac{\partial \hat{u}}{\partial \hat{x}}+\hat{w}\frac{\partial \hat{u}}{\partial \hat{z}}\right) = \frac{P_c}{L}\left(-\frac{\partial \hat{P}}{\partial \hat{x}}\right)+\frac{\mu q}{HL^2}\left(\frac{\partial \hat{\tau}_{xx}}{\partial \hat{x}}\right)+\frac{\mu q}{H^3}\left(\frac{\partial \hat{\tau}_{xz}}{\partial \hat{z}}\right) \tag{5.52}$$

$$\frac{\rho q^2}{HL^2}\left(\hat{u}\frac{\partial \hat{w}}{\partial \hat{x}}+\hat{w}\frac{\partial \hat{w}}{\partial \hat{z}}\right) = \frac{P_c}{H}\left(-\frac{\partial \hat{P}}{\partial \hat{z}}\right)+\frac{\mu q}{H^2 L}\left(\frac{\partial \hat{\tau}_{xz}}{\partial \hat{x}}+\frac{\partial \hat{\tau}_{zz}}{\partial \hat{z}}\right). \tag{5.53}$$

Now, by considering both equations simultaneously, we should rearrange the terms, by making the dominant term unity. As shear stresses dominate, and since $\mu q/H^3 \gg$ any other term, as $H \ll L$, Equations (5.52) and (5.53) are rewritten by dividing all terms by $\mu q/H^3$, which results in

$$\frac{\rho q H}{\mu L}\left(\hat{u}\frac{\partial \hat{u}}{\partial \hat{x}}+\hat{w}\frac{\partial \hat{u}}{\partial \hat{z}}\right) = -\frac{P_c}{\frac{\mu q L}{H^3}}\left(\frac{\partial \hat{P}}{\partial \hat{x}}\right)+\left(\frac{H}{L}\right)^2\frac{\partial \hat{\tau}_{xx}}{\partial \hat{x}}+\frac{\partial \hat{\tau}_{zx}}{\partial \hat{z}} \tag{5.54}$$

$$\frac{\rho q}{\mu}\left(\frac{H}{L}\right)^2\left(\hat{u}\frac{\partial \hat{w}}{\partial \hat{x}}+\hat{w}\frac{\partial \hat{w}}{\partial \hat{z}}\right) = -\frac{P_c}{\frac{\mu q}{H^2}}\left(\frac{\partial \hat{P}}{\partial \hat{z}}\right)+\left(\frac{H}{L}\right)\left(\frac{\partial \hat{\tau}_{xz}}{\partial \hat{x}}+\frac{\partial \hat{\tau}_{zz}}{\partial \hat{z}}\right). \tag{5.55}$$

One can now choose $P_c = \mu q L/H^3$, such that the coefficient in front of the pressure gradient term in Equation (5.54) is unity. Note that Equations (5.54) and (5.55) are scaled properly as the scaling term for pressure $P_c = \mu q L/H^3$ is the same in both Equations (5.54) and (5.55). Also, as $H \ll L$, one can ignore all terms that have H/L or their higher powers. Thus, the complicated equation of motion for flow in narrow gaps simplifies to

$$0 = -\frac{\partial \hat{P}}{\partial \hat{x}}+\frac{d\hat{\tau}_{xz}}{d\hat{z}} \tag{5.56}$$

$$0 = \frac{\partial \hat{P}}{\partial \hat{z}}. \tag{5.57}$$

■

5.4.1 Dimensionless Numbers Used in Composites Processing

After one obtains the nondimensional form of the equations, the coefficients in front of the terms can usually be reduced to one or more dimensionless numbers. This helps in identifying the role of various physical parameters in the process. From the Buckingham

Pi theorem, one can easily forecast how many dimensionless groups to expect [272]. The Buckingham Pi theorem states that the number of dimensionless groups we expect is equal to the number of variables in the system minus four if we encounter all M (mass), L (length), T (temperature) and t (time) units. If we encounter only three of the four units in the problem, then it will be minus three and so on. Temperature is always represented as a difference rather than an absolute value.

For example, the Fourier number measures the importance of transient effects in the absence of flow. How fast a composite will heat or cool due to conduction is directly proportional to the Fourier number, which is given by

$$\mathrm{Fo} = \frac{\alpha t_c}{L_c^2} \tag{5.58}$$

where L_c is the characteristic length in the direction of heat removal or heat addition by conduction, t_c is the characteristic time, and α is the thermal diffusivity and is given by

$$\alpha = \frac{k}{\rho C_p} \tag{5.59}$$

where ρ and C_p are the density and specific heat capacity of the composite, respectively. k is thermal conductivity of an isotropic composite.

Most commonly used dimensionless numbers in polymer and polymer composite processing are listed in Table 5.1.

Example 5.5: Importance of Fourier Number in Nonisothermal Processing

Consider a polymeric composite that one would like to cool from 200°C down to 70°C as shown in Figure 5.8. The top and bottom surfaces of the composite are kept in perfect thermal contact with an aluminum tool at $T_w = 25$°C. After how long should the engineer remove the part from the tool?

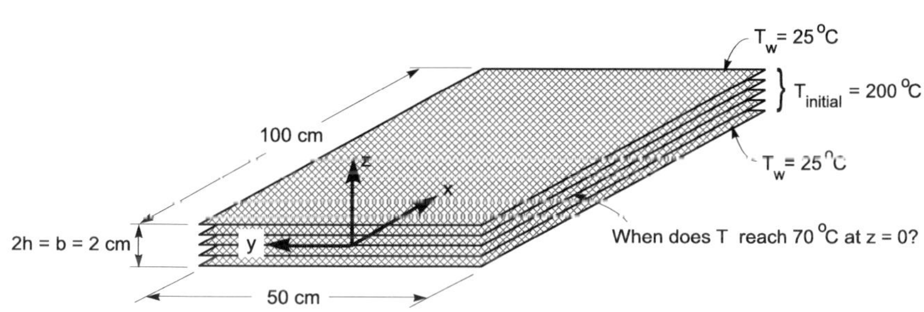

Figure 5.8: A polymeric composite cooled from 200°C to 70°C.

The thermal conductivity of the polymer is $k = 0.1\,\mathrm{W/m.C}$ and that of the S-glass used is $k = 10\,\mathrm{W/m.C}$ in both the longitudinal and transverse directions. The density of S-2 glass is $2500\,\mathrm{kg/m^3}$ and C_p is $0.41\,\mathrm{kJ/kg.C}$. For the polypropylene matrix, C_p is $2.0\,\mathrm{kJ/kg.C}$ and density is $900\,\mathrm{kg/m^3}$. The S-glass fraction in the composite is 50%.

Table 5.1: Dimensionless Groups Commonly Used in Polymer and Polymer Composites Processing

Dimensionless Number	Definition	Interpretation
Brinkman number	$\mathrm{Br} = \dfrac{\mu V^2}{k\Delta T}$	$\dfrac{\text{viscous dissipation}}{\text{conduction due to } \Delta T}$
Capillary number	$\mathrm{Ca} = \dfrac{\mu V}{\sigma}$	$\dfrac{\text{viscous forces}}{\text{surface tension}}$
Damkohler number	$\mathrm{Da} = \dfrac{\rho_r H_R h^2}{K\Delta T t_r}$	$\dfrac{(\partial T/\partial t) \text{ from reaction}}{(\partial T/\partial t) \text{ from conduction}}$
Deborah number	$\mathrm{De} = \dfrac{\lambda V}{R}$	$\dfrac{\text{elastic effects}}{\text{viscous effects}}$
Froude number	$\mathrm{Fr} = \dfrac{V^2}{gL}$	$\dfrac{\text{inertia}}{\text{gravity}}$
Gelling number	$\mathrm{G} = \dfrac{L}{t_r V}$	$\dfrac{\text{fill time}}{\text{reaction time}}$
Graetz number	$\mathrm{Gz} = \dfrac{VD^2}{\alpha L}$	$\dfrac{\text{flow direction convection}}{\text{transverse conduction}}$
Nahme-Griffith number	$\mathrm{Na} = \dfrac{V^2 \left\lvert \frac{\partial \mu}{\partial T} \right\rvert}{k}$	$\dfrac{\text{viscous dissipation}}{\text{heat required to alter the viscosity}}$
Nusselt number	$\mathrm{Nu} = \dfrac{hD}{k}$	$\dfrac{\text{convection at the boundary}}{\text{conduction of the fluid}}$
Pearson number	$\mathrm{Pn} = \dfrac{\mathrm{Na}}{\mathrm{Br}}$	$\dfrac{\text{heat required to alter viscosity}}{\text{heat conduction due to } \Delta T}$
Peclet number	$\mathrm{Pe} = \dfrac{\rho C_p L V}{k}$	$\dfrac{\text{heat dispersion}}{\text{heat conduction}}$
Prandtl number	$\mathrm{Pr} = \dfrac{C_p \mu}{k}$	$\dfrac{\text{momentum transfer}}{\text{heat transfer}}$
Reynolds number	$\mathrm{Re} = \dfrac{VL\rho}{\mu}$	$\dfrac{\text{inertia force}}{\text{viscous force}}$

Solution

Parameters:

$$\begin{aligned} \text{geometry}: &\quad W, L, h \\ \text{process}: &\quad \Delta T\ (= T_w - T_i), \\ \text{material}: &\quad k_{\text{fibers}}, k_{\text{resin}}, \rho_{\text{fibers}}, \rho_{\text{resin}}, C_{p,\text{fibers}}, C_{p,\text{resin}}, V_f. \end{aligned}$$

Governing equation:

$$\rho C_p \left(\frac{\partial T}{\partial t} + \mathbf{u} \cdot \nabla T \right) = \nabla \cdot \left(\underline{\underline{k}} \cdot \nabla T \right) + \underline{\underline{\tau}} : \nabla \mathbf{u} + \dot{R}. \tag{5.60}$$

In Equation (5.60), $\mathbf{u} \cdot \nabla T$ can be neglected as there is no flow, $\dot{R}$ is zero as there is no heat generation, and $\underline{\underline{\tau}} : \nabla \mathbf{u}$ vanishes as there is no dissipation. If we assume fibers are along the x direction, then the energy equation (5.60) simplifies to

$$\rho C_p \frac{\partial T}{\partial t} = \frac{\partial}{\partial x}\left(k_{xx} \frac{\partial T}{\partial x} \right) + \frac{\partial}{\partial y}\left(k_{yy} \frac{\partial T}{\partial y} \right) + \frac{\partial}{\partial z}\left(k_{zz} \frac{\partial T}{\partial z} \right). \tag{5.61}$$

For this case, $k_{xx} = k_{yy} = k_{zz} = k$. Choose characteristic values for independent (x, y, z, t) and dependent (T) parameters

$$\hat{x} = \frac{x}{L}, \quad \hat{y} = \frac{y}{W}, \quad \hat{z} = \frac{z}{h}, \quad \hat{t} = \frac{t}{t_c}, \quad \theta = \frac{T - T_i}{T_w - T_i} \tag{5.62}$$

Here t_c is a characteristic time value that we hope to uncover from the nondimensional version of the equation. Substitution of the above variables from Equation (5.62) into Equation (5.61) will result in the following nondimensional equation:

$$\frac{h^2}{\alpha t_c} \frac{\partial \theta}{\partial \hat{t}} = \left(\frac{h}{L} \right)^2 \frac{\partial^2 \theta}{\partial \hat{x}^2} + \left(\frac{h}{W} \right)^2 \frac{\partial^2 \theta}{\partial \hat{y}^2} + \frac{\partial^2 \theta}{\partial \hat{z}^2}. \tag{5.63}$$

Notice that $h^2/(\alpha t_c)$ is 1/Fo where Fo is the Fourier number, and the thermal diffusivity is $\alpha = k/(\rho C_p)$ where

$$k = (1 - V_f) k_{\text{resin}} + V_f k_{\text{fibers}} \tag{5.64}$$

and

$$\rho C_p = (1 - V_f)(\rho C_p)_{\text{resin}} + V_f (\rho C_p)_{\text{fibers}}. \tag{5.65}$$

As $h \ll L$, h/L and h/W are much smaller than unity, one can safely ignore the conduction in the x and y directions. Also one can choose $t_c = h^2/\alpha$, such that the coefficient in front of $\partial \theta / \partial \hat{t}$ will become unity. Thus, the nondimensional form becomes

$$\frac{\partial \theta}{\partial \hat{t}} = \frac{\partial^2 \theta}{\partial \hat{z}^2} \tag{5.66}$$

The initial condition and the two boundary conditions are given as follows:

$$\theta = 0 \quad \text{at } \hat{t} = 0 \tag{5.67}$$

$$\theta = 1 \quad \text{at } \hat{z} = 1 \tag{5.68}$$

$$\frac{\partial \theta}{\partial \hat{z}} = 0 \quad \text{at } \hat{z} = 0. \tag{5.69}$$

Note that the temperature distribution is symmetric about $\hat{z} = 0$ due to the geometry and the boundary conditions at $\hat{z} = \pm 1$. Hence, one could solve this problem in $-1 \leq \hat{z} \leq 1$ domain by using $\theta(-1) = 1$ and $\theta(1) = 1$. Or, alternatively, and as we did here, one can use the symmetry condition and solve the problem in $0 \leq \hat{z} \leq 1$ domain by using $\partial\theta/\partial\hat{z}(0) = 0$ and $\theta(1) = 1$.

Equation (5.66) is a diffusion equation. One can apply the separation of variables technique to solve it along with the boundary and initial conditions

$$\theta(\hat{z}, \hat{t}) = Z(\hat{z})T(\hat{t}). \tag{5.70}$$

Substitution of Equation (5.70) into Equation (5.66) yields

$$ZT' = Z''T. \tag{5.71}$$

Division of Equation (5.71) by ZT results in

$$\frac{T'}{T} = \frac{Z''}{Z}. \tag{5.72}$$

Since the left-hand side of Equation (5.72) is a function of $\hat{t}$ only, and the-right hand side of Equation (5.72) is a function of $\hat{z}$ only, Equation (5.72) can be true only if

$$\frac{T'}{T} = \frac{Z''}{Z} = \text{constant} = -\kappa^2. \tag{5.73}$$

By applying separation of variables, the original partial differential equation was converted into two ordinary differential equations:

$$\frac{Z''}{Z} = -\kappa^2 \qquad \text{and} \qquad \frac{T'}{T} = -\kappa^2,$$

or, in more familiar forms:

$$Z'' + \kappa^2 Z = 0, \tag{5.74}$$

$$T' + \kappa^2 T = 0. \tag{5.75}$$

The individual general solutions are given below

$$Z = \begin{cases} A\cos\kappa\hat{z} + B\sin\kappa\hat{z}, & \kappa \neq 0 \\ D + E\hat{z}, & \kappa = 0. \end{cases} \tag{5.76}$$

and

$$T = \begin{cases} Fe^{-\kappa^2\hat{t}}, & \kappa \neq 0 \\ G, & \kappa = 0. \end{cases} \tag{5.77}$$

Thus, the product solution is given as

$$\theta = H + I\hat{z} + (J\cos\kappa\hat{z} + K\sin\kappa\hat{z})e^{-\kappa^2\hat{t}}, \tag{5.78}$$

where DG was replaced by H, and so on. The unknown coefficients will be determined by using the boundary and initial conditions. Let us first apply the boundary condition given in Equation (5.69)

$$\begin{aligned}\frac{\partial\theta}{\partial\hat{z}}(\hat{z},\hat{t}) &= I + (-J\kappa\sin\kappa\hat{z} + K\kappa\cos\kappa\hat{z})e^{-\kappa^2\hat{t}},\\ \frac{\partial\theta}{\partial\hat{z}}(0,\hat{t}) &= 0 = I + (0 + K\kappa)e^{-\kappa^2\hat{t}}\end{aligned} \tag{5.79}$$

which reveals that $I = 0$ and $K = 0$ simplifying Equation (5.78) to

$$\theta = H + J\cos\kappa\hat{z}\, e^{-\kappa^2\hat{t}}. \tag{5.80}$$

Application of the other boundary condition, Equation (5.68), results in the following:

$$\theta(1,\hat{t}) = 1 = H + J\cos\kappa\, e^{-\kappa^2\hat{t}}. \tag{5.81}$$

Equation (5.81) can hold only if $H = 1$ and

$$\kappa = \frac{n\pi}{2} \tag{5.82}$$

where $n = 1, 3, 5, \ldots, \infty$. Using superposition gives

$$\theta(\hat{z},\hat{t}) = 1 + \sum_{n=1,3,5,\ldots}^{\infty} J_n \cos\frac{n\pi\hat{z}}{2}\, e^{-(n\pi/2)^2\hat{t}}. \tag{5.83}$$

Finally, one can use the initial condition, Equation (5.67):

$$\theta(\hat{z},0) = 0 = 1 + \sum_{n=1,3,5,\ldots}^{\infty} J_n \cos\frac{n\pi\hat{z}}{2}. \tag{5.84}$$

By using Fourier series expansion on Equation (5.84), one can determine J_n:

$$J_n = \frac{2}{1}\int_0^1 (-1)\cos\frac{n\pi\hat{z}}{2}\,d\hat{z} = -\frac{4}{n\pi}(-1)^{(n-1)/2} \tag{5.85}$$

for $n = 1, 3, 5, \ldots$. Hence, the final solution is given by

$$\theta(\hat{z},\hat{t}) = 1 - \sum_{n=1,3,5,\ldots}^{\infty} \frac{4}{n\pi}(-1)^{(n-1)/2} \cos\frac{n\pi\hat{z}}{2}\, e^{-(n\pi/2)^2\hat{t}}. \tag{5.86}$$

Note that if we want to know how long it takes to cool the composite from 200°C ($\theta = 0$) to 70°C [$\theta = (70 - 200)/(25 - 200) = 0.7429$] at the mid-point ($z = \hat{z} = 0$), one needs to evaluate and plot $\theta(\hat{z},\hat{t})$ numerically from Equation (5.86) and find out at what $\hat{t}$ value, $\theta(0,\hat{t}) = 0.7429$. In order to numerically evaluate Equation (5.86), a simple program can be written as shown below in FORTRAN, or any other programming languages such as C, BASIC, or mathematical solvers such as MATHEMATICA, MATLAB or Maple.

```
c------- A FORTRAN code to calculate the temperature ------------------
c---------------------------------------------------- starts here --
      open(unit=11,file='output.dat',status='unknown')
      pi = 4.0*atan(1.0)
      x=0.0
            do 100 i=1,200
            t=i*0.01
            theta=1.0
            err=1.0
            n=1
            previo = 999.0
                  do while(err.gt.0.00001)
                  theta=theta-( 4.0*(-1.0)**((n-1)/2)/(n*pi) )
     ?                  *cos(n*pi*x/2.0)*exp(-(n*pi/2.0)**2*t)
                  err=abs(theta-previo)
                  previo = theta
                  n=n+2
                  end do
100         write(11,500) t,theta
500         format(2f11.5)
      stop
      end
c------------------------------------------------------ ends here --
```

Figure 5.9 was prepared, and $\hat{t}$ was found to be 0.6484. This corresponds to $t = t_c\hat{t} = (h^2/\alpha)\hat{t} = (0.01^2/0.00000358)0.6484 = 18.1$ seconds. ■

Example 5.6: Re-solve Example 5.5 Using Numerical Methods

The solution to the previous example was found using an analytical method called "separation of variables." If an analytic solution can be obtained as in the above example, one can easily evaluate the effect of geometric, material and process parameters (such as mold dimensions, resin viscosity and boundary conditions) on the result. However, for problems with complicated differential equations or domains, application of analytical methods may be very tedious or may not be applicable at all. Keeping this in mind, we will now apply numerical methods to solve the same example, namely, the governing equation (5.66) with initial and boundary conditions (5.67)-(5.69).

Using your solution, plot the followings:

1. Nondimensional temperature at mid-point, $\theta(0,\hat{t})$ versus nondimensional time, $\hat{t}$ until the mid-temperature becomes 70°C, i.e., $T(z=0,t) = 70$°C.
2. Plot $T(z)$ versus z at different times.
3. t_{T70} versus k_{fiber} for a range of $k_{\text{fiber}} = 1 - 20\,$W/m.C where t_{T70} denotes the time at which the mid-temperature becomes $T(z=0,t) = 70$°C.

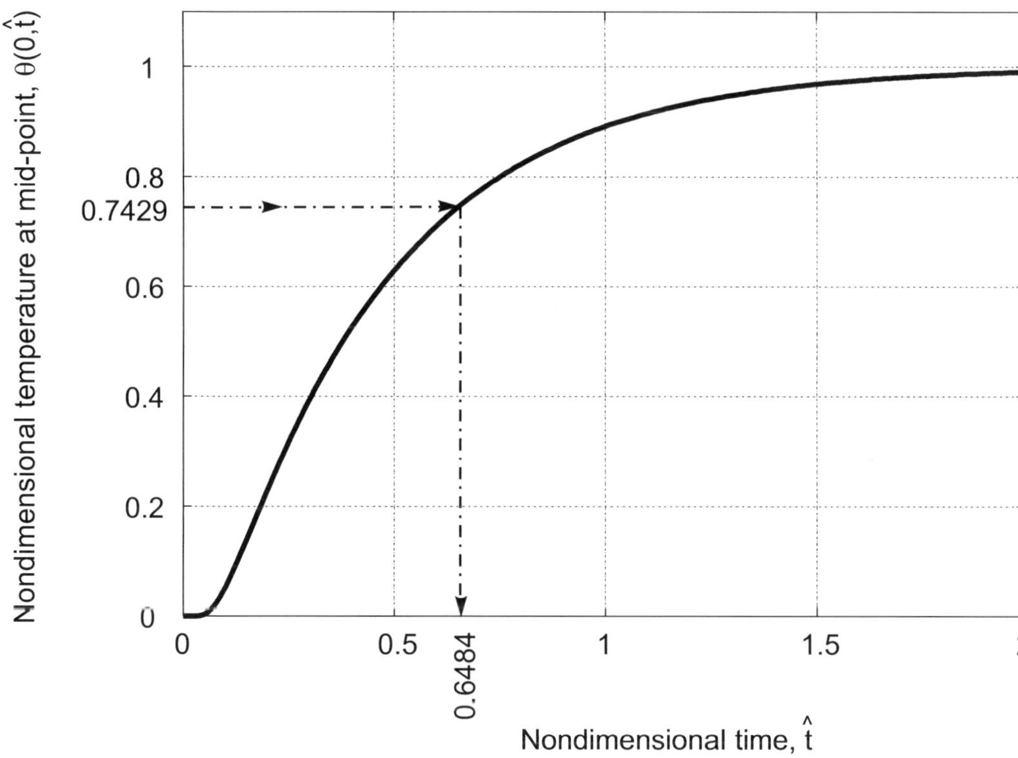

Figure 5.9: Nondimensional temperature versus nondimensional time at the center of the composite part. In order to cool the composite from 200°C ($\theta = 0$) to 70°C ($\theta = (70 - 200)/(25 - 200) = 0.7429$), the part should be kept in the mold for $\hat{t} = 0.6484$ which corresponds to 18.1 seconds.

Solution

Here, finite difference method (FDM) will be applied for the spatial derivatives, and Euler method will be applied for the time integration. One can discretize the $\hat{z}$ and $\hat{t}$ domains as

$$\hat{z}_i = -1 + (i-1)\Delta\hat{z} \tag{5.87}$$

$$\hat{t}_k = (k-1)\Delta\hat{t} \tag{5.88}$$

where $\Delta\hat{z} = 2h/(N-1)$ and $i = 1, 2, \ldots, N$; and $\Delta\hat{t}$ is the time step. The discretization is illustrated in Figure 5.10. Notice that we will solve the problem in $-1 \leq \hat{z} \leq 1$ with the boundary conditions $\theta(-1) = 1$ and $\theta(1) = 1$ (instead of Equations (5.68) and (5.69)) as explained in the previous example.

Equation (5.66) can be replaced with

$$\frac{\theta_i^{k+1} - \theta_i^k}{\Delta\hat{t}} = \frac{\theta_{i-1}^k - 2\theta_i^k + \theta_{i+1}^k}{(\Delta\hat{z})^2} \tag{5.89}$$

where subscripts $i-1$, i and $i+1$ denote the indexes for the spatial $\hat{z}$ discretization, and superscripts k and $k+1$ denote the indexes of the current time and next time steps for the time $\hat{t}$ discretization. Rearrangement of the terms yields

$$\theta_i^{k+1} = (1-2r)\theta_i^k + r\left(\theta_{i-1}^k + \theta_{i+1}^k\right) \tag{5.90}$$

where $r = \Delta\hat{t}/(\Delta\hat{z})^2$. Notice that if one knows θ_i^k for all i's, then θ_i^{k+1} can be calculated for all internal i indexes ($2 \le i \le N-1$), and $\theta_1^{k+1} = 1$ and $\theta_N^{k+1} = 1$ are known from the boundary conditions. This approach enables a time-marching solution of $\theta(\hat{z}, \hat{t})$.

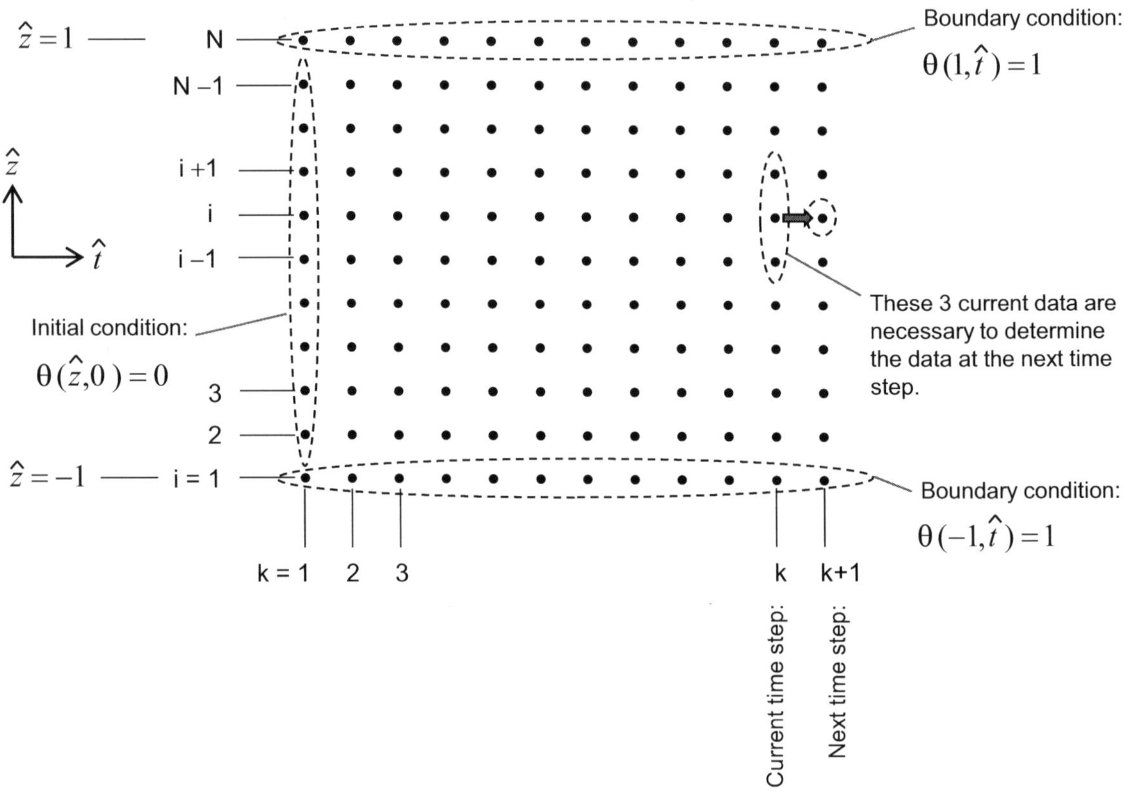

Figure 5.10: Discretization of $\hat{z}$ and $\hat{t}$ domains.

For a stable numerical solution, $r < 0.5$ must be selected [273].

A MATLAB code (named *example5_06.m*) has been coded and given in the Appendix. The reader is suggested to run it in MATLAB to animate the evolution of temperature profile. The results are shown in Figure 5.11. ■

Example 5.7: Role of Brinkman Number

A polymer suspension flows through a channel of thickness h and length L. The schematic is shown in Figure 5.12. The temperature of the polymer is T_i before it enters the mold. The mold wall temperature is kept at T_w. Formulate the energy equation that depicts the importance of viscous dissipation in heating of the suspension.

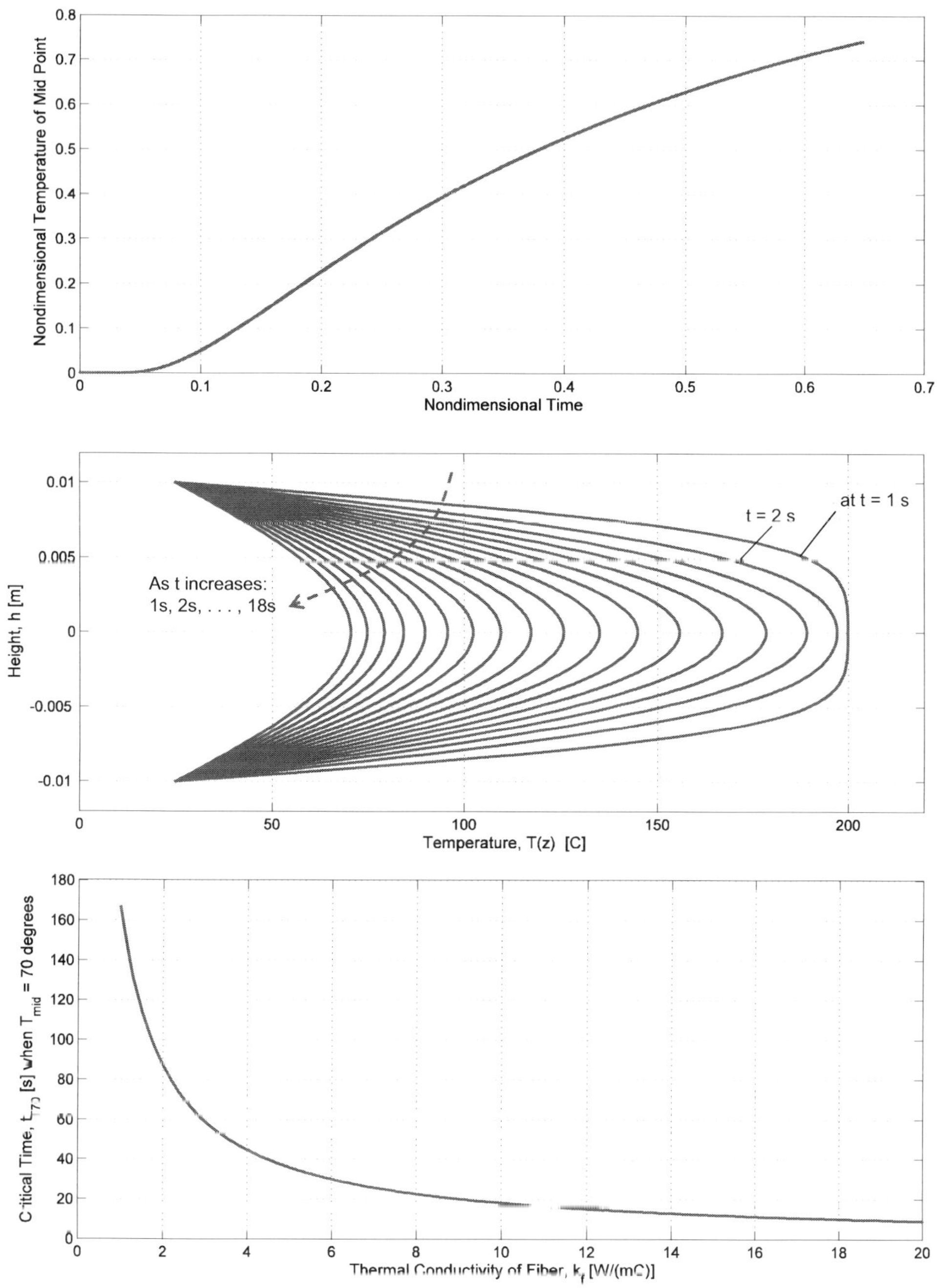

Figure 5.11: Results of numerical solution to Example 5.6.

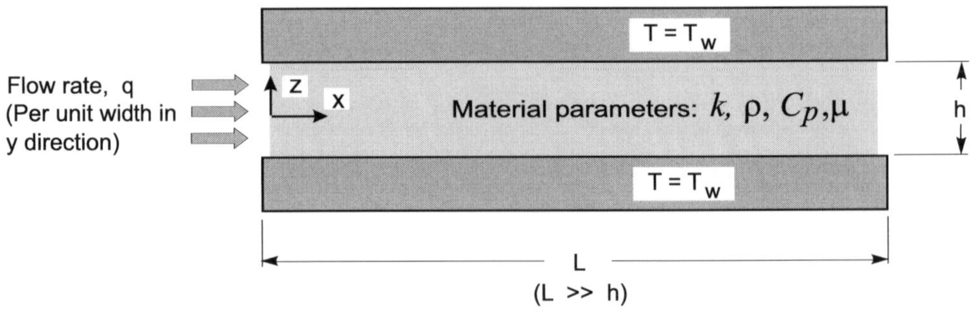

Figure 5.12: Nonisothermal flow of a polymer suspension through a channel.

Solution

Assuming that the flow is in the x direction, v and w components of the velocity vector $\mathbf{u}$ are zero, and one can ignore heat transfer in y-direction and write the energy equation in 2D as shown below,

$$\rho C_p \left(\frac{\partial T}{\partial t} + u \frac{\partial T}{\partial x} \right) = k \left(\frac{\partial^2 T}{\partial x^2} + \frac{\partial^2 T}{\partial z^2} \right) + \underline{\underline{\tau}} : \nabla \mathbf{u} + \dot{R}. \tag{5.91}$$

There is no internal energy generation, hence $\dot{R} = 0$. Also, as $h \ll L$

$$\frac{\partial^2 T}{\partial x^2} \ll \frac{\partial^2 T}{\partial z^2} \tag{5.92}$$

as can be seen from previous examples. Also the viscous dissipation term reduces to

$$\begin{aligned} \underline{\underline{\tau}} : \nabla \mathbf{u} &= \left(\tau_{xx} \frac{\partial u}{\partial x} + \tau_{xy} \frac{\partial u}{\partial y} + \tau_{xz} \frac{\partial u}{\partial z} + \tau_{yx} \frac{\partial v}{\partial x} + \tau_{yy} \frac{\partial v}{\partial y} \right. \\ &\quad \left. + \tau_{yz} \frac{\partial v}{\partial z} + \tau_{zx} \frac{\partial w}{\partial x} + \tau_{zy} \frac{\partial w}{\partial y} + \tau_{zz} \frac{\partial w}{\partial z} \right) \qquad (5.93) \\ &= \tau_{xz} \frac{\partial u}{\partial z} \qquad (5.94) \end{aligned}$$

as $v = 0$, $w = 0$, and as $h \ll L$, therefore, $\frac{\partial u}{\partial z} \gg \frac{\partial u}{\partial x}, \frac{\partial u}{\partial y}$, and $\tau_{xz} \gg \tau_{xx}, \tau_{xy}$ as seen from the example on flow in thin gaps. One could have arrived at the same conclusion using dimensionless analysis.

Using the following nondimensionalization of the variables:

$$\begin{aligned} \hat{u} &= \frac{u}{u_c} = \frac{u}{q/h} \\ \theta &= \frac{T - T_i}{\Delta T} \\ \hat{x} &= \frac{x}{L} \\ \hat{z} &= \frac{z}{h}. \end{aligned}$$

Assuming steady state in the temperature distribution (i.e., $\partial T/\partial t = 0$), Equation (5.91) takes the following form:

$$\rho C_p \left(u_c \frac{\Delta T}{L} \hat{u} \frac{\partial \theta}{\partial \hat{x}} \right) = \frac{k \Delta T}{h^2} \left(\frac{\partial^2 \theta}{\partial \hat{z}^2} \right) + \frac{\mu u_c^2}{h^2} \left(\frac{\partial \hat{u}}{\partial \hat{z}} \right)^2 \tag{5.95}$$

here $\Delta T = T_W - T_i$, and a Newtonian constitutive equation $\tau_{xz} = \mu \frac{\partial u}{\partial z}$ was assumed. We can select the conduction term as the one whose coefficient should be unity. Hence, division of Equation (5.95) by $k\Delta T/h^2$ results in

$$\left[\frac{u_c h}{\alpha} \left(\frac{h}{L} \right) \right] \hat{u} \frac{\partial \theta}{\partial \hat{x}} = \frac{\partial^2 \theta}{\partial \hat{z}^2} + \frac{\mu u_c^2}{k \Delta T} \left(\frac{\partial \hat{u}}{\partial \hat{z}} \right)^2 \tag{5.96}$$

$$\mathrm{Gz}\, \hat{u} \frac{\partial \theta}{\partial \hat{x}} = \frac{\partial^2 \theta}{\partial \hat{z}^2} + \mathrm{Br} \left(\frac{\partial \hat{u}}{\partial \hat{z}} \right)^2 \tag{5.97}$$

where the Graetz number $\mathrm{Gz} = u_c h^2/(\alpha L)$ and the Brinkman number $\mathrm{Br} = \mu u_c^2/(k\Delta T)$. The Graetz number measures the ratio of heat convection to heat conduction. The Brinkman number measures the ratio of viscous dissipation to heat conduction. If $\mathrm{Br} \ll 1$, then viscous dissipation is not important. If Gz is small and $\mathrm{Br} \gg 1$, then all heat generated due to viscous dissipation has to be conducted. If Gz and Br are large ($\gg 1$), then heat generated due to viscous dissipation is convected faster than it is conducted. ■

5.5 Customary Assumptions in Polymer Composite Processing

Certain assumptions can dramatically simplify a process model. However, inappropriate simplifications can also lead to process behavior not in sync with observations. Hence, it is important to justify the assumptions a priori or after the solution is obtained or by validating the assumption with experiments. We will list the common assumptions employed in one or more processes, depending on the phenomena of interest. It does not imply that all processes will employ the same assumptions listed. To model a phenomenon or a process, one should go through the exercise of justifying every assumption made to simplify the model [271].

5.5.1 Quasi-Steady State

Physically, if a variable such as velocity, temperature or pressure is in steady state, it indicates that, from a stationary reference frame, the variable does not experience any change with time at that spatial location. Mathematically, steady state can be described by setting $\partial()/\partial t = 0$ by placing the variable into the $()$. Quasi-steady state assumption refers to approximating an unsteady or time-dependent flow with small inertial effects as a succession of steady-state flows. So, even if the problem contains time-dependent boundary conditions, one can solve it as a steady-state problem at any instance of time t. By assuming the value of the time-dependent boundary condition, at that instant in time, to be a constant value, one can solve the steady-state governing equation.

The quasi-steady state approach is reasonable if the momentum transfer from the boundary conditions is immediate and the material inertia plays an insignificant role. The kinetic viscosity ν, usually governs the rate of momentum transfer. For polymers and polymer composites, this value is very high compared to that of water, and hence the rate of momentum transfer can almost always be approximated as immediate at any instant in time. Another way to describe this is that, when the viscosity is high, the inertial effects will be insignificant provided the process velocity is slow (low Reynolds number).

For example, if a thermoplastic polymer is squeezed between two parallel disks with a constant force F, how can we calculate the gap separation between the plates with time? A quasi-steady state assumption allows one to ignore the inertial terms in the momentum equation and solve the radial problem as a steady-state problem at any time t for the instantaneous gap height (which is a function of time). Other examples that use this assumption are the processes that simulate the mold-filling scenario. In these classes of problems, the solution domain in which one solves for the fluid pressure is bounded by the boundaries of the resin in the mold. The domain is bounded by either the mold boundaries if the resin has reached the mold walls, or the boundaries of the instantaneous resin front (also known as the flow front) if the resin has not reached the mold boundaries yet. After solving for the fluid pressure, one calculates the fluid velocities, advances the resin front based on the velocity field, and re-solves for the pressure in the new resin-filled domain. This procedure is continued until the mold is filled. Thus, the quasi-steady state solution of the pressure is dependent only on the instantaneous material properties, geometry of the domain and the boundary conditions, which may be functions of time.

Note that the justification of neglecting the transient terms in the equation of motion follows from the fact that the kinematic viscosity of polymers is high and as the speeds are low, the Reynolds number is low. However, the thermal diffusivity of polymers is low, so the heat transfer is much slower and one cannot always ignore the transient terms in the energy equation. The Prandtl number, which is the ratio of momentum transfer to energy transfer, is very high for polymers, of the order of 100 or higher. Thus, for non-isothermal processes, one usually cannot ignore the transient term in the energy conservation equation.

5.5.2 Fully Developed Region and Entrance Effects

When a system variable does not change along a certain direction, one assumes that the variable is "fully developed" with respect to that direction. For example, if the temperature profile does not change along the x direction, one can physically assume that the temperature is fully developed along the x direction. Mathematically, one can express this condition as $\partial T/\partial x = 0$.

"Fully developed flow and insignificant entrance effects" is a widely used assumption while developing "process models" in polymer and polymer composites processing. This assumption is again related to the rate of momentum transfer. The higher the kinematic viscosity, the faster the momentum transfer will be and the sooner the velocity profile will develop. For example, one can consider flow in a tube with a diameter D from a reservoir or from a pressure pot as shown in Figure 5.13. The entrance region is the region in which the flow profile is rearranging from plug flow to a parabolic profile for a viscous Newtonian fluid. Usually, this region is proportional to the diameter of the tube and the Reynolds number [272]

$$\frac{\delta}{D} \cong 0.054\,\mathrm{Re}. \tag{5.98}$$

If the Reynolds number is less than one, the entrance effects last for less than the tube diameter, and the assumption of fully developed flow will not influence the important physics or the solution.

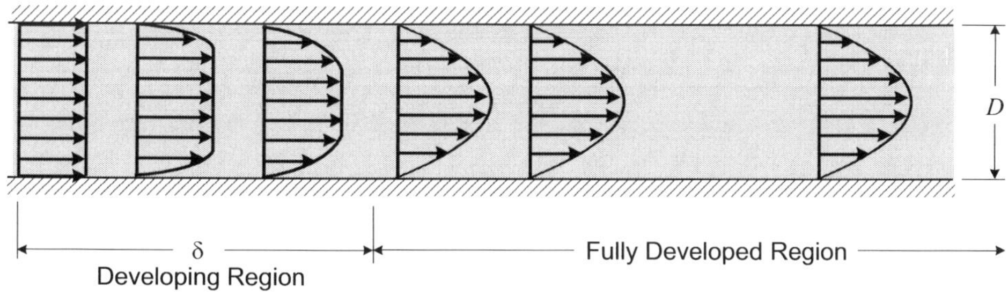

Figure 5.13: Development of velocity profile.

Similarly, if one wanted to assume a fully developed temperature profile along the flow direction, the following criterion needs to be met [274]:

$$\frac{\delta_t}{D} \cong (0.03 \text{ to } 0.04) \ \mathrm{Re}\,\mathrm{Pr} \tag{5.99}$$

where Re and Pr are the Reynolds and Prandtl numbers. In composites processing, this assumption is easy to justify for long tubes or long characteristic flow directions, but not for short tubes as Pr can be of the order of 1000 for most materials. Also, if the temperature boundary condition is changing with time, a fully developed temperature profile cannot be easily justified. The other important number here is the Graetz number, Gz. When it is small, one can assume that both the velocity and temperature profiles are developed in the direction of flow.

Fully developed flow is a common assumption used in polymer and composites processing due to low Reynolds number flows as compared to the fully developed temperature profile. One of the most common situations in which *fully developed* flow is used is in conjunction with the *lubrication approximation* that allows one to simplify the equations of motion considerably and retain the important physics of the flow in the process model.

5.5.3 Lubrication Approximation

Osborne Reynolds [275] coined the term "lubrication approximation" when he was addressing hydrodynamic lubrication analysis. The key requirements to apply this theory are that the flow should be viscous (the Reynolds number should be small) and flow should take place in thin narrow gaps. While processing composites, one is dealing with parts that may be meters in length and width but only millimeters in thickness. Hence, as they are also low Reynolds number flows, we can employ the lubrication analysis to simplify the model.

Let us review here how lubrication theory allows one to simplify the equations of motion for a Newtonian viscous material. The conditions that should be met before applying this analysis to a flow problem are

- The material should be incompressible.
- The flow should be isothermal.
- The Reynolds number should be less than one (that means the inertial forces should be smaller than the viscous shear forces).
- The gap height should be very small compared to the in-plane dimensions.
- If the gap height is not a constant, it could vary very slowly with the in-plane directions (for instance if the in-plane axes were x and y, dh/dx and $dh/dy \ll 1$).

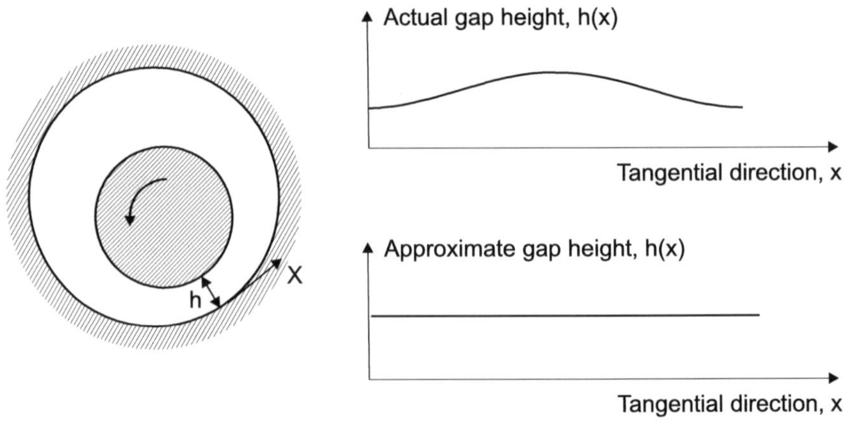

Figure 5.14: Approximation of flow in a thin gap using lubrication theory.

- The fluid does not slip at the walls of the surface.

A sketch where the lubrication approximation will be appropriate is shown in Figure 5.14.

5.5.4 Thin Shell Approximation

If the "no slip at the boundaries" condition is not satisfied, it may be difficult to justify the lubrication approximation. When you have full slip, it is possible to model the flow as plug flow through the cavity thickness. For example, when one considers Darcy's law for flow of thermoset resins through a network of fibers, one assumes a full slip boundary condition along the mold surfaces. In such cases, it is the square of the ratio of the thickness to the in-plane dimension multiplied by the ratio of the permeability in the thickness direction to that of the in-plane direction that dictates if one can ignore the flow in the thickness direction. Hence, if

$$\left(\frac{t}{L}\right)^2 \frac{K_t}{K_L} \ll 1 \tag{5.100}$$

one can ignore the flow in the thickness direction where t and L are the thickness and in-plane dimensions, and K_t and K_L are the permeability components along these directions, respectively.

5.6 Boundary Conditions for Flow Analysis

There are many types of boundary conditions that are encountered during processing of composite materials. In this section, we will discuss how one models some of them.

5.6.1 In Contact with a Solid Surface

When the material is in contact with a solid surface, the modeler has three options for the application of boundary condition for the velocity. They are (i) no slip (which implies that the material sticks to the surface), (ii) partial slip, and (iii) complete slip. Mathematically, one can express these conditions as a relationship between the velocity and the shear stress as shown below

$$u_x = -\beta \tau_{xy} \tag{5.101}$$

where u_x and τ_{xy} are the velocity and shear stress along the wall, and β is a weight coefficient indicating the amount of slip (see Figure 5.15). When $\beta = 0$, Equation (5.101) reduces to the no-slip boundary condition. A nonzero value for β implies partial slip. As β goes to infinity, the boundary condition tends to full slip (however, u_x is still finite because τ_{xy} approaches zero).

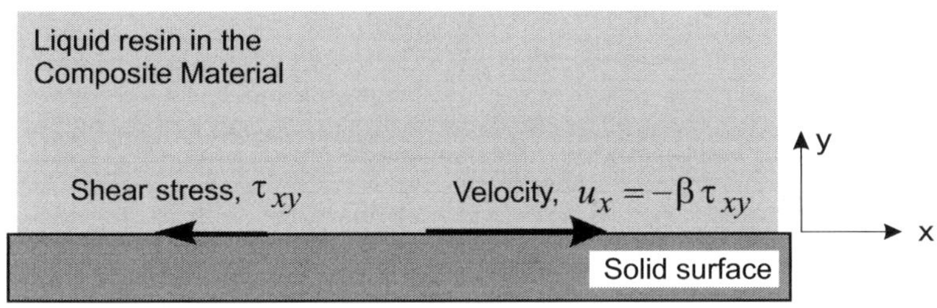

Figure 5.15: Boundary condition (BC) along a solid surface.

5.6.2 In Contact with Other Fluid Surfaces

When the composite material with the resin in liquid form or the fiber suspension is in contact with another fluid surface, one needs to specify the boundary condition at the interface between the two fluids. Usually, one matches the velocities and the shear stresses at the interface as shown in Figure 5.16.

$$\mathbf{u}_1 = \mathbf{u}_2 \tag{5.102}$$

$$(\tau_{nt})_1 = (\tau_{nt})_2 \tag{5.103}$$

where the subscripts denote the fluid domain.

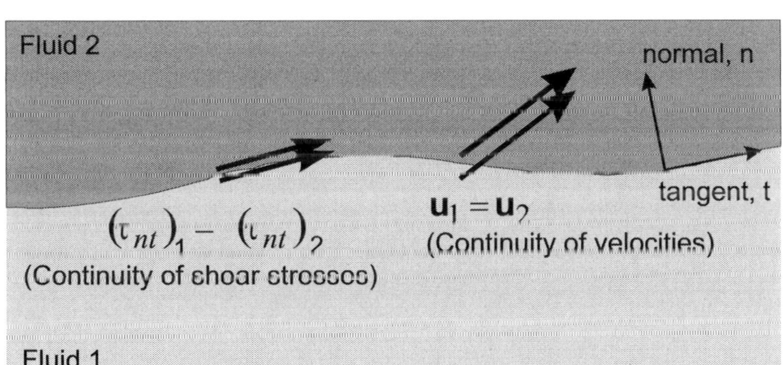

Figure 5.16: Boundary conditions along the interface of two fluids.

5.6.3 Free Surfaces

Sometimes, the viscous material in its liquid form will have a free surface. For example, in the extrusion process, once the material leaves the die, the mass balance requirement alone can cause the material to expand. When the material is in the die, one can use a no-slip or slip boundary condition, but as soon as the material is free, one requires a different boundary condition. Usually, "normal stress equal to zero" boundary condition is used.

When the resin is impregnating a preform, or in injection or compression molding where the flow front is pushing the air, one usually specifies a pressure boundary condition at the flow front which is usually the atmospheric pressure, $P = P_{\text{atm}}$ (or $P = P_{\text{vacuum}}$ if the vent is connected to a vacuum pump), as shown in Figure 5.17.

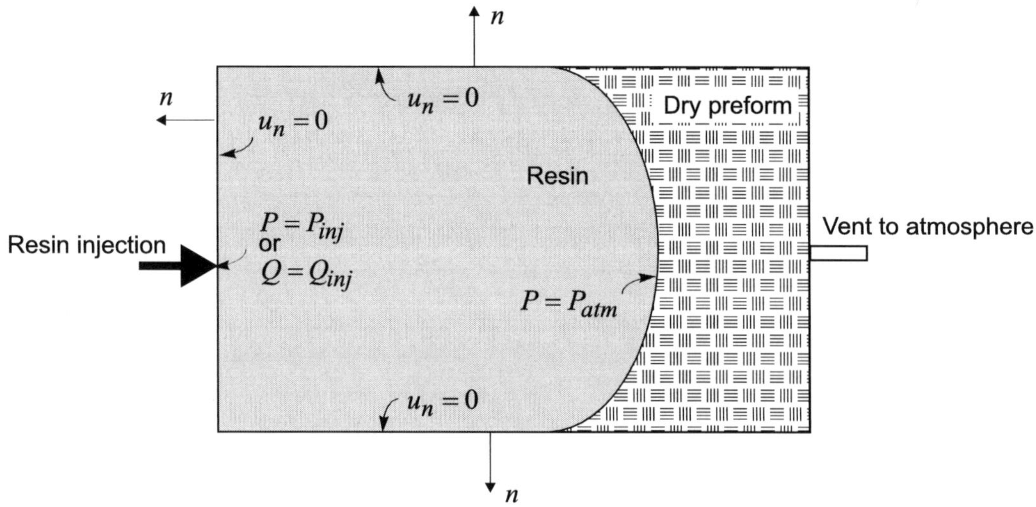

Figure 5.17: Boundary conditions along mold walls, on a free surface and at an injection point.

5.6.4 No Flow out of a Solid Surface

For dies and molds in which the resin is injected, if one uses a slip boundary condition along the wall, one usually specifies no flow out of the mold. Mathematically, this is represented as

$$u_n = 0 \tag{5.104}$$

where u_n is the velocity component normal to the mold wall as shown in Figure 5.17. If the polymer flow is modeled with Darcy's law ($\mathbf{u} = -\frac{1}{\mu}\mathbf{K} \cdot \nabla P$), then Equation (5.104) is written in terms of polymer pressure P as the primary unknown variable of the numerical solution. Thus, for a 2D flow

$$u_n = -\frac{1}{\mu}\left(K_{nn}\frac{\partial P}{\partial n} + K_{nt}\frac{\partial P}{\partial t}\right) = 0 \tag{5.105}$$

where n and t are the normal and tangential directions to the wall.

5.6.5 Specified Conditions

Sometimes, either the flow rate Q or the pressure P at the inlet is specified:

$$P = P_{\text{inlet}} \tag{5.106}$$

or

$$Q = Q_{\text{inlet}} \tag{5.107}$$

as shown in Figure 5.17. P_{inlet} and Q_{inlet} can be constant, or may vary with time. Usually, either one of the following two practices is applied (see Figure 5.18):

- $P_{\text{inlet}}(t)$ is set to a constant value, and $Q_{\text{inlet}}(t)$ decreases with time accordingly. This approach has a high flow rate initially due to a high initial pressure gradient, and it decreases with time.
- Q_{inlet} is set to a constant value and $P_{\text{inlet}}(t)$ increases with time as the resin continues to propagate through a porous fabric reinforcement; if $P_{\text{inlet}}(t)$ reaches a critical value, $P_{\text{inlet,allowed}} = P_{\text{limit}}$, the injection condition is switched from constant flow rate to constant pressure, thus $P_{\text{inlet}}(t)$ is set to a constant, $P_{\text{inlet,allowed}}$ and $Q_{\text{inlet}}(t)$ decreases with time accordingly. $P_{\text{inlet,allowed}}$ is usually the limiting pressure of the machine or the pressure which will ensure that the clamped mold parts won't open and leak or disrupt connection hoses between the injection machine and the mold. Also, $P_{\text{inlet,allowed}}$ should be low enough not to cause any fabric wash.

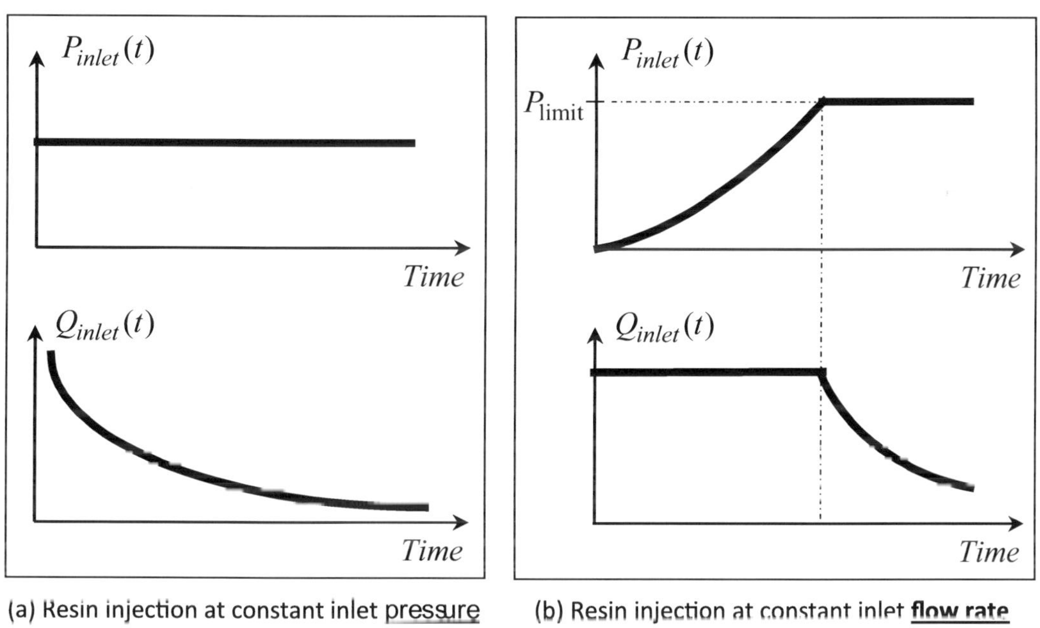

Figure 5.18: (a) Variation of the flow rate at the inlet when the resin is injected under constant fluid pressure. (b) When resin is injected under constant flow rate, the resin pressure at the inlet increases. One can switch the inlet condition to constant pressure at a pre-set pressure value during the injection.

5.6.6 Periodic Boundary Condition

At the microscale level, if one encounters a repetitive unit cell geometry, one can sometimes apply a periodic boundary condition to estimate the unit cell properties. For example, if one is interested in gauging the resistance (inverse of permeability) to flow of a preform that has this repetitive structure, one could do so in two steps. First, a pressure gradient is applied in the x direction, satisfying a periodicity in P and $\partial P/\partial x$ along the upper and lower edges of the unit cell as seen in Figure 5.19. Pressure and velocity fields are solved numerically within the empty channels of the fiber bed. By using 1D Darcy law in the x direction, one can calculate the effective permeability in the x direction, K_{xx}. A similar procedure can be repeated in the y direction, as shown in Figure 5.19.

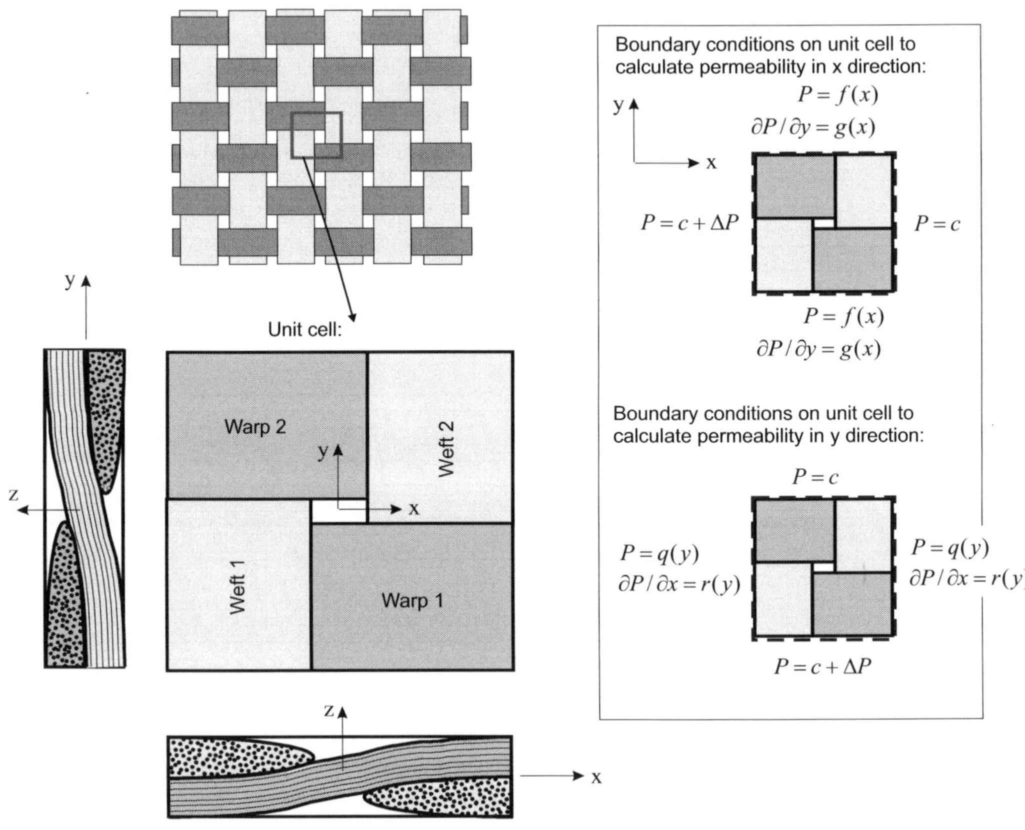

Figure 5.19: Boundary conditions on a repetitive unit cell geometry [251].

5.6.7 Temperature Boundary Conditions

The resin is either heated or cooled by the mold walls. One could specify one of the following as the boundary condition at the walls:

- **Fixed temperature (thermal contact)** boundary condition, $T = T_w$. This is used when the composite material and the mold walls are in perfect contact. Then, the temperature of the boundary of the composite is assumed to be the same as the wall temperature, T_w.

- **Fixed normal gradient** boundary condition, $\partial T/\partial n = c_1$. When the conductive heat flux, q, is known between the composite material and the mold walls, then one can use the 1D heat conduction equation $q = -k_n \partial T/\partial n$ where n is the normal direction to the wall, and k_n is the thermal conductivity of the composite in that direction. Thus, one specifies the temperature gradient at the boundary instead of specifying the temperature.

- **Relation between T_w and $\frac{\partial T}{\partial n}$ known**. When a fluid is used to heat or cool the mold, one can express the boundary condition as $h(T - T_\infty) = -k_n \frac{\partial T}{\partial n}$ at the wall. Here, T_∞ is the fluid temperature, and h is the convective heat transfer coefficient which is usually estimated from the fluid flow properties. This is done by relating the Nusselt number (Nu$=\frac{hL}{k}$) to the Reynolds and Prandtl numbers [274].

One can develop a general type of boundary condition as was done in Advani et al. [246] for a configuration as shown in Figure 5.20. Heating/cooling pipes are at a distance of a from the mold surface. The heating/cooling fluid flowing through these pipes is at a temperature of T_∞. The corresponding effective boundary condition can be written as [246]

$$\frac{\partial \theta}{\partial n} + C_{bc}\,\theta = 0 \tag{5.108}$$

where $\theta = T - T_\infty$, and the effective boundary condition constant C_{bc} is calculated as

$$C_{bc} = \frac{1}{k} \frac{1}{\left(\frac{1}{h} + \frac{a}{k_m}\right)}. \tag{5.109}$$

Here, k_m is the conductivity of the mold material, h is the heat transfer coefficient between heating/cooling fluid and the pipes, and k is given as

$$k = k_f \frac{(k_f + k_r) + V_f(k_f - k_r)}{(k_f + k_r) - V_f(k_f - k_r)} \tag{5.110}$$

where V_f is the fiber volume fraction of the preform, and k_f and k_r are the conductivity of the fiber preform and resin, respectively. When the mold is insulated well (not much heating or cooling), then C_{bc} approaches zero. The fixed temperature (perfect thermal contact) boundary condition corresponds to having a large value of C_{bc}. This formulation allows for the change in the mold surface temperature as experienced in practice. Due to the exothermic chemical cure reaction of polymeric resin, the temperature of the resin and hence the mold surface might increase away from the injection ports. Hence, this formulation allows the modeler flexibility in specifying the boundary condition. Usually the radiative boundary condition is not employed unless the temperature of the mold or that of the surroundings is very high.

5.7 Convection of Variables

Temperature, cure and fiber orientation will usually convect with the flow, and one must account for this phenomenon mathematically by using the material derivative instead of the partial derivative (see Chapter 3). So, instead of $\frac{\partial T}{\partial t}$, $\frac{\partial \alpha}{\partial t}$ and $\frac{\partial a_{ij}}{\partial t}$, one would use $\frac{DT}{Dt} = \frac{\partial T}{\partial t} + u\frac{\partial T}{\partial x} + v\frac{\partial T}{\partial y} + w\frac{\partial T}{\partial z}$ where u, v and w are components of the velocity of the moving fluid. Similarly, one would express any variable that is evolving during flow.

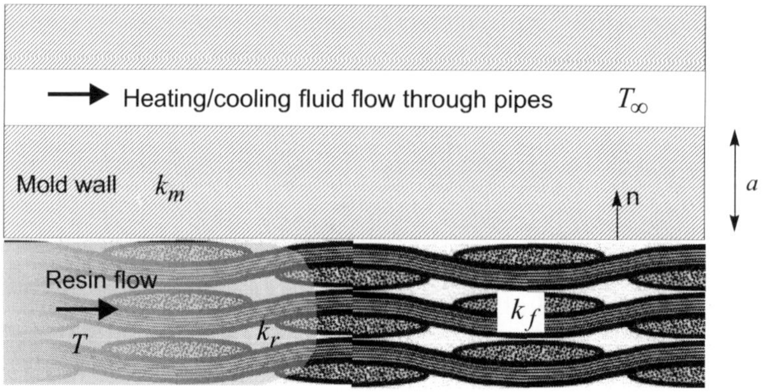

Figure 5.20: Heating/cooling of composite material by circulating a fluid flowing through a channel network machined inside the mold walls [276].

5.8 Process Models from Simplified Geometries

In composite manufacturing, one may fabricate simple geometries such as flat laminates to very complicated net shape structures such as an I-beam of a passenger van or the keel beam of a helicopter (see Figure 5.21). An ideal mathematical representation will stay true to the actual physical geometry of the part. For complex geometries, one must resort to numerical methods to meet this criterion. However, a simplified geometry model offers many advantages, including the following:

- Basic understanding of the process being modeled with the least investment in time and effort because it is possible to find analytic and close-form solutions.
- Ability to extract scaleable parameters from a simplified model.
- Ability to check the usefulness and validity of the phenomenological law or constitutive equation that is used in the process model.

Figure 5.21: Mold filling simulations for complex composite structures. Different gray tones indicate the location of resin front at different times during mold filling.

It is also possible to take simplified geometries and couple them to represent a complex geometry. This may geometrically produce an approximate solution, but will be able to address some of the complexities in the material modeling. Below, we will present results for Newtonian and power-law fluids for selected simple cases that were studied earlier in detail in Chapter 3.

Couette Flow

Couette flow is the flow in a rectangular channel induced by a moving upper boundary at a constant speed V. The in-plane dimension W in z direction is much larger than the thickness, h in y direction (see Figure 5.22), thus the flow in z direction is neglected. One expects $u_x = u_x(x, y)$, however by neglecting the end effects, we will consider the fully developed flow, thus $u_x = u_x(y)$ only. The no-slip boundary condition is assumed at the lower and upper solid walls. Hence, the resin velocity is zero and V at the lower and upper boundaries, respectively. The velocity profile is found to be

$$u_x(y) = \frac{V}{h}y \tag{5.111}$$

which is linear as shown in Figure 5.22.

Pressure Driven Flow in a Narrow Slit

This is the same as the previous Couette flow, except that both plates are fixed, but there is a pressure drop applied between the two ends of the channel as shown in Figure 5.23. The velocity profile is found to be

$$u_x(y) = \frac{P_e - P_i}{2\mu L}\left(y^2 - hy\right) \tag{5.112}$$

which is shown in the same figure for $P_e < P_i$.

Pressure Driven Flow in a Capillary or a Tube

This is the same as the previous pressure driven flow in a narrow slit, except that the flow is through a circular tube. The sketch and a typical velocity profile are shown in Figure 5.24. The velocity profile is found to be

$$u_x(r) = \frac{P_e - P_i}{4\mu L}\left(r^2 - R^2\right) \tag{5.113}$$

which is parabolic, and shown in the same figure for $P_e < P_i$.

One-Dimensional Flow in an Isotropic Porous Medium

In order to measure the permeability of a preform in the warp and the weft directions, two separate one-dimensional experiments are done in those directions. A fluid is injected into the mold from the left edge to fully saturate the fibrous preform as shown in Figure 5.25. If the injection is carried out under a constant flow rate Q, then the injection pressure rises linearly as follows,

$$P_{inj}(t) = \frac{\mu}{K_{xx}(1 - V_f)}\left(\frac{Q}{C_w C_h}\right)^2 t \tag{5.114}$$

where K_{xx} and V_f are the permeability in the flow direction and the fiber volume fraction of the preform. From the slope of the graph of pressure vs. time, one can calculate permeability as all other quantities are known.

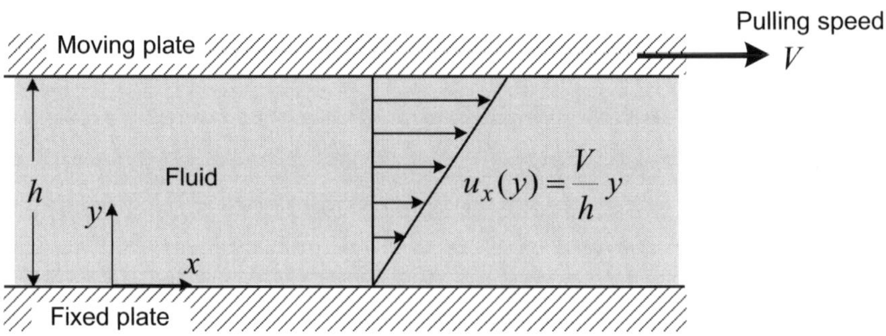

Figure 5.22: Velocity profile in a rectangular channel induced by a moving upper plate at a constant velocity V (Couette flow).

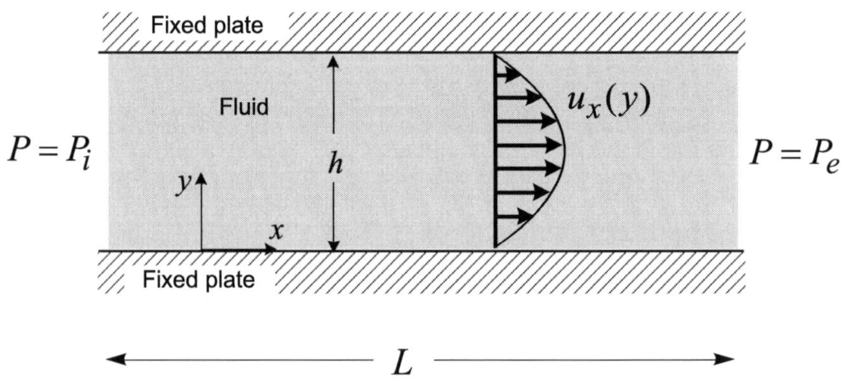

Figure 5.23: Velocity profile in a rectangular channel induced by a pressure drop in the direction of flow ($P_i > P_e$).

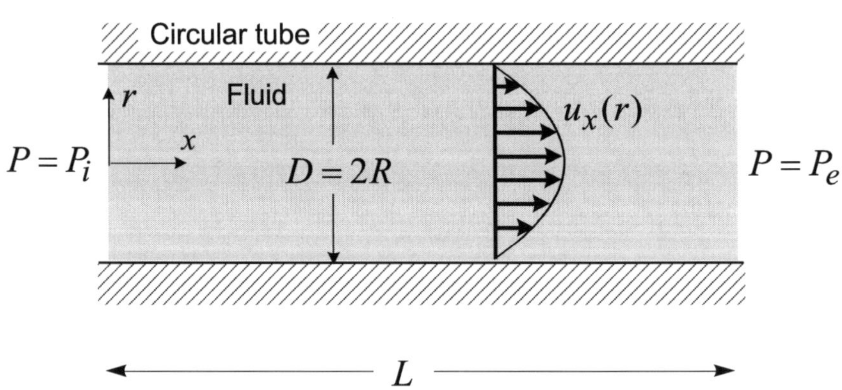

Figure 5.24: Velocity profile in a circular tube induced by a pressure drop in the direction of flow ($P_i > P_e$).

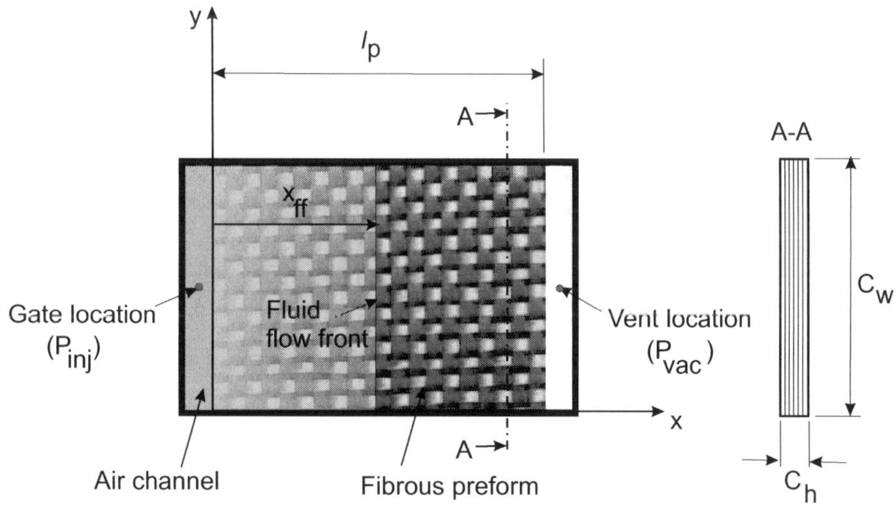

Figure 5.25: One-dimensional flow in x direction through an isotropic porous medium [81].

Radial Flow in an Anisotropic Porous Medium

In order to measure the permeability of a preform, a radial injection experiment is done from a tube at the center of the preform. The preform has a circular hole of radius R through which the fluid enters. This ensures radial flow in the plane as shown in Figure 5.26. If the preform is isotropic, then the flow front is circular. If it is anisotropic, then the flow front is elliptical. The aspect ratio of the ellipse (the ratio of the major and minor axes) is equal to the square root of the permeability components in those principal directions. If the fluid is injected at a constant flow rate Q, then the injection pressure rises logarithmically

$$P_{inj}(t) = \frac{\mu Q}{4\pi h\sqrt{K_{xx}K_{yy}}} \ln\left[1 + \frac{Qt}{\pi h R^2(1-V_f)}\right] \tag{5.115}$$

where K_{xx} and K_{yy} are the permeability components in the major and minor axes, and h is the mold cavity thickness.

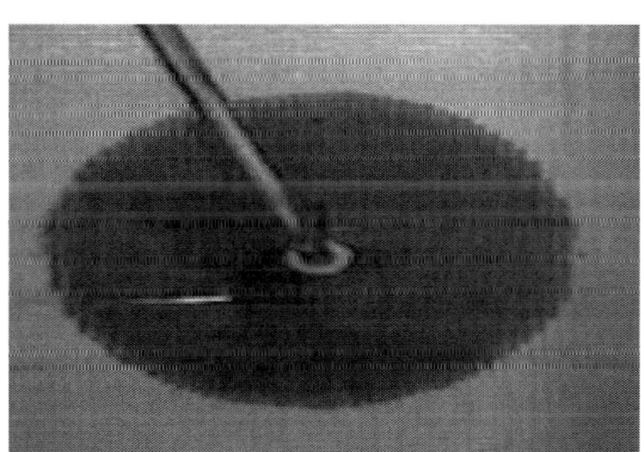

Figure 5.26: One-dimensional flow in radial direction in an isotropic porous medium.

Example 5.8: 1D Permeability Measurement Experiment

The experimental setup shown in Figure 5.25 was used to measure the permeability of a fabric preform. The parameters of the experiment are listed below:

- Resin viscosity, $\mu = 0.24\,\text{Pa.s}$
- Flow rate, $Q = 240\,\text{cc/min}$ (constant during the experiment)
- Mold thickness, $C_h = 0.005\,\text{m}$
- Mold width, $C_w = 0.200\,\text{m}$
- Mold length, $L_p = 1\,\text{m}$
- Number of layers in the fabric mat, $n = 7$
- Areal density of single layer fabric, $\rho_{sup} = 0.860\,\text{kg/m}^2$
- Density of glass fibers used in the fabric, $\rho_f = 2540\,\text{kg/m}^3$.

The following experimental data was collected using a pressure transducer at the inlet. Calculate the followings: (1) K_{xx} in the flow direction; (2) time to fill the mold cavity (neglect the inlet and exit channels); and (3) injection pressure at the end of mold filling. Make a comment about the validity of the experiment, and propose things to enhance the validity.

t [s]	0	5	10	15	20	25	30
P_{inj} [Pa]	0	11000	22400	33300	42600	53450	60850

Solution

One can rewrite Equation (5.114) as

$$K_{xx} = \frac{\mu}{(1 - V_f)} \left(\frac{Q}{C_w C_h} \right)^2 \frac{1}{\frac{dP_{inj}}{dt}}. \tag{5.116}$$

All of the parameters are known except V_f and dP_{inj}/dt. Fiber volume fraction of the preform is calculated as follows:

$$\begin{aligned} V_f &= \frac{\text{volume of fibers}}{\text{volume of the mold cavity}} \\ &= \frac{\text{mass of fibers}}{\text{mass of the mold cavity hypothetically occupied by 100\% fibers}} \\ &= \frac{n\rho_{sup}(L_p C_w)}{\rho_f (L_p C_w C_h)} = \frac{n\rho_{sup}}{\rho_f C_h} \\ &= \frac{(7)(0.860\,\text{kg/m}^2)}{(2540\,\text{kg/m}^3)(0.005\,\text{m})} = 0.474 = 47.4\%. \end{aligned} \tag{5.117}$$

To calculate dP_{inj}/dt, a linear curve (first order polynomial) is fit to the experimental data using least square method. One can use MATLAB's built-in function *polyfit* for this purpose. The simple MATLAB code (named as *example5_08.m*) is given in the Appendix, and its result is $dP_{inj}/dt = 2054.6\,\mathrm{Pa/s}$ (see Figure 5.27).

The flow rate should be converted to [m^3/s] before substituting in Equation (5.116) to be consistent with the other parameters' units, thus $Q = 240\,\mathrm{cc/min} = 240 \times 10^{-6}\mathrm{m}^3/(60\,\mathrm{s}) = 4 \times 10^{-6}\mathrm{m}^3/\mathrm{s}$.

Now, we can calculate K_{xx}:

$$K_{xx} = \frac{0.24\,\mathrm{Pa.s}}{(1-(0.474))}\left(\frac{4\times 10^{-6}\mathrm{m}^3/\mathrm{s}}{(0.200\,\mathrm{m})\,(0.005\,\mathrm{m})}\right)^2 \frac{1}{2054.6\,\mathrm{Pa/s}} = 3.55 \times 10^{-9}\mathrm{m}^2. \tag{5.118}$$

To calculate the time to fill the mold cavity, one can use the fact that at the end of mold filling, the empty volume of the mold cavity will be completely filled with resin at constant flow rate injection. The empty volume of the mold cavity is calculated as

$$\begin{aligned} \text{Empty volume} &= (L_p C_w C_h)\,(1 - V_f) \\ &= [(1\,\mathrm{m})\,(0.200\,\mathrm{m})\,(0.005\,\mathrm{m})]\;(1 - 0.474) \\ &= 526 \times 10^{-6}\mathrm{m}^3 = 526\,\mathrm{cc} \end{aligned} \tag{5.119}$$

where $1 - V_f$ is the porosity, ϕ of the fabric preform. Thus, the time to fill the mold cavity is calculated as

$$t_{fill} = \frac{\text{Empty volume}}{\text{Flow rate}} = \frac{526\,\mathrm{cc}}{4\,\mathrm{cc/s}} = 131.5\,\mathrm{s}. \tag{5.120}$$

The injection pressure at this final time will be

$$\begin{aligned} P_{inj}(t_{fill}) &= \frac{dP_{inj}}{dt} t_{fill} = (2054.6\,\mathrm{Pa/s})\,(131.5\,\mathrm{s}) \\ &= 270,180\,\mathrm{Pa} = 270\,\mathrm{kPa} = 2.70\,\mathrm{bars}. \end{aligned} \tag{5.121}$$

When Figure 5.27 is examined, it is seen that the injection pressure is not perfectly linear. This deviation from the linearity may be caused by any of the following reasons:

- The flow may not be perfectly 1D (although it is expected to be 1D for the validity of the Equation (5.116)). This may be caused by (1) racetracking channels between the preform and the mold wall, (2) irregular fabric structure and/or compaction, and (3) nonuniform mold thickness.
- Q may not be held constant (check the accuracy of your injection machine).
- The viscosity of the resin may be varying with time due to the changes in temperature and degree of cure.

Even if the injection pressure is almost linear, one should repeat the experiment using multiple specimens to calculate the mean and standard deviation of the permeability. Non-repeatable specimen preparation and placement in mold cavity may affect the permeability and thus cause significant variations. ■

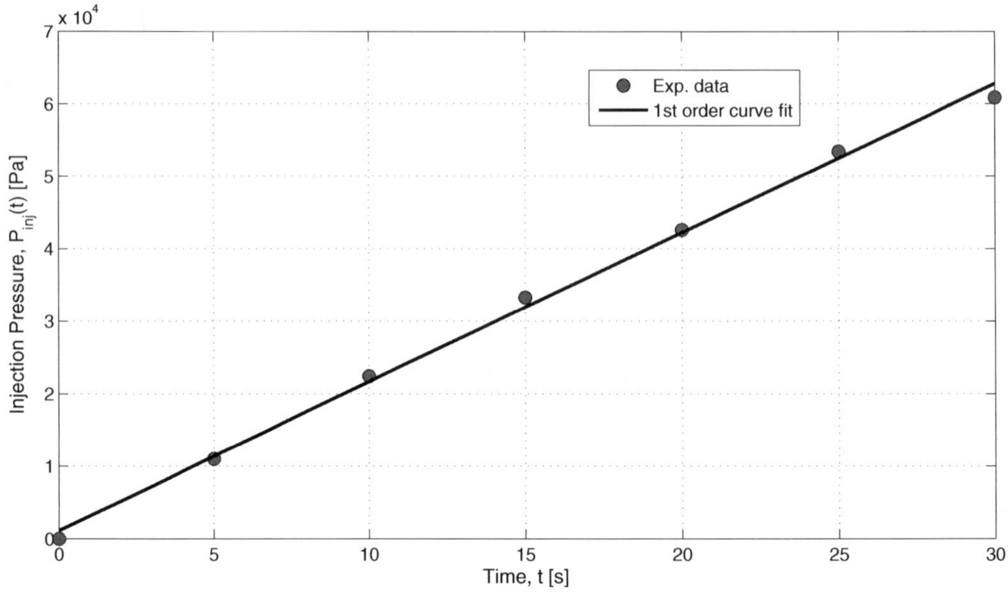

Figure 5.27: Injection pressure during a 1D permeability measurement experiment.

Example 5.9: Radial Permeability Measurement Experiment

A radial flow experiment is performed to measure the permeability of a fabric preform. The preform is made of random fabric, thus it is expected to have isotropic permeability in in-plane directions (i.e., $K_{xx} = K_{yy} \equiv K$ and $K_{xy} = 0$). The experiment is performed using transparent mold lids, and circular (i.e., not elliptical) flow front is observed at any time which validates this isotropic assumption. The fabric preform has a fiber volume fraction of $V_f = 0.30$. The other parameters of the experiment are listed below:

- Resin viscosity, $\mu = 0.24\,\text{Pa.s}$
- Flow rate, $Q = 240\,\text{cc/min}$ (constant during the experiment)
- Mold thickness, $C_h = 0.005\,\text{m}$
- Central hole radius, $R = R_1 = 0.006\,\text{m}$
- Mold radius = specimen's outer radius, $r_{mold} = 0.40\,\text{m}$.

The following experimental data is collected using a pressure transducer at the inlet: Calculate the following: (1) K; (2) the time to fill the mold cavity (neglect the inlet and exit channels); (3) injection pressure at the end of mold filling, (4) re-solve parts (2) and (3) if $R = 0.012\,\text{m}$ (i.e., the central hole size is doubled).

t [s]	0	5	10	15	20	25	30
P_{inj} [kPa]	0	125	147	160	169	176	182

Solution

Equation (5.115) does not allow the use of regular least square method (and thus MATLAB's built-in function *polyfit*) as the relation between P_{inj} and t is not polynomial. We can use the "overdetermined system of equations" [277] after rewriting Equation (5.115) in matrix form for all the experimental data as follows:

$$
\begin{aligned}
\left\{\begin{array}{c}
\ln\left(1+\frac{Qt_1}{\pi h R_1^2(1-V_f)}\right)\\
\ln\left(1+\frac{Qt_2}{\pi h R_1^2(1-V_f)}\right)\\
\vdots\\
\ln\left(1+\frac{Qt_7}{\pi h R_1^2(1-V_f)}\right)
\end{array}\right\}\{X\} &= \left\{\begin{array}{c} P_{inj,1}\\ P_{inj,2}\\ \vdots \\ P_{inj,7}\end{array}\right\}\\
\{C\}\{X\} &= \{P_{inj}\}
\end{aligned}
\tag{5.122}
$$

where $X = (\mu Q)/(4\pi h K)$ is an unknown scalar to be determined. To minimize the error between the experimental and curve-fit data, $C^T C X = C^T P_{inj}$ should be solved as the "overdetermined system of equations" method suggests [277]. The simple MATLAB code (named as *example5_09.m*) is given in the Appendix, and its result is $K = 4.805 \times 10^{-10}\,\text{m}^2$. Experimental and curve-fit injection pressures are shown in Figure 5.28.

The empty volume of the mold cavity is calculated as

$$
\begin{aligned}
\text{Empty volume} &= [\pi(r_{mold}^2 - R_1^2)\,h](1-V_f)\\
&= \left[\pi\left((0.40\,\text{m})^2 - (0.006\,\text{m})^2\right)(0.005\,\text{m})\right](1-0.30)\\
&= 1759\times 10^{-6}\text{m}^3 = 1759\,\text{cc}
\end{aligned}
\tag{5.123}
$$

The time to fill the mold cavity is calculated as

$$
t_{fill} = \frac{\text{Empty volume}}{\text{Flow rate}} = \frac{1759\,\text{cc}}{4\,\text{cc/s}} = 439.7\,\text{s}
\tag{5.124}
$$

since the flow rate is held constant during the injection.

The injection pressure at this final time will be 267.1 kPa using Equation (5.115); $P_{inj}(t)$ is plotted in Figure 5.28 until the mold is filled completely using the calculated permeability, K in Equation (5.115). The same equation is used with $R_2 = 0.012\,\text{m}$, and its result is shown in the same figure. Observe the significant decrease in the injection pressure as the inlet radius has been changed from 6 mm to 12 mm. ■

5.8.1 Model Construction Based on Simple Geometries

Example 5.10: Radial Flow of a Newtonian Resin through a Preform

Consider radial flow of a Newtonian resin of viscosity μ in a preform of isotropic permeability as shown in Figure 5.26. The resin is made to flow through a plastic tube of radius R_i and length L and impregnate the preform under a vacuum. Calculate the movement of the flow front with time.

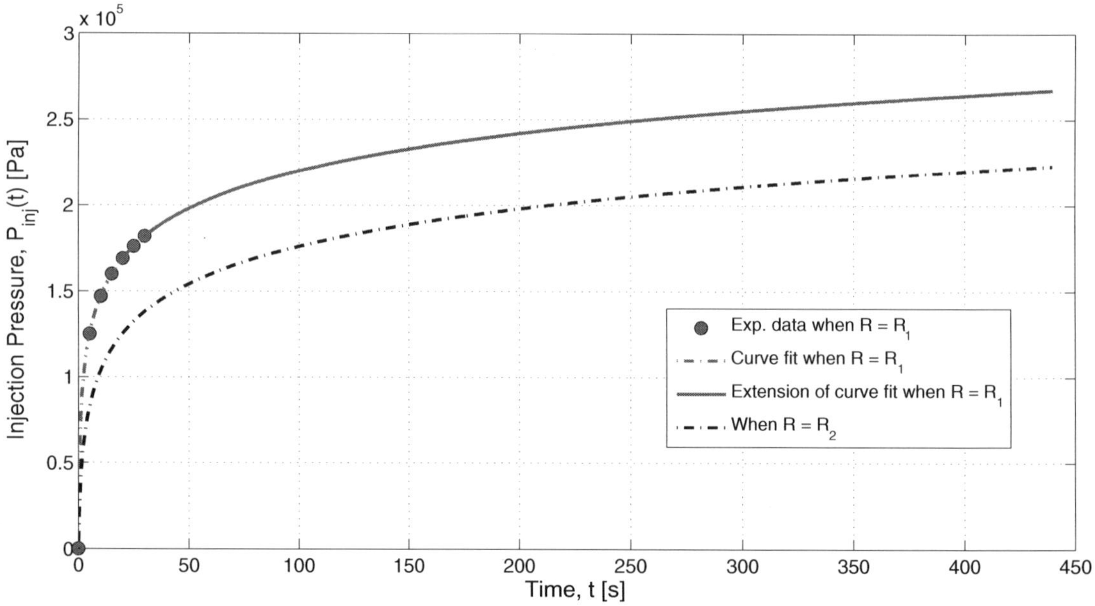

Figure 5.28: Injection pressure during a radial permeability measurement experiment.

Solution

From Figure 5.26, we can see that there are two simple geometry flows involved: flow through a tube and radial flow through the preform. At any time, the pressure difference between the resin container and the location of the flow front is constant and equal to $(P_{atm} - P_{vac})$. However, the pressure P_i at the inlet to the mold will vary with time, as the flow rate into the mold will vary with time. However the flow rate in the tube has to be equal to the flow rate inside the preform at all times. Also we will assume quasi-steady state for the flow. Hence, although $P_i \neq$ constant but $P_i(t)$ at any instant in time, we will assume that fully developed flow has been achieved at that time, t. Hence the flow rates in the two models will be as follows:

Model 1: Tube flow (see Figure 5.29)

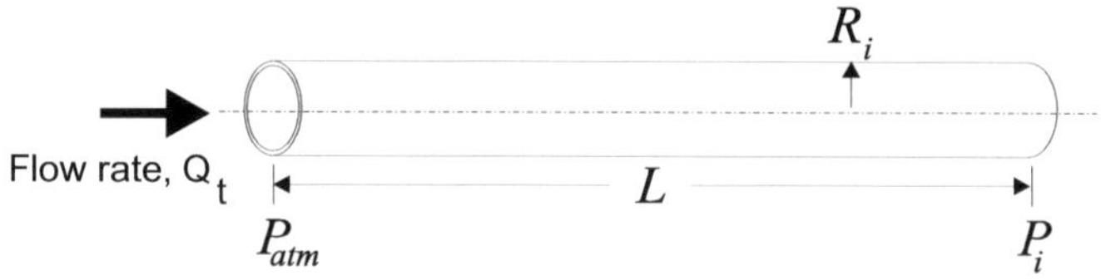

Figure 5.29: Tube flow.

$$Q_t = \frac{\pi R_i^4}{8\mu} \frac{P_{atm} - P_i}{L}. \tag{5.125}$$

Model 2: Radial flow (see Figure 5.30)

As shown in the figure, R_i is the inlet hole radius, $R(t)$ is the radius of the circular flow front at an instant of time, and r is the radial axis.

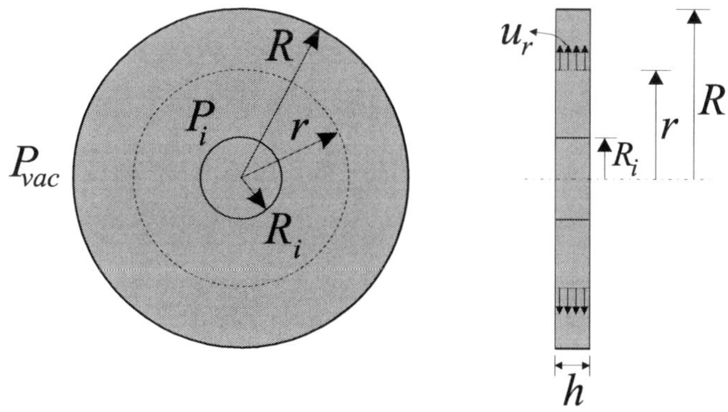

Figure 5.30: Radial flow through a preform.

$$\begin{aligned} Q_r &= \text{(Area of the flow front) (Volume averaged Darcy velocity)} \\ &= \text{(Area of the flow front) [(Velocity of the flow front) (porosity of the medium)]} \\ &= (A)[(\overline{u}_{r,f})(\phi)] \\ &= (2\pi Rh)\left[\left(\frac{dR}{dt}\right)(1-V_f)\right] \\ &= 2\pi h(1-V_f)R\frac{dR}{dt}. \end{aligned} \tag{5.126}$$

Although the flow rate, Q_r may vary with time, at any given time it should be the same for all r from $r = R_i$ to $r = R$ considering the conservation of mass (see Figure 5.30). Thus,

$$Q_r = (2\pi rh)\overline{u}_r. \tag{5.127}$$

From Darcy's law

$$\overline{u}_r = -\frac{K}{\mu}\frac{dP}{dr} \tag{5.128}$$

therefore

$$Q_r = -\frac{K}{\mu}\frac{dP}{dr}(2\pi rh). \tag{5.129}$$

Note that as Q_r is constant with respect to r, we can integrate Equation (5.129) with respect to r as follows:

$$\frac{Q_r\mu}{2\pi Kh}\int_{R_i}^{R}\frac{dr}{r} = -\int_{P_i}^{P_{vac}} dP. \tag{5.130}$$

Thus

$$Q_r = \frac{P_i - P_{vac}}{\ln(R/R_i)}\left(\frac{2\pi Kh}{\mu}\right). \tag{5.131}$$

Equating Equations (5.125) and (5.131), and assuming $P_{vac} = 0$, one can find P_i as a function of R, the location of the flow front

$$P_i = \frac{1}{1 + \dfrac{16KLh}{R_i^4 \ln(R/R_i)}} P_{atm}. \tag{5.132}$$

Equating Equation (5.125) to Equation (5.126) and substituting Equation (5.132) for P_i gives us a partial differential equation for R with respect to t:

$$2\pi R \frac{dR}{dt}(1 - V_f)h = \frac{\pi R_i^4}{8\mu L} \left[1 - \frac{1}{1 + \dfrac{16KLh}{R_i^4 \ln(R/R_i)}} \right] P_{atm}. \tag{5.133}$$

The initial condition is $R = R_i$ at $t = 0$, and one can apply a numerical method such as Euler method to determine R as a function of time. ■

Example 5.11: Radial Injection in VARTM Process

The schematic of the Vacuum Assisted Resin Transfer Molding (VARTM), which is also known as Vacuum Infusion (VI), is shown in Figure 5.31. The reader is referred to Chapter 8 for the details of a VARTM mold which contains one-sided mold part, fabric preform, peel-ply, resin distribution medium (a.k.a. flow mesh), vacuum bag, sealing tape and others. The resin is transferred from a tank to the mold through a tube. In Figure 5.31, the inlet tube is at the center of the circular mold, and the mold is vacuumed through its edge using a vacuum pump. The pressure of the resin is P_{atm} in the tank, and P_{vac} at the flow front. In the previous example, a mathematical model of this problem was developed, and a differential equation (Equation (5.133)) was obtained for the radius, $R(t)$ of the circular flow front. Notice that Equation (5.133) was obtained with the assumption of h = constant, which is not true in VI since the upper mold part is simply the vacuum bag; and it is decompacted as the resin pressure increases in the mold with time. For a more realistic model, the reader is referred to references [278, 279]. Here in this example, we will use Equation (5.133) and numerically solve $R(t)$ for the following typical VARTM parameters:

- Resin viscosity, $\mu = 0.1$ Pa.s
- Tube inner radius, $R_i = 0.006$ m
- Tube length, $L = 2.0$ m
- Mold thickness, $h = 0.004$ m
- Mold outer radius, $r_{mold} = 0.50$ m
- Fiber volume fraction of the fabric, $V_f = 0.35$
- Isotropic permeability of the fabric, $K = 6.0 \times 10^{-10}$ m^2
- Atmospheric pressure, $P_{atm} = 101$ kPa.

Plot $R(t)$ versus time. How will the result change if we use a different preform with the same fiber volume fraction but with a permeability of $K = 3.0 \times 10^{-10}$ m^2?

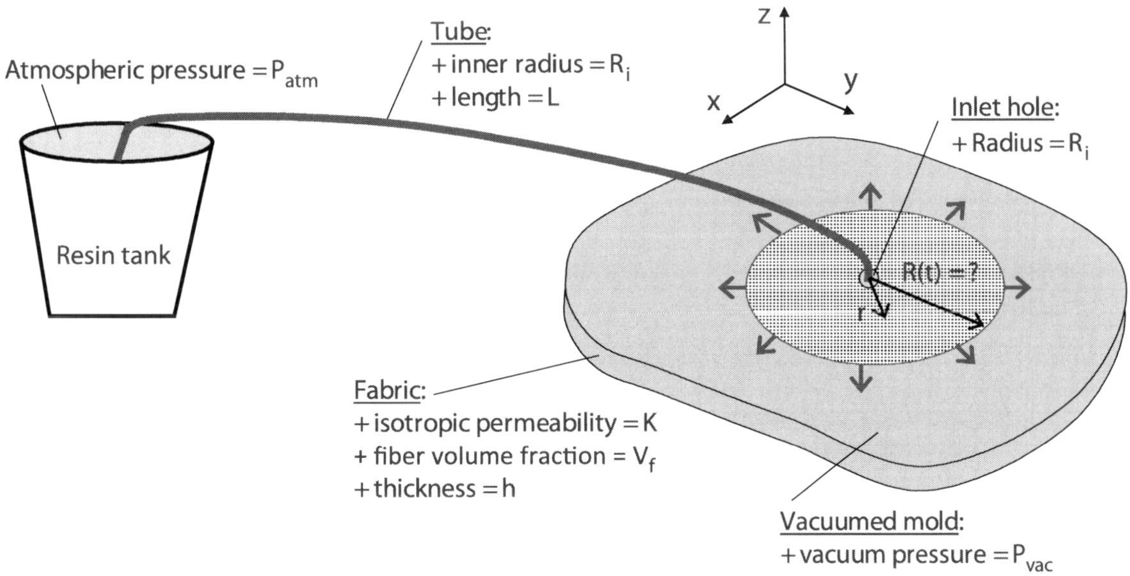

Figure 5.31: VARTM setup.

Solution

Equation (5.133) is reorganized to have a first order initial-value problem:

$$\text{O.D.E.:} \qquad \frac{dR}{dt} = \frac{R_i^4 P_{atm}}{16\mu L(1 - V_f)h} \frac{1}{R} \left[1 - \frac{1}{1 + \dfrac{16KLh}{R_i^4 \ln(R/R_i)}} \right] \tag{5.134}$$

$$\text{I.C.:} \qquad R(0) = R_i. \tag{5.135}$$

$R(t)$ will be calculated numerically using Euler method:

$$R_{j+1} = R_j + \frac{dR}{dt}\Delta t \tag{5.136}$$

where Δt is the time step which should be small enough for convergence of the numerical result. A MATLAB code (named *example5_11.m*) has been coded and given in the Appendix. The results are shown in Figure 5.32. The time to fill the mold cavity is 534 seconds. When the fabric permeability is one half of the original one, then the fill time is almost doubled: 1060 seconds.

5.9 Mathematical Tools for Simplification

Various mathematical tools can be used to simplify the governing equations and the boundary conditions, in search of a solution to the process model constructed. As discussed in Lee and Castro [271], we will present four different ways to simplify equations whenever possible.

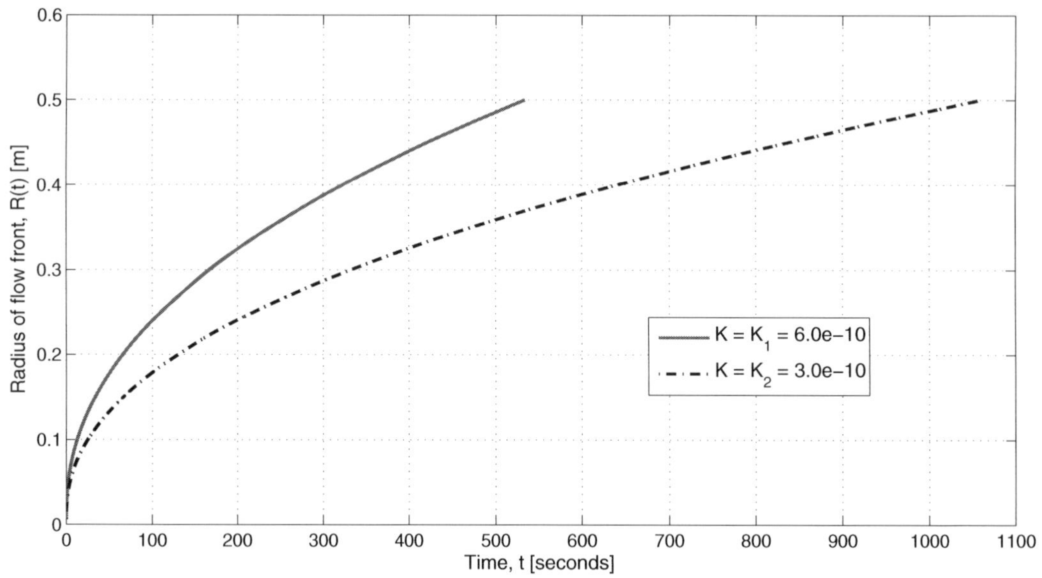

Figure 5.32: Radius of the circular resin flow front for Example 5.11.

5.9.1 Transformation of Coordinates

An example of how to transform a coordinate system will be studied below to demonstrate how one can simplify the calculations.

Example 5.12: Flow Through an Anisotropic Material

The flow of a liquid through a three-dimensional anisotropic material is shown in Figure 5.33. The three-dimensional Darcy's law is given by

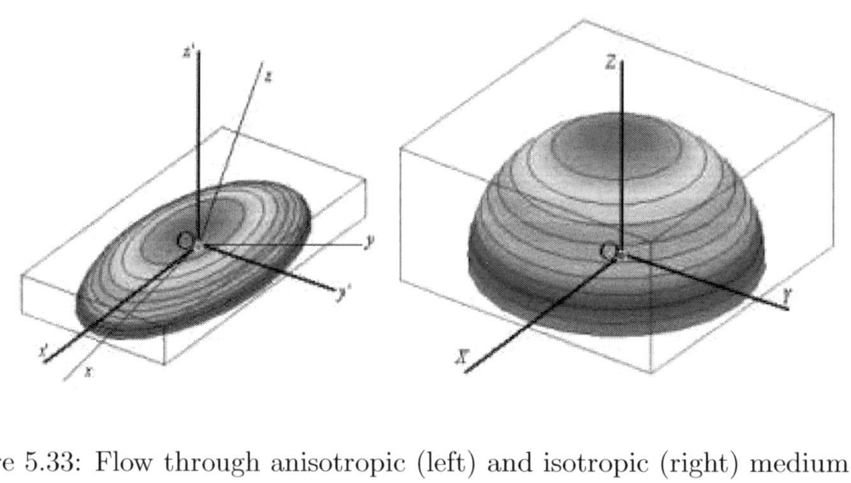

Figure 5.33: Flow through anisotropic (left) and isotropic (right) medium [255].

$$\begin{bmatrix} u_x \\ u_y \\ u_z \end{bmatrix} = -\frac{1}{\mu} \begin{bmatrix} k_{xx} & k_{xy} & k_{xz} \\ k_{yx} & k_{yy} & k_{yz} \\ k_{zx} & k_{zy} & k_{zz} \end{bmatrix} \begin{bmatrix} \dfrac{\partial p}{\partial x} \\ \dfrac{\partial p}{\partial y} \\ \dfrac{\partial p}{\partial z} \end{bmatrix} \tag{5.137}$$

The liquid is injected from a point source (denoted by O), and the shape of the flow front is that of the surface of an ellipsoid with major axes forming angles θ_{ij} with the axes of a general coordinate system x, y, z. The permeability tensor for this case is in the form used in Equation (5.137). After introducing a second coordinate system with axes x', y', z' coinciding with the major axes of the ellipsoid, the permeability tensor reduces to the form

$$\begin{bmatrix} K' \end{bmatrix} = \begin{bmatrix} k_1 & 0 & 0 \\ 0 & k_2 & 0 \\ 0 & 0 & k_3 \end{bmatrix} \tag{5.138}$$

where k_1, k_2, k_3 are the principal permeabilities of the fibrous material along the axes x', y', z', respectively. After the principal permeabilities k_1, k_2, k_3 are determined, the components of the permeability tensor $[K]$ in a general coordinate system can be obtained by the transformation:

$$[K] = C^{\mathrm{T}}[K']C \tag{5.139}$$

where C is the matrix of direction cosines of general coordinate axes x, y, z with respect to the principal axes x', y', z'. To convert the original anisotropic medium into an isotropic medium with reference permeability k, one can perform another transformation of coordinates to new independent variables X, Y, Z defined by the relations [94, 280]:

$$X = \left(\frac{k}{k_1}\right)^{1/2} x', \qquad Y = \left(\frac{k}{k_2}\right)^{1/2} y', \qquad Z = \left(\frac{k}{k_3}\right)^{1/2} z', \tag{5.140}$$

where k is a reference permeability. The volume element $dx'\,dy'\,dz'$ transforms into $[(k_1 k_2 k_3)^{1/2}/k^{3/2}]\;dX\,dY\,dZ$ and the reference permeability is selected such that

$$\frac{(k_1 k_2 k_3)^{1/2}}{k^{3/2}} = 1 \qquad \text{or} \qquad k = (k_1 k_2 k_3)^{1/3}. \tag{5.141}$$

The result of these transformations is that the general problem of point injection in an anisotropic medium is reduced to point injection in an isotropic medium with permeability $k = (k_1 k_2 k_3)^{1/3}$. The shape of the flow front in this case is that of a spherical surface (see the image on the right side of Figure 5.33). ■

Example 5.13: A Method to Determine 3D Permeability of Fibrous Reinforcements [255]

In a real experiment the injection is performed not from a point source but from a circular inlet with a finite diameter. The flow front transforms into an ellipse under the coordinate transformation mentioned in the previous example. Waitzenbock et al. [281] showed that for well-developed flow sufficiently far from the injection port, flow front shapes are the same for elliptical and circular inlet. This allows us to consider the flow far from the injection port and assume inlet original shape to be ellipse with major axes ratio $(k_1/k_2)^{1/2}$. After applying Equation (5.140), the shape of the inlet in the transformed isotropic domain is circular and the resulting setup is shown in Figure 5.34. Further analyses to determine the permeability tensor of the fabric is based on equating flow rates for: (1) injecting tube, (2) expanding hemisphere in isotropic medium, and (3) isotropic porous medium (applying Darcy's law):

$$Q = \frac{\pi b^4}{8\mu L}(P_0 - P_1) = \frac{\partial r}{\partial t} 2\pi r^2 \phi = \frac{k}{\mu}\frac{2\pi r b}{r - b} P_1 \tag{5.142}$$

where b: radius of the injecting tube; μ: viscosity of the liquid, considered constant; P_0: ambient pressure; P_1: pressure at the injection point; r: radius of the advancing spherical flow front; ϕ: porosity of the material; k: isotropic permeability, considered constant; L: length of injecting tube; and l: thickness of preform.

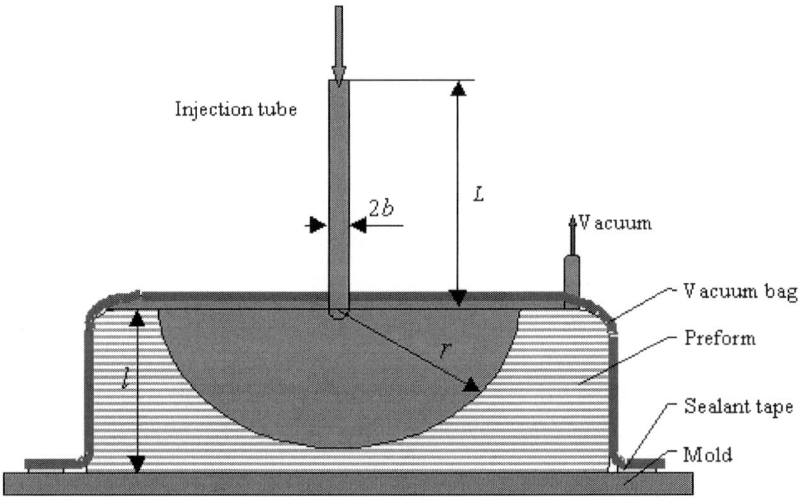

Figure 5.34: Expanding hemispherical flow in isotropic medium [255].

Equation (5.142) can be nondimensionalized by dividing all terms by $Q_C = \frac{\pi b^4 P_0}{8\mu L}$ and introducing the following notation: $\overline{Q} = \frac{Q_0}{Q_C}$, $\overline{t} = \frac{t}{t_C}$, $\overline{P} = \frac{P_1}{P_0}$, $\overline{r} = \frac{r}{l}$, $\overline{b} = \frac{b}{l}$, $t_C = \frac{16\mu L l^3 \phi}{b^4 P_0}$ and $C = \frac{16Lk}{b^3}$. The nondimensional form of (5.142) is:

$$1 - \overline{P} = \overline{r}^2 \frac{\partial \overline{r}}{\partial \overline{t}} = C \frac{\overline{r}\,\overline{P}}{(\overline{r} - \overline{b})}. \tag{5.143}$$

If the pressure P_0 is held constant, then Equation (5.143) can be solved to obtain the time $\overline{t}_{ff}$ required to fill a hemispherical region of radius r_{ff}

$$\int_0^{\overline{t}_{ff}} d\overline{t} = \int_{\overline{b}}^{\overline{r}_{ff}} \frac{\overline{r}(C\overline{r} + \overline{r} - \overline{b})}{C}\, d\overline{r}. \tag{5.144}$$

Performing the integration yields:

$$\bar{t}_{ff} = \frac{(1+C)(\bar{r}_{ff}^3 - \bar{b}^3)}{3C} - \frac{b(\bar{r}_{ff}^2 - b^2)}{2C}. \tag{5.145}$$

■

5.9.2 Superposition

Whenever the equations are linear, one can superimpose solutions. For example, when one has a pressure driven flow and drag flow occuring in a cylindrical geometry, one can add the solutions for the velocity profile to find the combined solution which is helical for this case.

Example 5.14: Pulling a Fiber Through a Concentric Cylinder

Find the force required to pull a fiber of diameter d from a concentric cylinder containing viscous polymer of diameter D as shown in Figure 5.35. The fiber is pulled with a velocity U_0 in the z direction. Assume $L \gg D$, d and that end effects are negligible. Only an insignificant amount of polymer coats the fiber. Assume the polymer to be Newtonian.

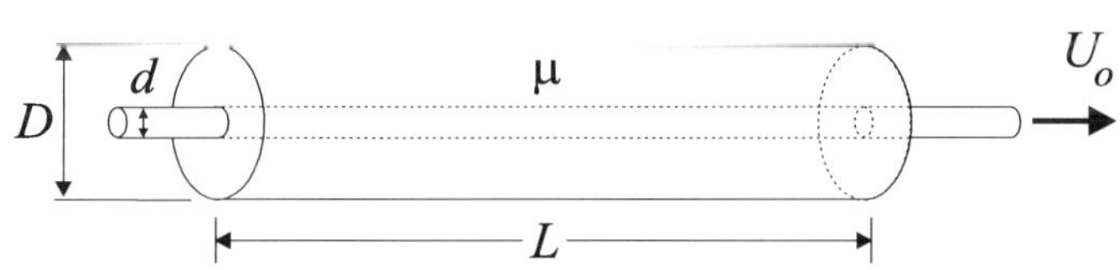

Figure 5.35: Fiber being pulled through a concentric cylinder.

Solution

The important physics to understand here is that we have a combination of drag-driven flow and pressure-driven flow. The pressure gradient is created because of flow being constrained in the z direction. If one neglects the end effects, the velocity profile, u_{zd} due to drag force will be

$$u_{zd} = U_0 \left[\frac{\ln(r/R_0)}{\ln\kappa}\right] \tag{5.146}$$

as shown in Figure 5.36. Here $\kappa = d/D$ and $R_0 = D/2$. Also Q_d, the flow rate due to drag flow will be

$$\begin{aligned} Q_d &= \int_{R_i}^{R_0} u_{zd}(r)(2\pi r\, dr) \\ &= \frac{2\pi U_0}{\ln\kappa}\int_{R_i}^{R_0} r \ln\left(\frac{r}{R_0}\right) dr \\ &= -2\pi R_0(R_0 - R_i)U_0\left[\frac{2\kappa^2\ln\kappa - \kappa^2 + 1}{4(1-\kappa)\ln\kappa}\right] \end{aligned} \tag{5.147}$$

where $R_i = d/2$. The velocity profile, u_{zp} and the flow rate Q_p due to the pressure-driven flow are given by

$$u_{zp} = \left[1 - \left(\frac{r}{R_0}\right)^2 + \frac{1-\kappa^2}{\ln(1/\kappa)}\ln(r/R_0)\right]\frac{R_0^2}{4\mu}\left(-\frac{\partial P}{\partial z}\right) \tag{5.148}$$

and

$$Q_p = \int_{R_i}^{R_0} u_{zp}(r)(2\pi r\, dr) = \left[1 - \kappa^4 - \frac{(1-\kappa^2)^2}{\ln(1/\kappa)}\right]\frac{\pi R_0^2}{8\mu}\left(-\frac{\partial P}{\partial z}\right). \tag{5.149}$$

The u_{zp} due to pressure-driven flow is shown in Figure 5.36. Hence, the velocity profile in the annulus is a superposition of drag flow and pressure-driven flow as shown in Figure 5.36.

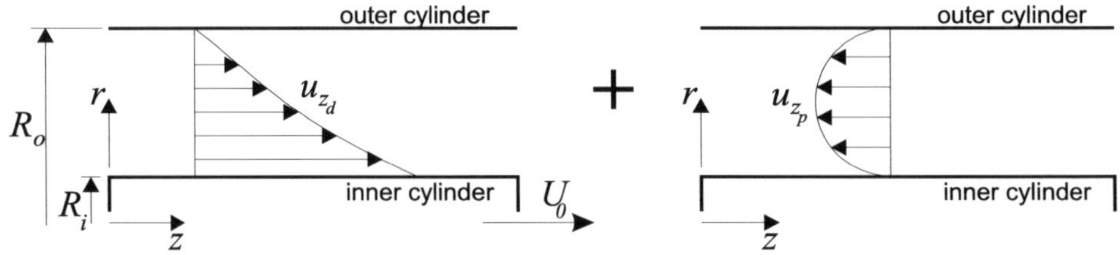

(a) drag flow (b) pressure-driven flow

Figure 5.36: Velocity profile due to drag- and pressure-driven flow.

Note that we do not know $-\frac{\partial P}{\partial z}$. However, we do know that total $Q = 0$. Thus, $Q = Q_d + Q_p = 0$. This allows one to solve for $-\frac{\partial P}{\partial z}$ and substitute in equation for u_{zp}. Hence, the final steady-state fully developed velocity profile will be

$$u_z = U_0\left[\frac{\ln(r/R_0)}{\ln\kappa}\right] + \left[1 - \left(\frac{r}{R_0}\right)^2 + \frac{1-\kappa^2}{\ln(1/\kappa)}\ln(r/R_0)\right]\frac{R_0^2}{4\mu}\left(-\frac{\partial P}{\partial z}\right) \tag{5.150}$$

To find the force, one can calculate the shear stress along the fiber surface (Figure 5.37):

$$F = (2\pi R_i L)\tau_{rz}|_{r=R_i}. \tag{5.151}$$

Figure 5.37: Shear stress along the fiber as it is pulled with velocity U_o from a concentric cylinder.

For Newtonian fluids

$$\tau_{rz} = \mu\frac{\partial u_z}{\partial r}. \tag{5.152}$$

Hence,

$$\begin{aligned} F &= (2\pi R_i L)\mu \frac{\partial u_z}{\partial r}\Big|_{r=R_i} \\ &= 2\pi L \left\{ \frac{\mu U_o}{\ln \kappa} + \frac{R_0^2}{4}\frac{\partial P}{\partial z}\left[2\left(\frac{R_i}{R_0}\right)^2 - \frac{1-\kappa^2}{\ln(1/\kappa)} \right] \right\}. \end{aligned} \tag{5.153}$$

Thus one can use the superposition principle to solve difficult problems more elegantly. ■

5.9.3 Decoupling of Equations

A detailed process model often will involve a coupled, nonlinear system of equations if there is more than one dependent variable. If the coupling between the equations and the dependent variables is strong, one has to solve all the variables simultaneously. This can be an effort and time intensive project and also difficult to prove that the mathematics will be stable under all conditions. However, if the equations are weakly coupled, one can solve them in succession.

For example, in nonisothermal models for injection molding, the momentum equations are coupled with the energy equation. The coupling will be strong if viscosity is a strong function of temperature (which makes the viscosity in the momentum equation a function of the temperature, the primary variable in the energy equation) and if there is a substantial amount of energy dissipation due to viscous heating (which makes the temperature depend on the viscosity and the shear stresses). The strength of the coupling is estimated by the Brinkman and Nahme-Griffith numbers (see Table 5.1). Large Brinkman and Nahme-Griffith numbers would imply that one must solve for momentum and energy equations simultaneously using an iterative or a direct solver. A small value for these numbers would signify that it is a justifiable approximation to solve the momentum equation first and then use the viscosity and the calculated shear stresses in the energy equation to solve for the temperature. Thus, the solution is decoupled. If the Brinkman number is small, it implies that one can solve the energy equation first (if there is not much convection) and then solve for the momentum equation.

Example 5.15: Nonisothermal Flow of a Viscous Suspension in a Channel

Consider nonisothermal flow of a viscous suspension in a channel (Figure 5.38). The conservation of mass, momentum and energy equations are in general coupled and nonlinear. The viscosity of the suspension is temperature dependent. Under what conditions can we decouple the equations and simplify the solutions for velocity, pressure and temperature? Assume flow is in the x direction only and $\eta = \eta_0 e^{-\xi(T-T_0)}$ where η_0 is the viscosity at reference temperature T_0.

Solution

There are four temperature scales here [93]:

- $\Delta T = (T_i - T_w)$, temperature difference between entering and wall temperature.
- The inverse of ξ, which has temperature units.

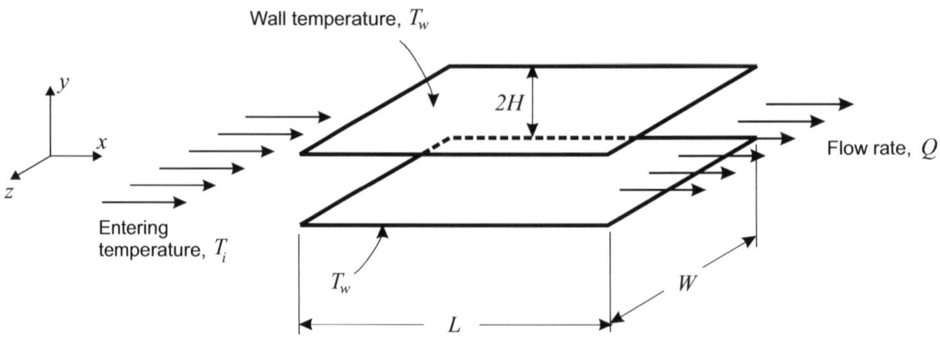

Figure 5.38: Nonisothermal flow between parallel plates.

- Temperature rise $\frac{\eta\left(\frac{Q}{WH}\right)^2}{k}$ due to the balance of dissipation and conduction across the thickness, H. Here k is the thermal conductivity of the suspension.
- Temperature rise $\frac{\eta QL}{\rho c_p W H^3}$ due to the balance of viscous dissipation and axial convection along the flow direction. Here ρ and c_p are the density and the heat capacity of the suspension, respectively.

Hence the three dimensionless numbers listed in Table 5.1 are important:

$$Gz = \frac{2\rho c_p QH}{kWL} \qquad Na = \frac{\eta Q^2 \xi}{4kW^2H^2} \qquad Br = \frac{\eta Q^2}{4kW^2H^2|\Delta T|}.$$

Now if Gz$\ll$ 1 and Na$\ll$ 1, this implies that transverse conduction dominates axial convection, so the temperature field is fully developed, and thus the entry temperature T_i is irrelevant and Br is of no consequence. Because Na$\ll$ 1, the effect of heat generation is too small to change the viscosity significantly. Hence, one can completely decouple the momentum and energy equations.

If Gz$\gg$ 1 and Na$\ll$ 1, the temperature field is underdeveloped and the transverse conduction is unimportant, one can assume adiabatic conditions and ignore T_i.

When Gz$\gg$ 1 and Na$\gg$ 1, the temperature field is underdeveloped and transverse conduction is important in a thin boundary layer by the wall.

When Gz$\ll$ 1 and Na$\gg$ 1, the temperature field is fully developed, but as Na$\gg$ 1, the heat generation does influence the viscosity; hence, the equations are coupled. ■

5.10 Solution Methods

After the governing equations and the corresponding boundary conditions are formulated and have been simplified to the satisfaction of the modeler, the next step is to select a solution method to predict how the dependent governing variables behave as a function of the independent variables. The solution method can range from a simple back-of-the-envelope calculation to very sophisticated numerical algorithms. The choice of the method will depend on

- The type of governing equation

- If it is an ordinary differential equation or a partial differential equation
- Whether the equation or the boundary conditions are linear or non-linear
- If the geometry is simple such as a pipe or a straight channel, or complex with inserts and ribs
- The number of variables that need to be solved, and if there is weak or strong coupling between them
- The nature and the stability of constitutive equations used.

One can broadly divide solution methods into two categories: closed-form solutions and numerical solutions. In most situations of practical interest, one will have to resort to a numerical solution, but an analytic or a closed-form solution should not be dismissed altogether. There are many situations in which the availability of a closed-form solution can be beneficial. We will first discuss the merits and situations for which one might want to seek a closed-form solution and then discuss briefly the merits and pitfalls of computer modeling with numerical methods.

5.10.1 Closed-Form Solutions

Analytic and closed-form solutions can be very useful in gaining insight into the physical processes occurring during composite manufacturing operations without unnecessary confusion caused by excessive geometric complexity or inclusion of detailed secondary influences in the equations. Simplified geometry models can reveal phenomena that can be easily missed by numerical solutions. Flow and heat transfer in a narrow gap provides a good example. Under certain combinations of the Graetz, Nahme-Griffith and Pearson numbers, the temperature and velocity profiles change rapidly inside the thin boundary layer near the wall [269]. A numerical solution will miss this layer unless the mesh or the grid is made extremely fine. Secondly, analytic solutions can be used as benchmarks for verification of sophisticated numerical solutions. Thirdly, closed-form solutions can allow the modeler to easily recognize the nondimensional parameters and the approach to scale the process from the laboratory to the production scale. Finally, analytic solutions are often easy to construct and hence inexpensive in terms of time and computer resources. This makes them attractive for use in design and optimization of the process and material parameters as one can carry out many solutions by varying the parameters quickly. Also, if one is interested in on-line control of the process, the availability of instant prediction of the variables for the next time step makes it possible to integrate the solution with the strategic or tactile controller of the process. This allows the controller to compare the predictive behavior with the actual behavior and make adjustments to the process parameters to drive it towards a successful outcome.

The analytic solution usually comes at the price of simplifying the geometry and/or the governing equations. A skillful modeler will make sure that the important physics is retained during these simplifications. It is usually easy to identify simplified geometric representations. For example, many molding operations involve flow through narrow confined channels [93]. Usually these channels will involve basic simple geometries. The six most common ones are listed below.

(a) Pipe or tube:

$$u_z = \left[1 - \left(\frac{r}{R}\right)^2\right] \frac{R^2}{4\mu} \left(-\frac{\partial P}{\partial z}\right) \tag{5.154}$$

$$\frac{\partial P}{\partial z} = \frac{\Delta P}{L} = \frac{P_e - P_i}{L} \tag{5.155}$$

$$Q = \frac{\pi R^4}{8\mu}\left(-\frac{\partial P}{\partial z}\right) \tag{5.156}$$

(b) Channel or parallel plates:

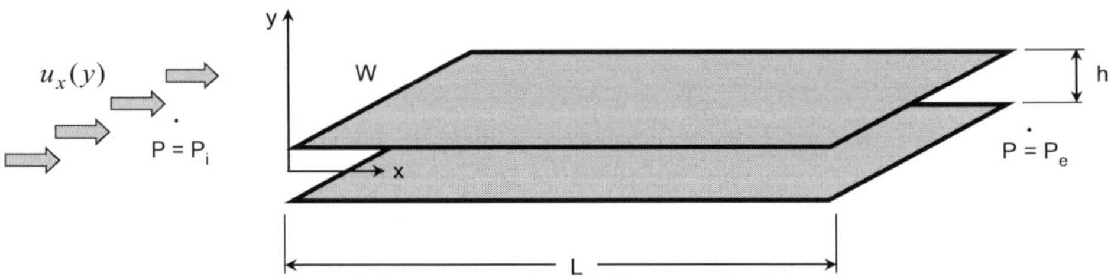

$$u_x = \left[1 - \left(\frac{2y}{h}\right)^2\right]\frac{h^2}{8\mu}\left(-\frac{\partial P}{\partial x}\right) \tag{5.157}$$

$$\frac{\partial P}{\partial x} = \frac{\Delta P}{L} = \frac{P_e - P_i}{L} \tag{5.158}$$

$$\frac{Q}{W} = \frac{h^3}{12\mu}\left(-\frac{\partial P}{\partial x}\right) \tag{5.159}$$

(c) Radial driven flow in a disk:

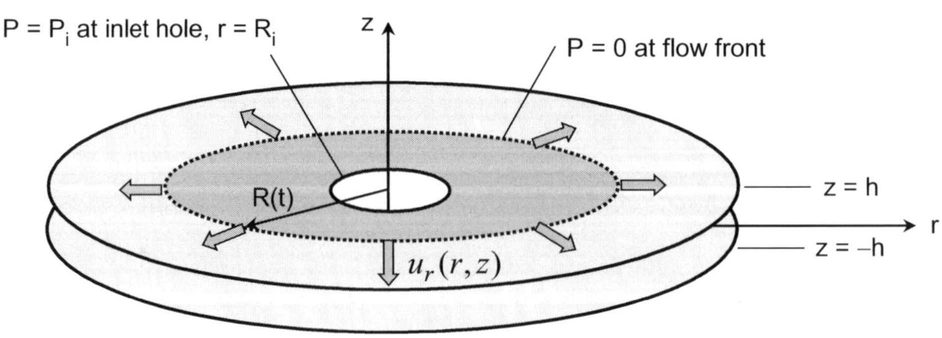

$$u_r = \frac{3Q}{8\pi h r}\left[1 - \left(\frac{z}{h}\right)^2\right] \tag{5.160}$$

$$P(r) = P_i + \frac{3\mu Q}{4\pi h^3} \ln\left(\frac{R_i}{r}\right) \tag{5.161}$$

$$Q(t) = \frac{4\pi h^3 P_i}{3\mu} \frac{1}{\ln[R(t)/R_i]}. \tag{5.162}$$

(d) Pressure driven flow in an annulus:

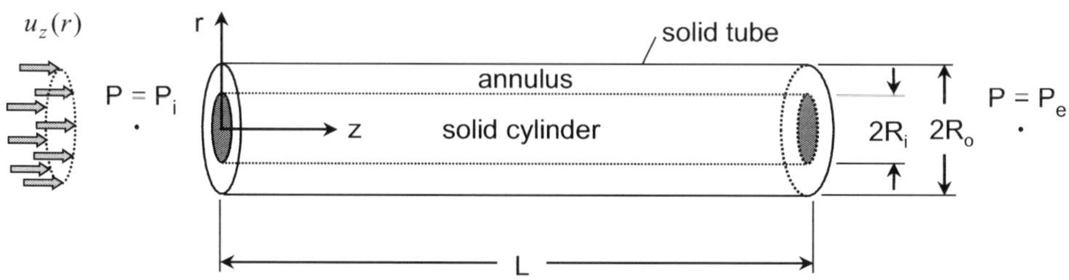

$$u_z = \left[1 - \left(\frac{r}{R_o}\right)^2 + \frac{1-\kappa^2}{\ln(1/\kappa)} \ln(r/R_o)\right] \frac{R_o^2}{4\mu} \left(-\frac{\partial P}{\partial z}\right) \tag{5.163}$$

$$Q = \left[1 - \kappa^4 - \frac{(1-\kappa^2)^2}{\ln(1/\kappa)}\right] \frac{\pi R_o^4}{8\mu} \left(-\frac{\partial P}{\partial z}\right) \tag{5.164}$$

where

$$\frac{\partial P}{\partial z} = \frac{\Delta P}{L} = \frac{P_e - P_i}{L} \tag{5.165}$$

$$\kappa = \frac{R_i}{R_o} \tag{5.166}$$

(e) Tangential annular drag flow between two concentric cylinders where the outer cylinder is rotating with angular velocity Ω:

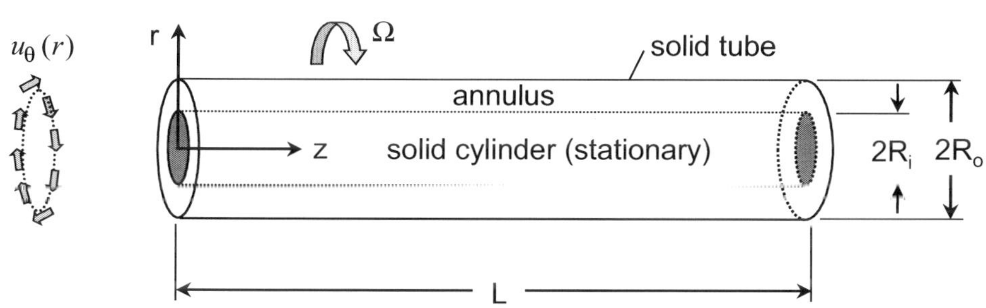

$$u_\theta = \Omega r \left[\frac{1 - \left(\frac{R_o}{r}\right)^2}{1 - \left(\frac{1}{\kappa}\right)^2}\right] \qquad \text{where} \quad \kappa = \frac{R_i}{R_o}. \tag{5.167}$$

(f) Axial annular drag flow between two concentric cylinder where the minor cylinder is being dragged with velocity u_0:

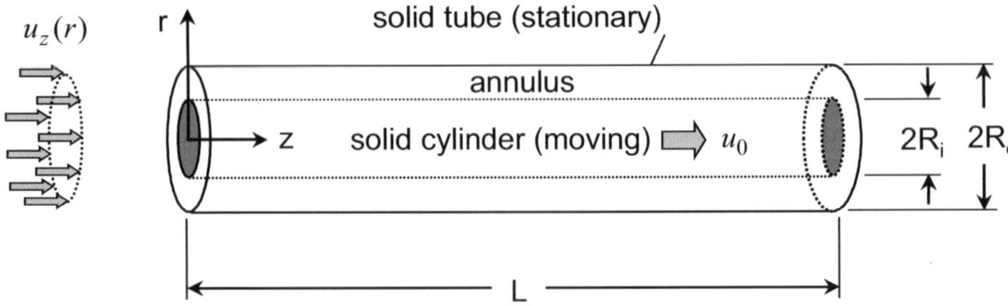

$$u_z = u_0 \left[\frac{\ln(r/R_o)}{\ln(\kappa)} \right] \tag{5.168}$$

$$Q = 2\pi R_o(R_o - R_i)u_o \left[\frac{2\kappa^2 \ln \kappa - \kappa^2 + 1}{4(1-\kappa)\ln \kappa} \right]. \tag{5.169}$$

The assumptions made are as follows:

- In (a): A long circular pipe of length L and radius R, where $R \ll L$.
- In (b): A long and wide rectangular channel of length L, width W and thickness h, where $h \ll L$ and $h \ll W$, which allows one to approximate the channel as two parallel flat plates.
- In (c): A concentric disk of inner radius R_i and outer radius R_o and height $h \ll (R_o - R_i)$, so that one can ignore inertia and elastic effects if present, and consider the radial flow as one-dimensional.
- In (d)-(f): A long, concentric annulus of length L, inner radius R_i and outer radius $R_o \ll L$. If the common additional assumption that $(R_o - R_i) \ll R_i$ is also made, the annulus can be unrolled into a channel of length L, width $\pi(R_o + R_i)$ and height of $(R_o - R_i)$.

One can increase the range of shapes that can be modeled if one of the dimensions varies slowly along one of the coordinate directions. For example, one can model a conical shape as a slowly varying pipe in the z direction (i.e., if $dR/dz \ll 1$), and a wedge shape can be represented by a slowly varying channel, etc. Such simplifications with obvious choice of the coordinate system can lead to closed form solutions as long as the governing equations and the boundary conditions are linear.

5.11 Numerical Methods

It is almost routine to see polymer composites being processed into complex shapes. It may be necessary to represent these shapes accurately to capture the important physics especially around inserts and corners during such composites manufacturing operations. This usually will require the governing equations to be solved numerically. Sometimes even the geometric simplification can be assumed, but the nonlinearity in the governing equations and boundary conditions will warrant a numerical solution. Finite difference, finite element, boundary element and spectral methods are some of the numerical methods that are used to obtain a solution. Each of these methods has its own particular strengths and weaknesses. We will not discuss the details of each method, but refer the reader to Chapters 6-8 of the computer modeling by Tucker [282] for background and demonstration of its applications. In this section, we will highlight some of the important characteristics of each method and state their strengths and weaknesses.

All numerical methods use discrete values at selected locations in space called nodes to represent the continuous solution of the field variables such as pressure, velocity and temperature over the geometry, or the domain of interest. Finite elements and spectral elements can represent virtually any geometry and there are many finite element and spectral element codes that use mesh generation modules to speed the development of the mesh, allowing the modeler to build the discrete representation of the domain. Mesh generators are very effective tools for two-dimensional domains, but building and verifying three-dimensional domains can still be time consuming. Boundary element methods can be very attractive for three-dimensional problems as in such methods only the boundary of the domain (which will be only a two-dimensional surface), and not the complete domain, needs to be discretized.

However, in the boundary element formulation, the differential equation must first be transformed into an integral equation over the domain boundary. The transformation is straightforward if one knows the kernel solution or Green's function for the differential equation. Without the kernel solution, there can be no boundary element solution. Usually, when there are nonlinearities in the material properties or the governing equations, either one does not have a kernel solution or one has to convert the nonlinearities into a body force term which requires some sort of discretization over the domain, eliminating the major advantage of boundary elements. While traditional finite difference methods can easily handle domains that are two-dimensional rectangles and three-dimensional boxes, they cannot model complicated shapes. Numerical grid generation and finite volume methods remove some of these obstacles, but are still far from being able to handle multiple connected domains. For example, finite difference may be able to handle a rectangular domain with one circular insert in it as it is a simply connected domain. But if it had two or more inserts, it will not be able to represent this with ease as finite element or spectral methods can. Hence, finite difference methods require formidable effort to model complex shapes, but they are very useful in developing solutions with new equations in simple geometries, as they take the least amount of time, and are very easy to implement.

Mesh refinement locally and globally is an issue that can be easily handled by finite elements and spectral methods. Finite element modeling discretizes the governing variable either linearly or with a second order polynomial over the element. So, by refining the mesh and making the elements smaller and smaller, one will be able to approach the continuous solution more accurately especially if the solution varies rapidly over the domain. Some problems, however, have regions where the field variables change rapidly combined with large regions where the field variables change slowly. Obtaining an accurate solution in such cases requires a fine mesh in the region where the variable is expected to change rapidly and a coarse mesh in the rest of the domain. Spectral methods usually use a coarse mesh but a very high polynomial to interpolate over the coarse element, and therefore can capture the rapidly varying variables with ease. Many composites processing models exhibit nonlinearity. Nonlinearity can enter governing equations through nonlinear terms in the equations such as a convective term in the momentum equation, through material properties that depend on field variables of the governing equations, such as temperature-dependent viscosity, through nonlinear constitutive equations or boundary conditions that may be nonlinear. Finite element methods are the most powerful methods for nonlinear problems, especially when solving the equations with the Newton-Raphson or one of its variant methods. The solution will usually converge rapidly, but requires additional programming to implement the method.

The overall recommendations once you have a model for the process at hand are as follows:

1. To solve a moving boundary problem, use the method most suited to handle moving boundaries.
2. For geometrically simple domains and to develop a quick solution for new equations, use finite difference methods.
3. There are many packages available that can solve general fluid flow and heat transfer problems, and one can modify the process model such that one can use the multi-purpose code to find the governing field variables. Some of these codes will allow you to input your own subroutines for viscosity and cure kinetics. Whenever possible, one should use this approach.

5.12 Validation

Almost all process models that are developed involve one or more assumptions, and many of them use constitutive or phenomenological equations to represent material behavior. Thus, it is important to do a reality check and see if the model predictions agree with experiments. Hence, there is a need for validation.

If a simulation of a model has been developed, it can provide many details such as the pressure and temperature distribution, fiber orientation state at every location, etc. However, one can only measure finite quantities from an experiment. Hence, the choices of which variables to measure and where to measure them become important considerations.

5.12.1 Various Approaches for Validation

There are many ways in which one can validate the model. One may choose to monitor the history of just one of the governing variables in the experiment, or all the field variables and compare them directly with the predictions. For example, one may monitor the temperature of the resin at the midpoint of the mold as a function of time in the injection molding process and compare it with the predictions from nonisothermal simulations. Another approach may be to indirectly compare experimental findings with predictions. For example, in curing of a thermoset composite, one may monitor the final state of the cure after the composite is demolded by performing DSC analysis on various samples that are cut from the fabricated composite, and compare the resulting cure distribution with the predictions of the model. So, although it does not validate the dynamics of the cure process, it allows one to compare the final cure state after demolding. A similar approach can be adopted for verification of the model for prediction of the cystallization state or fiber orientation state.

However, it is important to compare apples with apples when performing validation experiments. For example, let's consider the issue of validating the model for flow induced fiber orientation. To accomplish this, a simulation of injection mold filling is performed, and fiber orientation is predicted. The simulation is based on assumptions of no variations in fiber concentration and fiber length distribution and a constitutive rheological model. Next, the results are compared with the fiber orientation state measured from injection molded samples. This is not a good validation of the flow induced fiber orientation model as in the experiment there are many other parameters such as fountain flow, fiber length distribution, etc., that may have influenced the fiber orientation pattern which were not accounted for in the simulation. However, if one created a suspension of fibers in simple shear flow or other well-defined flows, and measured the orientation of the fibers in such flows, without complications introduced due to other assumptions, it will help validate the flow-induced fiber orientation model. Moreover, it will also elucidate fiber-fiber interactions, which could lead to refinement of the model.

Hence, the type of experiment is also an important factor in validation studies of the formulated process model. Process models usually consist of many phenomena and substeps with assumptions at every sublevel. If possible, one should try to validate every submodel independent of the overall model. For example, in filament winding, there are many submodels such as thermochemical, fiber motion, stress and void models. It will be useful to independently validate these models, instead of just comparing the final predicted mechanical properties with the measured properties. However, it is sometimes not feasible to validate every submodel, in which case, one may want to consider indirect validation of such submodels.

It is also crucial to validate the overall process model, in addition to validating individual submodels, as it sheds light on the accuracy to which one has modeled the interactions between the various submodels. One may validate every submodel of the process model and assume that the process model is validated, but there may be assumptions about the interactions between the submodels that may not have been verified. Thus, the validation of a process model is not complete until one also validates the final results with the predictions.

If the model predictions agree with the experiments, one has created a successful model that can be used for design, manufacturing and control situations instead of trial and error approaches. However, there will be many situations in which the predictions and the experimental results won't agree. In such cases, what should be the next step for the modeler?

The modeler should revisit his or her assumptions and make sure the validation experiments honor those assumptions. By this time, the modeler should have also verified that the simulation or the model has been verified with a closed-form solution or with mesh refinements. Next, the modeler should question the constitutive law that is being used to represent the phenomena. Sometimes, the modeler can learn from the experiments that will allow him or her to modify the law based on further physical insight. For example, if the model is using a power-law fluid to represent the material behavior, and the experiment involves very low shear rates, the experimental results and predictions won't match. This may be because the experiments involved very low shear rates, and we know that the power-law fluid model will not do a good job of capturing the physics. One may have to refine the model by using a Carreau fluid.

In the same way, we may have represented the fiber-fiber interaction with a constant parameter in a flow-induced fiber orientation model. However, the model experiments may show that the expected orientation behavior seems to be slightly off from the predictions. This may induce the modeler to think that maybe the interactions cannot be constant under all fiber orientation states. One would expect the fiber-fiber interactions to be higher when the orientation state is in complete disarray or random as compared to when all the fibers are aligned. Thus, the modeler can use this insight to improve the fiber orientation model.

Usually, models can be improved or refined at the expense of including more material parameters. For example, in the cases suggested above when we go from power-law to Carreau fluid, we need three parameters instead of two. Similarly, if we make the fiber interaction parameter dependent on orientation, we may need an additional equation that will allow us to determine this relationship. At other times, one may need to completely reformulate the constitutive law to accommodate the new phenomena being observed. For example, a random fiber preform clearly obeys Darcy's law as it has been validated many times. However, some of the stitched or tightly woven fabrics have two scales of pores: very small ones between fibers and large ones between fiber tows as shown in Figure 5.39. Thus, when the fluid impregnates these preforms, the pores between the fiber tows are filled at a much earlier time than the smaller pores between the fibers. This would require reformulation of the model as one has discovered that the saturation of the porous medium is not instantaneous as in random preforms.

Sometimes, the culprit may be the incorrect application of the boundary condition or an unknown boundary condition. For example, if one has a convective heat transfer boundary condition on the outside of the mold, one would have to specify a convective heat transfer coefficient which itself requires a constitutive law. One tip is to create validation experiments that do not use this boundary condition, but rather use specified temperature or heat flux.

Thus, this chapter has discussed tools and approaches that help build a process model and create a simulation. The next three chapters will focus on implementation of some of these tools in formulating models for polymer composite manufacturing processes.

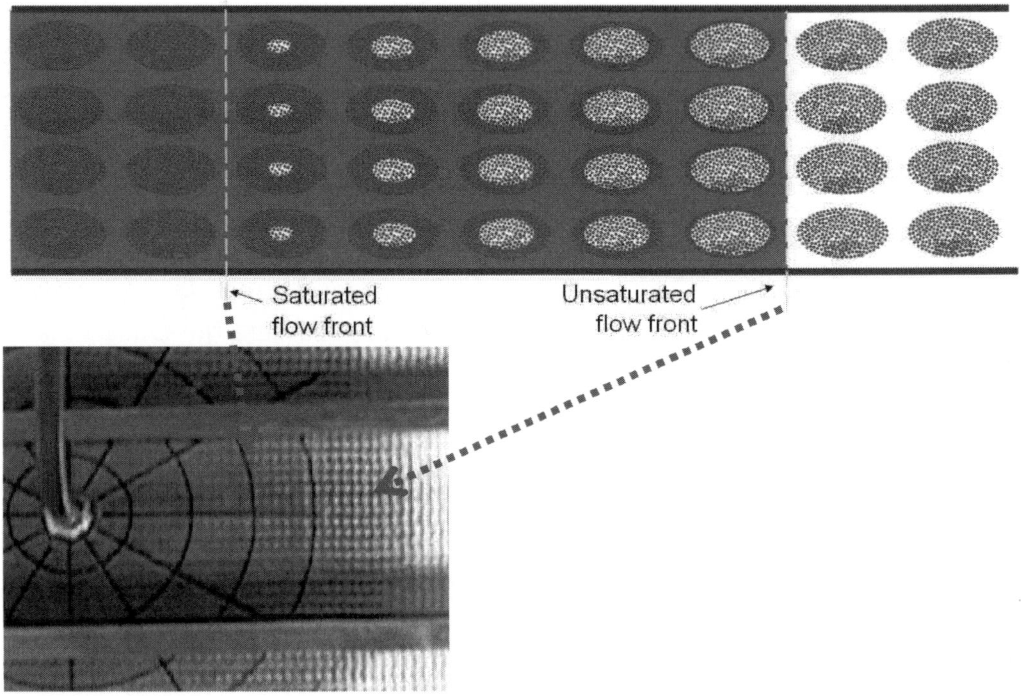

Figure 5.39: Schematic cross section of a woven fabric unit cell, demonstrating the dual length scale porous medium [283–285].

5.13 Exercises

5.13.1 Questions

1. What are the advantages of using a mathematical model and its analytical or numerical solution, compared to relying on a trial and error approach during composite manufacturing?

2. What are the main ingredients of a mathematical model of a manufacturing process?

3. What are the major steps in developing a model? In order to rely on the results of a model, what criteria should be satisfied?

4. Should models be accepted as representing the real process exactly?

5. Why do modelers usually make simplifying assumptions in their models instead of retaining all the physical details? How can a skilled modeler decide on which simplifications are acceptable?

6. Why do modelers need to verify and validate their models with controlled experiments? What is the difference between validation and verification?

7. What are the important considerations when comparing results from a model with the experiments?

8. Can a model be a good representation of a physical manufacturing process used in one production plant but not in a different plant? If so, do you need to adjust your model? How?

9. Is there a unique model for a particular process?

10. Does a higher number of parameters in the constitutive equation translate into greater accuracy of the model?

11. If there is more than one model available for a particular process, how would you determine which one to use?

12. Compare the advantages/disadvantages of sophisticated models without simplification that require numerical solution with simplified models with closed form solutions.

13. Explain what this means: "Virtual composite manufacturing scenarios based on models are very useful mold and process design tools." Give some examples.

14. Suppose that you are asked to design a mold for resin transfer molding (RTM) process for a given part geometry using a numerical mold filling simulation. (i) What material property data do you need in the model? (ii) Is the accuracy of that data important for your design? (iii) If you use this data from a database, is it necessary to question under what conditions the data was obtained?

15. For exactly the same part dimensions, can there be completely different mold and process designs? Explain this by considering the requirements of two different industries such as aerospace and automotive.

16. Suppose that two simulations are available for mold filling in the RTM, one is for 2D flow and the other one is 3D. What is the advantage of using a 2D model over 3D? Under what conditions should you use a 3D model?

17. When do you need to use constitutive equations for resin viscosity?

18. Considering the polymeric suspension flow between the extruder screw and barrel, answer the following three questions. (i) How can you model the geometry of the flow channel? (ii) What simplifications are possible in the conservation of momentum and energy equations? (iii) How does the velocity profile change if the resin material property a (that appears in $\mu = \mu_o e^{-a(T-T_{\rm ref})}$) and the temperature difference between the barrel and screw do change?

19. Why is dimensionless analysis very useful? How does one decide in choosing characteristic values? Can there be more than one characteristic length value in the same analysis? How does one decide what terms to drop in a governing differential equation?

20. What dimensionless numbers are commonly used in isothermal composite processing?

21. What are the major advantages/disadvantages of different numerical methods such as finite difference, finite element, finite volume and boundary element methods, to obtain solutions to mathematical models? How would you treat moving free boundaries such as free surface of a liquid composite material impregnating a mold containing fiber preforms?

22. Under what conditions (in terms of values of some dimensionless numbers), is quasi-steady state assumption acceptable?

23. If a quasi-steady state assumption is used in a model of the mold filling process, how would you solve the resin flow front location as a function of time?

24. What does "fully developed velocity" mean?

25. If the flow of a polymer through a circular tube has a Reynolds number of 1, what is the approximate length of the velocity developing region (in terms of tube diameter)? What is the approximate length of the thermal developing region (in terms of tube diameter) if the Prandtl number is 1000?

26. What do the Reynolds, Prandtl, Brinkman, Froude and Graetz numbers represent physically?

27. Under what conditions can the lubrication approximation be applicable?

28. Under what condition can one ignore the flow in the thickness direction within the thin shell approximation? Do you compare only the ratio of the dimensions in the thickness and in-plane directions, or are there additional considerations?

29. What are the boundary conditions, in mathematical form, if the liquid composite material has a contact with a solid surface with no/partial/full slip? Explain your answer in terms of the tangential and normal components of velocity.

30. If the liquid composite material has contact with another fluid, what is the corresponding boundary condition along the interface?

31. What boundary condition is used on the free surface as the liquid composite material (fiber suspension) propagates within a mold? Does your answer depend on whether the free surface is directly connected to a ventilation tube to outside air? For example, if the flow front traps some air not directly connected to a vent, what boundary condition can be applied on that free surface?

32. What different boundary conditions can be used at the injection ports?

33. When are periodic boundary conditions assumed? Give examples.

34. Explain what these thermal boundary conditions physically correspond to (i) fixed $T = T_w$, (ii) fixed normal gradient $\partial T/\partial n = c_1$, and (iii) difference of temperatures $T - T_{\text{surroundings}} = c_2$.

35. What are the mathematical tools to simplify governing differential equations and boundary conditions? Explain briefly.

5.13.2 Problems

1. Consider a large polymeric composite plate with a thickness of 2cm. Suppose that one would like to cool it from 300°C down to 70°C. The top and bottom surfaces of the composite are kept in perfect thermal contact with an aluminum tool at $T_w = 25$°C. Considering typical composite material data, how long should the part be kept in the mold?

2. Considering flow through a circular tube, what dimensions and properties of the problem do you need in order to calculate the length of the entrance region? When is this length significant?

3. For what values of Graetz number, Gz, are velocity and temperature profiles assumed to be developed in the direction of flow in a tube? Explain.

4. In flow modeling, a thin shell approximation is often used in order to simplify and speed up the computation. What are the criteria for this approximation to be valid? If the thickness of the composite part is much less than the in-plane dimensions, can you always use this approximation? Explain.

5. Consider Example 5.14. If there are two setups in which $\kappa_1 = d_1/D = 0.7$ and $\kappa_2 = d_1/D = 0.8$, find the ratio of required forces to pull the two cylinders. Leave your result in terms of μ, U_o, D and $\partial P/\partial z$ only.

6. What simplified flow problems can be useful for Newtonian fluids while studying flow of liquid composite material through narrow channels? What are the corresponding closed-form solutions for velocity profiles?

Chapter 6

Short Fiber Composites

6.1 Introduction

The polymer processing industry creates at least 30% of the objects that we see around us. There is a good chance that these objects were fabricated using one of the following three manufacturing processes: extrusion, injection molding or compression molding. They are all high-volume production processes. Extrusion is a continuous process, injection molding is an automated process and compression molding is a semiautomated process but can manufacture large and complex parts that cannot be easily accomplished by injection molding or extrusion. Most of the components fabricated are for nonstructural applications as the polymer structure can withstand only a limited amount of load and stress.

These methods were also being used to process food which had solid particles embedded in a viscous fluid. Hence, it was natural to extend these manufacturing methods to process fiber suspensions as well. Fiber suspensions refer to chopped fibers suspended in a polymer melt. Compression molding involved placing the material containing fibers and resin in a solid form called the *charge* inside the mold cavity. Usually, the heat conduction mode is employed to heat the charge by direct contact with the mold platen to melt the polymer and create a suspension of fibers in the polymer melt. This suspension is compressed by the platen to occupy the mold and take the form of the mold cavity. After solidification, the part is removed from the mold. Unlike compression molding, which uses a mold platen to melt and displace the material, the extrusion and injection molding process use a screw to melt and move the material. The function of the screw in both these processes is similar to some extent during preparation of the suspension. Examples of screws used are shown in Figure 6.1.

The screw by its rotation can generate shear flow which causes viscous dissipation that melts the polymer and mixes the fibers with the polymer melt at the same time. The screw also acts as a pump that can push the polymer melt suspension into a die during the extrusion process, or into a mold during the injection molding process. However, modifications need to be made to the screw material and design to ensure that it can withstand the abrasive environment created by the fibers. The channel depth of the screw is also usually increased to avoid excessive fiber length attrition, as fiber length plays a crucial role in load transfer.

The goal of the screw is to produce a well-mixed suspension suitable for forming operations. In extrusion, the suspension is pumped through an opening into a die to form a continuous product, whereas in injection molding the suspension enters a mold through one or more inlet gates and takes the shape of the mold. In extrusion, one does not need

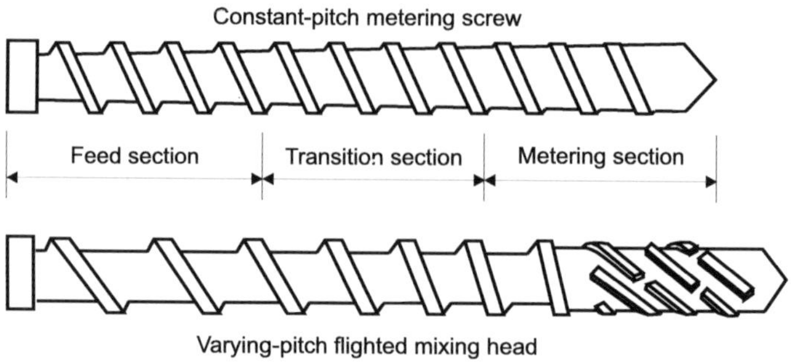

Figure 6.1: Examples of screws used in extrusion and injection molding for mixing and pumping the fiber suspension into a die or a mold [132].

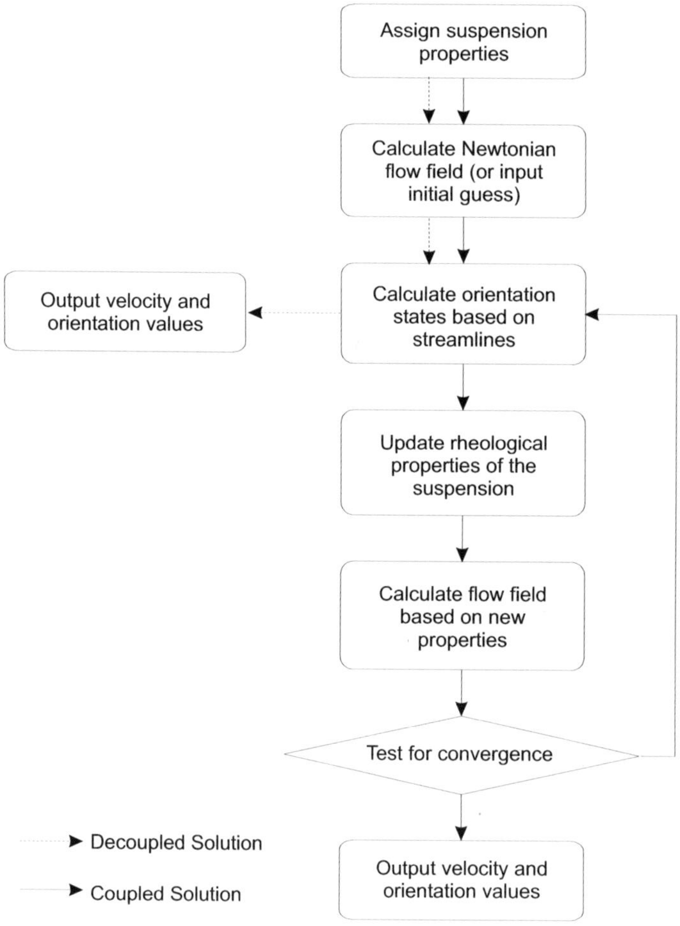

Figure 6.2: Decoupled and coupled solution approaches to describe the flow and fiber orientation in a fiber suspension [286, 287].

vents to let the displaced air out, whereas in the injection mold one must provide vents for the displaced air to exit. On the other hand, compression molding does not involve gates but requires the material that will fill the mold cavity to be preplaced in the mold and squeezed until the top and bottom mold platen meet. During this process, there is much less attrition of the fibers as the material is not subjected to large-scale deformations.

All the above processes involve flow of fibers along with the polymer melt to form the shape of the die or the mold. It is important to know how these suspensions flow into the mold or the die as the final properties of the product are determined by the flow history and the orientation state of the fibers when the polymer melt solidifies.

The solution methodology to calculate the flow and fiber orientation states is shown in Figure 6.2. There are two approaches: (i) decoupled and (ii) coupled. In the decoupled approach, the flow field is assumed to be unaffected by the presence of fibers.

In this chapter, we will first present the modeling of flow and heat transfer in the compression molding process. Next, we will briefly discuss the screw design in the extrusion process and present an analysis to calculate the pressure drop and flow rate relationship due to the presence of the die. Finally, in injection molding, we will focus on modeling the filling stage in the mold.

6.2 Compression Molding

Compression molding is one of the oldest technologies for processing polymers and hence was a natural candidate to manufacture near-net shaped, thin-walled composite parts at relatively low costs [288, 289].

6.2.1 Basic Processing Steps [1]

While there may be slight variations in this process, the basic steps can be defined as follows. First, during the material preparation stage, premeasured raw material, or "charge," is placed inside a preheated matched-die mold. The raw material can be either sheet molding compound (SMC), bulk molding compound (BMC), long fiber-reinforced thermoplastic sheets (LFTs) or glass-mat thermoplastics (GMTs). Polypropylene is usually the resin used in thermoplastic systems such as LFT and GMT [290–292]. SMC has been the widely used material because of its versatility [289, 293, 294]. SMC, which has been used since the early 1950s [290] is the generic name of a broad class of thermosetting resins reinforced by randomly oriented chopped fibers. Uncured SMC sheets are typically 4–6 mm thick with fibers 25 mm long. These are fiber bundles consisting of around 200 filaments each. Fiber volume fraction is 20-50%, while resin makes up 30% of the volume. The remainder of the bulk is comprised of mineral fillers such as calcium carbonate which are used to thicken the resin paste [26, 289]. Resin matrices used include unsaturated polyester, vinyl esters, epoxy, and phenolics [26, 289, 291, 293, 295]. Unlike SMCs which are in the form of sheets, BMCs are available as a gunk or soft log with typically low fiber contents (approximately 20%) along with a particulate mineral filler. There is a variation that uses chopped resin-coated fibers containing only the fiber reinforcement and no fillers [289]. Most of these material systems are reinforced with glass fibers. Limited compression molding is done with carbon-BMI and epoxy prepregs, and with carbon-epoxy BMCs and SMCs.

Once the *charge* is placed inside the mold cavity, then the two halves of the mold are closed to heat the resin paste and pressure is applied through a hydraulic press, causing the resin to flow and be squeezed into the shape of the mold cavity. The mold platens move towards each other until they meet. After the mold has been filled, the part is allowed to cure before it is cooled, then removed from the mold and possibly trimmed to remove flash [26, 289, 293].

There are slight variations in compression molding of thermoplastics. The obvious one is that no curing is necessary. Also, as thermoplastics are more viscous than thermosets, one usually requires higher pressures to make the material flow into the mold cavity. Thermoplastic composites are also used in compression molding operations such as stamping and forming where the flow is usually insignificant. In stamping, or rapid compression molding, the charge covers 95-98% of the mold cavity; hence, the flow is insignificant and the fiber orientation within the material is retained. Stretch compression molding, wherein the heated sheet is clamped in a frame and forced to stretch as it changes shape, is another variation of compression molding of thermoplastic composites [296].

Mold temperatures range between 135 and 160°C for thermoset materials [289, 293], whereas it has to be above the material's melt temperature if thermoplastics are used. Heat is usually supplied by pumping hot oil or by condensing saturated steam in heating lines drilled through the mold platens [293]. Press speeds are typically 5–10 mm/s while the applied pressure is in the range of 3.5–15 MPa [289, 293, 296]. Cure usually takes only 30–40 seconds and the entire cycle time can range from under a minute to 6 minutes, depending on the degree of automation incorporated into the process [289, 290, 293].

A recent variation of the compression molding process is the injection-compression molding process which combines conventional injection molding and compression molding to incorporate the advantages of both manufacturing processes. The resin is introduced into the cavity with the mold platen separated by a larger gap than the thickness of the part. Hence, the pressure requirement for injection is low. After injection is completed, the suspension is compressed to close the gap to the desired thickness of the part. The advantage of this process is that it can be done under low molding pressures which reduces residual stresses, sink marks, warpage, and density variations, and improves dimensional accuracy. However, the process has not been popular as the control required in processing is more complex and would require substantial investments in equipment to automate the process [297].

6.2.2 Applications [1]

Compression-molded parts are usually used in the automotive industries where high-volume production of moderately strong, stiff, corrosion resistant, lightweight parts with good surface finish and short processing times are needed [293, 294]. The process is particularly suitable for handling fiber-reinforced materials with high reinforcement loadings as longer fibers can be used in compression molding than in injection molding [295, 298]. Automotive body panels have been the primary application for compression molded components [26, 291, 293, 295]. For example, the front end of the Ford Galaxy and VW Sharan are compression molded from SMC while the Passat's is made from LFT. VW's Golf II has a GMT inner front end [290]. Pickup truck boxes and certain components of the tailgate are made from vinyl ester SMC. Several automatic transmission components are compression-molded from phenolic materials while compression-molded epoxy SMC is used in other more stressed components. Large floor components for a station wagon are also compression-molded from polypropylene GMT [26, 291, 293, 295]. The first structural application in automobiles was a leaf spring in 1979 made using continuous unidirectional glass fibers. The latest automotive use is for a composite wheel [289].

In aerospace applications, compression-molded components are usually used only in secondary structural or nonstructural parts as short fibers and low volume fractions do not meet the mechanical property requirements for primary structures. In commercial aircrafts, a number of covers, fairings, panels and housings are compression-molded from epoxy SMC to replace aluminum parts. Wings, control surfaces, bulkheads and nose cones are made from high content epoxy. Phenolic SMC and some thermoplastic materials with their superior high temperature and smoke generation properties are currently being evaluated for future aircraft applications [289].

Consumer industries also utilize compression-molded short fiber polymer composite components. For example, several office copiers utilize compression-molded end caps on the copier fusor roll. Embossing dies and counter dies are compression molded from phenolic rubber mats and from epoxy SMC. Even major elements of a centrifugal pump are compression molded from glass-reinforced epoxy or vinyl ester SMC [289]. Lamp housings have been reported to be compression molded as well [290].

6.2.3 Flow Modeling

Compression molding involves a modest amount of flow. Although the amount of flow may be small, it is critical to the quality of the part. The flow will control the orientation of the fibers, will influence the heat transfer and hence the curing of the composite, and will also influence the material properties if weld (knit) lines are formed (where flow fronts meet). It is beneficial to model the mold filling and curing behavior because, in addition to understanding the flow and cure during the operation, it can provide us with useful information such as knit line locations, velocity, pressure and temperature distributions, and the curing history.

The models developed for flow and heat transfer fall under two categories considering the thickness of the composite part: (i) thin parts and (ii) thick parts. Parts are classified as thin if their thickness is at least one to two orders of magnitude smaller than their in-plane dimensions as shown in Figure 6.3.

Thin parts comprise a large percentage of all compression-molded parts, since they can be cooled and cured quickly and meet the requirements of low cycle times. Thick parts may be necessary for some applications where cycle times may not be the driving factor, but the modeling approach followed is slightly different. We will first discuss thin cavity models.

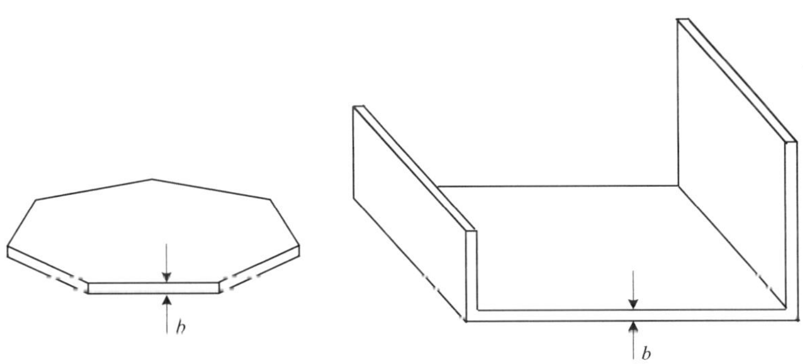

Figure 6.3: Two typical parts of thickness b, which is much smaller than the in-plane dimensions.

6.2.4 Thin Cavity Models

The first step in modeling a thin part is to use the concept of the lay-flat approximation. For example, if the part is a three-dimensional box, one can mathematically and conceptually "unfold" it in the x-y plane with the z direction being its thickness as shown in Figure 6.4. This approximation allows one to model any thin three-dimensional shape into an equivalent flat part, although the domain in the plane may be very complex.

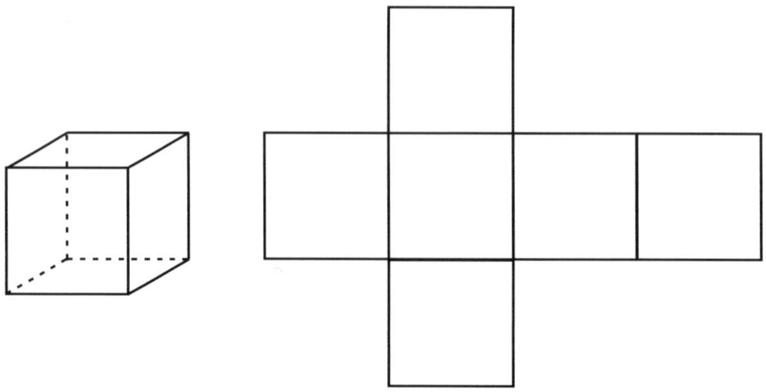

Figure 6.4: (a) A three-dimensional box, and (b) its mathematically and conceptually "unfolded" version.

The next step is to mathematically describe the material behavior to formulate the model. The simplest approximation is to assume the material to be Newtonian and isotropic. However, if the physics and experiments suggest something different, one could refine the material behavior by including nonlinearities. It is always useful to address the isothermal case first, as one can understand the important parameters if the complexities are kept to a minimum. For the moment, one can assume the viscosity to be a function of x, y and z. With these assumptions, one can now delve into the modeling.

A sketch of a typical compression molding flow is shown in Figure 6.5 with the coordinate system appropriately attached and some of the parameters such as the in-plane dimension L, height h and the closing speed s identified. For thin parts, $L \gg h$. As shown in Figure 6.5, the flow in the x-y directions is due to the squeezing motion of the upper half of the mold, which moves downward with a prescribed speed of s which is the same as $-\dot{h}$ or $-dh/dt$. The instantaneous thickness of the part is $h(x, y)$ and the velocity components of the material are denoted by u, v and w in the x, y and z directions, respectively.

Next, one can carry out an order of magnitude analysis to evaluate the magnitude of velocities in the x, y and z directions. The mass conservation equation for incompressible fluids may be stated as

$$\frac{\partial u}{\partial x} + \frac{\partial v}{\partial y} + \frac{\partial w}{\partial z} = 0. \tag{6.1}$$

Using this as the guide, and the characteristic dimensions in the in-plane directions (x and y) and thickness direction (z) as L and h, respectively, we find that

$$\frac{\partial u}{\partial x} \sim \frac{U_c}{L}, \qquad \frac{\partial v}{\partial y} \sim \frac{U_c}{L}, \qquad \text{and} \qquad \frac{\partial w}{\partial z} \sim \frac{s}{h} \tag{6.2}$$

where "$\sim$" denotes "of the order." We assume that the characteristic velocity, U_c, in the x and y directions is of the same order and the dimensions in the in-plane directions are of the same magnitude. Thus, substituting this order in the mass conservation equation, we

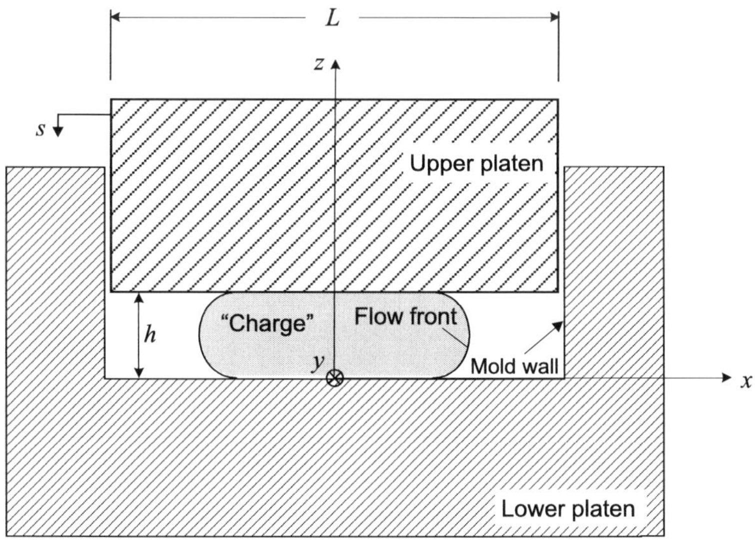

Figure 6.5: Coordinate system and nomenclature for compression mold filling analysis.

can show that

$$w \sim s, \qquad \text{and} \qquad U_c \text{ (in-plane velocities, } u \text{ and } v) \sim s\,\frac{L}{h}. \tag{6.3}$$

These relations will hold everywhere except very close to the edge of the flow domain and if $L \gg h$. Most of the flow will be planar and the velocity in the thickness direction w can be neglected. However, the velocities u and v are still functions of the z direction. Thus, to eliminate the z direction dependence, one can integrate the continuity equation in the z direction

$$\int_0^{h(x,y)} \frac{\partial u(x,y,z)}{\partial x}\,dz + \int_0^{h(x,y)} \frac{\partial v(x,y,z)}{\partial y}\,dz + \int_0^{h(x,y)} \frac{\partial w(x,y,z)}{\partial z}\,dz = 0 \tag{6.4}$$

which can be written as

$$\frac{\partial(\bar{u}h)}{\partial x} + \frac{\partial(\bar{v}h)}{\partial y} = -w\Big|_{z=0}^{z=h} = s \tag{6.5}$$

because $w = 0$ at $z = 0$ and $w = -s$ at $z = h$. Here $\bar{u}$ and $\bar{v}$ are the velocity components in the x and y directions averaged across the part thickness. Mathematically they can be written as

$$\bar{u}(x,y) = \frac{1}{h(x,y)} \int_0^{h(x,y)} u(x,y,z)\,dz \tag{6.6a}$$

$$\bar{v}(x,y) = \frac{1}{h(x,y)} \int_0^{h(x,y)} v(x,y,z)\,dz. \tag{6.6b}$$

Equation (6.5) is the continuity equation for a compression molding situation in which the lower mold platen is stationary and the top platen is moving down at the speed of s when in contact with the charge. Thus, this equation must be satisfied everywhere in the charge domain.

Next, we turn to equations of motion, which if we assume that the inertia and body forces are negligible, reduce to

$$\frac{\partial P}{\partial x} = \frac{\partial \tau_{xx}}{\partial x} + \frac{\partial \tau_{yx}}{\partial y} + \frac{\partial \tau_{zx}}{\partial z} \tag{6.7a}$$

$$\frac{\partial P}{\partial y} = \frac{\partial \tau_{xy}}{\partial x} + \frac{\partial \tau_{yy}}{\partial y} + \frac{\partial \tau_{zy}}{\partial z} \tag{6.7b}$$

$$\frac{\partial P}{\partial z} = \frac{\partial \tau_{xz}}{\partial x} + \frac{\partial \tau_{yz}}{\partial y} + \frac{\partial \tau_{zz}}{\partial z}. \tag{6.7c}$$

For thin parts, if one assumes no-slip of the material at the mold walls, the shear stresses (τ_{zx} and τ_{zy}) will dominate and one can adopt a lubrication approximation discussed in Chapter 5 to simplify the analysis. This results in the Hele-Shaw flow model. However, if one assumes a slip boundary condition at the mold walls, then the dominant stresses will be the in-plane stresses τ_{xx}, τ_{xy} and τ_{yy}. This leads to a lubricated squeezing flow model. These two cases represent limiting cases of thin compression molding. One could also introduce partial slip and obtain a more general solution, but first we will discuss the Hele-Shaw model.

6.2.5 Hele-Shaw Model

In the Hele-Shaw model, velocity may be a vector function of x, y and z, but at any (x, y) location the direction of the velocity vector is not a function of z as shown in Figure 6.6(a). This allows one to average the velocities u and v through the thickness and track them only as a function of x and y. This reduces the mathematical problem to two dimensions and reduces the number of calculations by an order of magnitude.

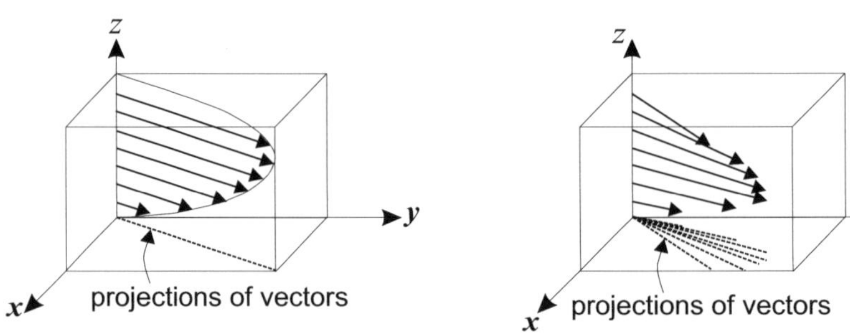

(a) (b)

Figure 6.6: (a) A typical Hele-Shaw velocity profile in which the *direction* of the velocity vector is not a function of z at a fixed (x, y) location. (b) A typical non-Hele-Shaw velocity profile: velocity vector direction changes with z.

In compression molding problems, if one assumes no-slip boundary conditions at the mold walls, an order of magnitude analysis will show that the velocity gradients in the thickness direction are much larger than in the in-plane direction. For example,

$$\frac{\partial u}{\partial x} \text{ and } \frac{\partial u}{\partial y} \sim \frac{U_c}{L} \tag{6.8a}$$

$$\frac{\partial u}{\partial z} \sim \frac{U_c}{h}. \tag{6.8b}$$

Thus, for an inelastic fluid

$$\frac{\partial \tau_{xx}}{\partial x} \sim \eta \frac{U_c}{L^2} \tag{6.9a}$$

$$\frac{\partial \tau_{zx}}{\partial z} \sim \eta \frac{U_c}{h^2}. \tag{6.9b}$$

As $h \ll L$, one can safely neglect Equation (6.9a) as compared to Equation (6.9b). This reduces the equations of motion to

$$\frac{\partial P}{\partial x} = \frac{\partial \tau_{zx}}{\partial z} \tag{6.10a}$$

$$\frac{\partial P}{\partial y} = \frac{\partial \tau_{zy}}{\partial z} \tag{6.10b}$$

$$\frac{\partial P}{\partial z} = 0. \tag{6.10c}$$

Equation (6.10c) implies that pressure varies within the x-y plane only. For isothermal Newtonian fluids, one can substitute the stress-strain rate relationship

$$\tau_{ij} = \mu \left(\frac{\partial u_i}{\partial x_j} + \frac{\partial u_j}{\partial x_i} \right) \tag{6.11}$$

into Equations (6.10a) and (6.10b) which results in

$$\frac{\partial P}{\partial x} = \mu \frac{\partial^2 u}{\partial z^2} \tag{6.12a}$$

$$\frac{\partial P}{\partial y} = \mu \frac{\partial^2 v}{\partial z^2}. \tag{6.12b}$$

Since P is not a function of z, Equations (6.12a) and (6.12b) can be integrated twice to obtain the velocity profile through the thickness

$$u = \frac{1}{2\mu} \frac{\partial P}{\partial x} z^2 + c_1 z + c_2 \tag{6.13a}$$

$$v = \frac{1}{2\mu} \frac{\partial P}{\partial y} z^2 + c_3 z + c_4. \tag{6.13b}$$

The boundary conditions to find the constants are no-slip at the mold platen walls,

$$u = 0, \quad v = 0 \text{ at } z = 0 \text{ and at } z = h. \tag{6.14}$$

This gives velocity distributions as

$$u = \frac{1}{2\mu} \frac{\partial P}{\partial x} h^2 \left[\left(\frac{z}{h}\right)^2 - \left(\frac{z}{h}\right) \right] \tag{6.15a}$$

$$v = \frac{1}{2\mu} \frac{\partial P}{\partial y} h^2 \left[\left(\frac{z}{h}\right)^2 - \left(\frac{z}{h}\right) \right]. \tag{6.15b}$$

Note that because of the very low Reynolds number in these flows, the parabolic velocity profile is achieved instantly. Hence, even though h is changing with time, the velocity profile can be assumed to be parabolic at every time step by assuming a quasi-steady state as discussed in the section on assumptions in Chapter 5. Note that u and v depend on x and y as well because $\partial P/\partial x$ and $\partial P/\partial y$ are functions of x and y.

Before we find an equation for P, note one striking feature of this velocity distribution: u and v are dependent on z but the ratio u/v is not, as made clear by taking the ratio of Equations (6.15a) and (6.15b):

$$\frac{u}{v} = \frac{\partial P/\partial x}{\partial P/\partial y} \neq f(z). \tag{6.16}$$

This form states that the magnitude of the velocity vector will vary across the cavity thickness but the direction will not. Thus, we have proved that this flow is a Hele-Shaw flow as shown in Figure 6.6(a). Heiber and Shen [299] have shown that even in a nonisothermal or non-Newtonian case, this type of flow has the characteristics of Hele-Shaw flow.

This simplifies the analysis to a great extent. Instead of calculating u and v at every x, y and z location, we can instead find the average velocity at each (x, y) location. These average velocities can be obtained by integrating Equations (6.6a) and (6.6b) to give [2, 3, 139, 300–303]

$$\bar{u} = -\frac{h^2}{12\mu}\frac{\partial P}{\partial x} \tag{6.17a}$$

$$\bar{v} = -\frac{h^2}{12\mu}\frac{\partial P}{\partial y}. \tag{6.17b}$$

For nonisothermal and non-Newtonian cases in which viscosity varies with z, [299–302] provide the general form

$$\bar{u} = -\frac{S}{h}\frac{\partial P}{\partial x} \tag{6.18a}$$

$$\bar{v} = -\frac{S}{h}\frac{\partial P}{\partial y} \tag{6.18b}$$

where S is a measure of the ease with which the suspension flows locally and is given by

$$S = \int_0^h \frac{(z-\lambda)^2}{\eta}\, dz. \tag{6.19}$$

Here $\eta = \eta(z)$ is the viscosity and λ is the value of z at which the shear stresses τ_{zx} and τ_{zy} are zero. For an isothermal case, S will be equal to $h^3/(12\eta)$.

To find the governing equation for the pressure distribution, one can substitute Equations (6.18a,b) into the integrated mass conservation equation (6.5). This results in

$$\frac{\partial}{\partial x}\left(S\frac{\partial P}{\partial x}\right) + \frac{\partial}{\partial y}\left(S\frac{\partial P}{\partial y}\right) = -s. \tag{6.20}$$

This equation is a Poisson equation and can be solved for $P(x, y)$, once the boundary conditions are specified. The physical boundary conditions will be that $P = 0$ at the free or moving boundary surface (assuming there is no back pressure in the mold) and $\partial P/\partial n = 0$

when the charge reaches the mold walls (assuming that there is no leakage of charge across the mold walls) where n is the normal direction in the x-y plane on the mold wall.

To solve Equation (6.20), one may have to resort to numerical methods such as finite difference or finite element, as the solution domain expands when the material is compressed. A procedure must also be chosen for tracking the moving flow front and the shape of the charge and for modifying the boundary conditions as portions of the charge front encounter the boundaries of the mold walls.

Once the pressure distribution is known, the entire velocity distribution may be found by rewriting Equations (6.18a,b) and (6.19) [299–302]

$$u(x, y, z) = \frac{\partial P(x, y)}{\partial x} \int_0^z \frac{z - \lambda}{\eta} dz \tag{6.21a}$$

$$v(x, y, z) = \frac{\partial P(x, y)}{\partial y} \int_0^z \frac{z - \lambda}{\eta} dz. \tag{6.21b}$$

If λ is not known a priori, one of the boundary conditions such as no-slip, i.e., $u = 0$, $v = 0$ at $z = h$ can be used to find λ.

The governing equation (6.20) can be solved to determine pressure and velocity distributions at any instant in time as we treat this problem as quasi-static in nature. The time-dependent nature of the pressure and velocity field enters into the problem because as the mold closes the redistribution of the charge takes place and the flow front position, where a boundary condition is applied, keeps changing. The usual procedure to predict the flow in compression molding can be outlined as follows [304]:

1. Specify initial conditions (location of the charge, boundaries of the charge, closing speed, etc.).
2. Solve for the pressure distribution at $t = 0$ using Equation (6.20).
3. From the pressure distribution, find the average velocities and advance the material to form the new boundaries of the charge.
4. Solve Equation (6.20) in the new domain with the new height h and the boundary conditions.
5. Repeat steps 3 and 4 until the charge touches all the mold walls or until the maximum force the compression molding machine can generate has been reached or until one achieves the final thickness in the z direction. To calculate the force required at any time t to move the platen at a speed s, one must integrate the pressure field at that t over the charge surface.

The two main assumptions that are used in this model are negligible w and insignificant in-plane stresses. These assumptions are valid everywhere in the domain except near the edges of the compressing material. Thus, one can gain tremendous simplification by sacrificing a little bit of accuracy at the edges. The Hele-Shaw model is always valid as long as the shear stress terms are larger than the in-plane stresses. This implies that

$$\frac{\partial \tau_{xz}}{\partial x} \sim \frac{\eta_w u_c}{h^2} \gg \frac{\partial \tau_{xx}}{\partial x} \sim \frac{\eta_c u_c}{L^2}. \tag{6.22}$$

In compression molding, usually a cold charge is molded in a hot mold so the viscosity of the material near the mold wall, η_w, where most of the shearing dominates, is lower, as compared to the viscosity near the middle, η_c, where the viscosity is higher as the material there is still not heated. Thus for the Hele-Shaw approximation to be valid, we must have

$$\frac{\eta_w L^2}{\eta_c h^2} \gg 1. \tag{6.23}$$

The viscosity could differ at the most by an order of magnitude from the surface to the center. However, usually if L is at least an order of magnitude larger than h, which is definitely true for a part such as an automobile hood, one can usually justify the use of the Hele-Shaw model.

6.2.6 Lubricated Squeeze Flow Model

If the charge or the mold surfaces are well lubricated, the charge material will slip along the upper and lower surfaces of the mold. This translates into a slip boundary condition. One could have a partial slip or complete slip. If one models a partial slip, an additional parameter to describe the degree of slip will be required. Here, in order to introduce the model, we will assume complete slip on both of the surfaces. Mathematically, this can be described as the shear stresses at the mold surface are zero:

$$\tau_{zx} = 0 \quad \text{and} \quad \tau_{zy} = 0 \qquad \text{at } z = 0 \text{ and } \; z = h. \tag{6.24}$$

It further assumes that there is no shearing across the thickness. This leads to u and v being only functions of x and y, and therefore $u = \bar{u}$ and $v = \bar{v}$. The equation of motion due to this assumption reduces to

$$\frac{\partial P}{\partial x} = \frac{\partial \tau_{xx}}{\partial x} + \frac{\partial \tau_{yx}}{\partial y} \tag{6.25a}$$

$$\frac{\partial P}{\partial y} = \frac{\partial \tau_{xy}}{\partial x} + \frac{\partial \tau_{yy}}{\partial y} \tag{6.25b}$$

with the boundary conditions that the velocity normal to the mold walls is zero and the total stress normal to the free flow front is zero.

Lubricated squeeze flow solutions can often be trivial, providing bi-axial extensional flow in the x and y directions. This implies that as you squeeze the material in the z direction, it elongates in the x and y directions equally, until it encounters a mold wall and then it elongates only in the direction it can as shown in Figure 6.7. The pressure inside the charge is usually constant. To find this pressure, one can use the boundary condition that the total stress normal to the boundary, σ_{nn}, is zero. If u_n is the velocity component normal to the free flow front boundary and n denotes the normal direction, it can be expressed as

$$\sigma_{nn} = 2\eta \frac{\partial u_n}{\partial n} - P = 0 \tag{6.26a}$$

$$\text{or,} \qquad 2\eta \frac{\partial u_n}{\partial n} = P. \tag{6.26b}$$

Here P is the pressure of the material inside the charge which is not equal to the pressure outside.

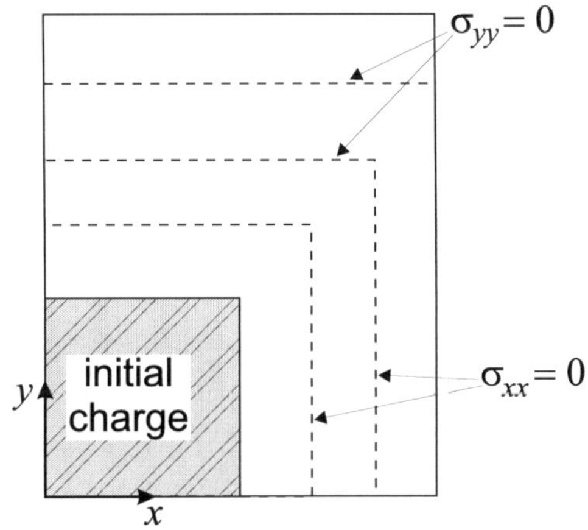

Figure 6.7: Biaxial elongation of a lubricated charge due to compression molding. The material elongates equally in the x and y directions until it encounters a mold wall.

Example 6.1: Lubricated Squeeze Flow

Consider a polymer charge being squeezed between two parallel disks as shown in Figure 6.8. The disk surfaces are lubricated so that there is a *slip boundary condition in the radial direction*. Assume that inertia and body forces are negligible. The thickness of the polymer charge is h and the radius of the disks, and hence the charge is R. The disks are compressed in a press with a closing speed of $s = -dh/dt = -\dot{h}$.

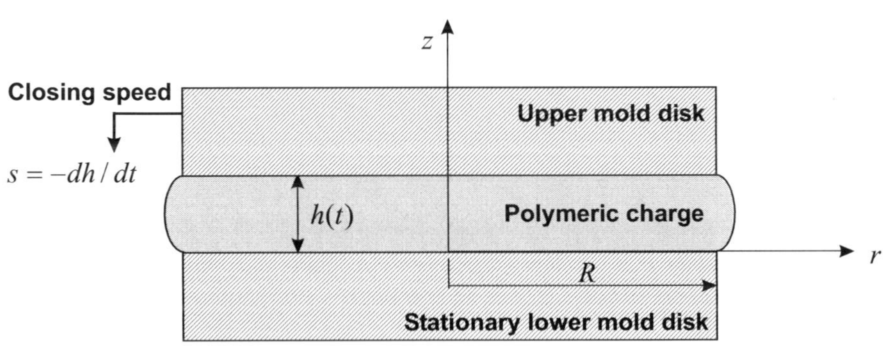

Figure 6.8: Schematic of a fiber reinforced polymer charge being squeezed between two parallel disks in compression molding.

Find (i) velocity distribution, (ii) stresses and the pressure in the fluid, (iii) the total force exerted on the upper disk by the fluid, and (iv) $h(t)$ if $h(0) = h_0$ and the total weight of the upper disk and the press is W.

Solution

(i) The symmetry of the problem requires that $v_\theta = 0$ and $\partial()/\partial\theta = 0$ for all the variables. The continuity equation in cylindrical coordinates reduces to

$$\frac{1}{r}\frac{\partial}{\partial r}(rv_r) + \frac{\partial v_z}{\partial z} = 0. \tag{6.27}$$

The boundary conditions are

$$\begin{aligned} v_z &= -s && \text{at } z = h && (6.28a)\\ \tau_{zr} &= 0 && \text{at } z = h && (6.28b)\\ v_z &= 0 && \text{at } z = 0 && (6.28c)\\ \tau_{zr} &= 0 && \text{at } z = 0 && (6.28d)\\ v_r &= 0 && \text{at } r = 0 && (6.28e)\\ \sigma_{rr} &= -P_a && \text{at } r = R && (6.28f) \end{aligned}$$

where P_a is the atmospheric pressure. Let's assume that $v_r = v_r(r)$ and $v_z = v_z(z)$ considering a thin charge, $R \gg h$. The assumption of $v_z = v_z(z)$ meets the boundary conditions (6.28a) and (6.28c). With these assumptions, the partial derivatives in Equation (6.27) reduce to ordinary derivatives, and hence Equation (6.27) can be rewritten as

$$\frac{1}{r}\frac{d}{dr}(rv_r) = -\frac{dv_z}{dz} = \text{constant} = c_1. \tag{6.29}$$

The term $\frac{1}{r}\frac{d}{dr}(rv_r)$ on the left-hand side of Equation (6.29) is a function of r only, and the term $-\frac{dv_z}{dz}$ is a function of z only. The equation indicates that these two terms are equal. This can be only true if these two terms are indeed equal to a constant, c_1. Integrating $-\frac{dv_z}{dz} = c_1$ with respect to z yields $v_z = -c_1 z + c_2$ where c_2 is another constant. Applying the boundary conditions (6.28a) and (6.28c) yields

$$v_z(z) = -\frac{sz}{h}. \tag{6.30}$$

By substituting $dv_z/dz = -s/h$ into Equation (6.29), we have the following:

$$\frac{d}{dr}(rv_r) = \frac{sr}{h}. \tag{6.31}$$

Integrating both sides of Equation (6.31) with respect to r, one can obtain

$$rv_r = \frac{sr^2}{2h} + c_3. \tag{6.32}$$

The constant c_3 is found to be zero by using boundary condition (6.28e); hence

$$v_r(r) = \frac{sr}{2h}. \tag{6.33}$$

Only the continuity equation (conservation of mass) was used to find the velocity field. That means the equation of motion (conservation of momentum) was not used for the velocity field calculation. So, the velocity field is the same whether the fluid is Newtonian or not. Once the velocity field is known, one can find the deformation field.

(ii) The components of the strain rate tensor for this flow field will be (see Table 3.5)

$$\dot{\gamma}_{rr} = 2\frac{\partial v_r}{\partial r} = \frac{s}{h} \tag{6.34a}$$

$$\dot{\gamma}_{\theta\theta} = 2\left(\frac{1}{r}\frac{\partial v_\theta}{\partial \theta} + \frac{v_r}{r}\right) = \frac{s}{h} \tag{6.34b}$$

$$\dot{\gamma}_{zz} = 2\frac{\partial v_z}{\partial z} = -\frac{2s}{h} \tag{6.34c}$$

$$\dot{\gamma}_{r\theta} = \dot{\gamma}_{\theta z} = \dot{\gamma}_{rz} = 0. \tag{6.34d}$$

Hence, the strain rate tensor,

$$\dot{\gamma} = \frac{s}{h}\begin{pmatrix} 1 & 0 & 0 \\ 0 & 1 & 0 \\ 0 & 0 & -2 \end{pmatrix} \tag{6.35}$$

is a biaxial elongation tensor. If the constitutive law between the stress and strain rate is known, one can also calculate the stress tensor. We will assume a Newtonian material behavior to evaluate the viscous stress tensor for the material. The stress tensor is given by

$$\tau = \mu\dot{\gamma} = \frac{\mu s}{h}\begin{pmatrix} 1 & 0 & 0 \\ 0 & 1 & 0 \\ 0 & 0 & -2 \end{pmatrix}. \tag{6.36}$$

What is the pressure distribution within the fluid (polymer charge)? One can use the stress tensor in the equation of motion in the r and z directions to evaluate that. The equation of motion in the r direction is as follows (assuming that the inertia terms are negligible compared to viscous terms):

$$\begin{aligned} 0 &= -\frac{\partial P}{\partial r} + \frac{1}{r}\frac{\partial}{\partial r}(r\tau_{rr}) + \frac{1}{r}\frac{\partial \tau_{r\theta}}{\partial \theta} - \frac{\tau_{\theta\theta}}{r} + \frac{\partial \tau_{rz}}{\partial z} \\ 0 &= -\frac{\partial P}{\partial r} + \frac{1}{r}\frac{\partial}{\partial r}\left(r\frac{\mu s}{h}\right) + \frac{1}{r}\frac{\partial (0)}{\partial \theta} - \frac{(\mu s)/h}{r} + \frac{\partial (0)}{\partial z} \\ 0 &= -\frac{\partial P}{\partial r} + \frac{\mu s}{rh} + 0 - \frac{\mu s}{rh} + 0 \end{aligned} \tag{6.37}$$

or, simply $\partial P/\partial r = 0$. The equation of motion in the z direction gives:

$$\begin{aligned} 0 &= -\frac{\partial P}{\partial z} + \frac{1}{r}\frac{\partial}{\partial r}(r\tau_{rz}) + \frac{1}{r}\frac{\partial \tau_{\theta z}}{\partial \theta} + \frac{\partial \tau_{zz}}{\partial z} \\ 0 &= -\frac{\partial P}{\partial z} + \frac{1}{r}\frac{\partial}{\partial r}(r(0)) + \frac{1}{r}\frac{\partial (0)}{\partial \theta} + \frac{\partial[(-2\mu s)/h]}{\partial z} \\ 0 &= -\frac{\partial P}{\partial z} + 0 + 0 + 0 \end{aligned} \tag{6.38}$$

or, simply $\partial P/\partial z = 0$. Hence, P is a function of neither r nor z. So, it is a constant $P = C$. In order to find C, we use the boundary condition (6.28f):

$$\begin{aligned} \sigma_{rr} &= \tau_{rr} - P \\ -P_a &= \frac{\mu s}{h} - P. \end{aligned} \tag{6.39}$$

Hence, the pressure P within the charge is

$$P = \frac{\mu s}{h} + P_a \tag{6.40}$$

which is higher than the pressure outside P_a. As seen from Equation (6.40), the higher the viscosity μ and closing speed s, the higher the charge pressure P.

(iii) σ_{zz} is the force per unit area exerted on the polymer charge. So, the force F on the upper disk is the integration of $-\sigma_{zz}$ over the disk area:

$$\begin{aligned} F &= -\int \sigma_{zz}\, dA = -\int_0^R \sigma_{zz} 2\pi r\, dr \\ &= -\int_0^R (\tau_{zz} - P)\, 2\pi r\, dr = -\int_0^R \left[\left(-\frac{2\mu s}{h}\right) - \left(\frac{\mu s}{h} + P_a\right)\right] 2\pi r\, dr \\ &= \int_0^R \left(\frac{3\mu s}{h} + P_a\right) 2\pi r\, dr \\ &= \pi R^2 \left(\frac{3\mu s}{h} + P_a\right). \end{aligned} \tag{6.41}$$

(iv) Force balance on the upper disk (in the z direction) is

$$\underbrace{\pi R^2 \left(\frac{3\mu s}{h} + P_a\right)}_{\text{force exerted by the polymer charge}} = W + \underbrace{\pi R^2 P_a}_{\text{force due to atmospheric pressure}} \tag{6.42}$$

Here, we neglect inertia forces. Figure 6.9 shows the force balance. Hence, $W = (3\pi R^2 \mu s)/h$. If a constant closing speed s is desired, then W should be increased with time by increasing the press force. One can see this mathematically due to the presence of $h(t)$ in the denominator. If W is kept constant, then the closing speed s will decrease with time. In this problem, W is specified as a constant, and we will calculate $h(t)$. The closing speed of the disk s is $-dh/dt$, hence

$$\frac{dh}{dt} = -\frac{Wh}{3\pi R^2 \mu} \tag{6.43a}$$

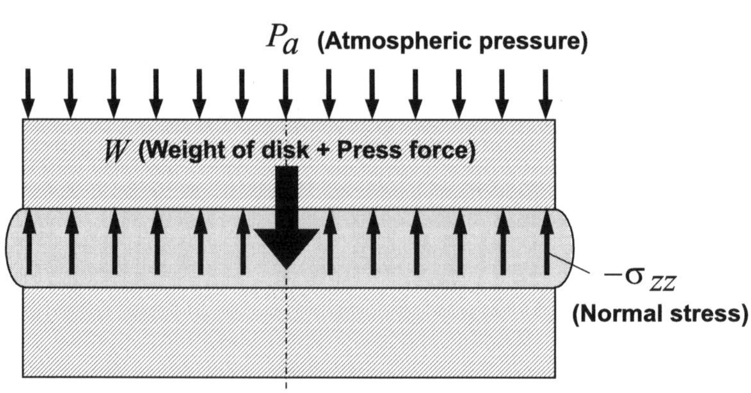

Figure 6.9: Force balance on the upper disk for lubricated squeeze flow Example 6.1.

$$\frac{dh}{h} = -\frac{W}{3\pi R^2 \mu}\, dt \tag{6.43b}$$

$$\ln h = -\left(\frac{W}{3\pi R^2 \mu}\right) t + c_1 \tag{6.43c}$$

$$h(t) = c_2\, e^{-\left[\left(\frac{W}{3\pi R^2 \mu}\right) t\right]}. \tag{6.43d}$$

In Equation (6.43b), integration of both sides with respect to t is carried out to obtain Equation (6.43c). The constant c_2 $(= e^{c_1})$ is found to be h_0 by using the initial condition $h(0) = h_0$. Hence,

$$h(t) = h_0\, e^{-\left[\left(\frac{W}{3\pi R^2 \mu}\right) t\right]}. \tag{6.44}$$

As shown in Figure 6.10, $h(t)$ decreases exponentially when a constant press force is applied.

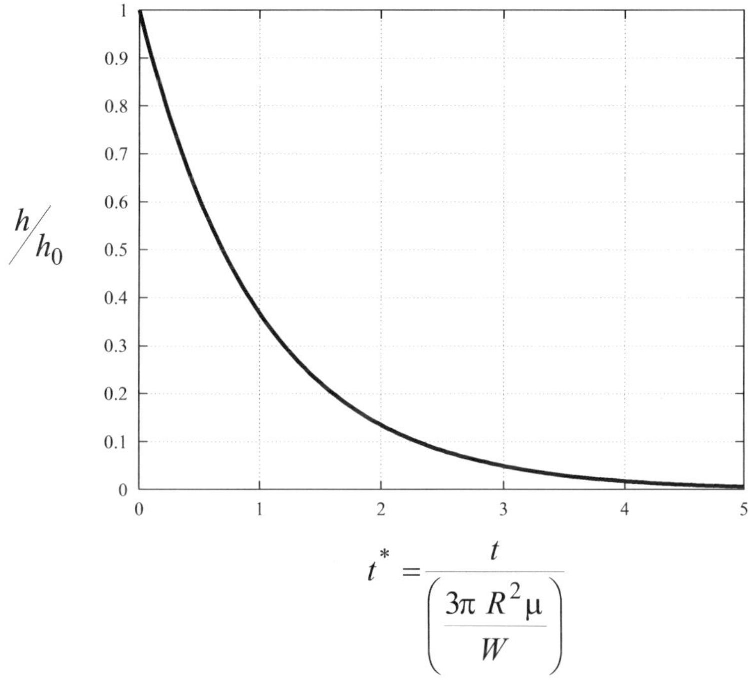

Figure 6.10: The mold gap thickness as a function of time for Example 6.1.

Obviously, this result breaks down at smaller values of h, as the assumptions of slip BC and no surface tension at the boundaries will no longer be valid. ■

6.2.7 Hele-Shaw Model with a Partial Slip Boundary Condition [2–4]

As we studied in Example 6.1, the lubricated squeezing model is based on the assumption that there is no shearing across the thickness ($\tau_{zx} = \tau_{zy} = 0$), and the prepreg completely slips across the mold cavity thickness.

The Hele-Shaw model has been found to give excellent results for thin, homogeneous polymer parts. The lubricated squeezing model provides unrealistic results for such cases. The lubricated squeeze flow model predicts that a rectangular charge will stay rectangular throughout the process until it encounters a mold wall. This type of behavior has been observed during the compression molding of thicker homogeneous parts, and parts molded from SMC. The two models can be thought of as providing two extremes, Hele-Shaw producing flow with zero slip at the mold boundaries, the lubricated squeezing flow model with complete slip boundary condition, producing "plug" flow through the mold thickness, as shown in Figure 6.11.

Barone and Caulk [3] proposed an alternative boundary condition at the mold walls to model the flow of SMC. They provided a model that lies somewhere between the extremes of the Hele-Shaw and lubricated squeezing models. They allowed for the partial slip of the molding compound at the wall, by specifying a slip velocity directly proportional to the shear stress at the boundary. For example:

$$u_x = \beta \left| \tau_{yx} \right| \tag{6.45}$$

at the mold walls. A schematic of the predicted shape of the velocity profile $v_r(z)$ is presented at the bottom of Figure 6.11.

In the following example, the proposed alternate boundary condition for the Hele-Shaw model will be investigated through an analytical case study.

Example 6.2: Hele-Shaw Model with Partial Slip Boundary Condition [4]

Investigate the flow induced by the compression molding of a cylindrical charge between circular plates which can undergo partial slip with respect to the mold walls. The plates are moving at a speed of $-\dot{h}$ towards each other, and are of radius R as shown in Figure 6.12.

Solution

The position of the charge flow front is defined by $\bar{R}$. The equations to be solved are as follows:

$$2\pi r \int_0^h v_r dz + \pi r^2 \dot{h} = 0 \tag{6.46}$$

$$\frac{\partial \tau_{rz}}{\partial z} = -\frac{dP}{dr} \tag{6.47}$$

Equation (6.46) was derived by integrating the continuity equation over the charge volume. Equation (6.47) is the momentum equation in the r direction with the assumption that shear stresses are higher than normal stresses. The boundary conditions are assumed to be

$$v_r = \beta \left| \tau_{rz} \right| \quad \text{at } z = h \tag{6.48a}$$

$$v_r = \beta \left| \tau_{rz} \right| \quad \text{at } z = -h \tag{6.48b}$$

$$\tau_{rz} = 0 \quad \text{at } z = 0 \tag{6.48c}$$

$$P = P_a \quad \text{at } r = \bar{R} \tag{6.48d}$$

where β is an arbitrary constant, defining the amount of slip occurring at the mold surfaces.

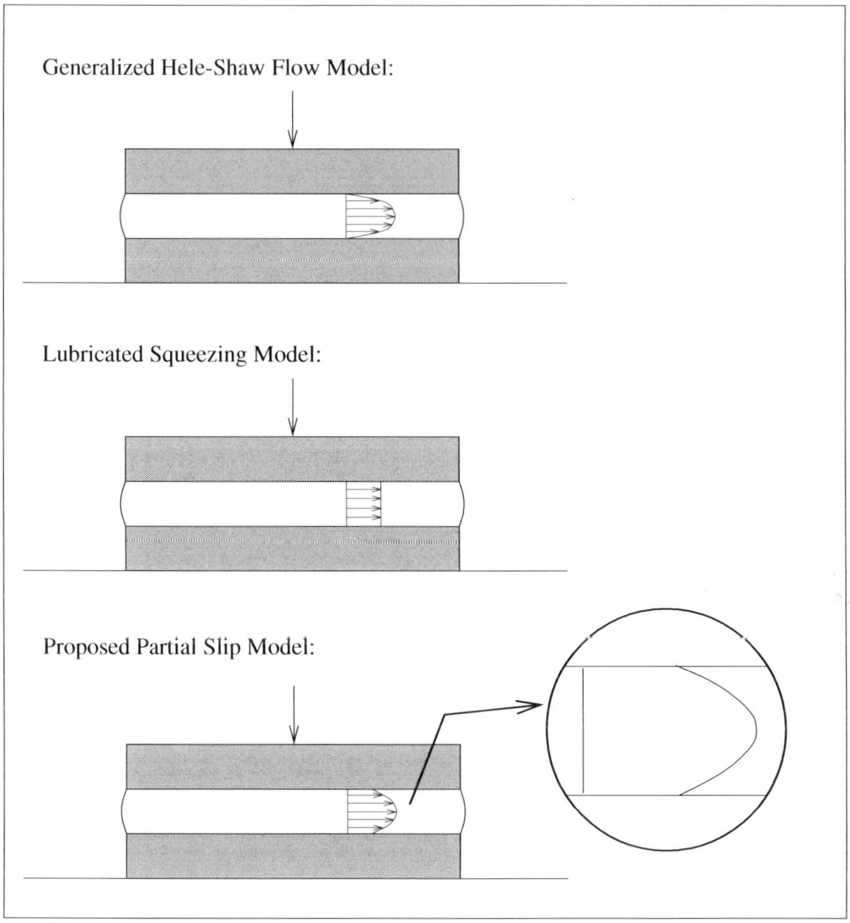

Figure 6.11: Velocity profiles $v_r(z)$ for various flow models [4].

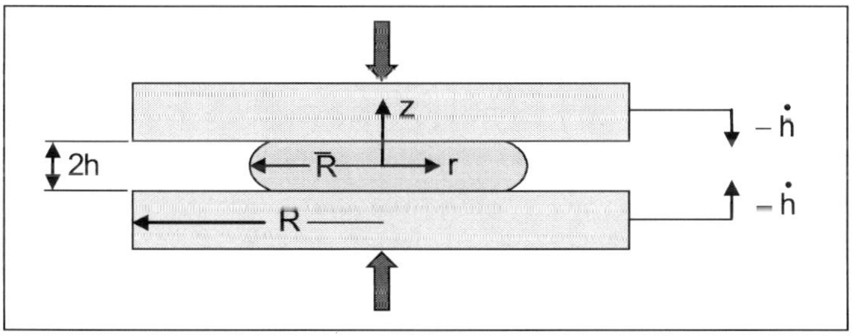

Figure 6.12: Axisymmetric squeeze flow of a cylindrical charge [4].

Here the problem will be solved for Newtonian fluids. The velocity profile is obtained by integrating Equation (6.47) with $\tau_{rz} = \eta \partial v_r / \partial z$ and the boundary conditions (6.48a)-(6.48c),

$$v_r = \frac{h^2}{2\eta}\left(-\frac{dP}{dr}\right)\left(1-(z/h)^2\right)+\beta\left(-\frac{dP}{dr}\right)h. \tag{6.49}$$

Then Equation (6.46) is used to find the pressure gradient:

$$\frac{dP}{dr} = \frac{r\dot{h}}{2h^2}\left(\frac{1}{\frac{h}{3\eta}+\beta}\right). \tag{6.50}$$

If β is given a value of zero, the velocity profile reduces to that of the normal Hele-Shaw solution.

Several velocity profiles have been plotted to examine the influence of the arbitrary coefficient β on the amount of slip occurring at the mold walls.

As the velocity profile varies linearly with $\dot{h}$, and scales linearly with radius r, all profiles have been calculated for a single $\dot{h}$ and radius. Calculated velocity profiles at different plate separations are shown in Figure 6.13. As the separation of the two plates decreases during the process, the average radial velocity through the thickness increases. Therefore, each velocity profile calculated is normalized to its centerline value, v_{max}. These plots demonstrate how the amount of slip at the mold boundaries develops as the process continues.

Three different sets of velocity profiles are plotted in Figure 6.13. In each plot, the velocity profiles have been calculated at h values of 0.1, 0.05, 0.01, 0.005 and 0.001 m. Each set of velocity profiles has been developed for a certain value of β in m/(Pa.s).

For Newtonian fluids, the ratio of the slip velocity at the wall to the centerline velocity increases as the separation of the mold wall decreases. The fluid is slipping more and more as the mold closes.

The pressure profiles were found by integrating $\frac{dP}{dr}$, applying the boundary condition $P = P_a$ at $r = \bar{R}$. For convenience, P_a has been set to zero at this point. For Newtonian fluids,

$$P = \frac{\dot{h}}{4h^2}\left(\frac{1}{\frac{h}{3\eta}+\beta}\right)(r^2-\bar{R}^2). \tag{6.51}$$

The expression for pressure distribution is useful to calculate the force required for compression of the charge.

The squeezing force required to move the circular disks towards each other will be calculated. This force was found by integrating the pressure at the mold surface from $r = 0$ to $r = \bar{R}$:

$$F = 2\pi \int_0^{\bar{R}} P(r)\, r\, dr. \tag{6.52}$$

The resultant force for a Newtonian fluid is calculated to be

$$F = -\frac{\pi \dot{h} \bar{R}^4}{8h^2}\left(\frac{1}{\frac{h}{3\eta}+\beta}\right). \tag{6.53}$$

Squeezing force will be calculated for a particular theoretical experiment. The initial diameter of the charge is $\bar{R}(0) = 0.3\,\text{m}$, and it has a half-thickness of $h(0) = 8\,\text{mm}$. The closing speed $-\dot{h}$ is 3 mm/sec, and the charge is squeezed to a thickness of 5 mm. The process takes 1.0 second to complete.

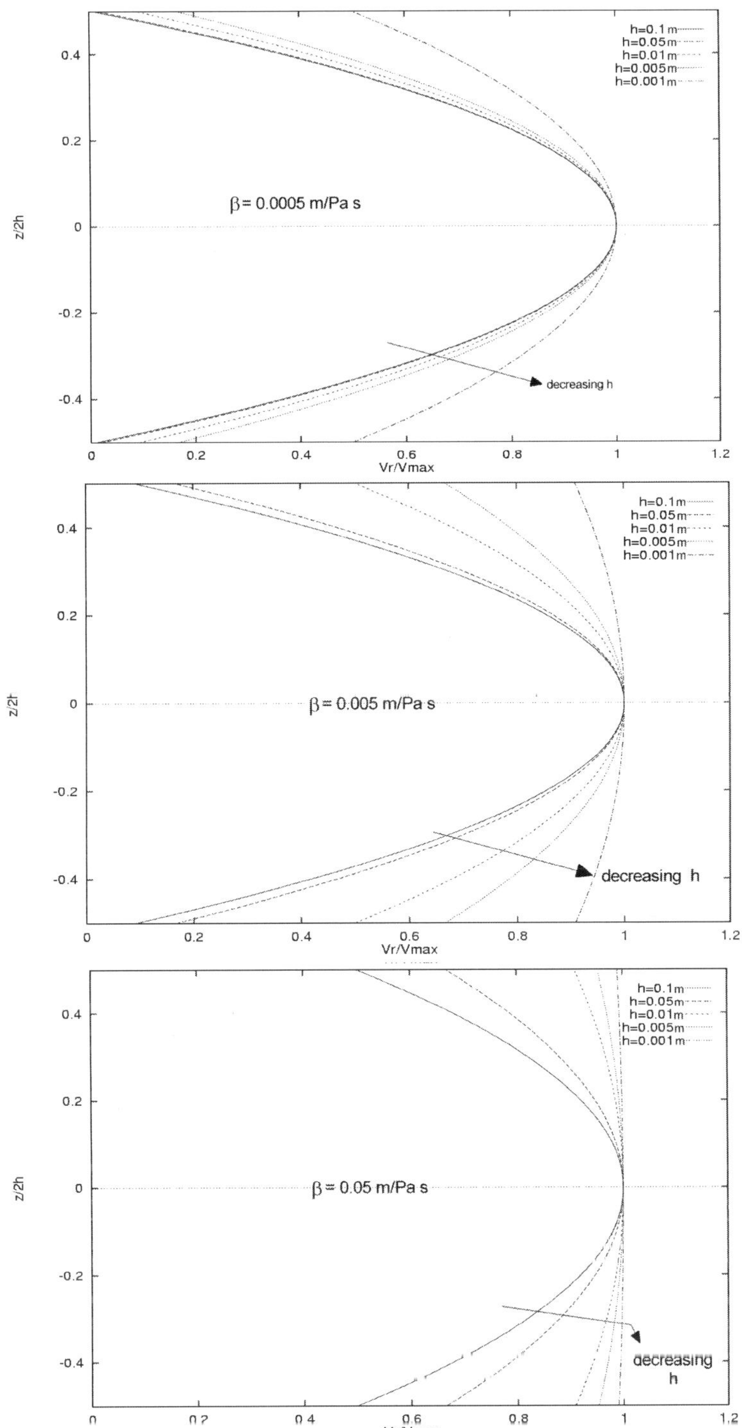

Figure 6.13: Normalized velocity profiles for compression molding flow of a Newtonian fluid with partial slip boundary conditions along the mold walls [4]. Top: $\beta = 0.0005$, Middle: $\beta = 0.005$, and Bottom: $\beta = 0.05$ in m/(Pa.s). In each plot, the profiles are shown for different values of h in meters.

Figure 6.14 presents several forces calculated for the experiment described above. Forces have been calculated for β values of 0.0, 0.0005, 0.005, and 0.05 m/(Pa.s). $\beta = 0.0$ corresponds to the Hele-Shaw flow model with no-slip boundary conditions. The force calculated using the lubricated squeezing model is also presented; however, this is very small, being 0.009% of the Hele-Shaw result at the end of the experiment.

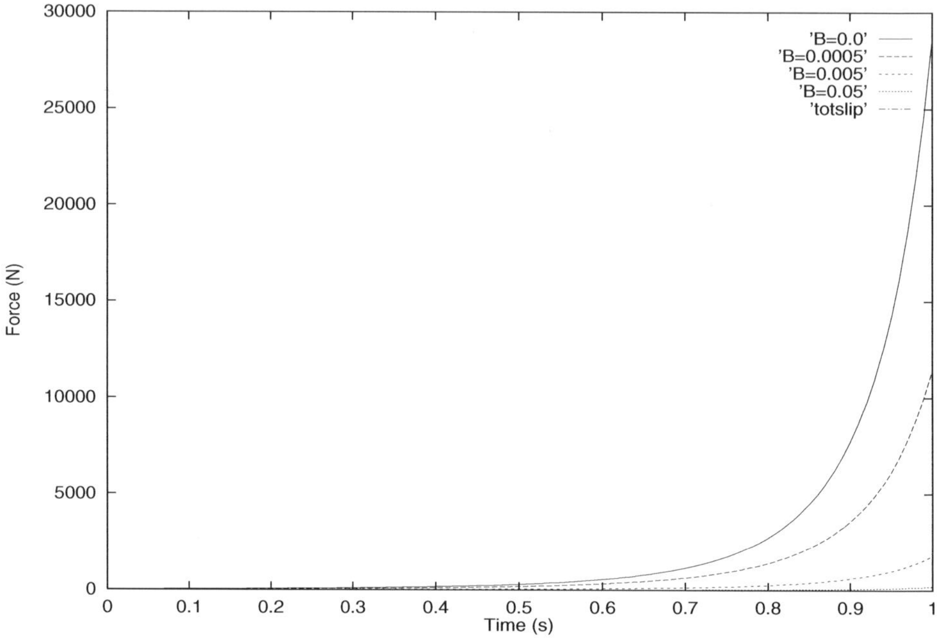

Figure 6.14: Force required to compression mold a Newtonian material with partial slip at the mold walls [4].

As β is increased from zero, the squeezing force approaches the lubricated squeeze flow model result. This is to be expected, because the velocity profile approaches "plug" flow as β increases. However, as β increases one would expect the Hele-Shaw model to break down as the in-plane stresses will become as important as or more important than the shear stresses. Thus, the results for the pressure distribution and the force are valid only at low values of β. ■

The successful application of the proposed alternate boundary condition depends on the determination of the arbitrary constant β. A method must be sought to measure this quantity, and verify if it indeed is a constant. If this can be done, this altered boundary condition will prove to be an invaluable tool for modeling large thickness and SMC parts.

6.2.8 Heat Transfer and Cure

In almost all compression molding processes, a cold charge is placed in a hot mold and the heat is provided to the charge by the mold walls. For thermoplastics, one needs to provide sufficient heat to melt the charge so it can flow and occupy the empty portion of the cavity before the cooling cycle is initiated to solidify the component. For thermosetting composites, one provides the heat also to initiate the curing of the composite. Curing is an exothermic reaction and will generate heat. If this heat is not extracted from the composite it could lead to degradation of the part. Hence, the thermal mold design is very crucial to the manufacturing of such components with the compression molding process.

It is also important to note that the filling and the thermal histories in molding are coupled through the dependence of viscosity on temperature. For example, if the flow is the Hele-Shaw type, the velocities at the center are much higher than the velocities near the walls. This will lead to cold material from the center migrating towards the hot mold surfaces as shown in Figure 6.15(a). On the other hand, if the resin viscosity changes dramatically with temperature and if there is sufficient slip near the surfaces, one could expect the material near the surfaces to get ahead of the material near the center. This would cause the material to flow towards the center as shown in Figure 6.15(b). Hence, the flow field is important to determine the temperature, the cure behavior in the case of thermoset, and the thermal design of the mold.

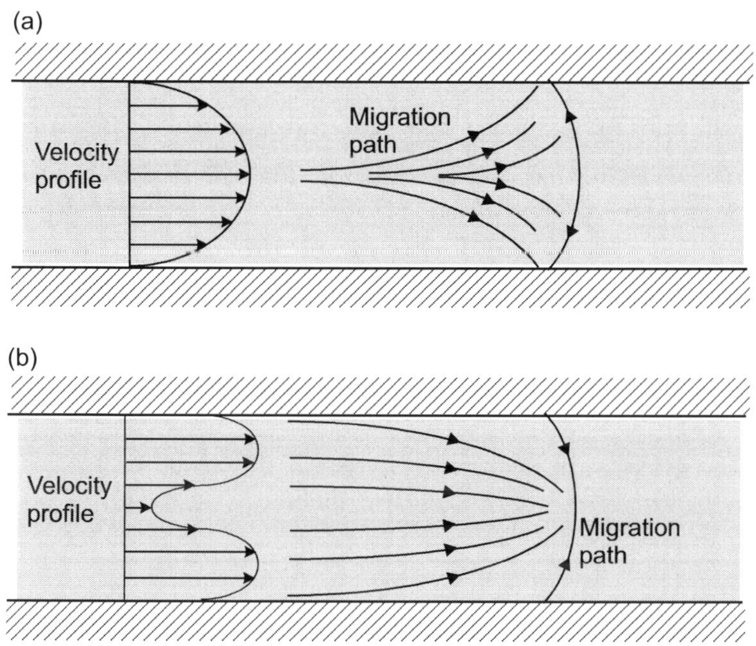

Figure 6.15: Velocity profile and the migration paths of cold material entering a hot mold. In (a), the no-slip velocity boundary condition leads to migration of the material from the center to the mold walls. In (b), the slip boundary condition at the walls and temperature dependence of viscosity cause the particles to migrate from the surface towards the center.

Example 6.3: Nonisothermal Lubricated Squeeze Flow [304]

Consider lubricated squeeze flow of a charge initially at temperature, T_0, placed in a hot mold at temperature T_w. The top mold platen contacts with the charge and is moving downwards compressing the material with constant speed s. Assume that viscous dissipation and conduction in the x-y plane are negligible. Find the temperature history as the charge flows to form the part. The thermal conductivity of the charge in the z direction is k, density is ρ, and heat capacity is c_p.

Solution

For lubricated squeeze flow, the velocity components u and v do not vary with z; therefore,

$$\frac{\partial u}{\partial z} = \frac{\partial v}{\partial z} = 0. \tag{6.54}$$

Differentiating the equation of continuity with respect to z for an incompressible fluid,

$$\frac{\partial}{\partial z}\left(\frac{\partial u}{\partial x} + \frac{\partial v}{\partial y} + \frac{\partial w}{\partial z}\right) = 0. \tag{6.55}$$

Interchanging the order of differentiation leads to

$$\frac{\partial}{\partial x}\left(\frac{\partial u}{\partial z}\right) + \frac{\partial}{\partial y}\left(\frac{\partial v}{\partial z}\right) + \frac{\partial^2 w}{\partial z^2} = 0. \tag{6.56}$$

Substitution of Equation (6.54) into Equation (6.56) results in

$$\frac{\partial^2 w}{\partial z^2} = 0. \tag{6.57}$$

The boundary conditions as seen from Figure 6.16 are $w = 0$ at $z = 0$ and $w = -s$ at $z = h$. This yields

$$w = -\frac{sz}{h}. \tag{6.58}$$

Thus, the motion of the material is uniform compression. If we scale z by introducing a new coordinate λ as

$$\lambda = \frac{z}{h(t)}, \tag{6.59}$$

then the fluid or the material particles will remain at the same position λ throughout the flow. However, the particles will move in the x-y plane.

The energy equation with no viscous dissipation or conduction in the x-y directions can be written as

$$\rho c_p \left(\frac{\partial T}{\partial t} + u\frac{\partial T}{\partial x} + v\frac{\partial T}{\partial y} + w\frac{\partial T}{\partial z}\right) = k\frac{\partial^2 T}{\partial z^2}. \tag{6.60}$$

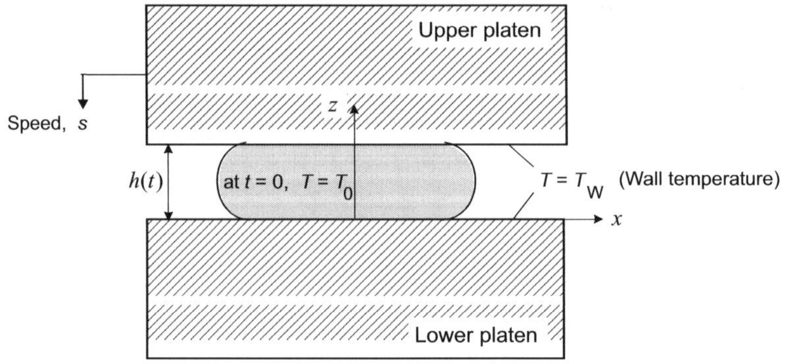

Figure 6.16: Temperature boundary conditions for nonisothermal compression molding of a charge.

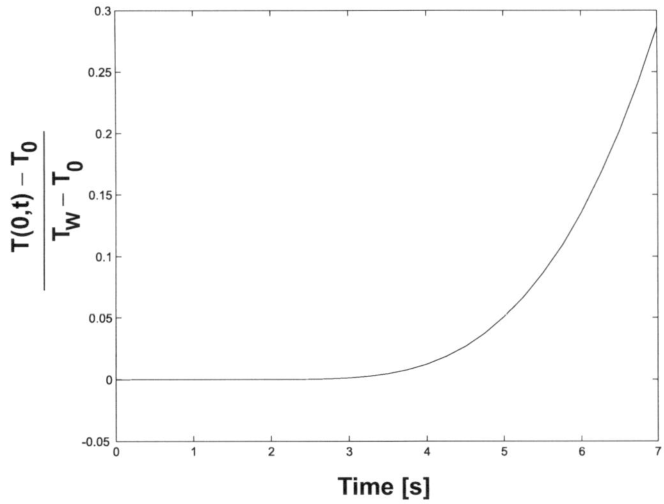

Figure 6.17: Nondimensional temperature at the center of the charge as a function of time with an initial thickness $h_i = 10$mm, final thickness $h_f = 3$mm, closing speed $s = 1$mm/s, and thermal diffusivity $\alpha = 1\text{mm}^2/\text{s}$.

As there is no x-y variation in the temperature or boundary conditions, $\partial T/\partial x$ and $\partial T/\partial y$ are zero. We can also assume that convection in the z direction is negligible for simplicity, although one can solve the equation by retaining this term. Using the transformation in Equation (6.59), Equation (6.60) can be rewritten as

$$\rho c_p \left(\frac{\partial T}{\partial t}\right) = \frac{k}{h^2(t)} \frac{\partial^2 T}{\partial \lambda^2}. \tag{6.61}$$

Here $\partial T/\partial t$ represents the time derivative at a fixed position λ. Equation (6.61) is similar to the conduction problem of a slab initially at temperature T_0, whose walls are exposed to T_w at $t = 0$ except the slab is shrinking across its thickness with time. The solution is given by

$$\frac{T - T_0}{T_w - T_0} = 1 + 2\sum_{n=0}^{\infty} \frac{(-1)^{n+1}}{a_n} \exp\left(\frac{-a_n^2 t^*(t)\alpha}{h_f^2}\right) \cos\left(a_n \lambda\right) \tag{6.62}$$

where α is the thermal diffusivity ($\alpha = \frac{k}{\rho c_p}$) and the coefficients a_n are

$$a_n = \pi\left(n + \frac{1}{2}\right) \tag{6.63}$$

and $t^*(t)$ is an integrated time

$$t^*(t) = \int_0^t h_f^2 \frac{dt'}{h^2(t')}. \tag{6.64}$$

Here t' is the dummy integration variable and h_f is the final thickness of the charge. The changing thickness of the sheet warps the time scale, accelerating the conduction of heat as the sheet becomes thinner, as illustrated in Figure 6.17. ■

6.2.9 Cure

In modeling the cure and coupling it with heat transfer, one must usually account for the dwell time, an initial period when no appreciable curing takes place due to the addition of inhibitors. Curing will be slow initially but then will rapidly accelerate and eventually level off. Isothermal curing behavior of unsaturated polyester is shown in Figure 6.18 by plotting degree of cure c versus time t. The degree of cure is defined as

$$c = \frac{q}{q_t} \tag{6.65}$$

where q is the cumulative amount of heat released from the beginning $t = 0$ to the current time t as compared with the total heat of reaction q_t. As listed in Chapter 4, there are many models to describe the cure kinetics. For example,

$$\frac{\partial c}{\partial t} = k_o e^{-(E/RT)} c^m (1-c)^n \tag{6.66}$$

where k_o, m and n are found by characterizing the resin experimentally.

Hence the rate of heat liberated per unit mass is given by

$$\dot{q} = q_t \frac{\partial c}{\partial t}. \tag{6.67}$$

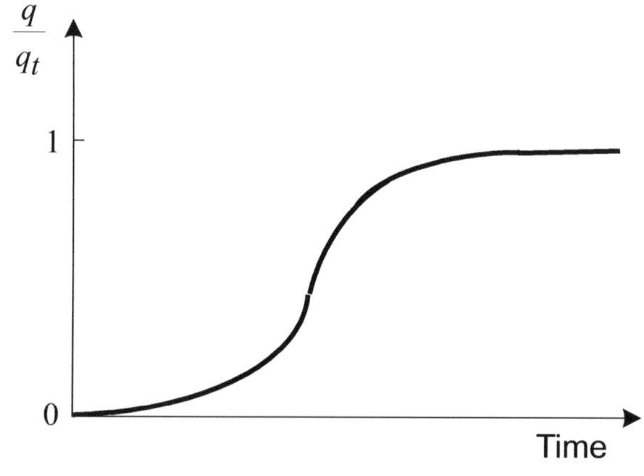

Figure 6.18: Schematic of isothermal curing behavior of unsaturated polyesters.

6.2.10 Coupling of Heat Transfer with Cure

As the cure depends on the temperature, and curing liberates heat into the composite and hence affects the temperature, the two phenomena are coupled. The heat liberated during cure can be modeled as volumetric internal heat generation; thus one can write the energy equation as

$$\rho c_p \frac{DT}{Dt} = k \nabla^2 T + \dot{S} \tag{6.68}$$

where $\dot{S} = q_t \partial c / \partial t$. At the same time, $\partial c / \partial t$ is dependent on cure and temperature as seen from Chapter 4 and Equation (6.66). Hence these equations need to be solved simultaneously if the curing has initiated. The viscosity of the material also dramatically increases with the cure rate, which makes it difficult for the material to flow. Many process engineers prefer to add inhibitors to the resin so that they can fill the part with low viscosity suspension, before the resin cure initiates. Therefore, the convective terms on the left-hand side of Equation (6.68) vanish, and it becomes a transient heat conduction problem with a space-dependent energy generation term.

Example 6.4: Importance of Damkohler Number

Nondimensionalize the energy equation to judge the relative magnitude of heat transfer due to conduction and reaction. Let us assume that conduction in only the z direction is important and there is no convection.

Solution

We will assume that the mold is saturated with resin when the resin cure initiates. As the in-plane dimensions are much longer than the thickness h in the z direction, nondimensionalization of the energy equation is carried out with conduction only in the z direction. Other variables can be nondimensionalized as follows:

$$T^* = \frac{T - T_0}{\Delta T_0} \tag{6.69}$$

$$t^* = \frac{t}{t_c} \tag{6.70}$$

$$z^* = \frac{z}{h} \tag{6.71}$$

where $\Delta T_0 = T_w - T_0$ (= difference between the wall and initial temperature) and $t_c = \rho c_p h^2 / k$ is the characteristic time for heat conduction. To make $\partial c / \partial t$ dimensionless and of the order unity, one must use the characteristic time of reaction, t_r instead of t_c. Thus

$$r^* = t_r \frac{\partial c}{\partial t} \tag{6.72}$$

The value of t_r will depend on the kinetic equation used. Also the adiabatic temperature rise just due to the heat released from the reaction can be estimated as follows:

$$\Delta T_r = \frac{q_t}{\rho c_p}. \tag{6.73}$$

One can now write the dimensionless form of the energy equation (6.68) with no convection and conduction neglected in the in-plane directions as

$$\frac{\partial T^*}{\partial t^*} = \frac{\partial^2 T^*}{\partial (z^*)^2} + \text{Da}\, r^*. \tag{6.74}$$

The dimensionless group Da is known as the Damkohler number and is given by

$$\text{Da} = \frac{t_c \Delta T_r}{t_r \Delta T_0} = \frac{h^2 q_t}{\alpha c_p t_r (T_w - T_0)}. \tag{6.75}$$

It represents the ratio of the rates of temperature rise due to reaction to conduction. Hence, when Da is large, the reaction kinetics dominate the heat transfer, and if Da is small then heat conduction dictates the heat transfer. ■

6.2.11 Fiber Orientation

When a charge with fibers is compressed, one would expect the fibers to rotate and move along with the charge. Should we then consider the anisotropy in the viscosity when modeling the compression effects or should we ignore it? It has been found that the change in viscosity of the material due to the change in fiber orientation is a minor effect in compression molding [305]. Hence, one can decouple the fiber orientation problem from the flow problem. So once the pressure field is solved, one can find the velocity and velocity gradients and use those in the constitutive equation to find fiber orientation. However, as fiber orientation is an initial-value problem, it does need an initial condition as will be seen from Equation (6.212) later in this chapter. In this equation, Da_{ij}/Dt is a function of a_{ij}, ω_{ij}, $\dot{\gamma}$ and C_I.

Thus, one needs to convect the orientation field as the material flows. Figure 6.19 shows the convection of fiber orientation during compression molding. The details of this are given in references [148, 150].

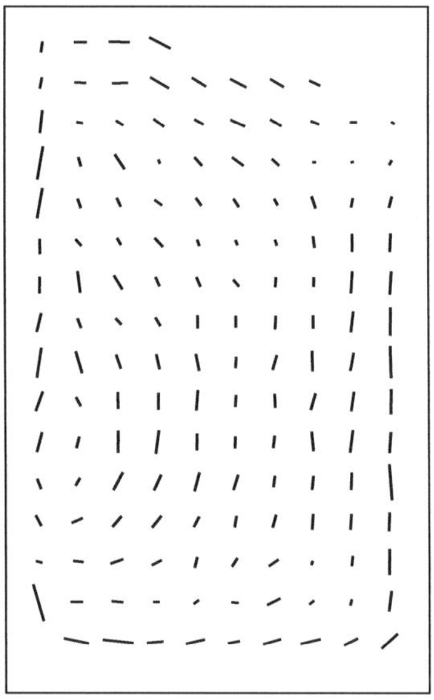

Figure 6.19: The convection of fiber orientation during compression molding [148]. The direction of the line shows the direction of most fibers and the length of the line shows the strength of the difference between the fibers in that direction as compared to the direction perpendicular to that direction. Thus, a dot would represent equal orientation distribution of fibers in all directions and a long line will denote that most fibers are oriented in that direction [148].

6.3 Extrusion

Extrusion is the most important forming method in polymer processing. More polymers and other materials are converted into useful objects by extrusion than any other method. An extruder is essentially a pump to displace viscous materials. It is also capable of other operations such as mixing along with pumping. The extruder can be fed with molten material or with solid polymer chips, beads or pellets. In the case of composite materials, the pellets have short fibers embedded in the polymer. A melting operation is achieved few diameters downstream of the inlet. The operation is called plasticating, and the extruder is known as a plasticating extruder. If molten material is fed to the extruder, it is known as melt extruder. If two dissimilar polymers are to be blended, or if the pellets contain short fibers, an additional operation of mixing is performed by the extruder.

The heart of the extruder is the screw. The details and the nomenclature for the screw are shown in Figure 6.20. Table 6.1 shows the various geometric relationships. The screw in the extruder is usually between 7.5 and 30 centimeters in diameter. The speed is usually in the range from 20 to 500 revolutions per minute with a working pressure between 50 and 1000 atmospheres. The flow rate could be as high as 2000 kg/hour. The length to diameter ratio could be anywhere from 5 to 100.

Table 6.1: Geometric Relationships and Operating Characteristics for an Extruder

Geometric Parameters and Relationships		
screw length	:	L
screw diameter	:	D
screw speed (revolution per second)	:	N
channel depth	:	H
helix angle	:	$\tan\theta = L_s/\pi D$
channel width	:	$W = L_s \cos\theta - e$
spiral channel length	:	$l = L/\sin\theta$
relative velocities	:	$V_z = \pi DN\cos\theta, \quad V_x = \pi DN\sin\theta$
Operating Characteristics		
flow rate	:	$Q = \dfrac{\pi DNWH\cos\theta}{2} - \dfrac{WH^3}{12\mu}\{\dfrac{P_2 - P_1}{L/\sin\theta}\}$
power	:	$\dot{\omega} = \left(\dfrac{4\mu V_x^2}{H} + \dfrac{\mu V_z^2}{H} + \dfrac{V_z W \Delta P}{2l}\right) lW$

The extruder can be divided into three zones, the solid conveying zone, the melting zone and the metering (melt conveying zone). The extruder operations can be grouped as follows: (1) melting, (2) mixing and devolatilization, and (3) pumping. The transport of the material inside the barrel is similar to a nut held in a wrench with the screw rotating in the nut. If a rotating screw is prevented from advancing, the nut will slide in the wrench as shown in Figure 6.21. In a similar manner the material slides between the screw and the barrel.

It is important to model the various operations in the extrusion process as it will prove useful in the design of the screw and help establish the operating conditions and scale it from laboratory scale to industrial scale. The objective of the model may be to focus on various aspects of the extrusion process. For example, if one wants to monitor how homogeneous the part produced is, the key issue would be to characterize and model the mixing operation. On the other hand, if one was interested in the pressure, and hence

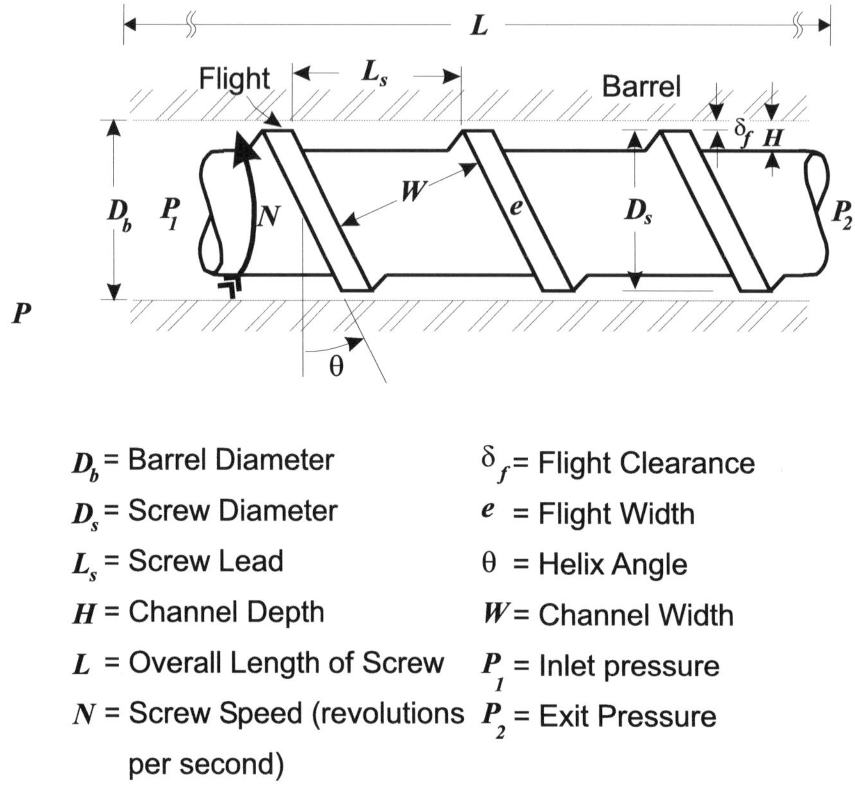

Figure 6.20: Nomenclature and dimensions of a typical extruder screw.

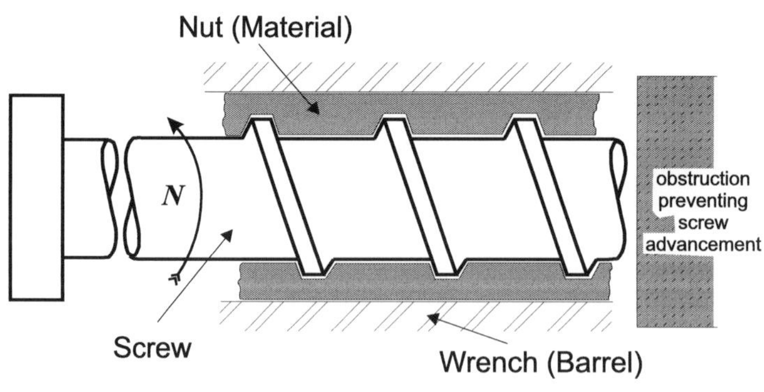

Figure 6.21: The principle of material advancement in an extruder can be compared to the sliding of the nut if a rotating screw is prevented from advancing.

power and the throughput relationship, one would focus on modeling the flow of the molten material through the screw and into the die. If one was interested in final dimensions of the components, then the die swell (a phenomenon that is common with any fluid once the material leaves the confined die shape and expands due to the surface tension and normal stress effects) would have to be modeled. Thus, based on the objective, the modeling for the same manufacturing process can address different issues. In this section, our goal will be to address the flow rate and pressure drop relationship.

6.3.1 Flow Modeling

An important choice in modeling the flow of the melt is the choice of the coordinate system. On first inclination, one may be tempted to use the spiral coordinate system, which can lead to ugly mathematics. A clever way to handle this would be to hold the screw stationary and rotate the barrel in the opposite direction [132]. Next, one must "unwrap the channel of the screw" and fix the coordinate system on the screw. This is schematically shown in Figure 6.22.

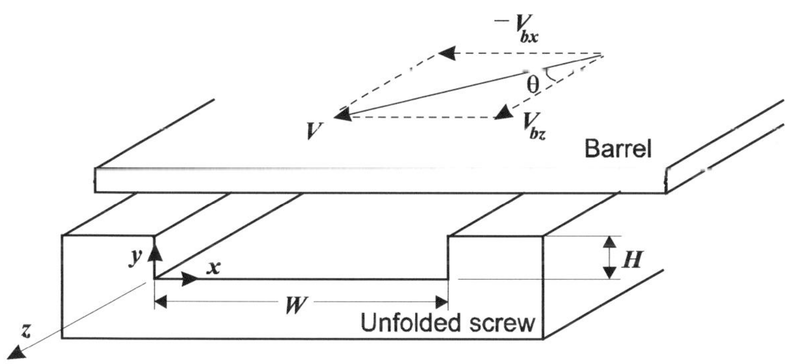

Figure 6.22: Schematic of the unwrapped extruder channel and the coordinate system on the screw. The boundary condition is $u_x = -V\sin(\theta) = -V_{bx}$ at $y = H$.

Next, we need to make simplifying assumptions to pose our problem correctly and uniquely. For our analysis, we can assume Newtonian fluid, although similar analysis has been conducted for non-Newtonian fluids. We will assume isothermal conditions. This allows us to solve the flow in the x and z directions separately and then add the solutions linearly as the fluid is Newtonian. There is no coupling with the temperature field.

The relative motion between the screw and the barrel becomes equivalent to steady motion of a plane at an angle θ to the helical axis z, which is the helical angle of the screw as shown in Figure 6.20. However, the drag flow is generated in the x and z directions. With inertia neglected, the equations of motion for Newtonian fluids reduce to:

$$0 = -\frac{\partial P}{\partial x} + \mu\left(\frac{\partial^2 u_x}{\partial x^2} + \frac{\partial^2 u_x}{\partial y^2} + \frac{\partial^2 u_x}{\partial z^2}\right) \quad (6.76)$$

$$0 = -\frac{\partial P}{\partial z} + \mu\left(\frac{\partial^2 u_z}{\partial x^2} + \frac{\partial^2 u_z}{\partial y^2} + \frac{\partial^2 u_z}{\partial z^2}\right). \quad (6.77)$$

As the flow is fully developed in the z direction, the last term in Equations (6.76) and (6.77) vanishes as $\partial u_x/\partial z = 0$ and $\partial u_z/\partial z = 0$. This also implies that u_x and u_z are not functions of z. Flow inside this channel is going to be a combination of pressure driven flow and drag flow. Hence, Equations (6.76) and (6.77) simplify to

$$\frac{\partial P}{\partial x} = \mu \left(\frac{\partial^2 u_x}{\partial x^2} + \frac{\partial^2 u_x}{\partial y^2} \right) \tag{6.78}$$

$$\frac{\partial P}{\partial z} = \mu \left(\frac{\partial^2 u_z}{\partial x^2} + \frac{\partial^2 u_z}{\partial y^2} \right). \tag{6.79}$$

The pressure gradient in the z direction is not a function of x or y, and u_x and u_z are not a function of z as the flow is fully developed in the z direction. Thus, in Equation (6.79), the pressure gradient has to be a constant along the z direction

$$\underbrace{\frac{\partial P}{\partial z}}_{\text{function of } z \text{ only}} = \underbrace{\mu \left(\frac{\partial^2 u_z}{\partial x^2} + \frac{\partial^2 u_z}{\partial y^2} \right)}_{\text{not a function of } z} = C. \tag{6.80}$$

Thus

$$\frac{\partial P}{\partial z} = C = \frac{\Delta P}{l} = \frac{P_o - P_i}{l} \tag{6.81}$$

where P_o is the pressure near the exit of the screw and P_i is the pressure at the inlet. l is the helical length of the screw and is given in Table 6.1. Now, one can solve

$$\frac{\Delta P}{\mu l} = \frac{\partial^2 u_z}{\partial x^2} + \frac{\partial^2 u_z}{\partial y^2} \tag{6.82}$$

with the boundary conditions

$$u_z = 0 \quad \text{at } x = 0 \text{ and } x = W \tag{6.83}$$

$$u_z = 0 \quad \text{at } y = 0 \text{ and } y = H \tag{6.84}$$

on all four walls. If $W \gg H$, one can ignore $\partial^2 u_z/\partial x^2$ as compared to $\partial^2 u_z/\partial y^2$ and the solution will be valid everywhere except near the walls at $x = 0$ and $x = W$. The complete solution for Equation (6.82) with boundary conditions (6.83) and (6.84) is given by an infinite (Fourier) series. However, the most useful result is the volumetric flow rate

$$Q = \int_0^H \int_0^W u_z \, dx dy = \frac{WH^3}{12\mu l} \Delta P \left(F_p\right) \tag{6.85}$$

where F_p is the shape factor given by the infinite series and is a function of the aspect ratio H/W [132]. The variation of this function with the aspect ratio is shown in Figure 6.23.

One can combine the drag flow in the z direction with the pressure driven flow in the same direction since the fluid is Newtonian and one is dealing with the isothermal case. Thus, the flow rate is

$$Q = \frac{1}{2} V_z W H F_d - \frac{WH^3}{12\mu} \frac{\Delta P}{l} F_p \tag{6.86}$$

where $V_z = V \cos\theta$ and l is the helical length as listed in Table 6.1. Note here the pressure flow is "backflow" in the direction opposite to that of the drag flow. For most extruders

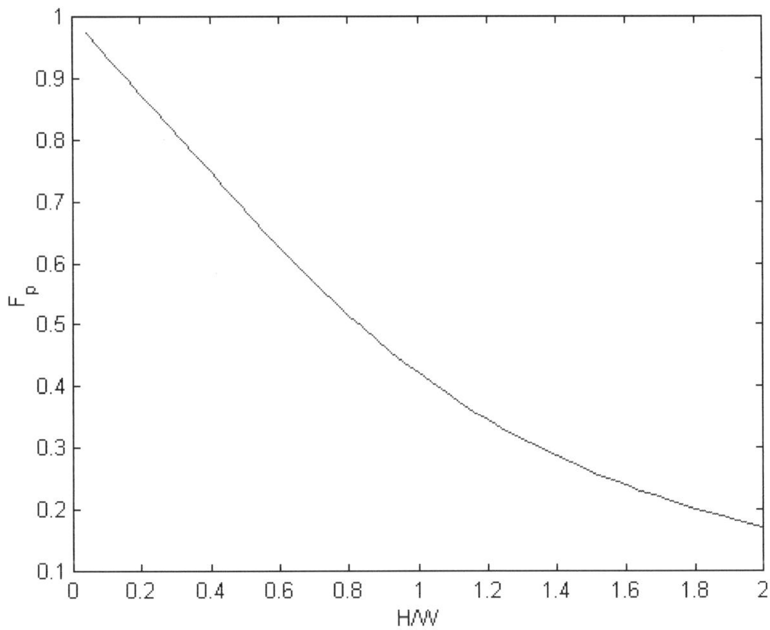

Figure 6.23: Influence of the aspect ratio of the channel on the flow rate in the extruder as expressed by the shape factor, F_p.

W/H is large; hence F_d and F_p are close to unity. The barrel velocity is $V = \pi DN$ and $V_x = V \sin\theta$ and $V_z = V \cos\theta$ and N is the screw speed in revolution per second.

There is also a transverse flow $u_x(x, y)$ due to drag flow in that direction induced by V_x. For large W/H, u_x will have weak dependence on x, except near the walls (the screw flights). The momentum equation can be rewritten as

$$0 = -\frac{\partial P}{\partial x} + \mu \frac{\partial^2 u_x}{\partial y^2} \tag{6.87}$$

subject to boundary conditions

$$u_x = 0 \quad \text{at } y = 0 \tag{6.88}$$

$$u_x = -V_x \quad \text{at } y = H. \tag{6.89}$$

The solution is

$$u_x = -V_x \frac{y}{H} - \frac{1}{2\mu} y(H - y) \frac{\partial P}{\partial x} \tag{6.90}$$

as $\partial P/\partial x$ is not a function of y. Even though the solution ignores the presence of the walls at $x = 0$ and $x = W$, the walls exert one effect that must be taken into account. That is, there is no net flow of material in the x direction because the screw flights prevent it. Therefore

$$\int_0^H u_x \, dy = 0 \tag{6.91}$$

which allows us to find

$$\frac{\partial P}{\partial x} = -\frac{6\mu V_x}{H^2}. \tag{6.92}$$

Hence

$$u_x = \frac{V_x y}{H}\left(2 - \frac{3y}{H}\right). \tag{6.93}$$

This approximation is valid in the middle of the screw, away from the walls. As there is no net flow, the extruder output flow rate Q is not affected. However, V_x is important from the viewpoint of mixing and power requirements. The velocity profile that will be created is shown schematically in Figure 6.24. This confinement and the velocity direction helps the fluid and the fibers to mix easily.

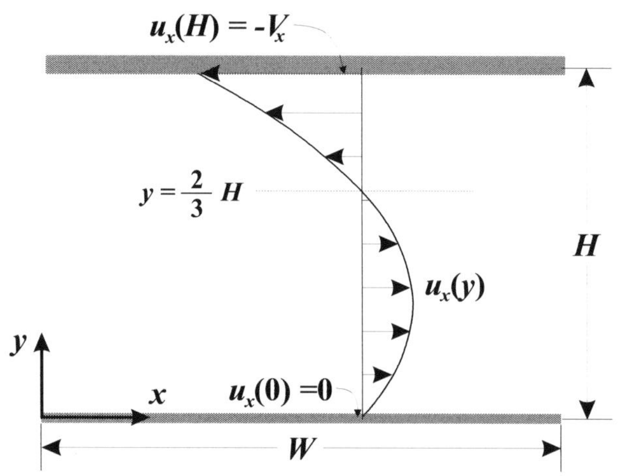

Figure 6.24: The velocity component u_x as a function of y inside the channel created by the extruder screw and barrel.

6.3.2 Calculation of Power Requirements [5]

The power needed for driving the screw in the melt zone is given by

$$\begin{aligned} \dot{\omega} &= \mathbf{F} \cdot \mathbf{U} \\ &= \int_0^l \int_0^W \left(\tau_{xy}\Big|_H u_x + \tau_{zy}\Big|_H u_z \right) dx\, dz. \end{aligned} \tag{6.94}$$

Here l is the length of the screw, $l = L/\sin\theta$, and W is the flight width. The shear stress terms are calculated as

$$\tau_{xy} = \mu\left(\frac{\partial u_x}{\partial y} + \frac{\partial u_y}{\partial x}\right) \tag{6.95}$$

$$\tau_{zy} = \mu\left(\frac{\partial u_z}{\partial y} + \frac{\partial u_y}{\partial z}\right) \tag{6.96}$$

The $\frac{\partial u_y}{\partial x}$ term in Equation (6.95) is zero since a uniform flow in the x direction is assumed, and the $\frac{\partial u_y}{\partial z}$ term in Equation (6.96) is negligible compared to the other term $\frac{\partial u_z}{\partial y}$ for large

W/H ratios. The velocity components are

$$u_x = \frac{V_x y}{H}\left(2 - \frac{3y}{H}\right) \tag{6.97}$$

$$u_z = \frac{V_z y}{H} - \frac{y}{2\mu}(H-y)\frac{\Delta P}{l}. \tag{6.98}$$

To obtain u_z component, Equation (6.82) was simplified to $\frac{\Delta P}{\mu l} = \frac{\partial^2 u_z}{\partial y^2}$ with the assumption of $\frac{\partial^2 u_z}{\partial x^2} << \frac{\partial^2 u_z}{\partial y^2}$, and integrating the equation in y and applying the boundary conditions given in Equation (6.84) and a pressure gradient of $\Delta P/l$ across the z direction. Hence the power required is

$$\dot{\omega} = \left(\frac{4\mu V_x^2}{H} + \frac{\mu V_z^2}{H} + \frac{V_z W \Delta P}{2l}\right) lW. \tag{6.99}$$

For basic performance, one needs to calculate Q, $\dot{\omega}$ and ΔP. Equations (6.86) and (6.99) are two of the three equations. We need a third equation to uniquely solve for Q, $\dot{\omega}$ and ΔP.

If there is no restriction at the end of the screw, $\Delta P = 0$. That implies that there is no mechanism to build up the pressure. Normally, there is a die downstream which establishes the pressure rise down the extruder axis. Hence the additional equation comes from pressure drop across the die associated with die output.

For Newtonian fluids

$$Q = \frac{K_G}{\mu}\Delta P \tag{6.100}$$

where K_G is related to die geometry and characteristics. To analyze this further, rewrite Equation (6.86) as

$$Q = AN - C\frac{\Delta P}{\mu} \tag{6.101}$$

where

$$A = \frac{1}{2} W H F_d \cos\theta \pi D \tag{6.102}$$

and

$$C = \frac{WH^3}{12l} F_p \tag{6.103}$$

where the channel length is given by $l = L/\sin\theta$, and L is the screw length, and N is the screw speed in revolution per second, then solving Equations (6.100) and (6.101) simultaneously results in

$$Q = \frac{AK_G}{C + K_G} N \tag{6.104}$$

$$\Delta P = \frac{\mu A}{C + K_G} N \tag{6.105}$$

as the flow rates in the screw section and die section must be the same. Note that Q does not depend on μ. Also, $Q_{\max} = AN$ as $K_G \to \infty$, and $\Delta P_{\max} = \mu AN/C$ as $K_G \to 0$.

6.3.3 Variable Channel Length [5]

Screws rarely have uniform geometry down the helical axis. The depth H varies in some way depending on the purpose of the extruder and the type of material it handles. As derived in [132], if channel depth is a function of z, but if the change is gradual, then u_z and u_x can still be considered independent of z. However, we can maintain $H(z)$ and $\partial P/\partial z$ as functions of z.

The simplest approximation in such situations is

$$u_z = V_z \frac{y}{H(z)} - \frac{1}{2\mu(z)} y \left[H(z) - y\right] \frac{\partial P}{\partial z}. \tag{6.106}$$

The flow rate Q, which is normal to the helical axis should still be a constant; hence

$$Q = \frac{1}{2} V_z W H(z) - \frac{W H^3(z)}{12\mu} \frac{\partial P}{\partial z} \tag{6.107}$$

and hence

$$\frac{\partial P}{\partial z} = \frac{6\mu V_z}{H^2(z)} - \frac{12\mu Q}{W H^3(z)}. \tag{6.108}$$

As $H(z)$ is a function of the screw dimensions, it is a known quantity. For example, consider a linear approximation

$$H(z) = H_o + \frac{H_l - H_o}{l} z \tag{6.109}$$

as shown in Figure 6.25. Now we can integrate Equation 6.108 as follows:

$$\int_0^l \frac{\partial P}{\partial z} dz = \int_0^l \frac{6\mu V_z}{H^2(z)} dz - \int_0^l \frac{12\mu Q}{W H^3(z)} dz. \tag{6.110}$$

Rearranging the integrated equation for flow rate results in

$$\begin{aligned} Q &= \frac{1}{2} V_z W \frac{H_o H_l}{\frac{1}{2}(H_o + H_l)} - \frac{W}{12\mu} \frac{(H_o H_l)^2}{\frac{1}{2}(H_o + H_l)} \frac{\Delta P}{l} \\ &= AN - C \frac{\Delta P}{\mu} \end{aligned} \tag{6.111}$$

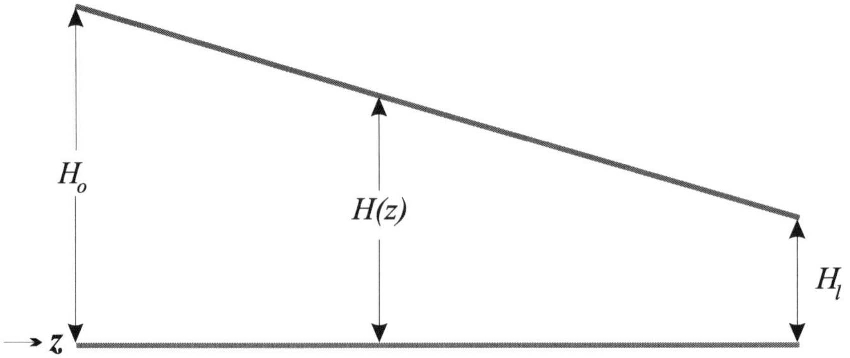

Figure 6.25: A linear variation in depth of the screw, H(z).

where

$$A = \frac{1}{2}\pi DW <H_1> \cos\theta \tag{6.112}$$

$$C = \frac{W <H>^3}{12l} \tag{6.113}$$

where

$$<H_1> = \frac{H_o H_l}{\frac{1}{2}(H_o + H_l)} \tag{6.114}$$

$$<H>^3 = \frac{(H_o H_l)^2}{\frac{1}{2}(H_o + H_l)}. \tag{6.115}$$

Example 6.5: Comparison of Results with (a) $H =$ constant and (b) $H = H(z)$

Calculate the pressure build-up, ΔP and flow rate, Q in the extruder channel by assuming that (a) the channel thickness, H is constant ($=$ the average channel thickness $= (H_o + H_l)/2$), and (b) $H(z)$ is decreasing linearly along the unwrapped channel direction z as illustrated in Figure 6.25. Use the following process parameters:

- Resin viscosity, $\mu = 120000$ Pa.s (assume that $\mu \neq \mu(z)$)
- Screw length, $L = 1.920$ m
- Screw diameter, $D = 0.080$ m
- Screw speed, $N = 1$ rev/s
- Inlet channel depth, $H_o = 0.015$ m
- Exit channel depth, $H_l = 0.002$ m
- Helix angle, $\theta = 20^o = (20/180)\pi$ rad
- Flight width, $e = 0.008$ m
- Die geometry characteristic, $K_G = 5.0 \times 10^{-7}\,\text{m}^3$ (see Equation (6.100)).

Solution

(a) Equations (6.104) and (6.105) were used in a MATLAB code (*example6_05.m* given in the Appendix) to calculate the pressure build-up, $\Delta P - 18.8$ MPa and the flow rate, $Q = 78.1 \times 10^{-6}\,\text{m}^3/\text{s} = 78.1$ cc/s. Note that, $H =$ constant $= (15 + 2)/2 = 8.5$ mm was assumed in this part, and $F_d = 1$ and $F_p = 1$ were taken considering that $H << W$.

(b) Assuming that the pressure build-up found in part (a), $\Delta P - 18.8$ MPa is valid here as well, the flow rate, $Q = 32.5 \times 10^{-6}\,\text{m}^3/\text{s} = 32.5$ cc/s was calculated using Equation (6.111) which is only 41.6 % of the answer calculated in part (a). This indicates that the assumption of $H =$ constant $= H_{\text{mid}}$ caused an overpredicted calculation of the flow rate compared to the more realistic solution when a linearly decreasing $H = H(z)$ was used in it. ■

6.3.4 Newtonian Adiabatic Analysis [5]

The heating of the polymer depends on the heat supplied by the heaters mounted on the barrels in the melting stage. The dissipative action of the polymer melt creates shearing forces that generate heat as well.

Let us assume for the sake of analysis that the barrel is insulated (hence adiabatic analysis) and no heat exchange takes place with the barrel in the molten polymer zone. This is strictly not true, but one can get a good understanding of the coupling between the momentum and energy equations to address the issue of temperature-dependent viscosity. This analysis follows the one presented in Tadmor and Gogos [132].

$$u_z = V_z \frac{y}{H} - \frac{1}{2\mu(z)} y(H - y) \frac{\partial P}{\partial z} \tag{6.116}$$

and

$$Q = AN - \frac{Cl}{\mu(z)} \frac{\partial P}{\partial z}. \tag{6.117}$$

Integrate with respect to z to obtain the pressure drop ΔP

$$\Delta P = \frac{AN - Q}{Cl} \int_0^l \mu(z)\, dz. \tag{6.118}$$

It is necessary to introduce the temperature rise into the equation

$$\frac{d\dot{\omega}}{dl} = \rho c_p Q \frac{dT}{dz} + Q\, dP \tag{6.119}$$

where the last term $Q\, dP$ in Equation (6.119) is neglected as it is small. From Equation (6.99)

$$\dot{\omega} = E\mu N^2 l + AN\Delta P \tag{6.120}$$

where the last term $AN\Delta P$ is neglected, and E is defined as

$$E = \frac{\pi^3 D^3 \sin\theta}{H} (1 + \sin^2\theta). \tag{6.121}$$

Taking the derivative of Equation (6.120) with respect to l, we have

$$\frac{d\dot{\omega}}{dl} = E\mu N^2. \tag{6.122}$$

By equating Equations (6.119) and (6.122), one obtains

$$\mu\, dz = \frac{\rho c_p Q}{EN^2} dT. \tag{6.123}$$

One can substitute Equation (6.123) into Equations (6.117) and (6.118), which results in

$$\Delta P = \frac{AN - Q}{Cl} \int_{T_0}^{T_l} \frac{\rho c_p Q}{EN^2} dT \tag{6.124}$$

and

$$Q = AN - \frac{EClN^2}{\rho c_p} \frac{\Delta P}{Q\Delta T}. \tag{6.125}$$

Q is now quadratic with N, and ΔT enters implicitly. Let us assume that the viscosity is dependent on temperature with the following form

$$\mu = a\, e^{-bT}. \tag{6.126}$$

Recalling that

$$\mu\, dz = \frac{\rho c_p Q}{EN^2}\, dT \tag{6.127}$$

from Equation (6.123), one can obtain

$$a\, e^{-bT}\, dz = \frac{\rho c_p Q}{EN^2}\, dT. \tag{6.128}$$

By multiplying both sides of Equation (6.128) by e^{bT} and integrating with respect to z, the following equation is obtained:

$$e^{b\Delta T} = \frac{b\mu_0 EN^2 l}{\rho c_p Q} + 1 \tag{6.129}$$

where $\mu_0 = a\, e^{-bT_0}$ is the entrance viscosity at $z = 0$ and ΔT is the temperature difference across the channel from $z = 0$ to $z = l$. If we now define a new variable $x = e^{b\Delta T}$ and consider that as our primary variable

$$Q = \frac{b\mu_0 EN^2 l}{\rho c_p (x-1)} \tag{6.130}$$

Now the die equation becomes

$$Q = \frac{K_G \Delta P}{\mu(z)} = \frac{K_G x \Delta P}{\mu_0}. \tag{6.131}$$

Q can also be written from the screw characteristic equation, Equation (6.125):

$$Q = AN - \frac{EClN^2}{\rho c_p Q} \frac{\Delta P}{\Delta T}. \tag{6.132}$$

Now as $e^{b\Delta T} = x$, one can write ΔT in terms of x and b

$$\Delta T = \frac{\ln x}{b} \tag{6.133}$$

and from Equation (6.131)

$$\frac{Q}{\Delta P} = \frac{K_G x}{\mu_0}. \tag{6.134}$$

By inserting Equations (6.133) and (6.134) into Equation (6.131), one can get

$$Q = AN - \frac{EClN^2 b V_0}{\rho c_p K_G x \ln x}. \tag{6.135}$$

By eliminating Q from Equations (6.130) and (6.135), one can get

$$x \ln x \left(N_2 - \frac{1}{x-1} \right) - N_1 = 0 \tag{6.136}$$

where $N_1 = C/K_G$ and $N_2 = (A\rho c_p)/(Nl\mu_0 bE)$. For any N_1 and N_2 one can obtain the solution for x (recall that $x = e^{b\Delta T}$) and hence find Q and ΔP.

More on the modeling of extrusion can be found in textbook written by Tadmor and Gogos on this topic [132].

Example 6.6: Plot the Contours for x as a Function of N_1 and N_2 using Equation (6.136)

Write a MATLAB code to plot the isocurves of x.

Solution

A MATLAB code, *example6_06.m*, was coded and given in the Appendix; the reader is suggested to run it to generate Figure 6.26. Notice that this code is very compact as it calculates N_2 for a given set of x and N_1 values, and then plot the results on N_2 versus N_1 graph. This figure can be used as a look-up table to calculate what x value satisfies Equation (6.136) for known values of N_1 and N_2. ■

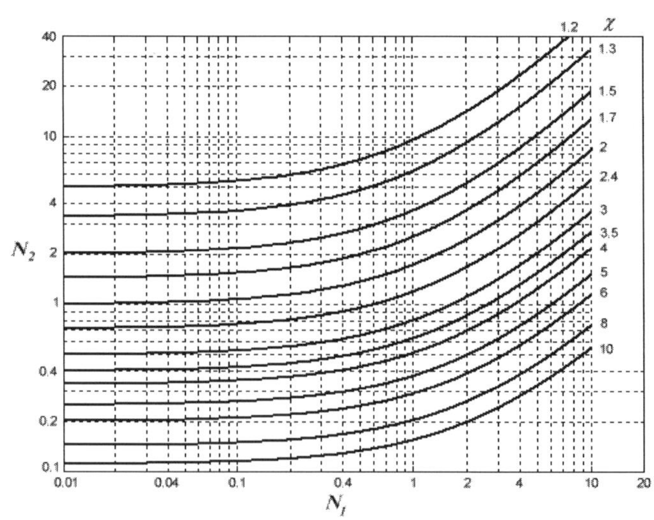

Figure 6.26: Variation of screw parameters (N_1 and N_2) for adiabatic case when the viscosity is temperature dependent [5]. Find x from N_1 and N_2, and use it in Equations (6.133)-(6.135) to calculate ΔT, Q and ΔP.

Example 6.7: Coupled Momentum and Energy Equations

Calculate the pressure build-up ΔP, temperature rise ΔT and flow rate Q in the extruder channel by using the coupled momentum and energy equations. Use the following material, geometric and process parameters:

- Screw length, $L = 1.920\,\text{m}$
- Screw diameter, $D = 0.080\,\text{m}$
- Screw speed, $N = 12\text{-}30\,\text{rpm} = 0.2\text{-}0.5\,\text{rev/s}$
- Inlet channel depth, $H_o = 0.020\,\text{m}$
- Exit channel depth, $H_l = 0.004\,\text{m}$
- Helix angle, $\theta = 20^o = (20/180)\pi\,\text{rad}$
- Flight width, $e = 0.008\,\text{m}$
- Die geometry characteristic, $K_G = 1.2 \times 10^{-8}\,\text{m}^3$ (see Equation (6.100))
- Fiber volume fraction, $V_f = 0.20$ (i.e., 20%)
- Density of fiber, $\rho_f = 2500\,\text{kg/m}^3$
- Density of matrix, $\rho_m = 1200\,\text{kg/m}^3$
- Thermal conductivity of fiber, $k_f = 1.20\,\text{W/(m}^o\text{K)}$
- Thermal conductivity of matrix, $k_m = 0.20\,\text{W/(m}^o\text{K)}$
- Specific heat capacity of fiber, $c_{pf} = 800\,\text{J/(kg}^o\text{K)}$
- Specific heat capacity of matrix, $c_{pm} = 700\,\text{J/(kg}^o\text{K)}$
- Inlet temperature, $T_o = 50 + 273.2 = 323.2\,^o\text{K}$
- Coefficient in viscosity equation, $a = 6.5 \times 10^5\,\text{Pa.s}$ (see Equation (6.126))
- Power term in viscosity equation, $b = 0.009\,^o\text{K}^{-1}$ (see Equation (6.126)).

Plot the results as a function of the rotational speed of the screw, N in the given range of 12-30 rpm.

Solution

Using the process parameters, the two terms $N_1 = C/K_G$ and $N_2 = (A\rho c_p)/(NlV_0bE)$ are calculated in a MATLAB code, *example6_07.m* (given in the Appendix), and a solution for x is found by solving the root of Equation (6.136) as shown in Figure 6.27. Then, Equations (6.135), (6.134) and (6.133) are used to calculate the flow rate Q, ΔP and ΔT, respectively. The readers are suggested to run *example6_07.m* to investigate their own process window so that a desired set of variables (Q, ΔP and ΔT) can be achieved. The results of this example are plotted in Figure 6.28 when the rotational speed of the screw, N, is set to a value in a range of 12-30 rpm. ■

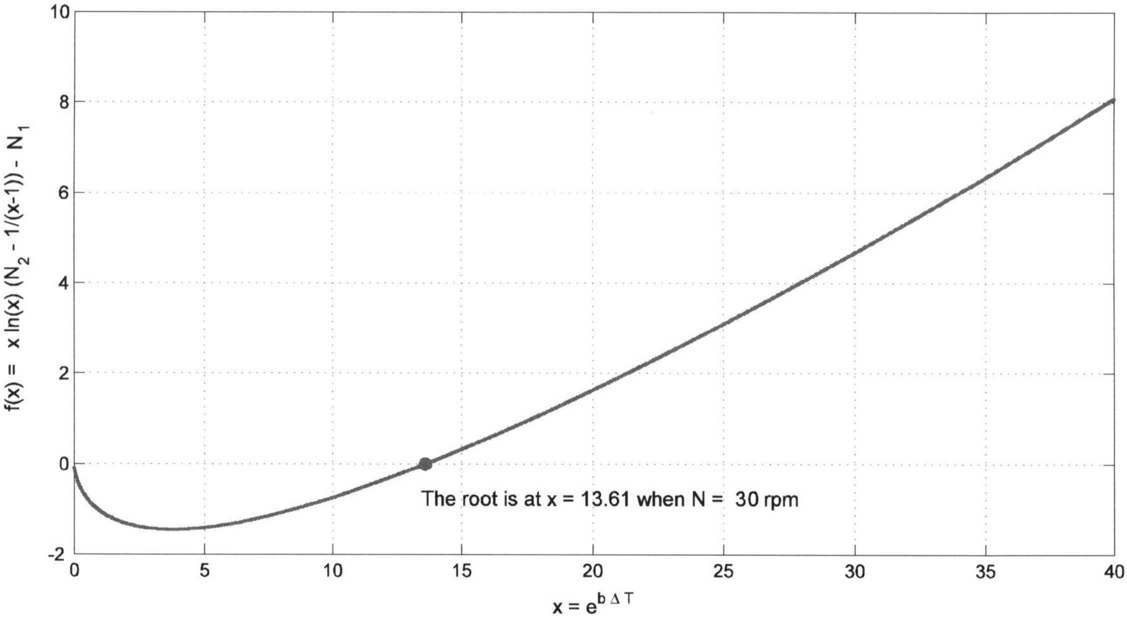

Figure 6.27: The root of Equation (6.136) when the screw speed is $N = 30\,\text{rpm}$.

6.4 Injection Molding

6.4.1 Process Description

Injection molding is a widely used manufacturing process for producing parts in a mold that has a cavity in the shape of the part being manufactured. In this process, solid pellets containing resin and fibers are fed through a hopper into the barrel, which contains the screw. It is similar in design to the extrusion process and is the heart of the injection molding machine. The basic steps of the process are schematically shown in Figure 6.29. During the melting stage, the heaters attached to the barrel heat the pellets. Simultaneously, the rotating screw pushes the raw material forward inducing friction and shear. The heat from viscous dissipation will melt the polymer and mix it uniformly with the short fibers, creating a fiber suspension. The accumulating suspension in the injection chamber relative to the screw moves forward gradually. After a sufficient amount of suspension is amassed, the screw moves axially forward acting as a pump and injecting the suspension into the cavity of the mold through an opening called a *gate*. The suspension, which is under high pressure due to the pumping action of the screw, quickly flows and fills the mold cavity. During the flow, the fibers move freely in the polymer melt. During flow, the fibers change the positions and orientations thus evolving the microstructure of the part. The heat from the polymer is conducted to the cold walls of the mold as soon as the suspension contacts the walls. The high pressure is maintained during the packing stage. In this stage, maintaining very high pressure offsets contraction due to cooling and solidification of the polymer. The injection pressure used is in the range of 10 to 100 MPa. During the cooling stage, the resin solidifies and freezes the orientation state of the fibers. Finally the mold is opened, and the part is ejected from the cavity. The complete cycle takes from a few seconds to less than a few minutes depending on the volume of the cavity to be filled. As a result, injection molding is well suited for high volume production parts.

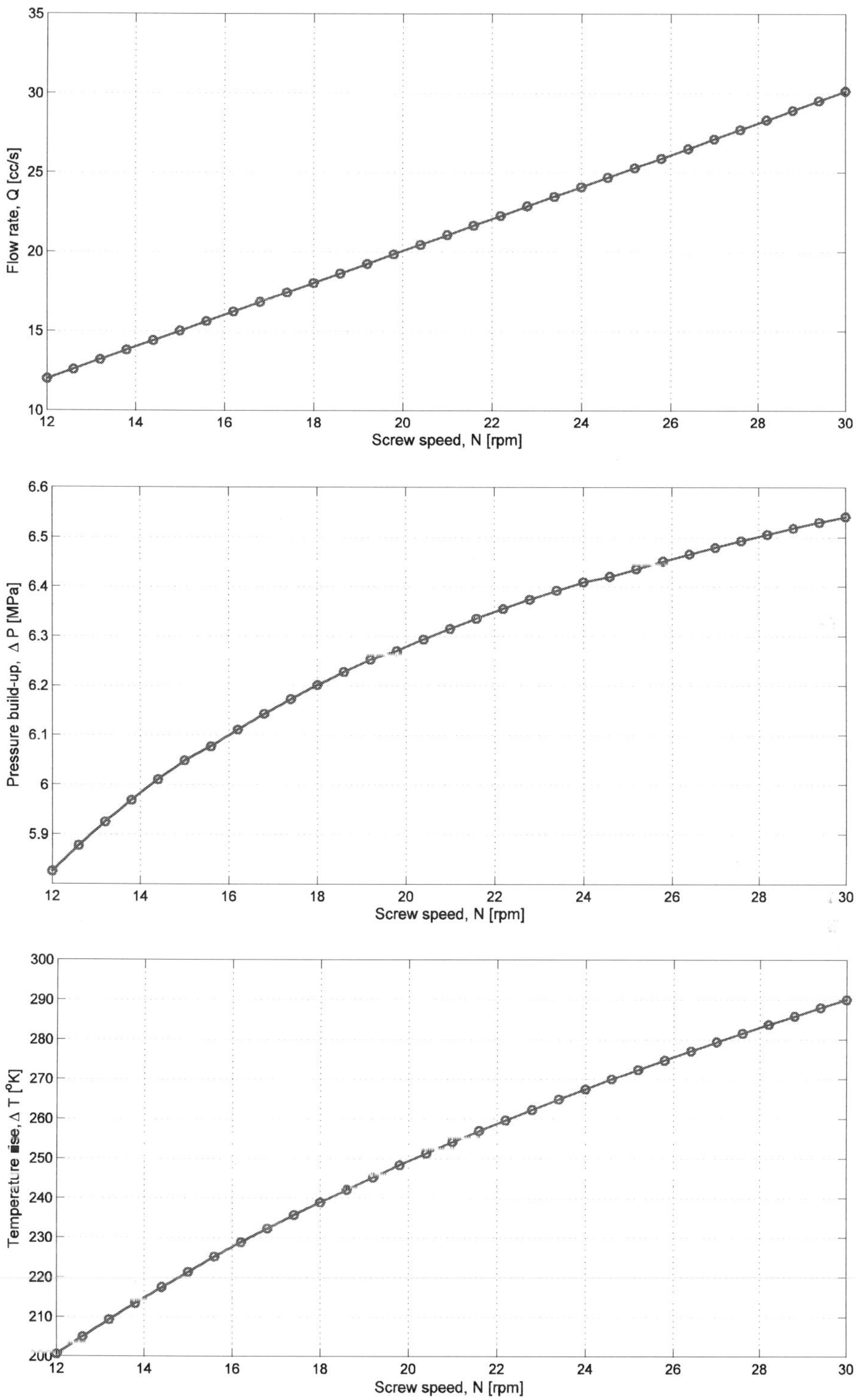

Figure 6.28: Flow rate Q, pressure build-up ΔP, and temperature rise ΔT in the composite when the screw rotates between 12 and 30 rpm.

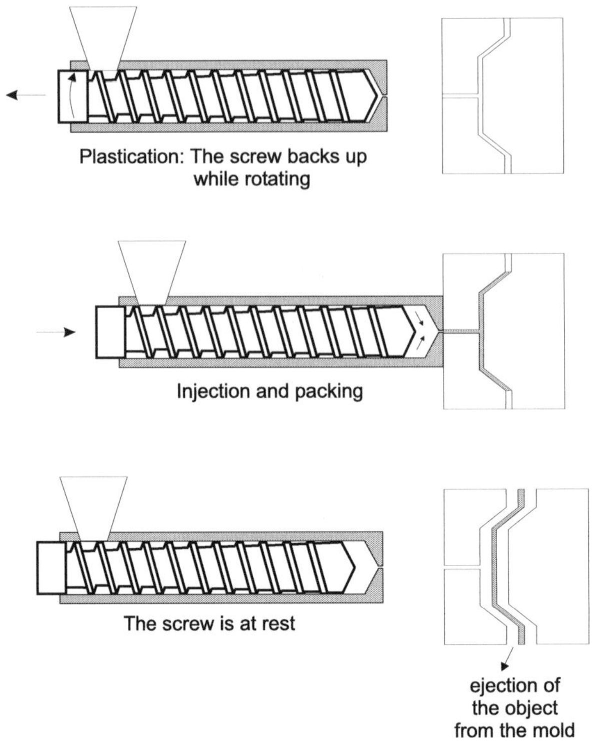

Figure 6.29: Basic steps in the injection molding process.

6.4.2 Materials

Both thermoplastic and thermoset polymers can be injection molded. Most often used thermoplastics are PE (polyethelene), PPS (polyphenylenesulfide) and PA (polyamide), which are usually glass-reinforced. High-performance thermoplastics such as PPS and PEEK (polyetheretherketone) might be carbon-reinforced. Unsaturated polyesters and phenolics are the most common thermosets used. The reinforcement is usually short glass fibers in the range of 0.5–5 mm in length. The raw material supplied to the barrel of the injection machine is usually in the form of pellets or powder.

6.4.3 Applications

One of the advantages of injection molding is that geometrically complex components with highly accurate dimensions can be economically manufactured in high volumes. Also, as the process is automated, the cost per part will decrease rapidly as the volume increases. It is not uncommon to make several hundred thousand to a million parts with a single mold. For example, most plastic household items are produced utilizing this process. Examples of short fiber reinforced articles are components for light machinery, equipment housings and sprockets, and a variety of automobile components including appearance parts such as tailgates, instrument clusters, lamp housings, oil pumps, water pumps, bumpers and air intake manifolds. This wide range of applications is constantly increasing. Recently as there is interest in use of bio-based materials, most of the natural materials such as wood flour, jute and hemp fibers can be easily integrated with polymers with an extruder or an injection molding machine to increase the application range to bio-based composite materials which is expected to create green products and reduce waste. Other possible applications include replacement of conventional structures consisting of numerous components (manufactured separately and assembled) with a single, lightweight, net shape composite part.

The number and variety of parts manufactured by injection molding is large. It includes items as small as paper clips to as large as automobile bumpers. Nonetheless, the applications are limited to non-load carrying components because of the poor mechanical properties of the injection molded parts caused by:

- Relatively low fiber volume fraction,
- Short length of reinforcement fibers,
- Not much control over the alignment of the fibers induced by the flow,
- Fiber degradation in the feed screw and injection gate.

Despite these limitations of the injection molding process, it has numerous advantages such as:

- Unlimited geometrical complexity of the manufactured parts,
- Size ranging from very small to very large,
- Excellent control of dimensional tolerances and surface quality,
- High volume production.

6.4.4 Critical Issues

The most important issues in injection molding arise during the filling and packing stages. During the filling stage, the cold mold must be filled completely with liquid resin suspension before it becomes solid (incomplete filling is called a "short shot"). At the same time, the shear rate and viscous dissipation should not exceed particular values to avoid resin degradation. The analysis of the filling stage is further complicated because of its non-isothermal and non-Newtonian character. Usually the mold surface is cold and the polymer melt containing the fibers is hot. Hence the polymer freezes at the mold surfaces, which leads to constriction of the flow path and thus it requires to increase in injection pressure to impregnate the suspension in the mold. Viscous heating can cause another complication. Because of high viscosity and small flow channels, the temperature of the resin will increase and may cause polymer degradation. In addition, the orientation of the fibers in the polymer resin will continuously change as the suspension flows into the mold. This can lead to significant variation in the fiber orientation state in the part and across the thickness which will influence the physical and mechanical properties of the part. Thus, it is important to understand the relationship between flow kinematics and fiber orientation. In summary, the mold filling stage is a nonisothermal, non-Newtonian transient flow of fiber suspensions in a complex geometry. To include all these effects, even in a simple mold, is a formidable task. In the next section, we will pose the problem with all the governing equations and the variables that need to be solved. However, it is beyond the scope of this book to provide a solution. We will provide selected results from research papers that have attempted this approach. In this chapter, our approach will be to address and solve selected aspects of the problem to develop a better understanding of some of the observed aspects.

Scientific understanding of the process, expressed in the form of reliable models, is of great help for example in the design of molds, which tend to be expensive components of the equipment. Once the mold is made, for it to produce over a million components, it is milled and treated to handle the wear and tear. A single mold can cost from several thousand to a million dollars, and take from several days to a few months to fabricate. After the mold is fabricated, an error in the mold design will lead to either incomplete filling or undesired flow and fiber orientation pattern. This will require one to retool the mold or scrap the mold and "redesign it for the next trial." This trial and error approach can prove expensive in time and money. Thus, it is time and cost effective to resort to process models and computer simulations to guide the mold design and also the injection machine characteristics such as maximum pressure required and the screw design. The process model and simulations can provide useful information such as

- Optimal number and positions of injection gates.
- Required filling pressures and clamping force to avoid *short shots* (incomplete filling) and *flashes* (excess material along the edges due to insufficient clamping pressure).
- Filling time: if the mold is filled too slowly, premature solidification may prevent complete filling; if the mold is filled too rapidly, thermal degradation may occur due to the heat released from viscous dissipation which is proportional to the square of the shear rate.
- Temperature distribution: a balanced temperature distribution is desirable to avoid excessive residual stresses that lead to warpage.

In order to obtain parts with desired mechanical properties, it is equally important to design the mold and control the process in such a way as to get favorable fiber orientation states. Thus, models that predict the evolution of the fiber orientation will also be useful. There is a simulation package available called Moldflow [306] which can predict fiber orientation during flow. It is based on the model we will discuss in the next section.

6.4.5 Model Formulation for Injection Molding

Most injection moldings are thin in one direction compared to their other dimensions. Hence, one can assume that the part being molded is thin and also flat for simplicity. The part may have otherwise arbitrary shape as long as there are no abrupt changes in thickness. Figure 6.30 shows a part that fits these requirements, along with the coordinate system used. Note that the cavity thickness, $2b$, may be a function of x and y. We will first restrict our formulation to a Newtonian isothermal case which will later be extended to non-Newtonian and nonisothermal cases.

Newtonian Isothermal Case

A thin cavity translates into velocity in the z direction being insignificant compared to the velocities in the x and y directions and this results in $\partial P/\partial z$ being close to zero. Hence, one can safely assume that P will be a function of x and y only and we need to be concerned about fluid motion in the x-y plane only. As with most composite processing situations, one can assume that inertia and gravity are negligible. For the isothermal Newtonian case, the equations of motion, N-S (Navier-Stokes) equations will reduce to

$$\frac{\partial P}{\partial x} = \mu \left(\frac{\partial^2 u}{\partial x^2} + \frac{\partial^2 u}{\partial y^2} + \frac{\partial^2 u}{\partial z^2} \right) \tag{6.137a}$$

$$\frac{\partial P}{\partial y} = \mu \left(\frac{\partial^2 v}{\partial x^2} + \frac{\partial^2 v}{\partial y^2} + \frac{\partial^2 v}{\partial z^2} \right). \tag{6.137b}$$

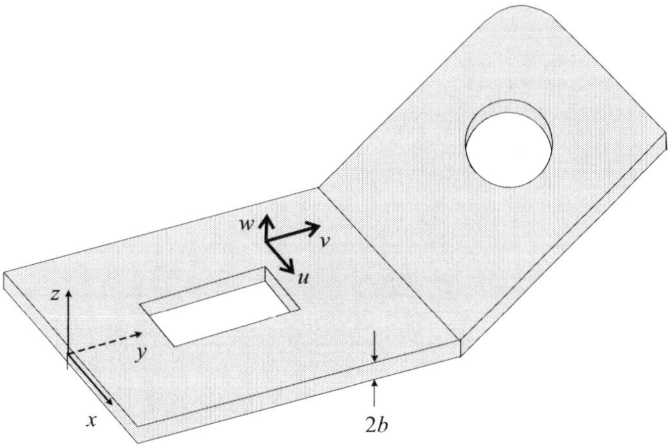

Figure 6.30: Typical dimensions of parts made using the injection molding process. u, v and w are the x, y and z components of the resin velocity during mold filling.

The velocity components u and v vary from zero at the mold walls to a maximum at the center of the thickness of the cavity along z. One may assume similar variation along the x-y plane. However, as the distance along the z direction is of the order of b, which is much smaller than the distances in the x-y plane, $\partial^2 u/\partial z^2$ and $\partial^2 v/\partial z^2$ will be much larger than the other terms on the right-hand side of Equation (6.137). This simplifies the equation of motion to

$$\frac{\partial P}{\partial x} = \mu \frac{\partial^2 u}{\partial z^2} \tag{6.138a}$$

$$\frac{\partial P}{\partial y} = \mu \frac{\partial^2 v}{\partial z^2}. \tag{6.138b}$$

Since P is not a function of z, each equation can be integrated twice to give the parabolic velocity profiles

$$u = \frac{1}{2\mu} \frac{\partial P}{\partial x} z^2 + c_1 z + c_2 \tag{6.139a}$$

$$v = \frac{1}{2\mu} \frac{\partial P}{\partial y} z^2 + c_3 z + c_4. \tag{6.139b}$$

The boundary conditions are that the velocity gradients are zero at the center due to symmetry and no-slip at the mold walls. Mathematically,

$$\frac{\partial u}{\partial z} = \frac{\partial v}{\partial z} = 0 \qquad \text{at } z = 0 \tag{6.140a}$$

$$u = v = 0 \qquad \text{at } z = b. \tag{6.140b}$$

Thus, the velocity profiles can be written as

$$u = \frac{1}{2\mu} \frac{\partial P}{\partial x} \left(z^2 - b^2\right) \tag{6.141a}$$

$$v = \frac{1}{2\mu} \frac{\partial P}{\partial y} \left(z^2 - b^2\right) \tag{6.141b}$$

This solution looks similar to flow between infinite parallel plates subjected to a pressure drop boundary condition. However, in this case, the pressure gradient terms, $\partial P/\partial x$ and $\partial P/\partial y$ are functions of x and y and are still unknown.

Also, as in compression molding, one can take the ratio of u/v and find that it is not a function of z, thus exhibiting the feature of Hele-Shaw flows. This allows one to average the velocities in the z direction at every x and y and simplifies the problem. These average velocities are

$$\bar{u}(x,y) = \frac{1}{b}\int_0^b u(x,y,z)\,dz \tag{6.142a}$$

$$\bar{v}(x,y) = \frac{1}{b}\int_0^b v(x,y,z)\,dz. \tag{6.142b}$$

Once $\bar{u}$ and $\bar{v}$ are known functions of x and y, and as one knows the velocity profile in the z direction from Equations (6.141a) and (6.141b), one can find the velocity as a function of x, y and z. We find that $\bar{u}$ and $\bar{v}$ for a part of thickness $2b$ depend on the pressure gradient as follows:

$$\bar{u} = -\frac{b^2}{3\mu}\frac{\partial P}{\partial x} \tag{6.143a}$$

$$\bar{v} = -\frac{b^2}{3\mu}\frac{\partial P}{\partial y} \tag{6.143b}$$

The incompressible continuity equation is

$$\frac{\partial u}{\partial x} + \frac{\partial v}{\partial y} + \frac{\partial w}{\partial z} = 0. \tag{6.144}$$

This contains the velocity component w which we have ignored so far in the analysis, so we may be tempted to do so here as well. However, although we know that w will be much smaller than u and v, this does not imply that $\partial w/\partial z$ will be smaller than $\partial u/\partial x$ or $\partial v/\partial y$. In fact, as the thickness is much smaller than the other two dimensions, one can safely argue that the $\partial w/\partial z$ term will not be smaller than the other two terms in the equation. Thus, one cannot just drop $\partial w/\partial z$. To address this issue, the general approach adopted is to not satisfy the continuity equation exactly at every location in the suspension. Instead, one can integrate the continuity in the z direction across the cavity thickness,

$$\int_{-b}^{b} \frac{\partial u}{\partial x}\,dz + \int_{-b}^{b} \frac{\partial v}{\partial y}\,dz + \int_{-b}^{b} \frac{\partial w}{\partial z}\,dz = 0. \tag{6.145}$$

This will ensure no material is lost. Rearranging the order of integration results in

$$\frac{\partial}{\partial x}\int_{-b}^{b} u\,dz + \frac{\partial}{\partial y}\int_{-b}^{b} v\,dz + \int_{-b}^{b} dw = 0. \tag{6.146}$$

The first two integrals are $2b\bar{u}$ and $2b\bar{v}$, while the third integral must equal zero (since w equals zero at $z = -b$ and at $z = b$), which results in

$$\frac{\partial}{\partial x}(b\bar{u}) + \frac{\partial}{\partial y}(b\bar{v}) = 0. \tag{6.147}$$

Substitution of Equation (6.143) into Equation (6.147) results in the governing equation to be solved for the pressure distribution

$$\frac{\partial}{\partial x}\left(-\frac{b^2}{3\mu}\frac{\partial P}{\partial x}\right)+\frac{\partial}{\partial y}\left(-\frac{b^2}{3\mu}\frac{\partial P}{\partial y}\right)=0. \tag{6.148}$$

If b is not a function of x and y (if the part is of uniform thickness, $2b$) and if the viscosity is constant, the equation can be further simplified to the Laplace equation

$$\frac{\partial^2 P}{\partial x^2}+\frac{\partial^2 P}{\partial y^2}=0. \tag{6.149}$$

Hence, by using simplifications and physics of the process, we have been able to reduce the problem from four variables (u, v, w, P) to one variable P. Once the pressure is found, one can calculate the velocities. The main issue now reduces to solving Equation (6.149).

To solve Equation (6.149), one needs to apply boundary conditions. As seen from Figure 6.31, at the flow front the polymer simply displaces air. If the mold cavity is well vented, this mathematically translates into a boundary condition of $P = 0$ at the flow front (assuming gauge pressure), where the air is being displaced by the suspension. The boundaries where the polymer touches the edge of the mold in the x-y plane can be mathematically represented as $\partial P/\partial n = 0$, where n denotes the normal to the boundary. Physically, this implies there is no flow through the mold walls (however, the tangential velocity component along the mold wall can be non-zero). The other important feature in modeling of such flows is that the flow front is in motion. When simulating mold filling, the approach is to start with the region near the gate. Solve for the pressure field, find the velocities once the pressures are known, advance the flow front and resolve Equation (6.149) for the new domain. This process is repeated until the complete mold is filled with the suspension. Again, in the solution of this problem, one assumes quasi-steady state. That is, at each time step, we assume that the process is steady and solve for Equation (6.149). This is a reasonable assumption as long as the Reynolds number of the fiber suspension is much less than one.

To solve these problems for isothermal and Newtonian cases in any arbitrary two-dimensional geometry, one has to resort to numerical methods. There is one commercial but many research software packages available that implement the model discussed above.

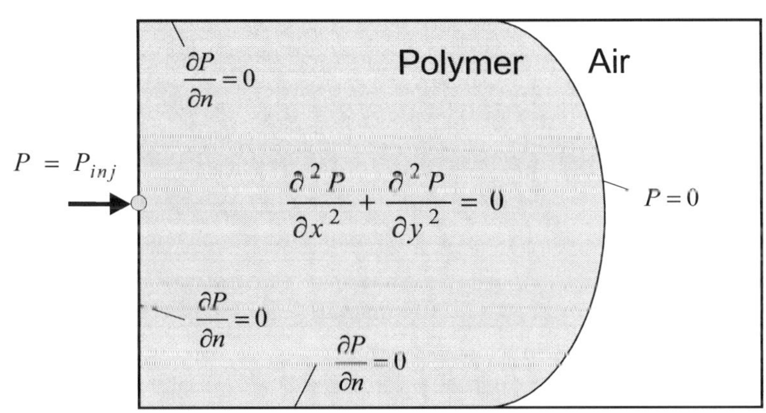

Figure 6.31: Boundary conditions on pressure for the domain containing the polymer suspension.

The commercial package used is called Moldflow [307]. However, we will present an example of injection molding of a center gated disk under isothermal and Newtonian conditions in which a closed form solution is possible.

Example 6.8: Mold Filling Time for Center Gated Disk [5]

Consider injection molding of a fiber suspension into a radial mold to make disks of radius R and thickness $2h$. The gate is located in the center of the disk and the suspension is injected under constant pressure (see Figure 6.32).

a) Find the time to fill the mold under the assumption that the suspension behaves as a Newtonian viscous fluid.

b) What will be the fill time under a constant flow rate injection at Q? Under this condition, what will be the maximum pressure?

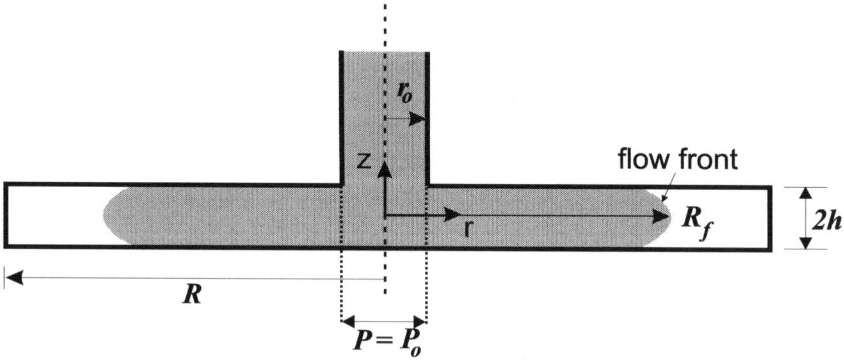

Figure 6.32: Injection molding of a center gated disk.

Solution

a) We need to make the following assumptions to simplify the solution to the problem:

- $h << R$.
- Entrance effects are negligible.
- Fiber orientation influence on viscosity is negligible.
- Quasi-steady state.
- Inertia effects are negligible.
- u_z and $u_\theta = 0$.

Thus, mass conservation dictates

$$\frac{1}{r}\frac{\partial}{\partial r}(ru_r) = 0 \tag{6.150}$$

Therefore,

$$u_r = \frac{C(z,t)}{r}. \tag{6.151}$$

Equation of motion in the r direction for a Newtonian fluid, when $h << R$ (only shear stresses being important) reduces to

$$\frac{\partial P}{\partial r} = \mu \frac{\partial^2 u_r}{\partial z^2} = \frac{\mu}{r}\frac{\partial^2 C}{\partial z^2}. \tag{6.152}$$

Since $P \neq P(z)$ and C is not a function of r,

$$\frac{r}{\mu}\frac{\partial P}{\partial r} = \frac{\partial^2 C}{\partial z^2} = A(t). \tag{6.153}$$

Integrating C twice and using the no slip boundary conditions at the mold walls, which translate into

$$C = 0 \text{ at } z = h \tag{6.154}$$

$$\partial C/\partial z = 0 \text{ at } z = 0 \tag{6.155}$$

results in

$$C = -\frac{Ah^2}{2}[1 - (\frac{z}{h})^2]. \tag{6.156}$$

Integrating P with respect to r in Equation (6.153) using the boundary condition of $P = P_0$ at $r = r_0$ results in

$$P - P_0 = A\mu \ln(\frac{r}{r_0}). \tag{6.157}$$

To find $A(t)$, one can use the B.C. at the flow front, R_f (see Figure 6.32), which assumes the pressure at the flow front to be atmospheric pressure ($P = P_{atm}$ at $r = R_f$). Let us assume that P_0 is gauge pressure; hence $P = 0$ at $r = R_f$ in terms of gauge pressure. Hence, Equation (6.157) will become

$$-P_0 = A\mu \ln(\frac{R_f}{r_0}). \tag{6.158}$$

As R_f is a function of time, A will also be a function of time. In order to find R_f as a function of time, first note that the flow rate Q is given by:

$$Q = 4\pi \int_0^h r u_r dz = 4\pi h R_f \frac{dR_f}{dt}. \tag{6.159}$$

Substitution of Equations (6.151) and (6.156) into Equation (6.159) and its integration results in

$$-\frac{4\pi h^3 A}{3} = 4\pi h R_f \frac{dR_f}{dt}. \tag{6.160}$$

Thus,

$$A = -R_f \frac{dR_f}{dt}(\frac{3}{h^2}). \tag{6.161}$$

Substitute Equation (6.161) in Equation (6.158) to get

$$P_0 = R_f \frac{dR_f}{dt}\mu(\frac{3}{h^2})\ln(\frac{R_f}{r_0}). \tag{6.162}$$

One can integrate R_f with respect to t to find how the flow front moves with time to fill the mold. When $R_f = R$, that value of time will give us the fill time.

If one defines

$$r^* = \frac{r_0}{R} \tag{6.163}$$

$$P^* = \frac{P_0}{3\frac{\mu R^2}{h^2}} \quad \text{(has units of 1/time)} \tag{6.164}$$

and

$$x^* = \frac{R_f}{r_0} \tag{6.165}$$

then Equation (6.162) can be integrated to give

$$x^{*2}\ln(x^*) - \frac{1}{2}(x^{*2} - 1) = \frac{2P^*t}{r^{*2}}. \tag{6.166}$$

The mold is filled when $R_f = R$, which translates into $x^* = \frac{1}{r^*}$. Substitution of this into Equation (6.166) gives t_{fill}, the time to fill the mold,

$$t_{fill} = \frac{1}{2P^*}[-\ln(r^*) - \frac{1}{2}(1 - r^{*2})]. \tag{6.167}$$

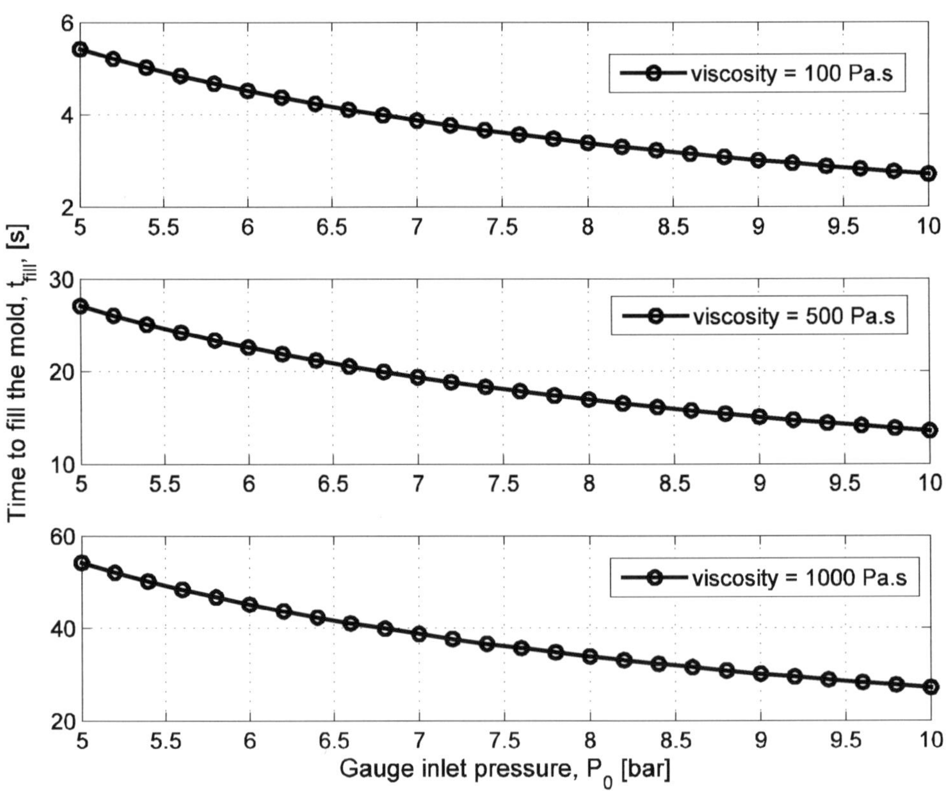

Figure 6.33: Mold filling time, t_{fill} at different constant gauge injection pressure, P_0 for μ =100, 500 and 1000 Pa.s.

For example, if $P_0 = 5$ bars, $r_o = 1$ cm, $R = 10$ cm, $h = 1$ mm and $\mu = 1000$ Poise ($= 100\,$Pa.s), then $t_{fill} = 5.4$ seconds. A MATLAB code, *example6_08a.m* (given in the Appendix), was used to plot the effect of resin viscosity and inlet pressure on the time to fill the mold, t_{fill} in Figure 6.33.

b) If the injection was done under constant flow rate, the fill time can be calculated from the information of flow rate and the volume of the disk that needs to be filled. The engineering quantity of interest for the constant flow rate case is the maximum pressure requirement.

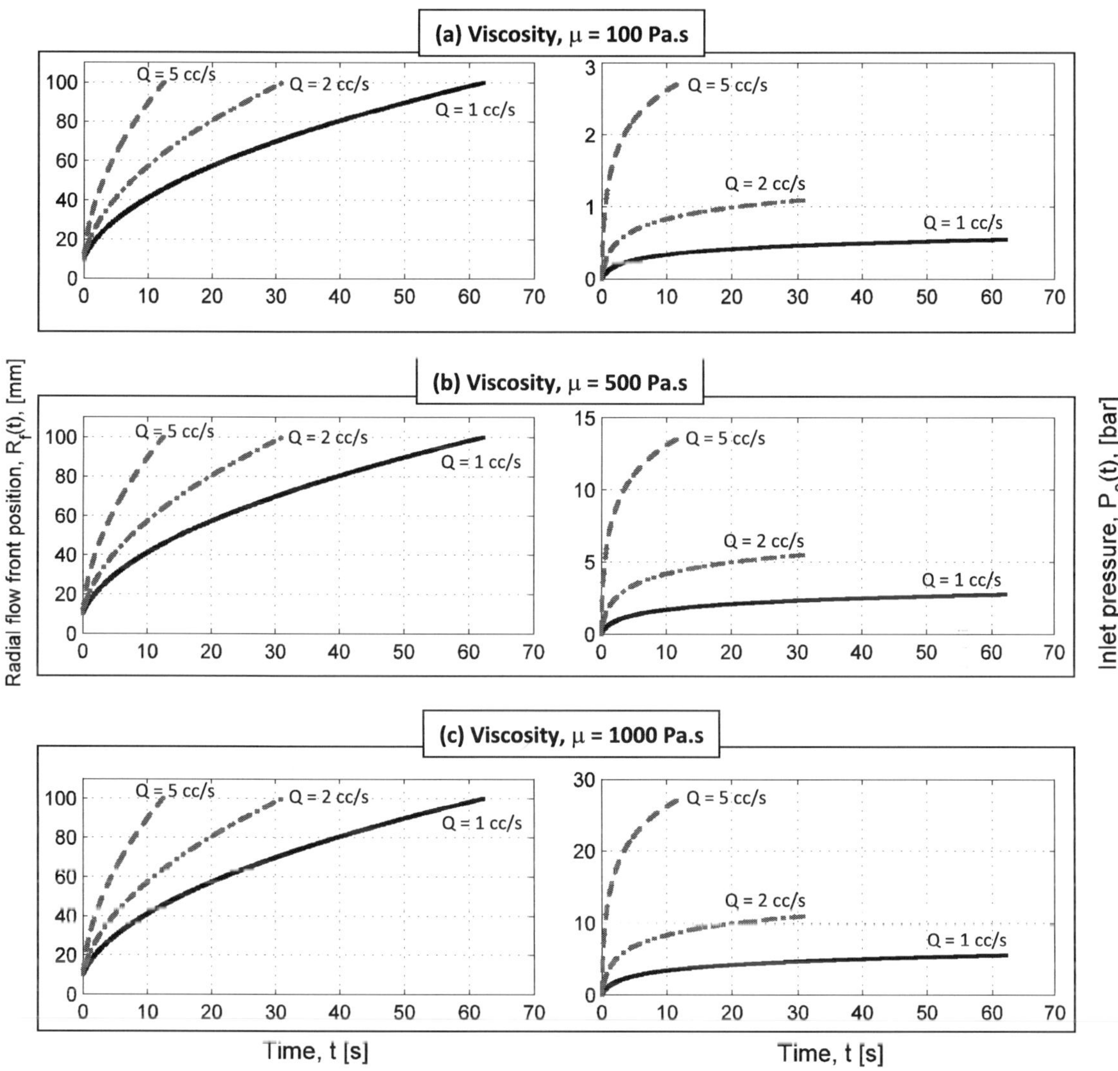

Figure 6.34: Radial flow front position, $R_f(t)$ and injection pressure, $P_0(t)$ at different constant flow rate, Q. Notice that $R_f(t)$ is not affected by the viscosity, μ, but the injection pressure increases as μ increases. $t_{fill} = 62.2\,$s, $31.1\,$s and $12.4\,$s for $Q = 1\,$cc/s, $2\,$cc/s and $5\,$cc/s, respectively.

For constant flow rate case,

$$t_{fill} = \frac{2\pi h(R^2 - r_0^2)}{Q} \tag{6.168}$$

where Q is the prescribed flow rate. Also from Equation (6.159)

$$R_f \frac{dR_f}{dt} = \frac{Q}{4\pi h}. \tag{6.169}$$

One can integrate this equation as Q is constant with B.C. that $R_f = r_0$ at $t = 0$ to give

$$R_f(t) = \sqrt{\frac{Q}{2\pi h}t + r_0^2}. \tag{6.170}$$

Thus,

$$P_0(t) = \frac{3Q\mu}{8\pi h^3}\ln(1 + \frac{Qt}{2\pi h r_0^2}) \tag{6.171}$$

As seen from Equation (6.171), the maximum pressure P_0 will be when $t = t_{fill}$.

Equations (6.170) and (6.171) were used in a MATLAB code, *example6_08b.m* (given in the Appendix) to plot $R_f(t)$ and $P_0(t)$ for $\mu = 100$, 500 and 1000 Pa.s and $Q = 1$, 2 and 5 cc/s in Figure 6.34.

See Ranganathan and Advani [308] to include the influence of fiber orientation on the flow velocity and the conditions under which it is important. ■

Example 6.9: Numerical Solution of Example 6.8

Solve Example 6.8 by using a numerical time integration method (here we will apply a straightforward method, Euler method), starting from Equation (6.162) instead of solving it analytically. Take $\mu = 100\,\text{Pa.s}$, $P_0 = 5\,\text{bars}$, $r_o = 1\,\text{cm}$, $R = 10\,\text{cm}$ and $h = 1\,\text{mm}$. In addition to calculating the time to fill the mold, t_{fill} also calculate $R_f(t)$ and $Q(t)$ and plot them versus time.

Solution

First, re-write Equation (6.162) in such a way that the time derivative term is on the left-hand side:

$$\frac{dR_f}{dt} = f(R_f, t) = \frac{P_0 h^2}{3\mu}\frac{1}{R_f \ln(\frac{R_f}{r_0})}. \tag{6.172}$$

The function, $f(R_f, t)$ is a general case, and in this case it is only a function of R_f as shown on the right-hand side of Equation (6.172). Euler method allows a discrete time integration:

$$R_f(i+1) = R_f(i) + f[R_f(i), t(i)]\Delta t \tag{6.173}$$

and

$$Q(i) = 4\pi h R_f(i) f[R_f(i), t(i)] \tag{6.174}$$

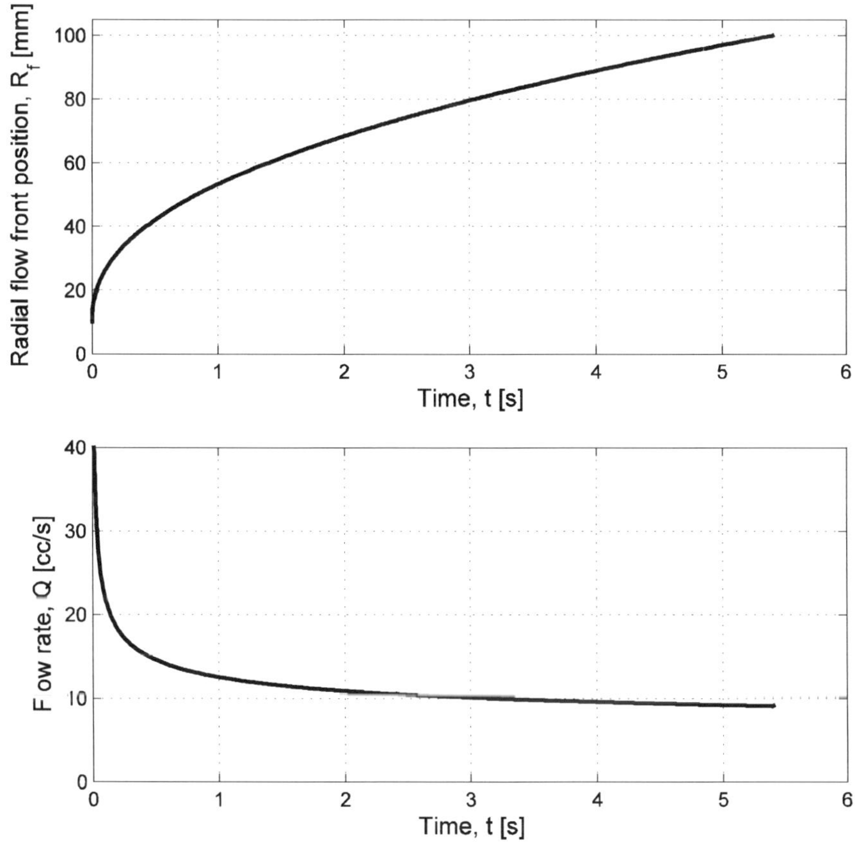

Figure 6.35: Numerically evaluated $R_f(t)$ and $Q(t)$ for a constant injection pressure of $P_0 =$ 5 bars and viscosity of $\mu = 100\,\text{Pa.s}$. Notice that dR/dt and Q decrease as t increases. The flow front reaches the end of the mold cavity (i.e., $R_f(t) = R = 100\,\text{mm}$) at $t = t_{fill} = 5.4\,\text{s}$ which agrees with the analytical solution found in the previous example.

where Δt is a discrete time increment, and the arguments i and $i+1$ correspond to the current and next time steps, respectively. The initial condition is

$$R_f(1) = r_0. \tag{6.175}$$

The function f and thus dR_f/dt is singular at $t = 0$. To avoid this singularity problem, the initial condition will be replaced with

$$R_f(1) = r_0(1+\epsilon) \tag{6.176}$$

where $\epsilon << 1$.

A MATLAB code, *example6_09.m* (given in the Appendix) was formulated, written and run to evaluate $R_f(t)$ and $Q(t)$, and the corresponding graphs are shown in Figure 6.35. The user is suggested to run the code and observe the convergence of the results at the selected Δt and ϵ values used. ■

Nonisothermal and Non-Newtonian Case

The cavity filling stage is rarely isothermal. One always has hot resin injected into a cold mold. Moreover, the viscosity is a function of the temperature. So, energy and momentum equations are coupled. Also, most polymer melts are shear thinning in nature, so one must also pay attention to this in engineering analysis of injection molding.

First, one can rewrite the energy equation in terms of temperature and heat generation. The heat is generated because of the viscous dissipation of the resin in the narrow cavity if the resin is a thermoplastic resin. The heat is generated due to cure-kinetic reaction if the material injected is a thermoset.

$$\underbrace{\rho c_p\left(\frac{\partial T}{\partial t}\right)}_{\text{transient}}+\underbrace{\rho c_p\left(u\,\frac{\partial T}{\partial x}+v\,\frac{\partial T}{\partial y}+w\,\frac{\partial T}{\partial z}\right)}_{\text{convection}} = \underbrace{\frac{\partial}{\partial x}\left(k_x\,\frac{\partial T}{\partial x}\right)+\frac{\partial}{\partial y}\left(k_y\,\frac{\partial T}{\partial y}\right)+\frac{\partial}{\partial z}\left(k_z\,\frac{\partial T}{\partial z}\right)}_{\text{conduction}}$$

$$+\underbrace{\eta\left[\left(\frac{\partial u}{\partial z}\right)^2+\left(\frac{\partial v}{\partial z}\right)^2\right]}_{\text{viscous dissipation}}+\underbrace{\dot{q}.}_{\text{cure kinetics}} \tag{6.177}$$

Note that we have retained terms of very high shearing in the viscous dissipation term. One may also ignore the conduction in the x and y directions. Using dimensionless analysis, one can easily show that the conduction in the z direction is at least two to three orders of magnitude higher than in the x and y directions (assuming the thermal conductivity is of the same order). For thermoplastics, $\dot{q} = 0$. For thermosets, usually the viscous dissipation term contribution is minimal, due to their low viscosity.

Almost all research text conveniently ignores the term $w\partial T/\partial z$, although by dimensionless analysis, you could prove that the unsteady term and all three convective terms are of the same order. Hence to be consistent, the $w\partial T/\partial z$ term should be retained. However, when we solve for flow, we do not solve for w, so it creates the problem of how to find w. So, the assumption is made that the convective contribution due to w, at least in the mid-region, is zero, as w is identically zero in that region. It is a reasonable engineering approximation. This assumption breaks down near the inlet gate or near the flow front where one will encounter entrance and exit effects, respectively.

In addition, the viscosity is temperature dependent, so one should account for its change in the momentum equation. As in the compression molding case, one does so in an average sense by integrating the viscosity through the thickness direction [309]

$$\frac{\partial}{\partial x}\left(S\,\frac{\partial P}{\partial x}\right)+\frac{\partial}{\partial y}\left(S\,\frac{\partial P}{\partial y}\right)=0 \tag{6.178}$$

where

$$S=\int_0^b \frac{z^2}{\eta}\,dz. \tag{6.179}$$

S is a measure of the fluidity, with b denoting the cavity half thickness in the z direction and $\eta(\dot{\gamma}, T, P)$ denoting the shear viscosity of the molten polymer. For example, in terms of the power-law model

$$\eta = m(T,P)\,\dot{\gamma}^{n-1} \tag{6.180}$$

where

$$m(T,P) = A\exp(\frac{T_a}{T})\,\exp(bP) \tag{6.181}$$

it follows that

$$S = \frac{1}{\Lambda}(\frac{\Lambda}{m_0 e^{bP}})^{\frac{1}{n}}\chi \tag{6.182}$$

where

$$\Lambda = \|\Delta P\| = \sqrt{\left(\frac{\partial P}{\partial x}\right)^2 + \left(\frac{\partial P}{\partial y}\right)^2} \tag{6.183}$$

and

$$\chi = \int_0^h z(\frac{z}{g(T)})^{\frac{1}{n}} dz \tag{6.184}$$

with

$$m(T, P) = m_0 g(T) e^{bP} \tag{6.185}$$

and

$$g(T) = \exp(\frac{T_a}{T} - \frac{T_a}{T_0}). \tag{6.186}$$

Note that if the power law index is $n = 1/4$, the fluidity index as expressed by Equation (6.182) will be proportional to the cube of the pressure gradient, which will make Equation (6.178) highly nonlinear. Under such conditions of coupled momentum and energy equation with nonlinearities present, it is difficult to solve such problems in an analytical manner. However, one can use dimensionless analysis to estimate the role of coupling and also the extent of heat dissipation.

Example 6.10: Flow of a Thermoplastic Suspension in a Thin Cavity

Consider the flow of a thermoplastic suspension in a thin cavity of height $2h$ as shown in Figure 6.36. The suspension is injected under a constant flow rate of Q in m^3/s. The surfaces of the mold are held at a constant temperature T_1. Assume that the viscosity does not change much with temperature and that steady state in flow and heat transfer has been achieved inside the long mold. Plot the T_{midplane} as a function of flow rate for a suspension of viscosity η.

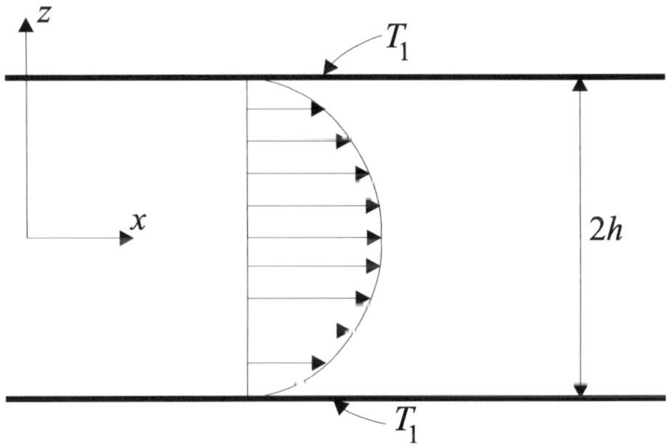

Figure 6.36: Flow of a thermoplastic suspension in a thin cavity of height $2h$.

Solution

As the viscosity, η is assumed to be not changing significantly with temperature, one can first solve the velocity distribution by using the equation of motion alone. Secondly, the calculated velocity distribution is used in the energy equation to solve the temperature profile.

Together with the assumptions made above, we will assume no-slip boundary conditions at the mold walls. These will lead to a Poiseuille flow. However, the flow rate Q is specified instead of the usual pressure drop ΔP along the channel of a length L. The equation of motion for this one-dimensional, steady-state flow is

$$0 = -\frac{\partial P}{\partial x} + \eta \frac{\partial u^2}{\partial z^2}. \tag{6.187}$$

For this fully developed flow, the pressure gradient in x direction is constant and can be written as $\partial P/\partial x = dP/dx = \Delta P/L = (P_e - P_i)/L$, where P_e and P_i are the exit and inlet pressures and L is the length of the channel. One might object to this by saying that the values of P_e, P_i and L were not specified in the problem. What we will do here is that we will first solve for $u(z)$ in terms of $\Delta P/L$, and then calculate the corresponding flow rate Q taking the integral of $u(z)$ across the thickness per width W. Then, we will rewrite u in terms of specified flow rate Q by eliminating the $\Delta P/L$ term. So, let's rewrite Equation (6.187) as

$$\frac{\partial^2 u}{\partial z^2} = \frac{\Delta P}{\eta L} \tag{6.188}$$

which can be integrated twice in z to give

$$u(z) = \frac{\Delta P}{2\eta L} z^2 + c_1 z + c_2. \tag{6.189}$$

Symmetry of the profile requires that $\partial u/\partial z = 0$ at $z = 0$ hence $c_1 = 0$. The no-slip boundary condition is $u = 0$ at the walls $z = \pm h$. This can be satisfied if $c_2 = -\Delta P h^2/(2\eta L)$. Hence, the velocity profile is given as

$$u(z) = \frac{\Delta P}{2\eta L}(z^2 - h^2). \tag{6.190}$$

The flow rate can be calculated as follows:

$$\begin{aligned} Q &= W \int_{-h}^{h} u(z)\, dz = 2W \int_0^h u(z)\, dz \\ &= 2W \int_0^h \frac{\Delta P}{2\eta L}(z^2 - h^2)\, dz \\ &= -\frac{\Delta P}{2\eta L}\frac{4Wh^3}{3} \end{aligned} \tag{6.191}$$

where W is the width of the channel. By replacing $\Delta P h^2/(2\eta L)$ term by $-3Q/(4Wh^3)$ in Equation (6.190), we have the velocity profile in terms of Q

$$u(z) = \frac{3}{4}\frac{Q}{Wh}\left[1 - \left(\frac{z}{h}\right)^2\right]. \tag{6.192}$$

One can introduce nondimensionalized variables as follows:

$$u^* = \frac{u}{Q/(Wh)}, \qquad z^* = z/h. \tag{6.193}$$

Equation (6.192) can be rewritten in nondimensional form as follows:

$$u^*(z^*) = \frac{3}{4}\left[1 - (z^*)^2\right]. \tag{6.194}$$

In order to solve the temperature profile, the energy equation will be used

$$\rho C_p u \frac{\partial T}{\partial x} = k\frac{\partial^2 T}{\partial z^2} + \eta\left(\frac{\partial u}{\partial z}\right)^2. \tag{6.195}$$

Assuming $\rho C_p u(\partial T/\partial x)$ is negligible compared to the two other terms, Equation (6.195) can be simplified to

$$\frac{\partial^2 T}{\partial z^2} = -\frac{\eta}{k}\left(\frac{\partial u}{\partial z}\right)^2. \tag{6.196}$$

By introducing a nondimensionalized temperature as follows:

$$\theta = \frac{T - T_1}{T_1}, \tag{6.197}$$

Equation (6.196) can be rewritten in nondimensional form

$$\frac{\partial^2 \theta}{\partial (z^*)^2} = -A\left(\frac{\partial u^*}{\partial z^*}\right)^2. \tag{6.198}$$

where the nondimensional coefficient A is given by

$$A = \frac{\eta}{kT_1}\left(\frac{Q}{Wh}\right)^2. \tag{6.199}$$

One can write $A = BQ^2$ where $B = \eta/(kT_1W^2h^2)$ to easily see the effect of Q on θ and hence T while plotting the mid-plane temperature versus flow rate.

One can evaluate $\partial u^*/\partial z^* = -3z^*/2$ from Equation (6.194) and then substitute it into Equation (6.198) to have

$$\frac{\partial^2 \theta}{\partial (z^*)^2} = -\frac{9A}{4}(z^*)^2. \tag{6.200}$$

By integrating both sides of Equation (6.200) twice with respect to z^*, one gets

$$\theta(z^*) = -\frac{3A}{16}(z^*)^4 + c_3 z^* + c_4. \tag{6.201}$$

The symmetry of the temperature profile about the mid-plane can be mathematically expressed as $\partial\theta/\partial z^* = 0$ at $z^* = 0$. This requires $c_3 = 0$. The other constant c_4 is evaluated by using the boundary condition $\theta = (T_1 - T_1)/T_1 = 0$ at $z^* = \pm 1$. Hence, the resulting profile is given by

$$\theta(z^*) = \frac{3A}{16}\left[1 - (z^*)^4\right]. \tag{6.202}$$

The nondimensional mid-plane temperature is

$$\theta_{\text{midplane}} \equiv \theta(z^* = 0) = \frac{3A}{16} = \frac{3B}{16} Q^2 \tag{6.203}$$

as a function of Q.

Equation (6.203) can be rewritten as follows:

$$T_{\text{midplane}} \equiv T(z = 0) = T_1 \left[1 + \left(\frac{3\eta}{16 k T_1 W^2 h^2} \right) Q^2 \right] \tag{6.204}$$

for the dimensional temperature at the mid-plane. Q and T_{midplane} versus P_i are shown in Figure 6.37 for the following material, geometric and process parameters:

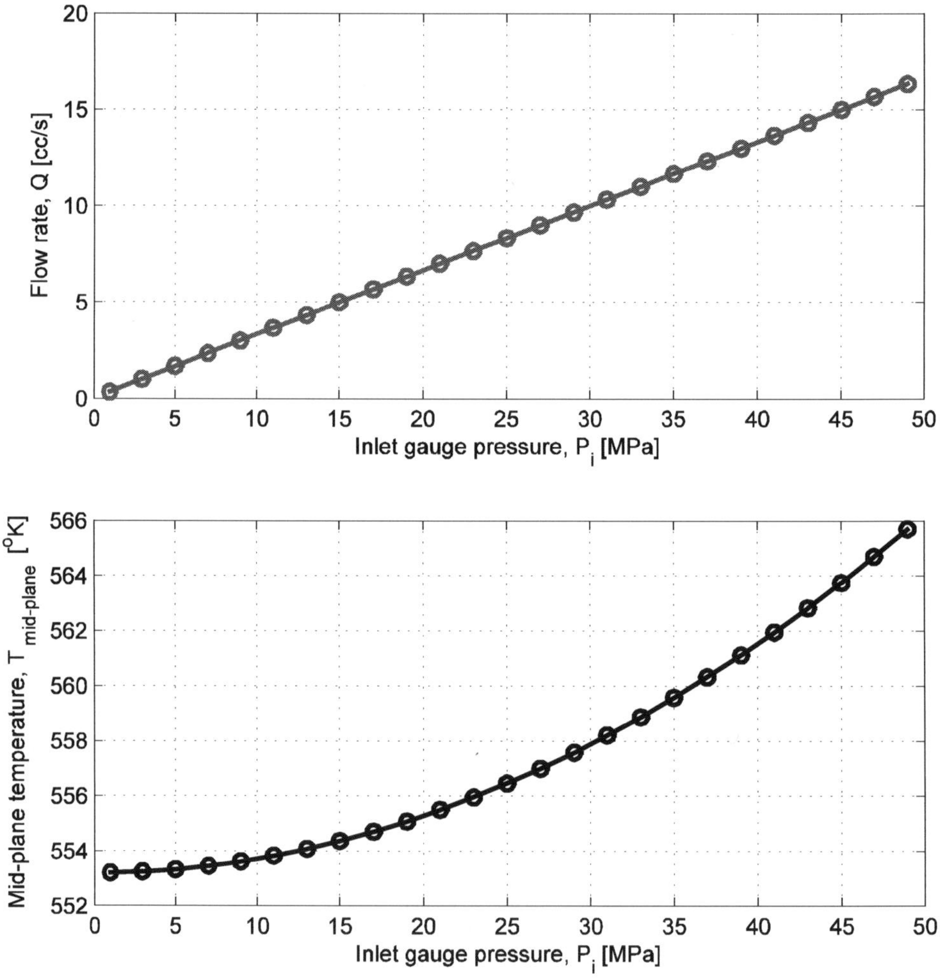

Figure 6.37: (a) Flow rate, Q and (b) mid-plane temperature, $T_{\text{midplane}} = T(z = 0)$, of thermoplastic material versus inlet gauge pressure, P_i. The material is injected at a flow rate of Q and the upper and lower mold walls are kept at a constant temperature T_1 which is 280 + 273.2 °K.

- Inlet temperature, $T_1 = 280\,°\mathrm{C} = 553.2\,°\mathrm{K}$
- Viscosity of the resin, $\eta = 50,000\,\mathrm{Pa.s}$
- Thickness of the channel, $h = 0.005\,\mathrm{m}$
- Width of the channel, $W = 0.1\,\mathrm{m}$
- Length of the channel, $L = 0.5\,\mathrm{m}$
- Thermal conductivity of composite, $k = 0.8\,\mathrm{W/(m°K)}$
- Inlet gauge pressure, P_i = 1-50 MPa.

A MATLAB code, *example6_10.m* (given in the Appendix), was used to calculate and plot Q and T_{midplane} in Figure 6.37. As seen from the figure, the flow rate increases linearly (see Equation (6.191)) and the mid-plane temperature increases quadratically (see Equation (6.204)) with inlet gauge pressure. ■

6.4.6 Fiber Orientation

As the molten polymer and the fibers are injected through a gate under high pressure (1 MPa to 10 MPa), they flow as a suspension into the mold cavity. As the suspension flows, the fibers are transported along with the resin and change orientation during the flow. The frozen-in orientation of the fibers determines the microstructure and properties of the composite. Hence, one of the key issues is to find the orientation of the fibers as they flow into the mold cavity. However, the flow and fiber orientation are coupled. We have already seen how the deformation of the suspension can influence fiber orientation state and how the flow is influenced by fiber orientation, as the fluid stresses will change with changes in fiber orientation as shown in the constitutive equations in Chapter 4, and will influence the viscosity of the suspension which influences the flow field.

Qualitatively, fibers align in the direction of shear and stretching. It has been observed that many injection molded parts exhibit a skin/core effect. That is, near the surface the fibers are aligned in the flow direction and in the core they can be oriented transverse to the flow direction [310]. The reason for this effect can be understood if one models the fiber orientation and the flow kinematics. Using simple fluid mechanics and the fact that shearing and stretching align fibers, we will explain this behavior with an example.

Example 6.10: Skin Core Effect in Injection Molding

Consider a circular disk being molded with short fiber suspensions from a gate in the center as shown in Figure 6.38. The disk is of radius r_0 and thickness $2h$, where $2h$ is much smaller than r_0. Assume that the suspension behaves as a power law fluid. Discuss qualitatively how the skin core structure is formed during injection molding of short fiber composites.

Solution

We will consider the isothermal case for simplicity. The suspension is injected through the center gate at a known flow rate. It makes perfect sense to choose a cylindrical coordinate system with its origin at the center of the disk, and $z = \pm h$ are the surfaces of the disk. As the flow would be radial, under the assumption of creeping flow and fully developed flow,

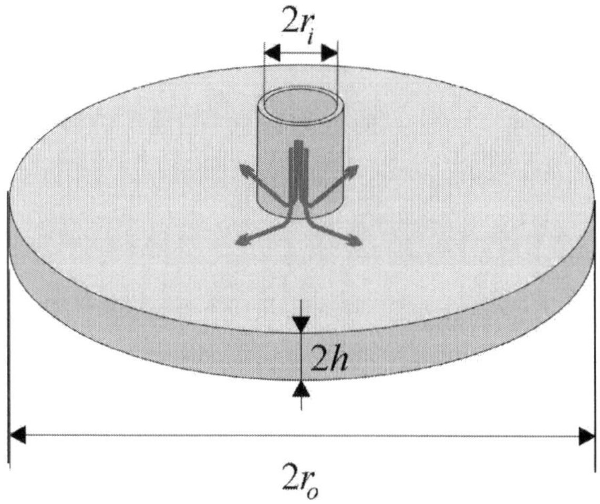

Figure 6.38: A circular disk being molded with short fiber suspensions injected through the gate in the center.

the only nonzero component of the velocity will be v_r. Except near the entrance and at the flow fronts, the velocity can be written as

$$u_r = \frac{C}{r}\left(h^{s+1} - z^{s+1}\right) \tag{6.205}$$

where C depends on the flow rate. The power-law index n is represented as $1/s$.

The rate of deformation tensor for this flow field has the form

$$\dot{\gamma}_{ij} = \begin{bmatrix} 2\dot{\epsilon}_0 & 0 & \dot{\gamma}_0 \\ 0 & -2\epsilon_0 & 0 \\ \dot{\gamma}_0 & 0 & 0 \end{bmatrix}. \tag{6.206}$$

Thus, the flow is a shearing/stretching flow. There is shearing of the flow along the walls at $\pm h$ due to the no-slip boundary condition along the walls of the mold. Along the mid-plane, away from the walls (close to $z = 0$) the flow will stretch or elongate in the tangential (θ) direction due to the radial nature of the flow. If $\dot{\gamma}$ is the shear rate, it can be easily found by using equation $\dot{\gamma}_0 = \partial u_r/\partial z + \partial u_z/\partial r$ from Table 3.5,

$$\dot{\gamma}_0 = \left|\frac{C}{r}(s+1)z^s\right|. \tag{6.207}$$

If we label the stretch or elongation rate by $\dot{\epsilon}_0$, one can use the definition of elongation rate from Chapter 3 to find the stretch rate

$$\dot{\epsilon}_0 = \frac{C}{r^2}\left(h^{s+1} - z^{s+1}\right). \tag{6.208}$$

Stretching will align the fibers in the transverse direction, while shearing will align them in the flow direction. The ratio

$$\frac{\dot{\gamma}_0}{\dot{\epsilon}_0} = \frac{r(s+1)z^s}{h^{s+1} - z^{s+1}} \tag{6.209}$$

can be thought of as the shear/stretch ratio. When it is close to zero, the fibers will be aligned in the transverse direction and when it is very large, the fibers will be aligned in the flow direction. Near the surfaces, the shearing dominates; hence, the fibers align in the flow direction and near the mid-plane ($z = 0$), the shearing is very low and the stretching is more prominent, thus inducing the fibers to align along the transverse direction. In between, the principal fiber orientation direction is determined by the shear/stretch ratio. Hence, a simple analysis like this explains the reason for the formation of the skin/core structure in short fiber injection molded parts.

Why is the thickness of the core different for different materials and flow rates? The gapwise velocity profile plays a key role in the determination. If the velocity profile is uniform across the thickness (power-law index tending to zero), then the stretch dominated region will be larger and one would expect a thicker core of transversely aligned fibers. The opposite is true if the velocity profile is more parabolic as is the case for Newtonian fluids. The temperature influence on the viscosity will also change the velocity profile and influence the skin/core distribution. Changes in the rheology of the suspension can also influence changes in the orientation. ■

The objective is to determine the velocity field and fiber orientation of flow of fiber suspensions not only through the thickness but also in the plane. There are two approaches that can be followed depending on whether the flow and orientation fields are coupled or de-coupled. If one considers a coupled approach, one cannot use the Hele-Shaw approximation and hence Equation (6.178) cannot be used to find the pressure and derive the velocity field. This is because in the coupled approach the shear and normal stresses are also functions of fiber orientation.

The general form of the constitutive equation to describe the stress tensor as a function of the strain rate tensor and the fiber orientation tensor is given according to [305] as:

$$\tau_{ij} = \eta_s \dot{\gamma}_{ij} + \eta_s c \dot{\gamma}_{kl} a_{ijkl} + B \left[\dot{\gamma}_{ik} a_{kj} + a_{ij} \dot{\gamma}_{kj} \right] + C \dot{\gamma}_{ij} + 2 F a_{ij} D_r \tag{6.210}$$

where $\dot{\gamma}_{ij} = (\partial u_j / \partial x_i) + (\partial u_i / \partial x_j)$ is the rate of deformation tensor, η_s is the viscosity of the solvent, c is the volume fraction of particles, A, B, C, F are material constants, D_r is the rotary diffusivity due to Brownian motion, a_{ij} and a_{ijkl} are second and fourth order tensors describing the orientation state of the fibers and defined as: $a_{ij} = \langle p_i p_j \rangle = \int p_i p_j \psi(p)\, dp$, and $a_{ijkl} = \langle p_i p_j p_k p_l \rangle = \int p_i p_j p_k p_l \psi(p)\, dp$, where $\psi(p)$ is a probability distribution function $P(\theta_1 \le \theta \le \theta_1 + d\theta, \phi_1 \le \phi \le \phi_1 + d\phi) \sin\theta_1\, d\theta\, d\phi$, $\vec{p}$ is a unit vector aligned with the axis of

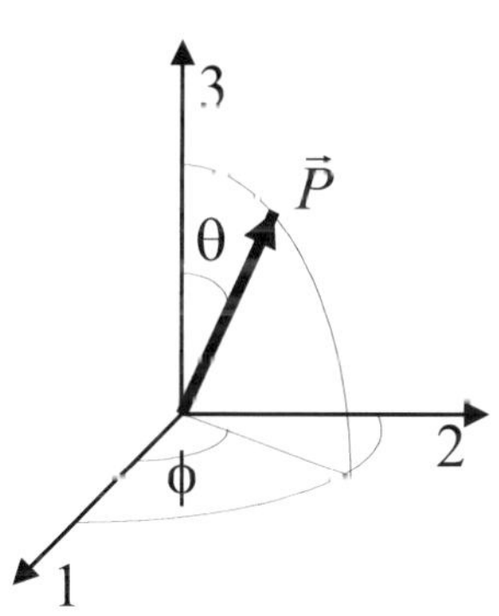

Figure 6.39: Definitions of θ, ϕ, and $\vec{p}$ to denote the direction of a short fiber.

symmetry of the fiber as shown in Figure 6.39. More details are given in Section 4.4. The rotary diffusivity term can be neglected [305] when the fibers are slender (which is usually the case in injection molding) and Equation (6.210) can be written in the form

$$\tau_{ij} = \eta_I \left[\dot{\gamma}_{ij} + N_p \dot{\gamma}_{kl} a_{ijkl} + N_s \left\{ \dot{\gamma}_{ik} a_{kj} + a_{ij} \dot{\gamma}_{kj} \right\} \right] \tag{6.211}$$

where N_p is a dimensionless parameter, called the particle number, and η_I contains isotropic contributions to viscosity from the solvent and the fibers.

In addition, one needs a constitutive equation to predict and calculate fiber orientation from flow. According to Advani and Tucker [150], the equation of change for the second order orientation tensor a_{ij} can be written as:

$$\frac{Da_{ij}}{Dt} = -\frac{1}{2}\left(\omega_{ik} a_{kj} - a_{ik}\omega_{kj}\right) + \frac{1}{2}\lambda\left(\dot{\gamma}_{ik} a_{kj} + a_{ik}\dot{\gamma}_{kj} - 2\dot{\gamma}_{kl} a_{ijkl}\right) + 2C_I \dot{\gamma}(\delta_{ij} - 3a_{ij}) \tag{6.212}$$

where $\omega_{kj} = (\partial u_j / \partial x_k) - (\partial u_k / \partial x_j)$ is the vorticity tensor, $\dot{\gamma} = \sqrt{\frac{1}{2}\dot{\gamma}_{ij}\dot{\gamma}_{ji}}$ is the scalar magnitude of the rate of strain tensor. C_I is interaction coefficient, measuring the intensity of interactions between the fibers in the suspension. In order to solve Equation (6.212), a closure approximation of the fourth order tensor a_{ijkl} in terms of the second order tensor a_{ij} is necessary. There are many such approximations available [311, 312]. Table 6.2 lists a few.

Table 6.2: Closure approximations for a_{ijkl} in terms of the second order tensor a_{ij}.

Linear closure (Hand [313])	$a_{ijkl} = -\frac{1}{35}(\delta_{ij}\delta_{kl} + \delta_{ik}\delta_{jl} + \delta_{il}\delta_{jk}) + \frac{1}{7}(a_{ij}\delta_{kl} + a_{ik}\delta_{jl} + a_{il}\delta_{jk} + a_{kl}\delta_{ij} + a_{jl}\delta_{ik} + a_{jk}\delta_{il})$
Quadratic closure (Leal and Hinch [314])	$a_{ijkl} = a_{ij}a_{kl}$
Hybrid closure (Advani and Tucker [315])	$a_{ijkl} = -\frac{1-f}{35}(\delta_{ij}\delta_{kl} + \delta_{ik}\delta_{jl} + \delta_{il}\delta_{jk}) + \frac{1-f}{7}(a_{ij}\delta_{kl} + a_{ik}\delta_{jl} + a_{il}\delta_{jk} + a_{kl}\delta_{ij} + a_{jl}\delta_{ik} + a_{jk}\delta_{il})$

Under certain assumptions about the fibers and the flow, one can formulate the complete set of governing equations to solve flow and fiber orientation simultaneously. We will describe the procedure here for an isothermal case. For a nonisothermal case, one must also include the energy equation such as Equation (6.177) in which the viscous dissipation term is expressed as $\tau : \nabla \mathbf{U}$ instead of just the velocity gradients multiplied by the viscosity. Other assumptions that are reasonable are that fibers are considered rigid axisymmetric particles that are smaller than any length scale of the flow field, the fibers are large enough to neglect Brownian motion, and inertia and body forces are ignored due to the highly viscous nature of the suspension. Thus using the continuity equation,

$$\nabla \cdot \mathbf{U} = 0 \tag{6.213}$$

and the equation of motion with inertia and body forces neglected

$$0 = -\nabla P + \nabla \cdot \tau \tag{6.214}$$

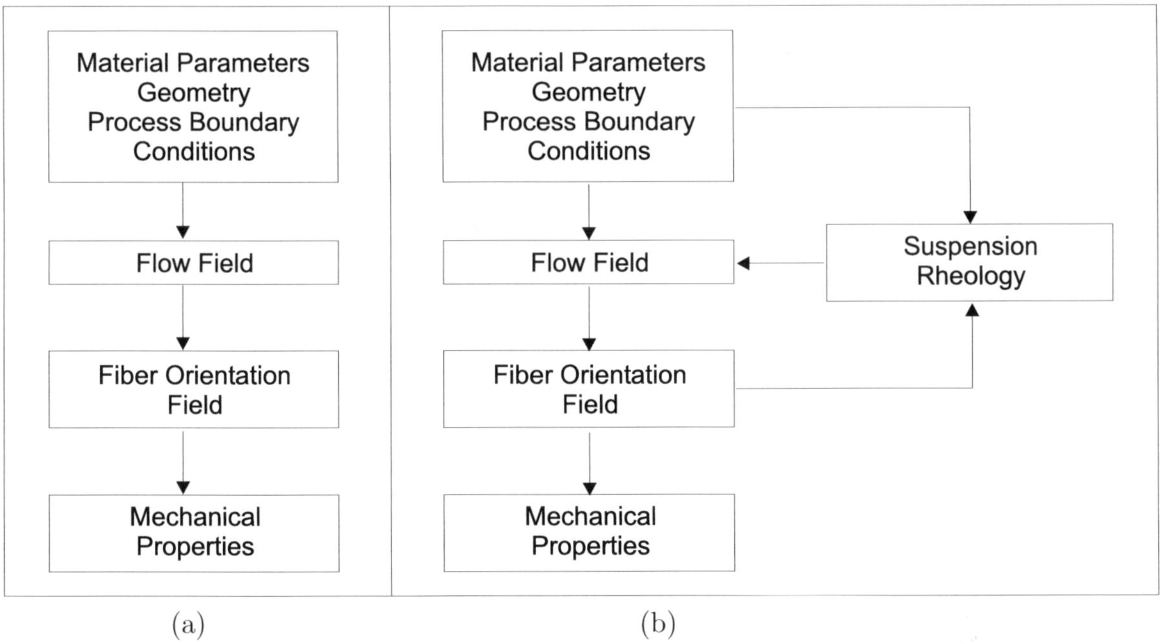

Figure 6.40: The approach to solve (a) decoupled and (b) coupled orientation field.

and combining the above two physical laws with the constitutive law for the stress and fiber orientation, one has to solve the complete set of equations expressed by Equations (6.211)-(6.214).

The dependent variables to be determined are pressure P, velocity $\mathbf{U}$, stress tensor τ_{ij} and orientation tensor a_{ij}. Because τ_{ij} and a_{ij} are symmetric and the trace of a_{ij} is equal to 1, the total number of unknowns is fifteen ($P, u, v, w,\ \tau_{11}, \tau_{12}, \tau_{13}, \tau_{22}, \tau_{23}, \tau_{33},\ a_{11}, a_{12}, a_{13}, a_{22}, a_{23}$).

As is clear that even considering the isothermal problem is quite involved with non-linear relationships and many dependent variables. In addition, as this is a moving boundary problem, one must continuously update the domain that is continuously being filled with the suspension. One also requires boundary and initial conditions to solve this problem. Hence, any simplifications are usually welcomed by the modeller.

If $N_p\delta^2 \ll 1$, where $\delta = p_z$ (z component of the vector $\mathbf{p}$) then one can safely decouple the orientation problem from the flow problem. This allows one to solve for the flow of the suspension independent of the fiber orientation state and then integrate the orientation field along the streamlines using Lagrangian description. One may use Equation (6.212) or the physical laws (Equations (6.213) and (6.214)) to solve for the pressure and velocity field first. And then a Lagrangian viewpoint is adopted [287]. The convective part of the material derivative in Equation (6.212) drops out which allows the system of 5 evolution equations to be integrated along the streamlines and the orientation state to be determined at various points. The flow charts depicting this are shown in Figures 6.40(a) and 6.40(b).

What does one do with the fiber orientation information? This information is quite useful in prediction of thermal or mechanical properties of a short fiber composite. For example, one can write the stiffness tensor in terms of the orientation tensors. For details, see Advani and Tucker [150]. Kim et al. [316] studied the fiber orientation using an X-ray photograph, image processing and Moldflow simulation for a composite part (see Figure 6.41).

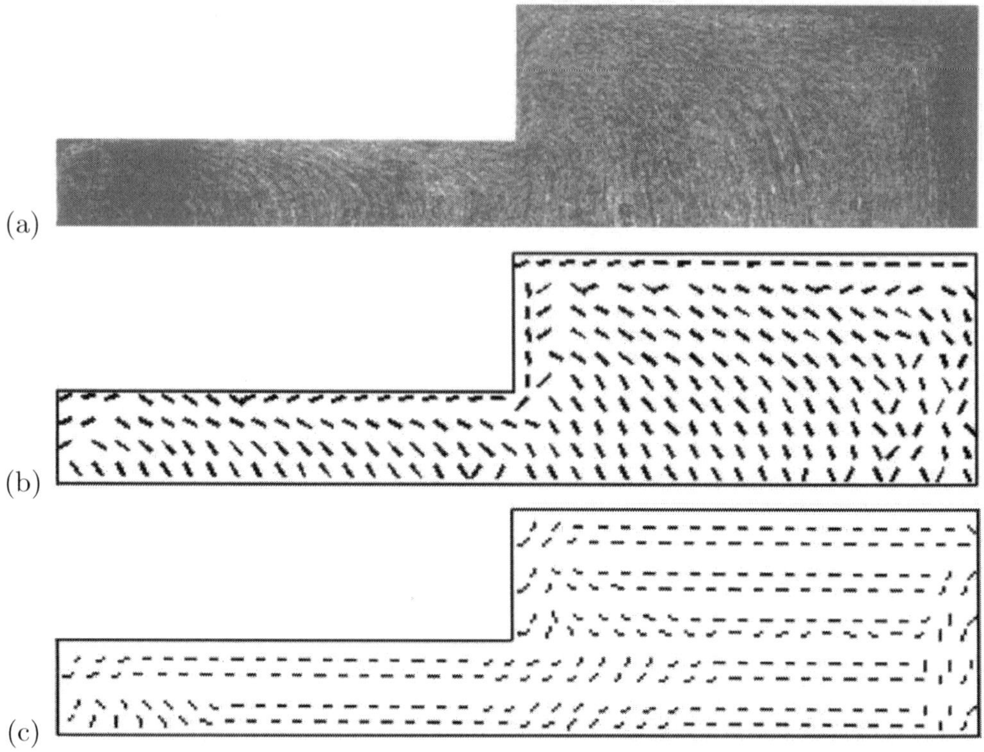

Figure 6.41: Fiber orientation in a composite part. (a) X-ray photograph; (b) result of image processing; and (c) simulated results using Moldflow (reprinted with permission from [316]).

In a recent study by Patcharaphun et al. [317], it has been shown how one can change processing conditions to influence the fiber orientation. As Figure 6.42 shows, a part is made by using two end gates connected to two synchronized injection units. The injection units can pull or push the resin during the packaging stage as many times one desires before solidification. They investigate the fiber orientation in the weldline region as a function of the number of push-pull strokes and the holding pressure differences between both injection units. Their experimental results (see Figure 6.43) compared well with the simulation package Moldflow using a 2.5-D model (based on Hele-Shaw approximation). Thus showing the usefulness of commercial software in helping in identifying the location of the weldline and the fiber orientation state within the weldline region.

6.5 Exercises

6.5.1 Questions

1. What is similar between the extrusion and injection molding processes?

2. What are the functions of the screw used in the extrusion and injection molding processes?

3. Why are vents needed in the mold during the injection molding process?

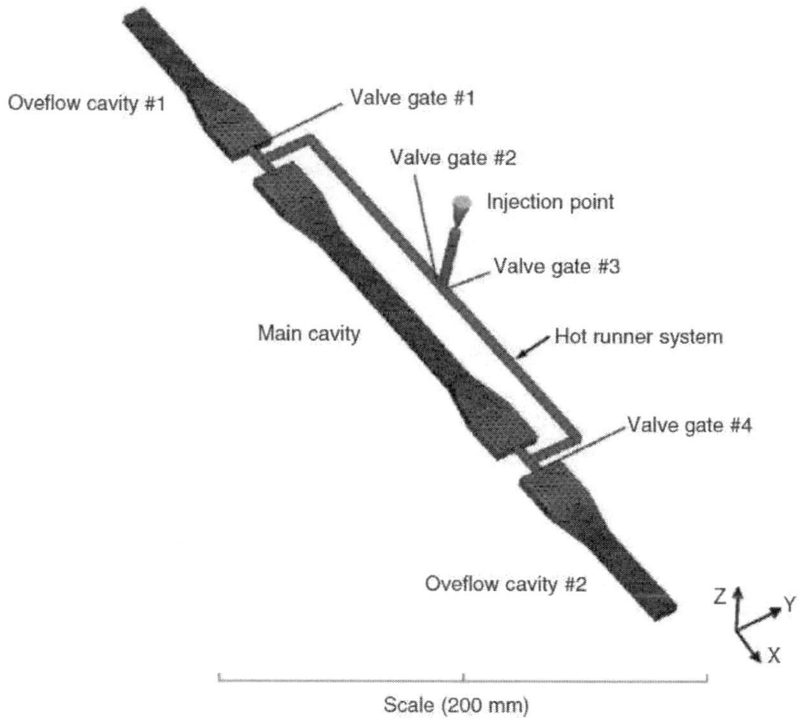

Figure 6.42: Two end gates connected to two synchronized injection units which can pull or push the resin (reprinted with permission from Patcharaphun et al. [317]).

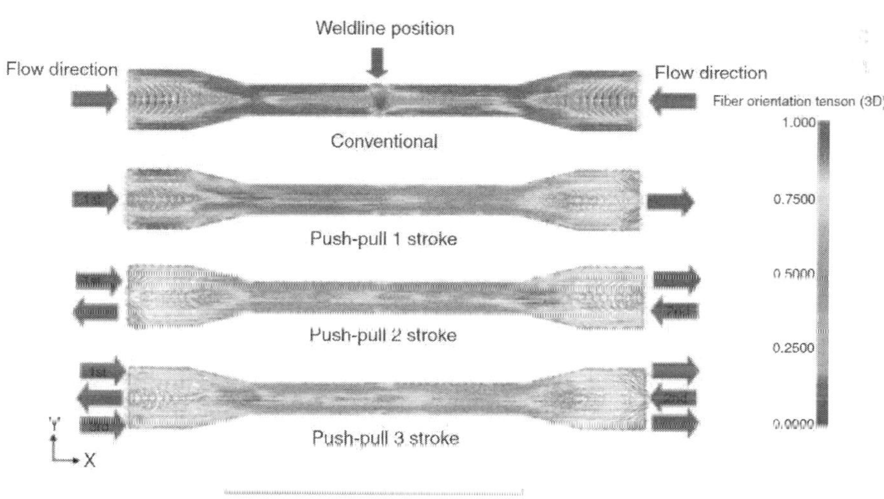

Figure 6.43: Simulation results of fiber orientation tensor at mid-plane layer of conventional and push-pull injection moldings (reprinted with permission from Patcharaphun et al. [317]). It shows how the skin core effect can be reduced by using a pull-push method of injection.

4. In the extrusion and injection molding processes, during the transportation of fibers and polymer melt, why is it important to predict how the suspension flows and what the fiber orientation will be? What is the difference in the two solution approaches: (i) decoupled and (ii) coupled?

5. What is "plasticating" in the extrusion process?

6. What are the three zones of an extruder?

7. What are the assumptions of the model that result in the following expression for the power needed for driving the screw in the melt zone?

$$\dot{w} = \left(\frac{4\mu V_x^2}{H} + \frac{\mu V_z^2}{H} + \frac{V_z B \Delta P}{2Z} \right) ZW.$$

Describe each term that appears above.

8. In order to understand the coupling between the momentum and energy equations to address the issue of temperature dependent viscosity, how would you model the heat transfer between the barrel and the screw in the extrusion process?

9. Describe the different stages of the compression molding process.

10. Why is it important to correctly premeasure the amount of raw materials, or charge, and then place it inside a preheated matched-die mold in the compression molding process? What will happen if the exact amount of charge is not placed inside the mold?

11. In the compression molding process, usually uncured SMC sheets are used. These are fiber bundles with uncured resin and mineral fillers such as calcium carbonate. What is the purpose of using mineral fillers?

12. What types of resin matrices are commonly used in the compression molding process?

13. What is the difference between the compression molding process using thermoplastic and thermoset matrices?

14. What are typical ranges of mold temperatures for thermoset materials in the compression molding process?

15. From which two processes was the injection-compression molding process recently developed? Describe this new process.

16. What is the major reason that the injection-compression molding process has not been popular, although it has some advantages such as improved dimensional accuracy and low molding pressure requirement?

17. In which high-volume manufacturing industry are compression molded parts usually used? Give examples of typical parts.

18. In aerospace applications, why are compression molded components usually used only in secondary structural or nonstructural parts, but not usually suitable for primary structures?

19. Why is it beneficial to model the mold filling and curing behavior in the compression molding process?

20. What are *weld lines* in the compression molding process? How do they affect the material properties of composite part being manufactured?

21. What is the concept of *lay-flat approximation* in compression molding process modeling? For what types of composite parts can it be applied? What is the benefit of using this approximation?

22. For thin and homogeneous polymer parts in the compression molding process, which one of the models is more realistic to apply: the Hele-Shaw model, or the lubricating squeezing model?

23. In the compression molding process, when a charge with fibers is compressed, the fibers change orientation and are convected along with the charge. Do the experiments show major or minor effects on the anisotropy in the viscosity of material due to the change in fiber orientation? Are the flow and fiber-orientation problems strongly coupled, or decoupled?

24. Describe the injection molding process by using all of the following words: solid pellets, resin, fibers, hopper, barrel, screw, melting stage, heaters, friction, shear, suspension, action of pump, gate, fill, mold, cavity, solidify, high pressure, cooling, ejection.

25. What types of thermoplastic and thermoset polymers are commonly used in injection molding?

26. What are the advantages of injection molding?

27. Give some examples of injection-molded parts from daily life.

28. Why do injection molded parts usually have poor mechanical properties, and hence are used as non-load carrying components?

29. Are small dimensional tolerances and good surface quality achievable by injection molding?

30. What are the critical issues in the injection molding process? Describe each of the following terms, what causes them, and what are the effects of them: short shot, flashes, resin degradation, difficult mathematical modeling and its solution, and variation in the fiber orientation and mechanical properties.

31. Why is it important to develop injection molding process models and computer simulations?

32. In the injection molding process model, what assumptions are used in order to obtain the governing differential equation: $\frac{\partial^2 P}{\partial x^2} + \frac{\partial^2 P}{\partial y^2} = 0$ which is a Laplace equation for the material pressure, P? What boundary conditions are usually used to solve this equation? Describe the solution procedure.

33. Describe the mathematical model and its solution for a nonisothermal and non-Newtonian flow of a suspension in the injection molding process.

34. As the molten polymer and the fibers are injected into a mold cavity under high pressure, do the fibers usually align in completely random directions, or in some preferential direction?

35. What is the skin/core effect?

36. Why is it important to know the fiber orientation distribution in a short fiber composite?

6.5.2 Fill in the Blanks

1. The screw acts as a pump by pushing the polymer melt suspension through a ____________ in the extrusion process, and inside a ____________ in the injection molding process.

2. There are different models of the extrusion process to address different issues. If one wants to monitor how homogeneous the part produced is, then the key issue would be to characterize and model the ____________ operation. On the other hand, if one is interested in the pressure and hence power requirement, then one would focus on modeling the ____________ of the molten material through the ____________ and into the ____________. If one is interested in the final dimensions of the components, then he/she would have to model the phenomenon called ____________ which is common with any fluid once the material leaves the confined die shape and expands due to ____________ and ____________ effects.

3. An important choice in modeling the flow of the melt is the choice of the coordinate system. On first inclination, one may be tempted to use the spiral coordinate system, which can lead to ugly mathematics. A clever way to handle this would be fix the ____________ and rotate the ____________ in the opposite direction. Next, unwrap the channel of the ____________ and fix the coordinate system on the ____________.

4. The relative motion between the screw and the barrel is equivalent to steady motion of a ____________ at an angle θ to ____________.

5. In extrusion process modeling, the flow could be assumed to be the combination of two flows: ____________ flow and ____________ flow if the fluid is assumed ____________ and the isothermal case is considered.

6. The heating of the polymer depends on the heat supplied by the ____________ mounted on the ____________ close to the melting stage. The ____________ action of the polymer melt creates ____________ forces that generate heat as well.

7. In compression molding, the raw material (charge) could be either ________________________ (SMC), ________________________ (BMC), ________________________ (LFTs), or ____________________ (GMTs).

8. If thermoplastic material is used in the compression molding process, then the mold temperature must be above the ________________________ temperature.

9. In the compression molding process, press speeds are typically ____________, the applied pressure is in the range of ____________, resin cure usually takes only ____________ and the entire cycle time can range from ____________ to ____________, depending on the degree of automation incorporated into the process.

10. For thin parts being manufactured in the compression molding process, if one assumes no-slip of the material at the mold walls, the shear stresses will dominate and one can adopt a lubrication approximation. This results in ____________ model. However, if one assumes a slip boundary conditions at the mold walls, then the dominant stresses will be the in-plane stresses. This leads to a ________________________ model. These two cases represent limiting cases of thin compression molding problems.

11. In the Hele-Shaw model, the ____________ of the velocity vector varies across the cavity thickness, but the ____________ does not.

12. In the Hele-Shaw model, because of the very ____________ Reynolds number in these flows, a ____________ velocity profile is achieved ____________. Hence, even though the gap height h is changing with time, the velocity profile can be assumed to be ____________ at every time step by assuming a ________________________.

13. In the Hele-Shaw model, the physical boundary conditions are as follows: ____________ at the free or moving boundary surface (assuming there is no back pressure in the mold), and ____________ when the charge reaches the mold walls (assuming that there is no leakage of flow across the mold walls) where n is the normal direction to the mold wall.

14. In the lubricating squeeze flow model of the compression molding process, as you squeeze the material in the z direction, the material elongates in the x and y directions ____________ , until it encounters a ____________ and then it elongates only in the direction that it can. The pressure inside the charge is usually ____________. To find this pressure, one can use the boundary condition that the total ____________ normal to the ____________ is ____________.

15. In the compression molding process, cold charge is placed in a ____________ mold and ____________ is provided to the charge by the mold walls. If the charge is ____________, one needs to provide sufficient heat to ____________ it so it can flow and occupy the empty portion of the cavity before the ____________ cycle is initiated to ____________ the component. For ____________ charges, one provides the heat to initiate the ____________ of the composite. Curing is an ____________ reaction and will generate ____________. If this heat is not extracted from the composite it could lead to ____________ of the part. Hence, the thermal mold design is very crucial to the manufacturing of such components by the compression molding process.

16. If the flow is the Hele-Shaw type, the velocities at the center are much ____________ than the velocities near the walls. This will lead to cold material from the ____________ migrating towards the hot ____________. On the other hand, if the resin viscosity changes dramatically with ____________ and if there is sufficient ____________ near the surfaces, one could expect the material near the ____________ to get ahead of the material near the ____________. This would cause the material to flow towards the ____________.

17. In the compression molding process, many process engineers prefer to add ____________ to the ____________ so that they can fill the part with low ____________ suspension, before the resin ____________ initiates.

18. In the compression molding process, when the Damkohler Number (Da) is large, the ____________ dominates, and if Da is small then ____________ dominates.

19. In the injection molding process, models and simulations can provide useful information such as (i) optimal number and positions of ____________, (ii) required filling pressures and (iii) clamping force to avoid ____________ and ____________, (iv) filling time to avoid premature ____________ due to slow filling, or thermal ____________ due to too fast filling, (v) temperature distribution to avoid excessive ____________ that might lead to ____________, and (vi) fiber orientation states to obtain parts with desired mechanical properties.

20. In the injection molding process, ____________ will align the fibers in the ____________ direction, while ____________ will align them in the ____________ direction. The ratio $\frac{\dot{\gamma}_0}{\dot{\epsilon}_0} = \frac{r(s+1)z^s}{b^{s+1} - z^{s+1}}$ can be thought of as the ____________/____________ ratio. In this equation, $\dot{\gamma}_0$ is ____________, $\dot{\epsilon}_0$ is ____________, r is ____________, s is ____________, z is ____________, and b is ____________. When this ratio is close to zero, the fibers will be aligned in the ____________ direction and when it is very large, the fibers will be aligned in the ____________ direction. Near the surfaces, the ____________ dominates; hence, the fibers align in the ____________ direction, and near the mid-plane ($z = 0$), the ____________ is very low and the ____________ is more prominent, thus inducing the fibers to align along the ____________ direction.

21. In the injection molding process, if the velocity profile is ____________ across the thickness, then the ____________ dominated region will be larger and one would expect a ____________ core of transversely aligned fibers. The opposite is true if the velocity profile is more ____________ as is the case for ____________ fluids. The temperature influence on the ____________ will also change the velocity profile and influence the ____________ distribution.

6.5.3 Problems

1. The approximated governing equation for the extrusion process was found to be

$$\frac{\Delta P}{\mu L} = \frac{\partial^2 u_z}{\partial x^2} + \frac{\partial^2 u_z}{\partial y^2}$$

with boundary conditions on the four walls: $u_z = 0$ at $x = \pm W/2$, and $u_z = 0$ at $y = \pm H/2$. The complete solution for the volumetric flow rate Q was found to be

$$Q = \int_0^H \int_0^W u_z \, dx dy = \frac{WH^3}{12\mu L} \Delta P \, (F_p)$$

where F_p is the shape factor which is a function of the aspect ratio H/W, and plotted in Figure 6.23. If $W \gg H$, one can ignore $\partial^2 u_z/\partial x^2$ as compared to $\partial^2 u_z/\partial y^2$ and the solution will be valid everywhere except near the walls. Calculate the volumetric flow rate Q by assuming $W \gg H$, and ignoring $\partial^2 u_z/\partial x^2$. Compare your result with the more accurate result for $W/H = 100$, 10, 5, and 2.

2. A polymer charge is squeezed between two parallel disks, and the disk surfaces are lubricated so that there is a full slip boundary condition in the radial direction, and inertia and body forces are neglected.

   (i) What is the velocity distribution as a function of thickness of the polymer charge h, at a closing speed of plates $s = -dh/dt = -\dot{h}$, and the radius of the disks and the charge being equal to R?

(ii) What is v_r if there is a partial slip with coefficient β between the walls and the material? Check your result by taking the limit as $\beta \to 0$ and then comparing your result with the result of part (i).

(iii) If the polymer charge is squeezed between two parallel rectangular plates instead of disks, would your answer to part (i) change away from the plate edges?

3. While answering part (i) of the previous problem, one student writes on a paper the following:

$$\text{Volume} = \pi r^2 s = v_r(2\pi r)h$$

and then simplifies that equation as $v_r = rs/(2h)$. Explain what each term corresponds to. What physical law was used? Which of the two models, Hele-Shaw or lubricating squeeze flow, was used? Is the material considered to be compressible, or incompressible? Is this solution valid for a Newtonian fluid as well as for a Non-Newtonian material?

4. If the material is Newtonian in the previous problems, then what is the required force F to compress the material by moving the upper plate? (i) If the closing speed of the disks, s is constant within a time interval, does the required force F stay constant, increase, or decrease with time? Express F mathematically. (ii) If the applied force is kept constant, and the initial thickness of the material is h_o, then how many seconds does it take to reduce the thickness to $0.75h_o$ and $0.50h_o$, respectively? You can express your answer in terms of F, R, P_a, and μ.

5. A polymer charge is squeezed between two parallel disks, and inertia and body forces are neglected. Consider (i) full slip, (ii) no-slip, and (iii) partial slip with $\beta = 0.02$ where $v_r = \beta\tau_{rz}$, and determine the required compression force on the upper plate to squeeze it to half its thickness. Express your answer in terms of the closing speed of plates $s = -dh/dt = -\dot{h}$, material viscosity μ, coefficient β and R_1 the radius of the disks.

Chapter 7

Advanced Thermoplastic Composite Manufacturing Processes

7.1 Introduction

To predict how a material will behave under deformation in any manufacturing operation will depend on the approach selected to describe its macroscopic behavior. For most thermoplastic materials containing continuous fibers, fiber preforms, or collimated fibers (long discontinuous fibers in an oriented fiber assembly), one can treat them as a homogeneous material with anisotropic viscosity or transversely isotropic viscosity with inextensibility in the fiber direction. However, at the microscopic level there are many phenomena taking place that are conglomerated into constitutive constants at the macroscopic level to allow for prediction of the deformation rate relationship with the applied force.

The microscopic phenomena that influence the constitutive equations at the macrolevel are resin percolation, interply (in between the plies) and intraply (within a ply) slip and shearing. Resin percolation and redistribution are functions of the fiber volume fraction and the load applied, whereas the shearing within the ply or in between the plies is limited to the resin rich areas. Experimental studies [141,318,319] suggest that this flow of the resin in between the fibers is a truly viscous flow phenomenon and can be modeled in that fashion. If the process time scales are large and viscosity is low as in the case of thermoset resins, shaping operations can squeeze excess resin out. However, as the resin viscosity is high and usually the process time scales are low for thermoplastics, if one does not understand the deformation behavior, the reinforcing fibers will not be able to adjust to the flow and buckling and distortion may result. The resin flow is schematically illustrated in Figure 7.1 within a unidirectional fiber bed and fabric tows of a mat, hence the need to understand the behavior of such materials.

Another change that the thermoplastic reinforced material undergoes is the level of crystallinity. The rate at which the material is cooled to below the glass transition temperature is the primary process variable, which governs the crystallization in the matrix. The presence of fibers also provides nucleation sites and can cause secondary crystallization.

In addition to influencing the crystallinity of the matrix, cooling can also cause internal stresses that can lead to microcracking. The internal stresses are caused due to four major factors: the mismatch of the coefficient of thermal expansion between the resin and the

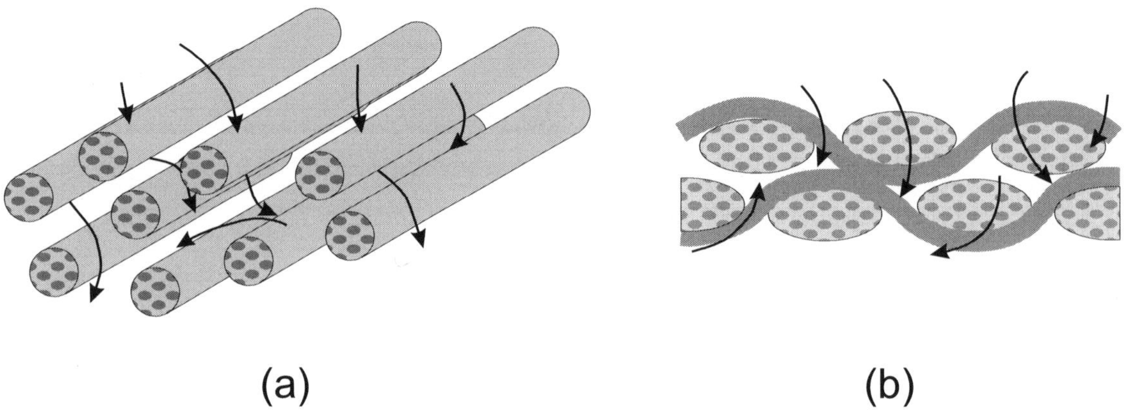

Figure 7.1: Schematic of resin flow through (a) unidirectional fiber mats and (b) woven or stitched fabrics.

fibers, temperature gradients between the laminates, the orientation of the fibers and the shape of the part to be manufactured. Usually as the composite is cooled from its processing temperature to below its glass transition temperature, the resin tries to shrink which is resisted by the stiff inextensible fibers. This places the fibers in compression and the resin in tension. Similar logic applies to the composite. The composite is cooled from outside to the center. This would shrink and solidify the outside first, and then the interior is constrained during its solidification. This puts the outside under compression and the interior under tension. Hence, controlled cooling is an important process parameter during manufacturing of such materials.

In summary, the inextensibility of the fibers coupled with the high viscosity of the resin limits the deformation processes that one can attempt for manufacturing operations to shape these materials into useful end products. The challenges are to be able to model this material and to ensure that stress build-up during cooling is minimum. We will consider modeling of three thermoplastic composites processing methods in this chapter that underpin the technology of advanced thermoplastics composites manufacturing. They are sheet forming, tape lay-up (also known as fiber tow placement) and thermoplastic pultrusion.

7.2 Composite Sheet Forming Processes

The basic operation in sheet forming is to deform sheets of material into curvilinear shapes with the help of tools or molds. Sheet forming with isotropic materials or unreinforced polymers can shear and deform in all directions equally well and allow for formation of complex structures with compound curvatures. Usually, the material used in advanced thermoplastic sheet forming is a polymer sheet preimpregnated with unidirectional continuous fibers. A schematic of a lamina, which is an assembly of unidirectional fibers and a polymeric matrix, is shown in Figure 7.2. This material is highly anisotropic and shows strong directional effects. Such a material cannot stretch in the fiber direction. Shearing in other directions has to compensate for inextensibility in the fiber direction to form the curvilinear shape. The deformation becomes even more complicated when there are several of these sheets stacked together in different directions to form the composite. The deformation behavior of such laminates of unidirectional fibers oriented in different directions and stacked together is similar to the use of woven or bidirectional fabrics impregnated in a polymer sheet and

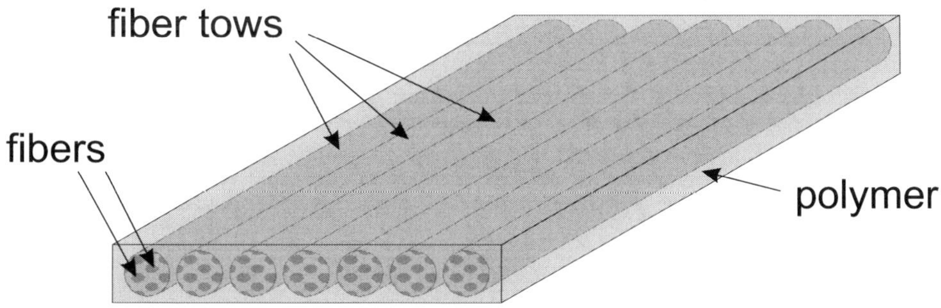

Figure 7.2: A composite lamina of unidirectional fibers impregnated in a polymeric matrix.

stacked together to form the laminate that is then deformed to contour to the shape of the tool or the mold. Long discontinuous fiber (LDF) material was introduced to accommodate the stretching mechanism, as the fibers were long and aligned but the discontinuous nature could create room for stretching.

There are four main composite forming processes: diaphragm forming, matched die forming, stretch forming and roll forming. Many details of these processes are given in chapters on sheet forming by O'Bradaigh [320] and Bhattacharrya [321]. Here, we will reiterate the important features of these processes in terms of the deformation mechanisms.

7.2.1 Diaphragm Forming

As shown in Figure 7.3, a laminate of the composite material is sandwiched and heated between two sheets of diaphragm, and a vacuum or a positive pressure is applied to form the composite. The edges of the material are not clamped as the inextensibility of the material can cause fiber wrinkling. The diaphragm material should be able to stretch without rupture at the forming temperature of the composite. It provides the composite, which is unrestrained, a guide to adhere to the shape of the tool and also consolidates the composite. This process allows for the formation of a wrinkle free composite with excellent thickness tolerance. However, one is limited to the deep draw ratios that can be achieved due to the inextensibility of the fibers and the inability to find diaphragms that are stiff enough to prevent the buckling of the laminate and also be flexible enough to stretch to large extents. Upilex-R material [322, 323] is one of the commonly used diaphragm materials.

The process modeling issues that can be addressed here are consolidation of laminates, rheology of the material in elongational flows, viscosity dependence on temperature, stretching/shearing motion of the material, type of boundary conditions to be applied between the diaphragm and the composite, etc.

7.2.2 Matched Die Forming

In this process, the composite laminate or sheet to be formed is held between two tool surfaces and stamped similar to sheet metal forming as shown in Figure 7.4. The advantages are that one can use the existing metal forming equipment with a few modifications and the process takes very little time. The disadvantages are that the process is nonisothermal as the heat has to be transmitted from the metal mold platen and can cause nonuniformity in the material flow. Also, as the tools are rigid, it can lead to localized high pressures instead of uniform pressure distribution. Thus, due to nonuniform pressure and temperature distri-

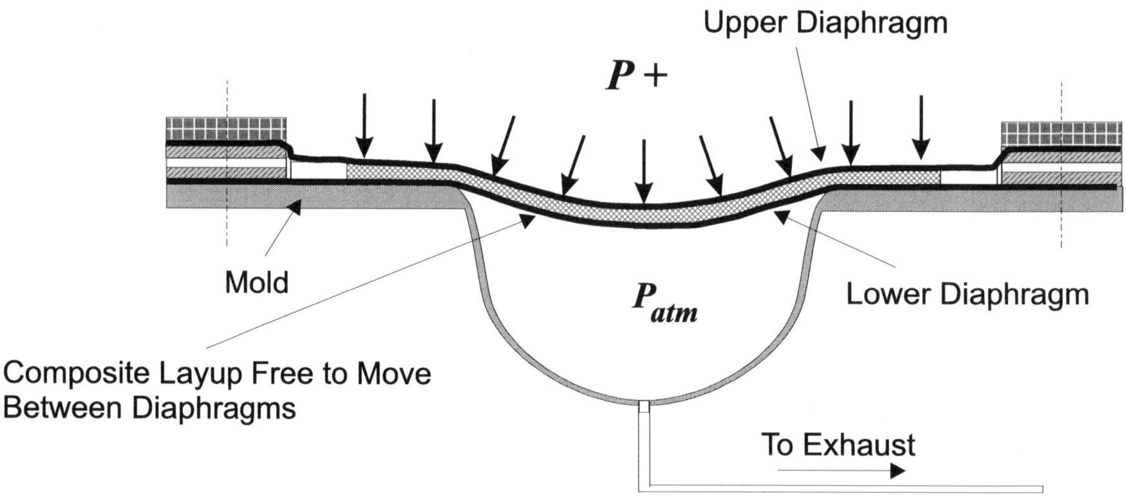

Figure 7.3: Schematic of diaphragm forming process of thermoplastic composite sheets [65].

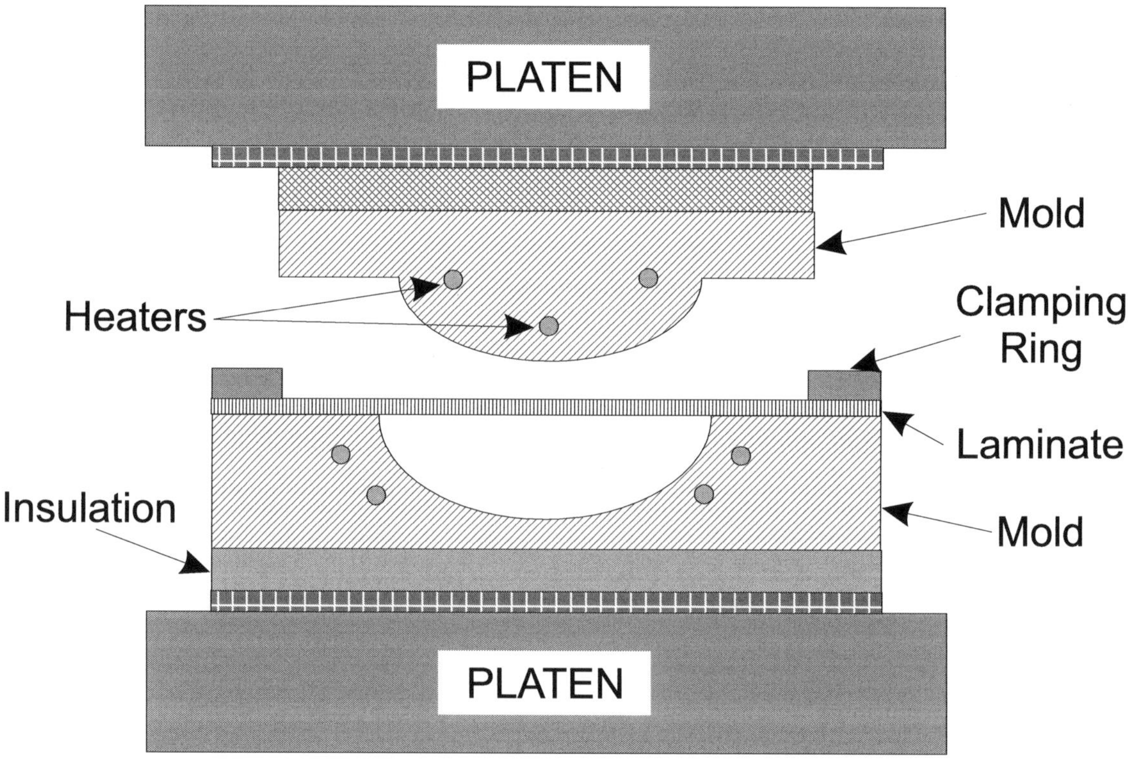

Figure 7.4: Schematic of matched die forming [145].

bution, the chances of failure are much higher than with diaphragm forming. Hence, most match die forming has been practiced on simple shapes with a single degree of curvature such as channel sections and bends.

The processing issues become more involved as one has to predict the nonuniformity of pressure due to the nonuniformity of the boundary conditions.

7.2.3 Stretch and Roll Forming

Stretch forming only can be performed with materials that can locally stretch. Continuous fiber thermoplastics cannot do this but LDF materials can. Hence, one can take formed I-beams or C-channels and stretch them by locally applying heat and pressure with a curved tool to stretch and then consolidate them as shown in Figure 7.5. Thus, one would have to characterize the rheology of such materials and examine its microscopic behavior and macroscopic transverse squeeze flow to model the flow of such materials.

Roll forming uses a series of rotating rollers to deform a composite sheet into a curvilinear shape as shown in Figure 7.6. This forming process is different from the others as the deformation is completed in a series of steps using a series of rollers. Again one must develop a model to describe the deformation of materials between two rollers to be able to predict the deformation.

For more details on stretch and roll forming, refer to Astrom [11].

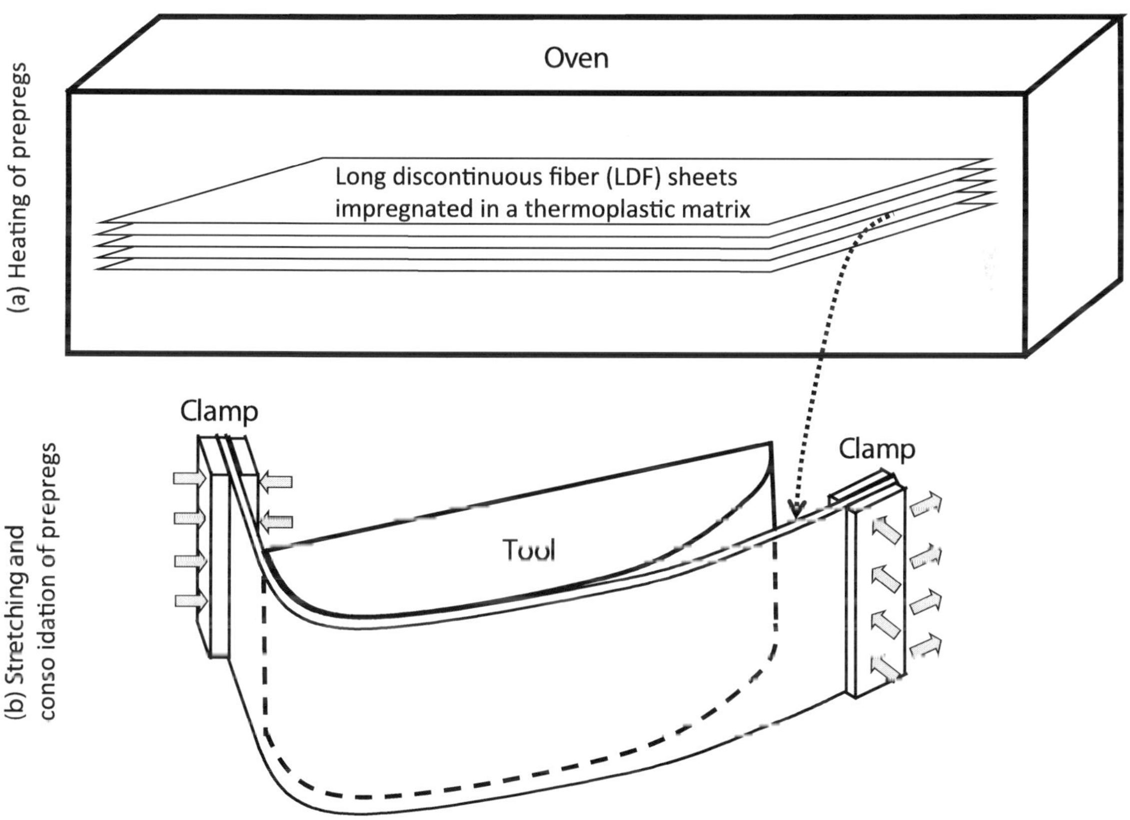

Figure 7.5: Stretch forming and consolidation of prepregs.

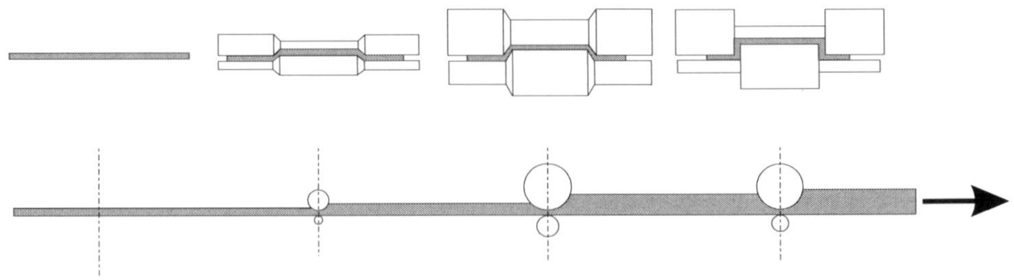

Figure 7.6: Roll forming process of fiber-reinforced materials [11].

7.2.4 Deformation Mechanisms

Deformation in shaping operations can be studied by isolating some of the mechanisms that occur. When consolidating the various sheets of the composite into a laminate, the dominant mechanism is consolidation in which the resin redistributes such that resin moves from resin-rich areas to resin-starved areas on the microscale as shown in the schematic in Figure 7.7.

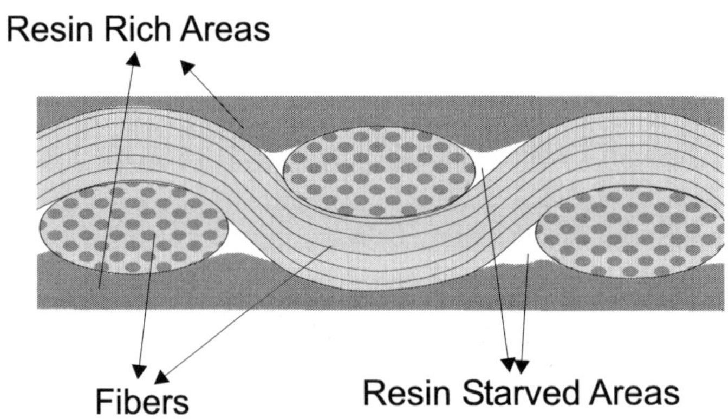

Figure 7.7: Schematic of flow of resin from resin-rich areas to resin-starved areas.

In processes such as matched die, there is also transverse flow in which the fibers and resin move as a unit together when squeezed as shown in Figure 7.8. One can characterize the rheology of squeezing flow of such materials as a single entity of a viscous material in order to model their behavior. When forming single curvature parts, in addition to transverse flow and resin redistribution, one also needs the individual plies to slip over one another in the laminate to form the part as the stretching mode along the fibers is unavailable due to the inextensibility of the fibers. Figures 7.9 and 7.10 show the phenomena of interply slip. Here, one would like to know the force required to pull a ply out from a stack of plies that have been consolidated. Rheological studies to characterize interply slip are usually conducted by flexural testing of heated laminates or by ply pull out experiments at elevated temperatures. Finally, if one is interested in forming double curvature parts, in addition to resin flow, transverse flow and interply slip, intraply slip and interply rotation must also occur.

Thus at the microscopic level within each sheet of the thermoplastic material with unidirectional fibers, the fibers have to move past one another to accommodate a complex curvature. The fibers can shear past one another. As each fiber is covered with a layer of

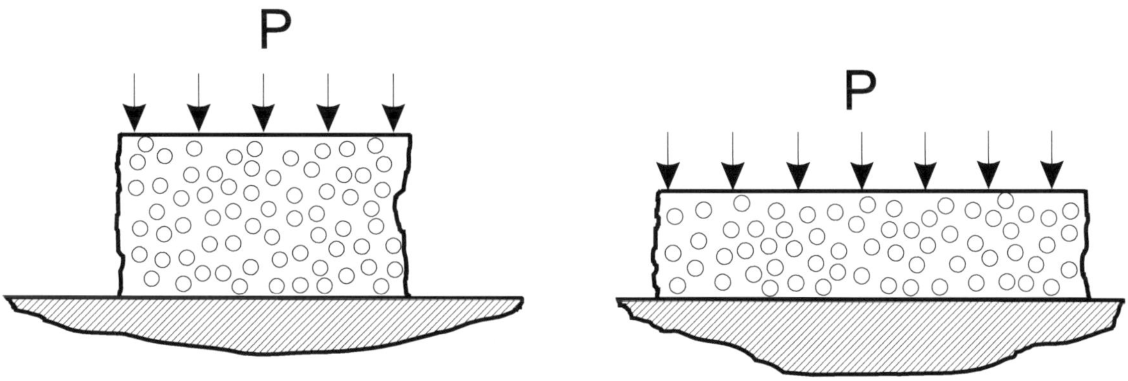

Figure 7.8: Transverse squeeze flow [320].

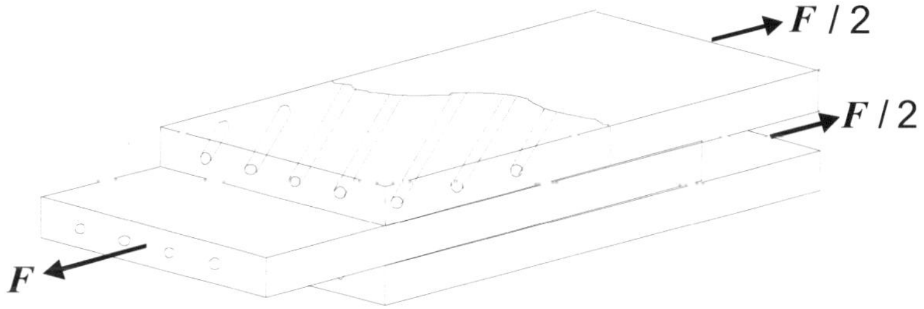

Figure 7.9: Interply shearing [320].

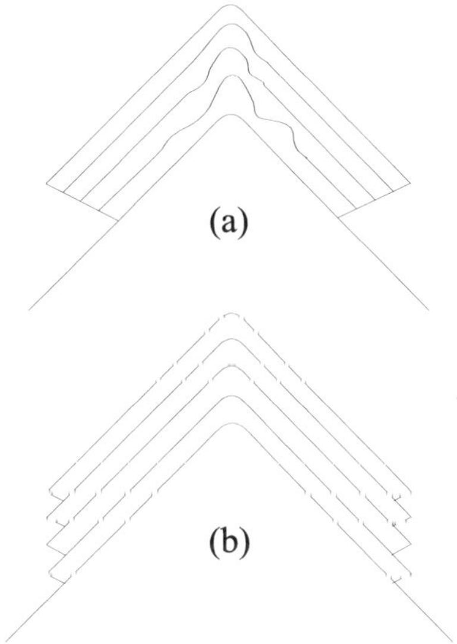

Figure 7.10: Interply deformation: (a) buckling of inner fibers and (b) interply slip [320].

resin, one can estimate the force needed for this activity. In addition, the fibers can also move in the transverse direction. These mechanisms are demonstrated in Figures 7.11(a) and (b).

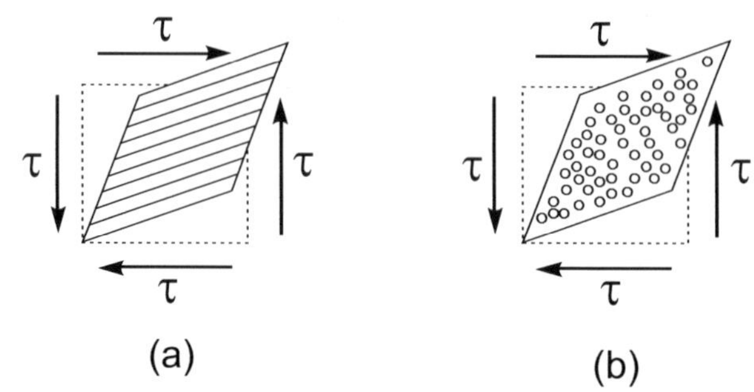

Figure 7.11: Shearing modes: (a) axial intraply, and (b) transverse intraply [320].

Example 7.1: Stretching of Unidirectional Composites [321, 324]

A composite laminate bar is extended in the x direction as shown in Figure 7.12.

The rate of extension of the length can be expressed as

$$V = \frac{dL}{dt}. \tag{7.1}$$

The strain rate in the x-direction may be written as

$$\dot{\epsilon} \equiv \frac{\epsilon}{dt} = \frac{\frac{dL}{L}}{dt} = \frac{\frac{dL}{dt}}{L}. \tag{7.2}$$

If at $t = 0$, the length is L_o,

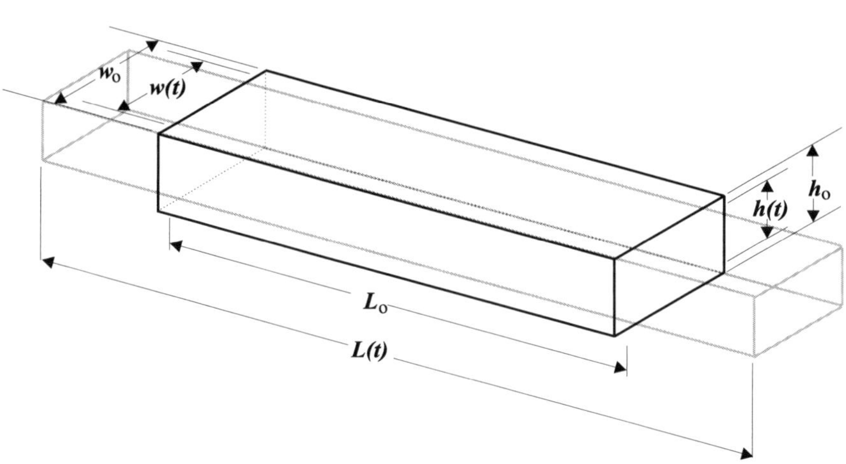

Figure 7.12: Elongation of a composite element.

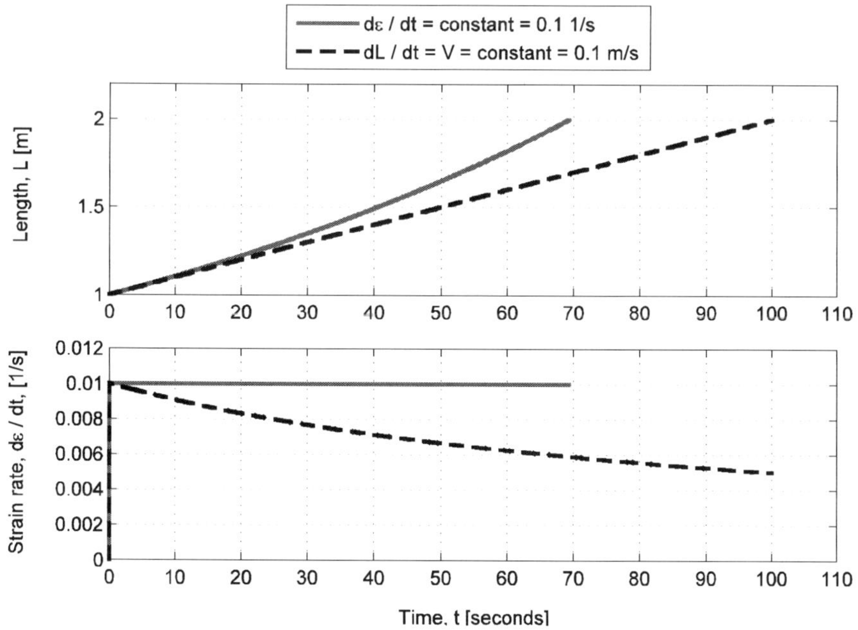

Figure 7.13: Plot of $L(t)$ and $\dot{\epsilon}(t)$ versus time for two different cases: $\dot{\epsilon}(t) = 0.01\,\mathrm{s}^{-1}$ and $V \equiv dL/dt = 0.01\,\mathrm{m/s}$ until $L = 2L_o = 2\,\mathrm{m}$.

(a) Find the rate at which one should extend L, such that $\dot{\epsilon} =$ constant.

(b) If one extends the length at a constant rate of $V =$ constant, how will $\dot{\epsilon}$ change with time?

(c) For $L_o = 1\,\mathrm{m}$, plot $L(t)$ and $\dot{\epsilon}(t)$ versus time for these two cases: (i) $\dot{\epsilon}(t) = 0.01\,\mathrm{s}^{-1}$ and (ii) $V \equiv dL/dt = 0.01\,\mathrm{m/s}$ until $L = 2L_o$.

Solution

(a) Integrating Equation (7.2) and using the initial condition $L = L_o$ at $t = 0$ gives

$$L(t) = L_o e^{\dot{\epsilon}t}. \tag{7.3}$$

Hence one has to extend L in an exponential manner to maintain the strain rate constant.

(b) If $V =$ constant, then

$$L = L_o + Vt. \tag{7.4}$$

Thus $dL/dt = V$, and one can use Equation (7.2) to obtain

$$\dot{\epsilon} = \left(\frac{L_o}{V} + t\right)^{-1}. \tag{7.5}$$

Thus $\dot{\epsilon}$ will reduce as time increases.

(c) A MATLAB code, *example7_01.m* (given in the Appendix) plots $L(t)$ and $\dot{\epsilon}(t)$ versus time in Figure 7.13.

The elongational viscosity is expressed as

$$\eta_\epsilon = \frac{\tau_{xx} - \tau_{yy}}{\dot{\gamma}}. \tag{7.6}$$

In this case, it will be given by

$$\eta_\epsilon = \frac{\tau_{xx}}{\dot{\epsilon}} \tag{7.7}$$

and the stress term is equal to

$$\tau_{xx} = \frac{\text{Force applied in } x \text{ direction}}{\text{Cross-sectional area in } x \text{ direction}}. \tag{7.8}$$

■

7.3 Pultrusion

Pultrusion is a continuous composites manufacturing process. The goal in pultrusion is to produce a consolidated profile with a constant cross-section by pulling impregnated reinforced material through a heated forming die. The schematic for the process is shown in Figure 7.14.

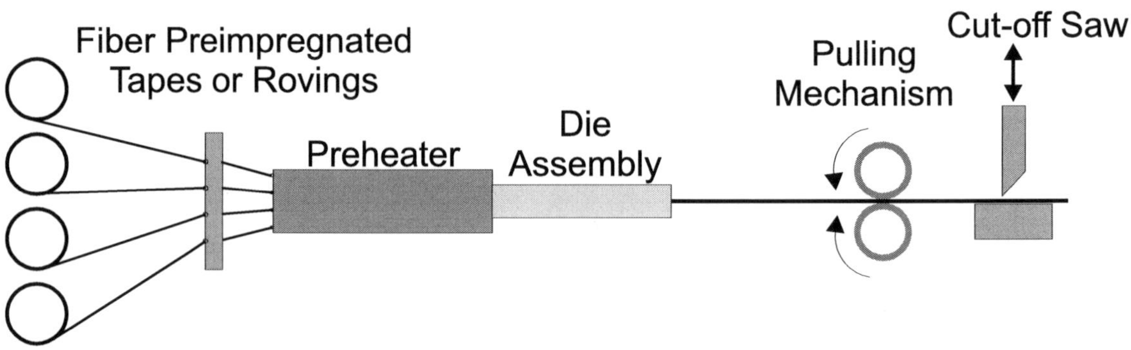

Figure 7.14: The schematic of the thermoplastic pultrusion process.

In this process, unidirectional preimpregnated tows, tapes or fiber rovings are guided through a preheater where their temperature is raised to the melting temperature of the polymer. Next, the heated polymer and fibers enter a die assembly. A die assembly consists of a series of dies which consolidate, form and cool the material. Usually, a significant taper is used in the entrance die to help consolidate the polymer with the fibers. Downstream, the cooling dies cool the formed product and influences the properties such as crystallinity. Also, faster cooling can allow one to locate the pulling mechanism closer to the die assembly. The pulling speeds can vary from a few millimeters per minute to a few meters per minute. A good design of the dies and heating and cooling mechanism can yield the maximum achievable speed.

The two important issues to model in pultrusion are the pressure profile in the tapered section of the die and the pulling force required to pull a profile through the die assembly. Hence one needs to be able to model flow as well as heat transfer through the die assembly to predict these important process parameters.

7.3.1 Thermoset versus Thermoplastic Pultrusion

Physically, thermoset pultrusion involves an additional station of a resin bath between fiber rovings and the preheater as shown in Figure 2.19 of Chapter 2. Despite this additional step, the thermoset and thermoplastic pultrusion processes are physically quite similar. They both pull fibers and resin through a converging die to form continuous composite sections. However, from an analysis viewpoint, the flow in thermoplastic pultrusion is modeled differently than in thermoset pultrusion. In thermosets, one uses Darcy's law to model resin impregnation and the heat transfer involves cure kinetics, whereas in thermoplastic pultrusion due to the high viscosity of the material, Darcy's law is not the correct model to calculate the relationship between flow rate and pressure drop. Also, due to the shear thinning nature of thermoplastics, one would expect the viscosity to vary inside the die. The heat transfer models do not usually involve cure but sometimes will consider the crystallization of the thermoplastics as it passes through the cooling die.

In this section, the focus will be on development of a model that can describe the flow dynamics for a thermoplastic matrix as the material passes through the tapering die. The goal will be to calculate the pressure drop as a function of the pulling speed and material parameters such as viscosity and fiber volume fraction.

7.3.2 Cell Model [6]

To determine the pressure drop, one must consider the tapered section of the die as shown in Figure 7.15.

If one assumes that most of the flow is parallel to the fibers, one can divide the tapered section of the die into self similar cells as shown in Figure 7.15. A cell model is formulated by considering a single tapered annulus of liquid resin surrounding each fiber. This assumption allows one to simplify the problem to a one-dimensional problem in the direction of the fiber movement. Also, due to the cylindrical nature of the fiber, we will assume that the annulus of the resin is also circular; thus one can choose the cylindrical coordinate system to pose the problem.

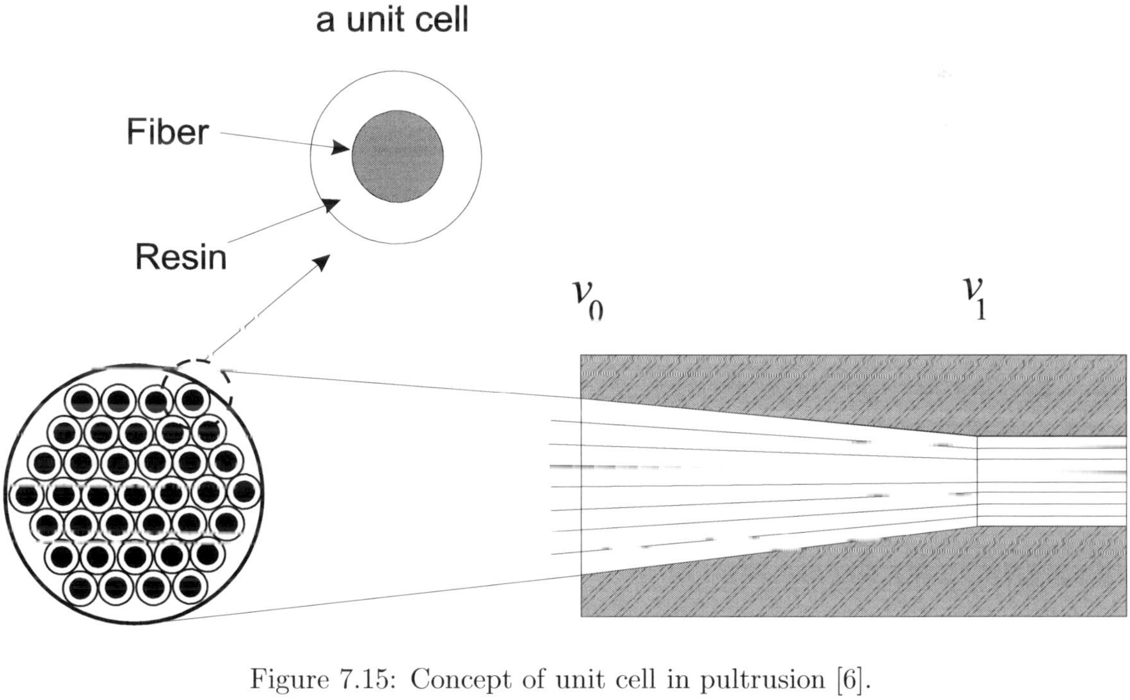

Figure 7.15: Concept of unit cell in pultrusion [6].

If one ignores gravity and inertia as these materials are very viscous and assumes symmetry in the θ direction, one can simplify the equation of motion in the axial direction to

$$0 = -\frac{\partial P}{\partial z} + \frac{1}{r}\frac{\partial}{\partial r}(r\tau_{rz}) + \frac{\partial}{\partial z}(\tau_{zz}). \tag{7.9}$$

Now as the cell dimensions in the radial direction are of the order of a few microns, whereas in the axial direction they could be 3 to 6 orders of magnitude larger, one can assume that $\partial\tau_{zz}/\partial z \ll \partial\tau_{rz}/\partial r$ and hence eliminate the last term in Equation (7.9). This assumption simplifies it to

$$\frac{\partial P}{\partial z} = \frac{1}{r}\frac{\partial}{\partial r}(r\tau_{rz}). \tag{7.10}$$

Now one needs an equation to describe the stress-strain rate relationship before one can proceed further. We will assume a constitutive equation which encompasses Newtonian as well as shear thinning fluid models:

$$\tau_{rz} = \eta\left(\frac{\partial u_r}{\partial z} + \frac{\partial u_z}{\partial r}\right) \tag{7.11}$$

where η can be a function of the shear rate.

As $R \ll L$ and $u_r \ll u_z$, $\partial u_r/\partial z \approx 0$ thus substituting Equation (7.11) into Equation (7.10), for Newtonian fluids we obtain

$$\frac{\partial P}{\partial z} = \frac{\eta}{r}\frac{\partial}{\partial r}\left(r\frac{\partial u_z}{\partial r}\right). \tag{7.12}$$

The pulling mechanism of the fibers in the axial direction induces a drag flow in that direction. However, due to the tapering geometry, the material experiences a pressure build-up. This makes the resin pressure higher near the exit of the die as compared to the entrance. Hence a pressure gradient is established in the opposing direction to drag flow, which may give rise to back flow. The boundary conditions for Equation (7.12) are

$$u_z = U \qquad \text{at } r = R_f \tag{7.13}$$

$$\frac{\partial u_z}{\partial r} = 0 \qquad \text{at } r = R. \tag{7.14}$$

Here R is the outer radius of the annulus and is a function of z due to the taper in the z direction, and U is the pulling speed with which the fibers are moving through the die. R_f is the radius of the fiber. The diameter of the resin annuli at the entrance and exit of the taper are defined by the fiber volume fractions v_0 at the entrance and v_1 at the exit. The zero velocity gradient in the r direction at the cell interface imposes the symmetry boundary condition of no flow in and out of each cell. At the end of the taper as shown in Figure 7.16, the velocity is uniform and no pressure gradient exists.

The flow pattern that satisfies all these boundary conditions is schematically shown in Figure 7.16 for two adjacent cells.

Notice that so far we have not utilized the continuity equation. We know that the flow rate of the resin has to be constant throughout the taper in order to satisfy the continuity equation. The flow rate at the end of the taper and in the uniform section of the die, where the velocity is uniform can be expressed as

$$Q = U\pi\left(R_e^2 - R_f^2\right) \tag{7.15}$$

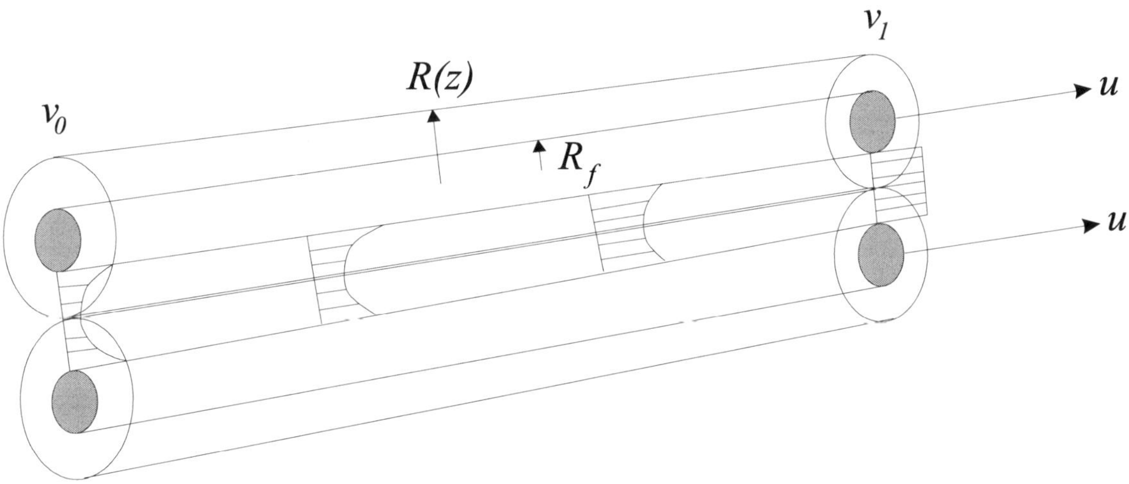

Figure 7.16: Flow profile in two adjacent cells [6].

where R_e is the radius of the annulus at the end of the taper. One should expect that at any axial position along z, the flow rate should be the same. Note that, $v_1 = \pi R_f^2/\pi R_e^2$ and the radius of the annulus at the inlet, R_i is related to initial fiber volume fraction of $v_0 = \pi R_f^2/\pi R_i^2$.

At any cross-section along z, the flow rate Q is the sum of plug flow induced by fiber drag and backflow generated due to the opposing pressure gradient as shown in Figure 7.17. As the resin is Newtonian, one can superimpose these flows

$$Q = Q_{\text{drag}} + Q_{\text{backflow}} = U\pi\left(R_e^2 - R_f^2\right). \tag{7.16}$$

The drag flow is uniform due to the slip boundary condition at the end of the annulus. This slip boundary condition is the same as symmetry or no flow out of the cell boundary. Thus

$$Q_{\text{drag}} = U\pi\left(R^2(z) - R_f^2\right) \tag{7.17}$$

and hence

$$Q_{\text{backflow}} = Q - Q_{\text{drag}} = U\pi\left(R_e^2 - R^2(z)\right). \tag{7.18}$$

at any cross-section along the axial direction, z.

One can also find the total flow rate at any cross section by integrating Equation (7.12) to find the velocity profile, u_z, and relating it to the flow rate as shown below:

$$Q = \int_0^{2\pi}\int_{R_f}^{R(z)} u_z r\, dr\, d\theta. \tag{7.19}$$

By equating Equation (7.16) and Equation (7.19), one can find details of the flow pattern and pressure gradient, once we have a rheological form that describes the viscosity η of the thermoplastic material.

Let us assume the fluid is Newtonian and that the viscosity does not vary appreciably with temperature. One can now integrate Equation (7.12) to obtain the velocity profile

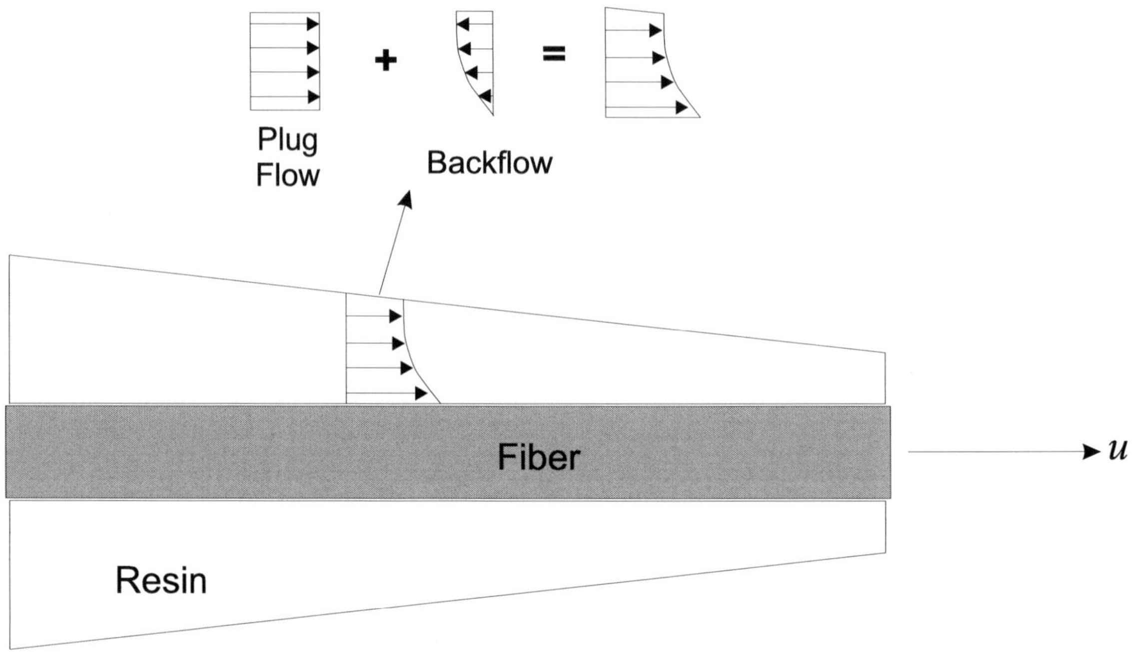

Figure 7.17: Velocity profile in the tapered section of the die. The velocity is considered to be the sum of plug flow and backflow [6].

which is

$$u_z = \frac{1}{4\mu}\frac{dP}{dz}\left[r^2 - R_f^2 + 2R^2\ln\left(\frac{R_f}{r}\right)\right] + U. \tag{7.20}$$

Note here that u_z varies with r which is between R_f and R. Substituting Equation (7.20) into Equation (7.19) and integrating it results in

$$\begin{aligned} Q(z) &= \underbrace{\frac{\pi}{8\mu}\frac{dP}{dz}R^4(z)\left[3 - 4\left(\frac{R_f}{R_e}\right)^2 + \left(\frac{R_f}{R_e}\right)^4 + 4\ln\left(\frac{R_f}{R_e}\right)\right]}_{\text{pressure driven flow}} \\ &\quad + \underbrace{U\pi\left(R^2(z) - R_f^2(z)\right).}_{\text{drag flow}} \end{aligned} \tag{7.21}$$

Thus equating the pressure driven flow in Equation (7.21) to Equation (7.18), the pressure driven flow or Q_{backflow} results in

$$\begin{aligned} Q_{\text{backflow}} &= U\pi\left(R_e^2 - R^2(z)\right) \\ &= \frac{\pi}{8\mu}\frac{dP}{dz}R^4(z)\left[3 - 4\left(\frac{R_f}{R_e}\right)^2 + \left(\frac{R_f}{R_e}\right)^4 + 4\ln\left(\frac{R_f}{R_e}\right)\right]. \end{aligned} \tag{7.22}$$

Inverting Equation (7.22) gives the pressure gradient along the tapered section of the die

$$\frac{dP}{dz} = \frac{8\mu U\pi\left(R_e^2 - R^2(z)\right)}{\pi R^4(z)\left[3 - 4\left(\frac{R_f}{R_e}\right)^2 + \left(\frac{R_f}{R_e}\right)^4 + 4\ln\left(\frac{R_f}{R_e}\right)\right]} \tag{7.23}$$

One can integrate Equation (7.23) with respect to z to find the pressure profile along the z direction. This pressure profile will be a function of the viscosity of the material, the pulling speed, the taper of the die and the final fiber volume fraction desired.

Example 7.2: Producing a Pultruded Profile

You have been asked to produce a pultruded profile with a final fiber volume fraction of 50%. The die is already made and has a linear taper and a length of 20 cm. The initial fiber volume fraction in the incoming rovings of fiber radius of 20 mm is 40%. The viscosity of the thermoplastic can be assumed to be Newtonian and is equal to 1,000,000 Poise (= 100,000 Pa.s). (a) Find the maximum speed with which you can pull the rovings through the die if the maximum gage pressure inside the die cannot exceed 1 MPa. (b) Calculate U for different die lengths, L between 0.1 and 0.5 m when the gage pressure is set to the maximum available value of 1 MPa.

Solution:

Given,

$$\begin{aligned} v_0 &= 0.40 \\ v_1 &= 0.50 \\ R_f &= 0.020 \text{ m} \\ L &= 0.2 \text{ m} \\ \mu &= 100,000 \text{ Pa.s} \\ P_{\max} &= 10^6 \text{ Pa} \\ R_i &= \sqrt{R_f^2/v_0} && (7.24) \\ R_e &= \sqrt{R_f^2/v_1} && (7.25) \\ R(z) &= (R_e - R_i)\frac{z}{L} + R_i. && (7.26) \end{aligned}$$

Substitute Equations (7.24), (7.25) and (7.26) into Equation (7.23), and integrate with respect to z to obtain the pressure profile

$$P(z) - P(0) = \int_0^z \frac{8\mu U\pi\left(R_e^2 - R^2(z')\right)}{\pi R^4(z')\left[3 - 4\left(\frac{R_f}{R_e}\right)^2 + \left(\frac{R_f}{R_e}\right)^4 + 4\ln\left(\frac{R_f}{R_e}\right)\right]}dz'. \tag{7.27}$$

Note that the maximum pressure occurs at $z = L$:

$$\begin{aligned} P(L) - P(0) &= P_{\max} \\ &= \int_0^L \frac{8\mu U\pi\left(R_e^2 - R^2(z')\right)}{\pi R^4(z')\left[3 - 4\left(\frac{R_f}{R_e}\right)^2 + \left(\frac{R_f}{R_e}\right)^4 + 4\ln\left(\frac{R_f}{R_e}\right)\right]}dz'. && (7.28) \end{aligned}$$

Inverting Equation (7.28), U is calculated as

$$U = \frac{P_{\max}\left[3 - 4\left(\frac{R_f}{R_e}\right)^2 + \left(\frac{R_f}{R_e}\right)^4 + 4\ln\left(\frac{R_f}{R_e}\right)\right]}{8\mu I} \tag{7.29}$$

where

$$I = \int_0^L \frac{R_e^2 - R^2(z')}{R^4(z')}\, dz'. \tag{7.30}$$

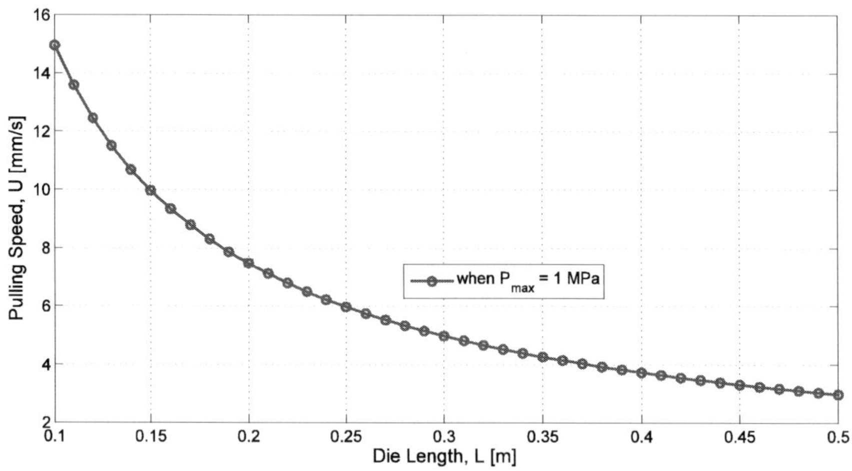

Figure 7.18: Use of mathematical solver MATLAB [325] to solve for U for different P_{max} values. Notice that $U = 7.5\,\mathrm{mm/s}$ when the die length is $L = 0.2\,\mathrm{m}$.

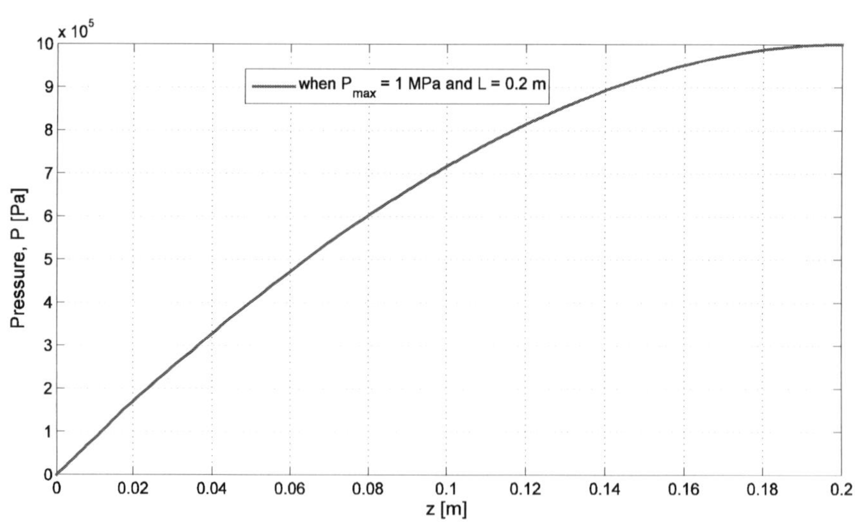

Figure 7.19: Pressure profile along the axial z direction in the tapering die during the pultrusion process. The rovings are pulled with a velocity of 7.5 mm/s.

One can use a mathematical solver program such as MATLAB [325] or Maple [326] to calculate the integral and solve for U. A MATLAB code (named *example7_02.m*) was written and is listed in the Appendix. It is suggested to the reader to run it in MATLAB to calculate U for the given set of parameters. The critical value of U is found to be 7.5 mm/s. If one pulls the rovings with this critical speed, then the pressure at the exit of the die will be 1 MPa. U versus L is shown in Figure 7.18. The pressure profile along the z axis is plotted in Figure 7.19. ■

Shear Thinning Thermoplastics

Most thermoplastics exhibit shear thinning behavior. As studied in Chapter 4, there are many different models to describe the shear thinning nature. Here we will consider only the power-law model, although it is quite straightforward to extend the analysis to Carreau or other detailed models for the resin.

The power-law model describes the viscosity as

$$\eta = m\,\dot{\gamma}^{n-1} \tag{7.31}$$

where m and n are curve fitting parameters and $\dot{\gamma}$ is the magnitude of the strain rate tensor. One can recover the Newtonian fluid model by choosing $m = \mu$ and $n = 1$. One can start with Equation (7.10), and integrate with respect to r using the boundary condition that at $r = R$, the shear stress is zero. This results in the shear stress profile

$$\tau_{rz} = \frac{1}{2}\frac{dP}{dz}\left(r - \frac{R^2}{r}\right). \tag{7.32}$$

For the power-law fluid model,

$$\tau_{rz} = \eta\,\frac{du_z}{dr} = m\dot{\gamma}^{n-1}\frac{du_z}{dr} = m\left(\frac{du_z}{dr}\right)^n. \tag{7.33}$$

The combination of Equations (7.32) and (7.33) results in

$$\frac{du_z}{dr} = -\left[\frac{1}{2m}\frac{dP}{dz}\left(\frac{R^2}{r} - r\right)\right]^{1/n}. \tag{7.34}$$

One must integrate u_z with respect to r to find the velocity profile. Once the velocity profile is known using the conservation of flow rate equation, one can obtain the expression for dP/dz along the fiber direction.

Example 7.3: Derive the Pressure Profile for a Shear Thinning Thermoplastic Composite Being Pultruded

Find the expression for dP/dz for a composite being pultruded through a die. The thermoplastic matrix in the composite exhibits shear thinning behavior which can be modeled by using the power-law model. Consider two cases, (a) the power-law index is 1/2, and (b) the power-law index is 1/3.

Solution (a):

Substituting $n = 1/2$ in Equation (7.34) and integrating with the boundary conditions listed in Equations (7.13) and (7.14) gives

$$u_z = -\left(\frac{1}{2m}\frac{dP}{dz}\right)^2 \left[-\frac{R^4}{r} - 2R^2 r + \frac{r^3}{3} + R^3\left(\frac{R}{R_f} + 2\frac{R_f}{R} - \frac{1}{3}\frac{R_f^3}{R^3}\right)\right] + U \quad (7.35)$$

if $\lambda = R_f/R$, then

$$u_z = -\left(\frac{1}{2m}\frac{dP}{dz}\right)^2 \left[-\frac{R^4}{r} - 2R^2 r + \frac{r^3}{3} + R^3\left(\frac{1}{\lambda} + 2\lambda - \frac{1}{3}\lambda^3\right)\right] + U \quad (7.36)$$

and using Equations (7.18) and (7.19)

$$Q_{\text{backflow}} = -2\pi\left(\frac{1}{2m}\frac{dP}{dz}\right)^2 R^5\left(-\frac{8}{5} + \frac{1}{2\lambda} + \frac{3\lambda}{2} - \frac{\lambda^3}{2} + \frac{\lambda^5}{10}\right). \quad (7.37)$$

Thus

$$U\pi\left(R_e^2 - R^2\right) = -2\pi\left(\frac{1}{2m}\frac{dP}{dz}\right)^2 R^5\left(-\frac{8}{5} + \frac{1}{2\lambda} + \frac{3\lambda}{2} - \frac{\lambda^3}{2} + \frac{\lambda^5}{10}\right). \quad (7.38)$$

One needs to invert Equation (7.38) to find dP/dz as a function of $R(z)$.

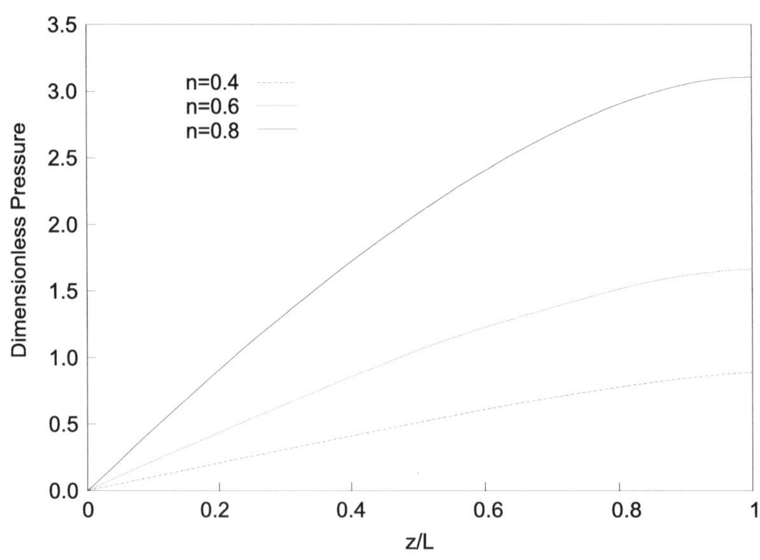

Figure 7.20: Effect of the power-law exponent on the pressure profile in thermoplastic pultrusion.

Solution (b):

For $n = 1/3$,

$$u_z = -\left(\frac{1}{2m}\frac{dP}{dz}\right)^3 \left[-\frac{R^6}{2r^2} + \frac{3}{2}R^2r^2 - \frac{r^4}{4} + 3R^4 \ln\left(\frac{\lambda R}{r}\right) + R^4\left(\frac{1}{2\lambda^2} - \frac{3}{2}\lambda^2 + \frac{1}{4}\lambda^4\right)\right] + U. \tag{7.39}$$

Thus

$$\begin{aligned} Q_{\text{backflow}} &= U\pi\left(R_e^2 - R^2\right) \\ &= -2\pi\left(\frac{1}{2m}\frac{dP}{dz}\right)^3 R^6 \left(2\ln(\lambda) + \frac{5}{6} + \frac{1}{4\lambda^2} - \frac{3\lambda^2}{2} + \frac{\lambda^4}{2} - \frac{\lambda^6}{12}\right). \end{aligned} \tag{7.40}$$

Again, one must invert Equation (7.40) to be able to integrate the pressure as a function of z. Figure 7.20 illustrates the effect of power-law exponent on the pressure profile using dimensionless variables. ■

7.4 Thermal Model

In the tapered die, one would expect heat generation by viscous dissipation. As one can calculate the velocity profile by using the concept of a cell model, it is possible to estimate the amount of heat generated due to viscous dissipation and its contribution to raising the temperature of the thermoplastic. The viscosity of thermoplastics is also a strong function of the temperature. For example, the viscosity of polypropylene reduces by 50% as the temperature increases from 190°C to 240°C. Such a variation has a significant influence on the pressure profile. Ideally, one may have to solve the equation of motion and energy equation simultaneously due to the two-way coupling between them. That is, the viscosity influences the temperature through viscous dissipation term in the energy equation and the temperature influences the viscosity in the equation of motion. However, here we will analyze and simplify the energy equation retaining the viscous dissipation term. In order to obtain a closed-form solution, we will assume that the viscosity is constant inside the unit cell.

7.4.1 Transient Heat Transfer Equation

The complete energy equation with heat generation in an anisotropic medium in cylindrical coordinates is given as

$$\begin{aligned} \rho c_p \frac{DT}{Dt} \;=\;& k_{rr}\frac{1}{r}\frac{\partial}{\partial r}\left(r\frac{\partial T}{\partial r}\right) + k_{\theta\theta}\frac{1}{r^2}\frac{\partial^2 T}{\partial \theta^2} + k_{zz}\frac{\partial^2 T}{\partial z^2} + (k_{r\theta} + k_{\theta r})\frac{1}{r}\frac{\partial^2 T}{\partial \theta\,\partial z} \\ &+ (k_{rz} + k_{zr})\frac{\partial^2 T}{\partial r\,\partial z} + (k_{\theta z} + k_{z\theta})\frac{1}{r}\frac{\partial^2 T}{\partial \theta\,\partial z} + \dot{q}(r, \theta, z, t). \end{aligned} \tag{7.41}$$

Here k_{ij}'s are components of the thermal conductivity tensor and $\dot{q}$ is the rate of internal energy generation per unit volume. In this case, it represents the energy generated due to viscous dissipation.

If one assumes that the principle axes coincide with the fiber direction, one can reduce the k_{ij} matrix to take on the diagonal form (i.e., $k_{ij} = 0$ when $i \neq j$). Also due to the axisymmetric nature of the problem, one can eliminate θ dependency. With these assumptions, the energy equation simplifies to

$$k_{rr}\,\frac{1}{r}\,\frac{\partial}{\partial r}\left(r\,\frac{\partial T}{\partial r}\right) + k_{zz}\,\frac{\partial^2 T}{\partial z^2} + \dot{q} \;=\; \rho c_p\,\frac{DT}{Dt} \tag{7.42}$$

where

$$\frac{DT}{Dt} \;=\; \frac{\partial T}{\partial t} + u_r\,\frac{\partial T}{\partial r} + u_z\,\frac{\partial T}{\partial z}. \tag{7.43}$$

Once the pultrusion has achieved a steady state within the die, one will not expect T to change with time; hence $\frac{\partial T}{\partial t} = 0$ is not an unreasonable assumption. Also, $u_r\,\frac{\partial T}{\partial r} \ll u_z\,\frac{\partial T}{\partial z}$ which will be the case here as $u_z = U$ (the pulling speed) whereas $u_r \approx 0$. Thus

$$\rho c_p\,\frac{DT}{Dt} \;=\; \rho c_p U\,\frac{\partial T}{\partial z}. \tag{7.44}$$

Hence,

$$k_{rr}\,\frac{1}{r}\,\frac{\partial}{\partial r}\left(r\,\frac{\partial T}{\partial r}\right) + k_{zz}\,\frac{\partial^2 T}{\partial z^2} + \dot{q} \;=\; \rho c_p U\,\frac{\partial T}{\partial z}. \tag{7.45}$$

Also the axial conduction term $k_{zz}\,\frac{\partial^2 T}{\partial z^2}$ could be neglected as L is much longer than R_d and one would expect most of the heat to be conducted radially. Thus, the final simplified form for the energy equation is

$$\underbrace{k_{rr}\,\frac{1}{r}\,\frac{\partial}{\partial r}\left(r\,\frac{\partial T}{\partial r}\right)}_{\substack{\text{conduction in}\\ \text{radial direction}}} + \underbrace{\dot{q}}_{\substack{\text{energy generation}\\ \text{due to viscous}\\ \text{dissipation}}} \;=\; \underbrace{\rho c_p U\,\frac{\partial T}{\partial z}}_{\substack{\text{convection of heat}\\ \text{in axial direction}\\ \text{due to the pulling}\\ \text{of the fibers}\\ \text{through the die}}}. \tag{7.46}$$

We need two boundary conditions in the r direction and one in the z direction to uniquely determine $T(r,z)$. The boundary conditions related to the temperature field in a pultrusion die can be modeled as

$$T(r,0) = T_{in} \tag{7.47a}$$

$$\left.\frac{\partial T}{\partial r}\right|_{r=0} = 0 \qquad \text{or} \qquad T(0,z) = \text{finite} \tag{7.47b}$$

$$T(R_d,z) = T_w. \tag{7.47c}$$

Here T_{in} and T_w are the preheater and die wall temperatures, respectively, R_d is the radius of the taper. Equation (7.46) can be solved using separation of variables

$$T(r,z) \;=\; \frac{2}{R_d^2}\sum_{n=1}^{\infty}\left[\frac{T_{in} - T_w - \frac{\dot{q}R_d^2}{4k_{rr}}}{\lambda_n}\,R_d\,J_1(\lambda_n R_d) + \frac{\dot{q}}{4k_{rr}}\int_0^{R_d} r^3\,J_0(\lambda_n r)\,dr\right]\frac{J_0(\lambda_n r)}{J_1^2(\lambda_n R_d)}\,e^{-\alpha\lambda^2 z/U} \tag{7.48}$$

where $\lambda_n R_d$ satisfies $J_0(\lambda_n R_d) = 0$. J_0 and J_1 are Bessel functions and α is the thermal diffusivity, $\alpha = (k_{rr})_c/(\rho c_p)_c$. Here $(k_{rr})_c$ and $(\rho c_p)_c$ are averaged composite properties determined using the rule of mixture

$$(k_{rr})_c = \frac{1}{\frac{1-V_f}{k_m} + \frac{V_f}{k_f}} \tag{7.49}$$

$$(\rho c_p)_c = (1 - V_f)\rho_m c_{pm} + V_f \rho_f c_{pf} \tag{7.50}$$

where the subscripts m and f represent matrix and fiber materials, respectively. The particular solution that will satisfy boundary conditions in Equations (7.47b) and (7.47c) is

$$T_s(r) = T_w + \frac{\dot{q}}{4\,(k_{rr})_c}\left(R_d^2 - r^2\right). \tag{7.51}$$

The overall temperature solution can be calculated by superimposing Equations (7.51) and (7.48).

Example 7.4: Finite Difference Method to Find Temperature Profile during Processing of Thermoplastics

Above, an analytical solution was found for the thermal model by using separation of variables method. This method is applicable when the domain is simple as in that case, which is a rectangular domain in rz plane (assuming that $dR/dz << 1$, i.e., the taper angle is small). As the domain becomes complicated, the method may not be applicable, or it may be tedious to apply it.

Here in this example, a numerical solution will be found to the same problem using finite difference method. The details of the method will be provided so one can adopt this method for similar problems in other composite manufacturing processes. The following material, geometric and process parameters are used for this example

- Inlet fiber volume fraction, $V_{f0} = 0.40$
- Exit fiber volume fraction, $V_{f1} = 0.50$
- Average fiber volume fraction, $V_f = (V_{f0} + V_{f1})/2 = 0.45$
- Radius of the fiber bundle (= mean radius of die), $R_f = 0.008\,\text{m}$
- Thermal conductivity of fibers, $k_f = 1.20\ \text{W/(m.°K)}$
- Thermal conductivity of matrix, $k_m = 0.20\ \text{W/(m.°K)}$
- Density of fibers, $\rho_f = 2500\,\text{kg/m}^3$
- Density of matrix, $\rho_m = 1200\,\text{kg/m}^3$
- Specific heat capacity of fibers, $c_{pf} = 800\,\text{J/(kg.°K)}$
- Specific heat capacity of matrix, $c_{pm} = 700\,\text{J/(kg.°K)}$
- Length of die, $L = 0.500\,\text{m}$
- Preheater temperature, $T_{in} = 20°\text{C} = 293.2°\text{K}$

- Die wall temperature, $T_w = 300°\text{C} = 573.2°\text{K}$
- Pulling speed, $U = 0.005\,\text{m/s}$
- Viscosity of thermoplastic polymer, $\mu = 400,000\,\text{Pa.s}$
- Estimated energy generation due to viscous dissipation, $\dot{q} = 200,000\,\text{W/m}^3$ (assumed to be uniform in the composite part).

(a) Plot the temperature distribution within the composite part as it is being pultruded.
(b) Observe the effect of decreasing the pulling speed, say to 60% of the previous value, i.e., re-solve part (a) with $U = 0.003\,\text{m/s}$.
(c) Re-solve part (a) with a higher preheater temperature, $T_{in} = 70°\text{C}$.
(d) Re-solve part (a) with no energy generation due to viscous dissipation, $\dot{q} = 0$.

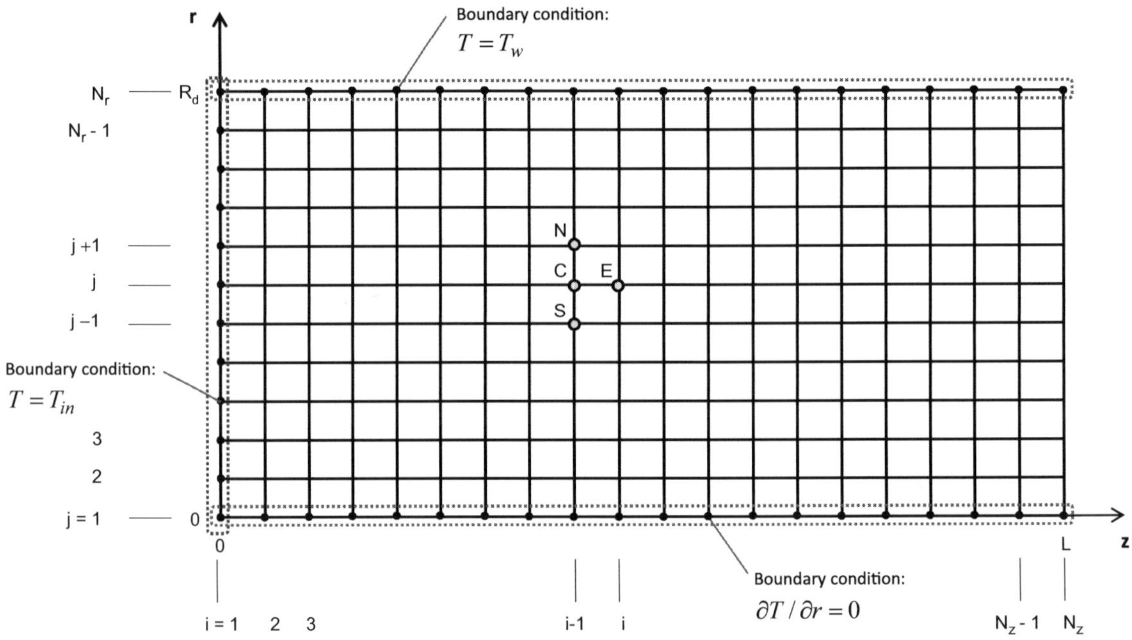

Figure 7.21: Discretized domain for the pultruded composite part and the boundary conditions. L and R_d are the length and average radius of the die through which the composite is pulled in z direction.

Solution:

In this solution, the radius of the taper, R_d is assumed to be constant along the die for the sake of simplicity; however, one can use a nonuniform mesh to solve a tapered domain. First, the continuous domain of $0 \leq r \leq R_d$ and $0 \leq z \leq L$ is discretized as shown in Figure 7.21.

The differential equation, Equation (7.46)

$$k_{rr}\left(\frac{1}{r}\frac{\partial T}{\partial r} + \frac{\partial^2 T}{\partial r^2}\right) + \dot{q} = \rho c_p U \frac{\partial T}{\partial z} \tag{7.52}$$

is discretized by using central finite difference in r direction and forward finite difference in z direction:

$$k_{rr}\left(\frac{1}{r_C}\frac{T_N - T_S}{2h_r} + \frac{T_N - 2T_C + T_S}{h_r^2}\right) + \dot{q} = \rho c_p U \frac{T_E - T_C}{h_z} \tag{7.53}$$

where the subscripts N (North), C (Central), S (South) and E (East) denote

$$N \equiv (i-1, j+1), \qquad C \equiv (i-1, j), \qquad S \equiv (i-1, j-1), \qquad E \equiv (i, j) \tag{7.54}$$

h_z and h_r are the increments in z and r directions, respectively. The boundary conditions, Equation (7.47a-c) correspond to

$$T_{1,j} = T_{in} \qquad \text{for } 1 \le j \le N_r \tag{7.55a}$$

$$T_{i,1} = \frac{1}{3}(4T_{i,2} - T_{i,3}) \qquad \text{for } 2 \le i \le N_z \tag{7.55b}$$

$$T_{i,N_r} = T_w \qquad \text{for } 2 \le i \le N_z \tag{7.55c}$$

where a second-order forward finite difference equation was used in Equation (7.55b).

One can apply a marching scheme in z-direction if Equation (7.53) is rewritten in such a way that the unknown T_E is isolated on the left-hand side and all the other known temperatures, T_N, T_C and T_S are on the right-hand side of the equation:

$$T_E = T_C + \frac{k_{rr}}{\rho c_p U}\left[\frac{1}{r_C}\frac{h_z}{2h_r}(T_N - T_S) + \frac{h_z}{h_r^2}(T_N - 2T_C + T_S)\right] + \frac{\dot{q} h_z}{\rho c_p U}. \tag{7.56}$$

T_E is calculated using Equation (7.56) for all the "internal nodes" ($2 \le i \le N_z - 1$ and $2 \le j \le N_r - 1$) as well as for the nodes at the right edge ($i = N_z$).

A MATLAB code, *example7_04.m*, was written and is listed in the Appendix. The temperature distributions for parts (a) through (d) are given in Figures 7.22 through 7.25, respectively.

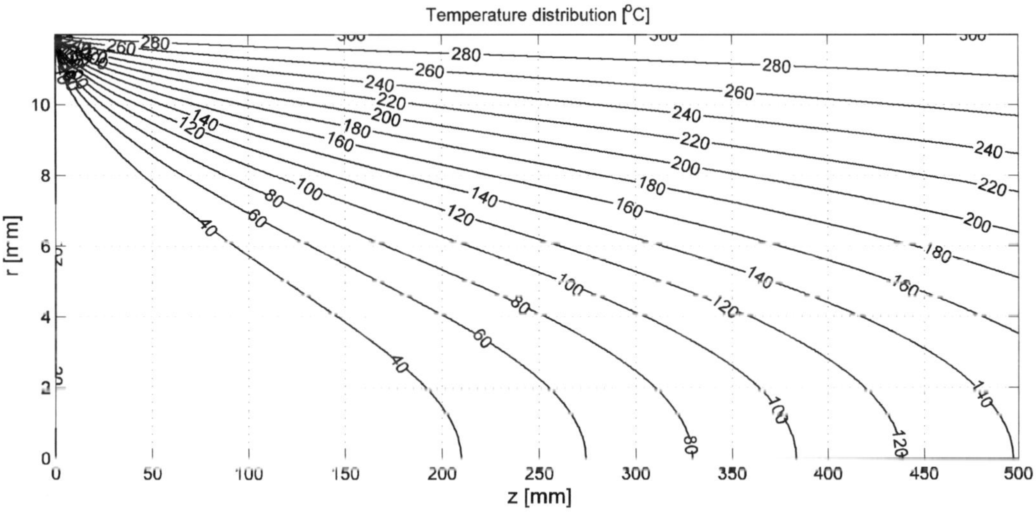

Figure 7.22: Temperature distribution for part (a) of Example 7.4. The exit temperature at $z = L$ ranges between 140.8°C at $r = 0$ and 300°C at $r = R_d$.

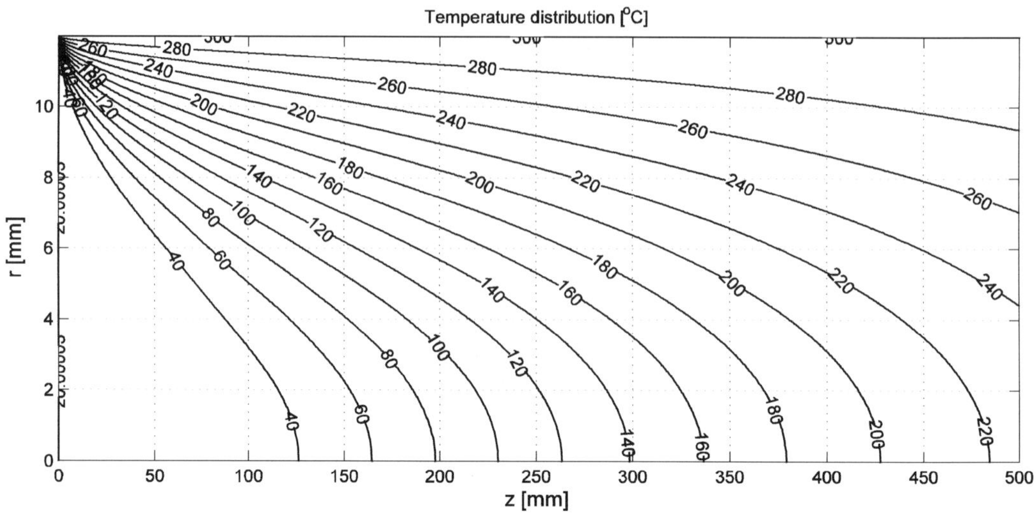

Figure 7.23: Temperature distribution for part (b) of Example 7.4 in which U is decreased to 3 mm/s. The exit temperature at $z = L$ ranges between 224.9°C at $r = 0$ and 300°C at $r = R_d$.

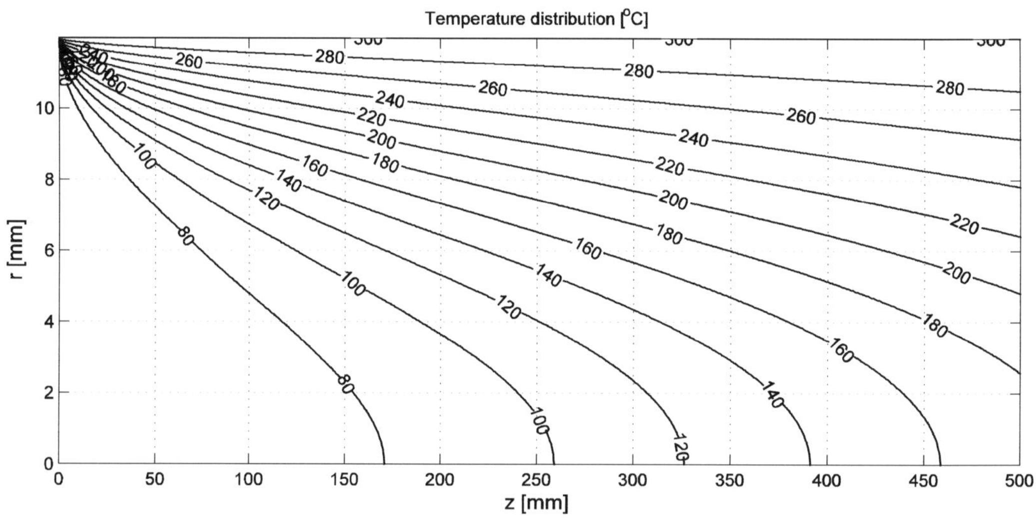

Figure 7.24: Temperature distribution for part (c) of Example 7.4 in which T_{in} is raised from 20°C to 70°C. The exit temperature at $z = L$ ranges between 171.5°C at $r = 0$ and 300°C at $r = R_d$.

Although the numerical solution takes several seconds of CPU time to calculate the results on a typical personal computer, it enables the user to design appropriate process, material and geometric parameters, not only the ones considered here but others such as die length, taper, material viscosity, etc., so that the temperature distribution can result in an adequate consolidation of the fibers as the thermoplastic viscosity reduces with temperature. If the temperature distribution is too low (or too high), the process, material and geometric

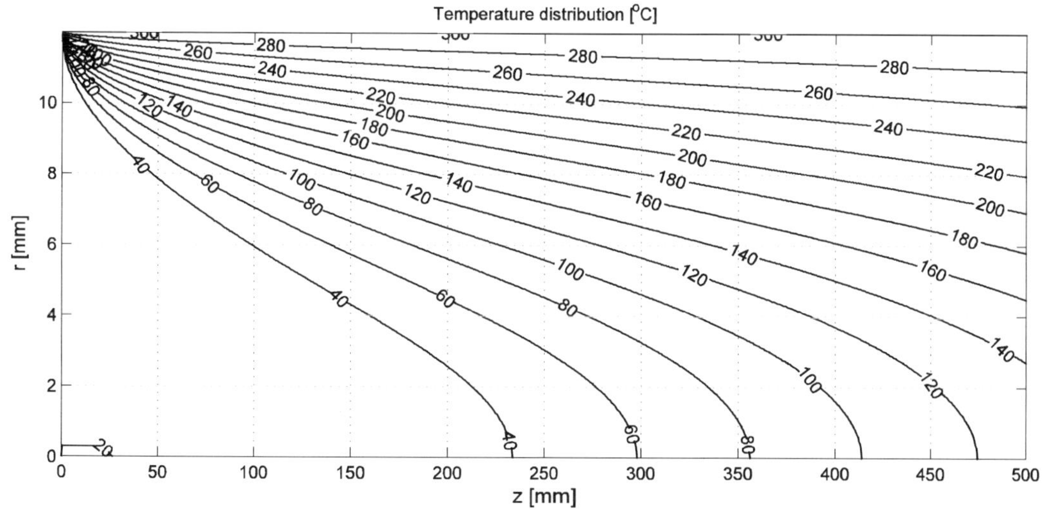

Figure 7.25: Temperature distribution for part (d) of Example 7.4 in which $\dot{q} = 0$. The exit temperature at $z = L$ ranges between 128.0°C at $r = 0$ and 300°C at $r = R_d$.

parameters (such as preheater temperature, wall temperature and pulling speed) should be adjusted and their effect on the results of this model estimated. ■

7.4.2 Viscous Dissipation

In the tapered section, one can assume that internal energy generation is due to viscous dissipation

$$\dot{q}(r, \theta, z, t) = \eta\phi \tag{7.57}$$

where η is the viscosity and ϕ is the dissipation function that describes the irreversible conversion of mechanical energy into thermal energy. For this one-dimensional, axisymmetric flow, the shearing of u_z in the radial direction is the only dominant term. Hence,

$$\phi = \left(\frac{\partial u_z}{\partial r}\right)^2. \tag{7.58}$$

One needs to find the average heat generated per unit volume of the cell to be used in the energy equation. This is done by using average shearing of u_z in the radial direction

$$\left(\frac{\partial u_z}{\partial r}\right)^2_{\text{average}} = \frac{1}{\pi R^2}\int_0^{2\pi}\int_{R_f}^{R}\left(\frac{\partial u_z}{\partial r}\right)^2 r\,dr\,d\theta \tag{7.59}$$

where R refers to the local radius of the cell model. Thus, dissipation will not be constant along the taper and will vary as R changes along z.

For Newtonian fluids

$$(\eta\phi)_{\text{average}} = \frac{1}{2\mu}\left(\frac{dP}{dz}\right)^2 R^2\left(-\frac{3}{4} - \ln\lambda - \frac{1}{4}\lambda^4 + \lambda^2\right) \tag{7.60}$$

where $\lambda = R_f/R$. Thus the viscous dissipation term varies with the square of the pressure gradient and the square of R. If one substitutes Equation (7.23) for dP/dz in Equation (7.60), one could plot how viscous dissipation changes with $R(z)$.

This viscous dissipation can then be included in Equation (7.46) and solved to evaluate the impact of the $\dot{q}$ term on the temperature profile. There is no closed-form solution for this, but one could easily develop a finite difference solution for this problem.

Example 7.5: Re-solve Example 7.4 by Using $\dot{q}(z)$ instead of Constant $\dot{q}$

In the previous example, an estimated energy generation term due to viscous dissipation, $\dot{q} = 200\,\mathrm{kW/m^3}$, was used in the energy equation (7.46), uniformly everywhere in the die domain. In this example, constant $\dot{q}$ will be replaced with

$$\dot{q}(z) = \frac{1}{2\mu}\left(\frac{dP}{dz}\right)^2 R^2 \left(-\frac{3}{4} - \ln\lambda - \frac{1}{4}\lambda^4 + \lambda^2\right) \tag{7.61}$$

where $\lambda = R_f/R$, and Equation (7.61) will be substituted in Equation (7.46).

Solution

The finite difference method will be applied exactly as in Example 7.4, except that the die radius R is calculated using $R(z) = (R_e - R_i)\frac{z}{L} + R_i$ and substituted in Equation (7.23) to calculate dP/dz and evaluate gauge pressure $P(z)$ (see Figure 7.26 for the graphs generated by running a MATLAB code, *example7_05.m*, which was given in the Appendix). Notice that `c4` term in this MATLAB code (which corresponds to $\dot{q}h_z/(\rho c_p U)$) is not constant, but varying along the discrete nodes in z direction. As seen from Figure 7.26, the calculated $\dot{q}(z)$ is maximum at the inlet and zero at the exit. The corresponding temperature distribution is shown in Figure 7.27.

The effect of $\dot{q}(z)$ on $T(r, z)$ should be investigated by comparing the temperature values in the entire domain for different cases; however, if we consider a particular point in the domain, say the right lower corner in the figures, which is the center of the circular part at the exit of the die. ■

Example 7.6: Nondimensionalization of Simplified Energy Equation in Thermoplastic Pultrusion

Nondimensionalize Equation (7.45).

Solution

If we use the following nondimensional variables,

$$\hat{z} = \frac{\alpha_{rr} z}{U R_d^2} \tag{7.62a}$$

$$\Theta = \frac{T - T_i}{T_w - T_i} \tag{7.62b}$$

$$\hat{r} = \frac{r}{R_d} \tag{7.62c}$$

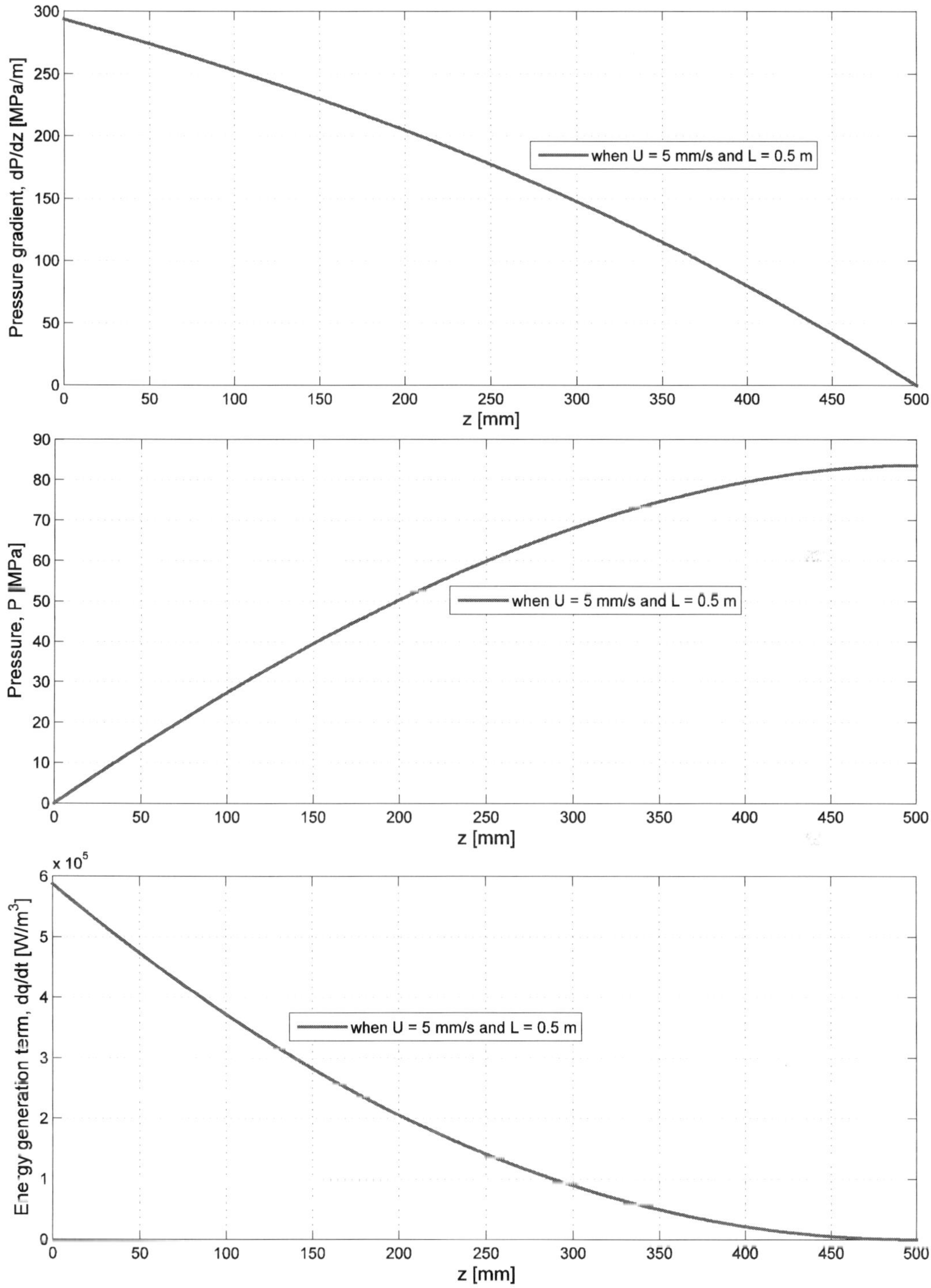

Figure 7.26: Plot of dP/dz, $P(z)$ and $\dot{q}(z)$ for Example 7.5.

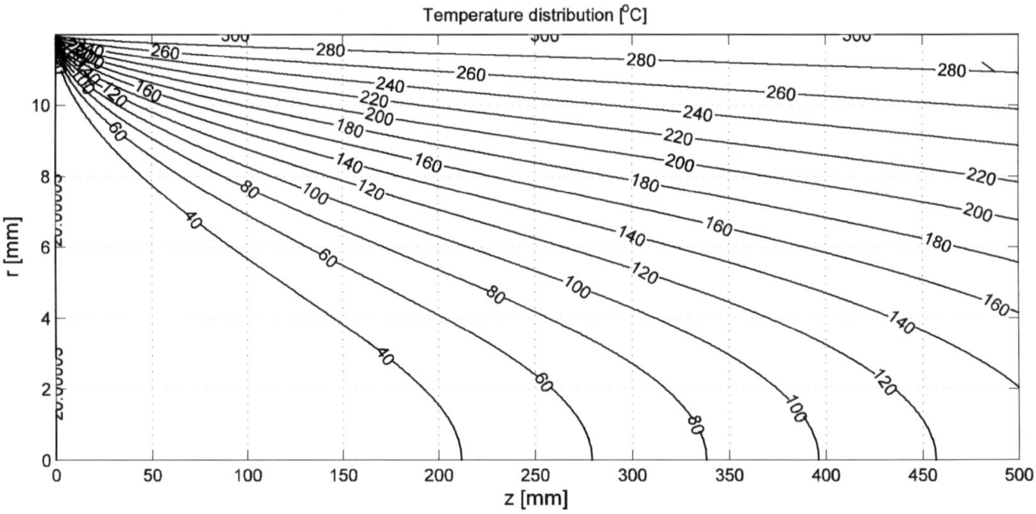

Figure 7.27: Temperature distribution for Example 7.5 in which $\dot{q} = \dot{q}(z)$. The exit temperature at $z = L$ ranges between 133.4°C at $r = 0$ and 300°C at $r = R_d$.

$$\hat{u}_z = \frac{u_z}{U} \tag{7.62d}$$

and choose

$$\text{Pe} = \text{Peclet number} = \frac{R_d U}{\alpha_{rr}} \tag{7.63a}$$

$$\text{Br} = \text{Brinkman number} = \frac{\eta U^2}{k_{rr}\Delta T}, \tag{7.63b}$$

we get

$$\frac{\partial \Theta}{\partial \hat{z}} = \frac{1}{\hat{r}} \frac{\partial}{\partial \hat{r}} \left(\hat{r} \frac{\partial \Theta}{\partial \hat{r}} \right) + \frac{k_{zz}}{k_{rr}} \frac{1}{\text{Pe}} \frac{\partial^2 \Theta}{\partial \hat{z}^2} + \text{Br} \left(\frac{R_d}{R} \right)^2 \left(\frac{\partial \hat{u}_z}{\partial \hat{r}} \right)^2. \tag{7.64}$$

Usually Pe $\approx 10^2$ to 10^3; hence, one can easily discard $\frac{k_{zz}}{k_{rr}} \frac{1}{\text{Pe}} \frac{\partial^2 \Theta}{\partial \hat{z}^2}$ term. The last term on the right-hand side of Equation (7.64) can make a significant contribution even for small values of Br number as there are many fibers and hence many cells inside the tapering die directly adding to the heat generation. ■

7.5 On-Line Consolidation of Thermoplastics

Consolidation in a composite manufacturing process is accomplished by applying pressure and heat to the system to squeeze the air, volatiles and excess resin out of the composite structure. It is a complex phenomenon with simultaneous transfer of mass, momentum and energy. The main goal of consolidation is to produce a monolithic structure by inducing intimate contact and adhesion between individual plies and minimizing void content inside the composite.

As seen in previous chapters, advanced composites are made from discrete plies. The plies consist of a tow of usually thousands of fiber strands partially or fully impregnated with resin. These plies are also referred to as tows or tapes. The approach in sheet forming in the previous section was to arrange and stack the plies in the required structure configuration and to apply heat and pressure to consolidate them to the desired geometry. Such processes are called post lay-up processes. On the other hand, one can consolidate the structure attaching one ply at a time. This process is known as in-situ or on-line consolidation.

On-line consolidation or in-situ consolidation refers to creation of a composite when thermoplastic prepregs or tapes are fused and compacted. This is done by locally heating and cooling the tapes as they are placed over a tool using a path trajectory to form the geometric shape of interest as shown in Figure 7.28. This form of in-situ processing allows one to place the reinforcing fibers in the desired directions without the constraint of following a near-geodesic path. The method can be potentially used to include even non-axisymmetric shapes. Two processes have emerged and been successfully used over the last ten years that fall under this category. They are thermoplastic filament winding and thermoplastic tow placement. The attractive features of these processes are that they easily lend themselves to automation and residual stresses are negligible.

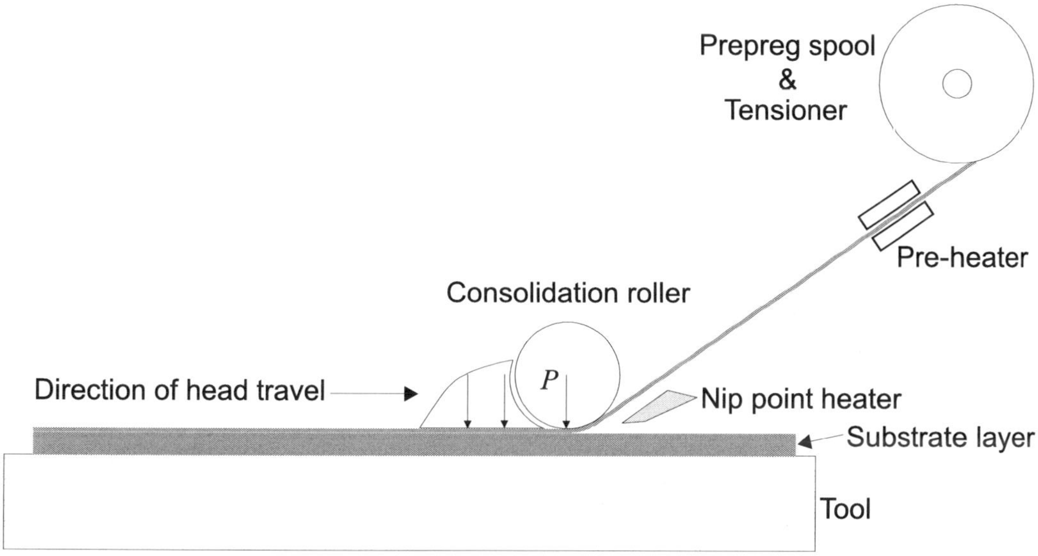

Figure 7.28: A thermoplastic fiber-reinforced tape being consolidated to the substrate layer with heating mechanism and consolidation roller.

As shown in Figure 7.28, there are five main components in this process. They are the preheater, tensioner, nip(contact)-point heater, consolidation roller and a tool for the placement of the fiber tows to form the composite.

The tensioner is used to mechanically maintain constant tension within the fiber tow from the moment it leaves the supply spool until it is wound on the tool or the previously placed fiber tows on the tool.

The tool, which is known as a mandrel in filament winding, is the surface on which the fiber tows are placed and wound to create the geometric form of the composite structure. In the case of filament winding, it is cylindrical in form, whereas in tow placement it can take various complex forms based on the geometry of the part to be manufactured. Usually metals such as stainless steel or aluminum are used to form the tool because of their durability and resistance to oxidation. The tool is usually heated to avoid creation of large thermal gradients between the heated tow and the mandrel. The large thermal mass of the tool can act as a heat sink, lower the matrix temperature of incoming tow and limit the inter-laminar flow of the molten polymer. Therefore, a heated tool is necessary to retain heat in previously wound layers. This is usually accomplished by blowing hot air into the hollow of the tool or, if higher temperatures are desired, resorting to electrically heated nitrogen torches.

A thermocouple is usually placed on the outer surface of the tool to provide the feedback to maintain the tool temperature to the desired value. A common problem in filament winding is the removal of the tool from the part. For example, if one is interested in making pipes, the process is to wind the fiber tows over a mandrel and then remove the mandrel after the desired OD of the pipes has been achieved. Usually a press or a collar fixture is used to slide the part from the mandrel. If the composite to be wound consists of carbon fibers and a thermoplastic matrix such as PEKK, the part can be easily removed upon cooling, as the thermal expansion coefficient of stainless steel is much higher than that of the composite.

The stainless steel compaction roller allows for the application of consolidation pressure to fuse and compact the incoming fiber tow with the previously placed fiber tow. Although the compaction roller may be preheated, it is generally not done, as due to low thermal mass and short contact times, it does not influence the temperature field to any significant degree. As the roller temperature is usually lower than the glass transition temperature of the thermoplastic, the incoming tow will tack on to the laminate rather than the roller. One can also monitor roller temperature if necessary with a sliding surface thermocouple. Once a layer has been wound on the part being created, the thickness of the composite part increases. This requires the compaction roller to move away from the tool while still continuing to apply constant pressure. One way to address this issue is to attach the roller to a sliding plate mounted vertically on the frame. The constant pressure is usually applied with the help of a pneumatic cylinder. The compaction force can range from 0 to 100 kN over the contact area.

The heater used in the preheating chamber must be easily controllable while supplying sufficient energy to raise the tow temperature to the desired level. Infrared strip heaters can do the job mounted equidistant from the tow along the top and the bottom surfaces. The heater power will be a function of tow speed, the material properties of the tow, and the temperature change one wants the tow to undergo.

Example 7.7: Power Requirement by a Heater

Find the heater power required to heat a tow from 30°C to 130°C. Assume that 20% of the heater energy is lost to surroundings.

Solution

$$\begin{aligned} Q_t &= \dot{m} c_{p_t} (T_{\text{exit}} - T_{\text{in}}) \\ &= \rho_t V_t w_t t_t c_{p_t} (T_{\text{exit}} - T_{\text{in}}) \end{aligned} \tag{7.65}$$

where

ρ_t	:	density of the tow
v_t	:	velocity of the tow
w_t	:	width of the tow
t_t	:	thickness of the tow
c_{p_t}	:	specific heat capacity of the tow
T_{in} and T_{exit}	:	incoming and exit temperatures of the tow in the preheater.

Hence, due to losses, the power required is

$$\text{Heater Power} = \frac{Q_t}{0.80}. \tag{7.66}$$

■

The requirements of the nip- or contact-point heater are the same as the preheater except the nip-point heaters are focused only at the point of contact of the incoming tow and the previously placed tows. The nip-point heater should induce rapid heating to bring the contact area to above the melting point of the resin instantly as the contact time is usually very short. A variety of heaters can be used for this purpose. Some examples are open torch, laser, hot gas, infrared and heated rollers, or hot shoes.

7.5.1 Introduction to Consolidation Model

There are several important issues during the consolidation process employed in composites manufacturing. The key process parameters that influence the extent of consolidation are the applied pressure, temperature, and time required for consolidation. Many phenomena such as intimate contact between two adjacent plies, resin flow across and in between the plies, adhesion, fiber deformation and movement, molecular interdiffusion, and diffusivity of air and volatiles in the molten matrix occur simultaneously during the consolidation process. There are complex interactions among all these phenomena. Thus, consolidation will be a function of process parameters such as the rate of application of pressure and temperature and also material parameters such as resin and fiber properties, the fiber volume fraction, and the state of the initial plies. The interactions between mass, momentum and energy transfer can be investigated by classifying them on macro-, micro- and molecular levels. The macrolevel analysis will allow us to establish the process windows of temperature, pressure, and time required to achieve complete consolidation. At the microlevel, one can investigate the effect of temperature and pressure on the size of individual voids and thus on the macroscopic density [327]. At the molecular level, one is concerned about intimate contact between two adjacent plies and diffusion of polymer chains across the ply interfaces [328, 329]. The process and material parameters affect phenomena at all scales, and the degree and mechanism of their interactions will depend on whether it is a post-consolidation or an on-line consolidation process [330]. The objective is to identify the mass, momentum, and energy transport mechanisms during the consolidation process so that one can establish their relative contribution and significance in the consolidation process. In on-line processes, the lay-up of plies and consolidation occur simultaneously.

7.5.2 Importance of Process Modeling

The phenomenon of consolidation during tow placement of composites may be mathematically modeled as the flow of a compressible viscous fluid. Such a model helps provide insight into the effect of the process parameters, such as the tow speed and the compaction force and about the variables of interest, such as the final void fraction and part thickness. Typically, the final void fraction of a composite is used as an estimate of the part quality. Therefore, the process inputs may be determined such that the quality requirement, for example $< 1\%$ void fraction, is satisfied. Also, the model may be used as part of an on-line control strategy for a tow placement process. Ideally, using a porosity sensor at the post-consolidation stage, the part porosity may be compared to the required value. If the quality requirement is not satisfied, the model may be used to correct the inputs appropriately to achieve the required porosity level. The process model also enables one to optimize the process in such a manner that the manufacturing speed is at the highest possible level given other constraints such as the machine limitations for compaction force and the desired final product quality. A typical situation is shown in Figure 7.29. The maximum compaction force the machine is capable of providing may be used as the limiting condition. Section A is taken at this level, and its projection on the processing speed-porosity plane is obtained. The maximum acceptable porosity can now be used to determine the maximum processing speed by following the line BCD. The line BC runs parallel to the processing speed axis at the required porosity value until it intersects the projection of section A. The line CD runs parallel to the porosity axis until it intersects the processing speed axis. The point D represents the highest processing speed that can be used to obtain a part of acceptable

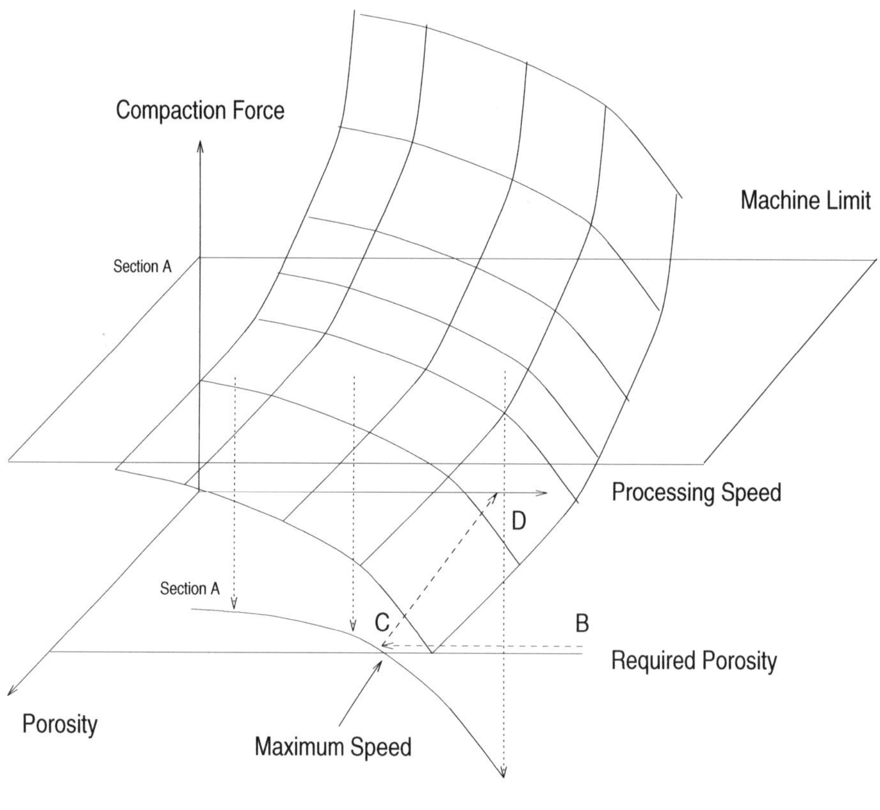

Figure 7.29: Obtaining the highest processing speed from a typical porosity-speed-force surface, taking into account machine limits and quality requirements for on-line consolidation [331].

quality. A processing window may be defined based on the maximum force and processing speed the machine is capable of operating at.

The process model (i) assists in development of the tow placement equipment, commonly called the thermoplastic tow placement head, (ii) assists in design choices, (iii) reduces the time to establish a working process window when the new process and equipment are used, and (iv) assists in process control.

7.5.3 Consolidation Process Model

Thermoplastic matrices in advanced composites applications are highly viscous compared to thermosets. Thermoset composites processing has been conventionally modeled as flow of resin through a network of fibers, which is described well by Darcy's law. This is valid for processes such as autoclave thermoset processing due to the low viscosity of thermosets prior to cure initiation. However, due to the high viscosity of thermoplastics, it has been observed that the fibers move with the resin rather than relative to the resin. Therefore, the approach best suited for thermoplastic composite consolidation is a squeeze flow model. The fibers, resin and any voids present in the tow are approximated as a homogeneous fluid with specified rheological properties. The rheological properties of this continuum depend on the temperature, the fiber volume fraction, and the void content.

7.5.4 Model Assumptions and Simplifications

The simplification of the geometry and the location of the consolidation of thermoplastic composites during tape lay-up or filament winding is shown in Figure 7.30. This process may be modeled as shown in Figure 7.31. A tow of a given height, h_i, and width, w_i, enters the region under the compaction roller. The coordinate system is defined such that x is the width direction, y is the direction of tow movement and z is the thickness direction. The tow speed, V, is specified. As the tow passes under the compaction roller, its height is reduced to h_f and its width changes to w_f. The consolidation force compresses the voids and changes the height and width of the tow. Since the tow dimension in the y direction is much larger than in the x and z dimensions, the y direction flow may be neglected. Due to the high matrix resin viscosity, the flow may be treated as creeping motion and the inertial effects can be neglected.

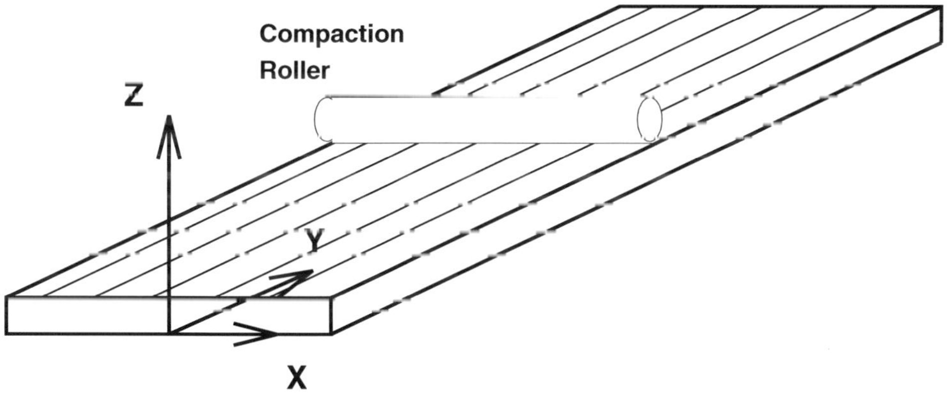

Figure 7.30: The coordinate axes for the consolidation process [327].

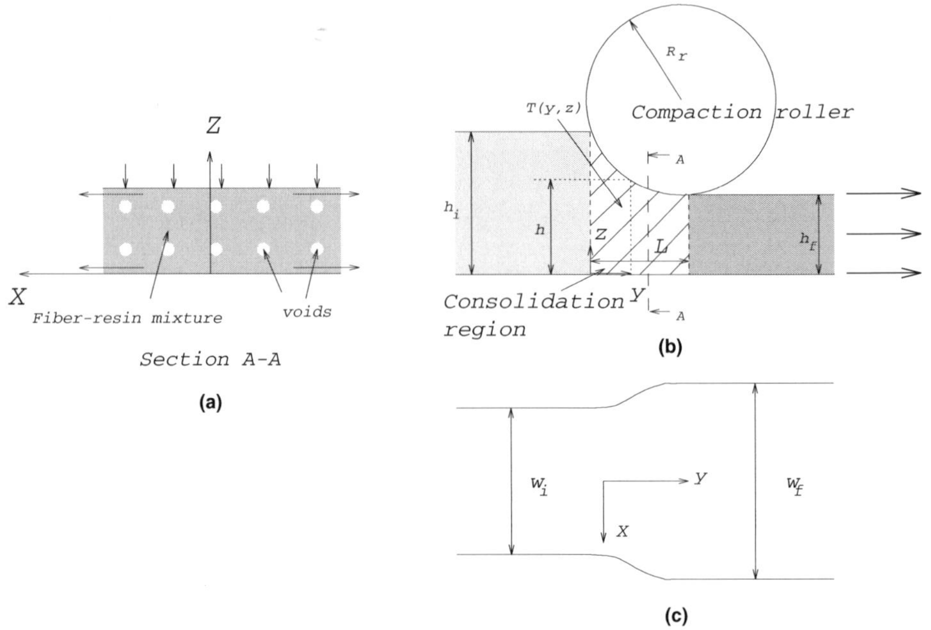

Figure 7.31: Schematic of the various views of the consolidation squeeze flow model with the input variables [327].

7.5.5 Governing Equations

The continuity and momentum equations that govern the motion of the fluid under the compaction roller, neglecting inertia and body force terms, may be written as,

$$\frac{\partial \rho}{\partial t} + \frac{\partial}{\partial x}(\rho v_x) + \frac{\partial}{\partial z}(\rho v_z) = 0 \tag{7.67a}$$

$$\frac{\partial P}{\partial x} = \frac{\partial}{\partial x}\left(\mu\left[2\frac{\partial v_x}{\partial x}\right]\right) + \frac{\partial}{\partial z}\left(\mu\left[\frac{\partial v_x}{\partial z} + \frac{\partial v_z}{\partial x}\right]\right) \tag{7.67b}$$

$$\frac{\partial P}{\partial z} = \frac{\partial}{\partial x}\left(\mu\left[\frac{\partial v_x}{\partial z} + \frac{\partial v_z}{\partial x}\right]\right) + \frac{\partial}{\partial z}\left(\mu\left[2\frac{\partial v_z}{\partial z}\right]\right) \tag{7.67c}$$

The density and viscosity of the continuum are given by ρ and μ, the pressure in the material by P, and the velocity by $\mathbf{v}$. In actual tows, the thickness ($0.006''$) is much smaller than the width ($0.25''$). Therefore, the dominant velocity gradient term in the momentum equations is the shear strain rate through the gap, $\partial v_x/\partial z$ and its derivative with respect to the thickness direction, z. Therefore, the momentum equations, Equations (7.67b) and (7.67c), may be simplified as,

$$\frac{\partial P}{\partial x} = \frac{\partial}{\partial z}\left(\mu\frac{\partial v_x}{\partial z}\right) \tag{7.68a}$$

$$\frac{\partial P}{\partial z} = 0. \tag{7.68b}$$

Here Newtonian constitutive law is assumed. It should be straightforward to extend it to shear thinning fluids. From Equation (7.68b), it is clear that $P = P(x)$. The density of the continuum is affected only by the local pressure. Therefore, the spatial variation of the density may also be approximated as a function of x only and Equation (7.67a) may be rewritten as,

$$\frac{\partial \rho}{\partial t} + \frac{\partial}{\partial x}(\rho v_x) + \rho\frac{\partial}{\partial z}(v_z) = 0. \tag{7.69}$$

Equation (7.69) may be integrated in the thickness direction from 0 to h, the instantaneous thickness of the domain

$$h\frac{\partial \rho}{\partial t} + \int_0^h \left[\frac{\partial}{\partial x}(\rho v_x)\right] dz + \rho(v_z)|_{z=0}^{z=h} = 0. \tag{7.70}$$

The velocity v_z at $z = 0$ is 0 and at $z = h$ is $\dot{h}$, the closing speed between the two bounding surfaces. Substituting these values and interchanging the order of integration and differentiation in Equation (7.70) results in

$$h\frac{\partial \rho}{\partial t} + \frac{\partial}{\partial x}\left(\rho\int_0^h v_x dz\right) + \rho\dot{h} = 0. \tag{7.71}$$

Equation (7.68a) may be integrated with respect to z to yield,

$$\mu\left[\frac{\partial v_x}{\partial z}\right] = z\frac{dP}{dx} + C_1(x). \tag{7.72}$$

Integrating Equation (7.72) once more with respect to z,

$$v_x(z) = v_x(0) + \frac{dP}{dx}\int_0^z \frac{\xi}{\mu} d\xi + C_1(x)\int_0^z \frac{1}{\mu} d\xi \tag{7.73}$$

where ξ is a dummy variable of integration. The value for v_x from Equation (7.73) may be substituted into Equation (7.71) to eliminate velocity as a variable and maintain pressure as the only variable to be solved

$$h\frac{\partial \rho}{\partial t} + \frac{\partial}{\partial x}\left(\rho\int_0^h \left[v_x(0) + \frac{dP}{dx}\int_0^z \frac{\xi}{\mu} d\xi + C_1(x)\int_0^z \frac{1}{\mu} d\xi\right] dz\right) + \rho\dot{h} = 0. \tag{7.74}$$

The values of $C_1(x)$ and $v_x(0)$ may be obtained from two velocity boundary conditions. Since we have a second-order equation for pressure, two pressure boundary conditions are also needed.

Let ρ^* be the nondimensionalized density of the continuum defined as,

$$\rho^* = \frac{\rho}{\rho_f} \tag{7.75}$$

where ρ is the density of the continuum and ρ_f is the density of the composite with no voids. Equation (7.74) can be rewritten as,

$$h\frac{\partial \rho^*}{\partial t} + \frac{\partial}{\partial x}\left(\rho^*\int_0^h \left[v_x(0) + \frac{dP}{dx}\int_0^z \frac{\xi}{\mu} d\xi + C_1(x)\int_0^z \frac{1}{\mu} d\xi\right] dz\right) + \rho^*\dot{h} = 0. \tag{7.76}$$

Equation (7.76) is the general form of the macroscopic governing equation for consolidation in which one is squeezing the voids out of the composite. This equation can be specialized depending on the particular problem that is to be solved (such as isothermal incompressible, isothermal compressible, etc.). However, first we must have a model to describe void mechanisms at the microscopic level.

Void Reduction Mechanisms and Models

Voids may be broadly characterized as interlaminar or intralaminar. Interlaminar voids will not be addressed in this work. Existing models for degree of intimate contact may be used to model them (for example see [332–336]). The focus here will be on the growth/collapse and transport of intralaminar voids. The various mechanisms of intralaminar void growth/collapse and transport are shown in Figure 7.32 schematically. Void coalescing may occur when the voids in a tow grow or move in such a manner that the boundaries of two or more bubbles meet; this leads to the formation of a bigger void of a different shape. Since our aim is to model void compression under the compaction roller, this mechanism is not of importance. The voids may migrate due to the pressure gradient in the fluid or other mechanical motions [337–339]). The transport of the voids along with the resin can be addressed here, but void migration relative to the resin has been neglected because the high viscosity of the resin impedes any such motion. In the region of interest, cooling of the tow takes place and should lead to reduction of the void size. This contraction can be taken into account using the ideal gas law. Void compression also takes place due to the applied compaction pressure. The effects of heat input and void compression are taken into account using a microscopic model. The effect of void transport is accounted for in the macroscopic flow model.

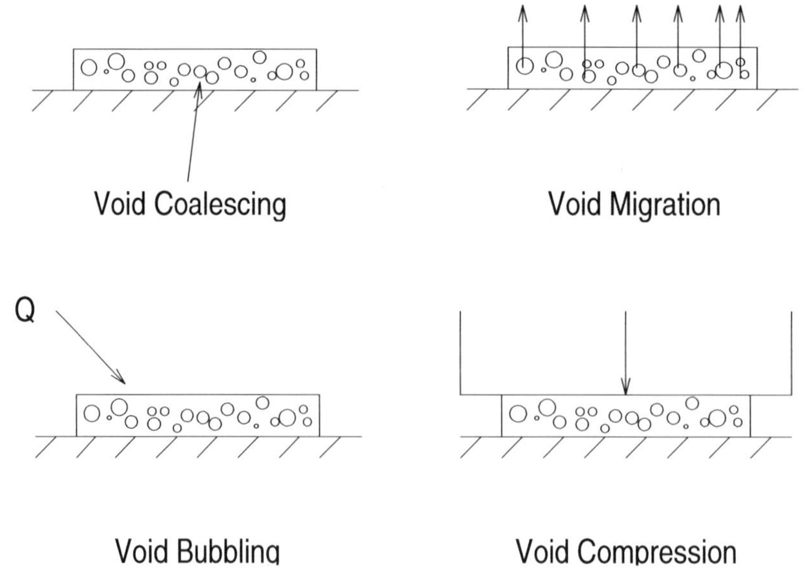

Figure 7.32: Different void mechanisms for intralaminar growth/collapse and transport.

Microscopic Model

A typical void at any location is approximated by a sphere of radius R, as shown in Figure 7.33. The void is assumed to be surrounded by a concentric spherical resin shell of outer radius S. The ratio of R and S is determined by the void fraction at the location under consideration. Void growth or collapse is governed by a balance between the pressure inside and outside the void, the surface tension and the resin viscosity [337]

$$4\mu\left(\frac{R^3}{S^3}-1\right)\frac{dR}{dt}+(P_g-P_f)R-2\sigma=0. \tag{7.77}$$

The initial radii of the void and the resin shell are R_o and S_o, respectively. P_g and P_f are the gas and resin pressure inside and outside the shell, respectively. The resin is assumed to be incompressible, therefore

$$S^3-R^3=S_o^3-R_o^3. \tag{7.78}$$

Also, from the ideal gas law,

$$\frac{P_{go}R_o^3}{T_o}=\frac{P_gR^3}{T} \tag{7.79}$$

where T_o and T are the initial and current temperatures of the void. The radius of the void and the shell of resin may be expressed in nondimensional form as,

$$S^* = \frac{S}{R_o} \tag{7.80}$$

$$R^* = \frac{R}{R_o} \tag{7.81}$$

Equation (7.77) may now be rewritten using Equations (7.78) and (7.79) as,

$$4\left(\frac{R^{*3}}{S_o^{*3}+R^{*3}-1}-1\right)\frac{dR^*}{dt}+\left(\frac{P_{go}}{R^{*3}}\frac{T}{T_o}-P_f\right)\frac{R^*}{\mu}-\frac{2\sigma}{\mu R_o}=0. \tag{7.82}$$

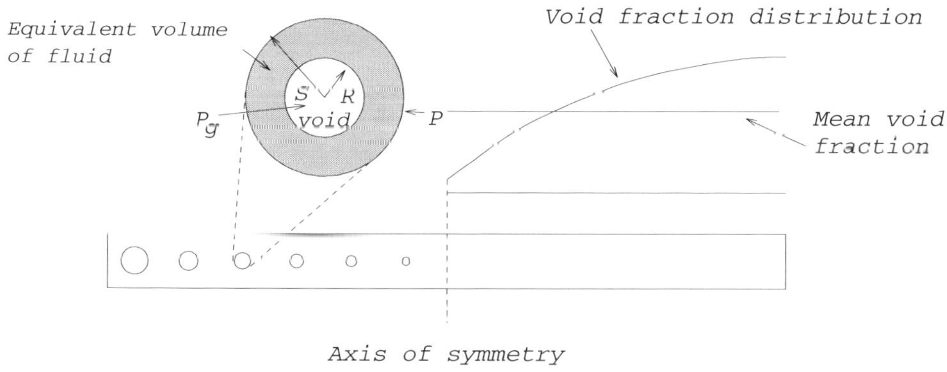

Figure 7.33: Microscopic void compression model used to determine the bubble radius as a function of the local pressure and temperature.

It can be shown that the rate of change of ρ^* with respect to time can be expressed in terms of the rate of change of the nondimensional radius of void as

$$\frac{\partial \rho^*}{\partial t} = \frac{-3R^{*^2}(S_o^{*^3} - 1)}{(S_o^{*^3} - 1 + R^{*^3})^2} \frac{dR^*}{dt}. \tag{7.83}$$

Thus the rate of change of the nondimensional density with respect to time may be obtained using Equations (7.82) and (7.83). The coupling between the microscopic model and the macroscopic pressure distribution is obtained by substituting Equation (7.83) in (7.76) to compute the pressure distribution in the domain. Equation (7.82) can be used to compute the individual change in the radius of the voids at various locations in the domain.

Example 7.8: Bulk Density and Voids

Derive the expression for change in bulk density due to void compression (Equation 7.83).

Solution

Initially, the composite material occupies a hollow spherical volume between $r = R_o$ and $r = S_o$. The mass of the composite material excluding the void in it, m_f is given as

$$m_f = \rho_f \frac{4\pi}{3}\left(S_o^3 - R_o^3\right) \tag{7.84}$$

where ρ_f is the density of the composite material excluding the void in it. At any time, the mass of the cell (composite material including the void) is given as

$$m = m_f + m_{\text{void}} = \rho \frac{4\pi}{3} S^3 \tag{7.85}$$

where ρ is the density of the composite material including the void and S is the radius of the cell at that time. If one neglects m_{void} relative to m_f, then

$$\rho \frac{4\pi}{3} S^3 = \rho_f \frac{4\pi}{3}\left(S_o^3 - R_o^3\right) \tag{7.86}$$

which reduces to

$$\frac{\rho}{\rho_f} = \rho^* = \frac{S_o^3 - R_o^3}{S^3}. \tag{7.87}$$

If one takes the time derivative of both sides of Equation (7.87), the following is obtained

$$\begin{aligned} \frac{\partial \rho^*}{\partial t} &= \frac{\partial}{\partial t}\left(\frac{S_o^3 - R_o^3}{S^3}\right) \\ &= -3\frac{S_o^3 - R_o^3}{S^4}\frac{dS}{dt}. \end{aligned} \tag{7.88}$$

Time derivative of Equation (7.78) allows one to express dS/dt in terms of dR/dt

$$\begin{aligned} \frac{d}{dt}(S^3) &= \frac{d}{dt}(S_o^3 - R_o^3 + R^3) \\ 3S^2\frac{dS}{dt} &= 3R^2\frac{dR}{dt} \\ \frac{dS}{dt} &= \frac{R^2}{S^2}\frac{dR}{dt}. \end{aligned} \tag{7.89}$$

By substituting for $\frac{dS}{dt}$ from Equation (7.89) into Equation (7.88), one can obtain the following:

$$\begin{aligned}\frac{\partial \rho^*}{\partial t} &= -3\,\frac{(S_o^3 - R_o^3)R^2}{S^6}\,\frac{dR}{dt} \\ &= -3\,\frac{(S_o^3 - R_o^3)R^2}{(S_o^3 - R_o^3 + R^3)^2}\,\frac{dR}{dt} \\ &= -3\,\frac{\left\{R_o^3\left[\left(\frac{S_o}{R_o}\right)^3 - \left(\frac{R_o}{R_o}\right)^3\right]\right\}\left\{R_o^2\left(\frac{R}{R_o}\right)^2\right\}}{\left\{R_o^3\left(\frac{S_o}{R_o}\right)^3 - R_o^3\left(\frac{R_o}{R_o}\right)^3 + R_o^3\left(\frac{R}{R_o}\right)^3\right\}^2}\,\frac{d(R_oR^*)}{dt} \\ &= \frac{-3{R^*}^2({S_o^*}^3 - 1)}{({S_o^*}^3 - 1 + {R^*}^3)^2}\,\frac{dR^*}{dt}\end{aligned} \tag{7.90}$$

which is Equation (7.83). ■

7.5.6 Boundary Conditions

Two velocity boundary conditions and two pressure boundary conditions are needed to obtain the solution for the governing equations. The velocity boundary conditions will be addressed first.

Velocity Boundary Conditions

There has been some speculation in the literature (e.g., [337]) about the type of velocity boundary conditions that needs to be imposed at the interface between the fluid and solid in processes such as compression molding. The problem under consideration is similar in nature to compression molding as far as the velocity boundary conditions are concerned. Traditionally, a no-slip boundary condition is imposed at the interface between a fluid and a solid wall. This implies that the friction between the fluid and the wall is much greater than the internal friction (due to the viscosity) in the fluid. However, in some cases a more appropriate boundary condition is partial or limited slip at a fluid-solid interface. Such a condition may be thought of as a friction factor boundary condition and expressed as,

$$\mu \left.\frac{\partial v}{\partial n}\right|_{\text{wall}} = K_{\text{wall}}\, v_{\text{wall}} \tag{7.91}$$

where K_{wall} is the friction factor and μ is the viscosity of the fluid.

The ratio K_{wall}/μ (where K_{wall} is the inverse of β in Equation (6.45)) determines the type of boundary condition that is imposed. If one expects the friction to be very high, *i.e.*, $K_{\text{wall}}/\mu \to \infty$, it is clear that v_{wall} must be zero such that the left-hand side is finite. This is equivalent to a no-slip boundary condition. On the other hand, if the friction is expected to be very low, *i.e.*, $K_{\text{wall}}/\mu \to 0$, the normal derivative of the velocity goes to zero at the wall. This is equivalent to a perfect slip boundary condition. Therefore, this friction factor boundary condition may be used to impose all types of velocity boundary conditions ranging from no-slip to perfect slip.

Pressure Boundary Conditions

We also need pressure boundary conditions to solve the consolidation problem uniquely. If the tow is assumed to be unrestricted along the width, i.e., there are no tows that have been laid adjacent to it, then the appropriate boundary condition would be to impose atmospheric pressure at the free surfaces. However, if the tow being considered has neighboring tows on one or both sides, then the boundary condition on the constrained edge needs to be modified. We do not expect any flow to take place out of the constrained edges and therefore need to impose a zero pressure gradient condition at these edges. The pressure boundary conditions may be summarized as follows:

$$P_{\text{unconstrained edge}} = P_{\text{atmospheric}} \tag{7.92}$$

$$\left(\frac{\partial P}{\partial x}\right)_{\text{constrained edge}} = 0. \tag{7.93}$$

While considering the tow that is being laid down, we apply the free surface boundary condition because the unconsolidated tow width is smaller than the consolidated tow width. However, for a tow that was placed previously, one applies the constrained boundary condition because it has tows along both edges of the width.

7.5.7 Rheology of the Composite

An important input to the consolidation process models described above is the viscosity of the composite. The composite consists of fibers, resins and voids. This combination may be treated as a continuum on length scales that are much greater than their individual dimensions. The viscosity of this continuum needs to be determined experimentally.

Previously, the viscosity of heterogeneous media has been measured using capillary rheometers, cone and plate rheometers and dynamic mechanical analysis. However, for application in a consolidation process model, the squeeze flow (transverse shear) viscosity of the composite is required. This is not obtainable using any of the above mentioned equipment. Therefore, the resistance of a composite laminate to squeezing should be used to estimate the transverse shear viscosity.

Example 7.9: Characterization of the Viscosity of AS4 Fibers with PEKK Resin

A constant load with a heating unit attached is used to characterize the viscosity of AS4 fibers with PEKK resin. The specimen, which is a ten layered consolidated AS4/PEKK laminate, was placed inside an oven using a dead weight as the source of the constant load. The composite and the dead weight were held at the desired temperature for 20 minutes. The dead weight was then placed on the specimen for a predetermined duration of time. The dimensions of the specimen before and after the dead weight being placed on it were measured to estimate the viscosity of the composite at different temperatures. Table 7.1 shows the results from these experiments. Find the temperature dependence of the viscosity of this system.

Solution

The change in dimensions of the laminate can be related to its viscosity assuming that the composite behaves as an isothermal incompressible Newtonian fluid. Thus,

Table 7.1: Viscosity of AS4/PEKK Composite at Various Temperatures [331]

Sample I.D.	L_i (m)	W_i (m)	h_i (m)	Temp (°C)	$\frac{L_f-L_i}{L_i}$ ×100%	$\frac{W_f-W_i}{W_i}$ ×100%	$\frac{h_i-h_f}{h_i}$ ×100%	Viscosity MPa.s	Average Viscosity MPa.s
A7	0.0508	0.0064	0.0014	350	0.96	32.39	15.06	12.21	
A8	0.0573	0.0051	0.0017		0.42	15.19	10.61	48.77	
A10	0.0510	0.0033	0.0015		0.15	54.16	22.84	44.92	50.46
A11	0.0509	0.0035	0.0014		0.30	27.51	10.53	95.94	
B1	0.0482	0.0065	0.0014	360	0.84	14.52	6.36	38.18	
B2	0.0487	0.0065	0.0014		0.81	17.05	13.93	13.62	
B3	0.0489	0.0064	0.0013		1.53	23.48	9.75	20.36	21.32
B4	0.0490	0.0065	0.0014		1.32	42.13	13.79	13.14	
D1	0.0494	0.0064	0.0013	380	1.26	23.45	8.88	22.30	
D2	0.0494	0.0065	0.0014		0.31	24.16	7.34	30.76	
D3	0.0494	0.0065	0.0014		1.47	29.78	12.52	15.36	18.76
D4	0.0495	0.0065	0.0014		0.92	48.65	20.66	6.64	
E1	0.0495	0.0064	0.0013	390	2.80	47.90	22.43	5.30	
E2	0.0495	0.0064	0.0013		2.62	41.79	20.12	6.32	
E3	0.0495	0.0064	0.0013		2.23	41.29	19.66	6.90	7.25
E4	0.0488	0.0065	0.0014		1.67	39.51	15.81	10.50	

Note: An average force of 266.72 N was applied for 60 seconds.

$$\mu = \frac{5}{8} \frac{Ft}{L_i W_i^3} \frac{h_i^2 h_f^5}{h_i^5 - h_f^5}. \tag{7.94}$$

Hence, at each temperature, one can find μ.

It is clear that the average viscosity of the composite decreases with an increase in temperature. The viscosity of AS4/PEKK can be approximated by the following functional form (see Chapter 4 for details):

$$\mu = A\,e^{\frac{B}{T}} \qquad \text{Pa.s} \tag{7.95}$$

where T is the temperature in K. By plotting viscosity values at 4 different temperatures, a best-fit line was found. Note that viscosity is plotted on a log scale due to the exponential nature of Equation 7.95. The values of A and B that best fit the experimental data were 1.28E-4 and 16548.6. The best-fit line and the experimental data are shown in Figure 7.34. ■

7.5.8 Model Solutions

In this section, sample solutions of the consolidation model described earlier will be presented. It is seen from Figure (7.31b) that the linear tow velocity with respect to the roller along with the geometric parameters, h_i, h_f and the roller radius, R_r, determine the closing speed $\dot{h}$ completely as a function of the instantaneous height.

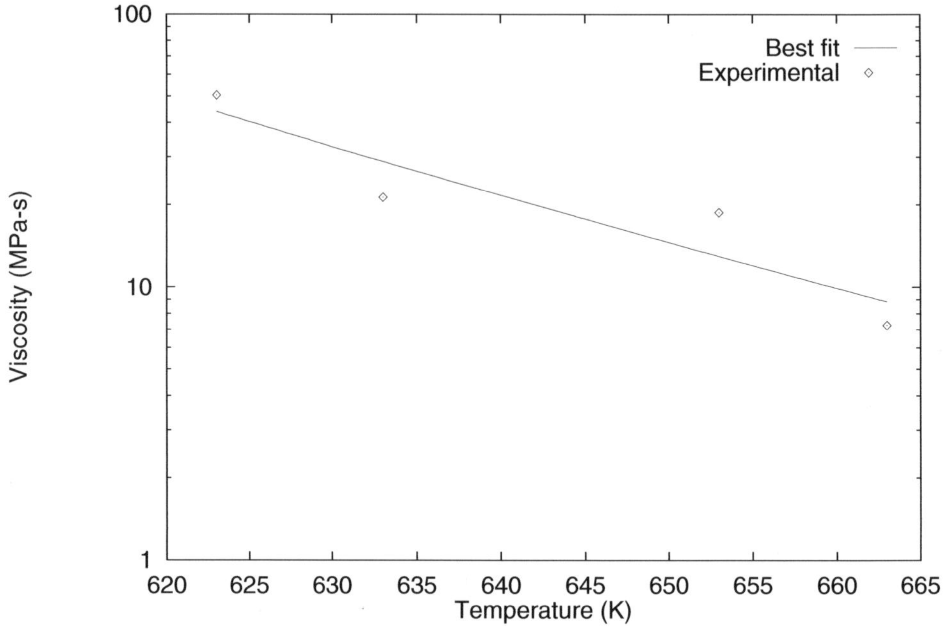

Figure 7.34: Viscosity of AS4/PEKK composite as a function of temperature [331].

From geometric considerations, the contact length L_c may be expressed as,

$$L_c = \left[R_r^2 - (R_r - h_i + h_f)^2\right]^{1/2} \tag{7.96}$$

The instantaneous height h may be expressed in terms of the position y as

$$h = R_r + h_f - \left[R_r^2 - (L_c - y)^2\right]^{1/2} \tag{7.97}$$

Differentiating both sides with respect to time and equating dy/dt to the tow speed V, we obtain

$$\dot{h} = \frac{dh}{dt} = -V\frac{\left[R_r^2 - (R_r - h + h_f)^2\right]^{1/2}}{(R_r - h + h_f)}. \tag{7.98}$$

Equation (7.98) is used to express the imposed closing speed as a function of the instantaneous height in all the computations.

The first solution we will present is for the simplest case of an isothermal incompressible fluid.

Isothermal Incompressible Fluid

Consider an isothermal incompressible Newtonian fluid with the following boundary conditions. The friction factors at $z = 0$ and $z = h$ are denoted by K_b and K_t in Equation (7.91). The susbcripts b and t characterize the bottom and top, respectively. The pressure at $x = w$ is zero and the pressure gradient $dP/dx = 0$ at $x = 0$. Equation (7.73) can be used to solve for the velocity profile

$$v_x(z^*) = -\frac{h^2}{\mu}\frac{dP}{dx}\left[-\frac{{z^*}^2}{2} + \left(\frac{1}{K_b^*} + z^*\right)\frac{1 + K_t^*/2}{1 + K_t^* + K_t^*/K_b^*}\right] \tag{7.99}$$

where K_t^* and K_b^* are K_t and K_b nondimensionalized with μ/h and z^* is z nondimensionalized with respect to h. Figure 7.35 shows the velocity profile for various values of K_t^* for a given value of K_b^*. It can be seen that the velocity profile changes from a symmetric parabolic profile for the case of no-slip at both walls to an asymmetric profile when no-slip is imposed on one wall and perfect slip is allowed at the other wall. Equation (7.76) can now be used to determine the pressure distribution in the domain

$$P = \frac{\mu\dot{h}}{h^3}\frac{w^2 - x^2}{2}\left[\frac{1}{6} - \left(\frac{1}{K_b^*} + \frac{1}{2}\right)\frac{(1 + K_t^*/2)}{(1 + K_t^* + K_t^*/K_b^*)}\right]^{-1}. \tag{7.100}$$

Figure 7.36 shows the pressure distribution in the domain for various friction factors. The pressure distribution is seen to be parabolic in nature independent of the type of velocity boundary condition that is applied. However, the pressure drop from the mid-plane to the outer edge is the lowest when a perfect slip condition is imposed on the top surface and the highest when a no-slip condition is imposed on the top surface, given that the bottom surface has a no-slip condition prescribed.

The change in width or the movement of the free surface may be tracked as a through-thickness average based on the velocity in the x direction at the free surface

$$\frac{dw}{dt} = \frac{1}{h}\left[\int_0^h v_x dz\right]_{x=w}. \tag{7.101}$$

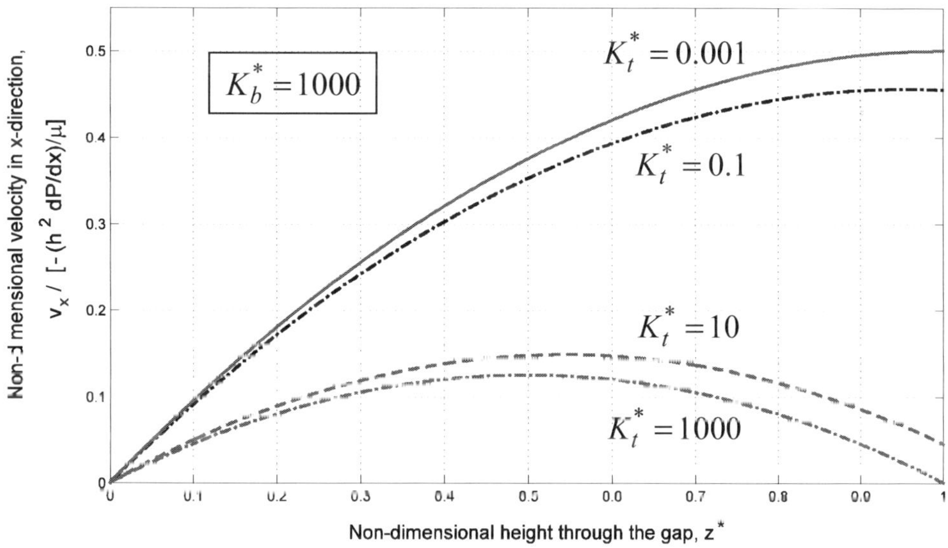

Figure 7.35: The velocity profile changes from a symmetric parabolic to an asymmetric parabolic shape as the top surface friction factor goes to zero with the bottom surface friction factor maintained constant at a high value (which models the no-slip boundary condition).

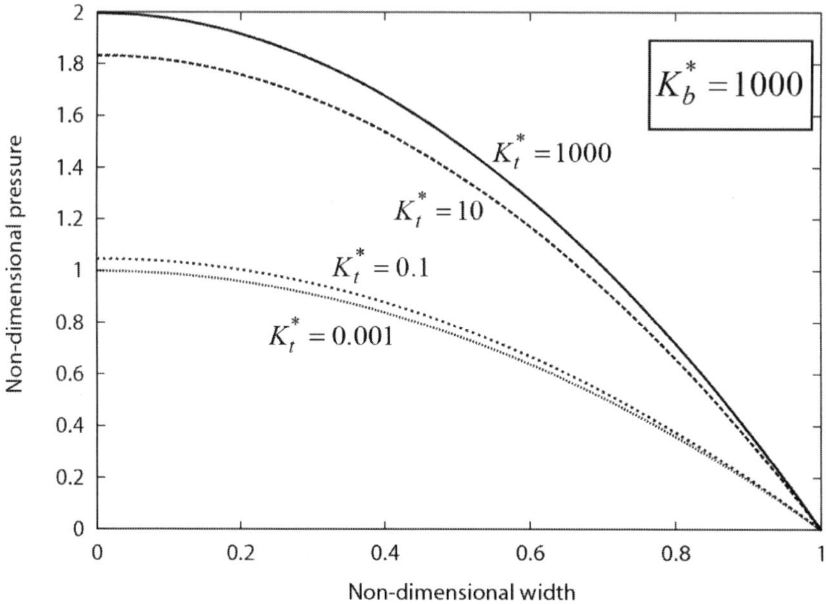

Figure 7.36: The pressure profiles maintain a parabolic shape. The pressure drop from the middle of the tow to the outer edge decreases with decreasing top surface friction factor. Pressure is nondimensionalized with respect to $\mu\dot{h}/h$ (see Equation (7.100)).

Substituting for v_x from Equation (7.99), it can be shown that,

$$\frac{dw}{dt} = -w\frac{\dot{h}}{h} \tag{7.102}$$

In other words,

$$w\,h = \text{constant.} \tag{7.103}$$

This result is not surprising because we are dealing with an incompressible fluid and expect the total volume to be conserved. This result is, in fact, a verification that we conserve volume in this formulation.

Example 7.10: Compaction Force

Find the compaction force required to compress the tow as a function of the final h_f shown in Figure 7.31.

Solution

The force required for compaction, F_c, for this incompressible material may be obtained by integrating the pressure distribution across the width and along the contact length

$$F_c = 2\int_0^{L_c}\int_0^{w} P(x,y)dxdy. \tag{7.104}$$

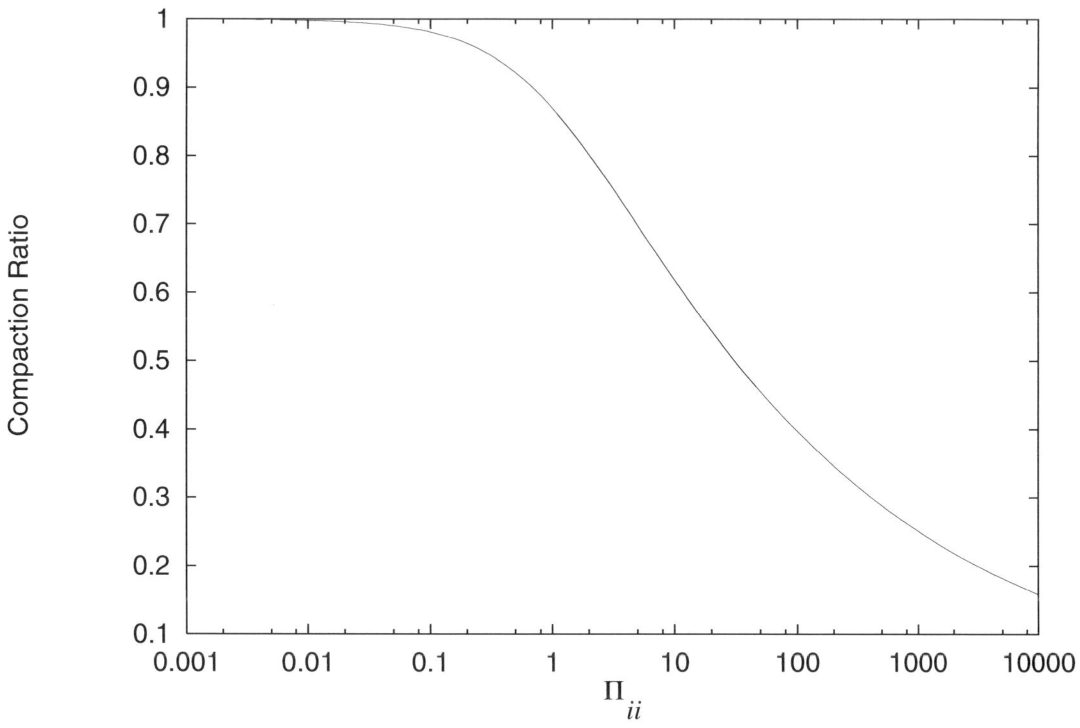

Figure 7.37: Variation of compaction ratio with Π_{ii}.

Substituting for $\dot{h}$ and the pressure solution from Equations (7.98) and (7.100), respectively, and carrying out the integrations, we obtain

$$1-\left(\frac{h_i}{h_f}\right)^5=\frac{15}{2}\frac{F_c}{\mu V}\frac{h_i^2}{w_i^3}\left[\frac{1}{6}-\left(\frac{1}{K_b^*}+\frac{1}{2}\right)\frac{(1+K_t^*/2)}{(1+K_t^*+K_t^*/K_b^*)}\right]. \tag{7.105}$$

The compaction ratio h_f/h_i is governed by a single nondimensional parameter, Π_{ii}, given by the right-hand side of Equation (7.105).

$$\Pi_{ii}=-\frac{15}{2}\frac{F_c}{\mu V}\frac{h_i^2}{w_i^3}\left[\frac{1}{6}-\left(\frac{1}{K_b^*}+\frac{1}{2}\right)\frac{(1+K_t^*/2)}{(1+K_t^*+K_t^*/K_b^*)}\right]. \tag{7.106}$$

Therefore, Equation (7.105) may be rewritten as

$$\frac{h_f}{h_i}=(1+\Pi_{ii})^{\ 1/5}. \tag{7.107}$$

Π_{ii} varies between zero and ∞ depending on the parameters of the problem. For example, for high compaction forces, low tow speeds or low material viscosity, the magnitude of Π_{ii} is large and thus h_f/h_i is a small number. Figure 7.37 shows the variation of the compaction ratio with the nondimensional parameter Π_{ii}. ■

Isothermal Compressible Fluid

In this section, we relax the assumption that the fluid is incompressible and solve the flow, pressure and density field in the domain of interest. Physically it implies that there are small voids in the material that can be squeezed out.

The formulation follows along the same lines as the incompressible case, but the terms with density gradients need to be retained in the solution. The governing equation for the pressure in the fluid field is given by

$$\frac{d}{dx}\left(\rho^* \frac{dP}{dx}\right) = -\frac{\mu}{h^3}\left(h\frac{\partial \rho^*}{\partial t} + \rho^* \dot{h}\right)\left[\left(\frac{1}{K_b^*} + \frac{1}{2}\right)\frac{(1 + K_t^*/2)}{(1 + K_t^* + K_t^*/K_b^*)} - \frac{1}{6}\right]^{-1}. \quad (7.108)$$

The velocity field is still given by Equation (7.99). Equation (7.108) needs to be solved numerically due to the dependence of the density on both time and position, x.

The solution methodology is shown in Figure 7.38. The initial density in the domain is assumed to be uniform. Equation (7.108) is solved with the pressure boundary conditions

$$\left(\frac{dP}{dx}\right)_{x=0} = 0 \quad (7.109)$$

$$P(x = w(t)) = 0 \quad (7.110)$$

using a central difference scheme for the spatial derivative of P. The velocity in the x direction is calculated from Equation (7.99). Based on the average velocity of the material in the x direction, the free surface and the other nodes are advanced to the new positions. The microscopic void compression equation, Equation (7.82), is solved using a Runge-Kutta algorithm at each node in the domain to obtain the updated void radii at all locations. The densities at these locations are updated, and Equation (7.108) is solved again. This procedure is repeated at every time step during the solution.

The model solution is illustrated for the case of a no-slip condition at the bottom surface, $z = 0$, and a perfect slip condition at the top surface, $z = h$, in order to limit the number of results to be presented. The initial void fraction is maintained at 0.05 (5%), and a tow speed of 0.05 m s^{-1} was used. The final void fraction distribution was calculated for the compaction ratio, h_f/h_i, of 0.99 to 0.95. A constant viscosity of 10^5 Pa.s was used for the medium. The initial pressure inside the voids was assumed to be atmospheric and the initial void radius was assumed to be 10 μm. The final void fraction distribution for these cases is shown in Figure 7.39. The final void fraction is seen to be highest at the outer edge of the tow because the pressure is the lowest at that point. As the compaction ratio decreases, i.e., the amount of compaction increases, the final void fraction decreases, which is consistent with our intuition.

Figure 7.38: Solution methodology.

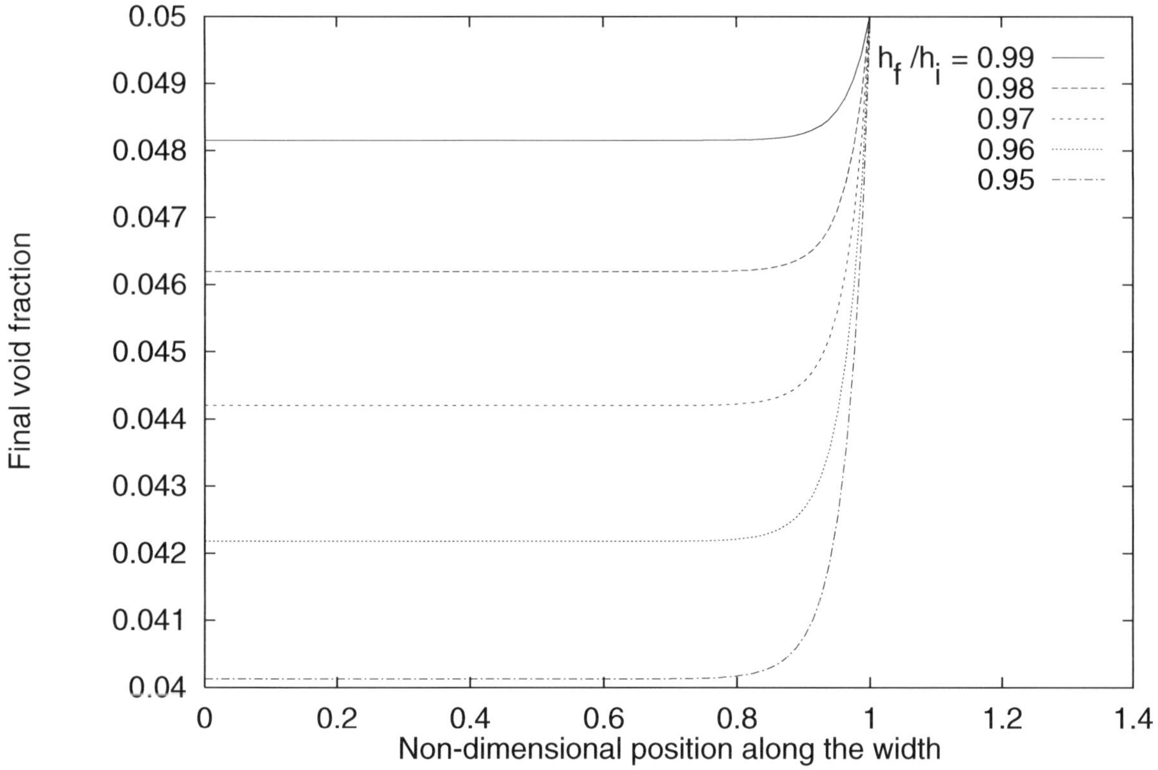

Figure 7.39: Final void fraction distribution in an isothermal compressible medium undergoing consolidation.

Nonisothermal Compressible Fluid

The most general case considers the continuum to be compressible and the domain to be nonisothermal. The viscosity of a continuum is a function of the temperature and therefore varies throughout the domain.

The values of $v_x(0)$ and $C_1(x)$ in Equation (7.73) may be evaluated from the velocity boundary conditions and the equation may be rewritten as

$$v_x(\xi) = \frac{dP}{dx}\left(F_1(\xi) - \frac{h + K_t F_1(h)}{K_b + K_t + K_b K_t F_2(h)}\left[1 + F_2(\xi)K_b\right]\right) \tag{7.111}$$

where $F_1(\xi)$ and $F_2(\xi)$ are defined by

$$F_1(\zeta) = \int_0^\xi \frac{z}{\mu}dz \tag{7.112}$$

$$F_2(\xi) = \int_0^\xi \frac{1}{\mu}dz. \tag{7.113}$$

The pressure solution is given by Equation (7.76). The solution procedure is similar to the previous case. However, the difference is that the temperature field needs to be known in this case to evaluate the viscosity of the material at various points in the domain. One could use Equation (7.95) for the temperature dependence of viscosity [327].

7.5.9 Inverse Problem of Force Control

The natural input to the problem being solved is the displacement rate given by $\dot{h}$. However, in practice, tow placement machines are run under a specified force rather than a displacement. Therefore, the results obtained using any of the above models except the isothermal incompressible model need to be inverted, and the force required for a certain compaction ratio is computed and expressed as the compaction ratio obtained for a given roller force.

7.5.10 Extended Consolidation Model

All the models that have been discussed have had free surfaces on the tow edge for the resin-fiber-void continuum to flow. This is appropriate for the topmost tow. However, tows beneath the top layer that are still at elevated temperatures ($> T_g$) also feel the effect of the compaction roller. These tows are constrained from the sides, and material cannot flow out in the width direction. Therefore, an appropriate boundary condition for them would be a no out-flow condition. This is depicted in Figure 7.40. The domain is divided into regions I and II. Region I denotes the topmost tow that is being placed. The void fraction in this tow reduces from the incoming value, ϕ_{v1} to an intermediate value, ϕ_{v2}. Region II denotes the layers beneath the topmost layer that are still at a temperature $> T_g$. The void fraction in this region reduces from ϕ_{v2} to the final void fraction ϕ_{v3}.

The pressure in region II is uniform and the continuity equation simplifies to

$$\rho h = \text{constant}. \tag{7.114}$$

The combined solution of the change in height of regions I and II needs to be carried out iteratively. The rates of change of the height of regions I and II are denoted by $\dot{h}_I$ and $\dot{h}_{II}$, respectively. These are related to the total change in height, $\dot{h}$ by

$$\dot{h} = \dot{h}_I + \dot{h}_{II} \tag{7.115}$$

$\dot{h}$ is known from the geometric parameters and the tow speed. An initial guess is used for $\dot{h}_I$ and ϕ_{v2} and the pressure distribution in region I is computed and averaged over the width of the tow. This average pressure is used as the driving pressure for the void compaction

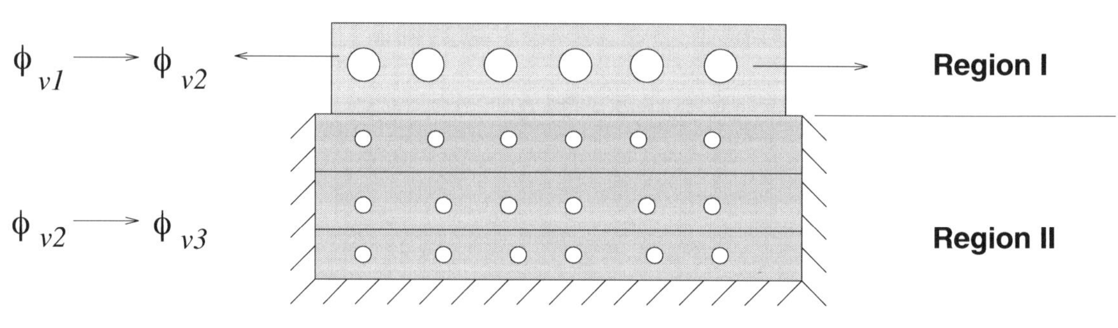

Figure 7.40: Extended consolidation model [331].

in region II. Based on the void compression model, the change in density of region II and the associated change in height may be computed. The sum of the changes in height is compared to the known total change and the guess for $\dot{h}_I$ is modified appropriately. This procedure is repeated until $\dot{h}$ is equal to the sum of $\dot{h}_I$ and $\dot{h}_{II}$. Figure 7.41 shows a typical result for void fraction changes in regions I and II.

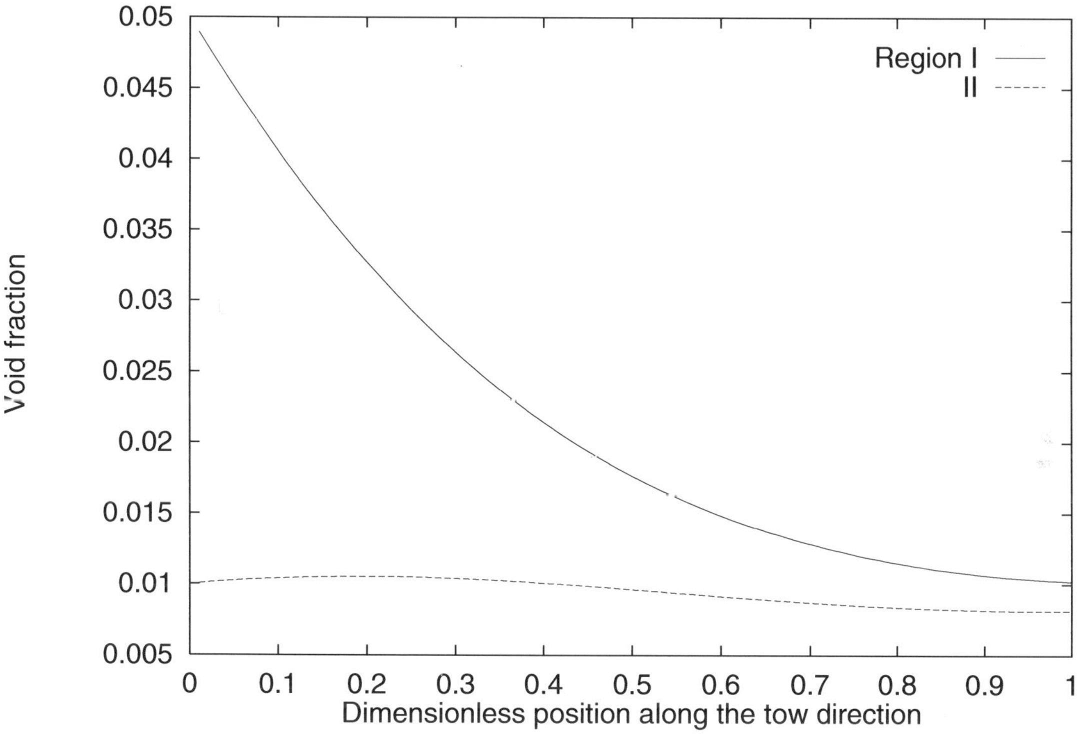

Figure 7.41: Variation in the void fraction in regions I and II as a function of the non-dimensional tow position under the compaction roller for a tow speed of 0.005 m/s.

Summary

A model for consolidation and void reduction during thermoplastic composites processing was discussed. This work is based on the assumption that the composite can be modeled as a compressible continuum. Thus the problem is solved as the squeeze flow of a compressible viscous fluid. After deriving a general formulation, three specific cases — isothermal incompressible, isothermal compressible and nonisothermal compressible — were solved. The final one is the most applicable model for the fiber tow placement manufacturing process. However, for an extended model of the consolidation process, the domain can be split into two regions, the topmost tow and the other tows beneath it that are at a temperature greater than T_g. Given the total thickness change in both regions, an iterative scheme can be used to estimate the individual changes in thickness in the two regions. Simultaneously the changes in their void fraction and the required compaction pressure to achieve the given thickness change can be computed.

This analysis demonstrates how one can start with a simple model and refine the model by relaxing the assumptions to move towards accurately describing the physics. Comparison with experiments can usually help determine if these refinements are necessary. Revisions and refinements are usually more time consuming in terms of model development and solution.

7.6 Exercises

7.6.1 Questions

1. Name at least three advanced thermoplastics composites manufacturing processes.

2. What are the microscopic phenomena that influence the constitutive equations at the macrolevel?

3. What are collimated fibers?

4. Why do the reinforcing fibers buckle and distort more when a thermoplastic resin impregnates than a thermoset resin?

5. What are the four key factors that cause internal stresses and hence can lead to microcracking in the composite part?

6. Describe the sheet forming process by using all of the following terms: deform, stretch, inextensibility, curvilinear shape, thermoplastic polymer, stacked fabrics, and pre-impregnated laminate.

7. What is the advantage of using long discontinuous fiber (LDF) material in the sheet forming process?

8. Name the four main composite sheet forming processes.

9. What are consolidation, interply slip, intraply slip and interply rotation?

10. What are the differences in the deformation mechanism of consolidation of fabric sheets between (i) single-curved and (ii) double-curved parts?

11. How can one calculate the force required to pull a ply out from a stack of plies that will be consolidated?

12. In the pultrusion process, (i) what types of reinforcing material are guided through a preheater where their temperature is raised to the melting temperature of the polymer? (ii) After the heater zone, which assembly do the heated polymer and fibers enter? What are the functions of this zone? (iii) What is the last zone?

13. Why is the cooling speed important in the pultrusion process?

14. What is a typical pulling speed range in the pultrusion process?

15. What are the two important issues to model in pultrusion?

16. What are the similarities and differences in the thermoset pultrusion and thermoplastic pultrusion processes?

17. Describe *viscous dissipation* in the pultrusion process.

18. What do the Peclet (Pe) and Brinkman (Br) numbers represent in the pultrusion process physically?

19. What are the key process parameters that influence the extent of consolidation?

20. What is the in-situ or on-line consolidation process? How does it differ from the regular consolidation process?

21. Give some examples of different types of heaters that are commonly used in the on-line consolidation process.

22. In the tow placement process, what properties of the composite part are used to estimate part quality?

23. How does a process model help engineers in the tow placement process?

24. How does void coalescing occur?

25. Why should there be a concern about the execution time of the numerical solution of a model for the tow placement process with on-line control?

7.6.2 Fill in the Blanks

1. Reinforced material can have different levels of crystallinity. The rate at which the material is ____________ to below the ____________ temperature, governs the crystallization in the matrix. Presence of fibers also provides ____________ sites and can cause secondary crystallization. Cooling rate can also cause internal ____________ that can lead to microcracking.

2. Usually as the composite is cooled, the resin tries to ____________ which is resisted by the stiff inextensible ____________. This places the fibers in ____________ and the resin in ____________.

3. The composite is cooled from the outside to the center. This would shrink and solidify the ____________ first, and then the ____________ is constrained during its solidification. This puts the outside under ____________ and the interior under ____________. Hence controlled cooling is an important process parameter during manufacturing of such materials.

4. In the diaphragm forming process, a laminate of the composite material is sandwiched and heated between two sheets of ____________ and a ____________ is used to form the composite. The edges of the material are not ____________ as the inextensibility of the material can cause fiber ____________.

5. In the matched die forming process, the composite laminate or sheet is held and ____________ between two metal tools surfaces as is accomplished in sheet metal forming. The process takes very ____________ time. Most match die forming has been practiced on ____________ shapes due to nonuniform ____________ and ____________ distributions.

6. The stretch forming process can be used only with materials that can ____________ stretch. ____________ fiber thermoplastics cannot do this but ____________ materials can.

7. The roll forming process uses a series of rotating ____________ to deform a composite ____________ into a ____________ shape.

8. When forming single curvature parts, individual plies ____________ over one another in the laminate. The stretching mode along the fibers is unavailable due to the ____________ of the fibers.

9. Pultrusion is a continuous composites manufacturing process. The goal in pultrusion is to produce a cured and ____________ profile with a constant ____________ by pulling ____________ reinforced material through a ____________ forming die.

10. Compared to thermoplastic pultrusion, thermoset pultrusion involves an additional station of a ____________ between ____________ and ____________.

11. In thermoset pultrusion, one uses ____________ law to model resin impregnation, whereas in thermoplastic pultrusion, due to the high ____________ of the material, this law is not the correct model.

12. The heat transfer model in thermoset pultrusion involves ____________. On the other hand, the heat transfer model in thermoplastic pultrusion focuses on ____________.

13. In pultrusion, if one assumes that most of the flow is parallel to the ____________, one can divide the ____________ section of the die into self similar cells. A cell model is formulated by considering a single tapered annulus of liquid ____________ surrounding each ____________. This assumption allows one to simplify the problem to ____________ problem in the direction of the fiber movement. Also, due to the cylindrical nature of the fiber, one can assume that the annulus of the resin is also ____________; thus one can choose the ____________ coordinate system to pose the problem.

14. In pultrusion, the ____________ mechanism of the fibers in the axial direction induces a ____________ flow in that direction. However, due to the ____________ geometry, the material experiences a pressure build-up. This makes the resin pressure higher near the ____________ of the die as compared to the ____________. Hence, a pressure gradient is established in the ____________ direction to drag flow, which may give rise to ____________ flow.

15. In pultrusion, at any cross-section along the axial direction of a tapered die, the flow rate Q is the sum of ____________ flow induced by the fiber ____________, and ____________ generated due to the opposing ____________.

16. In the tapered die of the pultrusion process, heat is generated by ____________________.

17. The viscosity of polypropylene (a thermoplastic resin) reduces by about ________% as the temperature increases from 190°C to 240°C. Such a variation has significant influence on the pressure profile in the pultrusion process. Ideally, one may have to solve the equation of ____________ and ____________ equation simultaneously due to the two-way ____________ between them. That is, the ____________ influences the ____________ through viscous ____________ term in the energy equation, and the ____________ influences the ____________ in the equation of motion.

18. There are five main components in on-line consolidation process. They are the ____________, ____________, ________________________, ________________________, and a ____________ for the placement of the fiber tows to form the composite.

19. The ____________ is used to mechanically maintain constant tension within the fiber tow from the moment it leaves the ____________ until it is wound on the tool or the previously placed fiber tows on the tool.

20. The tool, which is known as ___________ in filament winding and ____________________ in the tow placement manufacturing method, respectively, is the surface on which the fiber tows are placed and ___________ to create the geometric form of the composite structure. In the case of filament winding, it is ___________ in form, whereas in tow placement method it can take various complex forms based on the geometry of the part to be manufactured.

21. The stainless steel ____________________ allows for application of consolidation ___________ to fuse and compact the incoming _____________ with the previously placed _____________.

22. The heater used in the preheating chamber must be easily controllable while supplying sufficient energy to raise the _____________ temperature to the desired level. The heater power will be a function of the tow ___________, the material properties of the ___________, and the ___________ change one wants the tow to undergo.

23. In the consolidation process, the interactions between mass, momentum and energy transfer can be investigated by classifying them on ___________, ___________ and ___________ levels. The ___________ level analysis will allow us to establish the process windows of ___________, ___________, and ___________ required in order to achieve complete consolidation. At the ___________ level, one can investigate the effect of temperature and pressure on the size of individual ___________ and thus on the macroscopic ___________. At the ___________ level, one is concerned about intimate ___________ between two adjacent ___________ and ___________ of polymer chains across the ply interfaces.

24. In the consolidation process model, ___________, ___________ and any ___________ present in the tow are approximated as a homogeneous fluid with specified rheological properties. The rheological properties of this continuum depend on the ___________, the ____________________, and the ___________ content.

25. In the thermoplastic consolidation process model, due to the ___________ matrix resin viscosity, the flow may be treated as ___________ motion. In other words, the ___________ effects are neglected.

26. In the consolidation process, voids may migrate, along with resin, due to a pressure ___________. On the other hand, void migration relative to the resin is usually ___________ because of high ___________ of the thermoplastic resin.

27. In the consolidation process, cooling of the tow leads to ___________ of the void size. Void compression takes place due to the applied ___________ pressure. The effects of heat input and void compression are taken into account using a ___________ model. The effect of void transport is taken into account in the ___________ flow model.

28. In the microscopic void model of the consolidation process, a typical void is approximated by a ___________ surrounded by a concentric ___________ resin shell. The ratio of ___________ of these two ___________ is determined by the ___________ fraction at that location.

29. Void growth or collapse is governed by a balance between the ___________ inside and outside the void, the ____________________, and the resin ___________.

30. A mathematical model for consolidation and void reduction during thermoplastic composites processing can be developed based on the assumption that the composite can be modeled as a ____________________. Thus, the problem is solved as the ___________ flow of a ___________ and ___________ fluid.

7.6.3 Problems

1. Derive Equation (7.94).

2. If you are interested in obtaining a composite laminate of thickness 2 mm and you start with an initial thickness of 3 mm containing no voids, what will be the compaction force you will need for AS4/PEKK material at a temperature of 350°C? The initial width of the tow is 2 cm. Assume no-slip boundary condition and velocity of 1 cm/s.

3. Write a finite difference program to solve for compressible isothermal consolidation case as explained in the isothermal compressible section.

Chapter 8

Processing Advanced Thermoset Fiber Composites

8.1 Introduction

Thermoset resins are very conducive to processing, as their viscosity is reasonably low to allow easy impregnation of the empty spaces between the fiber preforms. The earliest composites manufacturing concentrated on processes such as *wet hand lay-up*, which is an open mold process. This process employed a one-sided male or female mold, in which a layer of resin, called the gel coat, is applied on the mold surface. The gel coat is usually about 0.5 mm in thickness, has good environmental resistance and produces a smooth and pigmented surface that covers the fiber preform. To manufacture a composite, a measured amount of the thermoset resin is poured over the gel-coated surface. The fiber preform is placed on top of the liquid resin and pressed by a hand-held roller. The resin percolates through the fibrous network due to the application of the pressure as shown in Figure 8.1.

After one layer of the fiber preform is satisfactorily impregnated and compacted, another layer is placed on top and the procedure is repeated until the desired lamination of the composite is complete. Although this process was one of the first techniques to be used in manufacturing composites, it is still widely used by the boat manufacturing industry due to its versatility and low capital investment. However, as the resin that is widely used in this open molding process contains styrene (which is emitted to the environment), there is a push to replace it with closed mold processes due to environmental concerns.

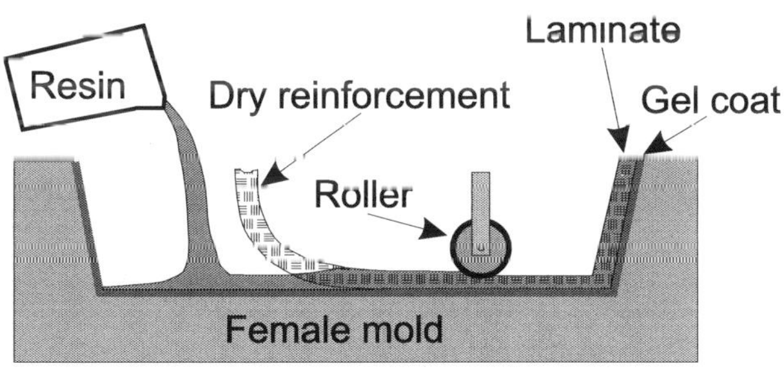

Figure 8.1: Schematic of the wet hand lay-up process.

All the processes we will discuss in this chapter evolved from the above process and were motivated by the need to introduce some or all of the following:

- Automation of the process to reduce labor costs,
- Improvement of the surface treatment,
- Elimination or reduction of emissions of styrene to the environment,
- Improvement of dimensional tolerances and/or increasing the fiber volume content in the composite.

The three processes we will analyze in this chapter are the autoclave molding process, liquid composites molding and filament winding.

8.2 Autoclave Molding

Autoclave processing is one of the oldest composites processing technologies. In this process, plies or prepregs or tapes (fibers that are preimpregnated with the resin) are stacked on the surface of the tool. They are then subjected to high pressure and temperature to allow the stacks to become a single coherent structure by forcing out air pockets and excess resin. The process is carried out in an autoclave, which is a large pressure vessel with an integral-heating element. One can think of an autoclave as an oven that can be pressurized. Not unlike a pressure vessel, it is usually constructed as a cylindrical tube with a door at one end. Its main function is to provide the heat and pressure necessary to consolidate and cure composite structures. A schematic is shown in Figure 8.2.

Autoclave processing is carried out under high pressure and temperature. The temperature and pressure requirements are set by the material systems used. For example, for high temperature thermoset resins such as polyamides and PMR-15, the temperatures required could be in the range of 300-400°C and pressure requirements may exceed 1 MPa. For such cases, one may need special autoclaves. Thermoplastic composites can also be processed in these autoclaves as most advanced thermoplastic resins such as PEKK and PEI melt at high temperatures and require higher pressures for consolidation.

Autoclave consolidation has been the champion of the aerospace industry for years. For aerospace applications, the parts can be very large, such as the case would be with

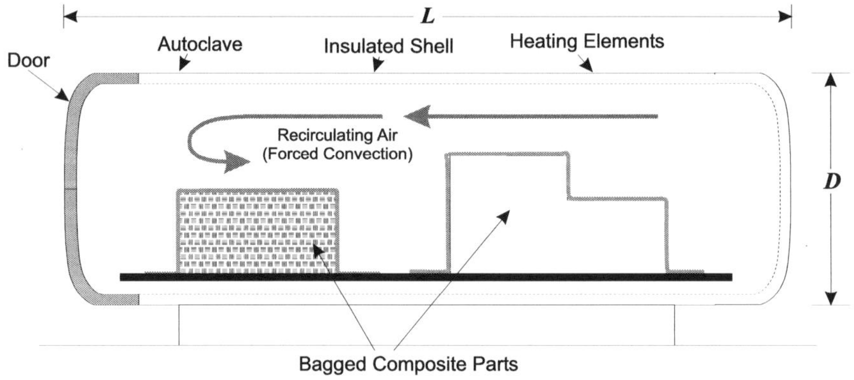

Figure 8.2: Schematic of an autoclave (redrawn from [78]).

an airplane wing; consequently, the autoclaves need to be even larger. The autoclave dimensions for some aerospace applications can be as large as 30 meters in diameter, and over 50 meters in length. This is equivalent to a multi-floor building. The largest existing autoclave in the U.S. is at the Boeing commercial aircraft plant in Seattle, WA — large enough to fit an entire length of a commercial airplane wing. The autoclaves are usually made of welded steel as they are subjected to high pressures and temperatures. Hence they are expensive pieces of equipment [78]. This is the main reason for reluctance to use this process in other industries besides aerospace, such as automotive and sporting.

However, in other industries where parts are small and not complex, the autoclave process can be cost effective. High volume production can be achieved by reducing cycle time by choosing materials that cure faster. Multiple parts could be cured simultaneously. One can address small curvatures and minor geometric complexities by placing the wet composite on a tool surface. As the resin heats up, the plies become less stiff and will adhere to the shape of the tool lying beneath it.

The advantages of the use of autoclaves include the ability to process a wide variety of materials and be able to obtain a very high fiber volume fraction in the composite due to its ability to apply high pressures. Hence, the autoclave has traditionally been used to process composite structures for the aerospace industry. The main disadvantages of the use of the autoclave are the high initial capital investment and the inability to use the pressure and temperature control effectively to produce void free composite structures. In addition, one cannot make complicated near net shaped structures as the mold is usually one sided. Also most of the cure cycles developed for aerospace structures are slow and can take hours before one can remove the part from the autoclave rendering it unsuitable for high volume production.

8.2.1 Part Preparation

The autoclave has been successfully used for many years to make composite parts which have no or negligible geometric complexity. Hence, the tool is usually a flat plate with very few curvatures and inserts. The tool material may be aluminum, steel, nickel, invar or even a composite depending on the application and the cost. A release film or a coating of release gel is usually applied on the tool surface for easy removal. The composite lay-up, consisting of prepregs or tapes, is placed on the tool surface sandwiched between release plies or fabrics. Cork or solid spacer plates are placed around the perimeter of the stacked plies, and they act as dams and limit resin flow in the lateral direction. The release film and release fabric are necessary for easy separation of the composite part from the tool surface and the vacuum bag. Bleeders are used to absorb excess resin in the thickness direction. The release fabric, in addition to assisting in releasing the composite from the tool, allows resin to flow into the bleeders. Peel plies provide surface texture and protect the part surface during handling. The breather material allows for uniform distribution of vacuum over the surface of the part. An inner vacuum bag is placed on top to seal the assembly along the tool surface and the perimeter dams. A caul plate covers the whole assembly. The role of the caul plate is to improve the surface finish of the part and improve dimensional tolerance by applying uniform pressure and minimizing ply movement. The caul plate may be semi-rigid and is made of thin metal, composite or rubber material or may be rigid for critical dimensional tolerance. The vacuum bag and breather material creates vacuum pressure inside the assembly. The driving force for consolidation is the difference between the autoclave pressure on the outside and the vacuum pressure inside the vacuum bag. A typical stacking order of layers is shown in Figure 8.3.

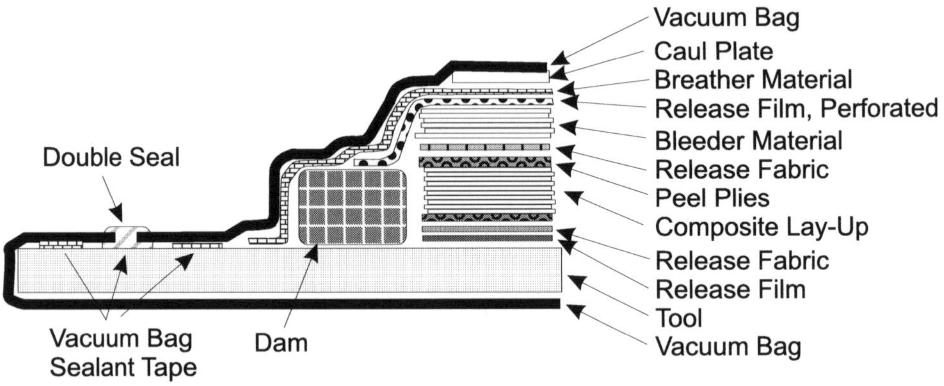

Figure 8.3: Schematic of stacking order in autoclave processing (redrawn from [340].)

8.2.2 Material and Process Parameters

The goal of consolidation is to force air pockets (or voids) and excess resin out of the structure. By squeezing these flaws out of the composite part, a single coherent structure can be made by inducing intimate contact between the individual plies. The final structure will be thinner than the original material placed into the autoclave, but the fiber volume fraction, V_f, will increase making it a stronger part. Moreover, due to consolidation, the interlaminar shear strength will also increase.

Typically, thermosets have been used in autoclave processing. For most thermoset resins, the heat of the autoclave is actually necessary to initiate the cure reaction. The typical temperature range for autoclave processing of thermosets is around 100-200°C, with pressures in the 600-700 kPa range. Thermoplastics can also be used, but as is typical with thermoplastics, the temperature and pressure requirements are much higher than those for thermosets. For example, thermoplastic matrix materials such as PEI or PEKK require processing temperatures in the 300-400°C range with pressures of the order of 1 MPa.

The typical inputs to the autoclave process are the raw material quality, attributes of part preparation and handling, and most importantly from the process modeling viewpoint, the pressure and temperature cycle imposed on the autoclave for composite manufacturing. The quality of the part manufactured is measured in terms of final thickness or dimensions, degree of cure or cross-linking (for thermosets) and void or porosity content. The prepregs used in the composite lay-up can be either impregnated (either partially or fully) tapes with smooth surfaces, or fully impregnated tapes with rough surfaces. It was found that fully impregnated tapes with smooth surfaces do not provide a path for the volatiles and air to escape from the part and can lead to a lower quality part. The material preparation can influence the part quality if the procedure of lay-up is not executed in a repeatable fashion. However, as a process modeler, the most effective way to influence the part quality is by controlling the pressure and the temperature cycle during the curing and consolidation of the composite part.

An example of a pressure and cure cycle is shown in Figure 8.4. There are typically three stages of autoclave processing. Stage I is the heating stage. At the same time, the pressure is ramped up inside the autoclave during this stage. The first pressure ramp and hold in stage I is for the viscosity to go down and for excess resin to flow vertically into the bleeder material. The second ramp and hold is for polymerization of the resin to initiate during which time the viscosity rises dramatically. Stage I ends when the selected processing pressure and temperature are reached. Stage II is essentially a "hold" stage. The

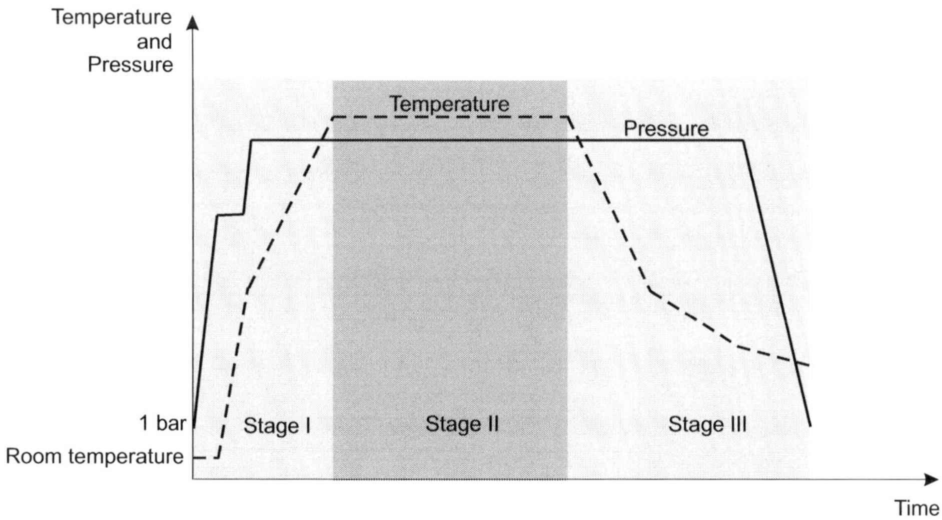

Figure 8.4: Typical autoclave temperature and pressure cycles during the three stages of autoclave processing.

autoclave is maintained at the processing pressure and temperature to allow consolidation and curing to occur during this stage. As the cure reaction proceeds and generates heat due to polymerization, in stage III, the temperature is lowered to allow the excess heat from the reaction to diffuse through the part but the pressure is maintained to prevent voids from growing. When the cure reaction is complete and the temperature of the part is lowered, the pressure is released and the part is removed from the autoclave. The vacuum is discontinued to allow for out-gassing of volatiles. Although the autoclave pressure is high, most of that pressure is borne by the fiber network.

Figure 8.5 schematically shows how the autoclave pressure redistributes between the fibers and the resin during the consolidation phase. Thus, at the end of consolidation, the resin pressure is low, the temperature is high and the volatile vapor pressure is high creating ideal conditions for void formation and growth. As resin pressure P_r tends to zero, no further resin will bleed, but it will be easy for volatiles to vaporize and form voids. Hence, it is important to maintain some pressure in the resin to arrest formation and growth of voids.

Example 8.1: Void Growth

Show that as the pressure in the resin reduces, the void growth rate of the nucleated voids increases.

Solution

Consider the sketch shown in Figure 8.6 of a nucleated void of radius R. To relate dR/dt, the rate of growth of this void, to the resin pressure, one can balance the stress at the interface between the resin outside the void and the air inside the void. One can write the following equation that equates the normal stress on either side of the interface location

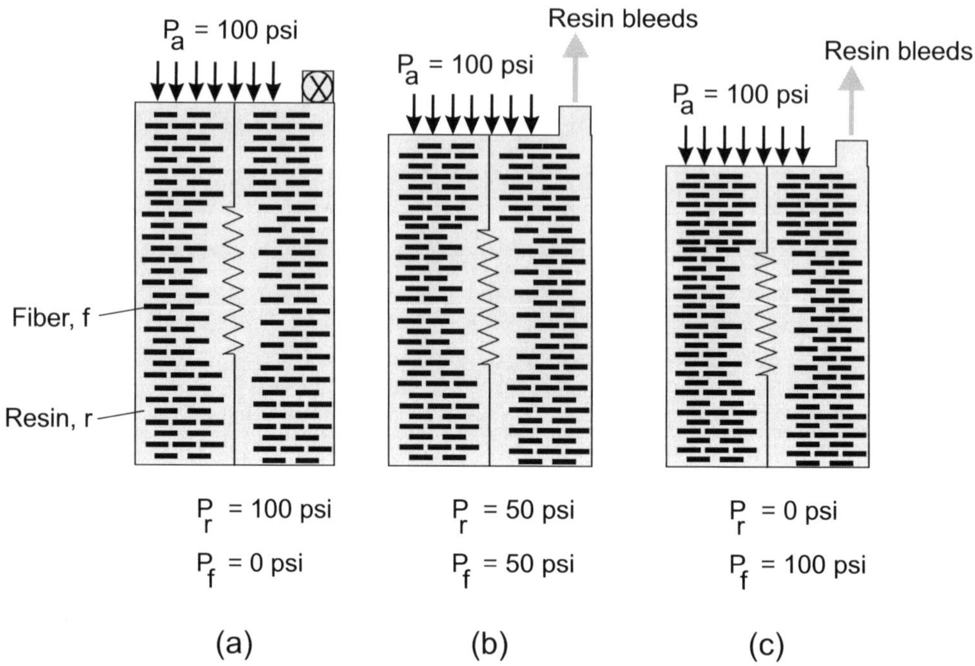

Figure 8.5: Redistribution of resin and fiber pressure as the network is compressed [90]. In (a), the fiber network is carrying no load. In (b), as the network is compressed partially, more and more load is being taken by the fibers, and in (c), all load is taken by the fibers.

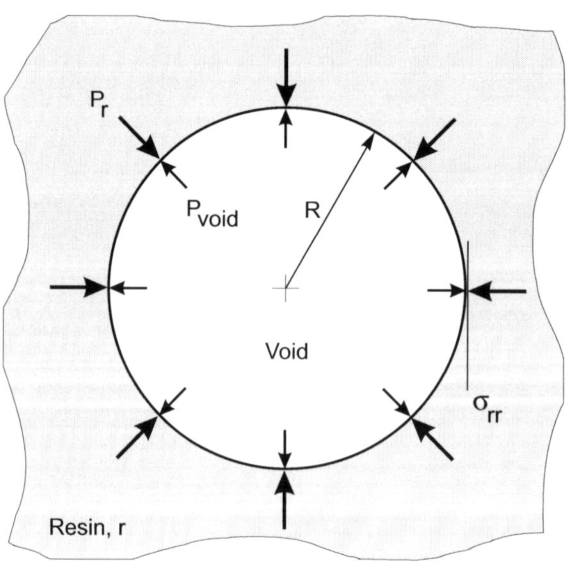

Figure 8.6: Schematic of force balance during void growth in autoclave processing.

$r = R$:

$$\sigma_{rr}\Big|_{r=R} = -P_{void} + \frac{2\alpha}{R} = -P_r + \tau_{rr} \tag{8.1}$$

where P_{void} is the void pressure, P_r is the resin pressure, R is the radius of the void, α is the surface tension of the resin, and τ_{rr} is the normal stress experienced by the resin. For a Newtonian fluid, low Reynolds flow and considering spherical symmetry, the normal stress is equal to $-4\mu\dot{R}/R$ where $\dot{R} = dR/dt$. Equation (8.1) can be rewritten as

$$(P_{\text{void}} - P_r) - \frac{2\alpha}{R} = \frac{4\mu\dot{R}}{R}. \tag{8.2}$$

From Equation (8.2), one can easily conclude that, as P_r decreases, $\dot{R}$ increases. Hence, it is important to have $(P_{\text{void}} - P_r) < 2\alpha/R$ in order to suppress the void growth. ■

Thus as resin pressure goes down, the rate of bubble growth could potentially increase. It is crucial to understand and predict the resin pressure inside the part during consolidation as it plays a dominant role in formation and suppression of voids. If resin pressure drops to zero, voids will form freely inside the composite since there is no back pressure to resist the void growth. Intuitively, one may want to increase the autoclave pressure to prevent this. However, as seen from Figure 8.4 applying too much pressure can cause consolidation to occur too quickly (in which case the pressure will be all borne by the fiber network), or cause regions to lose too much resin. If this happens too quickly, then resin pressure will be lost before the part has had a chance to fully cure initiating void growth. Also complete wet-out of the fibers may not occur which is detrimental to its mechanical properties. Hence it is important to control the resin pressure to encourage consolidation and allow total wet-out, while maintaining sufficient resin pressure until the part solidifies.

There are many combinations of pressure and temperature one could apply with time to change the cycle shown in Figure 8.4 to produce the part. However, one is limited by how fast one can change the pressure and temperature in the autoclave. At a given power, to find the time it takes to increase the temperature by 1°C, one would have to consider the mass and heat capacity of the autoclave. In general the MC_p term for the autoclave and the tooling can be much larger than that of the part where M and C_p are the mass and specific heat, respectively. Hence, large autoclaves are not well suited for rapid rates in heating or cooling due to their large thermal inertia.

Example 8.2: Heating Rate in Autoclave Made of Welded Steel

Consider an autoclave that is made of welded steel of thickness 25 mm. The autoclave is a cylindrical tube of diameter 2 meters and length 5 meters. The tooling and part thermal inertia can be considered negligible as compared to the autoclave inertia. The heater power supplied to the autoclave can vary from 10 to 50 kW. If one wants to heat the autoclave by 1°C, how long will it approximately take to do so at the power supply of 10 kW through 50 kW? Plot the effect of heater power on the time to heat the autoclave by 1°C. Also plot the heating rate as a function of the input power.

Solution

The schematic of the autoclave is shown in Figure 8.7. One can write the heat transfer to the autoclave as $Q = \dot{Q}\Delta t = MC_p\Delta T$. Here $\dot{Q}$ is the heater power, Δt is the time

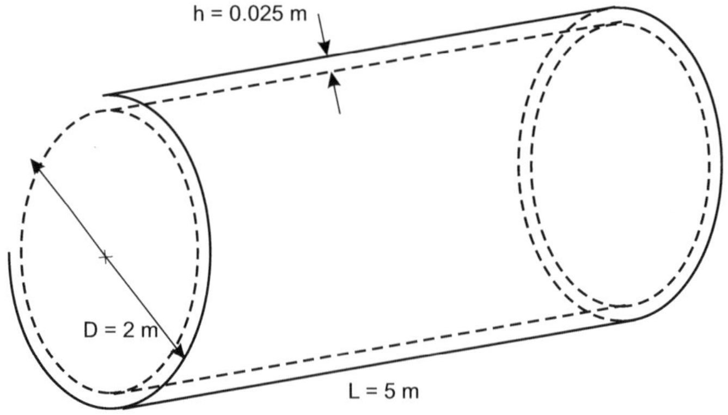

Figure 8.7: Dimensions of the autoclave for Example 8.2.

to increase the temperature of the autoclave by $\Delta T = 1°\text{C}$, C_p is the specific heat of the autoclave material, and M is the mass of the autoclave which can be calculated as

$$\begin{aligned} M &= M_{\text{tube}} + 2M_{\text{end plate}} \\ &= \rho\left(V_{\text{tube}} + 2V_{\text{end plate}}\right) \\ &= \rho\left[\pi DLh + 2\pi(D/2)^2 h\right] \\ &= 7,800\left[\pi(2)(5)(0.025) + 2\pi(1)^2(0.025)\right] \\ &= 7,351\,\text{kg}. \end{aligned} \tag{8.3}$$

Taking $C_p = 460\,\text{J/(kg°C)}$ for steel, one can find the time to heat the autoclave by 1°C

$$\begin{aligned} \Delta t &= MC_p\frac{\Delta T}{\dot{Q}} \\ &= (7351)(460)\frac{1°\text{C}}{\dot{Q}} \\ &= \frac{3.4{*}10^6}{\dot{Q}}\,\text{seconds} \end{aligned} \tag{8.4}$$

where $\dot{Q}$ is in Watts. Note that one Joule is equal to 1 Watt second. Figure 8.8 shows how long it takes to heat the autoclave of these dimensions by 1°C at various power inputs. For example, it takes 5 minutes and 40 seconds if the heater has a power of 10 kW, and only 1 minute and 8 seconds if the input power is 50 kW. The rate at which one can increase the autoclave temperature is plotted as a function of the power input in Figure 8.9. This is obtained by rewriting Equation 8.4 such that $\frac{\Delta T}{\Delta t}$ is on the left-hand side:

$$\frac{\Delta T}{\Delta t} = \frac{\dot{Q}}{MC_p} = \frac{\dot{Q}}{3.4{*}10^6}\,°\text{C/seconds} \tag{8.5}$$

Thus, the heating rate is directly proportional to power input. ■

For industrial autoclaves, which are generally heated by natural gas burners, it is not uncommon to have heating rates of about 2 to 3°C per minute.

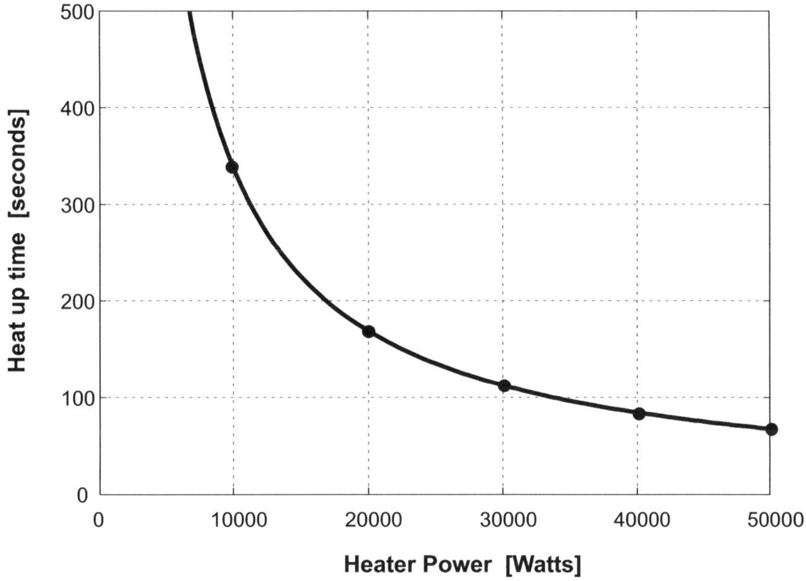

Figure 8.8: Time to heat the autoclave by 1°C at various power inputs.

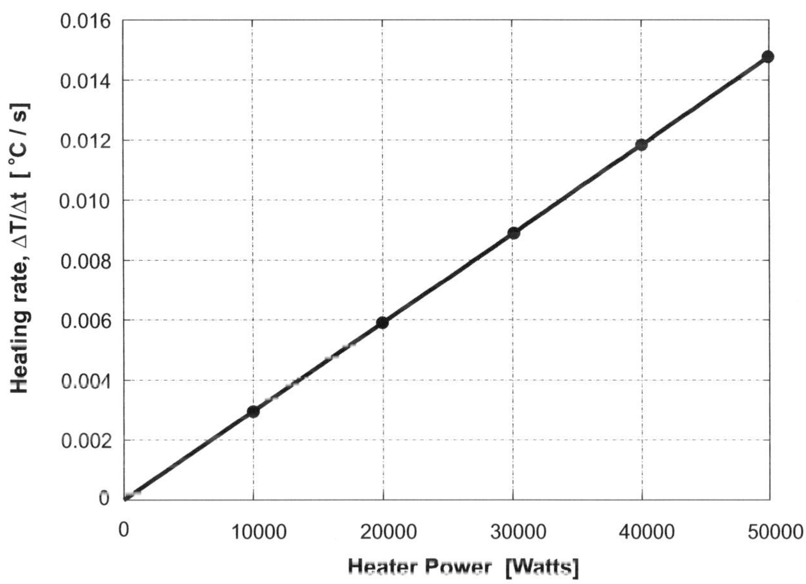

Figure 8.9: Heating rate (temperature rise) $\Delta T/\Delta t$ at various power inputs.

The most widely used method of pressurizing the autoclave is to use nitrogen gas. Air is unsuitable at high temperatures due to the risk of auto-ignition. However, as the approach to increase the pressure is to use a compressible medium and as the volume to be compressed is large, one can only expect sluggish rates of increase in pressure. However, there is sufficient flexibility in application of the pressure to the part. Usually, the parts are placed on one side of the tool and wrapped in a plastic bagging material. Pressure is then applied to the part, which presses it against the tool. Pulling a vacuum in the bag can intensify the pressure on the part. This way one can locally control the pressure more effectively.

The most important process parameters that influence the part quality manufactured by autoclave processing are pressure, temperature and time. The goal in autoclave processing has been to employ the science base of composite processing to develop the "most optimal pressure and temperature cycle" to manufacture void-free composite parts. In this section we will delineate the approach to model issues pertinent to autoclave processing using the scientific base. This is a necessary step before one can develop the optimal pressure and temperature cycle.

8.2.3 Processing Steps

There are three important steps involved in manufacturing of a composite part by autoclave processing. They are (i) material preparation, (ii) resin flow, consolidation and cure, and (iii) inspection and post-cure.

In the material preparation step, the precursor materials, which are usually in the form of plies of fabric or unidirectional prepregs, are cut and stacked to occupy the tool. Lay-up may be accomplished either manually or by an automated procedure such as automated tow placement or automated drape forming. The composite lay-up is placed on the tool surface along with the bagging and other ancillary materials. The arrangement of the tool, composite and ancillary material may vary slightly, but the general lay-up configuration is shown in Figure 8.10.

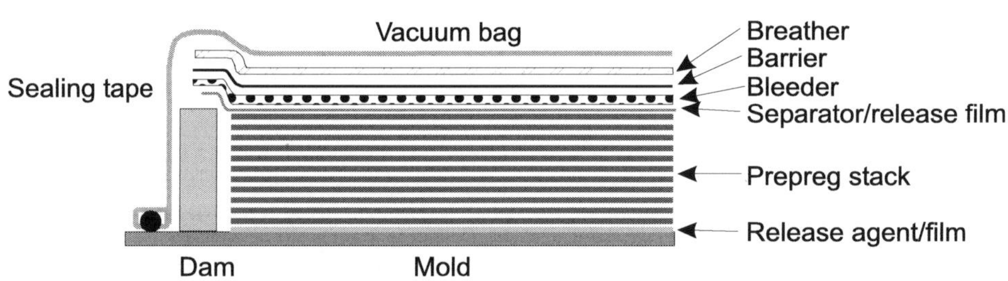

Figure 8.10: Typical vacuum bagging assembly.

The next step involves heat, resin flow, consolidation and cross-linking of the composite lay-up. In this step, the process parameters of pressure, temperature and time play an important role in the quality of the part that will be made, as this process is irreversible for thermosets. Insufficient pressurization can result in excess resin in the part. On the other hand, high pressures can lead to resin starved regions. The spatial curing distribution and the degree of cure is determined by the temperature cycle with time. Usually the focus is to understand the flow and heat transfer dynamics in this step to effectively model this step of autoclave processing. The final step of post processing usually will initiate with inspection and based on the findings may require either rework including further heat treatment to increase the degree of cure, trimming, machining or assembly.

8.2.4 Critical Issues

In this process there is ample room for error due to the number of manual procedures involved which then lead to flaws in the part and a very high rejection rate. Some of the errors involved in the material preparation step are [78]

- Failure to clean tool surface or apply release film which will cause imperfections on the composite surface or will make it difficult to separate the part from the tool surface.
- Failure to completely debulk (remove large air pockets in between the plies by adhering the layers to each other) the composite lay-up which can lead to void formation.
- Incorrect ply orientation or inaccurate placement which can lead to reduction in mechanical properties.
- Failure to place ancillary material correctly such as bleeder or breather material which can lead to resin build-up and low fiber volume fraction in the part.
- Improper sealing of the vacuum bag which can lead to loss of vacuum during cure and can result in void formation.

During the consolidation and curing step, an incorrect pressure and temperature profile can cause uneven cure, warpage, low fiber volume fraction, voids, resin degradation and insufficient consolidation.

The important issues for composites made by autoclave processing that are influenced by the cure and consolidation step are the thickness variation, void formation and growth and warpage of the part. The thickness variation is linked to the degree of consolidation of the plies during the process. The void formation and growth can be related to the moisture absorbed by the prepregs during manufacturing and storage. The warpage issue is linked to the residual stresses developed in the part and the differences in the coefficient of thermal expansion of the composite both along the fibers and in the transverse direction. The residual stresses are mainly caused by temperature gradients across the thickness of the part, which are influenced by the temperature control during processing. Process models can help the manufacturing engineer understand the influence of pressure and temperature on the quality of the composite manufactured. In the table below, critical issues involved in autoclave processing and the reasons for modeling them are summarized.

Table 8.1: Critical Issues in Autoclave Processing and Benefits of Modeling

Issue	**Possible Benefit of Modeling**
Resin Pressure	Minimize void growth
Resin Flow	Ensure completely wetted fibers and lamination of plies
Fiber Deformation/Consolidation	Consolidation of fiber network to maximize fiber volume fraction
Temperature	Avoid resin degradation and reduce temperature gradients
Cure Kinetics	Improve cycle time and reduce residual stresses

8.2.5 Flow Model for Autoclave Processing

The resin flow and resin pressure predictive model can be very complex as there is mass, momentum and energy transfer taking place simultaneously during the process. Below we will list assumptions that will simplify the model while retaining the important physics.

Assumptions

- The fibers in the composite can be considered as nonlinear elastic porous medium.
- The composite consists of incompressible resin and fibers (ρ_r and ρ_f are constant).
- The composite is void free; that is, the sum of the volume of the resin and the volume of the fibers is equal to the volume of the composite, $\text{Volume}_\text{r} + \text{Volume}_\text{f} = \text{Volume}_\text{c}$ hence the resin volume fraction, V_r and the fiber volume fraction, V_f add up to unity: $V_r + V_f = 1$.
- The applied pressure is distributed between the resin and the fiber network.
- Darcy's Law can describe the average resin flow between the fibers.
- Fibers act as a bending beam between multiple contact points and the load borne by the fibers depends on the fiber volume fraction.
- The permeability of the porous medium changes with V_f.
- V_f changes with only the transverse z direction, i.e., $V_f = V_f(z)$.
- The resin viscosity changes with temperature and cure, but it is expressed as a function of $\mu = \mu(t)$ only.

The composite domain is usually compressed in the z direction only. Hence, the fiber volume will change in that direction and so will the thickness as the part consolidates. Gutowski and Dillon [341] as shown in Figure 8.11 introduced a new variable, $\xi = w + z$, where w is the local displacement to account for the local deformation of the fiber network and thus convert a moving boundary problem to a fixed domain problem. Figure 8.11 shows the local deformation coordinate system. For a more detailed discussion of the local deformation network, please see Chapter 4 and Appendix 4A of Gutowski and Dillon [341].

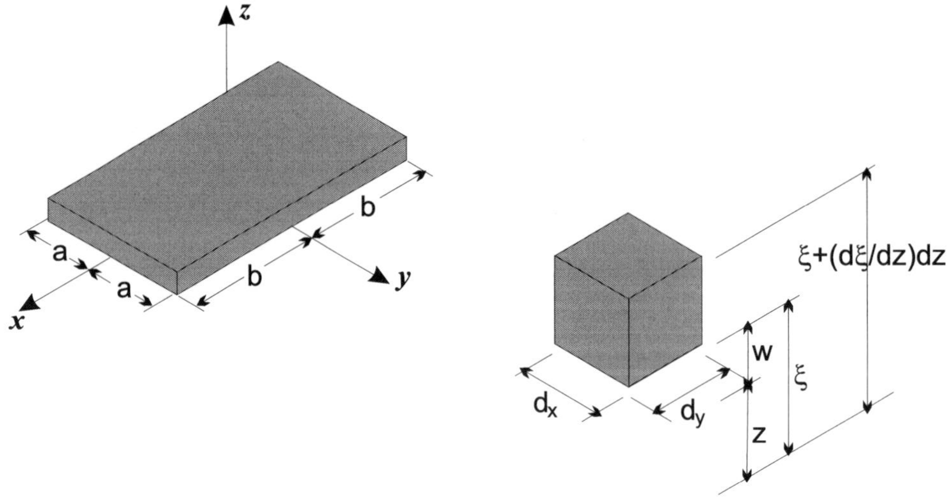

Figure 8.11: Schematic showing the geometry and the deforming coordinate system of a control volume.

Flow and Consolidation Model

First, we should ensure fiber and resin continuity. The fiber continuity allows one to relate the new coordinate to the z direction as follows:

$$\frac{V_o}{V_f} = \frac{\partial \xi}{\partial z}. \tag{8.6}$$

Here V_o is the initial fiber volume fraction and V_f is the fiber volume fraction during the consolidation stage. One can now use the control volume shown in Figure 8.11 and establish continuity of mass for the resin in the porous medium using the mass balance principle introduced in Chapter 3. This will result in

$$\frac{\partial}{\partial x}\left(U_x \frac{\partial \xi}{\partial z}\right) + \frac{\partial}{\partial y}\left(U_y \frac{\partial \xi}{\partial z}\right) + \frac{\partial U_z}{\partial z} + \frac{\partial}{\partial t}\left[(1 - V_f)\frac{\partial \xi}{\partial z}\right] = 0. \tag{8.7}$$

Replacing $\partial \xi / \partial z$ with V_o/V_f from Equation (8.6) results in

$$\frac{\partial}{\partial x}\left(U_x \frac{V_o}{V_f}\right) + \frac{\partial}{\partial y}\left(U_y \frac{V_o}{V_f}\right) + \frac{\partial U_z}{\partial z} + \frac{\partial}{\partial t}\left[(1 - V_f)\frac{V_o}{V_f}\right] = 0. \tag{8.8}$$

Using Darcy's law instead of the momentum equations, one can introduce the velocity and pressure gradient relationship through the permeability of the preform in the x, y, and z directions

$$U_x = -\frac{K_{xx}}{\mu}\frac{\partial P_r}{\partial x}; \qquad U_y = -\frac{K_{yy}}{\mu}\frac{\partial P_r}{\partial y}; \qquad U_z = -\frac{K_{zz}}{\mu}\frac{\partial P_r}{\partial z} \tag{8.9}$$

where K_{xx}, K_{yy}, K_{zz} are the permeabilities in the x, y, and z directions. It is assumed that x, y, and z are the principal axes of the permeability tensor. Substitution of velocities in terms of pressure gradients in the continuity equation and assuming that V_f does not change with x or y direction results in

$$\frac{K_{xx}}{V_f}\frac{\partial^2 P_r}{\partial x^2} + \frac{K_{yy}}{V_f}\frac{\partial^2 P_r}{\partial y^2} + \frac{1}{V_o^2}\frac{\partial}{\partial z}\left(V_f K_{zz}\frac{\partial P_r}{\partial z}\right) = \mu \frac{\partial}{\partial t}\left(\frac{1 - V_f}{V_f}\right). \tag{8.10}$$

P_r refers to the resin pressure and μ is the viscosity of the resin. From the force balance, one can relate the resin pressure and the fiber stress to the applied pressure, which is usually the autoclave pressure P_T, as follows:

$$P_T = P_r + \sigma_{zz}. \tag{8.11}$$

Gutowski et al. [342] developed a constitutive equation to describe the relationship between the stress taken on by the fibers as a function of fiber volume fraction. They used a conceptual model called the *Fiber in a Box* model (see Equation (4.119) and the related text in Chapter 4 for detail). The relationship is very nonlinear and can be expressed as

$$\sigma_{zz} = \sigma^f = A_s \frac{\sqrt{\dfrac{V_f}{V_o}} - 1}{\left(\sqrt{\dfrac{V_a}{V_f}} - 1\right)^4}. \tag{8.12}$$

In this relationship, A_s is a material or "spring" constant that will be a function of the elastic properties of the fibers and the fiber diameter. V_a is the maximum attainable fiber volume fraction, and V_o is the initial fiber volume fraction at which the stress taken on by the fibers is zero.

We also know that as the fiber bed compresses under the applied autoclave pressure, the transverse permeability, K_{zz}, will also reduce as the fiber volume fraction, V_f increases and the pore space between the fiber network reduces. Hence, one would need another constitutive equation to describe the relationship between the fiber volume fraction and the transverse permeability. Many such expressions have been developed. The best known and developed for isotropic porous media is known as the Kozeny-Carman equation [31] (Equation (8.13)). Other expressions that more specifically focus on fibrous porous media are the Gebart transverse permeability expression [247] (Equation (8.14)), the Bruschke-Advani model [248] (Equation (8.15)), and the modified Kozeny-Carman relation (Equation (8.16)). In these expressions, V_f is the fiber volume fraction, and V_a is the maximum attainable fiber volume fraction. One only needs to know how the transverse permeability in the z direction for these models changes with the fiber volume fraction. As we assume that the fiber volume fraction does not change in the plane of the composite, we do not expect the permeability to change in x and y directions. All the permeability relations are given below.

$$K_{zz} = \frac{r_f^2}{4p_{zz}} \frac{(1 - V_f)^3}{V_f^2} \tag{8.13}$$

$$K_{zz} = p_{zz} r_f^2 \left(\sqrt{\frac{V_a}{V_f}} - 1 \right)^{5/2} \tag{8.14}$$

$$\frac{K_{zz}}{r^2} = \frac{1}{3} \frac{(1 - L)^2}{L^3} \left(\frac{3L \tan^{-1} \sqrt{\frac{1+L}{1-L}}}{\sqrt{1 - L^2}} + \frac{L^2}{2} + 1 \right)^{-1} \quad \text{where } L^2 = \frac{4V_f}{\pi} \tag{8.15}$$

$$K_{zz} = \frac{r_f^2}{4k_c} \frac{\left(\sqrt{\frac{V_a}{V_f}} - 1 \right)^3}{\frac{V_a}{V_f} + 1} \tag{8.16}$$

Notice that all of these equations require an empirical constant such as p_{zz} or k_c except for the Bruschke-Advani model, which is based on flow across solid cylinders for a certain packing geometry and fiber volume fraction. r or r_f in these equations refers to the radius of the fiber.

Substitution of any of the permeability relationship with the fiber volume fraction in Equation (8.10) results in a partial differential equation for resin pressure in which V_f is also unknown. However, using Equations (8.10), (8.11) and (8.12) one can solve for V_f and P_r if the applied autoclave pressure is known. Another important quantity is the final fiber volume fraction after the compaction since this will give an idea of the mechanical properties of the panel, as well as the final consolidated thickness. These are related via

$$h = \frac{nA_W}{V_f \rho} \tag{8.17}$$

where h is the consolidated thickness in [m] in SI units, A_W is the weight of the fabric per unit area in [N/m^2], n is the number of layers, and ρ is the weight density of the fibers in [N/m^3].

Usually in autoclave processing, one strives to eliminate the flow in the in-plane direction (x and y directions) and focuses on bleeding the excess resin through the z direction into the bleeder cloth. Thus, one can easily show by dimensional analysis that if $h \ll a$ and b, one would expect the pressure gradient and the resin flow to be primarily in the z direction. This allows one to reduce Equation (8.10) to

$$\frac{\partial}{\partial z}\left(\frac{V_f}{V_o}\frac{K_{zz}}{\mu}\frac{\partial P_r}{\partial z}\right) + \frac{V_o}{V_f^2}\frac{\partial V_f}{\partial t} = 0. \tag{8.18}$$

Example 8.3: Find V_f as a Function of Time

Delineate the approach to obtain V_f as a function of time for a composite subjected to a constant autoclave pressure.

Solution

To solve Equation (8.18) and obtain V_f as a function of time, first assume μ to be constant and also the autoclave pressure, P_T to be constant. The approach is to define a new variable e as follows:

$$e = \frac{1 - V_f}{V_f} = \frac{\text{resin volume fraction}}{\text{fiber volume fraction}}. \tag{8.19}$$

Hence, $V_f = 1/(1+e)$ and $dV_f = -de/(1+e)^2$, and Equation (8.18) can be rewritten in terms of e as follows:

$$\frac{1}{V_o}\frac{\partial}{\partial z}\left(\frac{K_{zz}}{\mu(1+e)}\frac{\partial P_r}{\partial z}\right) + V_o(1+e)^2\left[\frac{-1}{(1+e)^2}\right]\frac{\partial e}{\partial t} = 0 \tag{8.20}$$

and simplification results in

$$\frac{1}{V_o^2}\frac{\partial}{\partial z}\left(\frac{K_{zz}}{\mu(1+e)}\frac{\partial P_r}{\partial z}\right) = \frac{\partial e}{\partial t}. \tag{8.21}$$

Now as $P_T = P_r + \sigma_{zz}$, and P_T = constant, differentiation results in $0 = \partial P_r + \partial \sigma_{zz}$, or rearranging, we have $\partial P_r = -\partial \sigma_{zz}$. Then, one can write $\frac{\partial P_r}{\partial z} = -\frac{\partial \sigma_{zz}}{\partial e}\frac{\partial e}{\partial z}$, and Equation (8.21) can be rewritten as

$$(1+e_o)^2\frac{\partial}{\partial z}\left(-\frac{K_{zz}}{\mu(1+e)}\frac{\partial \sigma_{zz}}{\partial e}\frac{\partial e}{\partial z}\right) = \frac{\partial e}{\partial t} \tag{8.22}$$

where $V_o = 1/(1+e_o)$.

One can solve the above equation for e, along with the equations for permeability and fiber stress constitutive equations. The initial condition is that the fiber volume fraction is V_o at $t = 0$ and two boundary conditions are necessary along the z direction. At $z = h$, the boundary condition would be that $P_r = 0$; therefore, σ_{zz} will be equal to P_T. Hence, one can find the fiber volume fraction using Equation (8.12) at $z = h$, which will allow one to calculate e at $z = h$. The second boundary condition is at the tool surface, that is at $z = 0$, there is no flow of the resin out of the tool surface. Mathematically, this can be expressed as $\partial P_r/\partial z = 0$ at $z = 0$. One can take the derivative of Equation (8.11) at $z = 0$, which gives $\partial P_T/\partial z = \partial P_r/\partial z + \partial \sigma_{zz}/\partial z$. This implies that $\partial \sigma_{zz}/\partial z = 0$ at $z = 0$ as the other two terms are zero as well. $\partial \sigma_{zz}/\partial z$ will be zero only if $\partial V_f/\partial z = 0$ at $z = 0$. One

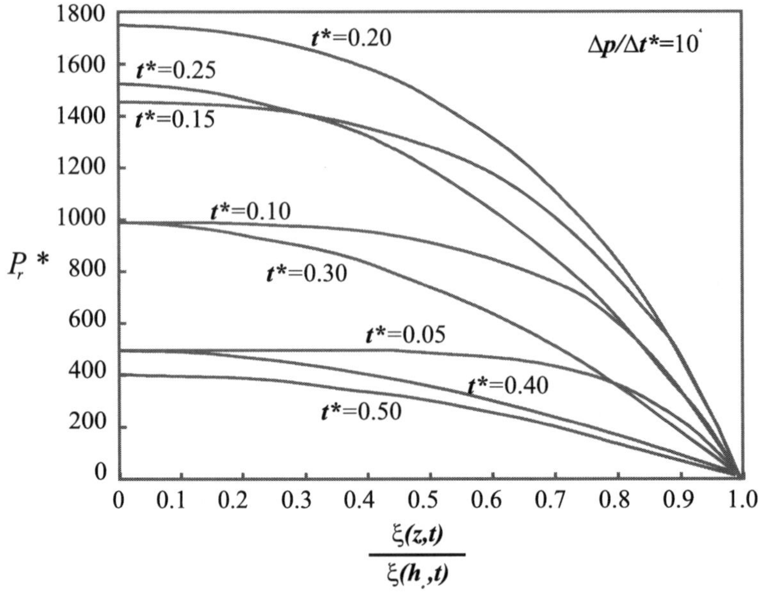

Figure 8.12: Resin pressure versus dimensionless thickness [341].

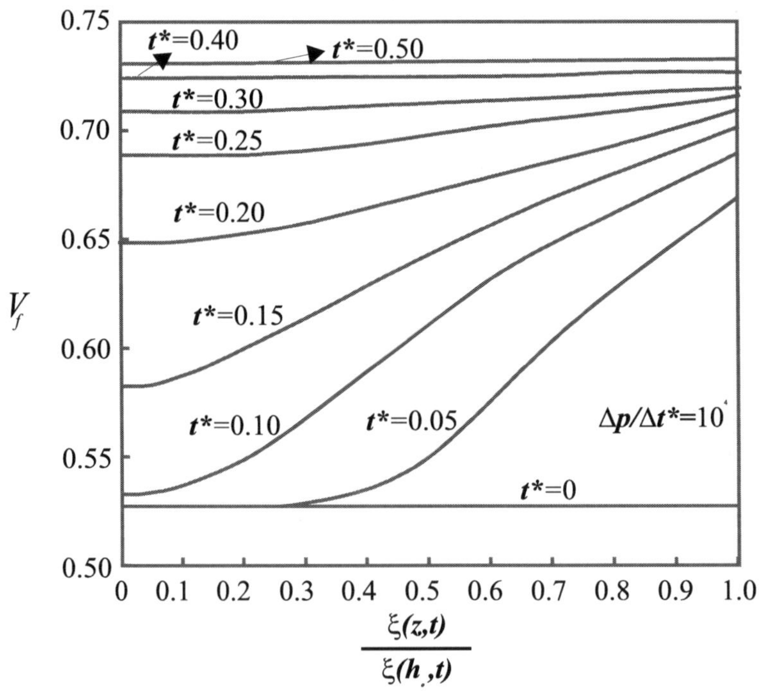

Figure 8.13: Fiber volume fraction versus dimensionless thickness [341].

can express gradient of V_f in terms of $\partial e/\partial z$. A typical solution of how the resin pressure and the fiber volume fraction change with time due to known applied time rate of change in pressure, dP/dt are shown in Figures 8.12 and 8.13, respectively. Note the significant variation in the fiber volume fraction as a function of thickness at early times, but as one approaches a steady state, V_f becomes more uniform. ∎

Example 8.4: Numerical Solution of V_f when the Autoclave Pressure Is Increased Linearly

Calculate the resin pressure $P_r(z,t)$, fiber stress $\sigma_{zz}(z,t)$ and fiber volume fraction $V_f(z,t)$ for a composite part subjected to a constant rise in the autoclave pressure, $dP_T/dt = 10,000$ Pa/s until the minimum fiber volume fraction within the part reaches 0.47 (from an initial value of $V_{f0} = 0.30$). Plot these three variables versus z at various times. Also plot these three variables versus time at three different locations, $z = 0$, $h/2$ and h. Use the following material, model and process parameters:

- Initial autoclave pressure, $P_{T0} = 0$
- Initial part thickness, $h = 0.02$ m
- Maximum attainable fiber volume fraction, $V_a - 0.70$
- Resin viscosity, $\mu = 0.40$ Pa.s

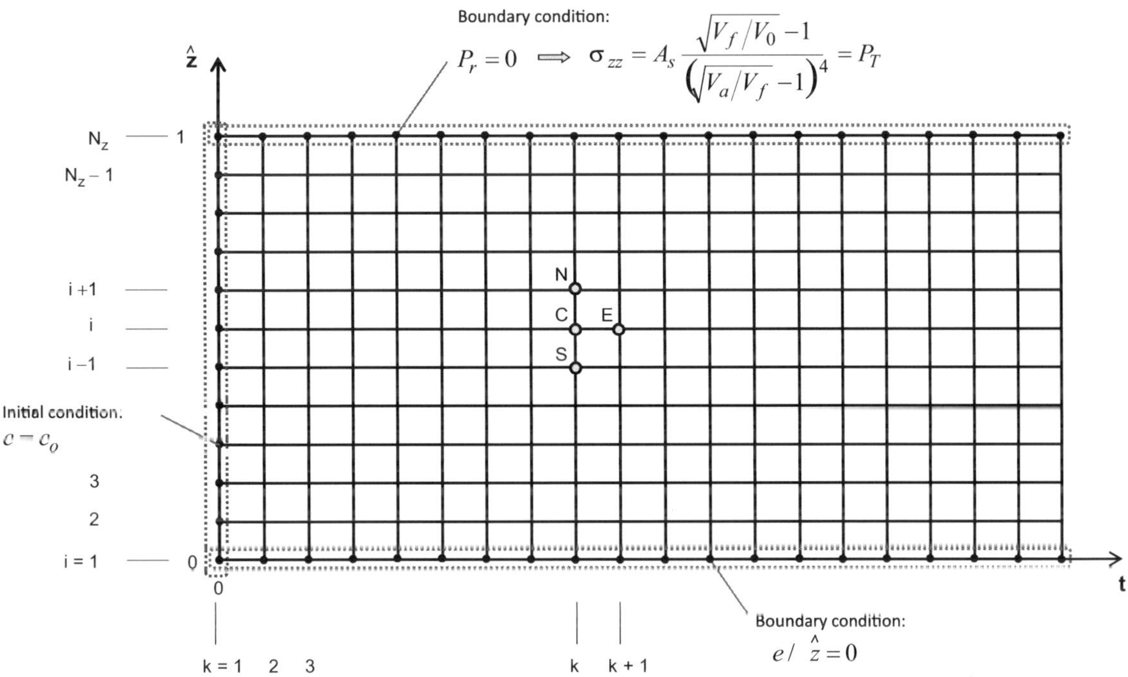

Figure 8.14: Discretized domain and the boundary conditions for a composite part being consolidated in an autoclave.

- Coefficient, $A_s = 280\,\text{Pa}$ in the "Fiber in a Box model," Equation (4.119) [342]
- Coefficient, $C = 5.0 \times 10^{-10}\,\text{m}^2$ in the Kozeny-Carman eq.: $K_{zz} = C(1 - V_f)^3/V_f^2$.

Notice that the coefficient, C corresponds to $r_f^2/(4p_{zz})$ in the Kozeny-Carman permeability [31], Equation (8.13).

Solution

For the solution of this example, as previously illustrated in Example 5.6, finite difference method (FDM) will be applied for the spatial derivatives, and Euler method will be applied for the time integration. The discretization of the spatial and time domains is shown in Figure 8.14 and formulated as follows

$$\hat{z}_i = (i-1)\Delta\hat{z} \qquad \text{where } i = 1, 2, \ldots, N_z \tag{8.23}$$

$$t_k = (k-1)\Delta t \qquad \text{where } k = 1, 2, \ldots \tag{8.24}$$

where $\hat{z} = z/h_k$ is the dimensionless z-axis, h_k is the instantaneous part thickness, $h(t)$ at $t = t_k$, $\Delta\hat{z} = 1/(N_z - 1)$ and Δt is the time step. As h_k is varying with time, one should update its value after each time step by using the relation between fiber volume fraction and part thickness, Equation (8.17):

$$h_k = C\,\frac{1}{V_{f,average}} \tag{8.25}$$

where $C = (nA_W)/\rho$ and it can be evaluated from the initial data.

Recall that $e = (1 - V_f)/V_f$ was defined in the previous example, and the differential equation on e was obtained in Equation (8.22) as $\partial e/\partial t = (1 + e_o)^2 \frac{\partial}{\partial z}\left(-\frac{K_{zz}}{\mu(1+e)}\frac{\partial \sigma_{zz}}{\partial z}\right)$. Note that $\frac{\partial \sigma_{zz}}{\partial e}\frac{\partial e}{\partial z}$ in Equation (8.22) was replaced by $\frac{\partial \sigma_{zz}}{\partial z}$ here. The discretized form of this differential equation

$$\begin{aligned} e_E = e_C - \frac{\Delta t(1+e_o)^2}{\mu}\frac{1}{h_k^2}\Bigg\{ & \left[\frac{\left(\frac{K_{zz}}{1+e}\right)_N - \left(\frac{K_{zz}}{1+e}\right)_S}{2\Delta\hat{z}}\right]\left[\frac{(\sigma_{zz})_N - (\sigma_{zz})_S}{2\Delta\hat{z}}\right] \\ & + \left(\frac{K_{zz}}{1+e}\right)_C\left[\frac{(\sigma_{zz})_N - 2(\sigma_{zz})_C + (\sigma_{zz})_S}{(\Delta\hat{z})^2}\right]\Bigg\} \end{aligned} \tag{8.26}$$

will be applied for the "inner" nodes ($2 \le i \le N_z - 1$ and $k \ge 2$) of the mesh shown in Figure 8.14. Here the subscripts N (North), C (Central), S (South) and E (East) denote

$$N \equiv (i+1, k), \qquad C \equiv (i, k), \qquad S \equiv (i-1, k), \qquad E \equiv (i, k+1). \tag{8.27}$$

Equation (8.13), $K_{zz} = C\,(1 - V_f)^3)/V_f^2$ is used to relate the permeability and fiber volume fraction in Kozeny-Carman model.

The boundary condition at the bottom surface is $\partial V_f/\partial\hat{z} = 0$ at $\hat{z} = 0$. In terms of the new variable, this boundary condition corresponds to

$$\partial e/\partial\hat{z} = 0 \qquad \text{for } i = 1. \tag{8.28}$$

The second order forward finite difference of this equation yields

$$e_1 = \frac{1}{3}(-e_3 + 4e_2). \tag{8.29}$$

The boundary condition at the top surface is $P_r = 0$ which corresponds to $P_T = \sigma_{zz}$ using Equation (8.11). The fiber stress, σ_{zz} is related to fiber volume fraction using Equation (8.12), and thus the boundary condition is written as

$$A_s \frac{\sqrt{\frac{V_f}{V_o}} - 1}{\left(\sqrt{\frac{V_a}{V_f}} - 1\right)^4} - P_T = 0 \qquad \text{for } i = N_z. \tag{8.30}$$

During the time marching scheme of the solution approach, one needs to evaluate V_f for a known value of the autoclave pressure, $P_T(t)$ at a particular instant. This involves

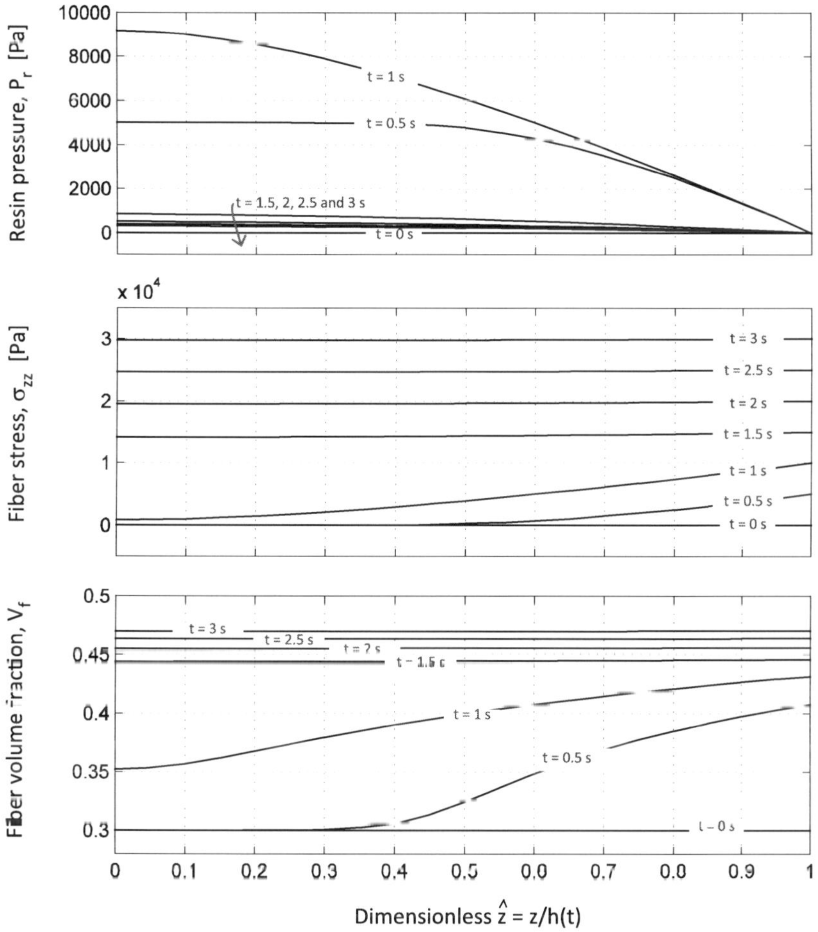

Figure 8.15: Resin pressure, fiber stress and fiber volume fraction at initial times when the autoclave pressure is increased linearly: $dP_T/dt = 10,000$ Pa/s.

root-finding of the nonlinear equation given above. One can apply a numerical method such as bi-section method, Newton's method or Secant method. However, here we will use Matlab's built-in function, *find.*

A Matlab code, *example8_04.m* was written and is listed in the Appendix. The variables of the process, resin pressure $P_r(z,t)$, fiber stress $\sigma_{zz}(z,t)$ and fiber volume fraction $V_f(z,t)$ were plotted in Figures 8.15 and 8.16.

The user of the code should use small enough time step, Δt for a stable solution. It is also suggested to re-run the code with smaller set of $\Delta\hat{z}$ and Δt and see if the results converge to the previous one. ■

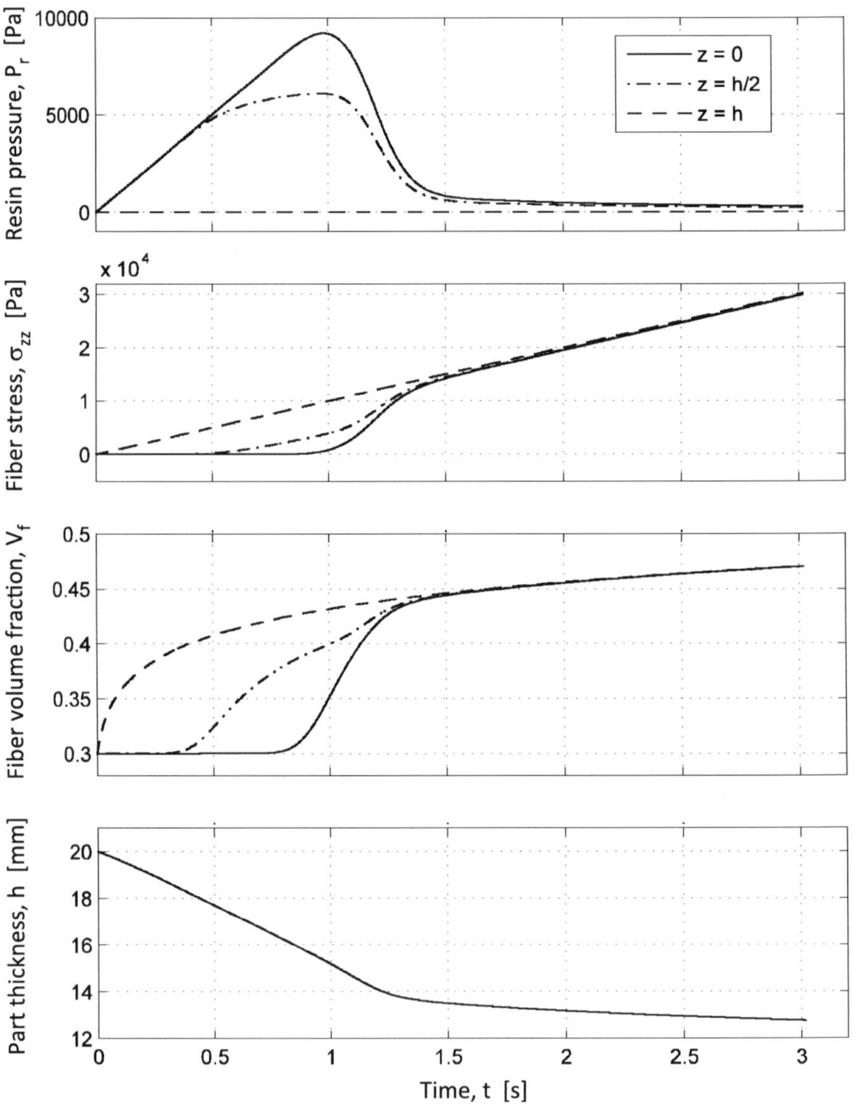

Figure 8.16: Resin pressure, fiber stress and fiber volume fraction at three different locations: $z = 0$, $h/2$ and h; and part thickness.

Heat Transfer and Cure

The hot autoclave walls and the heated air inside the autoclave transfers heat to the composite surfaces by radiation and convection, respectively, as shown in Figure 8.17. This heat is transferred to the interior of the composite by conduction and raises the internal energy of the composite and hence the temperature of the resin present in the composite. The raised temperature will initiate the curing of the resin which is an exothermic chemical reaction that cross-links the resin molecules and releases the heat. The heat released can be modeled as an energy generation term that further raises the temperature of the composite and needs to be conducted out of the interior of the composite as the curing reaction approaches completion. Thus, one needs to first increase the temperature of the autoclave and later reduce it to extract the heat of the reaction and cool the composite part.

To simplify the analysis, one can assume that the resin bleeding is complete and can use the energy conservation principle to write the governing equation to predict the temperature field inside the composite as a function of time. This temperature field will be a function of the thermophysical properties of the composite, the cure kinetics of the resin, the geometry of the part and the temperature of the autoclave. One could control the temperature history of the composite by changing the temperature of the autoclave. Mathematically, this translates into boundary conditions on the composite surface that may be of the first, second or third type as discussed in Chapter 3.

One can obtain the following governing equations for conservation of energy and resin cure state with temperature as a primary variable for a composite that is undergoing curing without any resin flow:

$$[V_f(\rho C_p)_f + (1 - V_f)(\rho C_p)_r] \frac{\partial T}{\partial t} = \frac{\partial}{\partial x}\left(k_x \frac{\partial T}{\partial x}\right) + \frac{\partial}{\partial y}\left(k_y \frac{\partial T}{\partial y}\right) + \frac{\partial}{\partial z}\left(k_z \frac{\partial T}{\partial z}\right) + (1 - V_f)\rho_r H_r \frac{dC^*}{dt}. \tag{8.31}$$

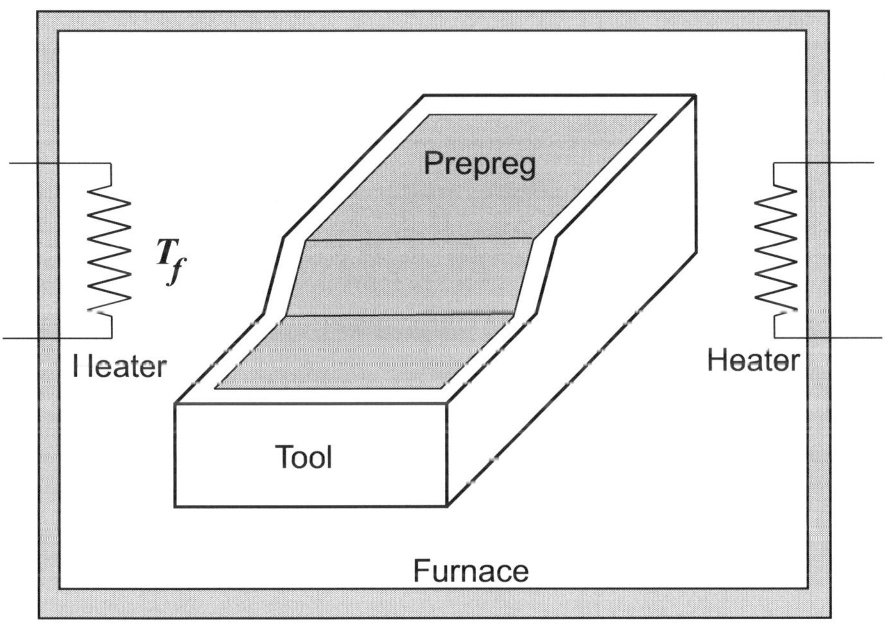

Figure 8.17: Curing of a composite in an autoclave.

Here ρ is the density, C_p is the heat capacity and V_f is the fiber volume fraction. The subscript r is for the resin and f is for the fibers. k_x, k_y and k_z are conductivities of the composite in the x, y and z directions, respectively. C^* is the degree of cure expressed by constitutive equation for that particular resin. Some examples are given by Equations (4.75), (4.80) and (4.81). For example,

$$\frac{dC^*}{dt} = k_0 \exp\left(-E/RT\right) C_{A0}(1 - C^*)(\Lambda - C^*) \tag{8.32}$$

By selecting the coordinate directions to coincide with the principal directions of the fiber preforms, one can write the governing equation as cast in the form of Equation (8.31). The composite is usually much longer in the in-plane directions (x and y directions) as compared to the thickness direction and, hence, one can usually eliminate the first and second terms on the right-hand side in the analysis (see Example 5.3 in Chapter 5 for details). The constitutive form of the cure kinetics equation is usually a function of the temperature and the current cure state (Equations (4.23) through (4.31)). One can solve Equations (8.31) and (8.32) simultaneously to obtain the temperature and cure history as a function of the imposed boundary and initial conditions. The initial condition for the composite temperature is usually room temperature, and for cure it is zero cure. The boundary conditions are related to the temperature of the autoclave. The simplest approximation is that the temperature of the two surfaces at $z = 0$ and at $z = h$ is equal to the autoclave temperature $T_w(t)$, and one can solve the coupled system of equations to obtain temperature and cure simultaneously. See references [343–345] for details.

Example 8.5: Temperature and Degree of Cure with Time

Initially, the temperature of a composite part is $T = T_o = 20°\text{C}$ and is placed in an autoclave which is held at a constant temperature of 120°C. Assume that the temperature of the top and the bottom surface attains the autoclave temperature instantly. Take the parameters $\Lambda = 1$, $k_0 = 2\,(\text{g/mole})/\text{s}$, $E/R = 1000\,°\text{K}$ and initial concentration of $C_{A0} = 0.5\,\text{mole/g}$ in the cure kinetics model, Equation (4.75) in Examples 4.5 and 4.6. Use the following material, geometry and process parameters:

- Average fiber volume fraction, $V_f = 0.40$
- Thermal conductivity of fiber, $k_f = 1.20\,\text{W}/(\text{m}°\text{K})$
- Thermal conductivity of matrix, $k_m = 0.20\,\text{W}/(\text{m}°\text{K})$
- Density of fiber, $\rho_f = 2500\,\text{kg/m}^3$
- Density of matrix, $\rho_m = 1200\,\text{kg/m}^3$
- Specific heat capacity of fiber, $C_{pf} = 800\,\text{J}/(\text{kg}°\text{K})$
- Specific heat capacity of matrix, $C_{pm} = 700\,\text{J}/(\text{kg}°\text{K})$
- Initial temperature, $T_{in} = 20°\text{C} = 293.2°\text{K}$
- Mold wall temperature, $T_o = 120°\text{C} = 393.2°\text{K}$
- Average 1D resin speed, $u = 0.0714\,\text{m/s}$

- Thickness of the part, $h = 0.020\,\text{m}$
- Constant to be used in cure kinetics, $H_r = 280,000\,\text{J/kg}$.

Assume that the part's in-plane dimensions are much larger than its thickness, thus the temperature is a function of z and t only: $T = T(z,t)$. Plot the distributions of temperature, T and degree of cure (also called the extent of reaction), C^* within the part ($0 \le z \le h$) until the minimum C^* within the part reaches 0.85.

Solution

With the assumption of $T = T(z,t)$ and k_z is constant, the conservation of energy, Equation (8.31) reduces to

$$\frac{\partial T}{\partial t} = \frac{1}{(\rho C_p)_c}\left[k_z \frac{\partial^2 T}{\partial z^2} + (1-V_f)\rho_r H_r \frac{dC^*}{dt}\right] \tag{8.33}$$

where

$$(\rho C_p)_c = V_f(\rho C_p)_f + (1-V_f)(\rho C_p)_r. \tag{8.34}$$

For the solution of this example, finite difference method (FDM) will be applied in the spatial direction, and Euler method will be applied for the time integration. The discretization of the spatial and time domains is shown in Figure 8.18 and formulated as follows

$$z_i = (i-1)\Delta z \qquad \text{where } i = 1,2,\ldots,N_z \tag{8.35}$$

$$t_j = (j-1)\Delta t \qquad \text{where } j = 1,2,\ldots \tag{8.36}$$

where $\Delta z = h/(N_z - 1)$ and Δt is the time step.

The discretized form of Equations (8.33) and (8.32) are given as

$$T_E = T_C + \frac{\Delta t}{(\rho C_p)_c}\left[k_z \frac{T_N - 2T_C + T_S}{\Delta z^2} + (1-V_f)\rho_r H_r \left(k_0 \exp\left(-E/RT_C\right) C_{A0}(1-C_C^*)(\Lambda - C_C^*)\right)\right] \tag{8.37}$$

$$C_E^* = C_C^* + \Delta t\left[k_0 \exp\left(-E/RT_C\right) C_{A0}(1-C_C^*)(\Lambda - C_C^*)\right] \tag{8.38}$$

and they will be applied for the "inner" nodes ($2 \le i \le N_z - 1$ and $j \ge 2$) of the mesh shown in Figure 8.18. To evaluate dC^*/dt, the cure kinetics model, Equation (8.32) is used. Here the subscripts N (North), C (Central), S (South) and E (East) denote

$$N \equiv (i+1,j), \qquad C \equiv (i,j), \qquad S = (i-1,j), \qquad E = (i,j+1). \tag{8.39}$$

During the time marching scheme of the solution, both variables, T and C^* will be updated simultaneously using the coupled Equations (8.37) and (8.38). A MATLAB code, *example8_05.m*, was written and is listed in the Appendix. The temperature and extent of reaction are plotted in Figures 8.19 and 8.20. ■

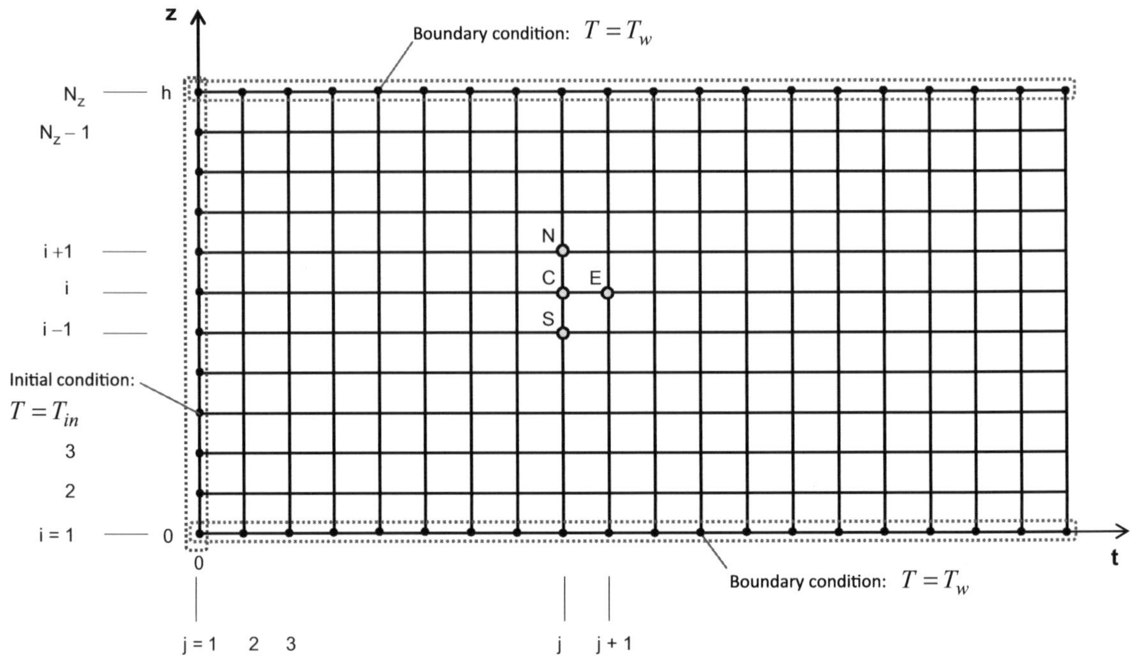

Figure 8.18: Discretized domain, initial and boundary conditions to solve temperature and degree of cure.

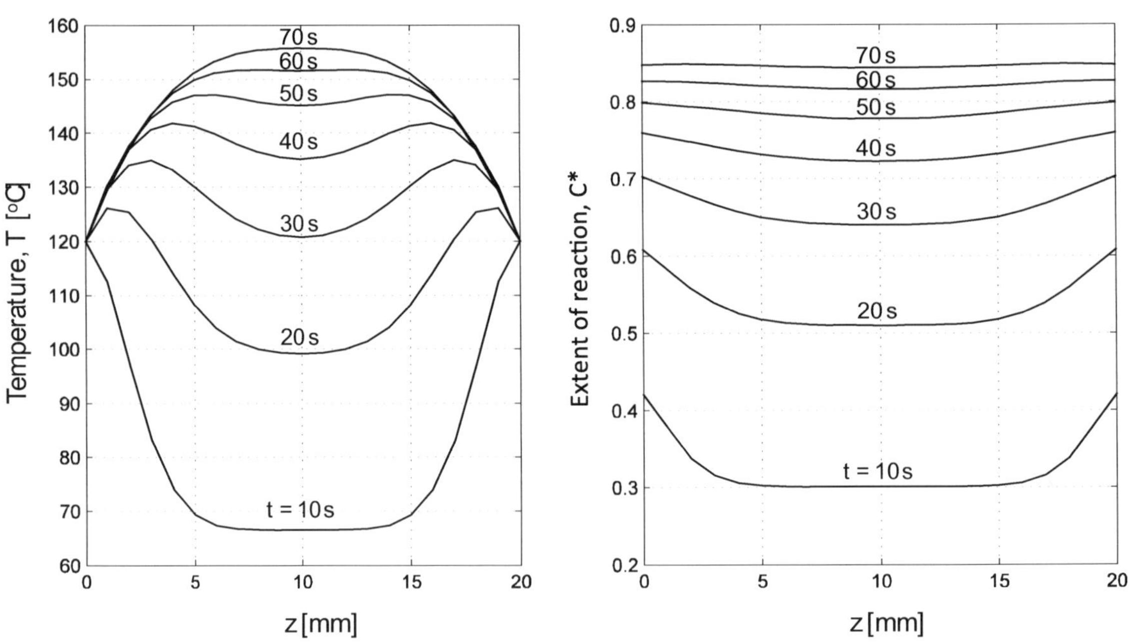

Figure 8.19: Temperature $T(z, t)$ and degree of cure $C^*(z, t)$ at $t = 10, \ldots, 70\,\text{s}$.

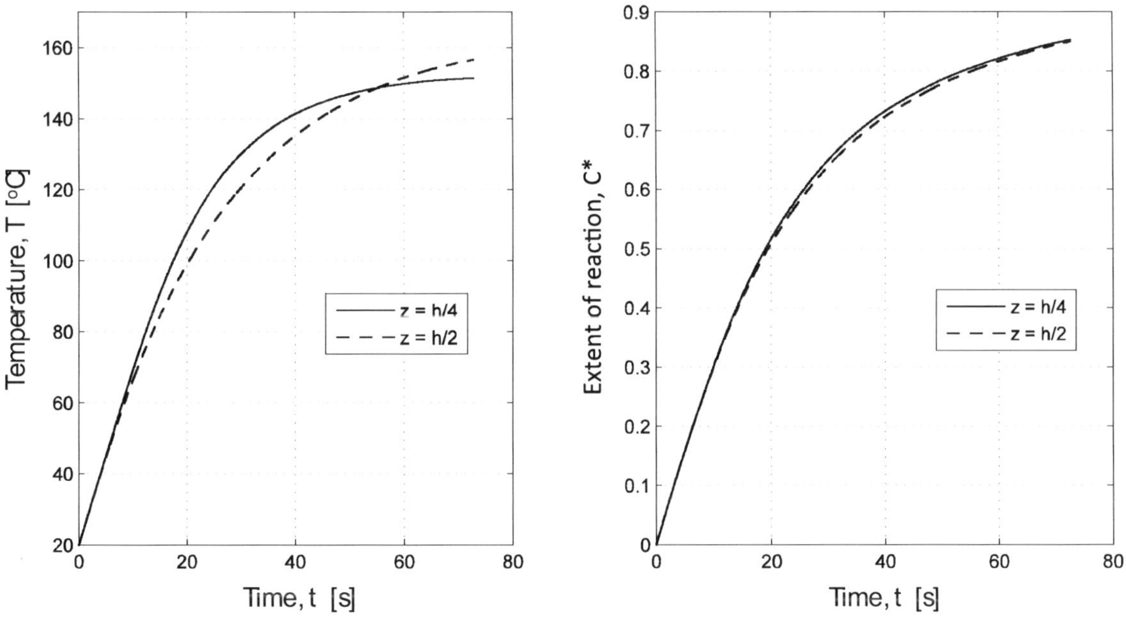

Figure 8.20: Temperature $T(z,t)$ and degree of cure $C^*(z,t)$ for the two inner locations ($z = h/4$ and $h/2$) until the minimum C^* reaches 0.85.

8.3 Liquid Composite Molding

Liquid Composite Molding (LCM) is a class of composite manufacturing processes in which a mold cavity containing a reinforcing preform is injected with liquid thermoset resin to fill the empty spaces in the mold. After the resin cures partially or completely, the composite part is demolded. LCM is used in almost all industries that benefit by switching from metals to composite parts. They include the aerospace, automotive, marine and civil industries. The reinforcing fabric preform is usually made of continuous glass or carbon fiber mats, and it can be dry or preimpregnated when it is placed into the mold cavity. The objective of the process is to manufacture composite parts by completely wetting the fabric preform with resin at the macro- and microscales with no regions that are resin starved. The following are the most commonly used LCM processes:

- Resin Transfer Molding (RTM)
- Vacuum Assisted Resin Transfer Molding (VARTM)
- Seeman's Composites Resin Infusion Molding Process (SCRIMP)
- Injection Compression Molding or Compression RTM
- Reinforced Reaction Injection Molding (RRIM)
- Structural Reaction Injection Molding (SRIM)
- RTM Light
- Resin Infusion Processes
- Resin Film Infusion

8.3.1 Similarities and Differences between Various LCM Processes

LCM processes can be broadly classified into two groups: matched mold processes and single-sided mold processes.

Matched Mold Processes

Matched mold LCM processes include RTM, Compression RTM and Injection Compression Molding. As the name implies, the molds have at least two rigid parts. If the composite part to be manufactured does not have a very intricate or complicated shape, then the mold usually consists of a two-part cavity for the preform. As the complexity of the shape or the size of the part increases, then the male and female halves might be constructed from smaller pieces in order to ease the demolding as shown in Figure 8.21.

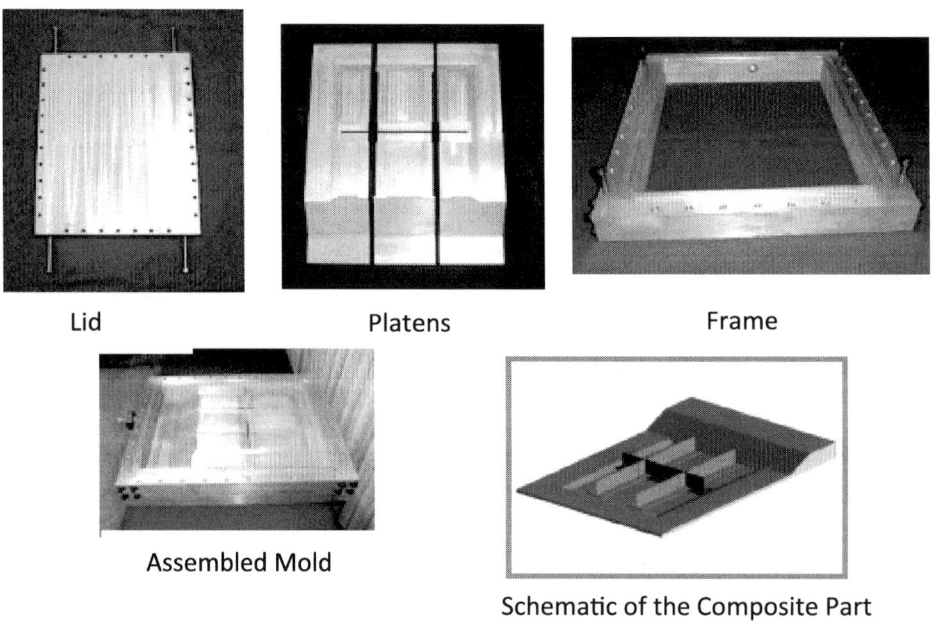

Figure 8.21: Various mold sections to manufacture the composite part depicted on the bottom right.

Under compressive clamping force, the mold parts completely enclose the fiber preform that is placed inside. By conforming the preform to the desired dimensions of the final part, a near-net-shaped part can be manufactured by this type of LCM process. However, process and design engineers should be careful in selecting the mold material. If the thermal expansion coefficients of the mold and composite materials are dissimilar, they will expand or contract at different rates during the heating/cooling stages of cure. This will usually lead to significant warpage or cracks in the composite part. Usually exact match of the thermal coefficients is not achievable, and design engineers have to adjust the mold cavity dimensions to compensate for the different expansion and contraction rates of the two materials. However, the key advantage is that this type of process allows one to obtain very tight tolerances which is rarely possible with one-sided molds.

RTM, the most widely used of the closed mold processes, was first introduced in the 1940s. A schematic of the RTM process is shown in Figure 8.22. One can describe the RTM process in five steps: (1) manufacturing of reinforcing preform, (2) draping of the preform over a tool surface, (3) compaction of the preform by closing and clamping of the mold, (4) filling the empty spaces between the fibers of preform with resin and cross-linking of the resin, and (5) demolding.

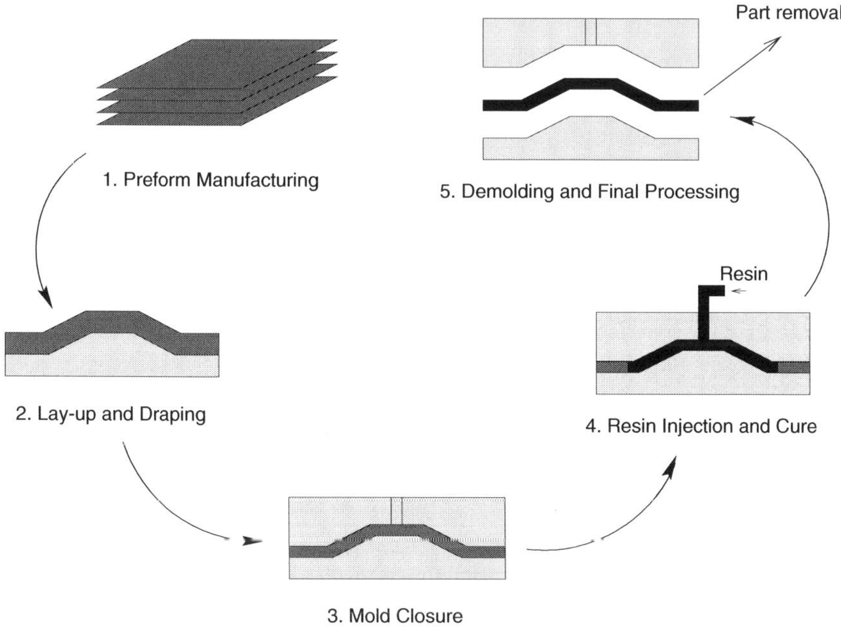

Figure 8.22: Schematic of the process steps in RTM.

The reinforcing fabric preform is usually manufactured by stacking individual fabric layers. During the draping stage, the fabric tows (bundles of fibers) undergo structural deformation if the mold has a doubly curved cavity, corners and ribs [346]. A tackifier adhesive (binder) is added to hold the individual fabric layers together in order to ease the handling and placement of fabrics, especially if the mold cavity is complicated or a large number of layers are stacked. During the compaction stage, the fibers mostly compact in the thickness direction. The draping and compaction can change the permeability of the reinforcing fabric very significantly and nonuniformly, and influence the resin injection stage [347, 348].

The RTM process is widely used due to its many **advantages**, including the following:

- Near-net-shaped composite parts with a good surface finish and close dimensional tolerances.
- Large parts with complicated shapes.
- Achievement of good mechanical properties by tailoring (i) the structure and the stacking order of reinforcing fabric, and (ii) the resin system.
- Use of metal inserts in order to attach the part to the main structure.
- Use of internal wood or foam cores to reduce cost and increase the stiffness of the part.
- Consistency in reproducibility of parts compared to wet hand lay-up.
- Fast production cycles compared to wet hand lay-up.

- Low mold clamping pressure compared to processes, such as injection compression molding.
- Much less exposure to volatile chemical vapor than wet hand lay-up.
- More automated (less labor skill needed) than wet hand lay-up.

Although the items listed above make the RTM process very promising for the manufacturing of composite parts, there are some **disadvantages** that need to be overcome, including the following:

- Inconsistency in reproducibility: inherent variations in preforming due to inconsistent fabric rolls from vendors, and inconsistencies in fabric cutting, stacking, pre-impregnating and placement into the mold cavity.
- Sensitivity to mold design, since improper gating and venting can cause incomplete filling of the empty spaces and result in macroscale dry spots (resin-starved regions) in the composite.
- Long production cycles compared to parts made from aluminum or steel in automotive industry.
- Higher clamping pressures than vacuum bagged VARTM and SCRIMP.
- Movement of preform (wash out) due to loose compaction and/or high resin pressure.

A variation of RTM involves drawing a vacuum in the cavity before the resin injection is initiated. By removing almost all of the air from the mold cavity, the chances of void formation are significantly reduced. It also reduces the required injection pressure and/or the time to fill the mold. However, vacuuming inside the mold sometimes can cause the resin to bubble if its vapor pressure is low.

A relatively new modification in matched mold processes is called injection compression molding. In this process, the mold is closed to nearly but not exactly its final thickness creating an empty space between the mold and the preform, after the preform is placed into the mold cavity. The amount of resin needed to fill the closed mold is injected into this empty space between the mold wall and the preform. Because the resistance to flow in the preform is orders of magnitude higher than the resistance offered by this empty space between the mold and the preform, the resin fills this space first. After the calculated resin is injected, the mold is gradually closed (compressed) to the desired dimensions pressuring the resin and squeezing it into the preform in the thickness direction as shown in Figure 8.23. Injection Compression Molding, also known as Compression RTM, allows one to reduce the injection pressure and fill time as compared to RTM but requires a press that can control the compression process [350].

Single-Sided Molding Processes

"Vacuum bagged" vacuum-assisted resin transfer molding (VARTM) and SCRIMP are single-sided LCM molding processes. They all belong to the category of resin infusion processes in which the resin is drawn into the mold with a vacuum and as there is no second rigid tool surface, the compaction of the fabric is also achieved through the use of a vacuum. As the name implies, these processes have only one rigid side of the mold. In "vacuum bagged" VARTM, after the preform is placed on the rigid mold part (male part in Figure 8.24, but in general it can be female as well), a vacuum exit tube and resin inlet tubes are placed around the preform. Then, a plastic film layer is placed on top of the preform and tubes, and sealed using "tacky" tape. The air inside is removed by using a vacuum pump drawing air through the vacuum exit tubes and clamping the resin inlet tubes. Once sufficient vacuum is created, the resin inlet tube is inserted into a resin pot, and the clamp is removed. Pressure differential of one atmosphere (or less) between the resin container and the interior of the mold pulls the resin into the mold. After the resin saturates the mold and reaches the vent, the resin injection is discontinued by clamping the tubes. Once the resin cures, the part is demolded, and the tubes and plastic film are discarded. Figure 8.24(a) shows a schematic of this process.

The major **advantages** of the single-sided LCM processes are (i) a very simple injection system, (ii) a simple mold, (iii) visual control of filling stage on the surface through the transparent plastic film, (iv) capability to manufacture parts of the order of a few meters such as train and bus compartment sides that can be up to 15 meters long, and (v) low capital investment.

However, there are many **disadvantages**, including (i) long fill time due to low injection pressure of 1 bar, (ii) low fiber volume fraction due to low compaction pressure of 1 bar, (iii) low quality surface finish and high dimensional tolerance, (iv) requirement for skilled labor during the bagging stage, (v) labor intensive as in autoclave processing, (vi) material wastage is much higher than other processes, and (vii) for thick parts, formation of voids through the thickness, as one can visualize only the resin close to the top layer adjacent to the plastic film.

In order to overcome the long fill time disadvantage of VARTM, The Seeman Corporation resin infusion molding process (SCRIMP) was developed and patented under their

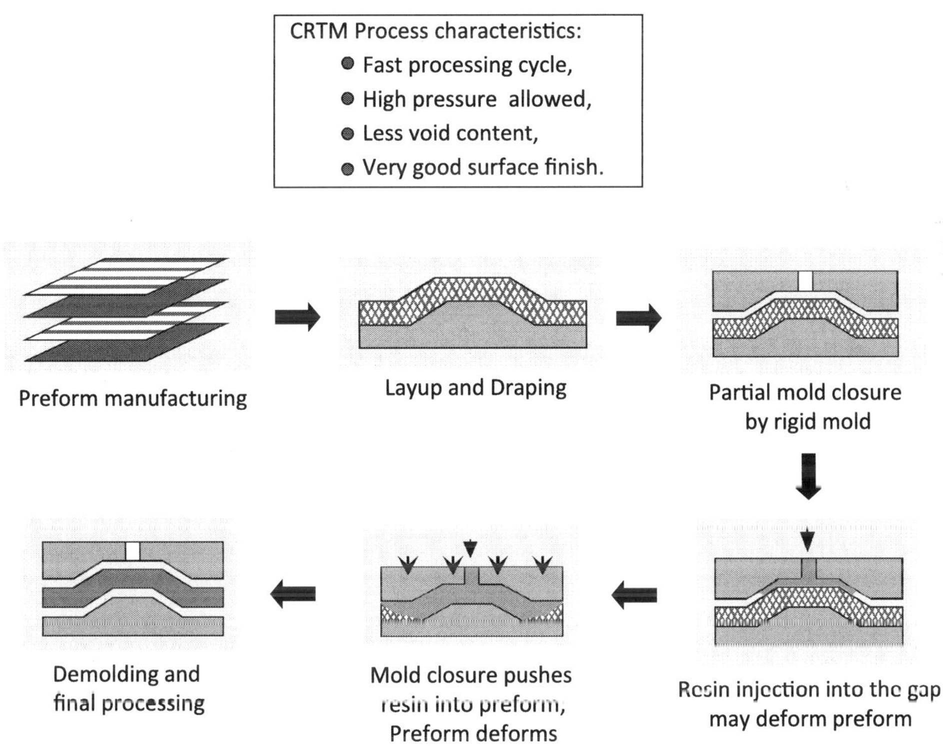

Figure 8.23: Schematic of Injection Compression Molding, also known as Compression RTM [349].

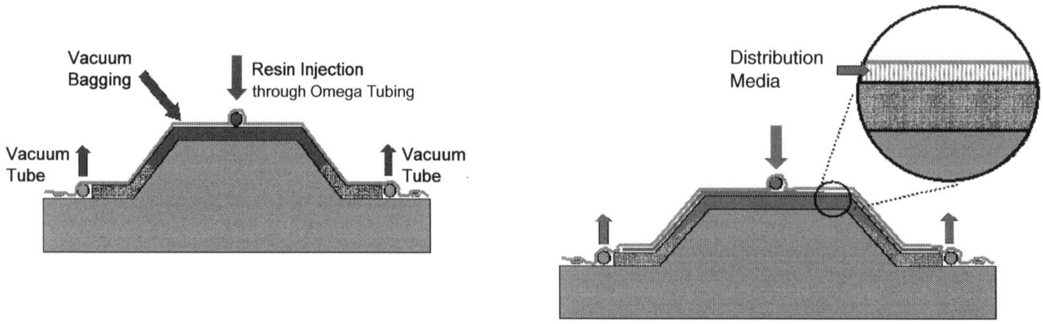

(a) (b)

Figure 8.24: (a) "Vacuum bagged" VARTM and (b) SCRIMP [87,351].

company's name. A very porous fabric, known as the distribution medium, is placed on top of the preform. Since it has very low resistance to resin flow, resin saturates this porous layer quickly and then flows into the reinforcing preform in the thickness direction. Other similar processes such as FASTRAC have been developed to improve the delivery approach to get the resin from the container to the surface of the preform quickly. In such approaches, channels are formed along one of the faces of the mold to distribute the resin along one of the faces as shown in Figure 8.25 [352]. The quick distribution of the resin is shown in Figure 8.26 [352,353].

A few other variations of these processes have been invented recently. For example, Fast Flooding Chamber Method [354,355] and the Vacuum Infusion Relaxation Method. Figure 8.27 shows the schematic of these processes. The main function of the chamber is to attract resin in regions difficult to fill. Thus, the chamber is used as a tool to manipulate the resin flow during filling.

8.3.2 Important Components of LCM Processes

The major ingredients of LCM processes are the fiber preform, resin system, mold and injection equipment and delivery system. We will discuss them briefly here. Details can be found in many books written about this process [80,246,356,357].

Fiber Reinforcements and Inserts

The most commonly used reinforcement fabric materials are fiberglass, carbon, and aramid (Kevlar). Proper fabric material and structure type (woven, stitched, knitted, braided, random, etc.) must be selected considering mechanical properties, surface finish and cost criteria. Strength and stiffness of composite parts depend on the reinforcing fabric. Resin acts as a matrix that holds the fibers together, and plays an important role against wear and corrosion of fibers.

Resin is expected to fill the empty spaces between fiber bundles and also the interior of each fiber bundle. Some of the terminology related with fiber reinforcement is listed below:

- *Fiber volume fraction* is the ratio of the fiber volume to the total volume of fabric. The fiber volume fraction of a preform in a closed mold depends locally on the mold

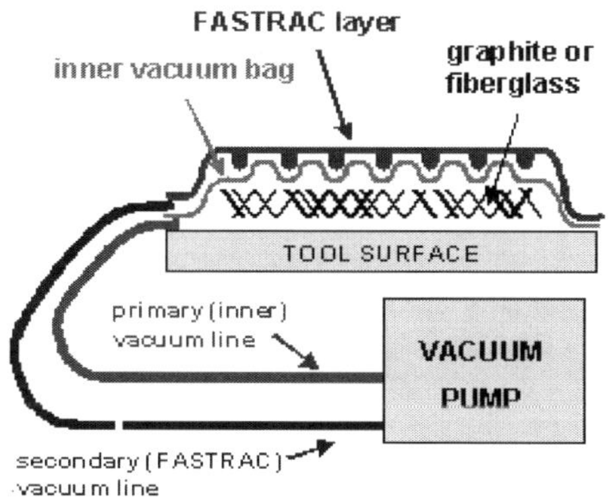

Figure 8.25: Schematic of the FASTRAC process [352].

gap thickness, number of fabric layers and in-plane compaction around corners and ribs if they are present.

- *Permeability* of a reinforcing preform, which is a measure of "inverse of resistance to resin flow," is dependent on fabric structure and fiber volume fraction. Permeability tensor, $\mathbf{K}$ and resin viscosity, μ are used to relate the resin velocity to the pressure gradient in Darcy's Law (an empirical momentum equation): $\mathbf{U} = -(\mathbf{K}/\mu) \cdot \nabla P$.

- *Wash* is the movement of fabric under high resin pressure.

- *Wettability* is a measure of how easily an interior of a fiber bundle is wetted as resin moves around the bundle.

- *Sizing* is a coating on the fiber surface in order to increase the wettability property and also to increase the bond between the fibers and the resin as it cures. Different sizings are used on the fiber surfaces to make them compatible with the selected resins.

An ideal fabric should have high wash resistance and wettability, and sizing on its fiber surfaces.

The structure of the fabric reinforcement can be tailored in order to achieve the desired mechanical properties such as tensile strength and stiffness. However, the structure affects how resin flows through it. Hence, while designing new fabric structures, not only mechanical properties, but also permeability and wettability properties must be considered

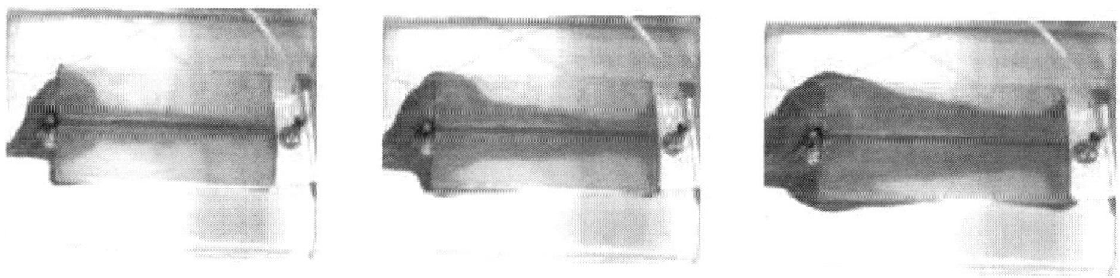

Figure 8.26: Resin flow in three early frames of FASTRAC [352, 353].

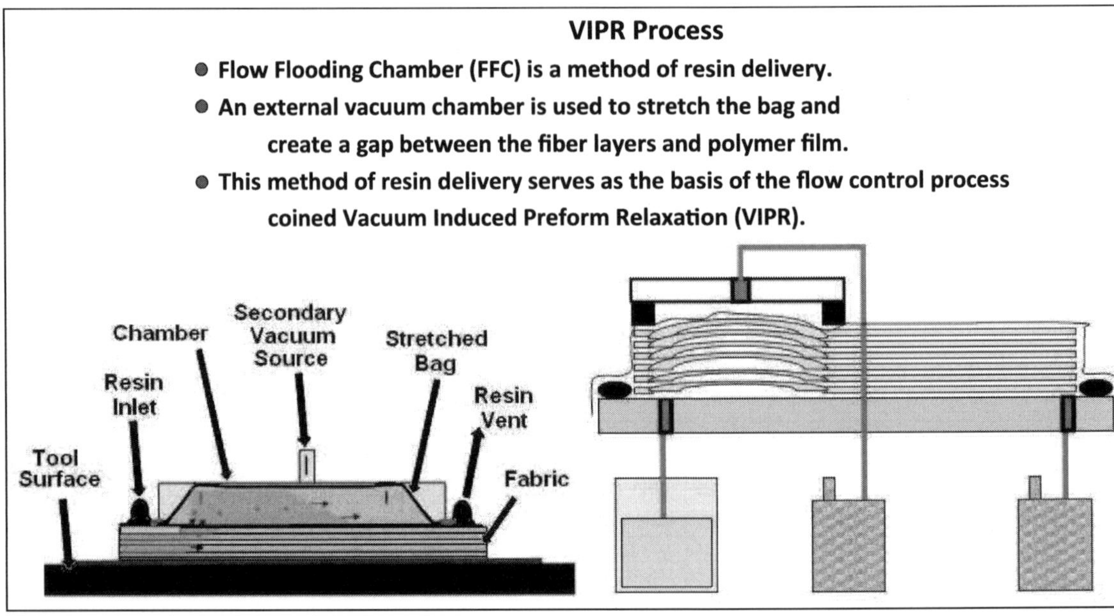

Figure 8.27: Schematic of the Fast Flooding Chamber Method and Vacuum Infusion Relaxation Method.

simultaneously. A fabric structure with a high resistance to resin flow (low permeability) will require higher injection pressure or filling time. In some industries such as aerospace, this might be acceptable as the mechanical properties are more important than the cycle time, but may not be viable for other high-volume production industries.

In order to ease the handling and placement of preform into the mold, a veil, or a tackifier adhesive (also called binder) is used by spot-glueing the adjacent fabric layers of preform to each other. Tackifiers are made by dissolving the resin in a solvent. A tackified preform allows one to place the preform into the mold easily and reduces the possibility of creating unplanned empty channels between the preform and the mold wall, where resin can flow easily due to low resistance to flow. This racetracking phenomenon can dramatically change the flow patterns as the resin impregnates the preform and can lead to unpredictable fill patterns. However, the disadvantage of using tackifier is that the permeability of the preform can reduce locally at the glued spots.

Between the reinforcing fabric and the resin, a strong bond should be created by treating the surface of fibers. This process is called *sizing*. The reasons why sizing should be applied are as follows:

- The major part of load on composite part will be directly transferred to the reinforcing fabric through the matrix without damaging the matrix due to the strong interface bond.
- The fibers will be lubricated against potential abrasive damage during handling and preforming operations such as weaving.
- Static electricity between the fibers will be minimized.

Cores such as foams and balsa wood are included inside large and thick parts. Cores are used to lower the cost, increase the moment of inertia, and reduce the weight of the part as cores have lower density than the rest of the composite part. If a runner network is machined into the cores, resin moves more rapidly in the mold through the core runners. Besides the advantages of cores, there are some disadvantages that design engineers must be cautious about. Cores might shrink or even collapse during the resin filling and curing stages. Resin pressure must be kept below the compressive strength of the cores. Also, during the heating and exothermic curing stages, the temperature of cores should not exceed a critical value which may otherwise deform the part.

As an alternative to cores, internal inflatable bladders might be used in thick parts such as tennis racquets and rotor blades [358]. The bladder is initially underinflated. After the preform is placed and the mold is closed, the bladder is inflated further.

Metal inserts, which are usually steel, can be placed inside the mold which allows the composite part to be attached to other parts of the structure. The surface of inserts must be processed to enable a good bond between the insert and the surrounding matrix. The thermal expansion coefficient of insert material should not be much different from the composite material. The insert material must also be resistant to corrosion.

Textile type preforms are manufactured by weaving, stitching or braiding fiber tows (fiber bundles). The tow cross-section is approximately elliptical, and it has dimensions of the order of millimeters. High strength composite parts such as the ones used in aerospace are made of very dense fiber tows with up to 48,000 fibers in each tow. "Chopped" and "continuous strand" random fabrics are manufactured from inexpensive and short E-glass fiber tows which contain about 100 fibers.

In hand preforming, also known as *cut-and-place preforming*, dry fabric layers are cut from rolls and placed into the mold cavity. Fabric stacking might consist of different types of fabric and/or varying orientation. For example, the laminate lay-up [0/90/0/90/0] has five layers of the same fabric, where the 1st, 3rd and 5th layers are placed in the 0° direction (warp direction), and the other two layers are in 90° direction (i.e., perpendicular to the warp direction). Hand preforming eliminates preforming instrumentation, but requires many manual hours. Consequently, reproducibility of parts is not consistent.

Stamping of fabrics with a thermoformable binder is a versatile and automated preforming process. A small amount of thermoformable binder is added to nonwoven glass fabric layers that are cut larger than the mold cavity surface area. Fabrics are heated around 150°C to soften the thermoformable binder in it. Fabric layers are then stamped in the mold cavity at lower temperatures as thermoformable binder stiffens and holds the shape. The excess fabric outside the mold cavity is then trimmed. This process has the advantage of preparing preforms with accurate dimensions which eliminates the tedious labor and skill required to cut and place preforms stacked by hand, and also eliminates the potential racetracking channels created between the preform and mold walls. Another advantage is that the preform has higher resistance against resin wash as it is stiffer. A drawback of this method is that some fabrics, such as directional and carbon fabrics, cannot take advantage of this method as they don't stretch during the stamping stage. Another disadvantage is that the fibers might buckle during the stamping stage.

Resin System

Resin forms the matrix of the composite part after it polymerizes. A resin system may consist of some or all of these ingredients: promoters, fillers, mold release agents, pigments and catalysts. Some of the most commonly used resins in LCM are polyester, epoxy, vinyl ester, urethane and nylon. Fillers such as clay, aluminum trihydrate and calcium carbonate are used to reduce cost. A good resin system should have the following properties:

- Low viscosity to ease injection.

- Effective wetting properties to wet the fabric.
- Uniform and consistent cure kinetics.
- Rapid cure after gelation.
- High transition temperature.
- High tensile strength to support the reinforcing fabric under load.
- Low thermal expansion coefficient.

Low resin viscosity is desirable as it will require low injection pressure under constant flow rate injection, or short filling time under constant pressure injection. If the resin viscosity exceeds a typical upper bound of 500 centiPoise (typically, 1 centiPoise is the viscosity of the water), it will not only require high injection pressure equipment, but might also "wash-out" the fabric by forcing it to move from its originally placed configuration. At the other extreme, injecting a resin with too low viscosity (typically lower than 50 centiPoise) is much easier, but it may fail to "wet-out" the interior of each individual fiber tow (fiber bundle) even if the resin impregnates the empty spaces between the fiber tows.

A thermosetting resin system can be injected either as a single component system or a multiple component system. In multiple component systems, the resin is premixed with the catalyst in a single container in the desired ratio and injected into the mold. This is a simple and reliable system, but one must use up the resin before it gels or have to clean the container before the next injection. The other option, the single component system, requires separate containers for the resin and the catalyst or the accelerator, which are mixed "on the fly" as they enter the mold. This allows the flexibility of changing the mix ratio either (i) to accelerate or decelerate the cure, or (ii) to hold the "gelation time" (also known as "pot life") constant at varying temperature of the manufacturing plant. However, this could introduce variability in the part quality due to uneven mix or cure.

Table 8.2 shows a comparison of the most commonly used LCM thermosetting resins from the viewpoint of performance, processing and cost [11].

Table 8.2: Thermoset Resins Commonly Used in LCM Processes [11]

Resin type	Performance	Processing	Cost	Aerospace industry	Automotive industry
Polyester	Low ↓	Easy ↓	Low ↓	√	√
Vinyl Ester				√	√
Urethane				√	√
Epoxy				√	√
Cyanata Ester				√	
Bismaleimide				√	
Phenolic				√	√
Polyimide	High	Difficult	High	√	

Resin cures due to the chemical reactions to create a matrix around the reinforcing fabric. Usually, the chemical reaction is a step- and/or chain-growth polymerization. The length of the cure cycle depends on the type of resin system and temperature of the heating system. During the cure cycle, the viscosity increases and then eventually gellation starts at T_{gel}. The cure kinetics of resin systems, especially the length of the cure cycle, can be adjusted by changing the amount of catalyst and accelerator that are added to the resin system. Some resin systems are designed for room temperature cure.

The strength of the matrix starts to build up after T_{gel}. The part is demolded when the matrix reaches a "green strength" such that the part will not deform or warp during or after being taken out of the mold. Wedges, mallets or hydraulic ejection system can be used to demold the part. Demolding the part before it reaches the maximum strength has advantages and disadvantages. The main advantage is that the production cycle is shortened by enabling early injection of the next part. Another advantage is that the damage due to exothermic resin cure reaction on the mold surface is reduced. The disadvantage is that the demolded part might warp or develop scratches on its surface. Hence, demolding should be done neither too early nor too late. Some trimming might be needed to remove excess matrix along the edges of the part.

Although most RTM applications use thermoset resin systems, research is underway to use thermoplastic resin for RTM due to the added attraction of recyclability. The research is focused on reducing the viscosity of cyclic thermoplastic resins to the order of thermoset resins to promote impregnation as discussed in Chapter 2.

Mold (Tool)

Mold material must be selected considering the production volume, the temperature bounds of the heating/cooling system, tolerances in part dimensions, and the quality of surface finish required. Steel, aluminum and copper are the most commonly used metals. Steel is preferred for high volume production because of its high resistance to wear and hence can last longer. Aluminum and copper are easier to cut during tool making. For low production volumes (less than several thousand parts), nonmetallic materials such as wood and reinforced polyesters or epoxy composites are preferred because of easy and rapid tool manufacturing and low cost.

Temperature fatigue, solvents, mold release agents and accidental scratches made on the manufacturing floor cause wear of the mold surfaces. A nickel shell on the mold cavity surface delays the wear of the surface and is usually applied to molds that need to produce large number of parts.

Release agent is sprayed on the mold cavity surface in order to ease separation of the composite part from the mold surface. An inflatable bladder mold half made of silicone rubber might be used, instead of a rigid mold half, to ease demolding of large parts with complicated shapes.

For VARTM and SCRIMP with one sided molds, the parts could be very large of the order of 10 to 50 meters especially when manufacturing wind blades.

The locations of gates (resin inlet ports) and vents (air and resin exit ports) must be properly designed such that the reinforcing fabric is completely wetted when resin reaches the last vent. Resin is injected into the mold cavity through the gates, and the air in the mold cavity is pushed out by the propagating resin through the vents. Mold design engineers use either their experience along with some trial and error mold fillings or mold filling simulations in order to properly place the gates and vents. Once the gate locations are decided, vents are placed at the last points of filling in order not to entrap air inside the mold. After the resin bleeds out from the last vent, a little extra time is allowed for complete wetting of the fibers. Then, all of the gates and vents are closed. At this time, resin might continue to percolate to wet the interior of the fiber tows. Unconventional techniques such as vibrating the mold or the inlet flow rate in order to wet the spaces between the fibers in a tow are used to eliminate microvoids [359].

Although the resin pressure in RTM is usually less than 20-30 bars, this pressure can cause resin leakage through the edges of the mold if the mold is not properly sealed. In order to prevent leakage, temperature and solvent resistant O-ring silicone rubbers are used along the perimeter of the mold. For resin infusion processes such as VARTM and SCRIMP, the maximum pressure is one bar; however, the mold here also needs to be sealed well by ensuring there is no leakage of air from outside into the mold which is usually sealed with a plastic bag.

Some resin systems require elevated temperatures to cure. Once the resin cures, the part is demolded much more easily if the mold is cooled down. In order to heat and then cool the mold, heating/cooling channels are machined inside the mold as shown in the computer model in Figure 8.28. Or, alternatively, the mold is placed into an oven for the heating stage.

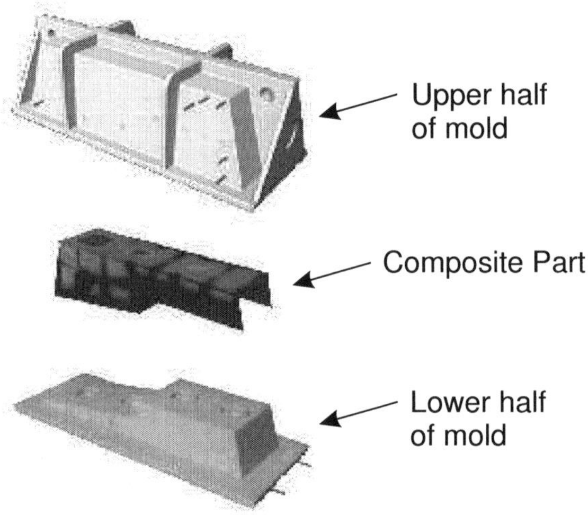

Figure 8.28: A finite element model of a complex RTM mold with heating and cooling channels and the composite part.

Injection and Resin Delivery System

Simple injection equipment can be used for single component resins, or batch mixed multiple component resin systems. However, expensive injection equipment with the capability of mixing the resin components is preferred for high production volumes of composite parts with multiple component resin systems. Ideal injection equipment should allow the manufacturing engineer (i) to change injection pressure or flow rate, (ii) to open and close gates for sequential gating by opening/closing gates, and (iii) to monitor the temperature, pressure, flow rate and cumulative amount of resin injected.

Most of the injection equipment is designed for either pressure controlled or flow rate controlled injections. Flow rate controlled injection equipment should allow the user to set an upper critical injection pressure to protect the mold from damage and bending. Once the critical pressure value is reached, either the flow rate may be reduced, or the injection may be continued at the critical pressure value.

Radius Engineering Inc. [360], Composite Integration [361], Magnum Venus Plastech Limited [362], and Liquid Control Corp. [363] are some injection equipment manufacturers for LCM. These manufacturers have two types of injection equipment: (i) pre-mixed multiple component resin systems, and (ii) single component systems, where one has separate containers for the resin and the catalyst and one can pre-select the mix ratio before the injection. Figure 8.29(a) shows an injection machine with simple controls from Radius Engineering Inc. [360]. An electric stepper motor drives a piston at a resin flow rate that the user defines. Resin cylinder has a pad heater to keep the resin at an elevated temperature, if desired. The control of the equipment can be done from a personal computer to adjust the process parameters such as flow rate, temperature and maximum pressure set points. Usually, the initial cost of pressure-controlled injection equipment is lower than the flow rate-controlled equipment. Figure 8.29(b) shows a pressure-controlled injection equipment from Composite Integration [361] which has two containers: one for the resin and one for

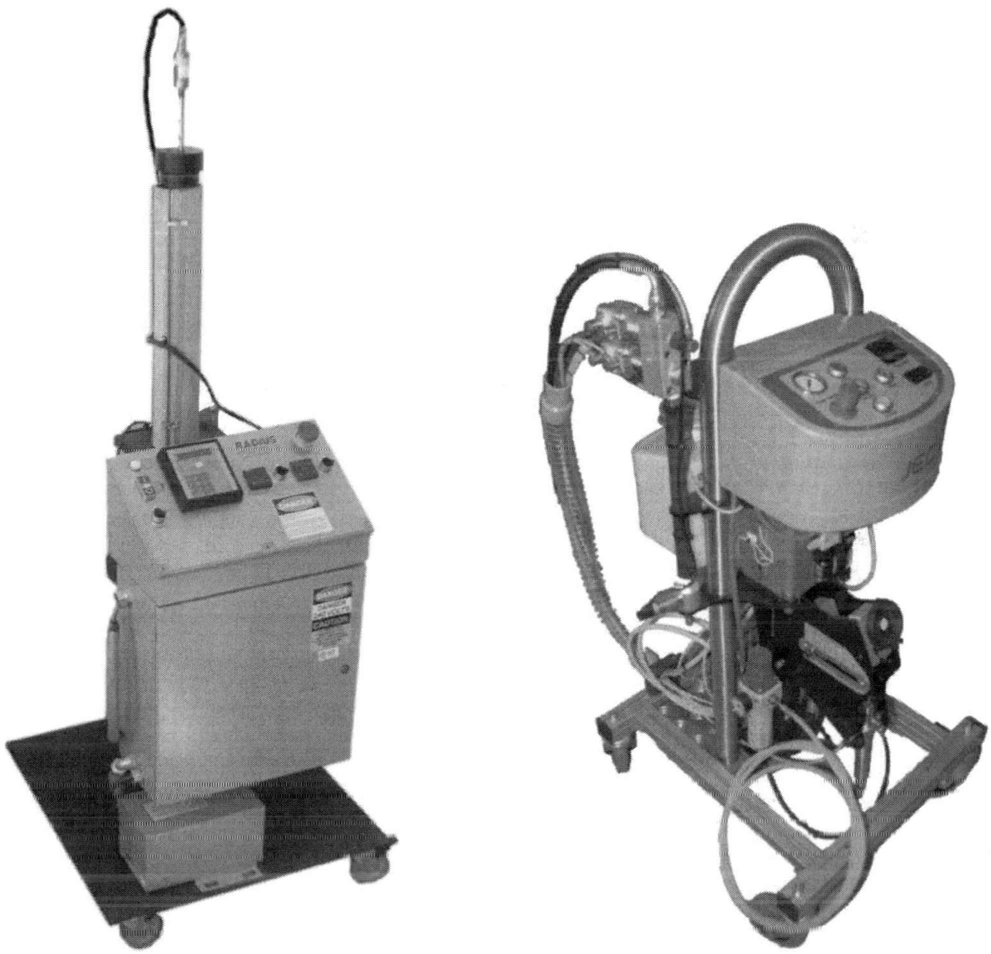

(a) (b)

Figure 8.29: (a) Flow-rate controlled RTM injection machine for pre-mixed multiple component resin systems from Radius Engineering Inc. [360] and (b) pressure-controlled RTM injection machine for single component systems from Composite Integration [361].

the catalyst, and they are mixed in-situ just before they enter into the outlet unit of the equipment.

When manufacturing small parts using RTM, resin is usually injected through a single gate. The resin may be delivered from a reservoir that is under constant pressure or injected under a constant flow rate or through a combination of the two as the mold is rigid and sealed. For larger parts made by RTM, runner systems are used to reduce the filling time or reduce injection pressure by having multiple tubes between the injection machine and the mold, instead of having a single tube and a single gate. However, while attempting to improve the process, the outcome could become worse. Design engineers must be vigilant about where the flow fronts of resin propagating from different gates meet because it could lead to entrapped air if the flow fronts do not merge in the vicinity of the vents.

For vacuum infusion processes, a vacuum pump is used that serves multiple purposes. It (i) draws the resin into the mold from a reservoir that is at room temperature and pressure, (ii) provides a good mold seal to draw most of the air out and (iii) compacts the preform in the mold as one of the surfaces is a flexible bag with vacuum pressure on the inside and atmospheric pressure on the outside. The resin in this process is usually introduced through an omega tube at one end and is vented at the other end as shown in Figure 8.24. As only one atmosphere is available, the resin flow is very slow and any mechanism to speed up the flow is highly desirable especially if one is interested in making large scale parts such as the wind blades or aerospace fuselages.

Hence SCRIMP was invented which introduced a delivery system called the flow distribution medium which is a flow enhancement layer placed on top of the fabrics as shown in Figure 8.24. This allows the resin to flood one of the faces of the preform which dramatically reduces the distance the resin has to travel to fill all the empty spaces between the fibers. However, one must be careful with the design of this delivery system as it could cause voids in the region close to the tool surface due to the transverse flow difference, i.e., the flow along the tool surface is much slower than the bag side where the highly permeable distribution medium facilitates the flow (also known as lead length). The filling time can be dramatically reduced by also introducing sequential injection lines along with the distribution media especially if the part to be infused is several meters. This introduces another decision about when to open the next injection line. Should one open it when one sees the resin on the bag side or should one wait until the resin on the tool side reaches the new injection location? Figure 8.30 and later Figures 8.37 and 8.38 show the effect of the lead length of the flow on void formation and entrapment.

8.3.3 Modeling Flow Issues in LCM

Once the mold is closed, even in one-sided mold processes, one cannot know or see if the resin has completely saturated the preform or if there are regions inside the composite with only "dry" fibers (uncovered by the resin). Sometimes even very small regions without resin can cause catastrophic failures of the composite part. Modeling of the flow will allow one to investigate the resin impregnation process and strategically design gates and vents and injection schemes to optimally fill the composite part without any dry spots. Almost all of the issues that influence flow of impregnating resin are related to the permeability in the mold. Higher permeability along the walls and corners of the mold due to unintentional gaps between the preform and the mold wall causes racetracking and changes the flow pattern dramatically. Draping the fabric over the tool surface can make the permeability change in magnitude and direction over doubly curved surfaces and introduce changes in the flow pattern of the impregnating resin. If the permeability of the fiber tows is distinctly less than the permeability between the fiber tows, it gives rise to a dual scale porous medium

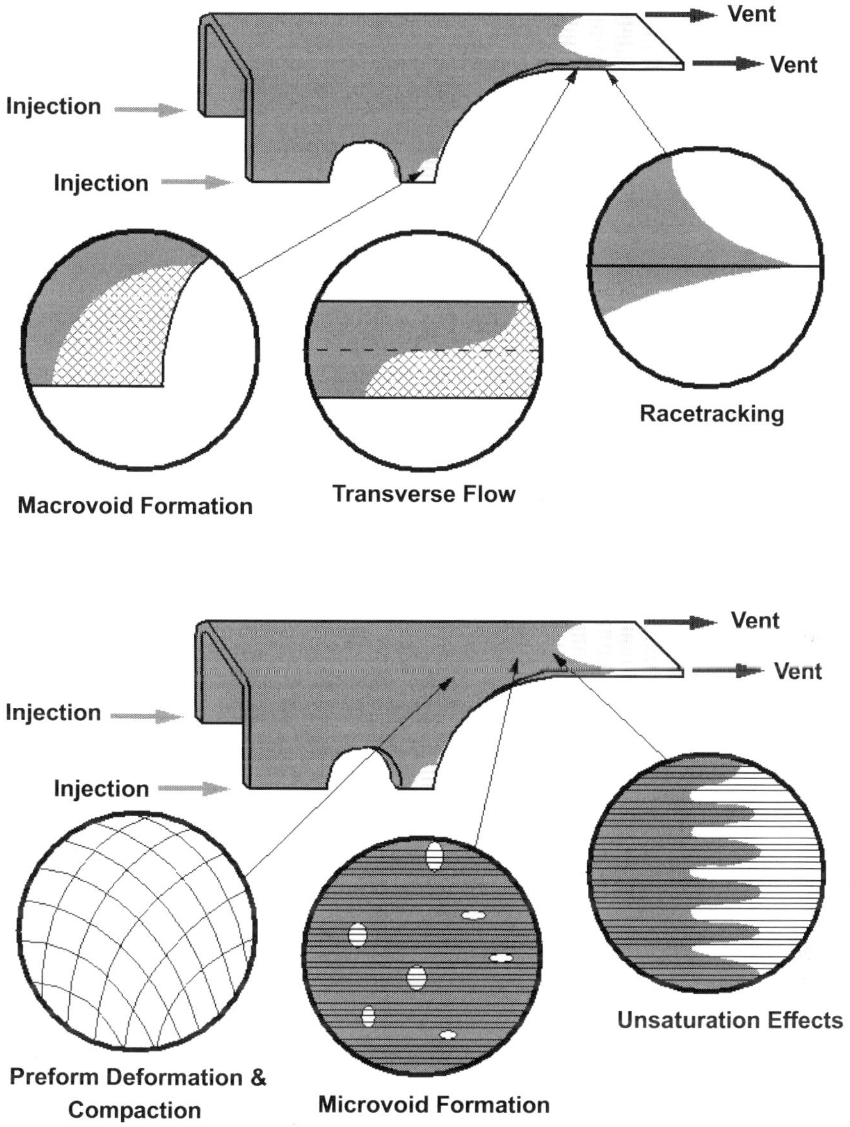

Figure 8.30: Flow issues in LCM filling process: racetracking, macrovoid formation, transverse flow, preform deformation due to draping and compaction, microvoid formation, and unsaturation effects [81].

flow in which the fiber tows may be filled (saturated) at a much slower rate than the flow in between the tows. In addition, if the flow slows down dramatically, surface tension effect of the resin can influence the microscopic flow pattern. These issues are schematically shown in Figure 8.30.

Figure 8.31 shows additional issues one must address when modeling vacuum infusion processes in Liquid Composite Molding due to use of one sided flexible mold which couples the fiber compaction with the flow and as only one atmospheric pressure can be exerted, the

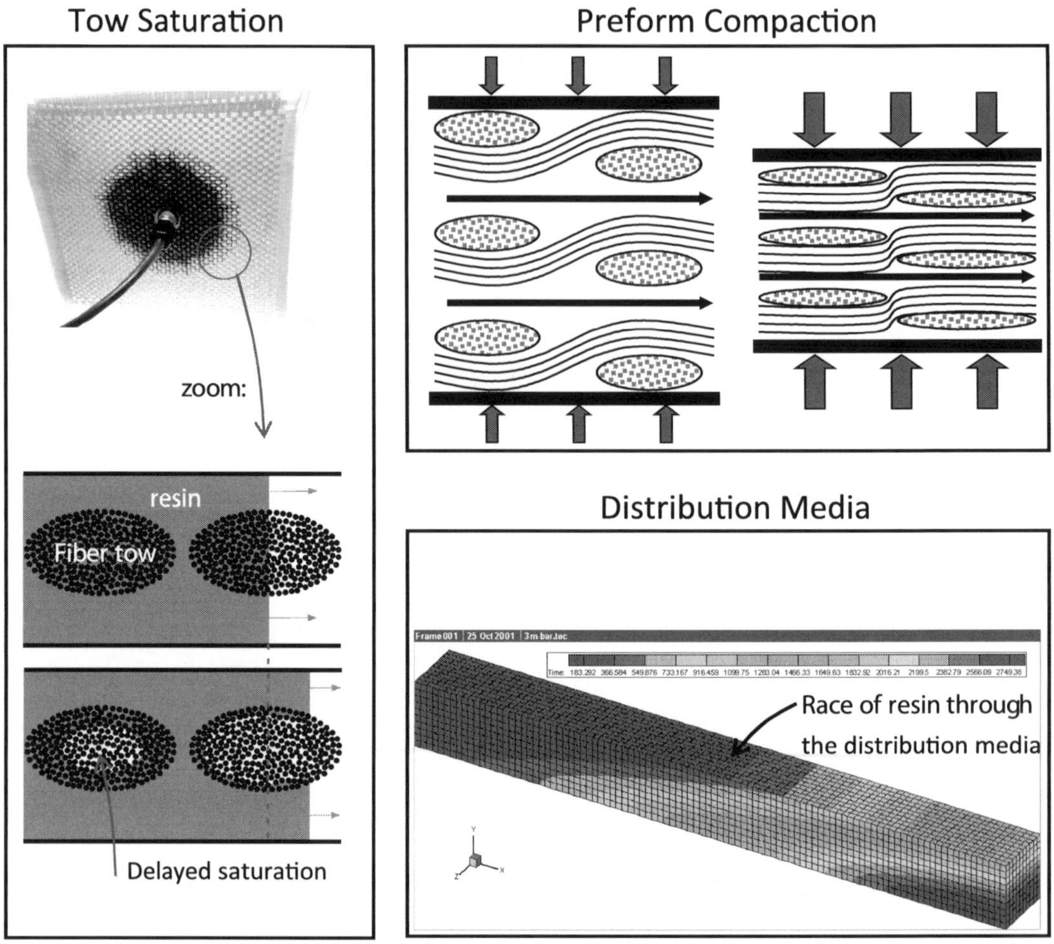

Figure 8.31: Important modeling issues for vacuum infusion processes in Liquid Composite Molding.

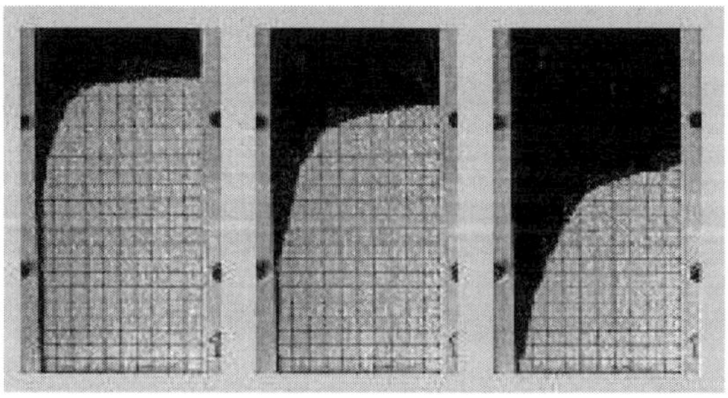

Figure 8.32: Experimental snapshots of racetracking along the left edge of a rectangular mold cavity [364, 365]. Resin enters the mold cavity from the upper edge by creating a line injection. Horizontal flow fronts would be developed if there was no racetracking.

resin impregnation in the fiber tows will be slow and delayed and it also becomes necessary to use a distribution medium to accelerate the filling. This requires one to couple the transverse flow with in-plane flow.

Thus, during the mold filling stage, these flow issues can influence the final quality of the manufactured part and the decision as to whether one can adopt the LCM process to manufacture a selected net-shape part. The flow issues arise due to significant changes in the flow pattern as a result of minor changes in the material or process parameters. Below, we discuss the most crucial ones that could be modeled and incorporated into simulations to understand and resolve such issues.

Racetracking

When the fiber preforms are placed in a mold, one may inadvertently create a small gap between the mold wall and the preform. This is mainly true for RTM type processes in which the mold is entirely from rigid tooling. The flow always uses the path of least resistance

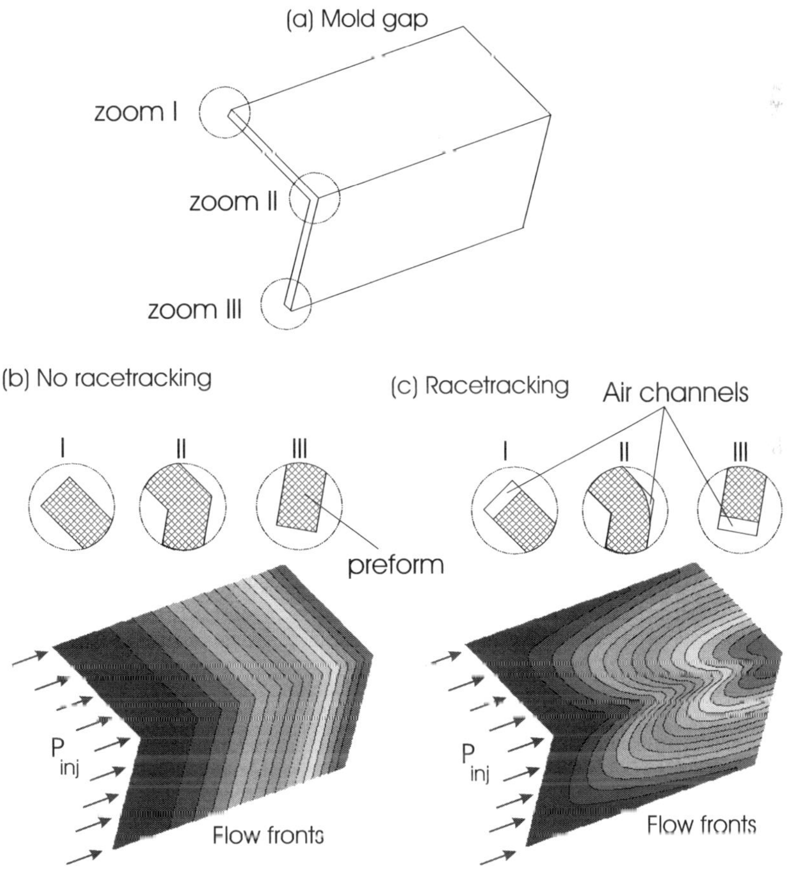

Figure 8.33: Simulated racetracking along the edges of a mold: (a) mold gap with possible regions for racetracking, (b) no low resistance areas; hence, simulation shows uniform filling pattern, and (c) racetracking along the edges changes the flow pattern significantly and can cause dry spots during manufacturing.

and will race along this edge. This phenomenon has been termed *racetracking* [81,364–370]. Figure 8.32 shows experimental snapshots of racetracking along an edge of a mold, and Figure 8.33 depicts the changes in the flow pattern due to racetracking around edges and a corner in a simulated mold filling example.

This racetracking is manifested not only along the mold wall edges and corners but also around ribs, T-joints and inserts which are usually present in complex net shaped structures. Figure 8.34 shows some examples of how racetracking is caused during molding. Tackified preforms could provide a good solution to reduce the possibility of racetracking channels or at least make such disturbances tractable. A predictable model of resin flow with racetracking along the edge can allow the process engineer to take advantage of such channels to impregnate the part. Hence modeling of such channels will be very useful.

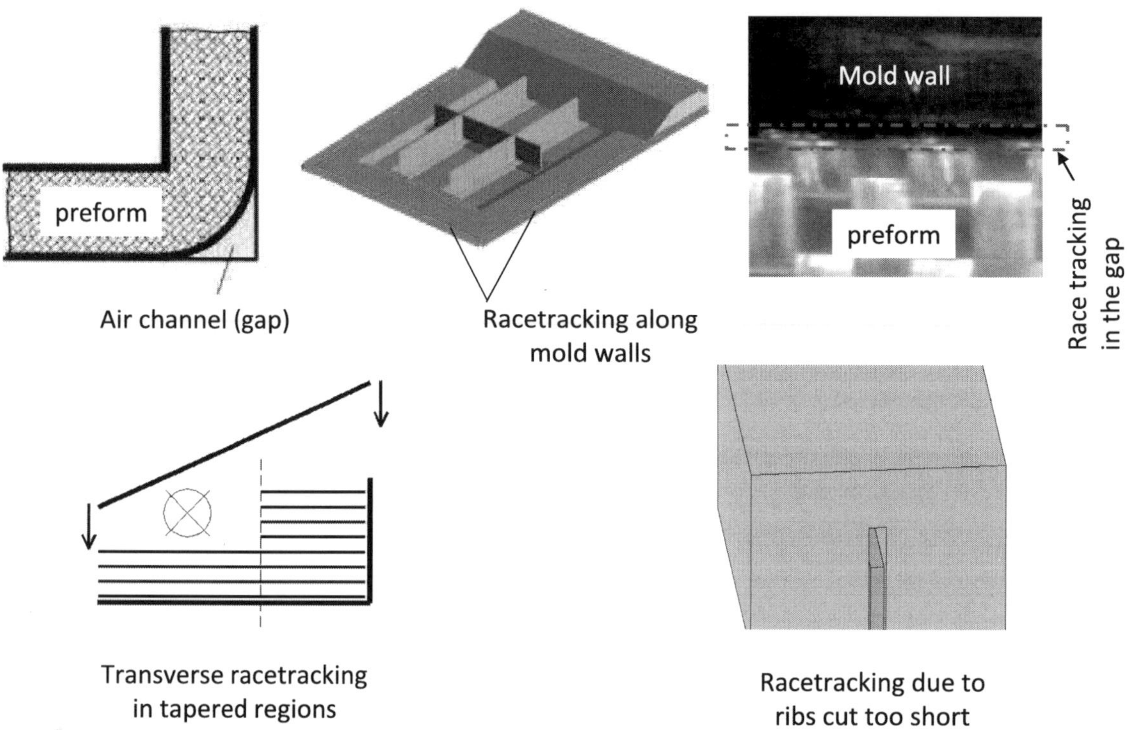

Figure 8.34: Racetracking is caused by air channels and other areas of high porosity set-up within a mold provide paths of least resistance to fluid flow, through which the fluid will flow faster [371].

Example 8.6: Permeability of a Racetracking Channel

Consider a racetracking channel as shown in Figure 8.35. The bulk permeability of the preform in the flow direction is $K_{\text{bulk}} = 1 \times 10^{-10}\,\text{m}^2$. The thickness of the mold is 1 millimeter and the channel width is, on an average, one millimeter. Estimate the permeability of the channel.

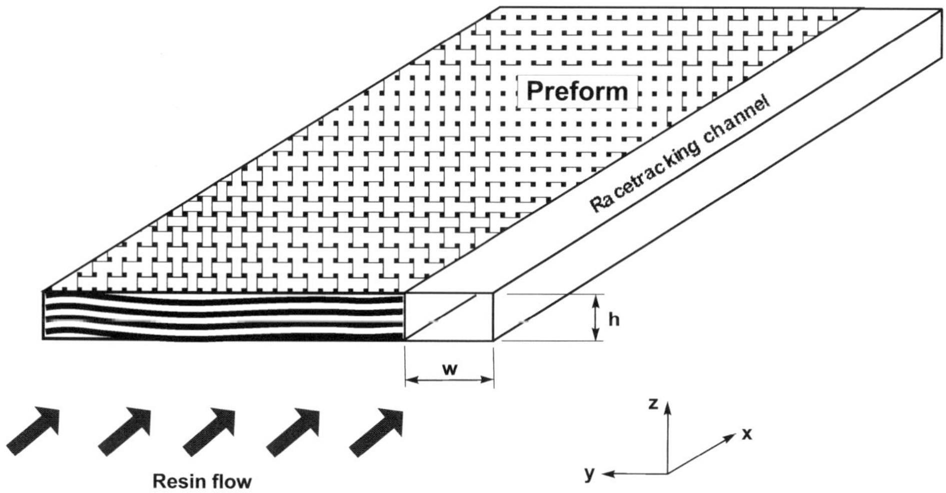

Figure 8.35: A schematic to estimate the permeability in a racetracking channel.

Solution

One can solve for fully developed flow through a rectangular channel to find the relationship between flow rate and pressure drop. For viscous resins and low Reynold's number flows, fully developed flow assumption is valid. By integrating the velocity profile in the rectangular channel area, one can find the flow rate; and dividing it by the area gives the average velocity. By comparing this average velocity with Darcy's law for one-dimensional flows, one can estimate the permeability of the channel. Two other assumptions are (i) the flow is in a channel with all solid boundaries (although in our case there is one porous boundary and one would expect slight loss of fluid in that direction) and (ii) most of the flow takes place in the flow direction, which is a reasonable assumption.

Consider a channel of height h and width w as shown in Figure 8.35. First, a simple and approximate answer will be attempted by assuming that the flow in the channel direction (which is the x direction) is a function of z only; that is, $u_x = u_x(z)$. This is a good assumption if $w \gg h$. Poiseuille flow will be induced under the applied external pressure gradient, as studied in Chapter 3. From Equations (3.83) and (3.84), the velocity and flow rate are written as follows:

$$u_x(z) = \frac{\Delta P}{\Delta x} \frac{1}{2\mu} (z^2 - hz) \tag{8.40}$$

and

$$Q_{\text{Poiseuille}} = w \int_0^h u_x(z)\, dz = -\frac{\Delta P}{\Delta x} \frac{wh^3}{12\mu} \tag{8.41}$$

where $\Delta P = P_{\text{exit}} - P_{\text{in}}$ and $\Delta x = L$ for 1D flow. One can also write the flow rate using Darcy's law:

$$Q_{\text{Darcy}} = \bar{u}_x(wh) = \left[-\frac{K_{\text{race}}}{\mu} \frac{\Delta P}{\Delta x} \right] (wh). \tag{8.42}$$

The permeability of the racetracking channel, K_{race} can be obtained by equating the two flow rates, $Q_{\text{Poiseuille}} = Q_{\text{Darcy}}$

$$K_{\text{race}} = \frac{h^2}{12}. \tag{8.43}$$

For a gap of $h = 0.001$ m, Equation (8.43) results in $K_{\text{race}} = 8.33 \times 10^{-8}\,\text{m}^2$.

To obtain a more accurate result, one can relax the assumption that $w \gg h$ as in most cases w will be less than or of the order of h. Now, as $w = O(h)$, and $u_x = u_x(y, z)$, one must solve the exact Stokes flow by applying no-slip boundary conditions on all four edges of the channel (still retaining the assumption that there is no porous wall) [81,132,372,373]. The corresponding flow rate is then calculated as [132, 365, 374]

$$Q = \int_0^h \int_0^w u_x(y,z)\,dy\,dz \;=\; -\frac{h^2}{12}\left[1 - \frac{192h}{\pi^5 w}\sum_{i=1,3,5,\ldots}^{\infty}\frac{\tanh\left(i\pi w/2h\right)}{i^5}\right]\frac{\Delta P}{\Delta x}\frac{wh}{\mu}. \quad (8.44)$$

The permeability of the channel K_{race} is obtained by equating Q in Equation (8.44) to one-dimensional Darcian flow rate, Q in Equation (8.42)

$$K_{\text{race}} = \frac{h^2}{12}\left[1 - \frac{192h}{\pi^5 w}\sum_{i=1,3,5,\ldots}^{\infty}\frac{\tanh\left(i\pi w/2h\right)}{i^5}\right]. \quad (8.45)$$

For channel dimensions of $h = w = 0.001$ m, it is calculated that $K_{\text{race}} = 3.51 * 10^{-8}\,\text{m}^2$, which is 42% of the value found by assuming $w \gg h$, which resulted in $K_{\text{race}} = \frac{h^2}{12}$. K_{race} is tabulated in Table 8.3 for different h and w values. The last column clearly shows that as the width of the channel, w increases by two or more orders of magnitude of the height of the channel, h, the permeability found using the exact Stokes equation approaches the simple one-dimensional channel flow solution which is independent of w.

Table 8.3: Permeability K_{race} of a Rectangular Racetracking Channel with a Height of h and a Width of w. All Four Walls of the Channel are Assumed to Be Nonporous.

h [m]	w [m]	K_{race} [m^2]	$\frac{K_{\text{race}}}{h^2/12}$
0.001	0.0001	7.81×10^{-10}	0.01
	0.0005	1.43×10^{-8}	0.17
	0.001	3.51×10^{-8}	0.42
	0.005	7.28×10^{-8}	0.87
	0.01	7.81×10^{-8}	0.94
	0.1	8.28×10^{-8}	0.99
0.002	0.0002	3.12×10^{-9}	0.01
	0.001	5.72×10^{-8}	0.17
	0.002	1.41×10^{-7}	0.42
	0.01	2.91×10^{-7}	0.87
	0.02	3.12×10^{-7}	0.94
	0.2	3.31×10^{-7}	0.99

Note that for h and $w = O(1)$mm, permeability of the racetracking channel, K_{race} is about two to three orders of magnitude larger than the permeability of a bulk fabric, K_{bulk} which may have a "typical" value between 1×10^{-11} and $1 \times 10^{-9}\,\text{m}^2$. ■

Macrovoid Formation

Macrovoids are formed when the air being displaced by the resin gets entrapped or when there is no sufficient resin pressure to overcome the resistance offered by the fiber preform and/or the viscosity of the resin. Figure 8.36 shows snapshots of a mold filling situation at different times in which the resin is impregnating the preform. The experiment clearly shows the formation of the dry spot or the macrovoid.

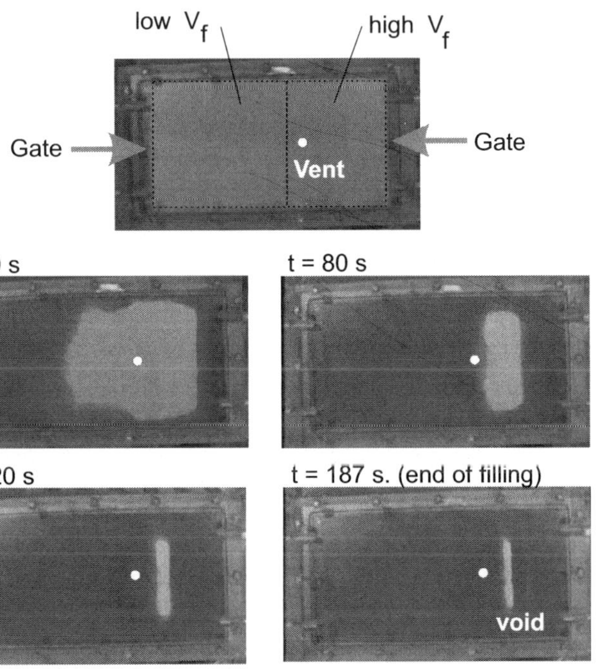

Figure 8.36: Snapshots of resin impregnation from two injection gates at different times in a mold containing fiber preforms with low and high fiber volume fraction V_f. Macrovoid or dry spot formation results due to incomplete filling [375].

These dry spots (voids) might significantly reduce the strength of the composite part depending on the location and the size of them. The size and location can usually be detected by nondestructive evaluation (NDE) methods such as ultrasound. Entrapment of dry spots is a critical issue as it will cause a high rejection rate. Flow modeling can definitely help in eliminating or reducing the size of the dryspot by identifying the last regions to fill, once the injection gate locations are specified. This will allow the mold designer to place the vent locations in the last regions to fill. The current trial and error methods allow the resin to bleed out of the vent for sufficient time in the hope that the void will get transported towards the vent and eliminated. This technique may be successful, if the void is within the proximity of the vent. However, flushing of the resin causes wastage and increases the cycle time.

Transverse Flow

As most composite structures are thin, one can ignore the flow in the thickness direction and assume an average value for it. This is a good assumption when the in-plane permeabil-

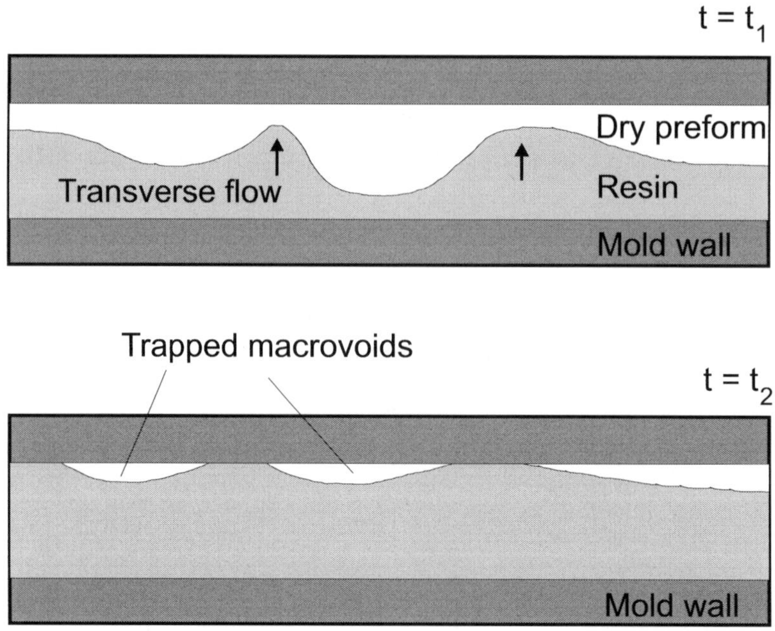

Figure 8.37: Transverse flow traps macrovoids in the part.

ities of the preform do not change by orders of magnitude through the thickness. However, transverse flow can be significant if the transverse permeability component of a fabric preform is much smaller than the in-plane components (see Example 5.12 in Chapter 5). Also, for processes such as SCRIMP and FASTRAC, the main goal in the process is to speed up the flow or reduce the pressure requirements by flooding the preform face with the resin by introducing a very high in-plane permeability material. This will introduce flow across the thickness of the composite which will be very nonuniform and may trap macrovoids as shown in Figure 8.37. In VARTM, the transverse flow is significant if $L_1 \gg L_2$ where L_1 and L_2 represent the distances between the injection gate and the flow fronts in the

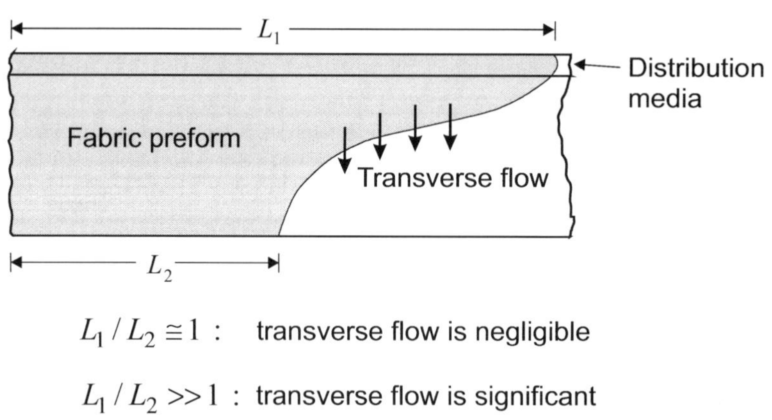

Figure 8.38: Transverse flow in VARTM/SCRIMP.

distribution medium and bulk preform, respectively, as shown in Figure 8.38.

To address this issue, one must address the flow in the transverse direction. Either analytic models [376, 377] can be developed or fully three-dimensional simulations need to be implemented [378–381]. One would also need to characterize the transverse permeability of the preform which is usually much more difficult to measure than measuring in-plane permeability components [382–384].

Dual Scale Preforms and Microvoids

Many fabrics consist of fiber tows (bundles of fibers) that are either woven or stitched. This usually will create two scales of permeabilities. The resistance to flow between the fiber tows may be different than the resistance within the tows. Two types of microvoids may be formed:

Inter-bundle microvoids: If the resin flows faster within the tows than in between the tows, it is likely to create microvoids between fiber tows when it gets wicked due to capillary action across a stitch as shown in Figure 8.39 [385].

Intra-bundle microvoids: However, if the flow is faster in between the fiber tows than within the tows, the resin will encircle itself along a stitch and entrap a microvoid within a fiber tow as shown in Figure 8.39. This issue is usually addressed on the manufacturing floor by slowing the flow down, allowing sufficient time for the regions within the tow to be impregnated.

As in most instances, the fiber tow regions are about two to three orders of magnitude less permeable than the regions in between the fiber tows. Modeling can help us understand the important parameters that influence this flow in between and within the tows which should prove useful in eliminating the microvoids. Hence, it may be important to include this phenomenon in flow modeling.

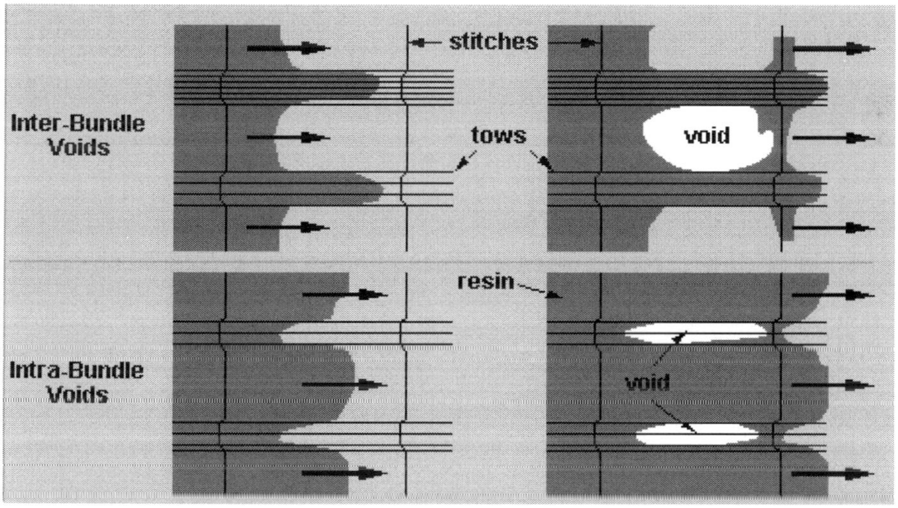

Figure 8.39: Microvoids are created because the permeability inside a fiber tow region is very different than the permeability of the region between the fiber tows in woven and stitched fabrics. (Top): Inter-bundle voids are created as resin races along the tows and encircles itself along a stitch. (Bottom): Resin flow is faster in between the tows which traps voids within the fiber tows [385].

Fiber Compaction and Flow

The flexibility of the vacuum bag in vacuum infusion processes introduce an additional aspect not present in RTM in which both the mold halves are rigid preventing any changes in the compaction of the preform once the mold is closed and sealed. In VARTM processes, the fluid pressure field that induces flow also modifies the local compaction state of the reinforcement as the flexible bag can deform. Therefore, as fluid pressure changes due to the movement of the flow front, so does the compaction pressure on the reinforcement, which leads to variations in porosity and permeability, which in turn will influence the fluid pressure. Thus, for all liquid molding processes that have a flexible bag, the flow modeling is coupled with preform deformation and needs to be addressed to predict not

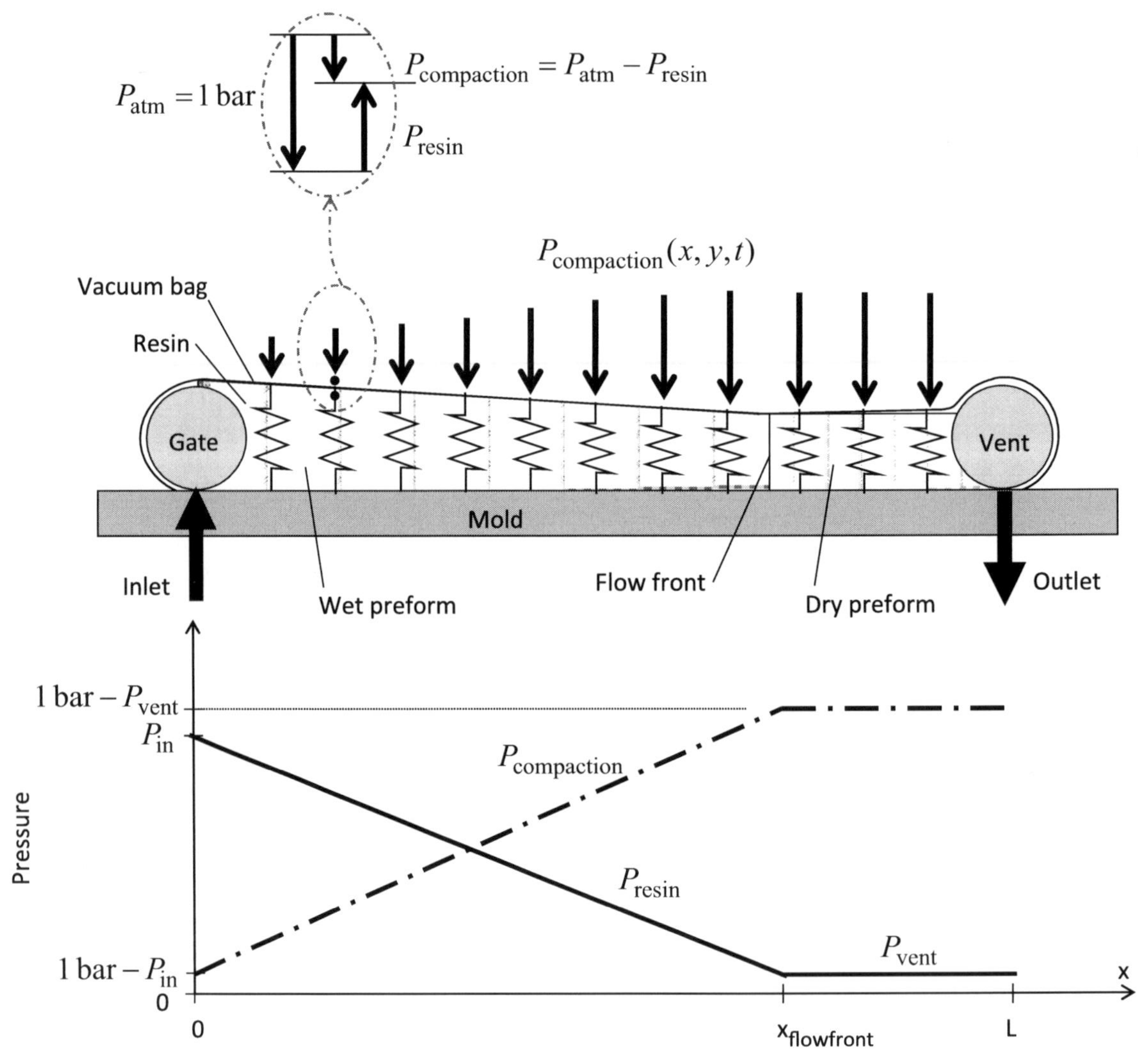

Figure 8.40: Fluid pressure changes compaction of the fabric in vacuum infusion processes that rely on flexible bag as one of the mold surfaces. Note that the linear resin and compaction pressures shown here are only valid for a 1D flow [386].

only fill times but also how this would influence the thickness and fiber volume fraction in the manufactured part. Figure 8.40 schematically shows how the two are coupled. In this figure and other parts of the text, the atmospheric pressure, P_{atm} was approximated as 1 bar ($= 100\,\text{kPa}$), however, it may slightly deviate from 1 bar depending on the location where the experiment is done; for example $P_{\text{atm}} = 1.01325$ bar at the sea level.

Heat Transfer and Cure

For LCM processes one usually aims to saturate the mold and fiber preforms with the resin before the curing reaction immobilizes the resin. Hence, it is important to fill the mold before resin cure reaches the threshold value where the viscosity starts to increase dramatically. (See Equations (4.23) through (4.31) in Chapter 4.) It is also important to know the time to demold the part which is equal to the time the part takes to reach its green strength (sufficient strength to retain its shape during and after de-molding). A very important issue is also the development of temperature and cure gradients across the thickness of the part. For thin parts of thickness less than 3 millimeters, this is less critical as one can remove the heat quickly by conduction through the surface of the part. For thicker parts, the conductive resistance increases and significant temperature gradients develop which will lead to build-up of residual stresses and eventually cause uneven shrinkage and warpage.

Example 8.7: Relationship Between Cooling Rate and Temperature Gradient Through the Thickness of a Composite

Thermal design criterion requires you to keep the temperature differential between the center and the surface of the composite to less than 10°C. You know that a cooling rate of 10 Watts/m^2 for a 3-mm thick composite was able to do so for an epoxy resin. Design and scaling requires you to quadruple the thickness of the composite. What is the maximum cooling rate you can impose if the thermal design criterion must also be met?

Solution

Under steady-state conditions (in which the temperature, $T = T(z)$ is linear in the thickness direction, z), one can show that

$$\begin{aligned} q_1 &= -k_t \frac{\partial T}{\partial n} \\ &= -k_t \frac{T_w - T_c}{h_1} = \frac{k_t}{h_1}(T_c - T_w) \end{aligned} \tag{8.46}$$

where q_1 is the cooling rate of the thin composite of thickness h_1 and k_t is the transverse conductivity of the composite. $(T_c - T_w)$ is the temperature difference between the center and wall. Thus, if the thickness is quadrupled, the cooling rate should be reduced to one-fourth to continue to maintain the thermal design criterion. Hence, the cooling rate for the thicker composite should not exceed 2.5 Watts/m^2. ■

The above example was a very crude estimate as we assumed steady state. Polymers and glass generally have a low thermal diffusivity and, hence, it takes them much longer to reach steady state as compared to the processing time. To obtain more accurate predictions, it is useful to solve the governing equations with numerical methods to obtain more accurate predictors. Example 5.5 in Chapter 5 shows how one could solve for the transient problem to obtain a better prediction of the temperature field.

In the last two decades, numerical simulation codes have been developed that incorporate the process physics of mold filling and cure and thus allow composite process design engineers to understand the physics and use the simulation as a design tool for prototype development instead of using trial and error manufacturing approaches. The engineers can change the process parameters in the simulation and evaluate the influence of the parameters on the simulation results in order to search for the optimal material and process parameters such as locations of the injection gates and vents, injection pressure or flow rate, cooling rate, resin system, fabric type and fiber volume fraction.

The focus of the next section is to model the impregnation of resin into the fabric preform, which is usually called the *filling* stage of LCM, and also develop the non-isothermal formulation with cure kinetics. The filling stage occurs in all LCM processes with slight variations. However, in most cases the resin propagates through a stationary fabric preform which is modeled as a porous medium. The filling model is based on Darcy's law which was introduced in Chapter 4.

8.3.4 Process Models

The objective of all LCM processes is to manufacture a composite part with no regions that contain voids or dry spots. Most applications, in addition, require a short production cycle and low manufacturing costs in order to be competitive with other manufacturing techniques and materials. To achieve these goals, the fabric, the resin system, the tool and the injection system should be selected such that the filling and the curing stages require short cycle times with low rejection rate due to dry spots.

Flow and cure modeling will allow one to control the process parameters and design the tool accordingly in order to achieve these goals. The process parameters that can be easily controlled and influence the process are location of the injection gates and vents and the flow rates or injection pressures to be applied at these locations. For nonisothermal injections, initial resin temperature and mold temperature will also play an important role. The curing process, as in the autoclave process, is decided by the mold wall temperatures and the reaction kinetics.

Permeability of the fabric and reaction kinetics will be important constitutive equations for realistic modeling and simulation of the process and must be characterized as outlined in Chapter 4.

Since the late 1980s, mathematical models for resin flow through reinforcing porous preforms and for the resin cure along with simulation codes based on the solution to these models have been developed. These models have been useful to manufacturing engineers in coupling tool design and process parameters for efficient manufacturing. In the next two subsections, we will formulate the flow, heat transfer and cure models. Many book chapters have also delineated the details [80, 246, 357, 379].

8.3.5 Resin Flow

As discussed in Chapter 4, Darcy studied liquid flow through a sand column, and put forth what is now known as Darcy's law. Darcy's law adequately models most LCM applications in which the Reynolds number is small. This allows us to ignore the inertial forces and assume a quasi-steady state process. The reinforcing fabric preform within the mold cavity is considered as porous medium, and the relationship between the resin pressure and its average velocity is given as follows [387, 388]:

$$\bar{\mathbf{u}} = -\frac{\mathbf{K}}{\mu} \cdot \nabla P \tag{8.47}$$

where $\bar{\mathbf{u}}$ is the volume averaged Darcy velocity, μ is the viscosity of the fluid, ∇P is the pressure gradient, and $\mathbf{K}$ is the permeability tensor of the preform. In expanded form, Darcy's law can be rewritten as

$$\begin{pmatrix} \bar{u}_x \\ \bar{u}_y \\ \bar{u}_z \end{pmatrix} = -\frac{1}{\mu} \begin{pmatrix} K_{xx} & K_{xy} & K_{xz} \\ K_{yx} & K_{yy} & K_{yz} \\ K_{zx} & K_{zy} & K_{zz} \end{pmatrix} \begin{pmatrix} \partial P/\partial x \\ \partial P/\partial y \\ \partial P/\partial z \end{pmatrix}, \tag{8.48}$$

where $\bar{u}_x$, $\bar{u}_y$ and $\bar{u}_z$ are the resin velocity components and K_{ij} are the permeability tensor components. The details on measurement and prediction of the permeability components were discussed in Chapters 4 and 5. Variations of RTM process such as SCRIMP involve three-dimensional resin flow, as the high permeability distribution medium placed on top of the preform creates a large transverse gradient flow through the thickness as shown in Figure 8.38. However, resin flow in most LCM applications can be approximated by two-dimensional flow because the transverse flow in the thickness direction is only of the order of a few thicknesses as seen from the example in Chapter 5 and the in-plane dimensions are usually two to three orders of magnitude larger than the thickness. Approximation of the flow in three-dimensional geometry with two-dimensional flow brings about significant savings in computational time and reduces the number of material parameters that one must measure or calculate. By ignoring the flow in the thickness direction, the two-dimensional form of Equation (8.48) can be rewritten as

$$\begin{pmatrix} \bar{u}_x \\ \bar{u}_y \end{pmatrix} = -\frac{1}{\mu} \begin{pmatrix} K_{xx} & K_{xy} \\ K_{yx} & K_{yy} \end{pmatrix} \begin{pmatrix} \partial P/\partial x \\ \partial P/\partial y \end{pmatrix}. \tag{8.49}$$

This assumption is a very good one as long as in-plane permeabilities of the preforms that are stacked in the thickness direction are about the same order of magnitude. Note that in SCRIMP, the permeability of the distribution medium may be two to three orders of magnitude larger than the preform permeability in the in-plane direction; therefore, the two-dimensional assumption is not suitable. Thus, for this process one has to be ready to accept a larger error in the calculations or resort to a fully three-dimensional flow model. The computational time for the three-dimensional model would increase by an order of magnitude and one would require data regarding the transverse permeability of the preform. Also, as we discussed the flow and fiber compaction are coupled and may prove crucial in some of the processes. We will address this issue in a later section.

In most LCM processes there is insignificant transverse flow; hence, they lend themselves to two-dimensional flow in the plane of the preform. Hence, we will restrict ourselves to discussing the solution of the mathematical model that will be based on a two-dimensional case for simplicity. The extension to three dimensions is straightforward and does not involve any conceptual problem.

The equation of mass conservation is obtained by the divergence of velocity as

$$\frac{\partial}{\partial x}(\bar{u}_x) + \frac{\partial}{\partial y}(\bar{u}_y) = 0. \tag{8.50}$$

We also assume here that the tows impregnate at the same rate as the space in between the tows. Substitution of the velocity terms from Equation (8.49) in terms of resin pressure P yields the following second order partial differential equation for the resin flow:

$$\frac{\partial}{\partial x}\left(\frac{K_{xx}}{\mu}\frac{\partial P}{\partial x}\right) + \frac{\partial}{\partial x}\left(\frac{K_{xy}}{\mu}\frac{\partial P}{\partial y}\right) + \frac{\partial}{\partial y}\left(\frac{K_{yx}}{\mu}\frac{\partial P}{\partial x}\right) + \frac{\partial}{\partial y}\left(\frac{K_{yy}}{\mu}\frac{\partial P}{\partial y}\right) = 0. \tag{8.51}$$

This governing equation is solved to calculate the resin pressure in the resin domain that has a moving free surface. The solution domain changes continuously. However, because of the "quasi-steady state" assumption, one solves for the pressure in the resin occupied domain at that instant in time. For open mold processes such as VARTM and SCRIMP, there is a dynamic distribution of pressure between the preform and the resin which may change the fiber volume fraction and the permeability due to the presence of the resin just as depicted conceptually in Figure 8.5 for autoclave processing. More details on this can be found in references [389–392].

With known material parameters of viscosity and permeability, which may vary with position, this governing differential equation with boundary conditions along the boundary of the resin domain has a unique solution. The boundary conditions are written for three segments of the boundary: (i) at injection gate(s), (ii) on the free surface (flow front), and (iii) along the mold wall. Usually, the injection system will inject the resin into the mold under constant pressure or constant flow rate:

$$P = P_{\text{inj}} \tag{8.52a}$$

or

$$A\left(-\frac{\mathbf{K}}{\mu}\cdot\nabla P\right)\cdot\mathbf{n} = -Q_{\text{inj}} \tag{8.52b}$$

where P_{inj} and Q_{inj} are the injection pressure and flow rate, respectively, and A is the injection cross-sectional area with an outward normal vector $\mathbf{n}$. Some injection equipment allows the users to vary the value of P_{inj} or Q_{inj} during the injection. This translates into the boundary conditions as replacing P_{inj} by $P_{\text{inj}}(t)$ and Q_{inj} by $Q_{\text{inj}}(t)$. This does not influence the solution technique as one solves the steady-state problem for the pressure at each time interval. The pressure of the free resin front is equal to the ventilation pressure, which is atmospheric or vacuum pressure (i.e., $P_{vent} = P_{atm}$ or $P_{vent} = P_{vacuum}$) depending on whether it is the RTM or VARTM process:

$$P_{\text{ff}} = \begin{cases} P_{\text{vent}}, & \text{if } P \text{ is absolute pressure,} \\ 0, & \text{if } P \text{ is gage pressure} \end{cases} \tag{8.53}$$

where P_{ff} and P_{vent} are the flow front pressure and vent pressure, respectively. Along the in-plane mold wall boundary, the normal component of velocity should be zero if there is no resin leakage:

$$\bar{u}_n = -\frac{1}{\mu}\left(K_{nn}\frac{\partial P}{\partial n} + K_{nt}\frac{\partial P}{\partial t}\right) = 0, \tag{8.54}$$

where n and t denote the directions normal and tangential to the mold wall, respectively. These three types of boundary conditions are illustrated in Figure 8.41.

It is possible that due to the presence of inserts or multiple gates, multiple resin fronts may result. When these fronts meet and are not linked to any vent, the air that they displace gets entrapped in the regions encompassed by the resin flow fronts. For such cases, the boundary condition given in Equation (8.53) can be modified to include the increase in the flow front pressure due to the increase in the void pressure. The entrapped air may be assumed to behave as an ideal gas satisfying

$$\frac{P_{\text{void}}V_{\text{void}}}{T_{\text{void}}} = \text{constant} \tag{8.55}$$

where P_{void}, T_{void} and V_{void} are the pressure, temperature and volume of the air enclosed [393]. However, this would require one to keep track of the void created by merging

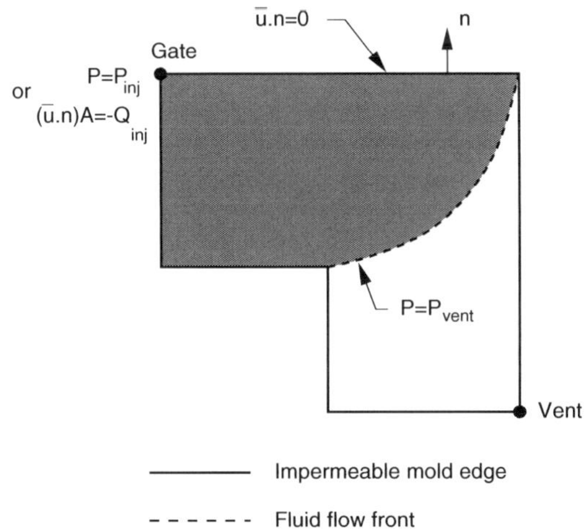

Figure 8.41: Boundary conditions for resin flow in a mold.

flow fronts and calculate its volume at each instant of filling as shown in Figure 8.42. In reality some of this air will dissolve in the resin, reducing the void pressure. To model this phenomenon, one may have to include a mass diffusion equation at the interface of the resin and the air, and characterize the diffusion coefficient. As this is usually not a significant factor in changing the flow front pattern, most modelers ignore this issue and assume that all air will dissolve in the resin and use Equation (8.53).

For preforms in which the fiber bundles impregnate at a slower rate than the flow in between the preforms as shown in Figure 8.43, one can model this effect by introducing a sink term in the governing equation. In order to model this phenomenon, a sink term is added to the mass conservation equation (see Equation (3.8) in Chapter 3): $\nabla \cdot \mathbf{U} = -s$. Or in expanded form, Equation (8.51) is replaced with the following:

$$\frac{\partial}{\partial x}\left(\frac{K_{xx}}{\mu}\frac{\partial P}{\partial x}\right)+\frac{\partial}{\partial x}\left(\frac{K_{xy}}{\mu}\frac{\partial P}{\partial y}\right)+\frac{\partial}{\partial y}\left(\frac{K_{yx}}{\mu}\frac{\partial P}{\partial x}\right)+\frac{\partial}{\partial y}\left(\frac{K_{yy}}{\mu}\frac{\partial P}{\partial y}\right)=s. \tag{8.56}$$

Here, s is the rate at which the fluid is impregnating inside the fiber tows, and should be proportional to the permeability of the fiber tows. For typical LCM fabrics, individual tow permeability is usually much lower than the overall fabric permeability. Besides the tow permeability, the value of s is also dependent on pressure outside the tows, the geometry of the tows and the density of these tows. For a circular tow, s can be derived as follows [395–398]:

$$s=\left[\frac{\epsilon_i}{\left(\frac{r_{2a}}{r_{1a}}\right)^2-1}\right]\frac{2r_{fa}(-\dot{r}_{fa})}{r_{1a}^2} \tag{8.57}$$

where

$$\frac{dr_{fa}}{dt}=\dot{r}_{fa}=-\frac{K_{it}(P_0+P_c-P_{\text{vac}})}{\epsilon_i\mu r_{fa}\ln\left(\frac{r_{1a}}{r_{fa}}\right)}. \tag{8.58}$$

Here, P_0, P_c, and P_{vac} are the free surface, capillary and vacuum pressures, respectively, ϵ_i is the porosity of the inner region (intratow), r_{fa} is the microfront radius inside the tow,

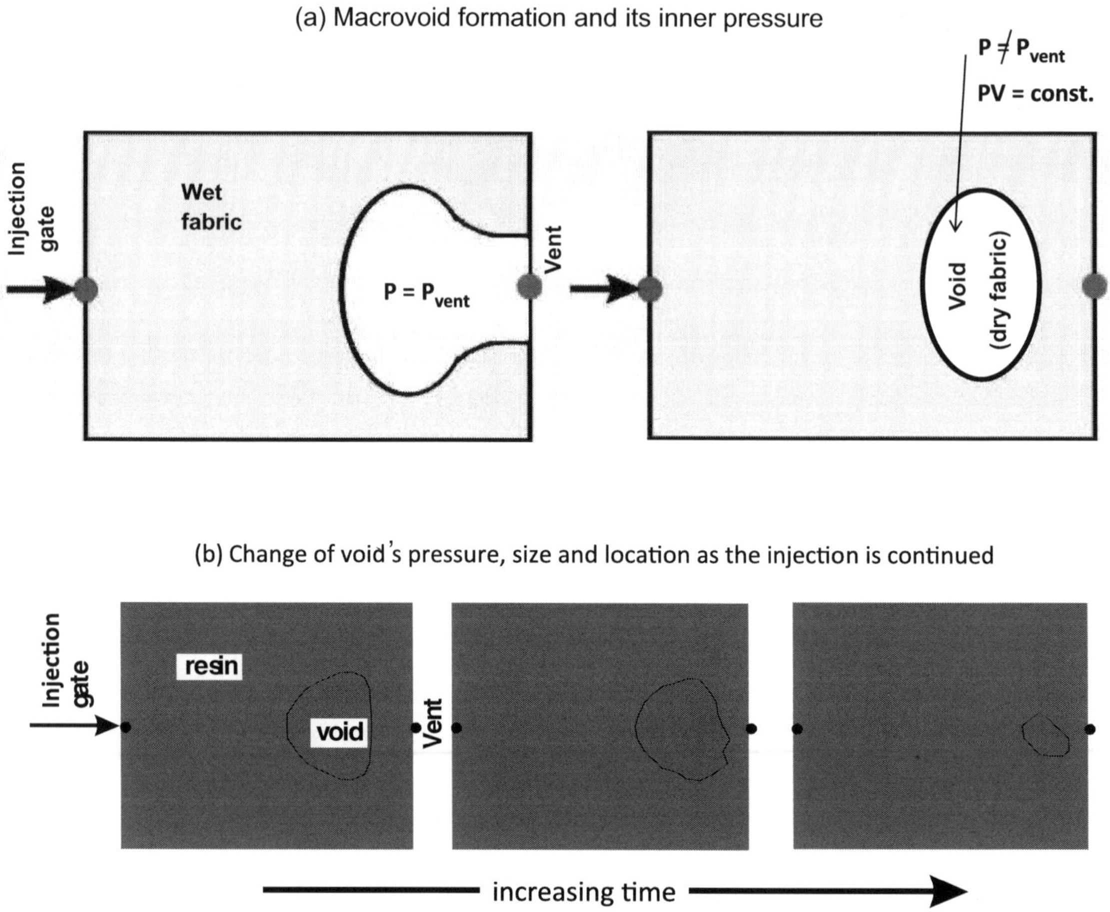

Figure 8.42: Applying boundary condition at the flow front not connected to the vent that can model air entrapment [394].

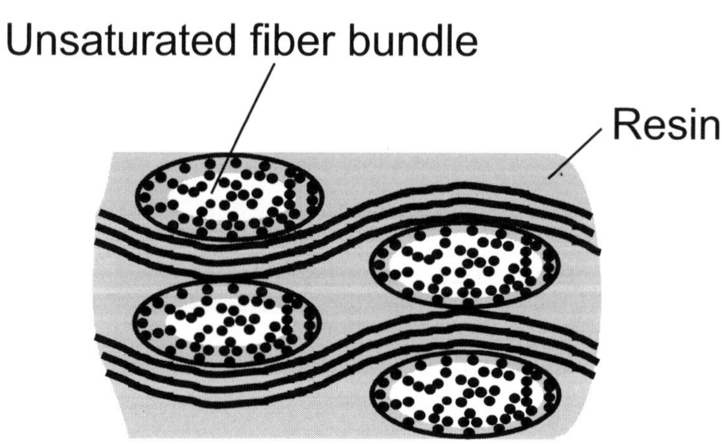

Figure 8.43: Fiber tows or bundles are impregnated much more slowly than the regions in between the fiber tows.

r_{1a} and r_{2a} are the tow and the cell (the circular computational domain with the tow and surrounding air channel) radii, K_{it} is the fiber tow permeability (see Figure 8.44).

Thus, one would have to solve Equations 8.51 and 8.56 simultaneously to solve for the macroflow front and the saturation of each tow as the mold filling proceeds. This usually will modify the pressure profile under constant injection conditions [399, 400].

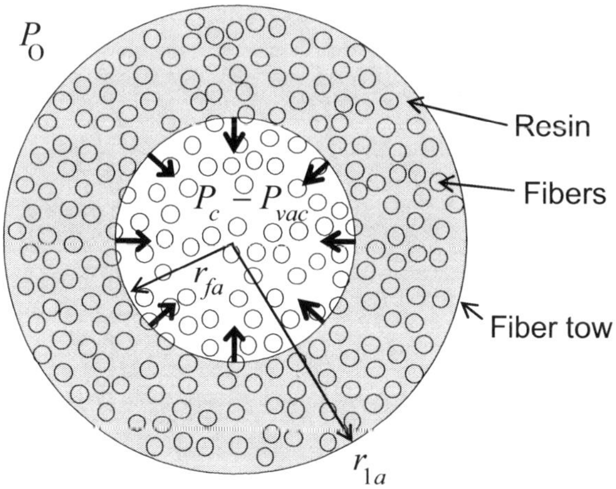

Figure 8.44: Parameters to model the sink effect during impregnation of a fiber tow with resin; the tow has a permeability of K_{it}.

Example 8.8: Influence of Constant Sink on Flow Front in One-Dimensional Flow

Consider one-dimensional flow in a preform.

(a) When one injects resin under a constant flow rate, one would expect to measure a linear rise in the pressure with time at the injection location. This is in accordance with Darcy's law and the assumption that the preform region that consists of spaces in between the fiber tows, and the fiber tows saturate instantly and at the same time (single scale preforms). Find the permeability of this single scale preform in the flow direction using the information of the injection pressure that one can gather with time.

(b) Now, assume that the preform is a dual scale preform, in which the spaces between the fiber tows saturate instantly but the fiber tows continue to absorb resin at a constant rate even after the flow front has gone past it. How would you expect the pressure curve to behave if the resin is injected under constant flow rate conditions?

Solution

(a) In Chapter 4, the relation between the permeability of a single scale preform and the injection pressure was derived in detail under the conditions when resin is injected at constant flow rate in one dimensional flow. We will rewrite the final equations for completeness of this problem:

$$P_{\text{inj}} = \frac{\mu}{\phi K_{xx}} \left(\frac{Q}{wh} \right)^2 t. \tag{8.59}$$

Thus,

$$K_{xx} = \frac{\mu}{\phi \dfrac{dP_{\text{inj}}}{dt}} \left(\frac{Q}{wh} \right)^2. \tag{8.60}$$

Here Q is the flow rate, w and h are the preform width and thickness, ϕ $(= 1 - V_f)$ is the porosity of the preform, $\frac{dP_{\text{inj}}}{dt}$ is time derivative of the injection pressure which is evaluated by using experimental data. If the fiber tows are saturated at the same time as the space in between the tows, then the injection pressure P_{inj} will increase linearly with time in one-dimensional flow (assuming line injection and no racetracking). Hence, experimentally measured $\frac{dP_{\text{inj}}}{dt}$ will be constant and Equation (8.60) will allow one to find K_{xx}.

(b) If the resin flow is in a fabric that exhibits dual scale, one-dimensional conservation of mass equation, $\frac{d\bar{u}_x}{dx} = 0$ is replaced with

$$\frac{d\bar{u}_x}{dx} = -s. \tag{8.61}$$

Here s corresponds to the sink term due to delayed saturation of fiber bundles compared to the empty spaces between the surrounding fiber tows. If s is a constant as stated in the problem, one can integrate both sides of Equation (8.61) with respect to x, and apply the boundary condition at the injection gate, $\bar{u}_x(x = 0) = Q/(wh)$, which then results in the following expression for the resin velocity as a function of x:

$$\bar{u}_x = -sx + \frac{Q}{wh}. \tag{8.62}$$

One-dimensional Darcy's law relates the resin velocity also to the pressure gradient through the resin viscosity and the preform permeability in the flow direction, K_{xx}

$$\bar{u}_x = -\frac{K_{xx}}{\mu}\frac{dP}{dx}. \tag{8.63}$$

Equating the two $\bar{u}_x$'s in Equations (8.62) and (8.63) yields $dP/dx = \frac{\mu}{K_{xx}}(sx - \frac{Q}{wh})$. One can obtain $P(x,t)$ by integrating both sides of this equation with respect to x, and applying the boundary condition at the injection gate, $P(x = 0, t) = P_{\text{inj}}(t)$

$$P(x,t) = \frac{\mu}{K_{xx}} \left[\frac{sx^2}{2} - \frac{Qx}{wh} + \frac{P_{\text{inj}}(t) K_{xx}}{\mu} \right]. \tag{8.64}$$

The boundary condition at the moving resin flow front, $x = x_f$, yields the following:

$$P(x = x_f, t) = 0 = \frac{\mu}{K_{xx}} \left[\frac{sx_f^2}{2} - \frac{Qx_f}{wh} + \frac{P_{\text{inj}}(t) K_{xx}}{\mu} \right] \tag{8.65}$$

assuming that $P(x = x_f, t) = P_{\text{vent}} = 0$ or considering that P is a gage pressure. One can rewrite Equation (8.65) in order to obtain a relation between K_{xx} and other process parameters and variables

$$K_{xx} = \frac{\mu}{P_{\text{inj}}(t)} \left[-\frac{sx_f^2(t)}{2} + \frac{Qx_f(t)}{wh} \right]. \tag{8.66}$$

One needs to write $x_f(t)$ explicitly in order to use Equation (8.66). This is shown below

$$\begin{aligned} x_f(t) &= \int_0^t v_f(t')\,dt' \\ &= \int_0^t \frac{\bar{u}_x}{\phi}\,dt' \\ &= \frac{1}{\phi}\int_0^t \left[-sx_f(t') + \frac{Q}{wh}\right] dt' \qquad \text{from Equation (8.62)} \\ &= \frac{Q}{wh\phi}\,t - \frac{s}{\phi}\int_0^t x_f(t')\,dt'. \end{aligned} \tag{8.67}$$

$x_f(t)$ appears on both hand sides of Equation (8.67). In order to write $x_f(t)$ explicitly, the Laplace transform of Equation (8.67) will be taken as follows:

$$\bar{x}_f(\rho) = \frac{A}{\rho^2} - \frac{B\bar{x}_f(\rho)}{\rho} \tag{8.68}$$

where $A = Q/(wh\phi)$ and $B = s/\phi$ were used for the sake of simplicity. In Equation (8.68), $\bar{x}_f(\rho)$ is the Laplace transform of $x_f(t)$, and ρ is the Laplace variable. Rearranging Equation (8.68) by using partial fractions, one can obtain the following:

$$\bar{x}_f(\rho) = \frac{A}{\rho(\rho+B)} = \frac{A}{B}\left(\frac{1}{\rho} - \frac{1}{\rho+B}\right). \tag{8.69}$$

One can now take the inverse Laplace transform of Equation (8.68) which results in the following:

$$\begin{aligned} x_f(t) &= \frac{A}{B}\left(1 - e^{-Bt}\right) \\ &= \frac{Q}{swh}\left(1 - e^{-st/\phi}\right). \end{aligned} \tag{8.70}$$

Finally, one can substitute this expression into Equation (8.66)

$$K_{xx} = \frac{\mu Q^2}{sw^2h^2P_{\text{inj}}(t)}\left[\left(1 - e^{-st/\phi}\right) - \frac{1}{2}\left(1 - e^{-st/\phi}\right)^2\right]. \tag{8.71}$$

Preform permeability K_{xx} along the flow direction is calculated by substituting injection pressure at any time (except at $t = 0$) into Equation (8.71). At $t = 0$, that expression is 0/0 and singular. As a good engineering practice, K_{xx} should be evaluated at several time values, and an average should be taken due to minor variabilities in s, K_{xx}, Q, and racetracking effects. Figure 8.45 illustrates some typical plots of $x_f(t)$ and $P_{\text{inj}}(t)$ with and without the effect of dual scale flow. In reality, s is not a constant but is coupled to the flow outside the tows because of the variation in resin pressure. The resulting equations are nonlinear and require numerical methods to solve them. ■

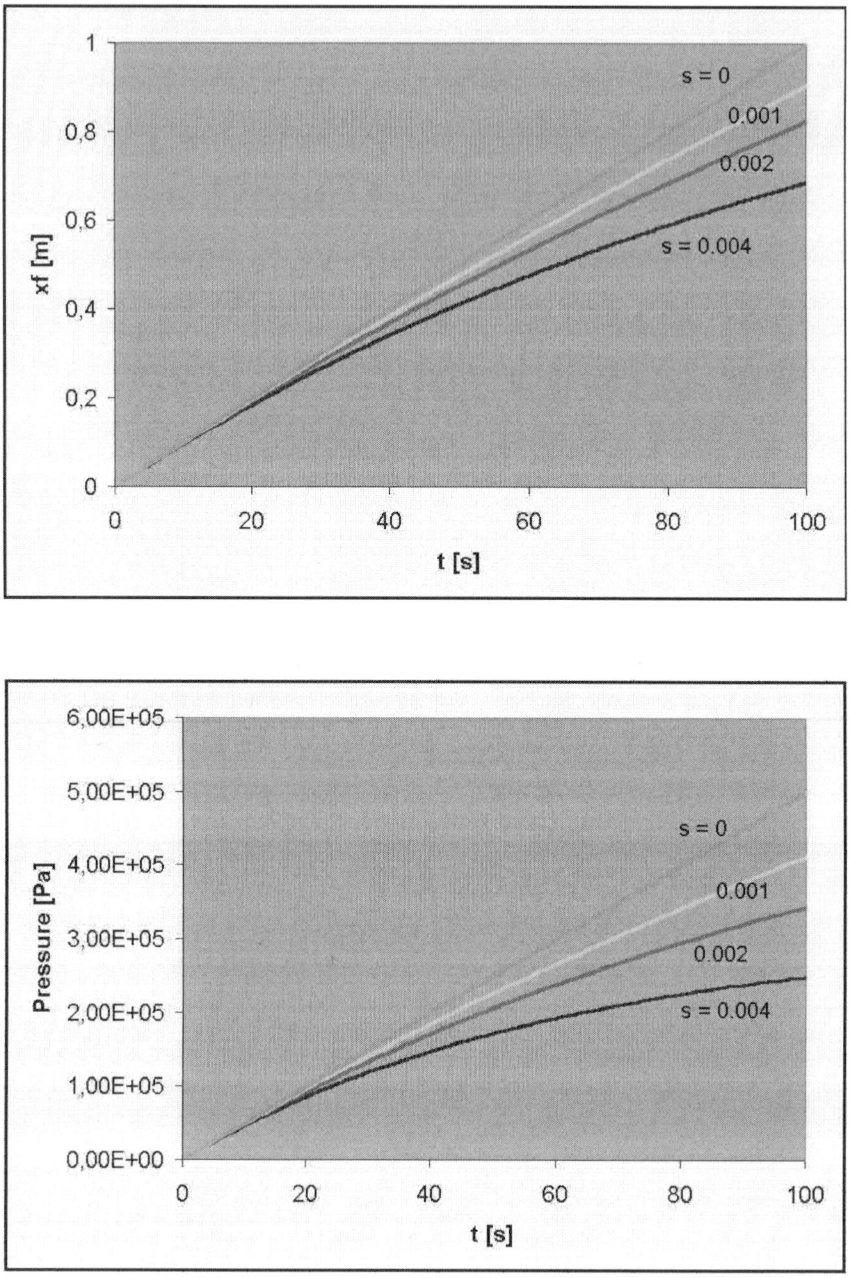

Figure 8.45: Flow front $x_f(t)$ and injection pressure $P_{\text{inj}}(t)$ as a function of time for four different values of sink term, $s = 0$, 0.001, 0.002 and 0.004 s^{-1}. Preform width $w = 0.1$ m, thickness $h = 0.01$ m, porosity $\phi = 0.5$, permeability $K_{xx} = 1 * 10^{-9}\,\text{m}^2$, resin viscosity $\mu = 0.1$ Pa.s, and flow rate $Q = 5 * 10^{-6}\,\text{m}^3/\text{s}$ were used in the calculations. As $s \to 0$, both $x_f(t)$ and $P_{\text{inj}}(t)$ become linear with time as one would expect for single scale porous media.

8.3.6 Heat Transfer and Cure

There are many resins, such as vinylester, that cure at room temperature. Hence, it may not be necessary to model nonisothermal mold filling for such resins. However, these are usually single component resins in which the second component to initiate the reaction is added. Hence, the cure will initiate with time. The goal in such cases is to fill the mold before the resin cure reaches a value where the viscosity of the resin starts to increase dramatically. However, as the resin continues to cure from the initiation of the process as demonstrated by Figure 8.46, the resin entering the mold will be at different degrees of cure; thus, the viscosities will be different and will be functions of time. Usually, this effect is ignored as the degree of cure during mold filling is minimal, and the increase in viscosity is usually of the order of the noise in the system and assumptions made in the model. Hence, unless we have a strong reason to believe that this is a primary influence, one can safely ignore the cure during the filling and focus on isothermal filling. The cure can then be separately handled in a similar form to autoclave cure, except the boundary conditions may be different.

For resins that need to be heated to initiate cure, usually warm or cold resin is injected into a hot mold containing heated preforms. In such cases the heat is transferred from the mold and the preforms to the resin as the resin flows into the mold. As the resin temperature increases, the resin viscosity decreases. However, higher temperature of the resin will initiate and accelerate its cure and as the degree of cure increases, the resin viscosity will increase. Usually the viscosity due to the cure can increase by orders of magnitude whereas the drop in viscosity due to increase in temperature for thermosets is only a few percent. To model the change in the viscosity due to temperature and cure kinetics in the flow equation, one must also characterize the viscosity dependence on these functions as discussed in Chapter 4.

The temperature and degree of cure will be nonuniform in the mold, as the heat dis-

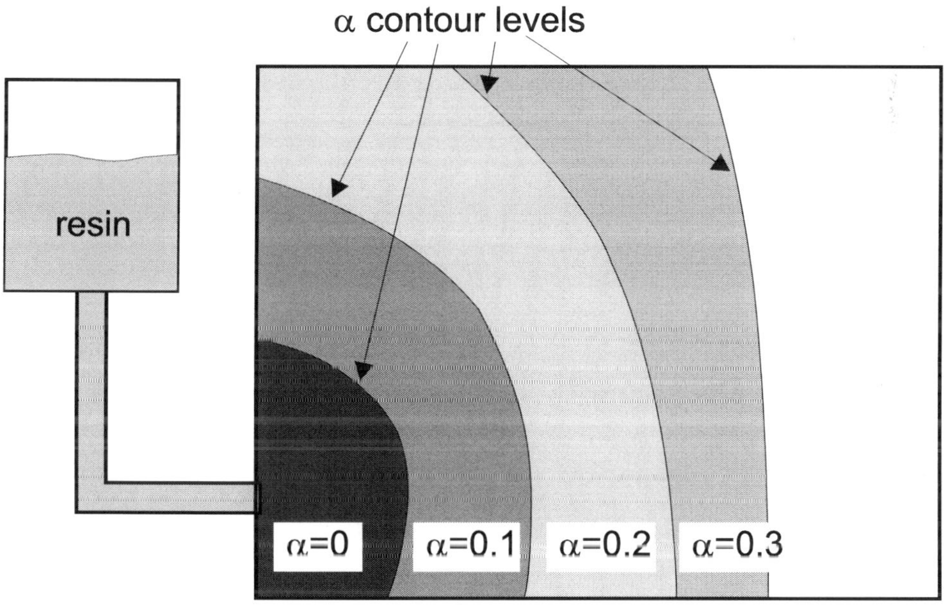

Figure 8.46: Schematic of resin cure during the mold filling stage. α is the degree of cure [379].

tribution is nonuniform due to convection of the resin and also due to conduction of the heat through the mold walls and the preform. Also, if one uses a multi-component injection system then the initial concentration of the catalyst may be different for different regions in the resin, initiating different rates of cure. The important processing step here is to design heating systems to prevent early gelation of resin and hence incomplete filling stage. For such situations, one would find the nonisothermal resin flow model along with the energy equation and cure kinetics equation very useful. The coupling between energy, flow and cure kinetics equations is through the cure and temperature dependent viscosity model.

One could describe heat transfer through fibrous porous media using different approaches and models [401]. One could either use a two-phase model in which one solves for the energy equation for fibers and resin separately and uses constitutive equations to express the energy transfer between the two phases or one could use the simpler local equilibrium model. In local equilibrium model, one focuses on the average temperature of a control volume containing both the resin and the fibers and assumes that the resin instantly equilibrates to the average temperature as soon as it covers the fibers [402]. This model is commonly employed due to its simplicity and the experimental results support the assumption especially for low Reynolds number flows. Thus, the local equilibrium model will be discussed here.

The energy equation for this model is given below [403]

$$\left[(1-V_f)\rho_r c_{p_r} + V_f \rho_f c_{pf}\right]\frac{\partial T}{\partial t} + \rho_r c_{p_r}\mathbf{u}\cdot\nabla T = \nabla\cdot\left[(\mathbf{k}_e+\mathbf{K}_D)\cdot\nabla T\right] + (1-V_f)\dot{s} + \mu\mathbf{u}\cdot\mathbf{K}^{-1}\cdot\mathbf{u} \tag{8.72}$$

where T is the average temperature of resin and fabric preform, t is the time, ρ and c_p are the density and heat capacity. The subscripts r and f denote the resin and fabric preform, respectively. V_f is the fiber volume fraction, $\dot{s}$ is the source term due to curing, $\mathbf{k}_e$ is effective conductivity which is expressed in terms of fiber and resin thermal conductivities κ_r and κ_f [401]. $\mathbf{K}_D$ is called the heat dispersion coefficient and quantifies the differences in heat dispersion not accounted for by the convection of heat. In accordance with Darcy's Law, it is assumed that the average velocity is in the flow direction and this is sufficient to find

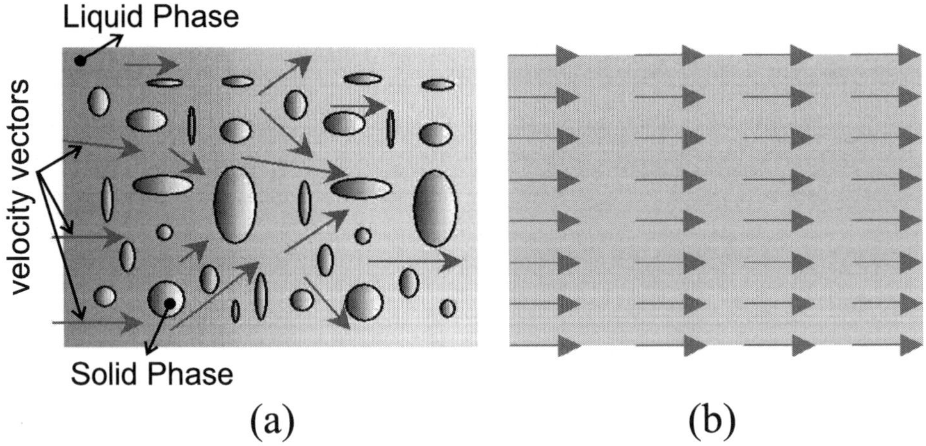

Figure 8.47: Schematic of heat dispersion effect; part (a) depicts the resin impregnation directions across the fibers. Part (b): due to the assumption of Darcy's law one can account for convection only in the flow direction. The difference in heat convection effect is captured by a dispersion term.

the pressure distribution. Figure 8.47 shows that the local velocity vector is in different directions, and convects the heat in other directions in addition to the flow direction. This influence is lumped in the second part of the first term on the right-hand side of the above equation, in which the dispersion is expressed as a conductive effect by the expression $\mathbf{K}_D \cdot \nabla T$.

A dimensional analysis shows that, for mold filling applications with a small Brinkman number, viscous dissipation is negligible. Further, for thin shell-like domains, heat conduction in the in-plane directions and heat convection in the transverse direction are also negligible [403]. However, heat dispersion may be significant [404,405]. It is difficult to evaluate the heat dispersion term $\mathbf{K}_D$ [406,407], and it is usually neglected in the models [408]. $\mathbf{K}_D$ is found to be proportional to the Peclet number. Hence, the higher the velocity of the flow, the larger will be the effect [405,406].

The first term on the left-hand side of Equation (8.72) is the change in the internal energy of the resin and the preform; the second term is due to fluid convection due to volume-averaged Darcy's velocity. Even if flow was considered two-dimensional, due to the presence of the first term on the right-hand side of Equation (8.72), the energy equation is three-dimensional. One can ignore the through thickness flow, but the dominant conduction mode is through the thickness direction; hence, one may ignore the heat conduction in the plane but must retain the conduction term through the thickness.

Equation (8.72) is a second order elliptic partial differential equation in spatial coordinates and first order in time. In order to solve it, boundary conditions on the entire spatial domain and an initial condition for temperature are needed. In some applications, the mold wall temperature is assumed to stay constant during the filling. However, this is not a very realistic assumption. Instead, a quasi-steady state approach yields the following boundary condition in the thickness direction [276,379]:

$$\frac{\partial T}{\partial n} + C_{bc}(T - T_\infty) = 0 \tag{8.73}$$

where n is the normal in the outward direction to the mold wall, and the boundary condition constant C_{bc} is written as follows:

$$C_{bc} = \frac{1}{k_t} \frac{1}{1/h_h + 1/h_m + a/k_m} \tag{8.74}$$

where h_h and h_m are the heat transfer coefficients between the heating fluid and the pipe, and between the mold and resin, respectively, k_m is the thermal conductivity of mold

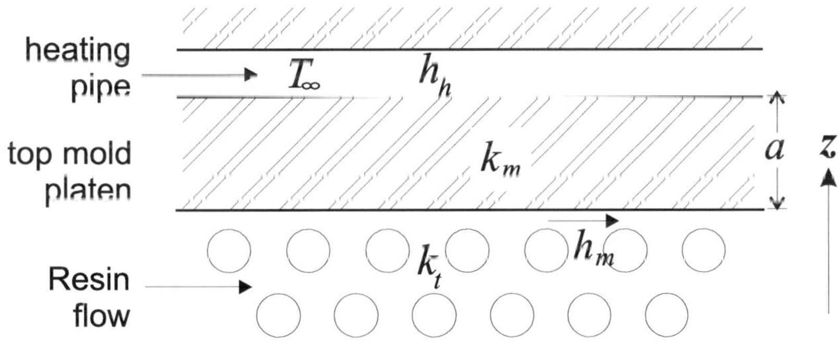

Figure 8.48: Boundary condition at the mold wall across the thickness of a composite [81].

material, k_t is the conductivity of the composite in the transverse direction, a is the distance between the mold wall and the heating pipe, and T_∞ is the temperature of the heating fluid as shown schematically in Figure 8.48.

The other boundary condition is usually at the injection gate where the temperature of the resin is usually prescribed. At the flow front, an energy balance allows one to account for the convection into the unsaturated region as shown in Figure 8.49.

The $\dot{s}$ term in the energy equation models the heat released by the exothermic reaction when the resin molecules are cross-linking as the resin cures. Hence, one needs a constitutive equation that can characterize cure and also provide the rate at which it generates energy.

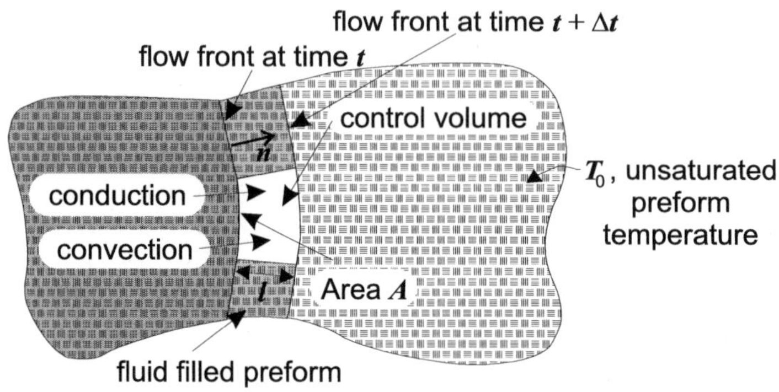

Figure 8.49: Energy balance at the flow front [404].

Example 8.9: Cold Resin Injected into a Hot Mold [406]

Consider flow of resin in the x direction. If cold resin continues to be injected into the hot mold (temperature of mold wall held at T_o) even after the preform is saturated and allowed to exit at one or more vents, the temperature distribution in the mold cavity will reach a steady state. Even if one does not flush the mold in this way, most of the resin near the vent will reach the steady state. Assuming steady state, find approximately the temperature distribution for slow, intermediate and fast resin injections. What dimensionless number governs this flow?

Solution

The energy equation, Equation (8.72), for steady state ignoring in-plane conduction reduces to

$$(\rho c_p)_r \langle u \rangle \frac{\partial \langle T \rangle}{\partial x} = k_{zz} \frac{\partial^2 \langle T \rangle}{\partial z^2} \tag{8.75}$$

with boundary conditions

$$\langle T \rangle = T_{in} \text{ at } x = 0 \tag{8.76}$$

$$\langle T \rangle = T_o \text{ at } z = \pm h. \tag{8.77}$$

Here $(\rho c_p)_r$ is the heat capacity of the resin, $\langle u \rangle$ is the Darcy's velocity in the flow direction (x direction) and $\langle T \rangle$ is the average temperature at a given location. Here k_{zz} is the conductivity of the medium in the thickness direction. Also, as thickness is much less than the length, one can ignore the conduction in the x direction.

When the filling speed is slow, the conduction across the thickness will dominate. At steady state, the mold wall temperature will diffuse through the thickness, and resin and preform will reach the mold wall temperature. The only exception will be the injection gate region of width h, where the temperature changes from T_{in} (from inlet resin temperature to T_o) as sketched in Figure 8.50 [406, 409].

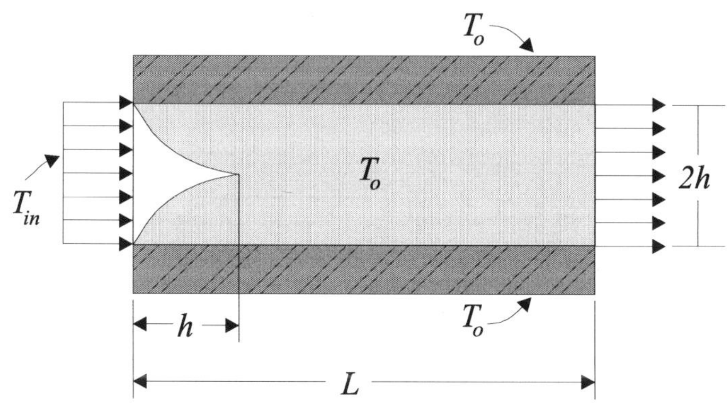

Figure 8.50: Schematic of steady-state temperature distribution at low injection speeds ($Gz << 1$). At steady state, $T = T_o$ in the shaded region. In the central white region, T changes between T_{in} and T_o.

The dimensionless number that governs the ratio of the convection in the inflow direction and conduction through the thickness direction is the Graetz number, Gz:

$$Gz = \frac{\langle u \rangle h^2}{\alpha_{zz} L} \qquad \text{where } \alpha_{zz} = \frac{k_{zz}}{(\rho c_p)_r}. \tag{8.78}$$

This can be obtained by nondimensionalization of Equation (8.75) by using the following nondimensional variables:

$$\Theta = \frac{\langle T \rangle - T_{in}}{T_o - T_{in}}, \qquad x' = \frac{x}{L}, \qquad z' = \frac{z}{h} \tag{8.79}$$

$$\frac{(\rho c_p)_r \langle u \rangle}{L} \frac{\partial \Theta}{\partial x'} = \frac{k_{zz}}{h^2} \frac{\partial^2 \Theta}{\partial z'^2} \tag{8.80}$$

$$\frac{h^2 \langle u \rangle}{\alpha_{zz} L} \frac{\partial \Theta}{\partial x'} = \frac{\partial^2 \Theta}{\partial z'^2} \tag{8.81}$$

$$Gz \frac{\partial \Theta}{\partial x'} = \frac{\partial^2 \Theta}{\partial z'^2}. \tag{8.82}$$

Thus when $Gz << 1$, conduction dominates as shown in Figure 8.50. One can easily solve Equation (8.81), by introducing a new variable

$$\xi = \frac{\hat{z}}{2\sqrt{\frac{\alpha_{zz} x}{\langle u \rangle}}} \tag{8.83}$$

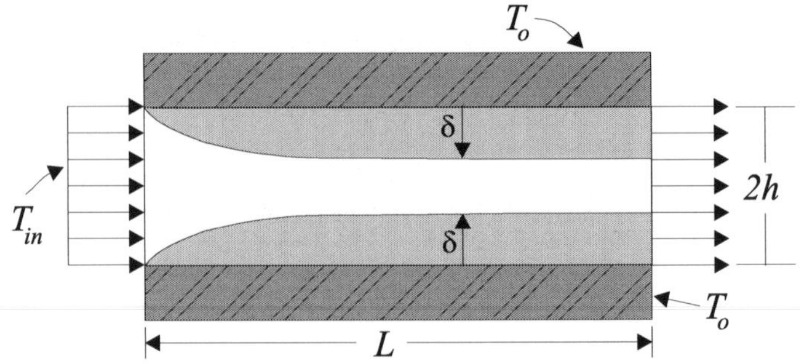

Figure 8.51: Schematic of steady-state temperature distribution at high injection speeds ($Gz >> 1$). At steady state, $T = T_{in}$ in the white unshaded region and changes from T_o to T_{in} in the shaded boundary layer.

where $\hat{z}$ is the distance from the wall towards the center in the thickness direction. This allows us to convert a partial differential equation into an ordinary differential equation by similarity method and the solution can be easily obtained:

$$\frac{\langle T\rangle - T_{in}}{T_o - T_{in}} = 1 - \text{erf}(\xi) \tag{8.84}$$

for $Gz >> 1$. Here erf represents the Gaussian error function. Thus for $Gz >> 1$ (large filling speeds), the thermal boundary layer increases with distance from the entrance as shown in Figure 8.51. The thickness of this boundary layer grows with the square root of the distance from the mold entrance and is given by

$$\frac{\delta}{h} = 4\sqrt{\frac{\alpha_{zz}/x}{\langle u\rangle}}. \tag{8.85}$$

Thus at high $Gz >> 1$, $\delta << h$. At intermediate speeds when $Gz \sim 1$, $\delta \sim h$ when $x = L$.
■

Example 8.10: Temperature Solution Using Equation (8.84)

In the previous example, a closed form solution, Equation (8.84) [406] was given for the temperature distribution when a cold resin is injected into a hot mold. Equation (8.84) is valid for $Gz >> 1$, and will be used here for the following material, geometry and process parameters:

- Average fiber volume fraction, $V_f = 0.40$
- Resin viscosity, $\mu = 0.25\,\text{Pa.s}$
- Thermal conductivity of fiber, $k_f = 1.20\,\text{W/(m°K)}$
- Thermal conductivity of matrix, $k_m = 0.20\,\text{W/(m°K)}$
- Density of fiber, $\rho_f = 2500\,\text{kg/m}^3$

- Density of matrix, $\rho_m = 1200\,\text{kg/m}^3$
- Specific heat capacity of fiber, $c_{pf} = 800\,\text{J/(kg}^\circ\text{K)}$
- Specific heat capacity of matrix, $c_{pm} = 700\,\text{J/(kg}^\circ\text{K)}$
- Inlet temperature, $T_{in} = 20^\circ\text{C} = 293.2^\circ\text{K}$
- Mold wall temperature, $T_o = 120^\circ\text{C} = 393.2^\circ\text{K}$
- Average 1D resin speed, $u = 0.0714\,\text{m/s}$
- Length of mold, $L = 0.500\,\text{m}$
- Half-thickness of the mold cavity, $h = 0.005\,\text{m}$.

Solution

The thermal conductivity, k_{zz} of the composite is determined using Equation (7.50),

$$k_{zz} = \frac{1}{\frac{1-0.40}{0.20\,\text{W/m}^\circ\text{K}} + \frac{0.40}{1.20\,\text{W/m}^\circ\text{K}}} = 0.30\,\text{W/m}^\circ\text{K} \tag{8.86}$$

based on the rule of mixture, and the thermal diffusivity is calculated using Equation (8.78)

$$\alpha_{zz} = \frac{k_{zz}}{(\rho c_p)_r} = \frac{0.30\,\text{W/(m}^\circ\text{K)}}{(1200\,\text{kg/m}^3)(700\,\text{J/(kg}^\circ\text{K))}} = 3.5714 \times 10^{-7}\text{m}^2/\text{s}. \tag{8.87}$$

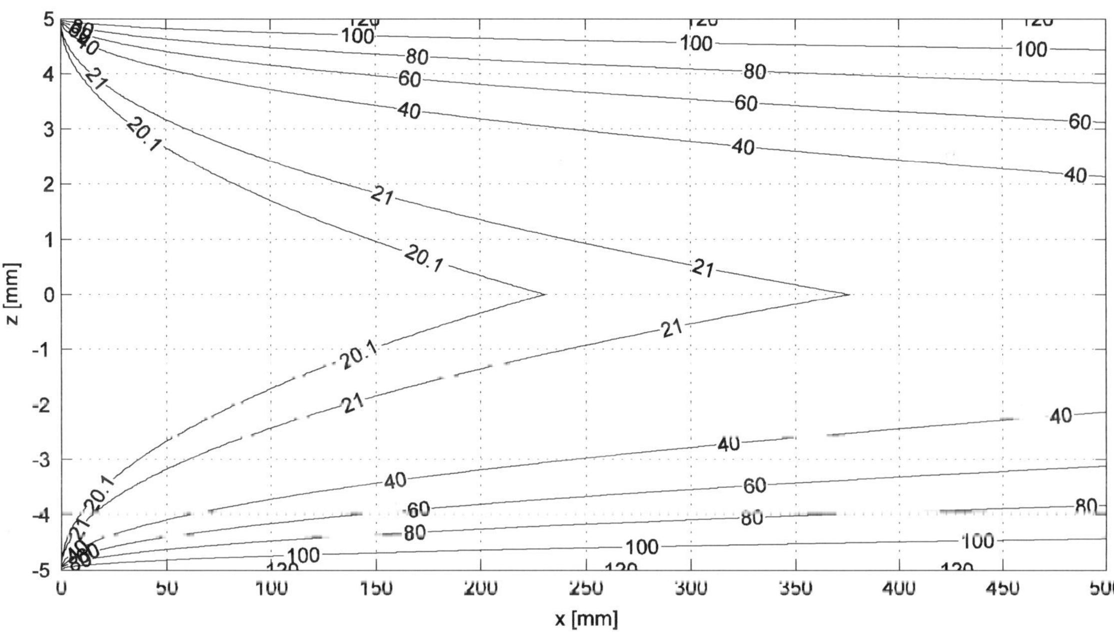

Figure 8.52: Isotherms of the composite temperature, $\langle T \rangle$ using a closed form solution, Equation (8.84) which is valid only for $Gz >> 1$.

The Graetz number, Gz, is calculated as

$$Gz = \frac{\langle u \rangle h^2}{\alpha_{zz} L} = \frac{(0.046\,\mathrm{m/s})(0.005\,\mathrm{m})^2}{(3.5714 \times 10^{-7}\mathrm{m^2/s})(0.500\,\mathrm{m})} = 10.0 \tag{8.88}$$

which satisfies the condition $Gz >> 1$.

A MATLAB code, *example8_10.m*, was written and is listed in the Appendix. The isocurves of the composite temperature, $\langle T \rangle$ were drawn at 20.1, 21, 40, 60, ..., 100 and 120°C in Figure 8.52. Notice a large entrance region where the temperature did not change from its inlet value, 20°C. ■

Example 8.11: Re-solve Example 8.10 Using Numerical Method

The Graetz number in the previous example was 10.0, and thus the closed form solution, Equation (8.84) could be used as it had been developed in Dessenberger and Tucker [406] by using similarity method for $Gz >> 1$. Now, you are asked to solve the same problem (Example 8.10) not only for Gz = 10, but also for 1 and 0.1 by setting the average 1D speed of resin to 71.4, 7.14 and 0.714 mm/s in these three different cases.

Solution

As previously illustrated in numerous examples, finite difference method (FDM) will be applied for the spatial derivatives. The discretization of the solution domain is shown in

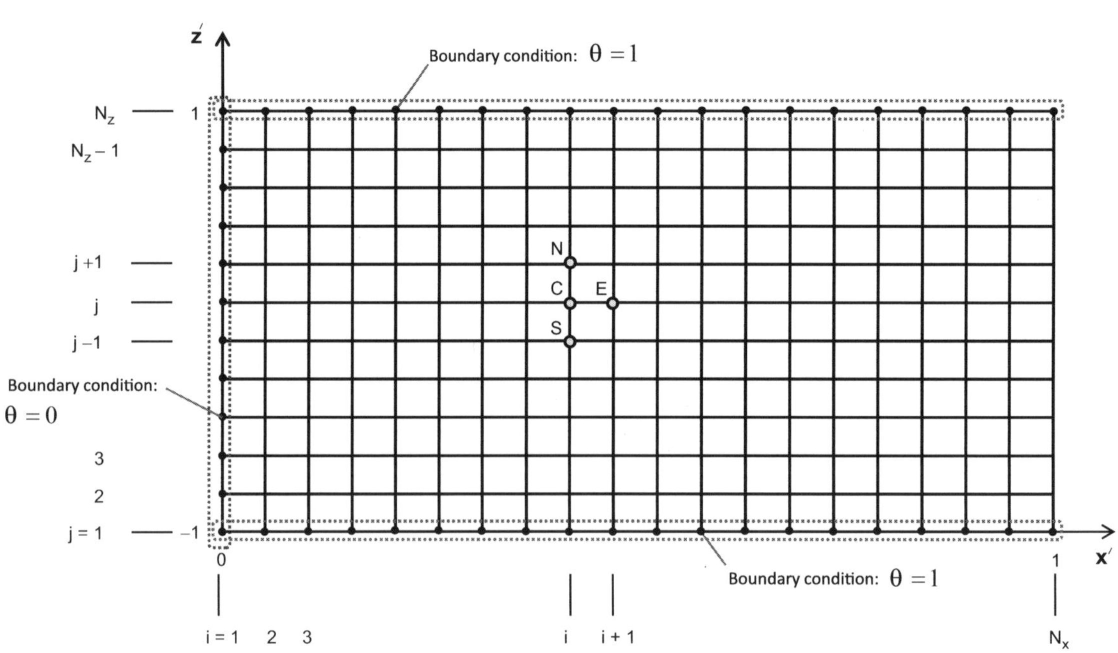

Figure 8.53: Discretized domain and the thermal boundary conditions for a composite part as the cold resin is injected into a hot mold. The nondimensional variables, Θ, x' and z' were defined in Equation (8.79).

Figure 8.53 and formulated as follows

$$x'_i = (i-1)\Delta x' \qquad \text{where } i = 1, 2, \ldots, N_x \tag{8.89}$$

$$z'_j = (j-1)\Delta z' - 1 \qquad \text{where } j = 1, 2, \ldots, N_z \tag{8.90}$$

where $\Delta x' = 1/(N_x - 1)$ and $\Delta z' = 2/(N_z - 1)$ are the increments in x' and z' directions.

The nondimensional temperature, Θ was defined in Equation (8.79), and the differential equation on it was obtained in Equation (8.82) as $Gz\, \partial\Theta/\partial x' = \partial^2\Theta/\partial z'^2$. The discretized form of this differential equation is given below

$$Gz\, \frac{\Theta_E - \Theta_C}{\Delta x'} = \frac{\Theta_N - 2\Theta_C + \Theta_S}{\Delta z'^2} \tag{8.91}$$

by using first order forward and second order central finite differences in x' and z' directions, respectively. Rearrangement of Equation (8.91) results in the following equation

$$\Theta_E = (1-2r)\Theta_C + r\,(\Theta_N + \Theta_S) \qquad \text{where } r = \frac{\Delta x'}{Gz\, \Delta z'^2} \tag{8.92}$$

will be applied for the "inner" nodes ($2 \leq i \leq N_x$ and $2 \leq j \leq N_z - 1$) of the mesh. Here the subscripts N (North), C (Central), S (South) and E (East) denote

$$N \equiv (i, j+1), \qquad C \equiv (i, j), \qquad S \equiv (i, j-1), \qquad E \equiv (i+1, j). \tag{8.93}$$

Equation (8.92) allows a x'-marching numerical scheme to obtain the solution starting with the boundary condition on the left edge of the domain:

$$\Theta_{1,j} = 0 \tag{8.94}$$

and applying the boundary conditions along the bottom and top edges:

$$\Theta_{i,1} = 1 \qquad \text{and} \qquad \Theta_{i,N_z} = 1. \tag{8.95}$$

A MATLAB code, *example8_11.m*, was written and is listed in the Appendix. The isocurves of the composite temperature, $\langle T \rangle$ were drawn at 20.1, 21, 40, 60, ..., 100 and 120°C in Figure 8.54. Notice the good agreement between the closed form solution found in the previous example and part (a) of this example. Also notice the effect of the Graetz number on the temperature distribution by examining parts (b) and (c) in the same figure.

As studied in numerous examples, for a stable numerical solution, r in Equation (8.92) was selected small enough (< 0.5); the user is suggested to re-run the same code with $r > 0.5$ and see how the solution could become unstable if the increments ($\Delta x'$ and $\Delta z'$) of the FDM are not synchronized appropriately. ■

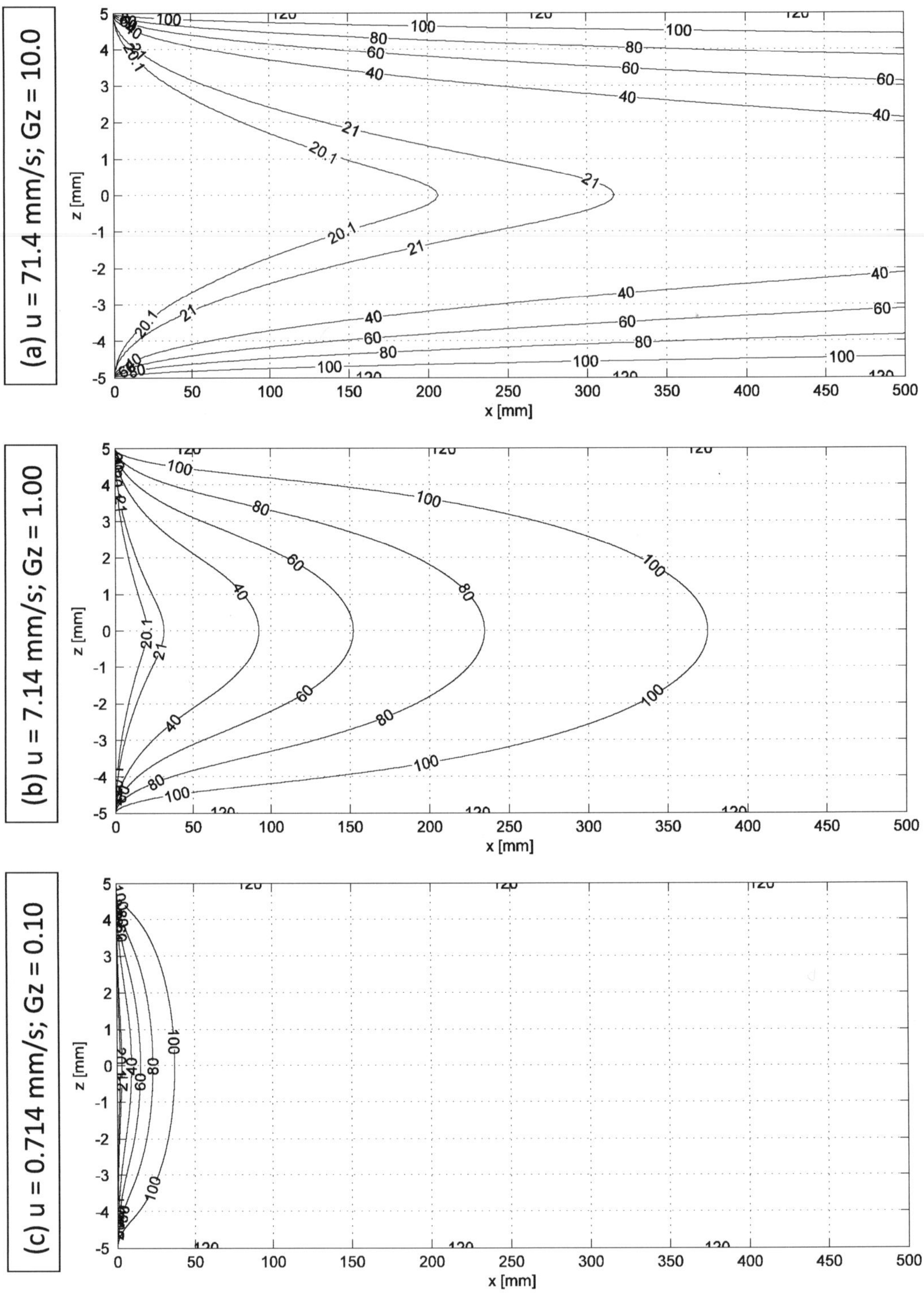

Figure 8.54: Isocurves of the composite temperature, $\langle T \rangle$ using FDM numerical method. Graetz number is 10, 1 and 0.1 in (a), (b) and (c), respectively, by setting the average 1D speed of resin to 71.4, 7.14 and 0.714 mm/s.

Resin Cure Models

There are many empirical models for resin cure as discussed in Chapter 4. Models for various resins are also presented in [276, 402, 410, 411]. One can assume that the heat generation due to the cure reaction, $\dot{s}$ is proportional to the rate of the reaction R_α

$$\dot{s} = R_\alpha E_\alpha \tag{8.96}$$

where E_α is the heat of reaction. R_α is assumed to be a function of temperature T, and degree of cure α. For example

$$R_\alpha = (k_1 + k_2\alpha^m)(1-\alpha)^n \tag{8.97}$$

where

$$k_1 = A_1 e^{E_1/RT} \tag{8.98a}$$

$$k_2 = A_2 e^{E_2/RT} \tag{8.98b}$$

where R is the universal gas constant. The resin material constants m, n, A_1, A_2, E_1 and E_2 are measured experimentally, as discussed to some extent in Chapter 4.

The cure convects with the resin as it flows to occupy the mold; hence, one can express the conservation of cure in mathematical form as follows:

$$R_\alpha = \frac{\partial \alpha}{\partial t} + \bar{u}_x \frac{\partial \alpha}{\partial x} + \bar{u}_y \frac{\partial \alpha}{\partial y}. \tag{8.99}$$

The right-hand side of Equation (8.99) represents the substantial derivative of the degree of cure and must be solved along with the flow model, the energy equation and the cure kinetics model. This system of equations is coupled through the viscosity dependence on the cure kinetics and temperature of the resin.

Temperature and Cure Dependent Viscosity

There are many empirical models available for viscosity. For thermoset resins that are undergoing nonisothermal process and cure, chemorheological models best describe the change in viscosity both as a function of the temperature and degree of cure [246, 379]

$$\mu = A_\mu e^{E_\mu/RT} \tag{8.100}$$

where the activation energy E_μ and the constant A_μ are given as follows:

$$E_\mu = a + b\alpha \tag{8.101a}$$

$$A_\mu = a_0 e^{-b_0\alpha} \tag{8.101b}$$

where the constants a, b, a_0 and b_0 are determined experimentally.

The temperature changes in the thickness direction as well as the in-plane directions. Hence, temperature dependent viscosity will change in all three directions. However, as the flow is modeled as two-dimensional, one must average the viscosity through the thickness as follows:

$$\frac{1}{\bar{\mu}} = \frac{1}{h}\int_0^h \frac{1}{\mu} dz \tag{8.102}$$

where h is the mold cavity thickness and $\bar{\mu}$ is the local in-plane viscosity.

The isothermal flow model of LCM processes can be solved analytically for simple one-dimensional flows with constant viscosity, permeability, fiber volume fraction and thickness. However, most practical situations where prediction will be useful involve complicated geometry with inserts and nonuniform thickness of the part, varying permeability across the region and multiple injection locations. Therefore, one has to resort to numerical solutions to address the mold filling process in LCM. In the next section, we briefly introduce the approach usually followed to accomplish this.

8.3.7 Numerical Simulation of Resin Flow in LCM Processes

A numerical simulation of the mold filling process can be developed by discretizing the governing partial differential equation with numerical methods such as finite difference, finite element or boundary elements. The advantage of numerical simulation is that it will help the process or manufacturing engineer to understand the flow behavior inside the mold, especially when the geometry of the part being manufactured is complex and has permeability and fiber volume fraction variations. This understanding can improve the tool design and placement of gates and vents in locations that help saturate the preform without any dry spots.

As the process engineer needs to run simulations on complex models multiple times with different choices for gates, it is important that the simulation is computationally efficient. One way to make moving boundary problems computationally efficient is to use whenever possible a simplification of the resin flow model to in-plane two dimensions from three dimensions and use a finite element/control volume (FE/CV) approach. The FE/CV approach does not require remeshing at each time step. It is based on the marker and cell method [300, 412, 413].

First, the part geometry is discretized as a thin shell by using triangular and/or quadrilateral elements. This allows one to model geometries in three-dimensional space as shown in Figure 8.55.

Material properties (permeability, fiber volume fraction and part thickness) can be assigned for each element individually, allowing for spatially nonuniform material properties and variable thickness of the part. A pressure profile is assumed between the nodes of an element

$$P^e = \sum_{i=1}^{n} N_i P_i \tag{8.103}$$

where n is the number of nodes of the element e, P^e is the pressure within the element e, P_i is the unknown nodal pressure and N_i is the interpolation function. The Galerkin finite element method can be used to convert the partial differential equation into a system of algebraic equations that minimize the error at the nodes and can be expressed as a linear system of equations as follows:

$$[[S^e]]\,[P] = [f] \tag{8.104}$$

where the components of the element stiffness matrix are given by

$$S^e_{ij} = \int_\Omega \left(\frac{K_{xx}}{\mu}\frac{\partial N_j}{\partial x}\frac{\partial N_i}{\partial x} + \frac{K_{yy}}{\mu}\frac{\partial N_j}{\partial y}\frac{\partial N_i}{\partial y} + \frac{K_{xy}}{\mu}\frac{\partial N_j}{\partial x}\frac{\partial N_i}{\partial y} + \frac{K_{xy}}{\mu}\frac{\partial N_j}{\partial y}\frac{\partial N_i}{\partial x} \right) d\Omega. \tag{8.105}$$

This set of governing equations can be solved at any instant during the filling process. The ith component of $[[S^e]][P]$ vector is the amount of mass generated per unit time at the ith node. Hence, the forcing vector $[f]$ on the right-hand side of Equation (8.104) has all zeros

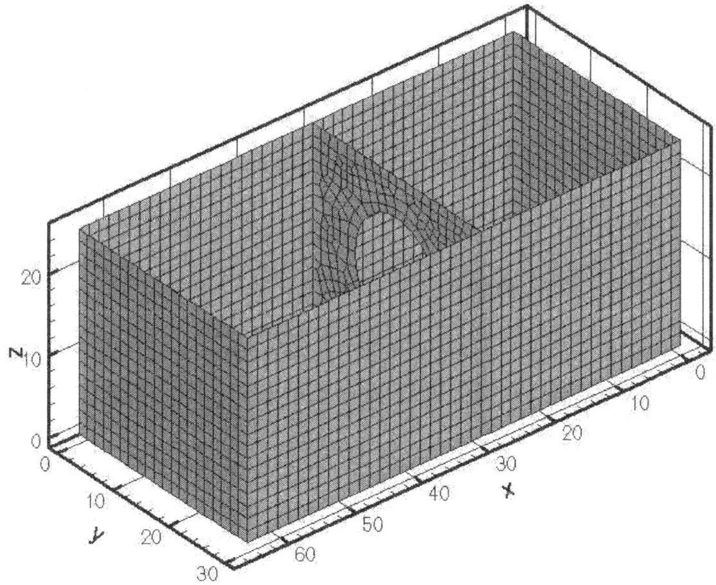

Figure 8.55: Mold geometry discretization in three-dimensional space with two-dimensional finite elements.

except at rows corresponding to the injection nodes. The value of the forcing vector at that row is related to the specified flow rate or pressure boundary condition. After pressures at the nodes are calculated by solving the set of algebraic equations described by Equation (8.104), the next step is to advance the flow front. The control volume approach is used to accomplish this as follows: first, the mold geometry is divided into control volumes, by associating one with every node. The control volume is bounded by the element centroids and the element mid-sides. Next, the flow rate between the control volumes is calculated by multiplying the average normal velocity with the area connecting the two control volumes. For example, the equation for the flow rate from the control volume associated with node i to the control volume associated with node j, q_{ij} is

$$q_{ij} = -\int_{s_{ij}} \left(\frac{h}{\mu} \mathbf{n} \cdot \mathbf{K} \cdot \nabla P \right) ds \tag{8.106}$$

where s_{ij} is the boundary between the two control volumes, h is the mold gap thickness at the boundary, and $\mathbf{n}$ is the normal vector to the boundary in the plane of the preform. Nodal fill factors are used to track the moving flow front. The fill factor for each node is defined as the fraction of its control volume occupied by the fluid as shown in Figure 8.56 [246, 379, 387]. In the calculation of the fill factor, only the porous control volume is considered, i.e., the volume of the fibers is not considered. For example, if the control volume around a node is $1\,\text{cm}^3$, and the local fiber volume fraction is 40%, then (i) fill factor $= 0$ means the preform is completely dry at that location, and (ii) fill factor $= 1$ means that the porous volume ($0.6\,\text{cm}^3$) is completely saturated with resin.

Pressures are calculated at full nodes only and empty nodes are ignored. Partially filled nodes are assumed to lie close to the flow front and the flow front boundary condition is applied there. The flow front is advanced at each time step by updating the fill factors

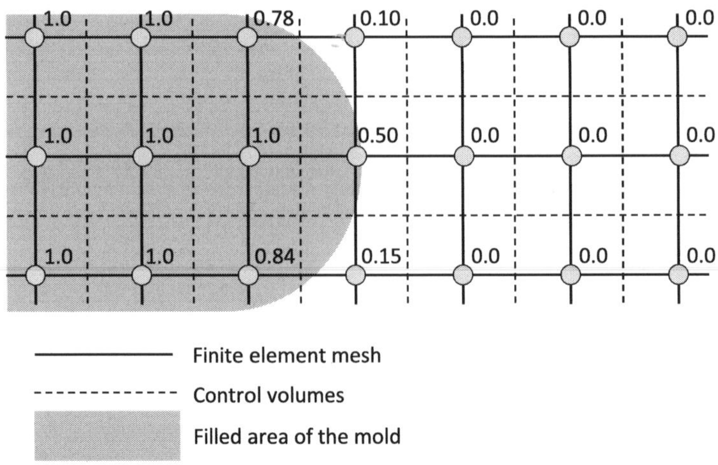

Figure 8.56: Nodal fill factors in FE/CV approach to mold filling.

of control volumes, using the flow rates between the connecting nodes, thus rigorously accounting for mass balance. With this technique, one can fill thin cavities with highly complex geometries in three-dimensional space and also account for the variation in the gap height of the part.

If the filling is nonisothermal or at elevated temperature, one must also solve the temperature equation during mold filling. If the flow analysis is carried out in 2D, each control volume used for the pressure analysis is rediscretized in the transverse z direction to generate a three-dimensional mesh for temperature and cure analysis. Equations (8.72) and (8.99) are solved by a method [379, 414] which is explicit in time t and the x-y plane but implicit in the z direction. The method does not require an excessive amount of memory considering the number of nodes in the third dimension, but the downside of the method is that it is only conditionally stable. If the time step is not chosen to be small enough, the numerical solutions will not converge. Since the source term in the energy equation is a nonlinear function of temperature, an iterative solution method is applied to evaluate the temperature distribution. The calculated temperature is then used to estimate the degree of cure. After the temperature and degree of cure are known within the three-dimensional resin domain, the averaged (or effective) resin viscosity can be calculated using Equations (8.100) and (8.102).

There are two disadvantages in conducting numerical simulations nonisothermally. First, the computational efficiency and speed go down drastically as one has to solve energy and flow equations simultaneously. Research in simplifying the nonisothermal analysis has reduced the time but still requires over two to three orders of magnitude of CPU time than solving the isothermal problem [379]. Secondly, more material parameters such as the dispersion coefficient, thermal conductivities, and resin cure kinetics are needed to obtain the results. There have not been many studies that have conducted rigorous experimental verification of nonisothermal simulations; hence, the sensitivity and accuracy of these models is untested. Further research in this area is essential if one expects to cure the resins during mold filling in an attempt to reduce cycle times.

For high-temperature resins, a good compromise would be to fill the mold under isothermal conditions and then invoke the cure kinetics. Numerically, it is very efficient as one does not have to deal with convective-diffusion problems which are notoriously difficult in converging and giving the correct result.

8.3.8 Case Studies

In the following two subsections, we will numerically solve several case studies in LCM processes:

- In the first set of examples (Examples 8.12-8.14 of Subsection 8.3.9), we will solve governing differential equation on resin pressure, P and the boundary conditions **at the end of mold filling** (i.e., assuming that the mold was already filled and all we are interested in are the pressure and velocity distributions at this final instant). **MATLAB** codes will be written based on Finite Difference Method (FDM) and the results will be discussed to investigate the effects of process parameters such as gate/vent locations and injection pressure.
- In the second set of examples (Examples 8.15-8.21 of Subsection 8.3.11), a robust numerical simulation program, **LIMS [415]** will be used to solve P not only at the end of mold filling, but also and more importantly, during the injection with a time-marching scheme (FEM + Control Volume Approach); and thus the propagation of flow front will be monitored virtually to investigate (i) if the mold can be filled completely with the selected locations of the gate(s)/vent(s), (ii) how long it takes to fill the mold cavity, and also (iii) design the process parameter window. Use of LIMS will also enable the user to investigate the effects of racetracking channels, lineal injection/ventilation, nonuniform material properties, and some control schemes such as sequential gating.

8.3.9 Numerical Solution of Pressure and Velocity Distributions at the End of Mold Filling Using Finite Difference Method

Example 8.12: Rectangular Mold with a Point Injection Gate

A rectangular mold cavity is shown in Figure 8.57; the material, geometry and process parameters are listed below:

- Injection pressure (gage), $P_{in} = 300{,}000\,\text{Pa}$
- Resin viscosity, $\mu = 0.10\,\text{Pa.s}$
- Fiber volume fraction, $V_f = 0.40$
- Preform permeability tensor, $\begin{bmatrix} K_{xx} & K_{xy} \\ K_{yx} & K_{yy} \end{bmatrix} = \begin{bmatrix} 1.0 \times 10^{-9} & 0 \\ 0 & 0.2 \times 10^{-9} \end{bmatrix} \text{m}^2$
- Inlet gate and exit port are square holes with a width and depth of 20 mm each.

Do the following:

- Obtain the governing differential equation on P and necessary boundary conditions.
- Discretize the solution domain.
- Apply FDM to express the unknown central nodal pressures in terms of neighbor nodal pressures.
- Write Darcy's Law which will be used to calculate the velocity components at each node once the pressures are already calculated.

- Write a MATLAB code to solve the governing equation and find pressure, P and velocity, $\mathbf{U}$ distributions.
- Plot P and $\mathbf{U}$ distributions and discuss them.
- Calculate the inlet and exit flow rates assuming that the injection is continued and the ventilation port is held open. Check if these two flow rates are equal to each other satisfying the mass balance.

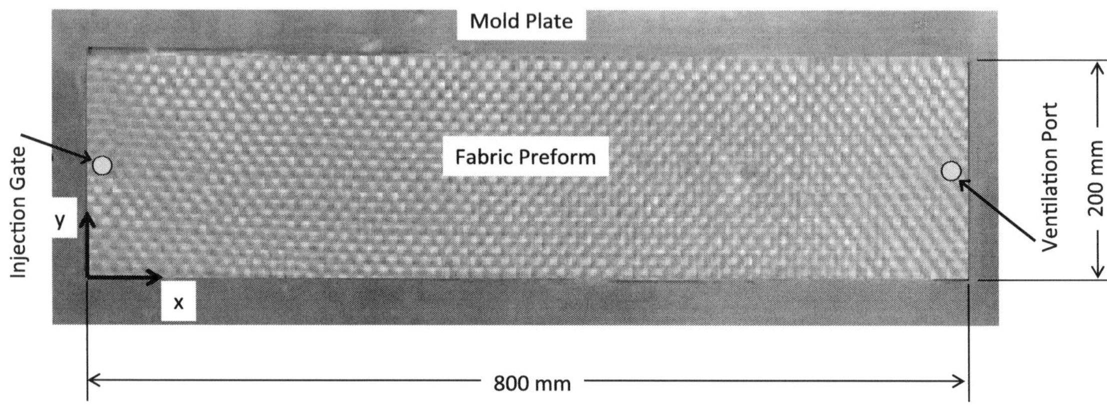

Figure 8.57: Top view of the mold cavity; the thickness of the cavity is 20 mm everywhere. The injection and ventilation holes are located at (0,100) and (800,100) where the coordinates are in millimeters.

Solution

For 2D incompressible and steady flow, the conservation of mass is written as

$$\begin{aligned} \nabla \cdot \mathbf{U} &= 0 \\ \frac{\partial u}{\partial x} + \frac{\partial v}{\partial y} &= 0 \\ \frac{\partial}{\partial x}\left(-\frac{K_{xx}}{\mu}\frac{\partial P}{\partial x}\right) + \frac{\partial}{\partial y}\left(-\frac{K_{yy}}{\mu}\frac{\partial P}{\partial y}\right) &= 0 \\ K_{xx}\frac{\partial^2 P}{\partial x^2} + K_{yy}\frac{\partial^2 P}{\partial y^2} &= 0 \qquad \text{(governing diff. eqn. on } P\text{)} \qquad (8.107) \end{aligned}$$

assuming that the resin viscosity and preform's permeability are constant within the mold cavity, and there is no sink effect (i.e., the fiber tows are saturated instantaneously as the flow front passes by). The velocity terms are written in terms of Darcy's Law: $u = -(K_{xx}/\mu)\partial P/\partial x$ and $v = -(K_{yy}/\mu)\partial P/\partial y$.

Along the nonpermeable mold walls, resin's normal velocity component should be zero considering the no-leakage condition (i.e., resin can neither enter nor exit the sealed mold wall; however, it can have a nonzero tangential velocity component); and at the inlet and exit points, the pressures are equal to the injection and ventilation pressures, respectively;

thus all the boundary conditions are given below:

$$\text{B.C. at } \underline{\text{left}} \text{ edge, } x = 0: \qquad u = 0 \Longrightarrow \frac{\partial P}{\partial x} = 0 \tag{8.108a}$$

$$\text{B.C. at } \underline{\text{right}} \text{ edge, } x = L_x = 0.8\,\text{m}: \qquad u = 0 \Longrightarrow \frac{\partial P}{\partial x} = 0 \tag{8.108b}$$

$$\text{B.C. at } \underline{\text{bottom}} \text{ edge, } y = 0: \qquad v = 0 \Longrightarrow \frac{\partial P}{\partial y} = 0 \tag{8.108c}$$

$$\text{B.C. at } \underline{\text{top}} \text{ edge, } y - L_y - 0.2\,\text{m}: \qquad v - 0 \longrightarrow \frac{\partial P}{\partial y} - 0 \tag{8.108d}$$

$$\text{B.C. at } \underline{\text{injection}} \text{ gate, } (x, y) = (0, 0.1)\,\text{m}: \qquad P = P_{in} = 300{,}000\,\text{Pa} \tag{8.108e}$$

$$\text{B.C. at } \underline{\text{ventilation}} \text{ port, } (x, y) = (0.8, 0.1)\,\text{m}: \qquad P = P_{vent} = 0. \tag{8.108f}$$

Notice that the pressures are gage pressures.

The mold cavity is discretized as shown in Figure 8.58. Equation (8.107) is discretized using second order central finite differences as follows

$$K_{xx}\frac{P_W - 2P_C + P_E}{h_x^2} + K_{yy}\frac{P_N - 2P_C + P_S}{h_y^2} = 0 \tag{8.109}$$

which yields the following expression for the unknown central pressure in terms of the neighbor pressures (see Figure 8.59):

$$P_C = \frac{\dfrac{K_{xx}}{h_x^2}(P_W + P_E) + \dfrac{K_{yy}}{h_y^2}(P_N + P_S)}{2\left(\dfrac{K_{xx}}{h_x^2} + \dfrac{K_{yy}}{h_y^2}\right)} \qquad \text{if } h_x \neq h_y, \tag{8.110a}$$

$$= \frac{K_{xx}}{2(K_{xx} + K_{yy})}(P_W + P_E) + \frac{K_{yy}}{2(K_{xx} + K_{yy})}(P_N + P_S) \qquad \text{if } h_x = h_y. \tag{8.110b}$$

The boundary conditions, Equations (8.108a)-(8.108d) are discretized using second order forward or backward finite differences as follows

$$\left.\frac{\partial P}{\partial x}\right|_{x=0} = 0 = \frac{1}{2h_x}(-3P_{1,j} + 4P_{2,j} - P_{3,j})$$

$$\Longrightarrow P_{1,j} = \frac{1}{3}(4P_{2,j} - P_{3,j}) \tag{8.111a}$$

$$\left.\frac{\partial P}{\partial x}\right|_{x=0.8\,\text{m}} = 0 = -\frac{1}{2h_x}(-3P_{N_x,j} + 4P_{N_x-1,j} - P_{N_x-2,j})$$

$$\Longrightarrow P_{N_x,j} = \frac{1}{3}(4P_{N_x-1,j} - P_{N_x-2,j}) \tag{8.111b}$$

$$\left.\frac{\partial P}{\partial y}\right|_{y=0} = 0 = \frac{1}{2h_y}(-3P_{i,1} + 4P_{i,2} - P_{i,3})$$

$$\Longrightarrow P_{i,1} = \frac{1}{3}(4P_{i,2} - P_{i,3}) \tag{8.111c}$$

$$\left.\frac{\partial P}{\partial y}\right|_{x=0.2\,\text{m}} = 0 = -\frac{1}{2h_y}(-3P_{i,N_y} + 4P_{i,N_y-1} - P_{i,N_y-2})$$

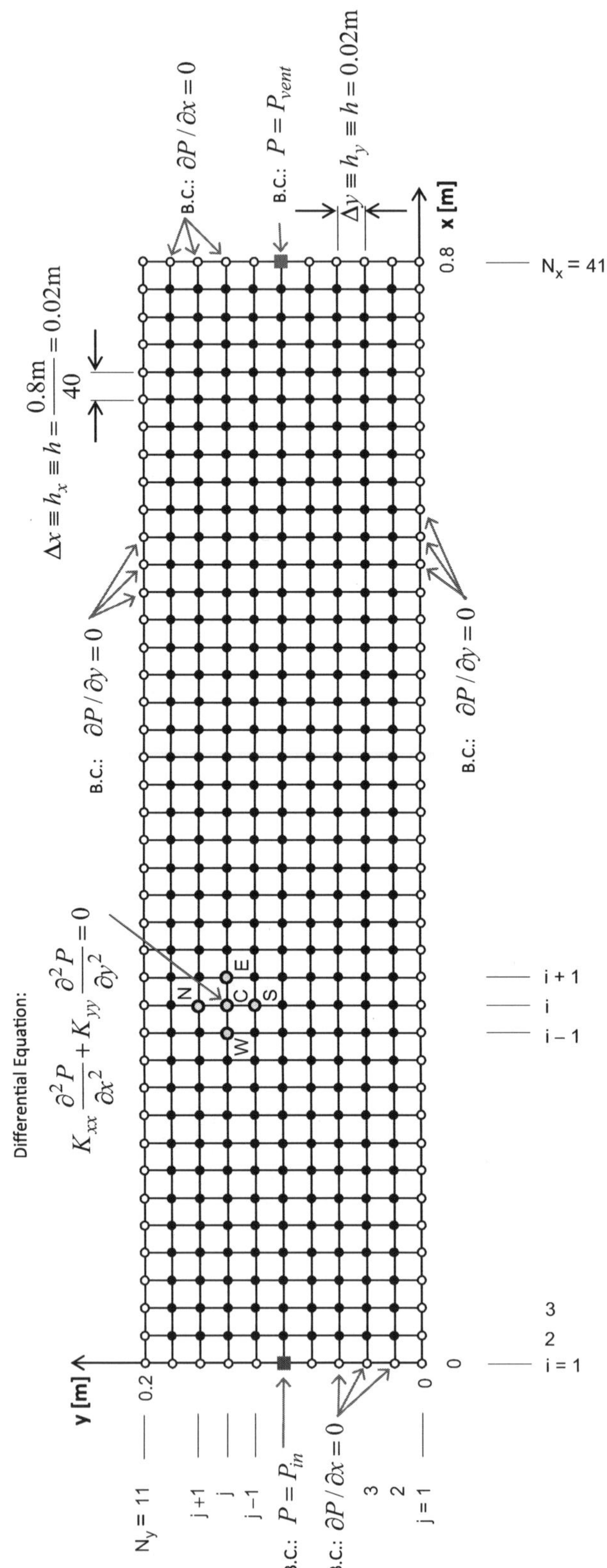

Figure 8.58: Discretized domain and the boundary conditions for the mold cavity shown in Figure 8.57.

$$\begin{aligned} \Longrightarrow P_{i,N_y} &= \frac{1}{3}\left(4P_{i,N_y-1} - P_{i,N_y-2}\right) && (8.111\text{d}) \\ P_{1,6} &= 300{,}000\,\text{Pa} && (8.111\text{e}) \\ P_{N_x,6} &= 0. && (8.111\text{f}) \end{aligned}$$

Equation (8.110) is used to solve the pressures at the internal nodes (shown with dark solid circles in Figure 8.58) in terms of the neighboring nodes which are also unknown. To solve the unknown pressures of the boundary nodes (shown with empty circles in Figure 8.58), Equations (8.111a)-(8.111f) are used; but, these equations also have unknown internal pressures on their right-hand sides. For this problem, we have 2 known (injection and ventilation) and 449 unknown ($41 \times 11 - 2$) nodal pressures. One can apply a direct method, such as inverse matrix or Gauss elimination to solve these 449 unknown P values as we have 449 linear equations on them: $[C]\{P\} = \{r\}$ where $[C]$ is the coefficient matrix of order 449 x 449 (most of its elements are zeros), $\{P\}$ is the unknown pressure vector of order 449 x 1 and $\{r\}$ is the known right-hand side vector of order 449 x 1 (all of its elements are zeros except only one node corresponding to the "East" of the injection gate, $(i, j) = (2, 6)$). However, constructing the coefficient matrix is tedious (especially if the solution domain is complex instead of having a rectangular domain as in this example). Thus, an iterative numerical method called Gauss-Seidel Iterative Method was applied in the MATLAB code, *example8_12.m* (listed in the Appendix). This iterative method involves two nested "for loops" (one for the index i and one for j) and iteration of P values until a convergence is achieved. This method guarantees a convergence no matter what the initial guess is due to the diagonally dominant coefficient matrix. With this choice of solution method, the coding is compact and straightforward.

Once the pressures are calculated, the next step is to calculate the velocity components at each nodal point by using Darcy's Law:

$$u = -\frac{K_{xx}}{\mu}\frac{\partial P}{\partial x} \tag{8.112a}$$

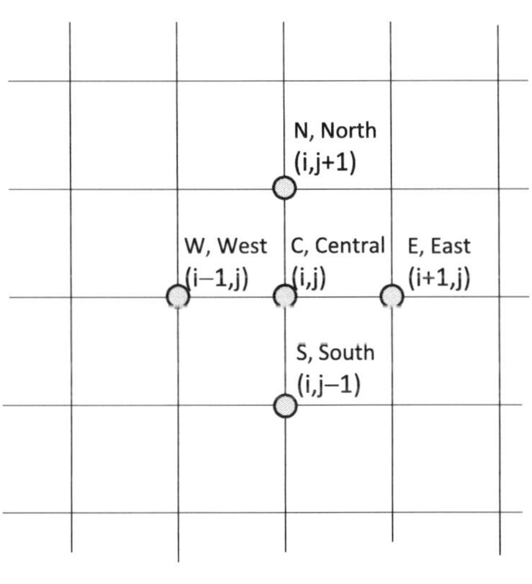

Figure 8.59: Nodal points to calculate the central pressure, P_C in terms of the four neighbor pressures, P_W, P_E, P_N and P_S (see Equation (8.110)).

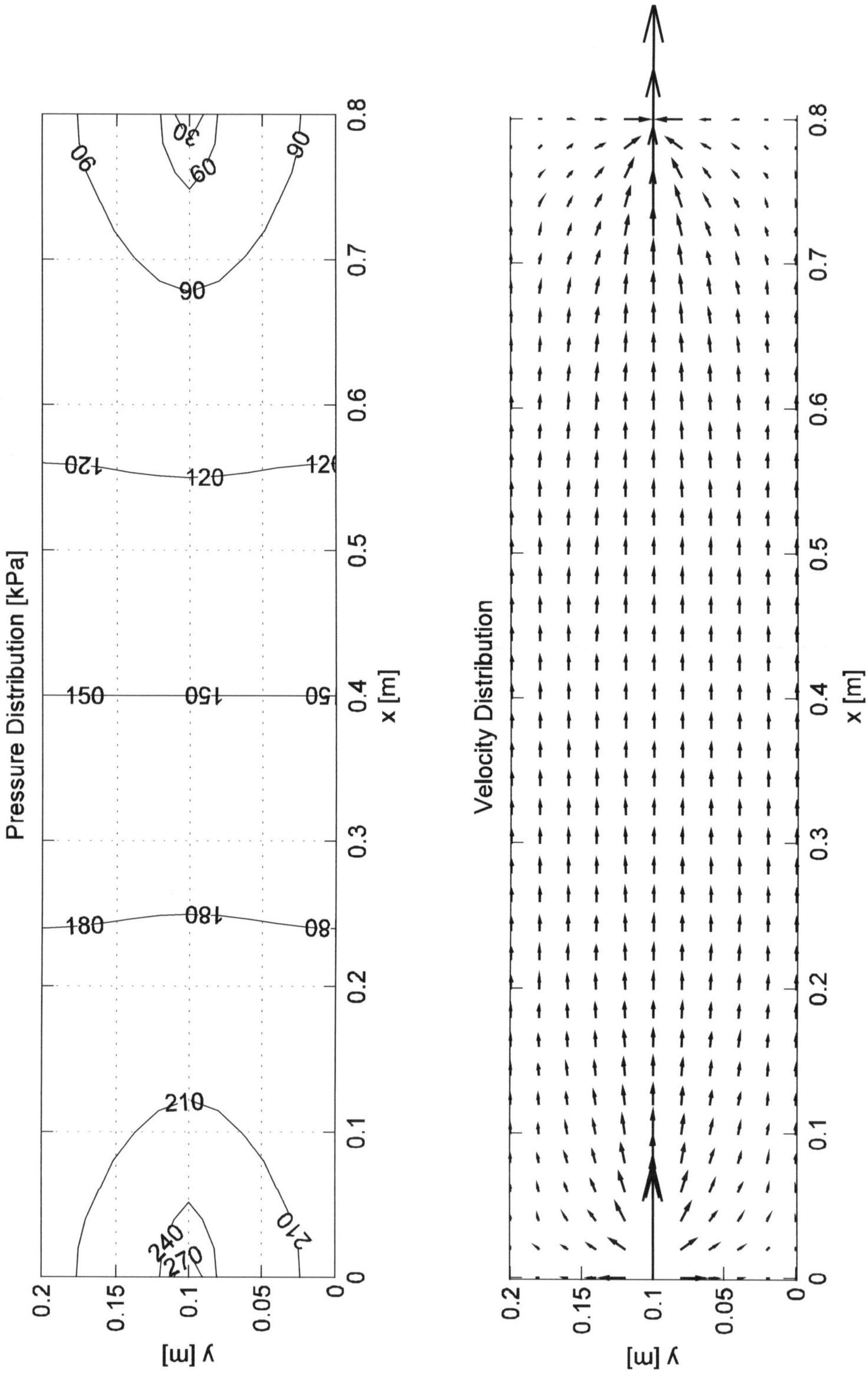

Figure 8.60: Pressure and average velocity distributions.

$$v = -\frac{K_{yy}}{\mu}\frac{\partial P}{\partial y} \tag{8.112b}$$

where the pressure gradients are calculated by using "central" finite differences at the internal nodes, and "forward" or "backward" finite differences at the boundary nodes (see the MATLAB code, *example8_12.m*).

Pressure and velocity distributions are plotted in Figure 8.60. The inlet and exit flow rates are calculated as $Q_{in} = Q_{vent} = 7.74 \times 10^{-6}\,\text{m}^3/\text{s} = 7.74\,\text{cc/s}$, thus the mass balance is satisfied.

This solution resulted in the distributions of resin pressure and velocity at the time of complete mold filling. However, we did not calculate how long it took to fill the mold cavity and we could not check if any macrosize dry region (void) would remain when the resin reached the ventilation port in the actual case. These can be calculated using a much more robust program, such as LIMS, as illustrated in the next subsection. ■

Example 8.13: L-Shaped Mold

An L-shaped mold is shown in Figure 8.61; the material, geometry and process parameters are listed below:

- Injection pressure, $P_{in} = 500,000\,\text{Pa}$
- Ventilation pressure, $P_{vent} = 100,000\,\text{Pa}$
- Resin viscosity, $\mu = 0.20\,\text{Pa.s}$
- Fiber volume fraction, $V_f = 0.40$
- Preform permeability: $K_{xx} = K_{yy} = 2 \times 10^{-9}\,\text{m}^2$ and $K_{xy} = K_{yx} = 0$
- Inlet gate and exit port are rectangular holes with a width of 50 mm and depth of 10 mm.

When the mold cavity is filled with resin completely, do the following:

- Draw P isocurves in xy plane.
- Draw the velocity field in xy plane.
- Calculate the pressure at $(x, y) = (0.3, 0.5)$.
- Compare the inlet and exit flow rates.

Use $\Delta x = \Delta y \equiv h = 0.050\,\text{m}$ in the mesh.

Solution

For the solution, see Appendix B. ■

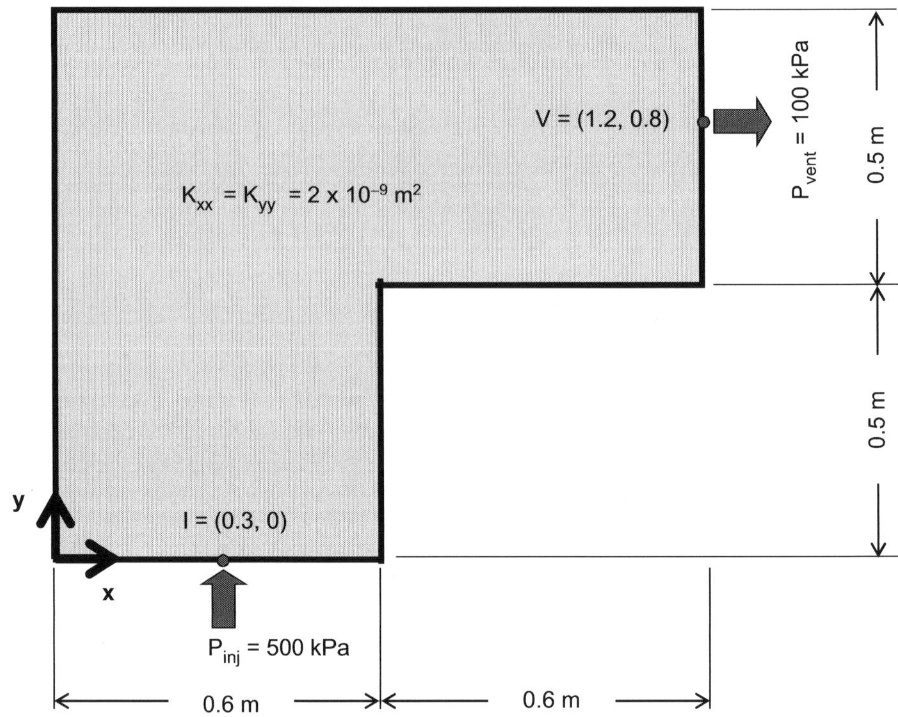

Figure 8.61: Top view of the mold cavity; the thickness of the cavity is 10 mm everywhere.

Example 8.14: Pressure Distribution in 1D Flow when $K_{xx} \neq$ constant

A polyester resin with a viscosity of $\mu = 0.20\,\text{Pa.s}$ is injected into a rectangular mold as shown in Figure 8.62. The mold thickness is constant, and two different but uniform fabric preforms are used in the left- and right-hand sides (their permeabilities are shown in the figure). There is no racetracking channel between the fabrics and the mold wall, and inlet/exit are linear gates and vents along the width as shown in the figure. Thus, one can assume that the flow is 1D ($v = 0$ and $w = 0$).

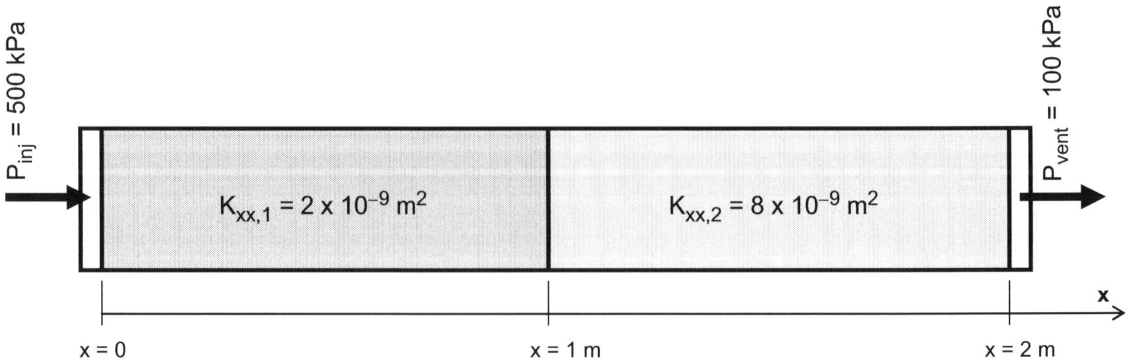

Figure 8.62: 1D flow in a mold in which two different preforms are used in the two ends.

Plot the resin pressure distribution, $P(x)$ when the mold cavity is filled completely.

Solution

Conservation of mass for 1D incompressible and steady flow is given as

$$\nabla \cdot \mathbf{U} = 0 \implies \frac{\partial u}{\partial x} = 0 \implies u \neq u(x) \implies u = \text{constant}. \tag{8.113}$$

Thus, u is constant everywhere ($0 \leq x \leq 2\,\text{m}$); and one can write

$$u_1 = u_2 = \text{constant} \tag{8.114}$$

where u_1 and u_2 are the velocities in the left- and right-half mold cavities. One could get the same result (Equation (8.114)) by using the fact that the flow rate must be constant in the two regions considering the mass balance:

$$u_1 A = u_2 A = \text{constant} \implies u_1 = u_2 = \text{constant} \tag{8.115}$$

where A is the cross-sectional area normal to the 1D flow direction, x. The velocities in Equation (8.114) can be written in terms pressure gradients using Darcy's law:

$$-\frac{K_{xx,1}}{\mu}\left(\frac{\partial P}{\partial x}\right)_1 = -\frac{K_{xx,2}}{\mu}\left(\frac{\partial P}{\partial x}\right)_2 \tag{8.116}$$

and, it can be rewritten as

$$\frac{\left(\frac{\partial P}{\partial x}\right)_1}{\left(\frac{\partial P}{\partial x}\right)_2} = \frac{\dfrac{P_{mid} - P_{in}}{x_{mid} - x_{in}}}{\dfrac{P_{exit} - P_{mid}}{x_{exit} - x_{mid}}} = \frac{K_{xx,2}}{K_{xx,1}} = \frac{8 \times 10^{-9}\,\text{m}^2}{2 \times 10^{-9}\,\text{m}^2} = 4. \tag{8.117}$$

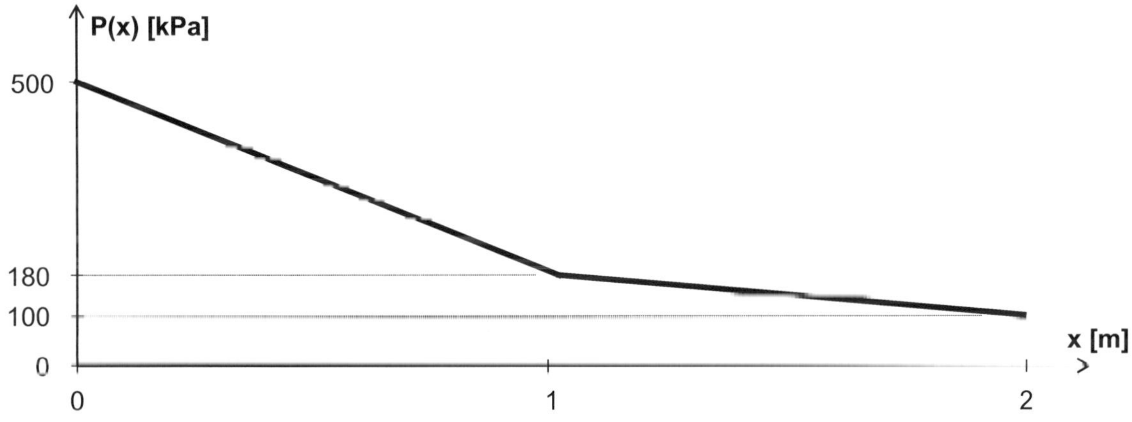

Figure 8.63: Pressure distribution in a one-dimensional mold with two different preforms placed in series.

Equation (8.117) is simplified to obtain the pressure at the mid-point:

$$P_{mid} = \frac{P_{in} + 4P_{exit}}{5} = \frac{(500\,\text{kPa}) + 4(100\,\text{kPa})}{5} = 180\,\text{kPa}. \tag{8.118}$$

The pressure distribution is shown in Figure 8.63. ■

8.3.10 Liquid Injection Molding Simulation (LIMS)

LIMS is a numerical simulation which describes the resin flow in a fibrous porous media for Liquid Composite Molding processes such as RTM and VARTM [415]. It can model the important physics encountered during mold filling of LCM processes such as (i) racetracking phenomenon, (ii) differential filling of fiber tows as compared to regions in between the tows for dual scale fabrics, (iii) accounting for drapability of the fabric over the tool surface during filling, (iv) entrapment of voids, (v) use of distribution media in VARTM and other Vacuum infusion processes, (vi) design of the process filling scheme, the tooling and the distribution medium lay-up and (vii) develop flow control algorithms in the process to counteract the ill effects of inherent variations from one part to the next due to the variability in the fabric permeability and the hand lay-up of the fabric. Figures 8.64-8.69 show some examples of these features. LIMS has been rigorously compared with experiments to validate various aspects of the process as shown in Figures 8.70 and 8.71. It was developed at the University of Delaware and now has a user interface which makes it easy to use within the windows environment as shown in Figure 8.72.

This section will introduce simple examples as a tutorial for simulating flow in liquid composite molding using LIMS along with the LIMS user interface.

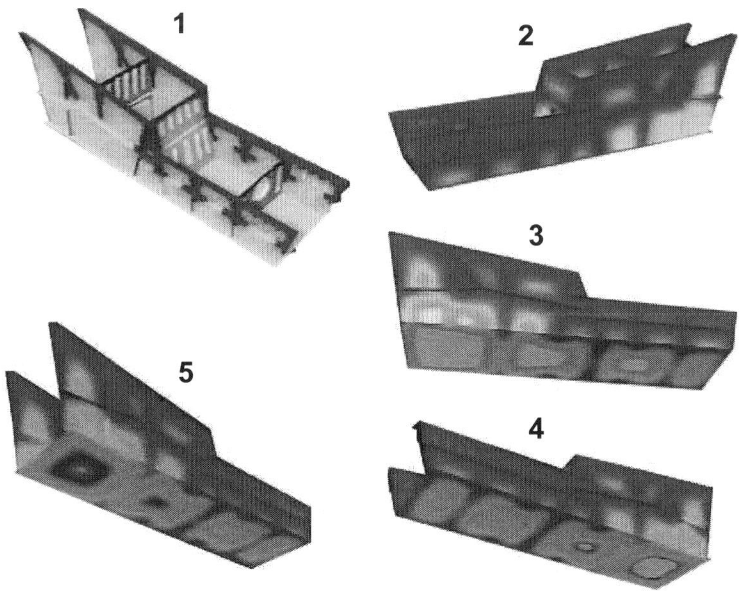

Figure 8.64: Resin flow simulation of a structure with LIMS in which racetracking has been addressed around edges by increasing the permeability of the edge elements. One can experiment and learn of how the mold fills by introducing weak, strong and no racetracking along the edges. The vents need to be placed at the last regions to fill. The injection gate is at the top edge of the structure.

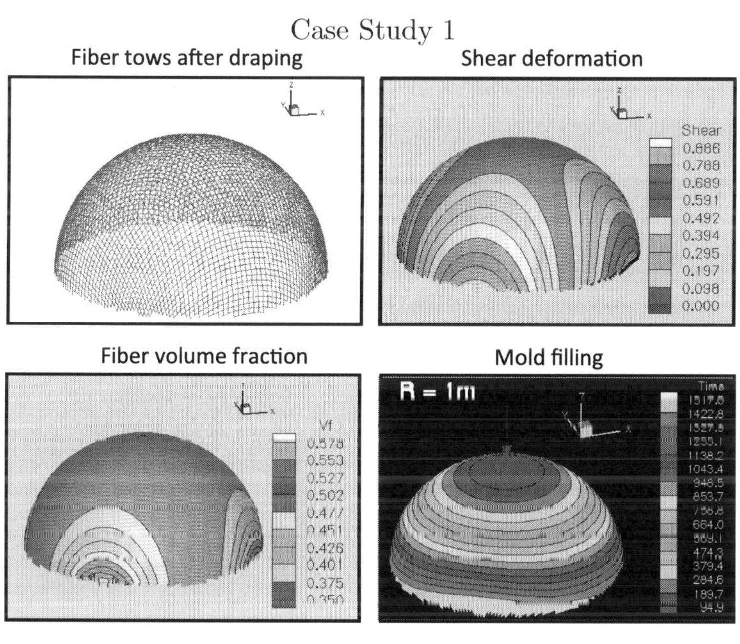

Figure 8.65: LIMS simulation of flow in a fiber preform draped over a compound curvature part to demonstrate the influence of draping on processing parameters such as flow pattern and fill times.

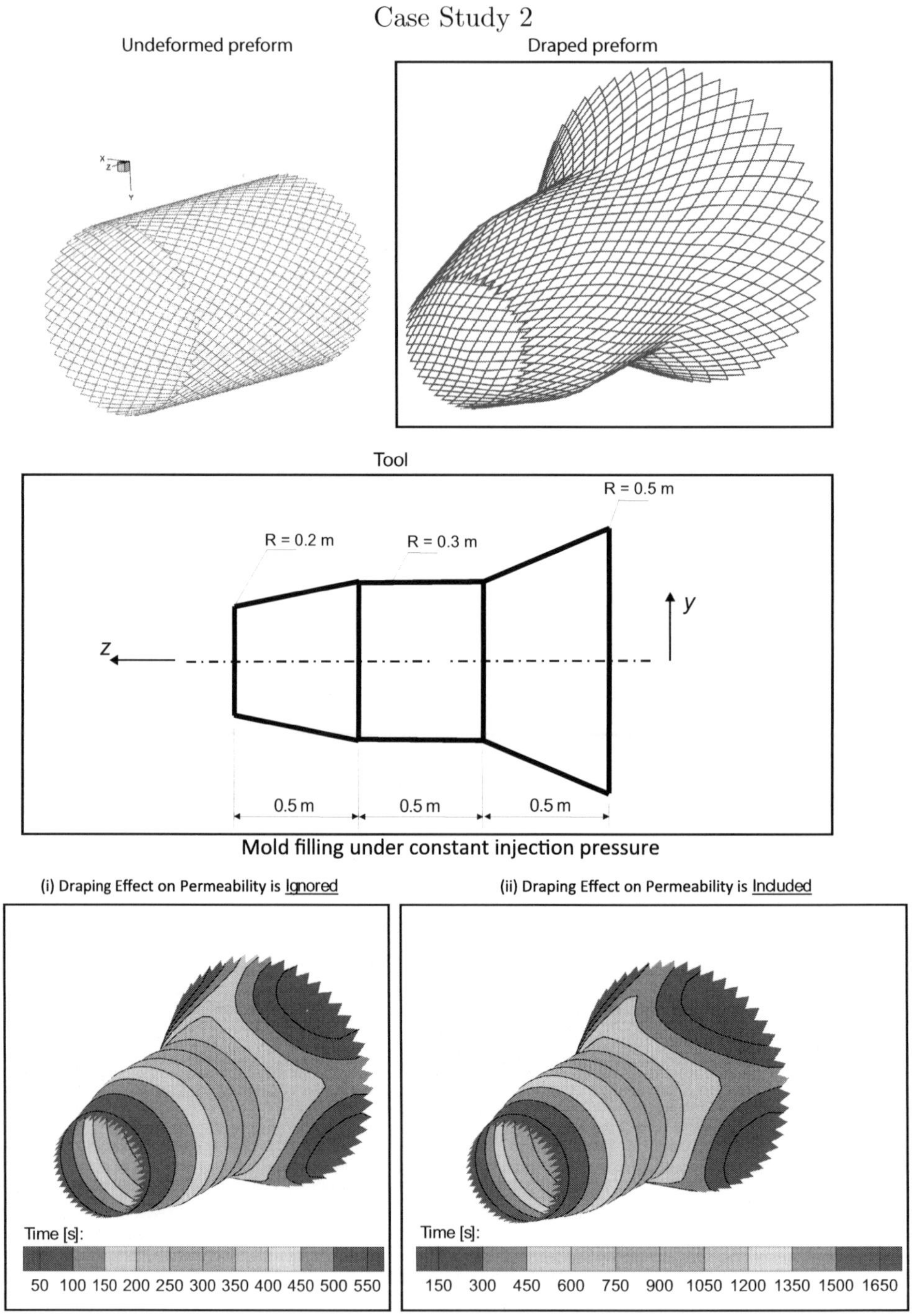

Figure 8.66: LIMS simulation of flow in a fiber preform draped over a compound curvature part to demonstrate the influence of draping on processing parameters such as flow pattern and fill times. DRAPE [346, 416, 417] is used to calculate the shear angle and fiber volume fraction of the deformed fabric preform due to the draping over the tool surface. DRAPE's results are used to calculate the permeability of the preform by interpolating permeability database of the preform; and the permeability is used as an input to LIMS.

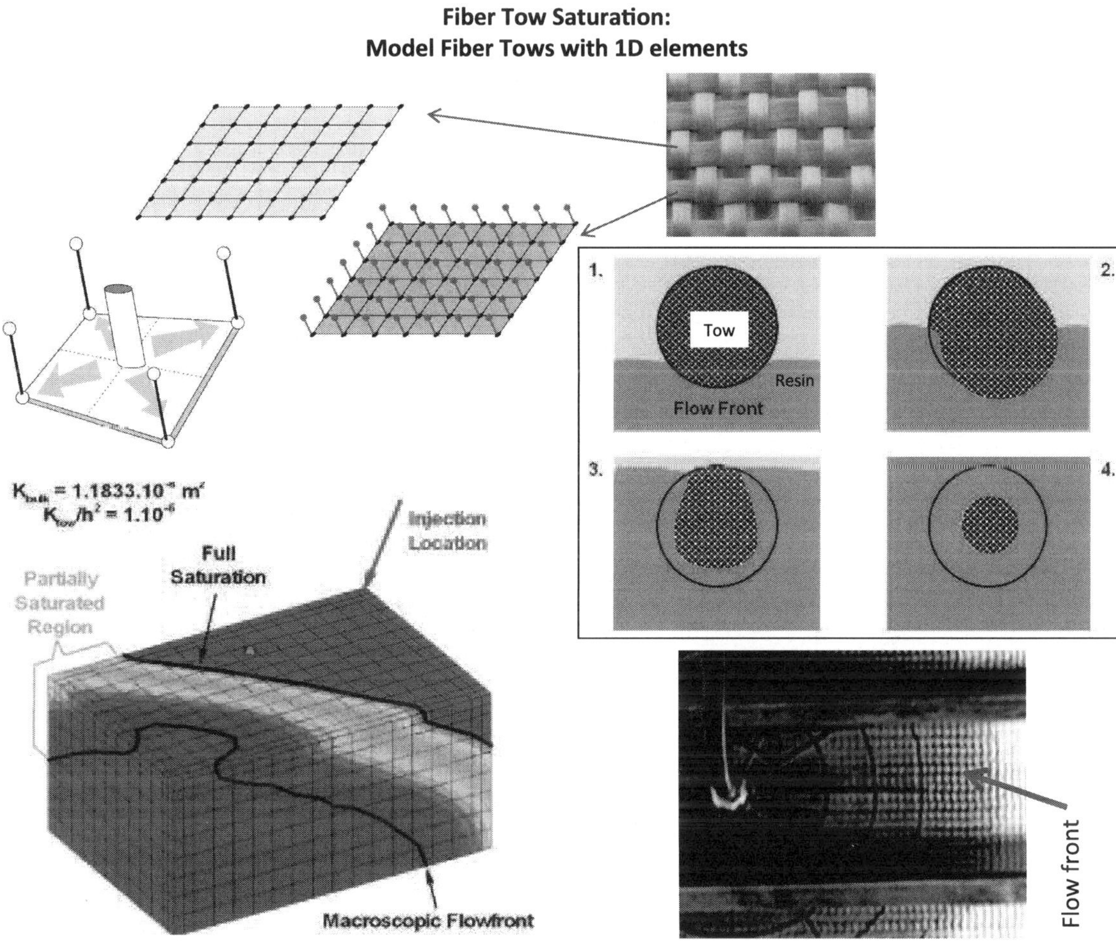

Figure 8.67: In LIMS one can also model the filling of fiber tows in addition to filling the space in between the fiber tows by attaching one-dimensional elements to every node of the mesh and assigning fiber tow properties to these elements. This way, one can separately track the filling of fiber tows along with the filling of the bulk flow for dual scale fabrics [415].

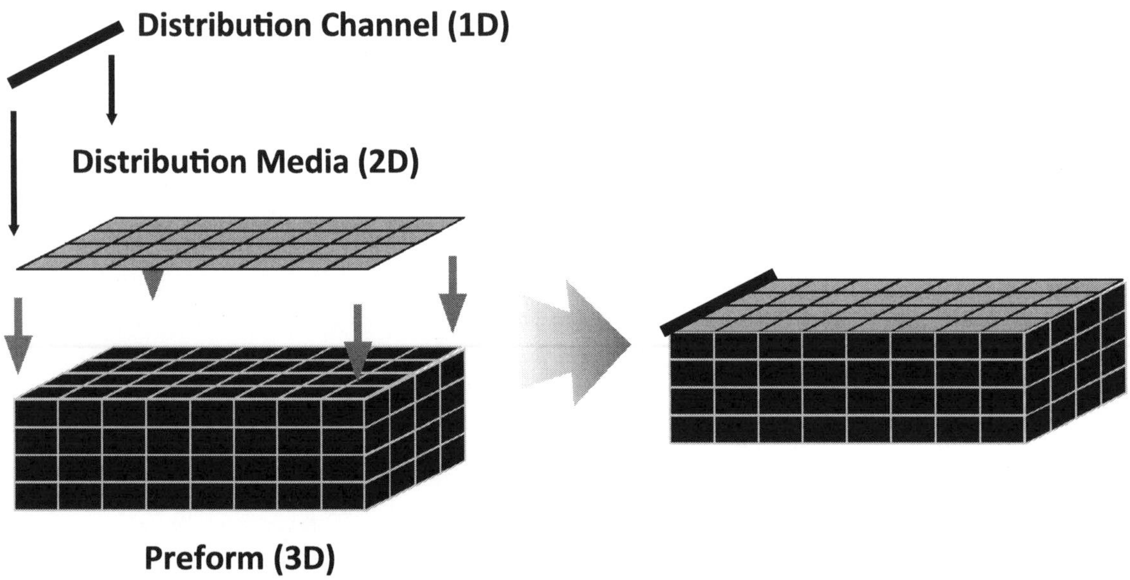

Figure 8.68: To overcome finite element aspect ratio issues, combination of 2D distribution media on top of the 3D part model can be integrated in LIMS. Distribution channels may be modeled as 1D elements. LimsUI allows setting of 1D element properties by specifying the tubing diameter [415].

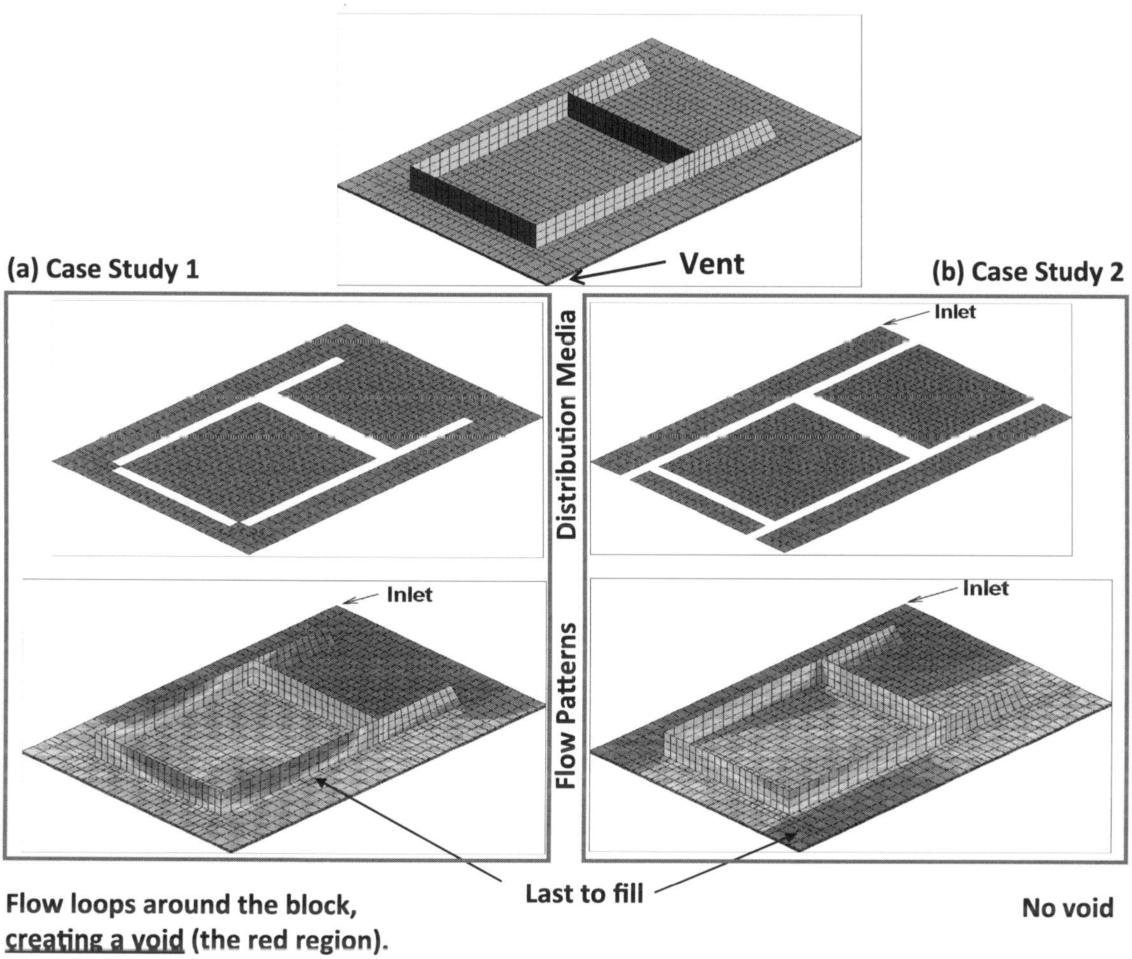

Figure 8.69: LIMS allows you to experiment with different Distribution Media (DM) designs in VARTM. Note how small changes in DM design can change the flow pattern significantly.

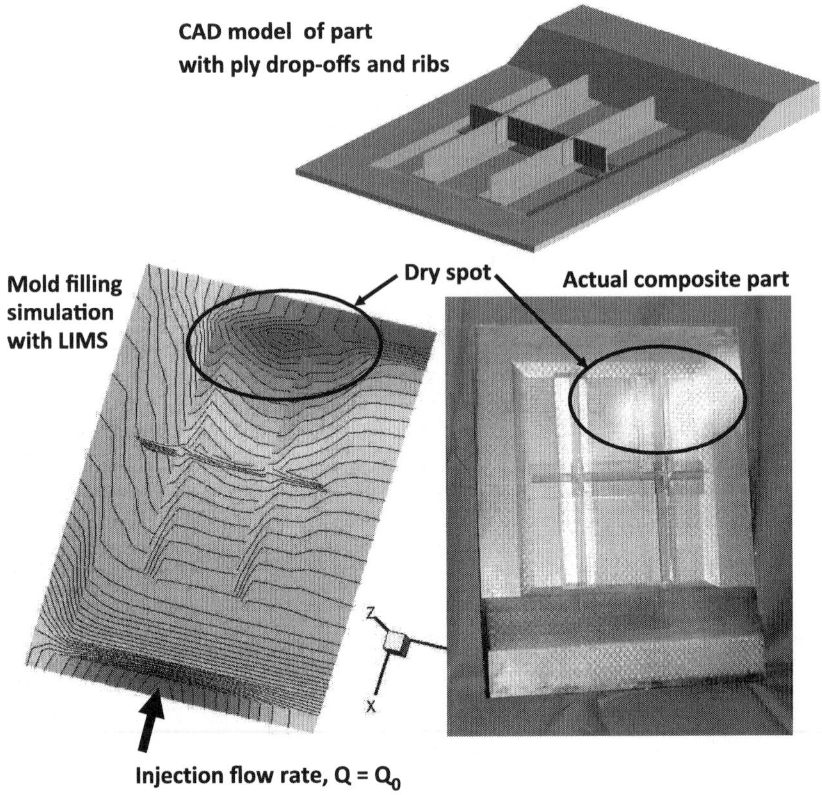

Figure 8.70: Resin flow in a 3D Complex Part: LIMS validation with experiment.

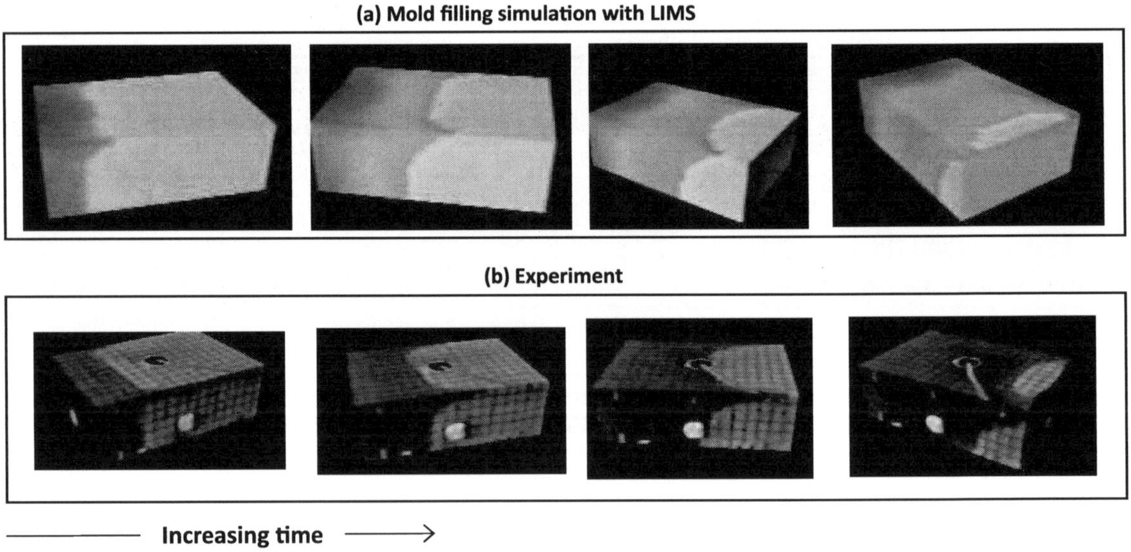

Figure 8.71: Validation of flow in a non-planar box mold: (a) the LIMS simulation, and (b) the experiment.

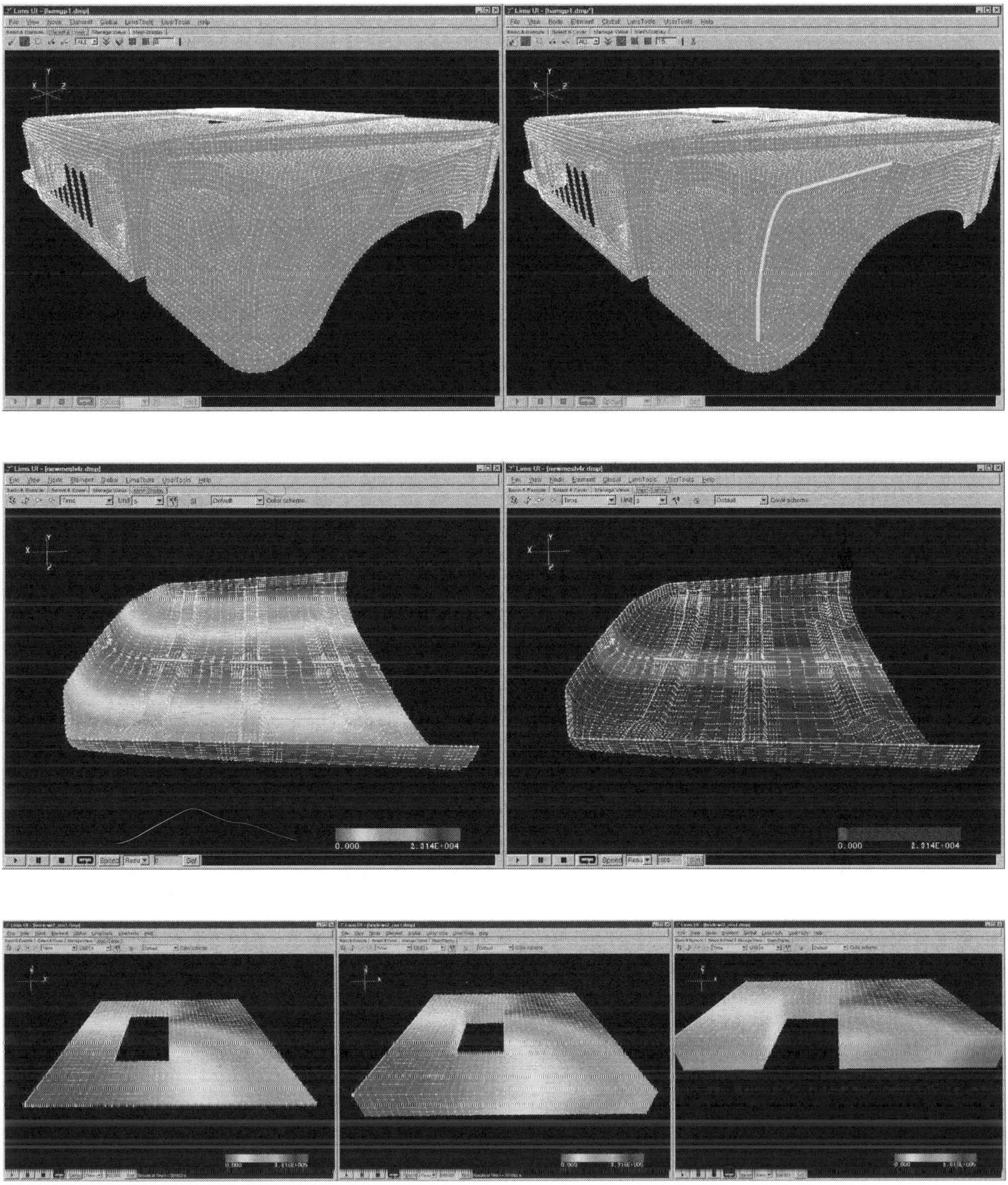

Figure 8.72: Example of LimsUI in action. (Top): Automated creation of the racetracking channel over geometric edge. (Middle): Standard result visualization – flow development as contour plots and as an animation. (Bottom): Post-processing of flow in thin plates: Standard contour plot display – through-the-thickness magnification – cross-section of the part.

8.3.11 Case Studies Using LIMS

The following examples will be used to illustrate how to use LIMS program for mold filling simulations. As mentioned earlier, this robust program can handle many issues of LCM processes such as

- Search for the optimal process parameter window (design of fabric preform and resin to be used, locations of injection/ventilation gates and their pressure or flow rate values) to achieve some process goals (short mold filling time, low injection pressure and elimination of macrovoids),
- Handling of nonuniform material properties (if the resin's viscosity and preform's fiber volume fraction and permeability vary with time and/or spatially),
- Investigation of the effects of process disturbances (such as racetracking channels and nonuniform material) on the process variables considering different "mold filling scenarios" that are likely to occur.

Example 8.15: Rectangular Mold with a Line Injection Gate

A rectangular mold cavity is shown in Figure 8.73; the material, geometry and process parameters are listed below:

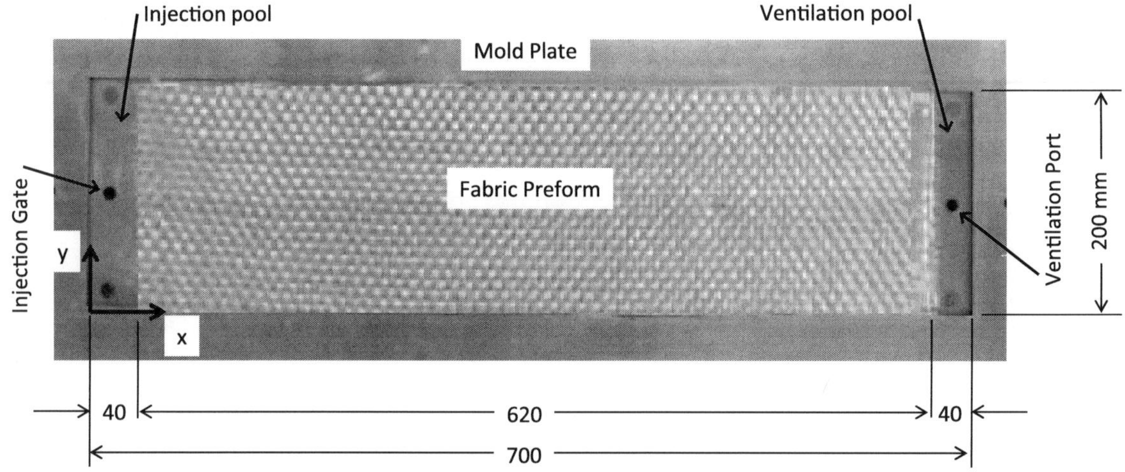

Figure 8.73: Top view of the mold cavity; the thickness of the cavity is 5 mm everywhere.

- Constant injection flow rate, $Q_{in} = 2 \times 10^{-6}\,\text{m}^3/\text{s} = 2\,\text{cc/s}$ from a single gate, Gate 1 located at $(x, y) = (0, 0.1)$ m
- Resin viscosity, $\mu = 0.10$ Pa.s
- Fiber volume fraction, $V_f = 0.40$
- Fabric preform's permeability tensor,

$$[\mathbf{K}_A] = \begin{bmatrix} K_{xx} & K_{xy} \\ K_{yx} & K_{yy} \end{bmatrix} = \begin{bmatrix} 4.0 \times 10^{-10} & 0 \\ 0 & 4.0 \times 10^{-10} \end{bmatrix} \text{m}^2$$

- Different regions of the mold cavity (see Figure 8.74):
  - A: uniform fabric preform (inner region),
  - B: the two edges of the fabric preform along the mold wall,
  - C: the two empty pools to allow lineal injection/ventilation; $K_{xx,\text{pool}} = K_{yy,\text{pool}} = 4.0 \times 10^{-7}\,\text{m}^2$ and $V_{f,\text{pool}} = 0$
- Scenarios for resin flow:
  - S1: $V_{f,B} = V_{f,A}$ and $[\mathbf{K}_B] = [\mathbf{K}_A]$ (i.e., no racetracking channel due to perfect fabric cutting and placement),
  - S2: $V_{f,B} = 0.50\,V_{f,A} = 0.20$ and $[\mathbf{K}_B] = 10\,[\mathbf{K}_A]$ (i.e., medium level of racetracking channel is expected),
  - S3: $V_{f,B} = 0.25\,V_{f,A} = 0.10$ and $[\mathbf{K}_B] = 100\,[\mathbf{K}_A]$ (i.e., high level of racetracking channel is expected)
- The mold cavity is initially vacuumed from the right end.

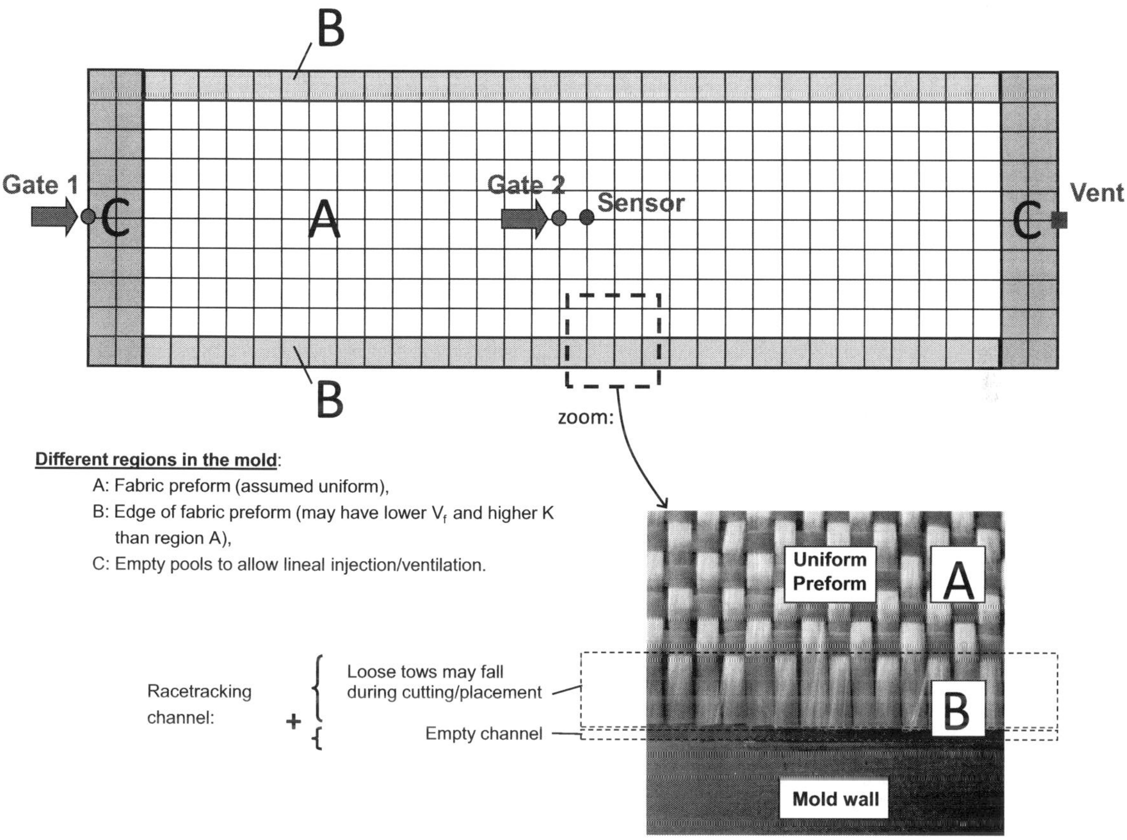

Figure 8.74: Different regions in the mold cavity.

Do the following:

- Perform and describe the procedure of the pre-processing (i.e., how to generate FEM mesh, input of material data and boundary conditions, and handle racetracking channels and the empty injection/ventilation pools),
- Describe the solution procedure (i.e., run LIMS program),
- Show how post-processing can be used to plot the results,
- Find the simulated mold filling time, $t_{fill,sim}$, in the three scenarios mentioned above; then, compare these values with the analytical solution,

$$t_{fill,an} = (\text{Volume of the empty cavity})/Q_{in} \tag{8.119}$$

  as the injection is carried out at constant flow rate,
- Make some concluding comments about the effect of racetracking channels on the flow front propagation, the design of vent location(s) and the maximum value of injection pressure needed.

Solution

One can use LIMS in one of the two following ways:

1. **LimsUI** (User Interface for the Liquid Injection Molding Simulation) to
   - Input data (material, geometry and process parameters),
   - Run the simulation,
   - View the results.

   The mesh file can be either
   - Built by a script for LEGO (included with LIMS), or
   - Converted from any commercial CAD software (such as I-DEAS or ABAQUS) to LIMS's "DMP" input file format.

   The former choice has two advantages: (i) it does not use any external program, and (ii) it allows one to create regions in the mesh if one wants to simulate different resin flow scenarios in LimsUI.
2. Prepare the FEM mesh using an external program such as PATRAN; convert it to LIMS's "DMP" input file format; attach the material and process data to the prepared file; write a LIMS script file to be run on the **Command Window** of LIMS; store the results in an output file; and eventually view the results using another external program such as Microsoft Excel, Tecplot or MATLAB.

Here, we will use the former choice because it has the advantage of using only one single program from the beginning to the end. The user is suggested to follow the tutorial and help menu of the User's Manual of LimsUI.

Figure 8.75 shows the mesh with uniform quadrilateral elements. The nodes and elements of the mesh are detailed in Figure 8.76. Note that the mesh is a coarse one since we will be using the "Demo" version of LimsUI (which is limited to a maximum of 2000 nodes). The unlimited version of LimsUI and LIMS can easily handle tens of thousands of nodes at a very reasonably low CPU time of computation and is available for licensing through the University of Delaware.

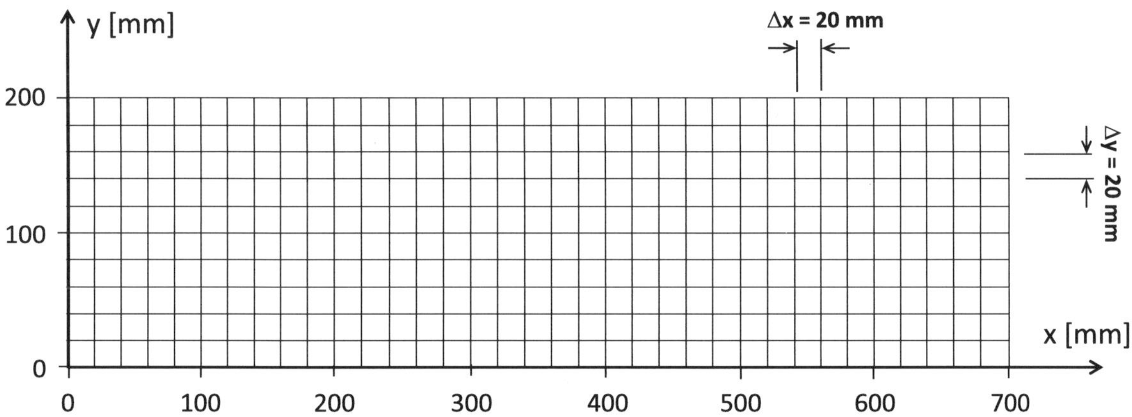

Figure 8.75: The FEM mesh to represent the mold cavity. Although the dimensions are given in millimeters here, the coordinates of the nodes will be in meters when used in LIMS.

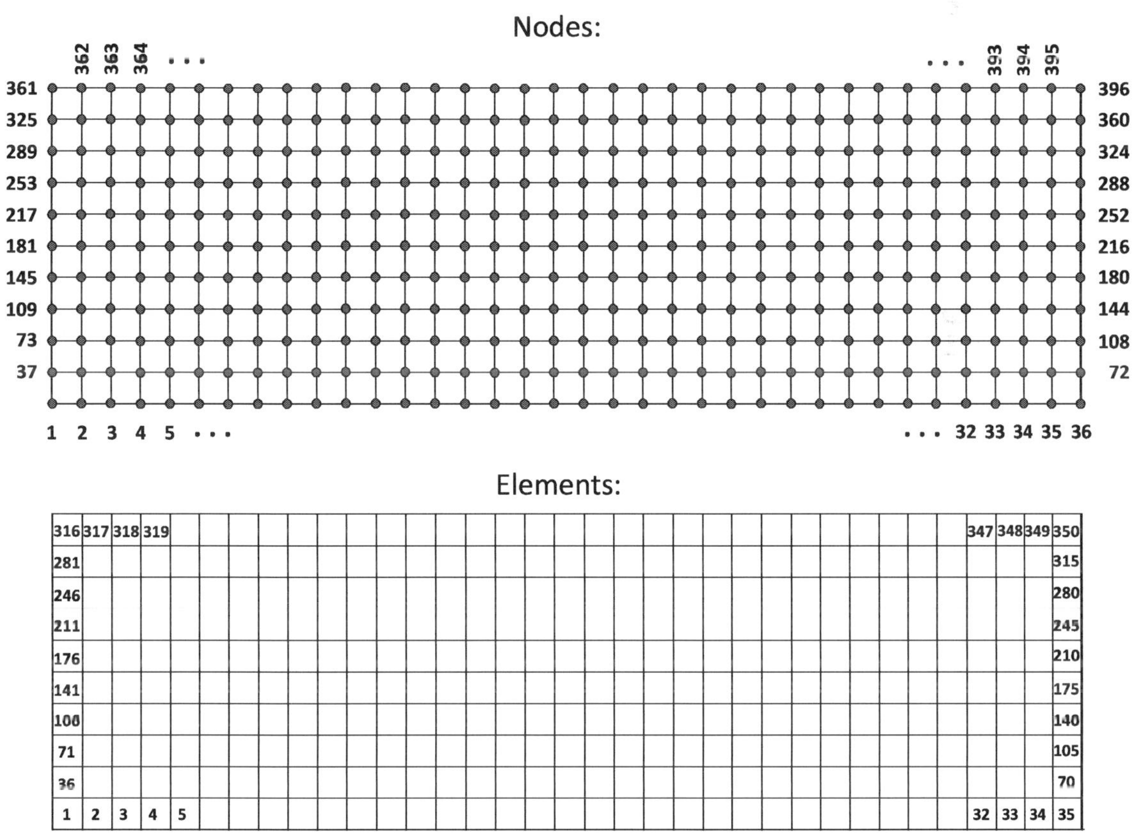

Figure 8.76: Nodes and elements used in the FEM solution of the mold filling.

LIMS input file, "**rectanglemold.dmp**" is listed below:

```
#Lims dump File (.dmp) generated by LimsUI
...............................................
Number of nodes : 396
 Index        x             y             z
===============================================
     1        0.000000      0.000000      0.000000
     2        0.020000      0.000000      0.000000
     3        0.040000      0.000000      0.000000
 ..............................................
   394        0.660000      0.200000      0.000000
   395        0.680000      0.200000      0.000000
   396        0.700000      0.200000      0.000000
Number of elements : 350
 Index  NNOD  N1    N2    N3    N4  ...  h        Vf          Kxx        Kxy   Kyy     ...
==========================================================================================
     1   4     1     2    38    37       0.005    0.400000    4e-010     0     4e-010
     2   4     2     3    39    38       0.005    0.400000    4e-010     0     4e-010
     3   4     3     4    40    39       0.005    0.400000    4e-010     0     4e-010
 ........................................................................................
   348   4   357   358   394   393       0.005    0.400000    4e-010     0     4e-010
   349   4   358   359   395   394       0.005    0.400000    4e-010     0     4e-010
   350   4   359   360   396   395       0.005    0.400000    4e-010     0     4e-010
Resin Viscosity model NEWTON
Viscosity :              0.1
```

A brief explanation of major entries will be given below; the user is referred to the *LIMS User's Manual* for a more complete index and a help menu.

- The node and element numbers must be continuous starting from 1. In this example, there are 396 nodes and 350 elements (see Figure 8.76).
- Each node number is followed by (x, y, z) coordinates of that node in meters.
- Each element number is followed by
  - Number of nodes that the element has (3 for a triangular element, 4 for a quadrilateral element, and so on).
  - Node-connectivity (i.e., the nodes, N_1, N_2,... of the element in either CW or CCW direction, consistently applied to all the elements). For example, observe that the connectivity of element 1 is [1 2 38 37] (see Figure 8.76).
  - Thickness of the mold cavity is the dimension normal to the 2D shell element used here; for 3D analysis refer to the *LIMS User's Manual*.
  - Permeability components of the preform in [m^2] at the location of element.
- Resin viscosity is in [Pa.s]; and viscosity model is denoted as the Newtonian viscosity model in this example. For other viscosity models, see the *LIMS User's Manual*.

Follow these steps for the simulation of resin flow using LimsUI:

- Start LimsUI by clicking on the "**limsui.exe**" icon; a user-interface window of LIMS will open.
- From the "File" menu, select "Open," and locate the mesh file "rectanglemold.dmp." The mesh will load (see Figure 8.77). One can zoom in a region of interest of the mesh or change the view by using the "Manage Views" menu.
- Using the "Select and Cover" menu, assign the gate(s) as follows:
  - "Select Node"
  - Click on the node where you wish to locate the gate(s).
  - Using "Node" menu click on "Set Gate."
  - Select the type of gate(s); for this example it is constant "FlowRate," specify flow rate as $2\,\mathrm{cm}^3/\mathrm{s}$ (see Figure 8.77).
- If no node is assigned as a vent, it is meant to have a perfect vacuum in the mold cavity; and this is the default case in LIMS. Considering this, we will not assign any vent in this example. However, for a different example, one can use "`SETGATE i,5,p`" command of LIMS in a script file to assign a vent where `i` and `p` are the node number and ventilation pressure, respectively.
- In order to assign the material properties (K and V_f) of the inlet/exit resin pools and racetracking channels, the elements of the regions B and C need to be modified. Using the "Select and Cover" menu, follow these steps:
  - "Select Element"
  - Click on a group of elements to define a section (area).
  - The right button on the mouse allows one to open "Material Properties" window and specify (i) "Volume Fraction" of fibers or porosity; and (ii) "Permeability" components.
- Run the simulation of resin flow by selecting the following sequence from the menu: "Basic and Execute" → "File" → "Run Simulation" (see Figure 8.78).
- The results will be stored in a default file name "basename_res1.dmp" and it will be viewed in LimsUI window as shown in Figure 8.78.
- Example 8.18 in Appendix C illustrates sequential simulations of the three different scenarios using a single LIMS script file. This script file is called on the command window of LIMS. One can track the injection pressure or flow rate storing the data in an output file, and then viewing it using a program such as Microsoft Excel or MATLAB (see Figure 8.79).

Some comments about the results:

- As expected, the flow is almost 1D (except in the inlet pool) when there is no race-tracking channel in S1.
- The patterns of flow front are greatly affected and become 2D as $[\mathbf{K}_B] >> [\mathbf{K}_A]$ in S2 and especially S3.

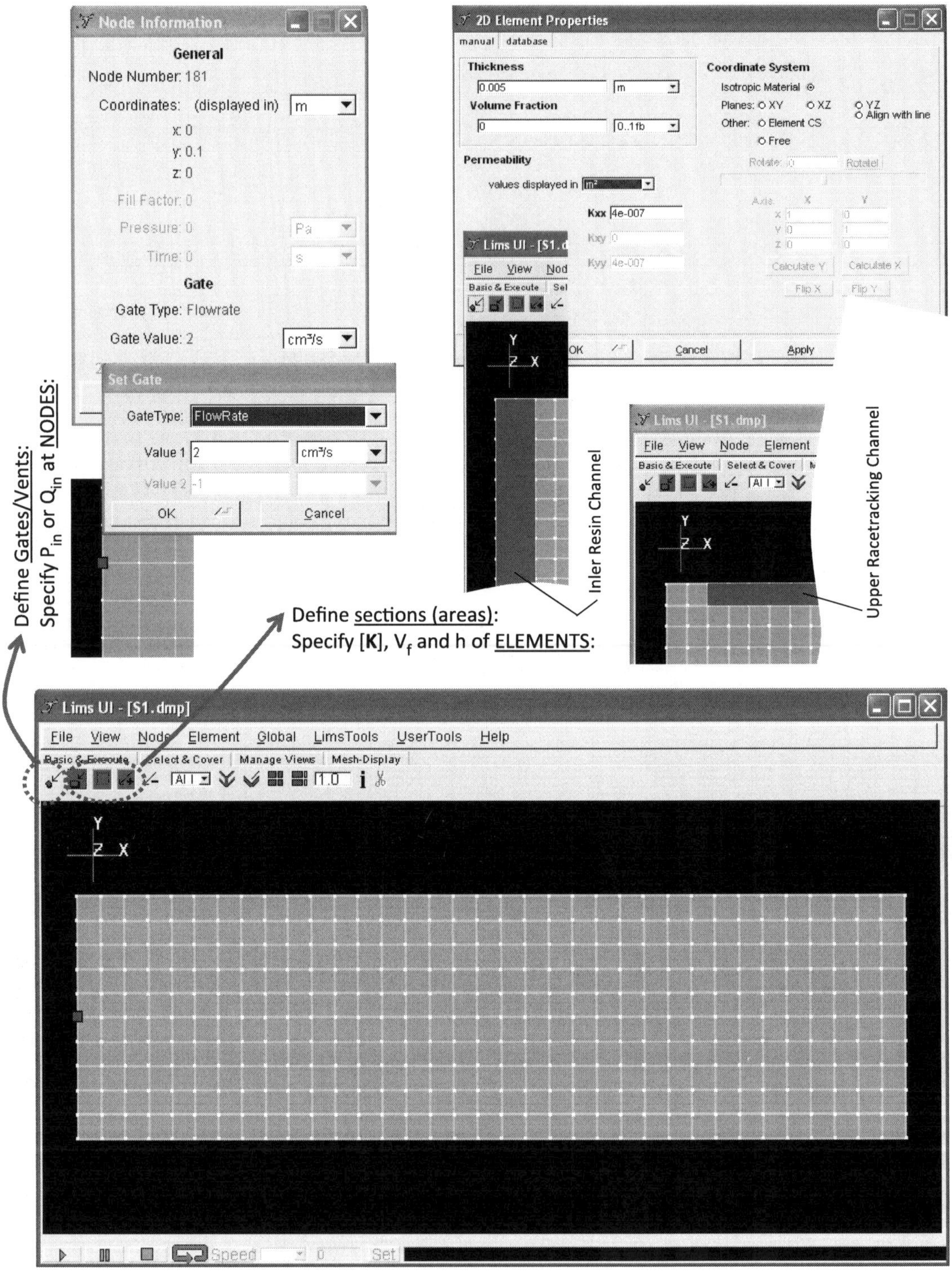

Figure 8.77: Assigning material properties in different sections of the mesh and defining the boundary conditions at the gate(s) by using LimsUI.

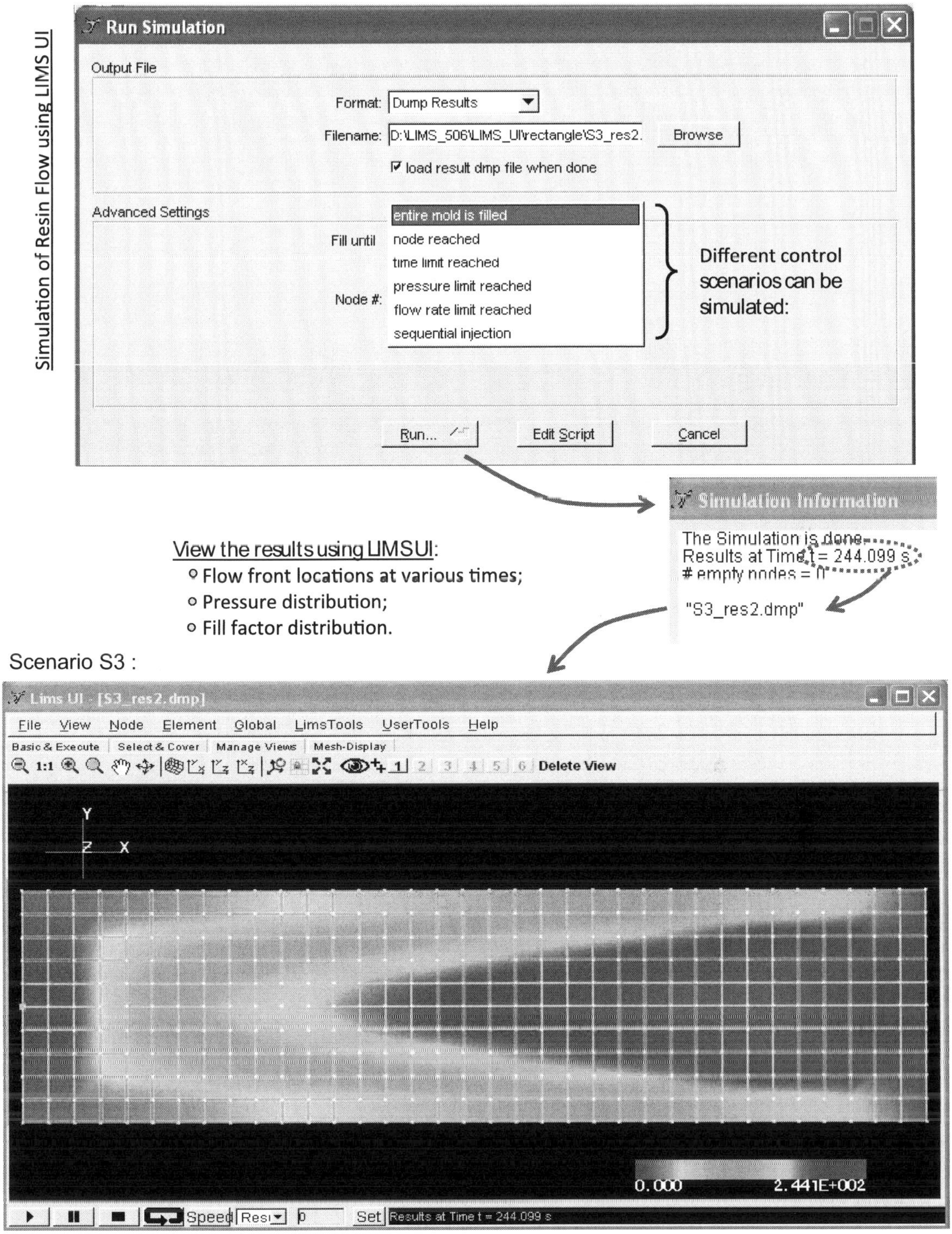

Figure 8.78: (Top): Window during the simulation run. (Bottom): flow front locations during filling as a function of time with LimsUI.

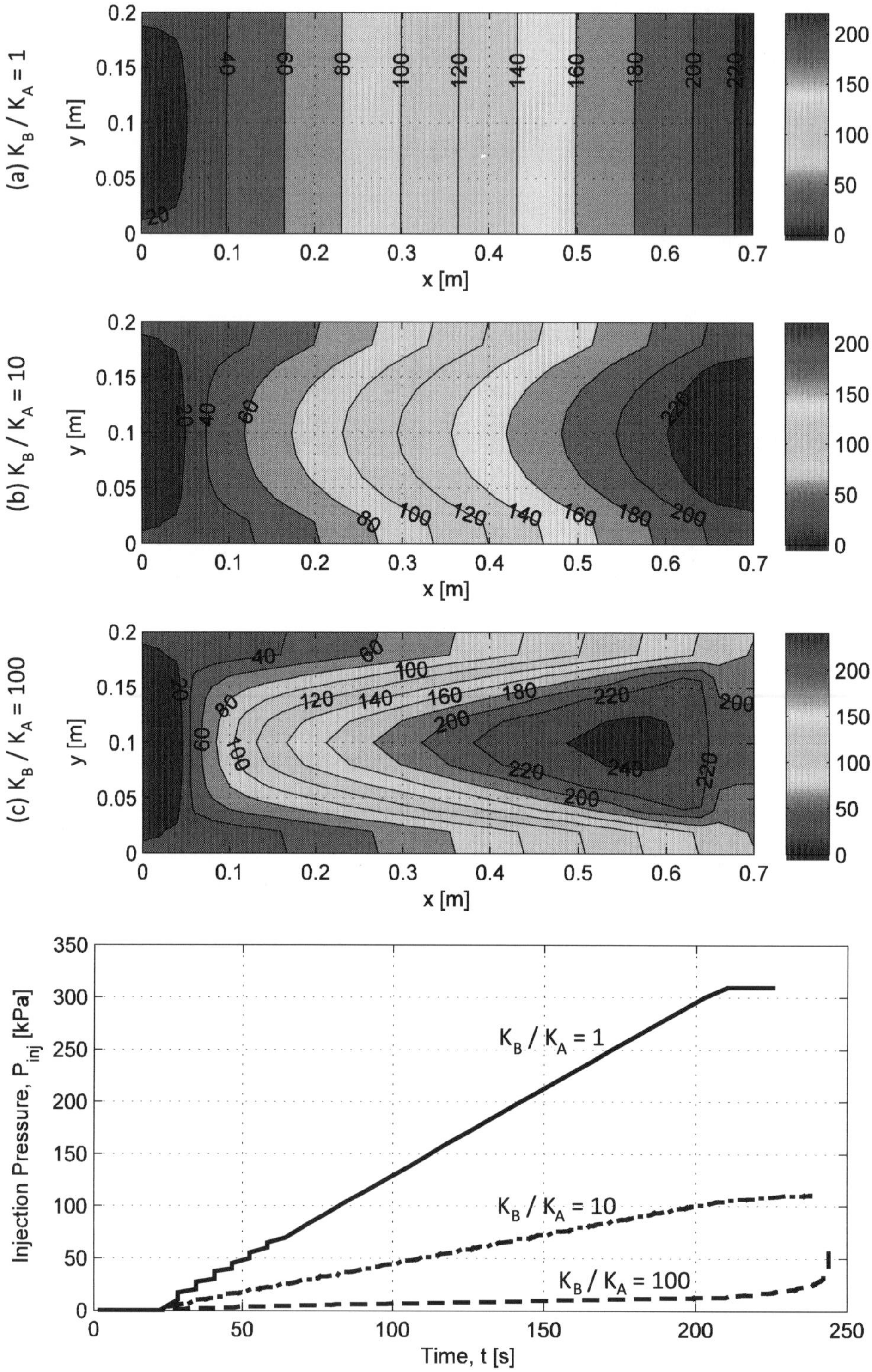

Figure 8.79: Simulated flow front propagation and injection pressure for $K_B/K_A = 1$, 10 and 100 by using LIMS program.

- The simulated $t_{fill,sim}$ is listed in the following table for the three scenarios. The analytical time to fill, $t_{fill,an}$, was calculated using Equation (8.119) in a MATLAB code, *example8_15_analytical.m* (listed in the Appendix) and the results are also listed in the same table. There is an excellent agreement even at the coarse resolution of the FEM mesh used here. The slight variation in t_{fill} for different scenarios is due to the three different $V_{f,B}$ values (0.40, 0.20 and 0.10 in S1, S2 and S3, respectively) which affects the empty volumes of two B regions.

Table 8.4: Mold Filling Time Using LIMS Simulations and an Analytical Formula

Scenario	$t_{fill,sim}$ [s] by LIMS	$t_{fill,an} =$ [s]	Error [%]
S1	225.4931	225.5000	0.0031
S2	237.8988	237.9000	0.0005
S3	244.0999	244.1000	0.0000

- The effect of $[\mathbf{K}_B]$ on the P_{inj} is very clearly seen in Figure 8.79. The maximum injection pressure, which occurs at $t = t_{fill}$, is 310.1, 110.9 and 62.6 kPa in S1, S2 and S3, respectively. Clearly, the racetracking channels helped reducing the injection pressure greatly. One can conclude that racetracking channels should be desired from this point of view; however, it should not be forgotten that the last point to be filled is also affected. Thus, the location(s) of ventilation port(s) should be carefully designed if one tries to add intentional racetracking channels. Otherwise, the disadvantages of racetracking channels (such as controlling the clamps of multiple vents, being unsure about how long the early bleeding from the vents should be allowed, the corresponding waste of resin during the bleeding, and perhaps the most importantly any potential macrosize voids remaining in the composite part) could overshadow the advantages of racetracking channels.

- The simulations were done assuming that the mold cavity was fully vacuumed before the start of injection, so that no ventilation port had to be placed on the mesh as a boundary condition. In reality, this corresponds with the fact that the vacuuming and the ventilation are applied at the last point to fill. If this is not the case, the vent location should be added in either the input or script file. This will be illustrated in one of the examples later in the Appendix. ■

Example 8.16: Semicircular Part with Flanges in VARTM Process

A prototype composite part and its mold used in the VARTM process are shown in Figure 8.80. In order to design a process parameter window suitable for high volume manufacturing, a design engineer would like to adjust the curing agents of a polyester so that the pot life (gellation time) is approximately 15 minutes. The region B is an isotropic fabric preform with a low permeability. The region A is the same preform with a distribution medium (a.k.a. flow mesh) placed on it; thus, its effective in-plane permeability is higher than the region B. The resin is injected from the two point gates (G1 and G2) using inlet connectors with short omega-tubes underneath. The engineer experiments with different sizes and placement locations for the distribution media (one of them is shown in Figure 8.81 schematically) so that the resin races in the outer peripheral region and then slows down when it starts flowing in the inner region shown with the darker color in the same figure (where there is only fabric preform, i.e., no distribution medium exists). The engineer's aim is to complete the mold filling in approximately 11-12 minutes; and then wait for the remaining 3-4 minutes to obtain a full saturation by filling any possible microvoids between dry fibers. During this extra time, the two injection gates and the ventilation port are still held open.

The material, geometry and process parameters are listed below:

- Pressure differential between inlet and exit, $P_{in} - P_{vent} = 100{,}000\,\mathrm{Pa}$
- Resin viscosity, $\mu = 0.10\,\mathrm{Pa.s}$
- Part thickness, $h = 0.005\,\mathrm{m}$ (assume constant)
- Fiber volume fraction, $V_f = 0.40$

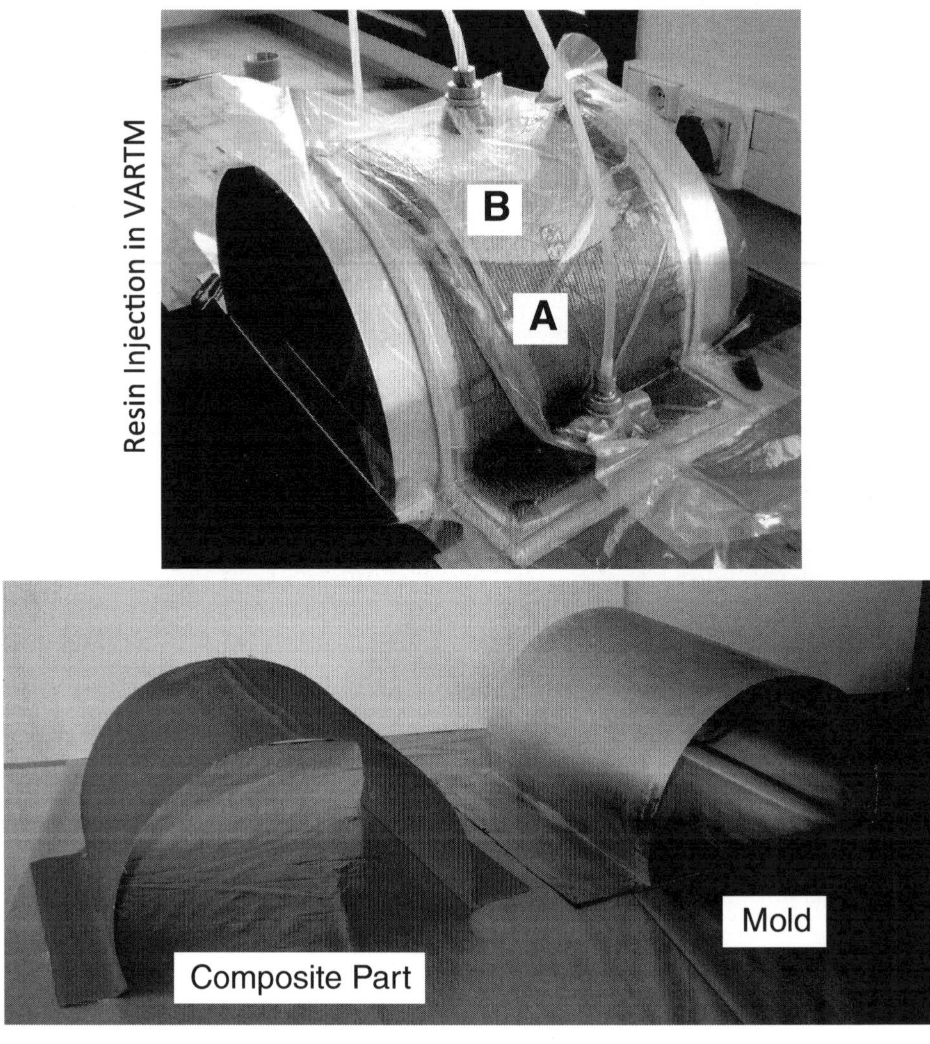

Figure 8.80: Composite part and mold in the VARTM process.

- Fabric preform's permeability tensor,

$$[\mathbf{K}_B] = \begin{bmatrix} K_{xx} & K_{xy} \\ K_{yx} & K_{yy} \end{bmatrix} = \begin{bmatrix} 0.19 \times 10^{-10} & 0 \\ 0 & 0.19 \times 10^{-10} \end{bmatrix} \text{m}^2$$

- Effective permeability tensor for the "fabric preform + distribution medium" assembly,

$$[\mathbf{K}_A] = \begin{bmatrix} 5.3 \times 10^{-10} & 0 \\ 0 & 5.3 \times 10^{-10} \end{bmatrix} \text{m}^2.$$

Although, in this example it is assumed that the part thickness is constant during the injection, it is not true in the actual case since the resin and compaction pressures change with time in VARTM as experimentally shown in [279, 418]. If the user wishes to include

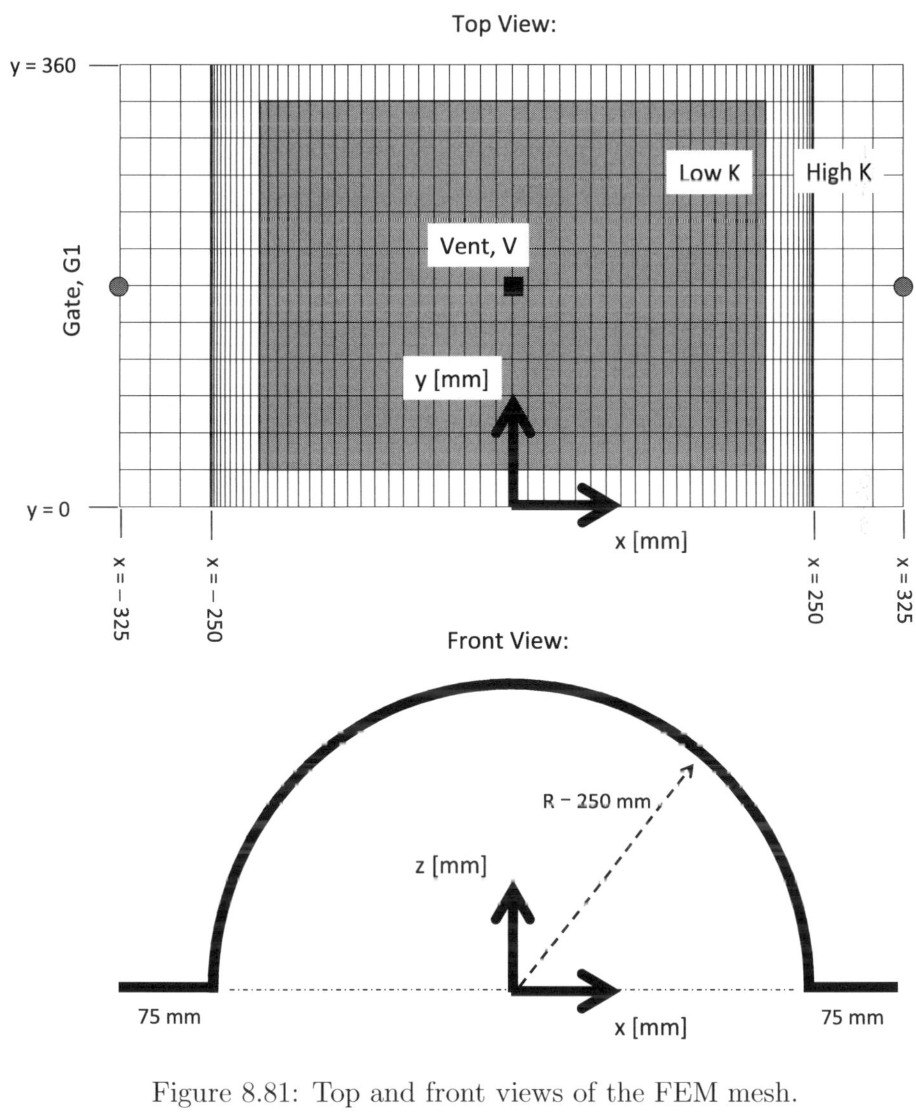

Figure 8.81: Top and front views of the FEM mesh.

this effect and get more accurate mold filling simulation, LIMS can appropriately handle this effect simply by updating the distributions of thickness, fiber volume fraction and permeability after each time step. Each FEM element may have a different and time-varying h, V_f and K; and the initial material data can be easily updated via the `SCALEPERM` and `SETTHICK` commands in a script file (see the LIMS examples in the Appendix which execute the script files for different control schemes). However, one needs a broad material characterization database on compaction/decompaction and permeability as experimentally obtained in [279, 418] for three types of fabric.

Solution

The mesh of the mold cavity is stored in "**semicirclemold.dmp**" file by using the converter program "convert2dmp.exe" to convert the mesh from any commercial CAD software to LIMS's input format. The nodes and elements of the mesh are detailed in Figure 8.82.

LIMS input file, "**semicirclemold.dmp**" is listed below:

```
#Lims dump File (.dmp) generated by LimsUI
...............................................
Number of nodes : 871
  Index        x             y             z
===============================================
      1     -0.325000     0.000000      0.000000
      2     -0.300000     0.000000      0.000000
      3     -0.275000     0.000000      0.000000
    ............................................
    869      0.275000     0.360000      0.000000
    870      0.300000     0.360000      0.000000
    871      0.325000     0.360000      0.000000
Number of elements : 792
  Index  NNOD   N1    N2    N3    N4  ...  h        Vf          Kxx        Kxy    Kyy  ...
=========================================================================================
      1    4     1     2    69    68      0.005    0.400000    5.3e-010    0   5.3e-010
      2    4     2     3    70    69      0.005    0.400000    5.3e-010    0   5.3e-010
      3    4     3     4    71    70      0.005    0.400000    5.3e-010    0   5.3e-010
   ....................................................................................
    790    4   801   802   869   868      0.005    0.400000    5.3e-010    0   5.3e-010
    791    4   802   803   870   869      0.005    0.400000    5.3e-010    0   5.3e-010
    792    4   803   804   871   870      0.005    0.400000    5.3e-010    0   5.3e-010
Resin Viscosity model NEWTON
Viscosity :             0.1
```

Notice that the permeability of all FEM elements are assigned as $K_{xx} = K_{yy} = 5.3 \times 10^{-10}\,\mathrm{m}^2$ which is the permeability of the region A. The elements of region B are modified in LimsUI as shown in Figure 8.83 by following steps similar to the previous example. The flow front locations at various times are shown in Figure 8.84 using the post-processing feature of LimsUI. The injection flow rate $Q_{in}(t)$ is also shown in the same figure. As expected, the resin fills the outer region A very fast (less than in 200 seconds); then it arrives the ventilation port at $t_{fill} = 671.0\,\mathrm{s} = 11.2$ minutes. This mold filling time is within the desired interval of 11-12 minutes. The injection flow rate is high initially, and decreases as the pressure gradient decreases with time. ■

In Appendix C, additional examples are presented that use LIMS to illustrate different features such as (i) changing the boundary conditions during filling, (ii) sensor data evaluation, and (iii) sequential gating. In some of these simulations, preparation of script files and their execution in LIMS program is also shown. The examples presented in Appendix C are

- Example 8.18: Sequential Simulations of Scenarios S1-S3 of Example 8.15 with a Script File.
- Example 8.19: Re-solve Example 8.15 with a Constraint of $P_{in} \leq 200\,\text{kPa}$.
- Example 8.20: Re-solve Example 8.15 with a Sequential Gating Control Scheme.

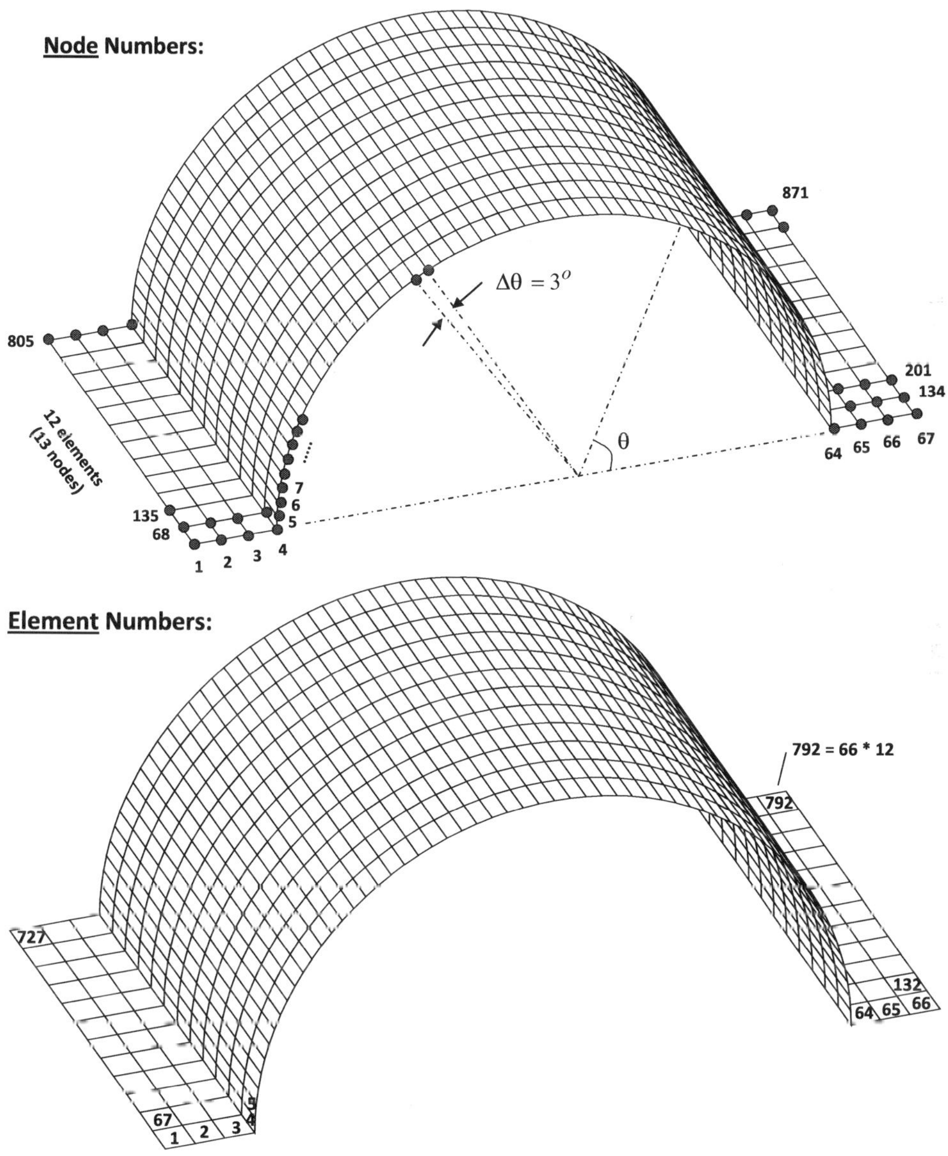

Figure 8.82: Nodes and elements used in the FEM mesh. For brevity, not all nodes and elements are shown.

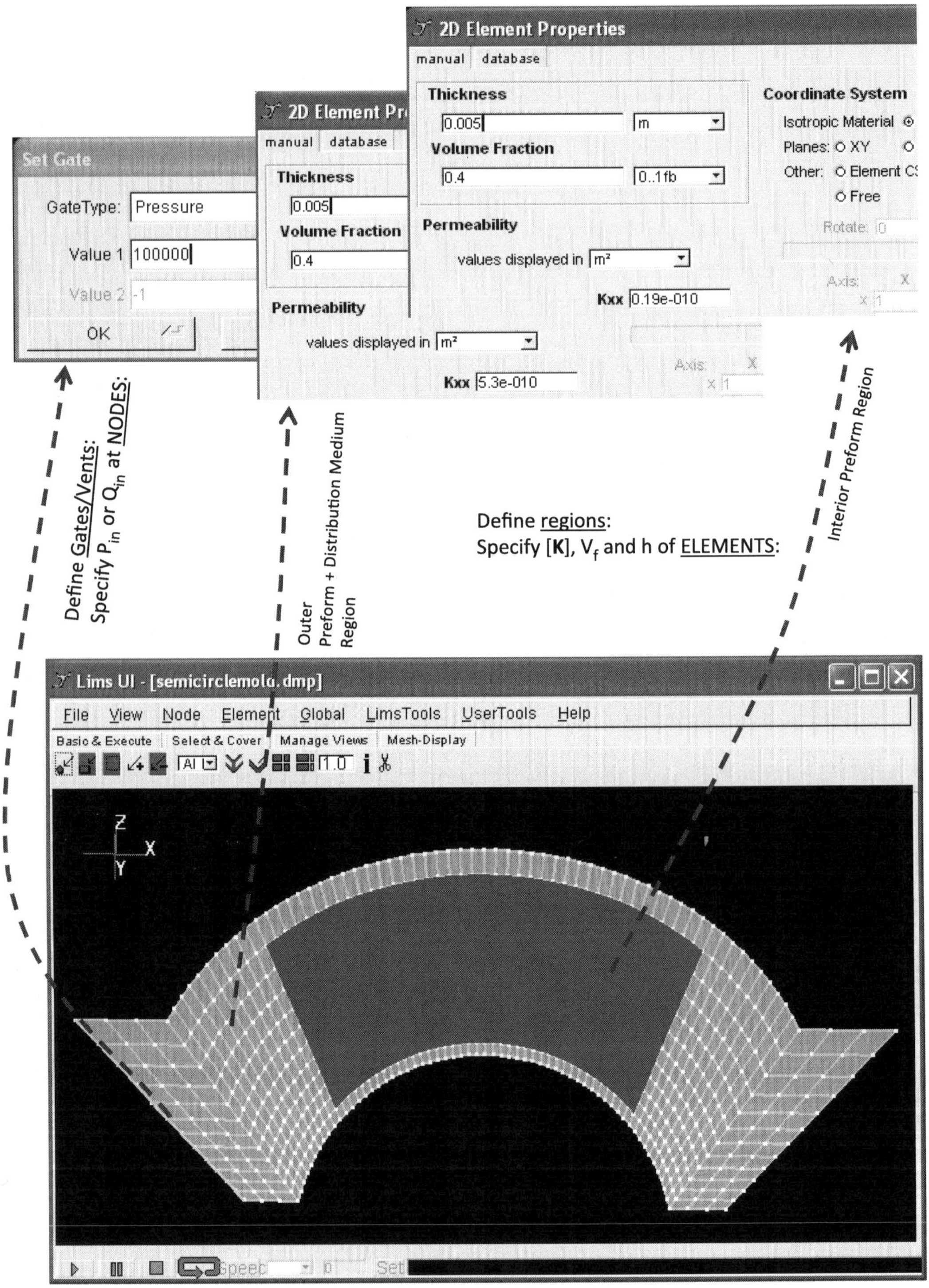

Figure 8.83: Assigning material properties in different regions of the mesh and defining the boundary conditions at the gates/vents by using LimsUI.

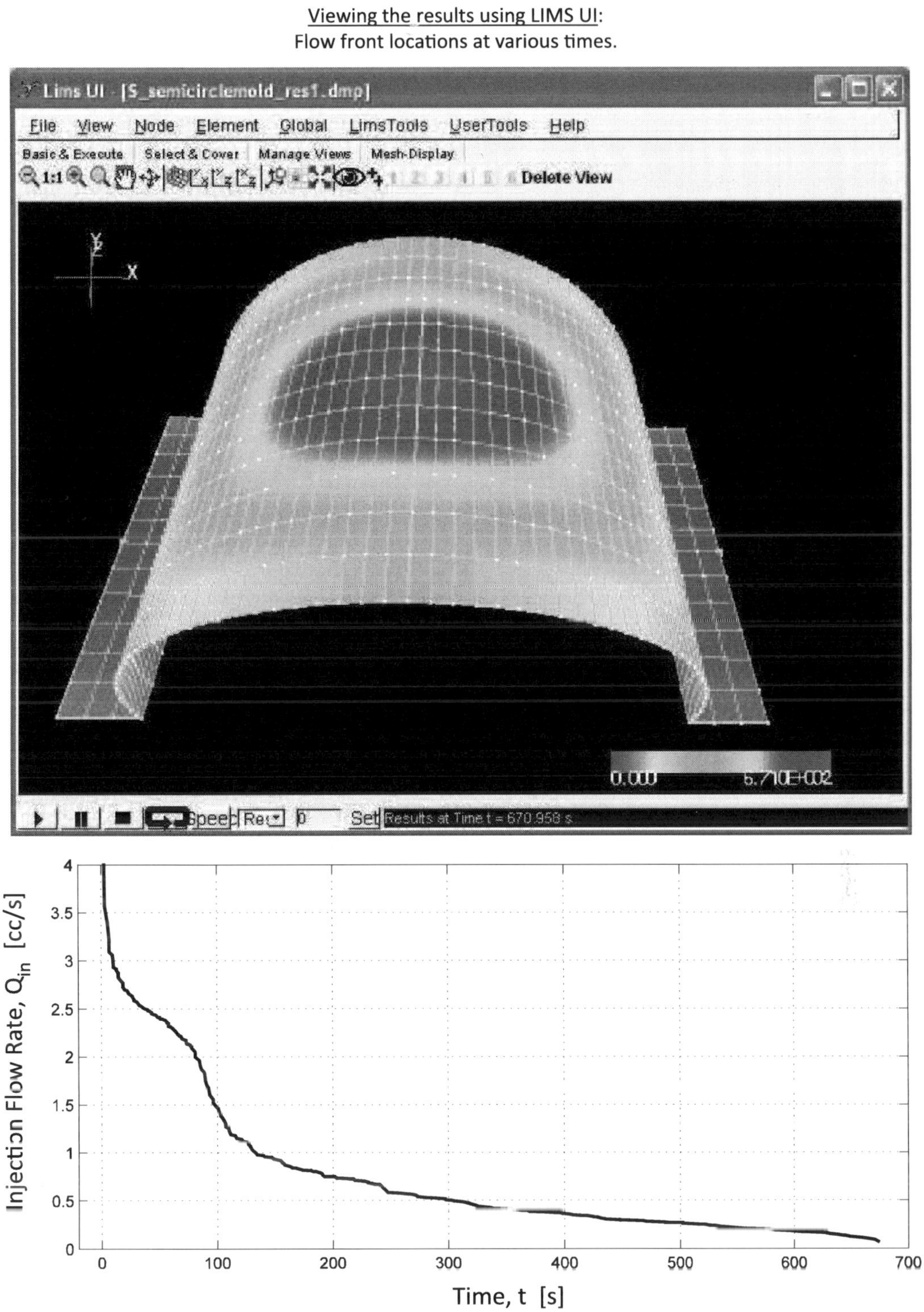

Figure 8.84: Viewing the flow front propagation by using LimsUI, and simulated injection flow rate at the gate.

- Example 8.21: 1D Permeability Measurement Experiment.
- Example 8.22: Lineal Injection under Constant Injection Pressure.
- Example 8.23: Radial Flow in a Hollow Disk.
- Example 8.24: Box Mold in RTM Process.

8.4 Filament Winding of Thermosetting Matrix Composites

8.4.1 Introduction

In thermoplastic filament winding, more commonly known as thermoplastic tape lay-up, a tape of continuous fibers is preimpregnated with thermoplastic resin. In this process, a relatively thin tape is consolidated on a substrate under the application of heat and pressure (Figure 8.85).

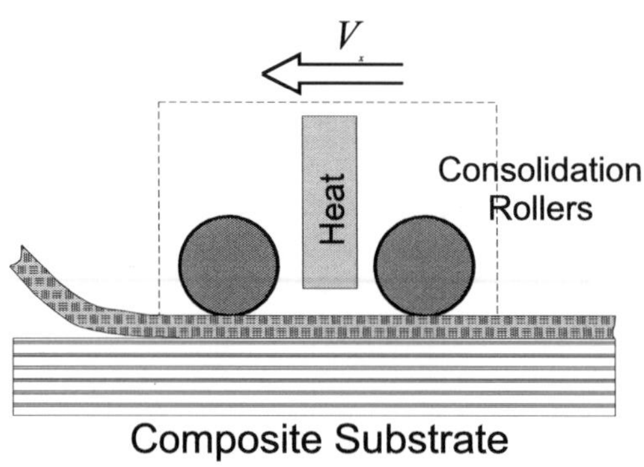

Figure 8.85: In-situ thermoplastic tape lay-up process.

In most cases, the feed tape, the heater (gas, induction, laser, etc.) and the consolidation rollers/shoes are traversed over the substrate at a predefined path and velocity.

Thermoset filament winding is a process of winding strands of reinforcement fibers embedded in a thermoset resin matrix onto a tool (mandrel), then curing the resin and removing the mandrel. In simple terms, a continuous strand (rope-like) reinforcement which may be a glass or a carbon strand is saturated in a resin bath, then wound over the mandrel like a thread over a spool by going across from one end to another and back. This winding is repeated until the desired thickness is achieved. Figure 8.86 illustrates the process [419].

One can precisely place the fibers on the mandrel as it is a machine controlled process and hence can be accomplished at high speed. The major development costs are the filament winding machine and the mandrel. However, as the process is automated, the labor cost is low and repeatability is high. High-speed, precise positioning of fibers in predescribed patterns and automation are the attractive features of the filament winding method for manufacturing composites. The fiber strands are either in the form of continuous resin-impregnated rovings or tows that are wound over a rotating male mandrel. The mandrel may be cylindrical, spherical, rectangular, or any shape that does not have a concave curvature. The reinforcement fiber may be wound either in adjacent bands or in repeating bands, which eventually may cover the entire mandrel surface. The winding tension, fiber angle, resin content and fiber material are important process and material variables that

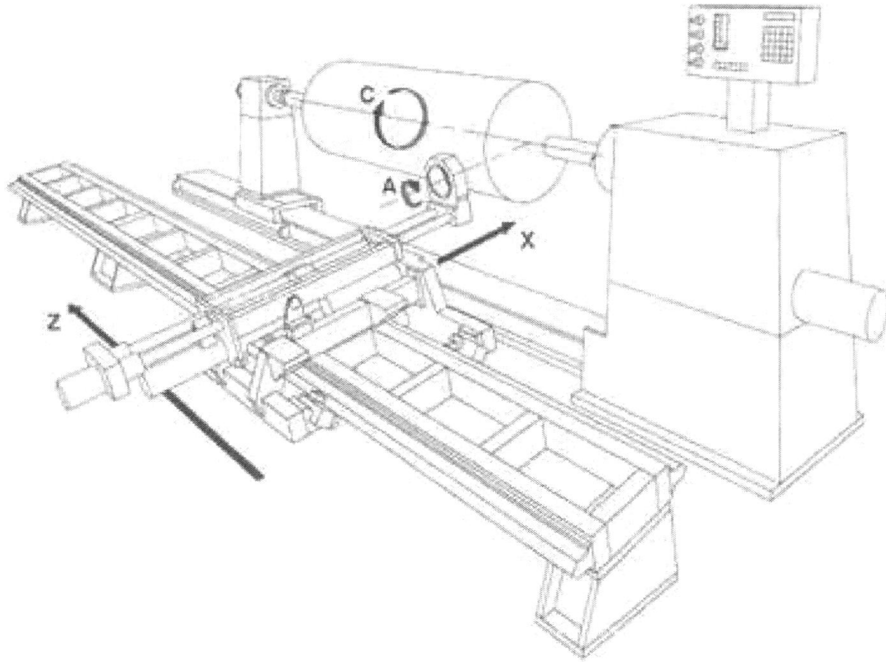

Figure 8.86: Thermoset filament winding process [419].

can be varied in each layer of reinforcement until the desired wall thickness is obtained. A good filament winding process model can enable the designer to engineer components by placing the exact mechanical characteristics where needed within the part and allow the design to be optimized for weight, stress concentrations and fatigue.

The thermoset filament winding process is primarily used for the manufacture of cylindrical and hollow composite parts. Figure 8.87 shows the mandrel and the composite part being wound. The tension in the filament can usually provide the consolidation pressure for fiber motion and to percolate the resin through it. The composite along with the mandrel may be cured in an autoclave if the curing needs to be initiated at a higher temperature.

Some of the key benefits of the process include machine controlled fiber placement, ease of automation, and the potential for high fiber volumes. Accurate, repeatable fiber placement from layer to layer and from part to part are other attractive features from the viewpoint of manufacturing. The capability to use continuous fibers over the length of a component (no joints or splices) and ability to easily orient fibers in the load direction are other useful features of filament winding that can influence the use of this process in applications. Complex shapes can first be wound as a preform and then molded to near net shape using filament winding. Filament winding also allows for customization of properties such as torsional stiffness, axial stiffness, tensile strength, proportional elastic limit, hardness, abrasion and wear resistance.

During this process, the following physical phenomena occur:

- *Thermochemical reaction and heat transfer:* The resin cures due to the cross-linking exothermic reactions of the thermoset matrix, and the heat generated is conducted to the ambient.

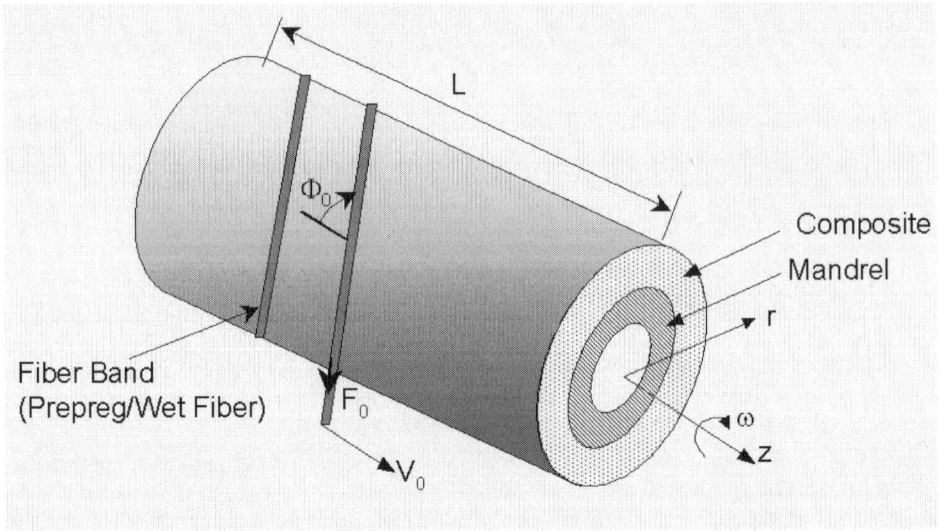

Figure 8.87: Schematic of filament winding process [88].

- *Resin slippage:* Since there is uncured resin between the fibers, the fibers slip from their original position under tension.
- *Void formation:* Voids are formed due to the air trapped between the bands of fibers.
- *Thermal expansion effects and the development of stresses and strains:* The mandrel and the composite expand due to the changes in temperature which creates thermal stresses and strains due to mismatch in thermal expansion. This adds to the stresses developed due to the fiber tension.

8.4.2 Process Models

For filament winding of thermosets, process models have been developed to (i) calculate fiber motion and tension in the fiber filaments during winding, (ii) predict temperatures and cure during winding and in the autoclave, (iii) estimate void growth and final fiber volume fraction, and (iv) evaluate residual stresses. Springer and co-workers [420–423] have developed extensive formulations that include thermochemical, fiber motion, stress, and void submodels. Their models have been experimentally validated for prepreg thermoset composite materials.

The thermochemical submodel provides temperature, viscosity and degree of cure during winding and when the part is cured in the autoclave. The inputs needed for this model are thermophysical properties, constitutive equations that describe the cure behavior and the viscosity change with temperature and cure along with the boundary and initial conditions. The fiber motion or fiber seepage submodel predicts the fiber position and the fiber volume fraction within the cylinder along with fiber tension during the winding process.

The stress submodel calculates the stresses and strains experienced by the composite. Fiber tension, thermal expansion or contraction and chemical changes cause the stress and strain development. Usually, the stresses and strains due to each of the causes are calculated separately and then added to predict the total stress and strain in the composite. The details on this submodel may be found in references [420–423].

Voids may form during winding when the resin percolates through the fibers. The voids may also be introduced in the prepreg that is wound over the cylinder and modest amount of flow during the winding may not be able to flush it out. Void nucleation is not modeled, only changes in void size is considered by Springer et al. [420–423]. All voids are considered to be spherical and contain a vapor of known composition (usually initial concentration of water and air). The volume of the void is considered to change because vapor is transported in and out of the void across the void-composite interface due to the pressure differential between the resin pressure outside and the vapor pressure inside the void. Figure 8.88 shows the interrelationships of the various filament winding submodels. They have been reproduced from Mantell and Cohen [424].

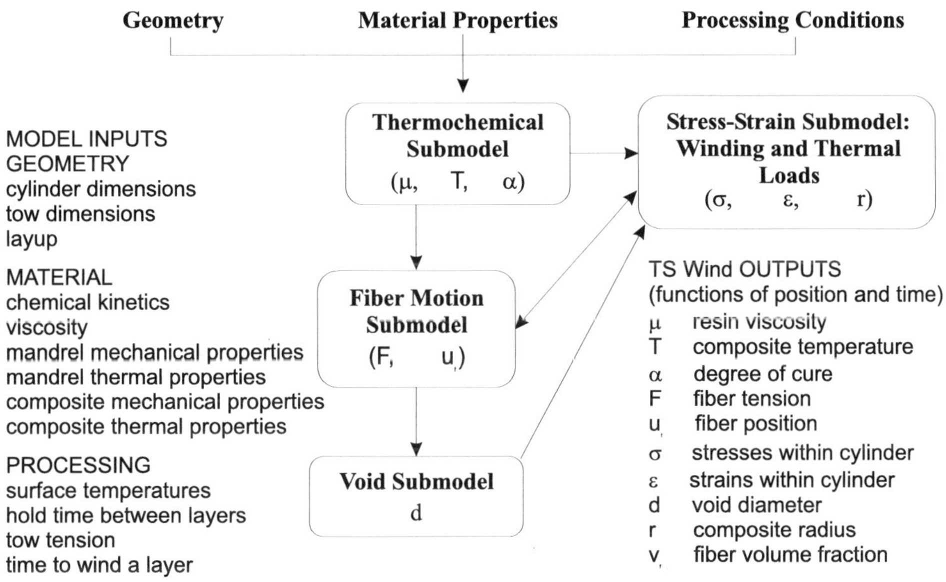

Figure 8.88: Interrelationship of submodels for filament winding with thermosetting matrix materials [424].

In this section, we will consider only the *thermochemical reaction* and the fiber motion or the *resin seepage* submodel. These two models are coupled via the resin viscosity, which is temperature and cure dependent. It is assumed that the winding of the mandrel is instantaneous. The modeling approach will be described with the use of a simple example. Significant work in the modeling of the filament winding process for thermosetting matrix composites has been undertaken by Mantell and Cohen [424], and the models have been verified with experiments. The following process submodels are adopted from their work. The goal here is to introduce modeling and improve the understanding of the modeling process and not solve comprehensively for all the details.

Thermochemical Model

In this process, heat transfer and cure kinetics play an important role, whereas resin flow is considered to be negligible. Hence, one can focus on the energy transport and ignore resin motion. From the energy balance equation introduced in Chapter 3, we can simplify the equation by neglecting the convective terms and ignoring the variations in the azimuth direction due to cylindrical symmetry. So, terms that are retained are an increase in the internal energy of the composite due to the increase in the temperature on the left-hand side which is a result of the energy generated by the curing of the resin (last term on the right-hand side of the equation). Part of the heat is lost to the ambient by conduction in the radial as well as axial directions (first and second terms on the right-hand side)

$$\rho C \frac{\partial T}{\partial t} = \frac{1}{r} \frac{\partial}{\partial r} \left(k_{rr} r \frac{\partial T}{\partial r} \right) + \frac{\partial}{\partial z} \left(k_{zz} \frac{\partial T}{\partial z} \right) + \rho_r (1 - V_f) H_u \frac{d\alpha}{dt} \tag{8.120}$$

where k_{rr} and k_{zz} are the composite conductivities in the radial and axial directions, respectively. ρ and C are the density and heat capacities of the composite. One could use simple rule of mixtures to find these values knowing the density and the heat capacities for the resin and the fibers and the fiber volume fraction. H_u is the heat of reaction for the resin. $d\alpha/dt$ is the rate of cure, where α is the degree of cure. One would need an equation that can describe the cure-kinetic reaction. The cure-kinetic reaction is usually a function of temperature. Hence the cure kinetics is coupled with the energy equation and one must solve them simultaneously

$$\frac{d\alpha}{dt} = F(\alpha, T, t). \tag{8.121}$$

Hence our independent variables are t, r and z and our dependent variables are α and T. The initial conditions for cure and temperature need to be specified (usually α is zero at $t = 0$). As the temperature equation is a partial differential equation, boundary and initial conditions are required to solve for the temperature. Temperature, heat flux or a convective boundary condition is usually specified on the composite outside diameter and on the mandrel inner diameter. In Figure 8.89, constant temperature boundary conditions are applied in the radial and axial directions. Here h_m and h are the thickness of the mandrel and the composite, respectively.

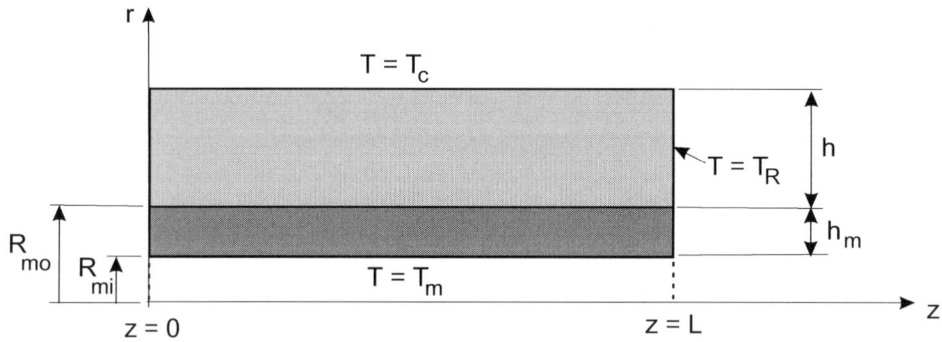

Figure 8.89: Thermal boundary conditions of constant temperature.

Fiber Motion or Resin Seepage Submodel

Darcy's law as applied to the fiber governs the movement of fiber through the resin. In liquid molding processes, we considered the fiber network stationary and the resin moves relative to the fiber network because of the pressure gradient created. In autoclave processing, the resin was made to bleed across a network of fibers that was being compressed. In this process, the fibers or the porous network formed by the fibers are forced through the resin due to the applied tension as seen in Figure 8.90.

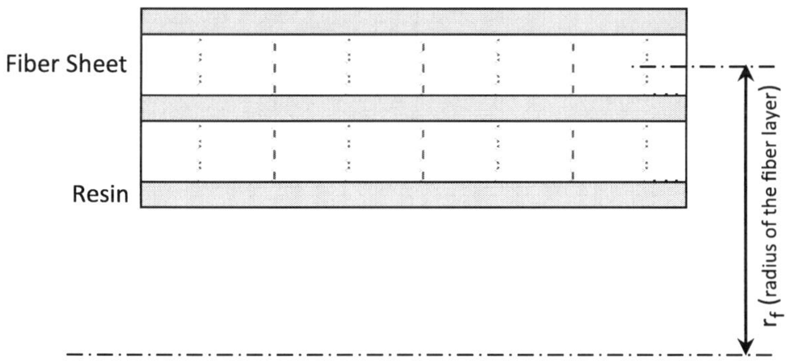

Figure 8.90: Schematic of fiber motion through the resin due to the applied tension.

Flow of the resin, expansion of the mandrel and expansion of the composite affect the fiber position. However, in the fiber motion model we will consider fiber movement only due to the flow of the resin. As seen in Figure 8.90, each composite layer is idealized as a fiber sheet surrounded by thermoset resin. The motion of the fibers can be denoted by x_f. Then, Darcy's law gives the rate at which the fibers move relative to the resin

$$\frac{dx_f}{dt} = -\frac{K_{rr}}{\mu}\frac{dP}{dr}. \tag{8.122}$$

This model assumes that the resin flows transverse to the fibers in the radial direction. Another critical assumption is that the resin carries the entire load and none is carried by the fibers, unlike in autoclave processing. The consolidation pressure is provided by the component of the tension in the filament over the distance r_f, as recast below:

$$\frac{dx_f}{dt} = -\frac{K_{rr}}{\mu}\frac{\sigma_f}{r_f}\sin^2\Phi_o. \tag{8.123}$$

Here dx_f/dt is the rate of radial movement of the fibers, σ_f is the tension in the fiber, K_{rr} is the permeability of the fibrous network in the radial direction, since the fibers are moving through and relative to the resin of viscosity, μ. Φ_o is the angle at which the fiber is wound onto the mandrel and r_f is the radius of the fiber layer. The fiber motion submodel is coupled with the thermochemical submodel through the resin viscosity, μ, which is temperature dependent

$$\mu = G(T,t). \tag{8.124}$$

The radius of the fiber layer, r_f, is given by the following relation:

$$r_f = r_f^0 + x_f \tag{8.125}$$

where r_f^0 is the initial radius and x_f is the displacement of the fiber layer. Under this displacement, the fiber deforms and the fiber tension changes. The equation governing the rate at which the tension, σ_f, is changing and is given by:

$$\frac{d\sigma_f}{dt} = -\frac{E_f}{\mu}\frac{K_{rr}\sin^4\Phi_o}{r_f^2}\sigma_f \tag{8.126}$$

where E_f is the elongational modulus of the fiber material. Hence, we have a set of coupled nonlinear equations for the fiber displacement and the fiber tension, which have to be integrated to obtain the displacement for every layer. The initial conditions for the above equations are

$$x_f = 0 \qquad (8.127)$$

$$\sigma_f = \frac{F_o}{A_f} \qquad (8.128)$$

$$r_f = r_f^o \qquad (8.129)$$

at $t = 0$. Here F_o is the winding tension and r_f^o is the initial radius of the fiber layer. We will conclude this section by demonstrating the approach to calculate the temperature, cure kinetic and fiber motion by coupling the thermochemical and the fiber seepage submodels.

Example 8.17: A Cylindrical Composite Part Wound on an Aluminum Mandrel

A sample problem based on the work of Calius et al. [421] is used here for the purposes of demonstrating the approach to carry out the analysis. A cylindrical composite part is wound on an aluminum mandrel that is 1/4″ thick, 6″ outside diameter by using 18 layers of Fiberite T300/976 (Thornel T-300 fibers/Fiberite 976) prepreg at an angle of 40 degrees, giving it a thickness of 5/32″. Each band of prepreg was wound with tension of 6 lbf and has a width of 0.125″. After winding, the cylinder is cured in an autoclave and subjected to a cure cycle that lasts 6 hours. This cure cycle has a ramp section of one hour to 350°F (177°C) from room temperature, followed by a dwell time of 4 hours and then a cooling period of an hour, back to room temperature. The geometry and the cure cycle are illustrated in Figure 8.91. Describe the approach to solve for fiber tension with time during the curing process. ■

Thermochemical Model Solution

By using the following nondimensional variables,

$$T^* = \frac{T - T_{\text{init}}}{T_{\text{air,max}} - T_{\text{init}}}, \qquad z^* = \frac{z}{L}, \qquad r^* = \frac{r}{\Delta t_{\text{mandrel}} + \Delta t_{\text{composite}}} = \frac{r}{\Delta t_{\text{tot}}}, \qquad t^* = \frac{t}{t_c} \qquad (8.130)$$

the equations governing the heat transfer and curing reaction in the composite and the mandrel were nondimensionalized as follows:

$$\frac{\partial T^*}{\partial t^*} = \frac{k t_c}{\rho C_p \Delta t_{\text{tot}}^2}\left[\frac{1}{r^*}\frac{\partial}{\partial r^*}\left(r^*\frac{\partial T^*}{\partial r^*}\right) + \left(\frac{\Delta t_{\text{tot}}}{L}\right)^2 \frac{\partial^2 T^*}{\partial (z^*)^2}\right] + \frac{\rho_r V_r H_u}{\rho C_p (T_{\max} - T_{\text{init}})}\frac{d\alpha}{dt^*}. \qquad (8.131)$$

Using the documented values of the thermal capacity and heat conductivity of the composite and aluminum, it was found that the time scale of heat conduction in the composite (45 s) was much slower than the conduction time scale in the mandrel. In addition, since the composite layer is thin compared to the length of the part, the terms for the conduction in the z direction can be neglected.

$$\left(\frac{\Delta t_{\text{tot}}}{L}\right)^2 \ll 1, \qquad t_c = \frac{\rho C_p \Delta t_{\text{tot}}^2}{k}, \qquad \beta = \frac{\rho_r V_r H_u}{\rho C_p (T_{\max} - T_{init})}, \qquad t_{c,\text{Al}} \ll t_{c,\text{composite}}. \qquad (8.132)$$

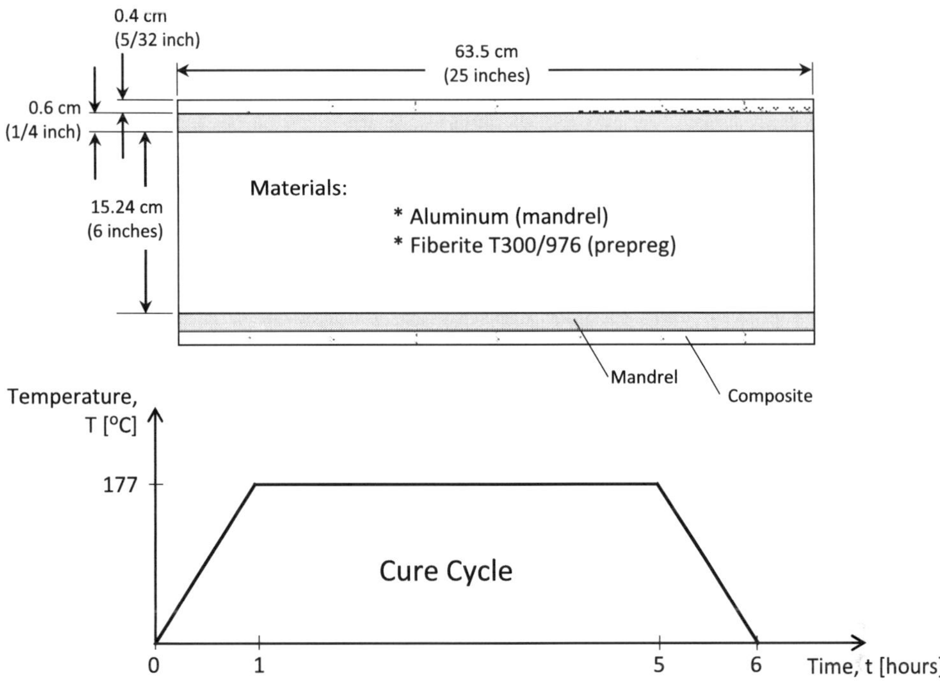

Figure 8.91: Sample composite part and cure cycle used for the model solution [421].

This leads to the following simplified thermochemical equation in nondimensional form:

$$\frac{\partial T^*}{\partial t^*} = \frac{1}{r^*}\frac{\partial}{\partial r^*}\left(r^*\frac{\partial T^*}{\partial r^*}\right) + \beta\frac{d\alpha}{dt^*}. \tag{8.133}$$

The cure kinetics of the epoxy resin and the relation for resin viscosity can be recast as follows:

$$\frac{d\alpha}{dt} = \left(K_1 + K_2\alpha^{1.03}\right)\left(B - \alpha^{1.22}\right) \tag{8.134}$$

$$K_{1,2} = A_{1,2}\exp\left(\frac{-\Delta E_{1,2}}{RT}\right) \tag{8.135}$$

$$\frac{d\alpha}{dt^*} = t_c\frac{d\alpha}{dt} \tag{8.136}$$

$$\mu = \mu_\infty \exp\left(\frac{U}{RT} + \kappa\alpha\right). \tag{8.137}$$

Since the conduction time scale in the mandrel is much smaller than that of the composite, the conduction in the mandrel can be assumed to be in quasi-steady state and the transients there can be neglected. In addition, the chief mechanism of heat transfer from the autoclave to the composite part is the convective heating or cooling. This leads to the following boundary conditions, where the conductive resistance of the mandrel and the convective heat resistance between the part and the autoclave atmosphere are used to derive a mixed type of boundary conditions:

$$\text{inside surface, composite:} \quad \frac{T_{\text{air}} - T_{\text{composite,inside}}}{R_{\text{conv.,in}} - R_{\text{cond.,mandrel}}} = -k\frac{\partial T}{\partial r} \quad \text{at } r = r_{\text{in}} \tag{8.138}$$

$$\text{outside surface, composite:} \qquad \frac{T_{\text{air}} - T_{\text{composite,out}}}{R_{\text{conv.,out}}} = k\frac{\partial T}{\partial r} \quad \text{at } r = r_{\text{out}} \tag{8.139}$$

$$R_{\text{conv1}} = \frac{1}{2\pi r_{\text{in,mandrel}} h}, \qquad R_{\text{cond}} = \frac{\ln\left(\dfrac{r_{\text{in,composite}}}{r_{\text{in,mandrel}}}\right)}{2\pi k_{\text{mandrel}}}, \qquad R_{\text{conv2}} = \frac{1}{2\pi r_{\text{out,composite}} h}. \tag{8.140}$$

Here k is the thermal conductivity of the composite in the radial direction. These boundary conditions can be nondimensionalized to give the following conditions:

$$\text{inside surface, composite:} \qquad \frac{T^*_{\text{air}} - T^*_{\text{composite,inside}}}{R_{\text{conv.,in}} - R_{\text{cond.,mandrel}}} = -\frac{k}{\Delta t_{\text{tot}}}\frac{\partial T^*}{\partial r^*} \quad \text{at } r^* = r^*_{\text{in}} \tag{8.141}$$

$$\text{outside surface, composite:} \qquad \frac{T^*_{\text{air}} - T^*_{\text{composite,out}}}{R_{\text{conv.,out}}} = \frac{k}{\Delta t_{\text{tot}}}\frac{\partial T^*}{\partial r^*} \quad \text{at } r^* = r^*_{\text{out}} \tag{8.142}$$

The initial conditions are given by:

$$\text{initial condition:} \qquad T = T_{\text{room}}, \quad \text{and} \quad \alpha = \alpha_{\text{init}}. \tag{8.143}$$

The governing PDE may be discretized using finite differences to give the following scheme:

$$(1 + 2v + \mu)T_i^{n+1^*} = T_i^{n*} + (v + \mu)T_{i+1}^{n+1^*} + vT_{i-1}^{n+1^*} + \beta\Delta t^*\left(\frac{d\alpha}{dt^*}\right)^n. \tag{8.144}$$

$$v = \frac{\Delta t^*}{\Delta r^{*2}}, \qquad \mu = \frac{\Delta t^*}{r^*[i]\Delta r^*}, \qquad \forall \;\; i = 1 \ldots r \text{ steps} \quad \text{and } j = 0 \ldots t \text{ steps} \tag{8.145}$$

$$\text{B.C.'s:} \qquad T_0^{n+1^*} = \frac{\dfrac{T_1^{n+1^*}}{R} + \dfrac{T_{\text{air}}^{n+1^*}}{R_{\text{conv.,in}} + R_{\text{cond.,mandrel}}}}{\dfrac{1}{R} + \dfrac{1}{R_{\text{conv.,in}} + R_{\text{cond.,mandrel}}}}, \qquad T^{n+1^*}_N = \frac{\dfrac{T_{N-1}^{n+1^*}}{R} + \dfrac{T_{\text{air}}^{n+1^*}}{R + R_{\text{conv.,out}}}}{\dfrac{1}{R} + \dfrac{1}{R_{\text{conv.,out}}}} \tag{8.146}$$

where $R = k/(\Delta t_{\text{tot}}\Delta r^*)$. The numerical scheme selected to solve this problem is fully implicit since the explicit scheme has an inherent limitation in time-step size which can increase the computational time substantially.

Fiber Motion Submodel: Solution

The governing ODEs for the motion of the fiber through the resin, while the resin is curing, can be integrated using the viscosity data from the thermochemical model to obtain the following equations for the fiber motion and fiber tension:

$$x_f^{n+1} = x_f^n - \frac{K_{rr}}{\mu^{n+1}}\frac{\sigma_f^{n+1}}{r_f^{n+1}}\sin^2\Phi_o\,\Delta t \tag{8.147}$$

$$\sigma_f^{n+1} = \sigma_f^n\left(1 - \frac{E_f}{\mu^{n+1}}\frac{K_{rr}\sin^4\Phi_o}{r_f^{n+1^2}}\Delta t\right) \tag{8.148}$$

$$r_f^{n+1} = r_f^0 + x_f^{n+1} \tag{8.149}$$

$$\text{initial conditions:} \qquad x_f^0 = 0, \qquad \sigma_f^0 = \frac{F_{\text{tension}}}{A_f}, \qquad r_f^0 = R_f. \tag{8.150}$$

The discretized thermochemical model can be solved using an iterative solver for the temperature and the cure rate. The viscosity can be determined from the calculated values of the temperature and cure. These viscosity values can then be used to solve for the fiber motion and tension. The fibers tend to move downwards through the resin pool, until the resin gels and solidifies. As the fibers move down, they become slightly slack and the fiber tension decreases. Hence, fiber tension also decreases with time.

However, other phenomena — the development of stresses and strains, thermal expansion of the mandrel and composite with temperature rise — have not been modeled here. These phenomena will definitely have a significant effect on the fiber movement. Hence, though the presented results are reflective of the process physics, a complete picture can be obtained only by including all these effects.

8.5 Summary and Outlook

In this chapter, we introduced the modeling approach to three processes in which the resin moves relative to the fibers. Each process has its own complexities although the general approach is to adopt Darcy's law to describe the impregnation of the fluid into the porous network created by the fibrous preform. The fibrous network is represented by its permeability. Accurate modeling and characterization of this quantity is the key to success in modeling of the process. However, in autoclave, VARTM and filament winding there is a dynamic change in the permeability due to the rearrangement of the fibrous media as the load on the fiber network changes during the impregnation process.

Hence, the challenges in this modeling efforts are description of the compacting effect and its influence on the preform permeability and also more accurate methods to characterize the preform permeability such that one can design the preform to ease the processing needs. Use of these process simulations in an effective manner in optimization and control of the process is currently being pursued and with the use of sensors, one can see the manufacturing of composites moving into the era of intelligent processing which will allow for statistical variation of material properties and also for human error. This would naturally lead one into the realm of automation of the process, which should be the final goal of any modeling and simulation effort.

8.6 Exercises

8.6.1 Questions

1. Describe the wet hand lay up process by using these words and terms: tool (mold) surface, gel coat, reinforcement material, resin, and rollers.

2. What industry widely uses the wet hand lay-up process?

3. What are the advantages and disadvantages of the wet hand lay-up process?

4. List three common processes used to make advanced thermoset fiber composites.

5. What process has traditionally been used to manufacture very large composite structures in the aerospace industry?

6. Describe the autoclave process by using these words and terms: preimpregnated, plies, prepregs, tapes stack, tool surface, high pressure and temperature, air pockets, excess resin, autoclave, large pressure vessel, heating element, consolidate and cure.

7. Which one requires higher pressure and temperature during processing, thermoset or thermoplastic composites? Why?

8. What are the main advantages and disadvantages of the autoclave process as compared to other processes?

9. Describe a typical stacking order of layers in an autoclave process. Explain the purpose of the different layers.

10. What is the role of consolidation in the autoclave process?

11. What are the three most important process parameters in the autoclave process that influence the part quality?

12. What are the typical ranges of temperatures and pressures exercised in autoclave processing?

13. What is the drawback to using fully impregnated tapes with smooth surfaces in the autoclave process?

14. How many stages are there in autoclave processing? Describe each stage.

15. Increasing the autoclave pressure prevents possible void growth. This suggests one should use as high autoclave pressure as possible. However, there is an important reason why the autoclave pressure is not increased to maximum pressure. What is that reason?

16. Why is air not suitable for pressurizing the autoclave at high temperatures? What gas is commonly used for this purpose?

17. List the common errors involved in the material and cure preparation step in the autoclave process.

18. What are the resulting effects of incorrect autoclave pressure and temperature profile on the composite part?

19. Why is the mathematical modeling of flow and consolidation in autoclave processing important?

20. Under what conditions during autoclave processing would one expect the pressure gradient and the resin flow to be primarily in the thickness direction rather than in the in-plane directions? What is the corresponding simplified partial differential equation to relate the fiber volume fraction to the pressure?

21. List the most commonly used liquid composite molding (LCM) processes.

22. What are the two groups in which you can classify LCM processes?

23. Describe the five stages of the resin transfer molding (RTM) process.

24. What are the major advantages and disadvantages of RTM compared to other processes?

25. What are the main differences between RTM and SCRIMP?

26. Among RTM and vacuum bagged VARTM, which process should be selected if small dimensional tolerances are required? Compare these two processes in terms of initial investment and labor cost.

27. Describe the following for a reinforcing fabric: fiber volume fraction, permeability, wettability and sizing.

28. What are the common fibers and fabrics used in LCM?

29. What are the major ingredients of a resin system?

30. List the desired behavior of a good resin system.

31. What factors determine the types of fabric and resin system to be used in LCM? Give examples.

32. What materials are used for the manufacturing of molds (tools)? Describe their advantages and disadvantages.

33. Why is the design of a mold important?

34. What are the undesired effects of improper design of gate and vent locations?

35. Why do some molds need heating/cooling channels machined into them?

36. Suppose that you have a flow rate controlled injection system for RTM. In order to reduce the fill time and hence the cycle time, should you set the injection flow rate to the maximum? Explain.

37. Why are the mathematical models of resin flow and cure and the simulation codes based on these models important?

38. What are the major issues in LCM?

39. What is fiber washout? What causes it?

40. What is racetracking? What causes it? Is it always undesired? What are the major consequences of racetracking on the resin flow?

41. Describe the injection compression molding process. What must be calculated carefully before injecting the resin?

42. Under what conditions in LCM can resin flow be modeled using Darcy's law?

43. In LCM, when can the resin flow be modeled in two dimensions instead of three dimensions? Can the heat transfer model also be simplified into two dimensions?

44. When and how are the resin flow and cure models coupled?

45. When is the modeling of resin flow inside a dual scale porous medium important? What modification should be made to the mass conservation equation if dual scale flow is to be considered?

46. Although a unique numerical solution of resin pressure distribution can be obtained for a completely filled mold under a given set of gate and vent boundary conditions, why do the simulation codes apply a time marching scheme to solve the pressure and flow front location?

47. In order to measure the permeability of a preform at a given fiber volume fraction, how is a one-dimensional experiment performed? What variables and parameters are recorded in each experiment? What relationship is used? What causes error in the measured value?

48. What is the heat dispersion coefficient? Under what conditions can it be neglected?

49. What physical phenomena occur during the filament winding process? Briefly explain them.

8.6.2 Fill in the Blanks

1. All the advanced thermoset fiber composite processes evolved from the wet hand lay-up process in which the goal was either to ____________ the process, to reduce ____________ cost, to improve ____________ treatment, eliminate or reduce emissions of ____________ to the environment, improve dimensional ____________ and/or increase the ______________________ content in the composite.

2. In order to manufacture a thermoset composite part as large as 30 meters in diameter and over 50 meters in length, ____________ process is used. ____________ industry uses this process for years to manufacture large parts.

3. The largest existing autoclave in the U.S. is at the ____________ commercial aircraft plant in ____________.

4. Autoclaves are usually made of ____________ steel as they are subjected to high pressure and temperature.

5. In autoclave process lay-up stacking, ____________ are used to absorb excess resin in the thickness direction. The ____________ material allows for uniform distribution of vacuum over the surface of the part. A ______________________ covers the whole stacking assembly. The role of this plate is to improve the ______________________ of the part and improve ______________________ by applying uniform pressure and minimizing ply movement.

6. The goal of autoclave consolidation is to force ____________ and excess ____________ out of the structure.

7. There are typically three stages of autoclave processing. In Stage I, ____________ and ____________ are ramped up inside the autoclave. In Stage II, the autoclave is maintained at the processing pressure and temperature to allow ____________ and ____________. As the cure reaction proceeds and generates heat due to polymerization, in stage III, the ____________ is lowered to allow the excess ____________ from the reaction to diffuse through the part but the ____________ is maintained, to prevent voids from ____________.

8. The most widely used gas for pressurizing the autoclave is ____________. Air is unsuitable at high temperatures due to the risk of ____________________.

9. The goal in autoclave processing is to develop the most optimal ____________ and ____________ cycles to manufacture ____________ free composite part.

10. The warpage of composite parts that are autoclaved is linked to the ____________________ developed in the part and the differences in the coefficient of ____________________ of the composite both along the fibers and in the transverse directions.

11. In the autoclave process, void growth can be minimized by modeling resin ____________ and hence controlling the autoclave ____________.

12. The resin flow model of the autoclave process enables complete ____________ of fibers.

13. The temperature model of the autoclave process allows the engineers to reduce temperature ____________ within the composite and hence to avoid resin degradation.

14. In order to overcome the long fill time disadvantage of VARTM, ____________ process was developed and patented under the ____________ Corporation's name. In this process, a very ____________ fabric is placed on top of the preform. Since it has very ____________ to resin flow, resin saturates this layer quickly and then flows into the reinforcing preform in the ____________ direction.

15. ____________ of a reinforcing preform is a measure of resistance to resin flow. It is neither a scalar, nor a vector, but it is a ____________.

16. ____________ is a measure of how easily the interior of a fiber bundle is wetted as resin moves around the bundle.

17. ____________ is a coating on the fiber surface to increase the wettability property and also to increase the bond between the ____________ and the ____________.

18. Cores such as ____________ and ____________ are included inside large and thick parts to lower the cost, increase the moment of inertia, and reduce the weight of the part.

19. Alternative to cores, internal inflatable ____________ might be used in thick parts such as tennis racquets and rotor blades.

20. ____________ and ____________ fabrics cannot be used in stamping of fabrics with thermoformable binder process as they don't ____________ during the stamping stage.

21. Most of the injection equipment is designed for either ____________ controlled, or ____________ controlled injections.

22. Among the most commonly used thermoset resins in LCM, ____________ has the lowest performance, easiest processing and lowest cost. On the other extreme end, ____________ has the highest performance, most difficult processing and highest cost.

23. The part is demolded when the matrix reaches a ____________ strength such that the part will not deform or warp after being taken out of the mold.

24. Microvoids are created because the ____________ inside a fiber tow region is very different than the ____________ of the region between the fiber tows in woven and stitched fabrics. ____________ voids are created as resin races along the tows and encircles itself along a stitch. When resin flow is faster in between the tows, then ____________ voids are trapped in between the fiber bundles.

8.6.3 Problems

1. Assume that (i) the resin used in an autoclave is Newtonian, (ii) the Reynolds number is low, and (iii) the voids are spherical. If P is the void pressure, P_r is the resin pressure, R is the radius of the void, and α is the surface tension, find an expression for these process variables so that the void growth will be suppressed. If you can vary the temperature and pressure within the autoclave, how will it affect the void growth? Would you set the pressure to be maximum for autoclave control? Explain why.

2. Equation (8.2) relates the void growth to viscosity, void radius, surface tension, pressures of void and the resin. Can you express $R(t)$ in closed form? How would you solve $R(t)$ numerically, with an initial condition $R(0) = R_o$, and what data will you need?

3. Consider an autoclave that is made of welded steel of thickness 20 mm. The autoclave is a cylindrical tube of diameter 0.5 m and length of 1 m. The tooling and part thermal inertia can be considered negligible as compared to the autoclave inertia. How long will it take to heat the autoclave from room temperature of 25°C to 350°C, if the heater power supplied to the autoclave is 20 kW?

4. By referring to Figures 8.12 and 8.13, answer the following questions: (a) It takes t_1 minutes for the mid-point of the composite to reach a fiber volume fraction of about 0.61 from the initial value of 0.53. How long will it take for the entire composite to be above fiber volume fraction of about 0.70? (b) If it takes t_2 minutes for the resin pressure at the top layer to reach its maximum value P_{max}, at roughly what time(s) the resin pressure at this location reaches $P_{max}/2$? Explain why there is more than one unique time values to your answer.

5. Consider a resin flow racetracking along a rectangular channel of height of $h = 5$ mm and a width of $w = 2.5$ mm. If all four walls of the channel are assumed to be nonporous, calculate the permeability K_{race} of the channel to be used as a parameter in one-dimensional Darcy's law.

6. Thermal design criteria require you to keep the temperature differential between the center and the surface of the composite to less than 20°C. You know that a cooling rate of 5 Watts/m^2 for a 6 mm thick composite was able to do so for an epoxy resin. For a second part of 8 mm thickness, what is the maximum cooling rate you can impose to meet the thermal design criteria?

7. Resolve Example 8.7 assuming that the transient temperature is important. See Examples 5.5 and 5.6 on the details for solving transient thermal problems.

8. Suppose that a resin is injected from the left edge of a rectangular mold cavity from a line source at a constant pressure of $P = P_{vent} + \Delta P$. There is a line vent at the right edge of the mold. Assume that there is no racetracking and a uniform preform. Hence the flow is one-dimensional. The mold has a length of $L = 1$ m, a thickness

of $h = 0.01$ m and a width of $w = 0.2$ m. The preform has a porosity of 0.5 and a uniform permeability of $2 * 10^{-9}\,\text{m}^2$ along the resin flow direction, resin viscosity is $\mu = 0.2$ Pa.s. (i) Calculate the flow rate when the mold is half full for $\Delta P = 1$ bar. (ii) If the length of the mold is changed to $L = 2$ m, and viscosity is reduced to $\mu = 0.15$ Pa.s, what will be the flow rate when the mold is 50% full?

9. In the previous problem, if the air inside the mold is initially at 1 bar, but the vent is kept closed, what will be the steady-state void size if $\Delta P = 10$ bar assuming that the entrapped air inside the void doesn't dissolve into the resin and behaves as an ideal gas?

10. Considering an unsteady state one-dimensional permeability experiment, plot the flow front $x_f(t)$ and injection pressure $P_{\text{inj}}(t)$ as a function of time at two different values of sink term, $s = 0$ and $0.002\ \text{s}^{-1}$ if the preform width $w = 0.2$ m, thickness $h = 0.005$ m, porosity $\phi = 0.4$, permeability $K_{xx} = 1{*}10^{-10}\,\text{m}^2$, resin viscosity $\mu = 0.15$ Pa.s, and flow rate $Q = 4{*}10^{-6}\,\text{m}^3$.

11. For $Gz >> 1$ in RTM applications, the steady-state temperature distribution is given by $\frac{\langle T\rangle - T_{in}}{T_o - T_{in}} = 1 - \text{erf}(\xi)$ where $\xi = \frac{\hat{z}}{2\sqrt{\frac{\alpha_{zz}x}{\langle u\rangle}}}$ and $\hat{z}$ is the distance from the wall towards the center in the thickness direction. If $T_{in} = 20°$C and $T_o = 150°$C, plot the composite temperature, T, as a function of z along the thickness direction at a downstream location (far away from the injection ports).

Appendix A

MATLAB Files

```
%%%%%%%%%%%%%%%%%%%%%%%%%%%%% Matlab code for Example 4.2 %%%%%%%%%%%%%%%%%%%%%%%%%%%%%%
clear all, close all
R           = 8.3145;
MU0         = [ 7.93e-14  1.06e-06];
U           = [ 9.08e+04  3.76e+04];
K           = [14.1       18.8      ];
ALPHA(1,:) = linspace(0,0.2,1001);
ALPHA(2,:) = linspace(0,0.2,1001);
T           = [350 375 400];

for i = 1:2
    mu0   = MU0(i); u = U(i); k = K(i);
    alpha = ALPHA(i,:);
for j = 1:3
    t = T(j)
    mu(j,:) = mu0*exp(u/(R*t)+k*alpha);
end
figure(i)
subplot(1,3,1), plot(alpha,mu(1,:),'b-', 'linewidth',2), grid on%, legend('T = 350^oK',0)
subplot(1,3,2), plot(alpha,mu(2,:),'b-', 'linewidth',2), grid on%, legend('T = 375^oK',0)
subplot(1,3,3), plot(alpha,mu(3,:),'b-', 'linewidth',2), grid on%, legend('T = 400^oK',0)
end
%%%%%%%%%%%%%%%%%%%%%%%%%%%%%%%%%%%%%%%%%%%%%%%%%%%%%%%%%%%%%%%%%%%%%%%%%%%%%%%%%%%%%%%%%%
```

```
%%%%%%%%%%%%%% Matlab code for Example 4.5 %%%%%%%%%%%%%%%%%%
clear all, close all, clc
R      = 8.3145;
CA0    = 0.5;
k0     = 2;
E      = 1000 * R;
LAMBDA = 1;
T      = 400;
C(1)   = 0;
t(1)   = 0;
dt     = 0.01;
i      = 1;

while  C(end) < 0.8
dCdt   = k0 * exp(-E/(R*T)) * CA0 * (1-C(i)) * (LAMBDA-C(i));
C(i+1) = C(i) + dCdt*dt;
t(i+1) = t(i) + dt;
i      =    i + 1;
end

figure(1), plot(t,C,'b-','linewidth',2)
xlabel('Time, t [seconds]'), ylabel('Extent of reaction, C^*')
grid, legend('T = 400^oK',0)
%%%%%%%%%%%%%%%%%%%%%%%%%%%%%%%%%%%%%%%%%%%%%%%%%%%%%%%%%%%%%%
```

```
%%%%%%%%%%%%%%% Matlab code for Example 4.6 %%%%%%%%%%%%%%%%
clear all, close all, clc
R = 8.3145;
CA0 = 0.5;
k0 = 2;
E = 1000 * R;
LAMBDA = 1;
T = 400;
C(1) = 0;
t(1) = 0;
dt = 0.01;
i =1;
while C(end) < 0.8
dCdt = k0 * exp(-E/(R*T)) * CA0 * (1-C(i)) * (LAMBDA-C(i));
C(i+1) = C(i) + dCdt*dt;
t(i+1) = t(i) + dt;
i = i + 1;
end
figure(1), plot(t,C,b-,linewidth,2)
xlabel(Time, t [seconds]), ylabel(Extent of reaction, C^*)
grid, legend(T = 400^oK,0)
%%%%%%%%%%%%%%%%%%%%%%%%%%%%%%%%%%%%%%%%%%%%%%%%%%%%%%%%%%%%%%%
```

```
%%%%%%%% Matlab code for Example 5.6 %%%%%%%%%%%%%%%%%%%%%%%%%%
clc, clear all, close all
%%% INPUT:
k_r     =   0.1;   % in [W/(m C)]
k_f     =  10.0;
Rho_r   =  900;    % in [kg / m^3]
Rho_f   = 2500;
C_p_r   = 2000;    % in [J / (kg C)]
C_p_f   =  410;
Vf      =   0.50;  % fiber volume fraction
h_thick =   0.010; % half thickness in [m]
T_w     =  25;     % temperature in [C]
T_i     =  200;

RhoCp_r = Rho_r * C_p_r;
RhoCp_f = Rho_f * C_p_f;
k       = (1-Vf)*k_r   + Vf*k_f;
RhoCp   = (1-Vf)*RhoCp_r + Vf*RhoCp_f;
alpha   = k/(RhoCp);              % thermal diffusivity
t_c     = h_thick^2 / alpha;      % reference time
DeltaT  = (T_w - T_i);

r          = 0.1; % = Delta t^ / (Delta z^) ^2
N          = 101; % Number of increments in z direction
%%% I.C.:
theta      = zeros(1,N);
theta(1)   = 1;
theta(end) = 1;
theta_next = theta;
dz_non     = (1-(-1))/(N-1);
dt_non     = r * (dz_non)^2;
dt         = dt_non * t_c;
z_non      = linspace(-1,1,N);

t_non  = 0.0;
t      = 0.0;
k_time = 1;
mid    = (N+1)/2;

TIME_non(k_time)    = 0;
TIME(k_time)        = 0;
THETA(k_time)       = 0;
TEMPERATURE(k_time) = T_i;

figure(1)
while theta(mid) < (70 - T_i)/(T_w - T_i)

for i = 2:N-1
theta_next(i) = r*theta(i-1) + (1-2*r)*theta(i) + r*theta(i+1);
end

theta                = theta_next;
temperature          = theta * DeltaT + T_i;
t_non                = t_non + dt_non;
t                    = t     + dt;
k_time               = k_time + 1;
```

```
TIME_non(k_time)     = t_non;
TIME(k_time)         = t;
THETA(k_time)        = theta(mid);
TEMPERATURE(k_time) = temperature(mid);

    if (mod(k_time,round(1/dt)) == 1)
    text_plot = ['time [seconds] = ',num2str(t,3)];
    plot(temperature,z_non*h_thick,'linewidth',2); grid on
    axis([0 1.1*T_i -1.2*h_thick 1.2*h_thick])
    text(0.55*T_i,0.95*h_thick,text_plot);
    xlabel('Temperature, T(z)  [C]')
    ylabel('Height, h [m]')
    pause(0.10)
    else
    end
end

t_non_critical = t_non(end); % non-dime. time when T_mid = 70 C

figure(2)
plot(TIME_non,THETA,'linewidth',2)
grid
xlabel('Nondimensional Time')
ylabel('Nondimensional Temperature of Mid Point')

%%%% Investigate the effect of k_f
%%%% INPUT:
k_r     =    0.1;   % in [W/(m C)]
k_f     =   linspace(1,20,191);
Rho_r   =  900;     % in [kg / m^3]
Rho_f   = 2500;
C_p_r   = 2000;     % in [J / (kg C)]
C_p_f   =  410;
Vf      =    0.50;  % fiber volume fraction
h_thick =    0.010; % half thickness in [m]
T_w     =   25;     % temperature in [C]
T_i     =  200;

RhoCp_r = Rho_r * C_p_r;
RhoCp_f = Rho_f * C_p_f;
k       = (1-Vf)*k_r   + Vf*k_f;
RhoCp   = (1-Vf)*RhoCp_r + Vf*RhoCp_f;
alpha   = k./(RhoCp);              % thermal diffusivity
t_c     = h_thick^2 ./ alpha;      % reference time
DeltaT  = (T_w - T_i);

t_critical      = t_non_critical*t_c;
figure(3)
plot(k_f,t_critical,'b-','linewidth',2)
grid
xlabel('Thermal Conductivity of Fiber, k_f [W/(mC)]')
ylabel('Critical Time, t_{T70} [s] when T_{mid} = 70 degrees')
%%%%%%%%%%%%%%%%%%%%%%%%%%%%%%%%%%%%%%%%%%%%%%%%%%%%%%%%%%%%%%%
```

```
%%%%%%%%%% Matlab code for Example 5.8 %%%%%%%%
clear all, close all, clc
t    = 0:5:30;
P    = [0 11000 22400 33300 42600 53450 60850];
C    = polyfit(t,P,1);
dPdt = C(1)
T    = linspace(t(1),t(end),1000);
Pfit = polyval(C,T);
plot(t,P,'bo', T,Pfit,'k-'), grid on
xlabel('Time, t [s]')
ylabel('Injection Pressure, P_{inj}(t) [Pa]')
legend('Exp. data','1st order curve fit',0)

mu   = 0.24;
Cw   = 0.200;
Ch   = 0.005;
Vf   = 0.474;
Q    = 4e-6;
Kxx  = (mu/(1-Vf))*(Q/(Cw*Ch))^2*(1/dPdt)
%%%%%%%%%%%%%%%%%%%%%%%%%%%%%%%%%%%%%%%%%%%%%%%%
```

```
%%%%%%%% Matlab code for Example 5.9 %%%%%%%%%%%%%%%%%%%%%%%%%%%%%%%%%%%
clear all, close all, clc
mu    = 0.24;
 h    = 0.005;
Vf    = 0.30;
Q     = 4e-6;
R1    = 0.006;
R2    = 0.012;
rmold = 0.40;
t     = [0:5:30]';
P     = [0 125 147 160 169 176 182]' * 1000;
C     = [log(1+(Q*t)./(pi*h*R1^2*(1-Vf)))];
X     = (C'*C)\(C'*P);
K     = ((mu*Q)/(4*pi*h))/X

EmptyVolume1 = (pi*(rmold^2 - R1^2))*h*(1-Vf)
t_fill       = EmptyVolume1/Q
T1           = linspace(0,      t(end),1000);
T2           = linspace(t(end),t_fill,1000);
T3           = linspace(0      ,t_fill,1000);
PP1          = (mu*Q)/(4*pi*h*K) * log(1+(Q*T1)./(pi*h*R1^2*(1-Vf)));
PP2          = (mu*Q)/(4*pi*h*K) * log(1+(Q*T2)./(pi*h*R1^2*(1-Vf)));
PP3          = (mu*Q)/(4*pi*h*K) * log(1+(Q*T3)./(pi*h*R2^2*(1-Vf)));
plot(t,P,'ro', T1,PP1,'r-.', T2,PP2,'b-', T3,PP3,'k-.')
grid, xlabel('Time, t [s]'), ylabel('Injection Pressure, P_{inj}(t) [Pa]')
legend('Exp. data when R = R_1', ...
       'Curve fit when R = R_1', ...
       'Extension of curve fit when R = R_1', ...
       'When R = R_2',0)
%%%%%%%%%%%%%%%%%%%%%%%%%%%%%%%%%%%%%%%%%%%%%%%%%%%%%%%%%%%%%%%%%%%%%%%%%%%%%%%
```

```
%%%%%%%%%%% Matlab code for Example 5.11 %%%%%%%%%%%%%%%%%%%%%%%%%%%
clc, clear all, close all
mu    = 0.1;
Ri    = 0.006;
L     = 2.0;
h     = 0.004;
rmold = 0.50;
Vf    = 0.35;
k     - [6.0e-10 3.0e-10];
Patm  = 101000;
str   = 'bk';

for icase = 1:2
clear R
clear t
K    = k(icase);
R(1) = Ri;
t(1) = 0;
dt   = 0.001;
i=1;

while  R(end) < rmold
    if R(end) > 0.01, dt = 0.01; end
    if R(end) > 0.10, dt = 0.10; end
    a1   = (Ri^4*Patm)/(16*mu*L*(1-Vf)*h)*(1/R(i));
    a2   = 1 - 1/(1+(16*K*L*h)/(Ri^4*log(R(i)/Ri)) );
    dRdt = a1*a2;
    R(i+1) = R(i) + dRdt*dt;
    t(i+1) = t(i) + dt;
    i      = i+1;
end

t_fill(icase) = t(end)
figure(1)
plot(t,R,str(icase)), grid on, hold on
xlabel('Time, t [seconds]'), ylabel('Radius of flow front, R(t) [m]')
end
legend('K = K_1 = 6.0e-10','K = K_2 = 3.0e-10',0)
%%%%%%%%%%%%%%%%%%%%%%%%%%%%%%%%%%%%%%%%%%%%%%%%%%%%%%%%%%%%%%%%%%%%%
```

```
%%%%%%%%%% Matlab code for Example 6.5 %%%%%%%%%%%%%%%%%%%%%%%%%%%%%%%%%%%%%%%%
clear all, close all, clc
mu    = 120000;             % viscosity of polymer, [Pa.s]
L     = 1.920;              % screw length, [m]
D     = 0.080;              % screw diameter, [m]
N     = 1;                  % screw speed, [rev/s]
Ho    = 0.015;              % inlet channel depth, [m]
Hl    = 0.002;              % exit  channel depth, [m]
theta = (20/180)*pi;        % helix angle, [rad]
KG    = 5.0e-7;             % related to die geometry and characteristics, [m^3]
e     = 0.008;              % flight width, [m]

Ls    = tan(theta)*(pi*D);  % screw lead, [m]
W     = Ls*cos(theta) - e;  % channel width, [m]
l     = L/sin(theta);       % spiral channel length, [m]
Vz    = pi*D*N*cos(theta);  % relative velocity in z direction, [m/s]
Vx    = pi*D*N*sin(theta);  % relative velocity in z direction, [m/s]

disp('Assume constant channel thickness:')
H     = 0.5*(Ho+Hl);
Fp    = 1;
Fd    = 1;
A     = 0.5*W*H*Fd*cos(theta)*pi*D;
C     = (W*H^3)/(12*l)*Fp;
Q     = (A*KG)/(C+KG)*N
DP    = (mu*A)/(C+KG)*N

disp('Consider varying channel thickness:')
slope = atan((Ho-Hl)/l) * (180/pi)      % slope of the channel, [degree]
H1    = (Ho*Hl)  /(0.5*(Ho+Hl));
H3    = (Ho*Hl)^2/(0.5*(Ho+Hl));
A     = 0.5*pi*D*W*H1*cos(theta);
C     = (W*H3)/(12*l);
DP    = DP                              % Pressure built up, [Pa]
Q     = A*N-C*DP/mu
%%%%%%%%%%%%%%%%%%%%%%%%%%%%%%%%%%%%%%%%%%%%%%%%%%%%%%%%%%%%%%%%%%%%%%%%%%%%%%%
```

```
%%%%%%%%% Matlab code for Example 6.6 %%%%%%%%%%%%%
clear all, close all, clc

X = [1.2 1.3 1.5 1.7 2 2.4 3 3.5 4 5 6 8 10];

for i = 1:length(X)
x  = X(i);
N1 = logspace(log10(0.01),log10(10),1001);
N2 = N1./(x*log(x)) + 1/(x-1);
loglog(N1,N2,'k-','linewidth',2), grid on, hold on
axis([0.01 20 0.09 40])
xlabel('N_1'), ylabel('N_2')
end
hold off
%%%%%%%%%%%%%%%%%%%%%%%%%%%%%%%%%%%%%%%%%%%%%%%%%%%%%
```

```
%%%%%%%%% Matlab code for Example 6.7 %%%%%%%%%%%%%%%%%%%%%%%%%%%%%%%%%%%%%%%%%%
clear all, close all, clc
L      = 1.920;               % screw length, [m]
D      = 0.080;               % screw diameter, [m]
Nscrew= 0.20:0.01:0.50;       % screw speed, [rev/s] = [rpm/60]
Ho     = 0.020;               % inlet channel depth, [m]
Hl     = 0.004;               % exit  channel depth, [m]
theta = (20/180)*pi;          % helix angle, [rad]
KG     = 1.2e-8;              % related to die geometry and characteristics, [m^3]
e      = 0.008;               % flight width, [m]

Vf     = 0.20;         % Fiber volume fraction
rhof   = 2500;         % density of fiber,  [kg/m3]
rhom   = 1200;         % density of matrix, [kg/m3]
kf     = 1.20;         % thermal conductivity of fiber,  [W/(m K)]
km     = 0.20;         % thermal conductivity of matrix, [W/(m K)]
cf     =  800;         % specific heat capacity of fiber,  [J/(kg K)]
cm     =  700;         % specific heat capacity of matrix, [J/(kg K)]

a      = 6.5e5;        % where mu = a exp(-b T), [Pa.s]
b      = 0.009;        %                       , [1/oK]

To     = 50 + 273.2;  % inlet temperature, [oK]
Vo     = a*exp(-b*To);% viscosity at the inlet, [Pa.s]

for j = 1:length(Nscrew)
N      = Nscrew(j);
Ls     = tan(theta)*(pi*D);  % screw lead, [m]
W      = Ls*cos(theta) - e;  % channel width, [m]
l      = L/sin(theta);       % spiral channel length, [m]
Vz     = pi*D*N*cos(theta);  % relative velocity in z direction, [m/s]
Vx     = pi*D*N*sin(theta);  % relative velocity in z direction, [m/s]

k      =  1/((1-Vf)/km  + Vf/kf);       % thermal conductivity of composite
rho_c = (1-Vf)*rhom*cm + Vf*rhof*cf;   % (density * specific heat) of composite

disp('Consider varying channel thickness:')
H1     = (Ho*Hl)  /(0.5*(Ho+Hl));
H      = H1;
H3     = (Ho*Hl)^2/(0.5*(Ho+Hl));
A      = 0.5*pi*D*W*H1*cos(theta);
C      = (W*H3)/(12*l);
E      = (pi^3 * D^3 * sin(theta))/H * (1+(sin(theta))^2);
N1     = C/KG
N2     = (A*rho_c)/(N*l*Vo*b*E)

X      = linspace(0.01,40,4000);
F      = X.*log(X) .* (N2 - 1./ (X-1)) - N1;
```

```
figure(1)
plot(X,F), grid on, hold on
xlabel('x = e^{b \Delta T}')
ylabel('f(x) = x ln(x) (N_2 - 1/(x-1)) - N_1')

for i = 2:length(F)
if sign(F(i))*sign(F(1)) < 0, break, end
end
x     = X(i)
str   = ['The root is at x = ',num2str(x),' when N = ',num2str(N*60),' rpm'];
plot(x,F(i),'ro','linewidth',2), text(7,0.85*max(F),str), hold off
pause(0.2)

Q(j)  = A*N - (E*C*l*N^2*b*Vo)/(rho_c*KG*x*log(x));  % flow rate, [m^3/s]
DP(j) = (Q(j)*Vo)/(KG*x);                            % pressure build up, [Pa]
DT(j) = log(x)/b;                                    % temperature rise, [oK]
Te(j) = To + DT(j);                                  % exit temperature, [oK]
Ve(j) = a*exp(-b*Te(j));                             % exit viscosity, [Pa.s]
end

figure(2), plot(Nscrew*60,Q*1.0e6,'b-o','linewidth',2), grid on
xlabel('Screw speed, N [rpm]'), ylabel('Flow rate, Q [cc/s]')

figure(3), plot(Nscrew*60,DP*1.0e-6,'b-o','linewidth',2), grid on
xlabel('Screw speed, N [rpm]'), ylabel('Pressure build-up, \Delta P [MPa]')

figure(4), plot(Nscrew*60,DT,'b-o','linewidth',2), grid on
xlabel('Screw speed, N [rpm]'), ylabel('Temperature rise, \Delta T [^oK]')
%%%%%%%%%%%%%%%%%%%%%%%%%%%%%%%%%%%%%%%%%%%%%%%%%%%%%%%%%%%%%%%%%%%%%%%%%%%%%%%%%%%%%%%%%
```

```
%%%%%%%%%% Matlab code for Example 6.8 (a) %%%%%%%%%%%%%%%%%%%%%%%%%%
clear all, close all, clc

r0     = 0.010;               % radius of central gate, [m]
R      = 0.100;               % outer radius of disk, [m]
h      = 0.001;               % half of the thickness, [m]
MU     = [100 500 1000];      % viscosity of the resin, [Pa.s]
P0     = [5:0.2:10] * 1.0e5; % gage inlet pressure, [Pa]

for i = 1:length(MU)
mu     = MU(i);
rs     = r0/R;
Ps     = P0/(3*mu*R^2/h^2);

tfill = 1./(2*Ps) * (-log(rs)-(1-rs^2)/2);  % time to fill the mold

subplot(3,1,i), plot(P0/100000,tfill,'k-o','linewidth',2)
str = ['viscosity = ',num2str(mu),' Pa.s'];
grid on, legend(str,0)
end
xlabel('Gauge inlet pressure, P_0 [bar]')
ylabel('Time to fill the mold, t_{fill}, [s]')
%%%%%%%%%%%%%%%%%%%%%%%%%%%%%%%%%%%%%%%%%%%%%%%%%%%%%%%%%%%%%%%%%%%%%%
```

```
%%%%%%%%% Matlab code for Example 6.8 (b) %%%%%%%%%%%%%%%%%%%%%%%%%%%
clear all, close all, clc

r0      = 0.010;              % radius of central gate, [m]
R       = 0.100;              % outer radius of disk, [m]
h       = 0.001;              % half of the thickness, [m]
MU      = [100 500 1000];     % viscosity of the resin, [Pa.s]
QQ      = [1 2 5]  * 1.0e-6;  % gage inlet pressure, [Pa]

for i = 1:length(MU)
for j = 1:length(QQ)
mu      = MU(i);
Q       = QQ(j);

clear t Rf P0
t(1)    = 0;
Rf(1)   = r0;
P0(1)   = 0;
k       = 1;
dt      = 0.01;
str_cur = ['k- ';'b-.';'r--'];

while Rf < R
t(k+1)  = t(k) + dt;
Rf(k+1) = sqrt(Q*t(k+1)/(2*pi*h) + r0^2);
P0(k+1) = (3*Q*mu)/(8*pi*h^3)*log(1+(Q*t(k+1))/(2*pi*h*r0^2));
k       = k+1;
end

figure(1)
subplot(3,1,i), plot(t,P0/100000,str_cur(j,:),'linewidth',2), hold on
str_title  = ['Viscosity = ',num2str(mu),' Pa.s'];
grid on, title(str_title)
xlabel('Time, t [s]')
ylabel('Inlet pressure, P_{0}(t), [bar]')

figure(2)
subplot(3,1,i), plot(t,Rf*1000,str_cur(j,:),'linewidth',2), hold on
str_title  = ['Viscosity = ',num2str(mu),' Pa.s'];
grid on, title(str_title)
xlabel('Time, t [s]')
ylabel('Radial flow front position, R_{f}(t), [mm]')

end
end
%%%%%%%%%%%%%%%%%%%%%%%%%%%%%%%%%%%%%%%%%%%%%%%%%%%%%%%%%%%%%%%%%%%%%%
```

```
%%%%%%%%%% Matlab code for Example 6.9 %%%%%%%%%%%%%%%%%%%%%%%%%%%%%
clear all, close all, clc

r0    = 0.010;              % radius of central gate, [m]
R     = 0.100;              % outer radius of disk, [m]
h     = 0.001;              % half of the thickness, [m]
mu    = 100;                % viscosity of the resin, [Pa.s]
P0    = 5 * 1.0e5;          % gage inlet pressure, [Pa]

epsil = 0.001;

Rf(1) = r0 * (1 + epsil);
t(1)  = 0;
Q(1)  = 0;
i     = 1;

while Rf < R
        if     Rf(end) < 1.1*r0, dt = 0.00001;
        elseif Rf(end) < 1.5*r0, dt = 0.0001;
        elseif Rf(end) < 2.0*r0, dt = 0.001;
        elseif Rf(end) < 5.0*r0, dt = 0.01;
        end
f       = (P0*h^2)/(3*mu) * (1/(Rf(i)*log(Rf(i)/r0)));
Q(i)    = 4*pi*h*Rf(i)*f;
Rf(i+1) = Rf(i) + f*dt;
t(i+1)  = t(i) + dt;
i       = i+1;
end
Q(i)    = 4*pi*h*Rf(i)*f;

figure(1)
subplot(2,1,1), plot(t,Rf*1000,'k-','linewidth',2)
grid on
xlabel('Time, t [s]')
ylabel('Radial flow front position, R_f [mm]')

subplot(2,1,2), plot(t(2:end),Q(2:end)*1e6,'k-','linewidth',2)
grid on
xlabel('Time, t [s]')
ylabel('Flow rate, Q [cc/s]')
%%%%%%%%%%%%%%%%%%%%%%%%%%%%%%%%%%%%%%%%%%%%%%%%%%%%%%%%%%%%%%%%%%%%%%%%
```

```
%%%%%%%%%% Matlab code for Example 6.10 %%%%%%%%%%%%%%%%%%%%%%%%%%%%%%%%%%
clear all, close all, clc

T1     = 280 + 273.2;          % inlet temperature, [oK]
mu     = 50000;                % viscosity of the resin, [Pa.s]
h      = 0.005;                % thickness of the channel, [m]
W      = 0.100;                % width of the channel, [m]
L      = 0.5;                  % length of the channel, [m]
k      = 0.8;                  % thermal conductivity of composite, [W/(m oK)]
Pi     = [1:2:50] * 1.0e6;     % inlet gauge pressure, [Pa]
Pe     = 0.0e6;                % exit  pressure, [Pa]
DP     = Pe - Pi;

Q      = -DP/(2*mu*L) * ((4*W*h^3)/3);                % flow rate, [m^3/s]

Tmid   = T1 * ( 1 + ( (3*mu)/(16*k*T1*W^2*h^2)*Q.^2) );

figure(1)
subplot(2,1,1), plot(Pi*1.0e-6,Q*1.0e6,'b-o','linewidth',2)
grid on
xlabel('Inlet gauge pressure, P_i [MPa]')
ylabel('Flow rate, Q [cc/s]')

subplot(2,1,2), plot(Pi*1.0e-6,Tmid,'k-o','linewidth',2)
grid on
xlabel('Inlet gauge pressure, P_i [MPa]')
ylabel('Mid-plane temperature, T_{mid-plane} [^oK]')
%%%%%%%%%%%%%%%%%%%%%%%%%%%%%%%%%%%%%%%%%%%%%%%%%%%%%%%%%%%%%%%%%%%%%%%%%%%%
```

```
%%%%%%%%%%%%% Matlab code for Example 7.1 %%%%%%%%%%%%%%%%%%%%%%%%%%%%%%%%%%%%%%%%
clear all, close all, clc
L0          = 1;     % original length, [m]

%%% part (a), constant strain rate:
epsdot      = 0.01;   % strain rate, [1/s]
dt          = 0.01;
La(1)       = L0;
ta(1)        = 0;
k           = 1;
while La(end) < 2*L0
k           = k+1;
ta(k)       = ta(k-1) + dt;
La(k)       = L0*exp(epsdot*ta(k));  % length, [m]
EPSDOTa(k) = epsdot;                 % strain rate, [1/s]
end

%%% part (b), constant elongation speed:
V           = 0.01;   % elongation speed, [m/s]
dt          = 0.01;
Lb(1)       = L0;
tb(1)       = 0;
k           = 1;
while Lb(end) < 2*L0
k           = k+1;
tb(k)       = tb(k-1) + dt;
Lb(k)       = L0 + V*tb(k);     % length, [m]
EPSDOTb(k) = 1/((L0/V)+tb(k));           % strain rate, [1/s]
end

figure(1)
subplot(2,1,1), plot(ta,La,'b-',tb,Lb,'k--','linewidth',2)
xlabel('Time, t [seconds]'), ylabel('Length, L [m]'), grid on
legend('d\epsilon / dt = constant = 0.1 1/s','dL / dt = V = constant = 0.1 m/s',0)

subplot(2,1,2), plot(ta,EPSDOTa,'b-',tb,EPSDOTb,'k--','linewidth',2)
xlabel('Time, t [seconds]'), ylabel('Strain rate, d\epsilon / dt, [1/s]')
grid on
legend('d\epsilon / dt = constant = 0.1 1/s','dL / dt = V = constant = 0.1 m/s',0)
%%%%%%%%%%%%%%%%%%%%%%%%%%%%%%%%%%%%%%%%%%%%%%%%%%%%%%%%%%%%%%%%%%%%%%%%%%%%%%%%%%
```

```
%%%%%%%%% Matlab code for Example 7.4 %%%%%%%%%%%%%%%%%%%%%%%%%%%%%%%%%%%%%%%%%%%%%
clear all, close all, clc
v0      = 0.40;         % inlet fiber volume fraction
v1      = 0.50;         % exit  fiber volume fraction
Rf      = 0.008;        % radius of the fiber bundle, [m]
Vf      = (v0+v1)/2;    % average fiber volume fraction
mu      = 400000;       % resin viscosity, [Pa.s]
kf      = 1.20;         % thermal conductivity of fiber,  [W/(m K)]
km      = 0.20;         % thermal conductivity of matrix, [W/(m K)]
rhof    = 2500;         % density of fiber,  [kg/m3]
rhom    = 1200;         % density of matrix, [kg/m3]
cf      =  800;         % specific heat capacity of fiber,  [J/(kg K)]
cm      =  700;         % specific heat capacity of matrix, [J/(kg K)]
L       = 0.500;        % length of die, [m]
Tin     =  20 + 273.2;  % preheater temperature, [K]
Tw      = 300 + 273.2;  % die wall temperature,  [K]
U       = 0.005;        % pulling speed, [m/s]

qdot    = 200000;       % energy generation due to viscous dissipation, [W/m3]

Ri      = sqrt(Rf^2/v0);            % inlet radius of the die, [m]
Re      = sqrt(Rf^2/v1);            % exit  radius of the die, [m]
z       = linspace(0,L,1001);       % z array, [m]
dz      = z(2)-z(1);                % Delta z, [m]
R       = (Re-Ri)*(z/L)+Ri;         % radius of the die, [m]
Rd      = mean(R);                  % average radius of the die, [m]

krr     =  1/((1-Vf)/km + Vf/kf);       % thermal conductivity of composite
rho_c   = (1-Vf)*rhom*cm + Vf*rhof*cf;  % (density * specific heat) of composite

Nr      =  41;                 % number of nodes in r
Nz      = 1001;                % number of nodes in z
hr      = Rd/(Nr-1);           % Delta_r
hz      = L /(Nz-1);           % Delta_z
r       = linspace(0,Rd,Nr);   % r array
z       = linspace(0,L ,Nz);   % z array
T       = Tin * ones(Nz,Nr);   % initial assignment for T matrix
T(:,Nr) = Tw;                  % boundary condition along r = Rd

c1     = krr/(rho_c*U);
c2     = hz/(2*hr);
c3     = hz/(hr^2);
c4     = (qdot*hz)/(rho_c * U);
```

```
for i = 2:Nz
   for j = 2:Nr-1
   T(i,j)=  T(i-1,j)                                                   ...
          + c1* (   (1/r(j))*c2*(T(i-1,j+1)            -T(i-1,j-1)  ) ...
                  +           c3*(T(i-1,j+1)-2*T(i-1,j)+T(i-1,j-1)) ) ...
          + c4;
   end
   T(i,1) = (1/3)*(4*T(i,2)-T(i,3));
end

TC          = T - 273.2;
T_corner    = TC(end,1)

T_physical = TC';
T_physical = T_physical(end:-1:1,:);
[Z,R]      = meshgrid(0:hz:L, Rd:-hr:0);

figure(1)
[pp,qq] = contour(Z*1000,R*1000,T_physical,[20:20:300]); clabel(pp,qq)
title('Temperature distribution [^oC]'), xlabel('z [mm]'), ylabel('r [mm]'), grid
%%%%%%%%%%%%%%%%%%%%%%%%%%%%%%%%%%%%%%%%%%%%%%%%%%%%%%%%%%%%%%%%%%%%%%%%%%%%%%%%%%%
```

```
%%%%%%%%% Matlab code for Example 7.5 %%%%%%%%%%%%%%%%%%%%%%%%%%%%%%%%%%%%%%%%%%%
clear all, close all, clc
v0      = 0.40;         % inlet fiber volume fraction
v1      = 0.50;         % exit  fiber volume fraction
Rf      = 0.008;        % radius of the fiber bundle, [m]
Vf      = (v0+v1)/2;    % average fiber volume fraction
mu      = 400000;       % resin viscosity, [Pa.s]
kf      = 1.20;         % thermal conductivity of fiber,  [W/(m K)]
km      = 0.20;         % thermal conductivity of matrix, [W/(m K)]
rhof    = 2500;         % density of fiber,  [kg/m3]
rhom    = 1200;         % density of matrix, [kg/m3]
cf      =  800;         % specific heat capacity of fiber,  [J/(kg K)]
cm      =  700;         % specific heat capacity of matrix, [J/(kg K)]
L       = 0.500;        % length of die, [m]
Tin     =  20 + 273.2; % preheater temperature, [K]
Tw      = 300 + 273.2; % die wall temperature,   [K]
U       = 0.005;        % pulling speed, [m/s]

Ri      = sqrt(Rf^2/v0);             % inlet radius of the die, [m]
Re      = sqrt(Rf^2/v1);             % exit  radius of the die, [m]
z       = linspace(0,L,1001);        % z array, [m]
dz      = z(2)-z(1);                 % Delta z, [m]
R       = (Re-Ri)*(z/L)+Ri;          % radius of the die, [m]
Rd      = mean(R);                   % average radius of the die, [m]

s       = 3 - 4*(Rf/Re)^2 + (Rf/Re)^4 + 4*log(Rf/Re);
dPdz    = (8*mu*U*pi*(Re^2-R.^2))./(pi*R.^4*s);
lamd    = Rf ./ R;
QDOT    = 1/(2*mu) * dPdz.^2 .* R.^2 .* (-0.75 - log(lamd) - 0.25*lamd.^4 + lamd.^2);

P(1) = 0;
for i = 2:length(z)
P(i) = P(i-1) + dPdz(i-1)*dz;
end

figure(11), plot(z*1000,dPdz*1.0e-6,'b-','linewidth',2), grid on
xlabel('z [mm]'), ylabel('Pressure gradient, dP/dz [MPa/m]')
legend('when U = 5 mm/s and L = 0.5 m')

figure(12), plot(z*1000,P*1.0e-6,'b-','linewidth',2), grid on
xlabel('z [mm]'), ylabel('Pressure, P [MPa]')
legend('when U = 5 mm/s and L = 0.5 m')

figure(13), plot(z*1000,QDOT,'b-','linewidth',2), grid on
xlabel('z [mm]'), ylabel('Energy generation term, dq/dt [W/m^3]')
legend('when U = 5 mm/s and L = 0.5 m')

krr     =  1/((1-Vf)/km + Vf/kf);       % thermal conductivity of composite
rho_c   = (1-Vf)*rhom*cm + Vf*rhof*cf;  % (density * specific heat) of composite
```

```
Nr      =  41;                  % number of nodes in r
Nz      = 1001;                 % number of nodes in z
hr      = Rd/(Nr-1);            % Delta_r
hz      = L /(Nz-1);            % Delta_z
r       = linspace(0,Rd,Nr);    % r array
z       = linspace(0,L ,Nz);    % z array
T       = Tin * ones(Nz,Nr);    % initial assignment for T matrix
T(:,Nr) = Tw;                   % boundary condition along r = Rd

c1      = krr/(rho_c*U);
c2      = hz/(2*hr);
c3      = hz/(hr^2);

for i = 2:Nz
qdot   = QDOT(i);
c4     = (qdot*hz)/(rho_c * U);
   for j = 2:Nr-1
   T(i,j)=  T(i-1,j)                                              ...
         + c1* (   (1/r(j))*c2*(T(i-1,j+1)           -T(i-1,j-1)  ) ...
                 +          c3*(T(i-1,j+1)-2*T(i-1,j)+T(i-1,j-1)) ) ...
         + c4;
   end
   T(i,1) = (1/3)*(4*T(i,2)-T(i,3));
end

TC          = T - 273.2;
T_corner    = TC(end,1)

T_physical = TC';
T_physical = T_physical(end:-1:1,:);
[Z,R]      = meshgrid(0:hz:L, Rd:-hr:0);

figure(1)
[pp,qq] = contour(Z*1000,R*1000,T_physical,[20:20:300]); clabel(pp,qq)
title('Temperature distribution [^oC]'), xlabel('z [mm]'), ylabel('r [mm]'), grid
%%%%%%%%%%%%%%%%%%%%%%%%%%%%%%%%%%%%%%%%%%%%%%%%%%%%%%%%%%%%%%%%%%%%%%%%%%%%%%%%
```

```
%%%%%%%%% Matlab code for Example 8.4 %%%%%%%%%%%%%%%%%%%%%%%%%%%%%%%%%%%%%%%%%
clear all, close all, clc
Vfo      = 0.30;        % initial fiber volume fraction
mu       = 0.40;        % resin viscosity, [Pa.s]
C        = 5.0e-10;     % coefficient [m^2] in Kozeny-Carman permeability:
                        %                    Kzz = C (1-Vf)^3/Vf^2
Va       = 0.70;        % maximum attainable fiber volume fraction
As       = 280;         % coefficient [Pa] in Fiber in a Box model:
                        %           Sigma_zz = As (sqrt(Vf/Vo)-1)/(sqrt(Va/Vf)-1)^4
PTo      = 0.0e5;       % initial autoclave pressure, [Pa]
dPTdt    = 10000;       % slope of the applied pressure ramp, [Pa/s]
h        = 0.02;        % initial part thickness, [m]
Nz       = 21;                              % # of nodes in z'
z        = linspace(0,1,Nz);                % z' array, [m]
dz       = z(2)-z(1);                       % Delta z', [m]
DT       = [0.00001 0.00002 0.00005];       % Delta t,  [s]
Vf       = Vfo*ones(Nz,1);                  % fiber volume fraction
PT(1)    = PTo;                             % autoclave pressure, [Pa]
Pr       = PTo*ones(Nz,1);                  % resin pressure, [Pa]
e        = (1-Vf)./Vf;
e0       = e(1);
Kzz      = C*(1-Vf).^3./Vf.^2;              % permeability, [m^2]
Sigmazz  = As*(sqrt(Vf./Vfo)-1)./(sqrt(Va./Vf)-1).^4;
t(1)     = 0;                               % time, [s]
k        = 1;                               % time index
Pr_0(1)  = Pr(1);
Pr_mid(1)= Pr(round((Nz+1)/2));
Pr_h(1)  = Pr(end);
Sigmazz_0(1)   = Sigmazz(1);
Sigmazz_mid(1)= Sigmazz(round((Nz+1)/2));
Sigmazz_h(1)   = Sigmazz(end);
Vf_0(1)  = Vf(1);
Vf_mid(1)= Vf(round((Nz+1)/2));
Vf_h(1)  = Vf(end);
H(1)     = h;
e_next   = e;
Vf_next  = Vf;
VF       = Vfo:0.0001:Va-0.0001;
SIGMAZZ  = As*(sqrt(VF./Vfo)-1)./(sqrt(Va./VF)-1).^4;
VFstop   = 0.47;

while min(min(Vf)) < VFstop
    if t(end) < 0.5, dt = DT(1); end
    if t(end) > 0.5, dt = DT(2); end
    if t(end) > 1.0, dt = DT(3); end
    if abs(mod(t(end),0.5)   0.5) < 0.000000001 |  mod(t(end),0.5) < 0.000000001
    figure(1)
    subplot(311), plot(z,Pr,'k-','linewidth',1), hold on
    xlabel('Dimensionless z/h(t)'), ylabel('Resin pressure, P_r [Pa]'), grid on
    subplot(312), plot(z,Sigmazz,'k-','linewidth',1), hold on
    xlabel('Dimensionless z/h(t)'), ylabel('Fiber stress, \sigma_{zz} [Pa]'), grid on
    subplot(313), plot(z,Vf,'k-','linewidth',1), hold on
    xlabel('Dimensionless z/h(t)'), ylabel('Fiber volume fraction, V_f'), grid on
    pause(0.1)
    else
    end
```

```
h              = H(end);
t(    k+1)     = t(k) + dt;
PT(k+1)        = PT(k)+ dPTdt*dt;
e_next(1)      = (1/3)*(-e(3)+4*e(2));        % B.C. at z = 0
c              = VF(find(abs(SIGMAZZ-PT(end)) == min(abs(SIGMAZZ-PT(end)))));
e_next( end)   = (1-c)/c;                     % B.C. at z = h
for i = 2:Nz-1
c1             = - ( (Kzz(i+1)/(1+e(i+1))) - (Kzz(i-1)/(1+e(i-1))))/(2*dz  );
c2             =   ( Sigmazz(i+1)          - Sigmazz(i-1)          )/(2*dz  );
c3             = -    Kzz(i)/(1+e(i));
c4             =   ( Sigmazz(i+1) - 2*Sigmazz(i) + Sigmazz(i-1)    )/(  dz^2);
e_next(i)      =   e(i) + (dt*(1+e0^2)/mu) * (1/h^2) * (c1*c2+c3*c4);
end
Vf_next        = 1./(1+e_next);
fprintf('Time = %10.5f s   min.Vf = %7.4f  max.Vf = %7.4f \n',t(end),min(Vf),max(Vf))
Kzz_next       = C*(1-Vf_next.^3)./Vf_next.^2;
Sigmazz_next   = As*(sqrt(Vf_next./Vfo)-1)./(sqrt(Va./Vf_next)-1).^4;
Pr_next        = PT(k+1) - Sigmazz_next;
Pr_next(end)   = 0;
Sigmazz_next(end) = PT(k+1);
k              = k+1;
Pr_0(k)        = Pr(1);
Pr_mid(k)      = Pr(round((Nz+1)/2));
Pr_h(k)        = Pr(end);
Vf_0(k)        = Vf(1);
Vf_mid(k)      = Vf(round((Nz+1)/2));
Vf_h(k)        = Vf(end);
Sigmazz_0(k)   = Sigmazz(1);
Sigmazz_mid(k)= Sigmazz(round((Nz+1)/2));
Sigmazz_h(k)   = Sigmazz(end);
e              = e_next;
Vf             = Vf_next;
Pr             = Pr_next;
Kzz            = Kzz_next;
Sigmazz        = Sigmazz_next;
H(k)           = (H(1)*Vfo) * (1/mean(Vf));
end

figure(2)
subplot(411), plot(t,Pr_0,'k-', t,Pr_mid,'k-.', t,Pr_h,'k--','linewidth',1)
grid on, xlabel('Time, t [s]'), ylabel('Resin pressure, P_r [Pa]')
legend('z = 0', 'z = h/2', 'z = h', 0)
subplot(412), plot(t,Sigmazz_0,'k-', t,Sigmazz_mid,'k-.', ...
                  t,Sigmazz_h,'k--','linewidth',1)
grid on, xlabel('Time, t [s]'), ylabel('Fiber stress, \sigma_{zz} [Pa]')
subplot(413), plot(t,Vf_0,'k-', t,Vf_mid,'k-.', t,Vf_h,'k--','linewidth',1)
grid on, xlabel('Time, t [s]'), ylabel('Fiber volume fraction, V_f')
subplot(414), plot(t,H*1000,'k-','linewidth',1)
grid on, xlabel('Time, t [s]'), ylabel('Part thickness, h [mm]')
%%%%%%%%%%%%%%%%%%%%%%%%%%%%%%%%%%%%%%%%%%%%%%%%%%%%%%%%%%%%%%%%%%%%%%%%%%%%%%%%%%%%%%%%%%%%%%%%%%%%%%%%%%%%%%%%%%%%%%%%%%%%%%%%%%%%%%%%%%%%%%%%%%%%%%%%%%%%%%
```

```
%%%%%%%%%%%%% Matlab code for Example 8.5 %%%%%%%%%%%%%%%%%%%%%%%%%%%%%%%%%%%%%
clear all, close all, clc
Vf       = 0.40;        % average fiber volume fraction
kf       = 1.20;        % thermal conductivity of fiber,  [W/(m K)]
km       = 0.20;        % thermal conductivity of matrix, [W/(m K)]
rhof     = 2500;        % density of fiber,  [kg/m3]
rhom     = 1200;        % density of matrix, [kg/m3]
cf       =  800;        % specific heat capacity of fiber,  [J/(kg K)]
cm       =  700;        % specific heat capacity of matrix, [J/(kg K)]
Tin      =  20 + 273.2; % initial temperature, [K]
Tw       = 120 + 273.2; % mold wall temperature,  [K]
h        = 0.020;       % thickness of the part, [m]
R        = 8.3145;      % constant to be used in cure kinetics
CA0      = 0.5;         % constant to be used in cure kinetics
k0       = 2;           % constant to be used in cure kinetics
E        = 1000 * R;    % constant to be used in cure kinetics
LAMBDA   = 1;           % constant to be used in cure kinetics
Hr       = 280000;      % constant to be used in cure kinetics, [J/kg]
k        =  1/((1-Vf)/km + Vf/kf);            % thermal conductivity of composite
rho_c    = Vf*(rhof*cf) + (1-Vf)*(rhom*cm);  % (density * specific heat) of composite
Nz       = 21;                                % # of nodes in z
z        = linspace(0,h,Nz);                  % z' array, [m]
dz       = z(2)-z(1);                         % Delta z', [m]

T        = Tin * ones(size(z));
T(1)     = Tw;
T(end)   = Tw;
C        = zeros(size(z));
To(1)    = T(1                    );
Tq(1)    = T((round(Nz-1)*0.25)+1);
Tm(1)    = T((round(Nz-1)*0.50)+1);
Co(1)    = C(1                    );
Cq(1)    = C((round(Nz-1)*0.25)+1);
Cm(1)    = C((round(Nz-1)*0.50)+1);
t(1)     = 0;
dt       = 0.01;
j        = 1;
Cnext    = C;
Tnext    = T;
```

```
while min(C) < 0.85

for i = 2:Nz-1
dCdt       = k0 * exp(-E/(R*T(i))) * CA0 * (1-C(i)) * (LAMBDA-C(i));
dTdt       = (1/rho_c)*(k*((T(i+1)-2*T(i)+T(i-1))/(dz^2))+(1-Vf)*rhom*Hr*dCdt);
Cnext(i)   = C(i) + dCdt*dt;
Tnext(i)   = T(i) + dTdt*dt;
end
Cs         = C(end-1) + (C(end-1)-C(end-2));
Cnext(1)   = Cs;
Cnext(end)= Cs;
C          = Cnext;
T          = Tnext;
t(j+1)     = t(j) + dt;
j          = j + 1;
To(j)      = T(1                      );
Tq(j)      = T((round(Nz-1)*0.25)+1);
Tm(j)      = T((round(Nz-1)*0.50)+1);
Co(j)      = C(1                      );
Cq(j)      = C((round(Nz-1)*0.25)+1);
Cm(j)      = C((round(Nz-1)*0.50)+1);
    if abs(mod(t(end),10) - 10) < 1.0e-8 |  mod(t(end),10) < 1.0e-8
    figure(1)
    subplot(121), plot(z*1000,T-273.2,'k-','linewidth',1), grid on, hold on
    xlabel('z [mm]'), ylabel('Temperature, T [^oC]')
    subplot(122), plot(z*1000,C,'k-','linewidth',1), grid on, hold on
    xlabel('z [mm]'), ylabel('Extent of reaction, C^*')
    pause(0.1)
    else
    end
end

figure(2)
subplot(121), plot(t,Tq-273.2,'k-',t,Tm-273.2,'k--','linewidth',1)
grid on, legend('z = h/4','z = h/2',0)
xlabel('Time, t [s]'), ylabel('Temperature, T [^oC]')
subplot(122), plot(t,Cq,'k-',t,Cm,'k--','linewidth',1), grid on, hold on
xlabel('Time, t [s]'), ylabel('Extent of reaction, C^*')
grid on, legend('z = h/4','z = h/2',0)
%%%%%%%%%%%%%%%%%%%%%%%%%%%%%%%%%%%%%%%%%%%%%%%%%%%%%%%%%%%%%%%%%%%%%%%%%%%%
```

```
%%%%%%%%% Matlab code for Example 8.10 %%%%%%%%%%%%%%%%%%%%%%%%%%%%%%%%%%%%%%
clear all, close all, clc
Vf       = 0.40;         % average fiber volume fraction
mu       = 0.25;         % resin viscosity, [Pa.s]
kf       = 1.20;         % thermal conductivity of fiber,  [W/(m K)]
km       = 0.20;         % thermal conductivity of matrix, [W/(m K)]
rhof     = 2500;         % density of fiber,  [kg/m3]
rhom     = 1200;         % density of matrix, [kg/m3]
cf       =  800;         % specific heat capacity of fiber,  [J/(kg K)]
cm       =  700;         % specific heat capacity of matrix, [J/(kg K)]
Tin      =  20 + 273.2;  % inlet temperature, [K]
To       = 120 + 273.2;  % mold wall temperature,  [K]
u        = 0.0714;       % average 1D speed, [m/s]
L        = 0.500;        % length of die, [m]
h        = 0.005;        % half-thickness of the mold cavity, [m]

k        =  1/((1-Vf)/km + Vf/kf);       % thermal conductivity of composite
rho_c    = rhom*cm;                      % (density * specific heat) of resin
alpha    = k/rho_c;                      % thermal diffusivity, m^2/s
Gz       = (u*h^2)/(alpha*L);            % Graetz number
Nz       =  81;                          % # of nodes in z'
z        = linspace(-1,1,Nz);            % z' array, [m]
dz       = z(2)-z(1);                    % Delta z', [m]
r        = 0.2;
dx       = r*(Gz*dz^2);
Nx       = round(1/dx) + 1;              % # of nodes in x'
x        = linspace( 0,1,Nx);            % x' array, [m]

Tasy         = zeros(Nx,Nz);             % asymptotic temp. array
Tasy(1,: )   = Tin;                      % inlet  B.C.
Tasy(:,1 )   = To;                       % bottom B.C.
Tasy(:,Nz)   = To;                       % top    B.C.

for i = 2:Nx
for j = 2:Nz-1
    if z(j) < 0, zhat = (1+z(j))*h;
    else
                 zhat = (1-z(j))*h;
    end
    ksi     = zhat/(2*sqrt(alpha*(x(i)*L)/u));
 Tasy(i,j) = (1-erf(ksi)) * (To-Tin) + Tin;
end
end

TTC          = Tasy - 273.2;
TT_physical  = TTC';
TT_physical  = TT_physical(end:-1:1,:);
[X,Z]        = meshgrid(linspace(0,1,Nx)*L,linspace(1,-1,Nz)*h);
Tiso         = Tin + 0.001* (To  - Tin) - 273.2;
figure(1)
[pp,qq] = contour(X*1000,Z*1000,TT_physical,[Tiso,21,40:20:120]); clabel(pp,qq)
xlabel('x [mm]'), ylabel('z [mm]'), grid
%%%%%%%%%%%%%%%%%%%%%%%%%%%%%%%%%%%%%%%%%%%%%%%%%%%%%%%%%%%%%%%%%%%%%%%%%%%%%%%%
```

```
%%%%%%%%% Matlab code for Example 8.11 %%%%%%%%%%%%%%%%%%%%%%%%%%%%%%%%%%%%%%%%%
clear all, close all, clc
Vf       = 0.40;          % average fiber volume fraction
mu       = 0.25;          % resin viscosity, [Pa.s]
kf       = 1.20;          % thermal conductivity of fiber,  [W/(m K)]
km       = 0.20;          % thermal conductivity of matrix, [W/(m K)]
rhof     = 2500;          % density of fiber,  [kg/m3]
rhom     = 1200;          % density of matrix, [kg/m3]
cf       =  800;          % specific heat capacity of fiber,  [J/(kg K)]
cm       =  700;          % specific heat capacity of matrix, [J/(kg K)]
Tin      =  20 + 273.2;   % inlet temperature, [K]
To       = 120 + 273.2;   % mold wall temperature,  [K]
u        = 0.0714;        % average 1D speed, [m/s]
L        = 0.500;         % length of die, [m]
h        = 0.005;         % half-thickness of the mold cavity, [m]

U        = [u 0.10*u 0.01*u];
for icase=1:1
u        = U(icase);
k        = 1/((1-Vf)/km + Vf/kf);          % thermal conductivity of composite
rho_c    = rhom*cm;                        % (density * specific heat) of resin
alpha    = k/rho_c;                        % thermal diffusivity, m^2/s
Gz       = (u*h^2)/(alpha*L);              % Graetz number
Nz       =  81;                            % # of nodes in z'
z        = linspace(-1,1,Nz);              % z' array, [m]
dz       = z(2)-z(1);                      % Delta z', [m]
r        = 0.20;                           % the ratio, dx/(Gz dz^2); use < 0.5
dx       = r*(Gz*dz^2);
Nx       = round(1/dx) + 1;                % # of nodes in x'
x        = linspace( 0,1,Nx);              % x' array, [m]

TH       = zeros(Nx,Nz);                   % theta array
TH(1,: ) = (Tin - Tin)/(To - Tin);         % inlet  B.C.
TH(:,1 ) = (To  - Tin)/(To - Tin);         % bottom B.C.
TH(:,Nz) = (To  - Tin)/(To - Tin);         % top    B.C.

for i = 1:Nx-1
for j = 2:Nz-1
 TH(i+1,j)=  (1-2*r)*TH(i,j) +  r * (TH(i,j+1) + TH(i,j-1));
end
end

T            = TH   * (To - Tin) + Tin;
TC           = T - 273.2;
T_physical   = TC';
T_physical   = T_physical(end:-1:1,:);
[X,Z]        = meshgrid(linspace(0,1,Nx)*L,linspace(1,-1,Nz)*h);
Tiso         = Tin + 0.001* (To  - Tin) - 273.2;
figure(icase)
[pp,qq] = contour(X*1000,Z*1000,T_physical,[Tiso,21,40:20:120]); clabel(pp,qq)
xlabel('x [mm]'), ylabel('z [mm]'), grid
pause(0.1)
end
%%%%%%%%%%%%%%%%%%%%%%%%%%%%%%%%%%%%%%%%%%%%%%%%%%%%%%%%%%%%%%%%%%%%%%%%%%%%%%%%
```

```
%%%%%%%%%%%%% Matlab code for Example 8.12 %%%%%%%%%%%%%%%%%%%%%%%%%%%%%%
%%%%%%%%%%%%% Solution to resin pressure within a rectangular domain
clear all, close all, clc
Kxx         =  1.0e-9;      % permeability in x direction
Kyy         =  0.2e-9;      % permeability in y direction
Lx          =  0.8;         % length of mold
Ly          =  0.2;         % width  of mold
Lz          =  0.02;        % depth  of mold = inlet depth
Nx          =  41;          % # of nodes in x
Ny          =  11;          % # of nodes in y
hx          =  Lx /(Nx-1); % dx
hy          =  Ly /(Ny-1); % dy
mu          =  0.1;         % viscosity of resin

if abs(hx-hy) > eps, stop, else h = hx; end
p          = zeros(Nx,Ny);
p_prev     = p;

c1         = Kxx/(2*(Kxx+Kyy));
c2         = Kyy/(2*(Kxx+Kyy));
err        = 1000;

while (err > 0.01)

for  i = 2:Nx-1
for  j = 2:Ny-1
p(i,j) = c1*(p(i-1,j)+p(i+1,j)) + c2*(p(i,j-1)+p(i,j+1));
end
end

p(1 ,: ) = (4*p(2   ,:)-p(3   ,:))/3;     % B.C. at left   edge
p(Nx,: ) = (4*p(Nx-1,:)-p(Nx-2,:))/3;     % B.C. at right  edge
p(: ,1 ) = (4*p(:,2   )-p(:,3   ))/3;     % B.C. at bottom edge
p(: ,Ny) = (4*p(:,Ny-1)-p(:,Ny-2))/3;     % B.C. at top    edge
p(1 , 6) = 300000;                        % B.C. at injection   node
p(41, 6) =      0;                        % B.C. at ventilation node

err      = norm(p-p_prev);
p_prev   = p;
end

p_physical = p';
p_physical = p_physical(end:-1:1,:);
[X,Y]      = meshgrid(0:h:Lx, Ly:-h:0);

figure(1)
[pp,qq] = contour(X,Y,p_physical*0.001,[0:30:300]); clabel(pp,qq)
title('Pressure distribution [kPa]'),xlabel('x [m]'),ylabel('y [m]'),grid

delPdelx = zeros(Nx,Ny);
delPdely = zeros(Nx,Ny);
u        = zeros(Nx,Ny);
v        = zeros(Nx,Ny);
```

```
   for i = 2:Nx-1
   for j = 2:Ny-1
          delPdelx(i,j)  = (p(i+1,j  )-p(i-1,j  ))/(2*h);
          delPdely(i,j)  = (p(i  ,j+1)-p(i  ,j-1))/(2*h);
   end
   end

   for i = 2:Nx-1
        delPdelx(i,1 ) = (p(i+1,1   )-p(i-1,1   ))/(2*h);
        delPdelx(i,Ny) = (p(i+1,Ny  )-p(i-1,Ny  ))/(2*h);
   end

   for j = 2:Ny-1
        delPdely(1 ,j) = (p(1   ,j+1)-p(1   ,j-1))/(2*h);
        delPdely(Nx,j) = (p(Nx  ,j+1)-p(Nx  ,j-1))/(2*h);
   end

   for i = 2:Nx-1
        delPdely(i,1 ) = (-p(i, 1+2)+4*p(i, 1+1)-3*p(i,   1))/(2*h);
        delPdely(i,Ny) =-(-p(i,Ny-2)+4*p(i,Ny-1)-3*p(i,Ny  ))/(2*h);
   end

   for j = 2:Ny-1
        delPdelx(1 ,j) = (-p( 1+2,j)+4*p( 1+1,j)-3*p(   1,j))/(2*h);
        delPdelx(Nx,j) =-(-p(Nx-2,j)+4*p(Nx-1,j)-3*p(Nx  ,j))/(2*h);
   end

u          = -(Kxx/mu)*delPdelx;
v          = -(Kyy/mu)*delPdely;
u_physical = u';
u_physical = u_physical(end:-1:1,:);
v_physical = v';
v_physical = v_physical(end:-1:1,:);

figure(3)
quiver(X,Y,u_physical,v_physical,3), hold off, axis image
title('Velocity Distribution'), xlabel('x [m]'), ylabel('y [m]')

Qinlet  = u( 1,6)*(h*Lz)
Qexit   = u(41,6)*(h*Lz)
%%%%%%%%%%%%%%%%%%%%%%%%%%%%%%%%%%%%%%%%%%%%%%%%%%%%%%%%%%%%%%%%%%%%%%
```

```
%%%%%%%%%%%%% Matlab code for Example 8.13 %%%%%%%%%%%%%%%%%%%%%%%%%%%%%%%%%%%
%%%%%%%%%%%%% Solution to resin pressure within a L-shaped domain
clear all, close all, clc
Kxx         =  2.0e-9;     % permeability in x direction
Kyy         =  2.0e-9;     % permeability in y direction
Lx          =  1.2;        % length of mold
Ly          =  1.0;        % width  of mold
Lz          =  0.01;       % depth  of mold = inlet depth
Nx          =  25;         % # of nodes in x
Ny          =  21;         % # of nodes in y
hx          =  Lx /(Nx-1); % dx
hy          =  Ly /(Ny-1); % dy
mu          =  0.2;        % viscosity of resin

if abs(hx-hy) > eps, stop, else h = hx; end
p           = zeros(Nx,Ny);
p_prev      = p;
c1          = Kxx/(2*(Kxx+Kyy));
c2          = Kyy/(2*(Kxx+Kyy));
Area        = Lz*h;
err         = inf;

while(err > 1e-5)

%%% Left half of inner points:
for i=2:12
for j=2:Ny-1
p(i,j) = c1*(p(i-1,j)+p(i+1,j)) + c2*(p(i,j-1)+p(i,j+1));
end
end

%%% Right side of inner points:
for i=13:Nx-1
for j=12:Ny-1
p(i,j) = c1*(p(i-1,j)+p(i+1,j)) + c2*(p(i,j-1)+p(i,j+1));
end
end

p(1    ,:    ) = (4*p(2    ,:    )-p(3    ,:    ))/3;    % B.C. at left         edge
p(13   , 1:11) = (4*p(12   , 1:11)-p(11   , 1:11))/3;    % B.C. at right-bottom edge
p(Nx   ,11:21) = (4*p(Nx-1 ,11:21)-p(Nx-2,11:21))/3;     % B.C. at right-upper  edge
p(:    ,Ny   ) = (4*p(:    ,Ny-1 )-p(:    ,Ny-2 ))/3;    % B.C. at top          edge
p(1:13 ,1    ) = (4*p(1:13 ,2    )-p(1:13 ,3    ))/3;    % B.C. at bottom       edge
p(13:Nx,11   ) = (4*p(13:Nx,12   )-p(13:Nx,13   ))/3;    % B.C. at mid-horiz.   edge
p(7    ,1    ) = 500000;                                 % B.C. at injection    node
p(Nx   ,17   ) = 100000;                                 % B.C. at ventilation  node
p(13   ,11   ) = (4*p(12   ,12   )-p(11   ,13   ))/3;    % B.C. at kink         node
err            = norm(p-p_prev);
p_prev         = p;
end
```

```
p(14:Nx,1:10 ) = -inf;
p_physical     = p';
p_physical     = p_physical(end:-1:1,:);

figure(1)
[X,Y]          = meshgrid(0:h:Lx,Ly:-h:0);
[pp,q]         = contour(X,Y,p_physical*0.001);clabel(pp,q)
title('Pressure Distribution [kPa]'), xlabel('x(m)'), ylabel('y(m)'), grid on
figure(2)
[X,Y]          = meshgrid(0:h:Lx,Ly:-h:0);
[pp,q]         = contourf(X,Y,p_physical*0.001);clabel(pp,q),colorbar
title('Pressure Distribution [kPa]'), xlabel('x(m)'), ylabel('y(m)'), grid on

delPdelx = zeros(Nx,Ny);
delPdely = zeros(Nx,Ny);
u        = zeros(Nx,Ny);
v        = zeros(Nx,Ny);

    for i=2:12
    for j=2:Ny-1
          delPdelx(i,j)  = (p(i+1,j  )-p(i-1,j  ))/(2*h);
          delPdely(i,j)  = (p(i  ,j+1)-p(i  ,j-1))/(2*h);
    end
    end

    for i=13:Nx-1
    for j=12:Ny-1
          delPdelx(i,j)  = (p(i+1,j  )-p(i-1,j  ))/(2*h);
          delPdely(i,j)  = (p(i  ,j+1)-p(i  ,j-1))/(2*h);
    end
    end

    for j = 2:20
        delPdelx(1,j) = (-p(3,j)+4*p(2,j)-3*p(1,j))/(2*h);
        delPdely(1,j) = ( p(1,j+1)-p(1,j-1)       )/(2*h);
    end

    for i = 2:24
        delPdelx(i,21) = ( p(i+1,21)-p(i-1,21)        )/(2*h);
        delPdely(i,21) =-(-p(i,19)+4*p(i,20)-3*p(i,21))/(2*h);
    end

    for j = 12:20
        delPdelx(25,j) =-(-p(23,j)+4*p(24,j)-3*p(25,j))/(2*h);
        delPdely(25,j) = ( p(25,j+1)-p(25,j-1)        )/(2*h);
    end

    for i = 14:24
        delPdelx(i,11) = ( p(i+1,11)-p(i-1,11)        )/(2*h);
        delPdely(i,11) = (-p(i,13)+4*p(i,12)-3*p(i,11))/(2*h);
    end
```

```
    for j = 2:10
        delPdelx(13,j) =-(-p(11,j)+4*p(12,j)-3*p(13,j))/(2*h);
        delPdely(13,j) = ( p(13,j+1)-p(13,j-1)          )/(2*h);
    end

    for i = 2:12
        delPdelx(i,1) = ( p(i+1,1)-p(i-1,1)          )/(2*h);
        delPdely(i,1) = (-p(i ,3)+4*p(i,2)-3*p(i,1))/(2*h);
    end

u          = -(Kxx/mu)*delPdelx;
v          = -(Kyy/mu)*delPdely;
u_physical = u';  u_physical = u_physical(end:-1:1,:);
v_physical = v';  v_physical = v_physical(end:-1:1,:);

figure(3)
quiver(X,Y,u_physical,v_physical,3), hold off, axis image
title('Velocity Distribution'), xlabel('x [m]'), ylabel('y [m]')

Qinlet  = v( 7, 1)*Area    % inlet flow rate, [m^3/s]
Qexit   = u(25,17)*Area    % exit  flow rate, [m^3/s]
P_03_05 = p(7,11)          % pressure at (x,y) = (0.3,0.5), [Pa]
%%%%%%%%%%%%%%%%%%%%%%%%%%%%%%%%%%%%%%%%%%%%%%%%%%%%%%%%%%%%%%%%%%%%%%%%%%%%%%%%%%
```

```
%%%%%%%%% Matlab code for Example 8.15 %%%%%%%%%%%%%%%%%%%%%%%%%%%%%%%%%%%%%%
%%%%%%%%% FEM MESH GENERATOR FOR A RECTANGLULAR DOMAIN
clear all, close all, clc,
Lx = 0.700; Nx = 35; hx = Lx / (Nx);
Ly = 0.200; Ny = 10; hy = Ly / (Ny);

H   = 0.005000;
Vf  = 0.400000;
Kxx = 4.0e-010;
Kxy = 0;
Kyy = 4.0e-010;
mu  = 0.10;

Number_nodes = (Nx+1) * (Ny+1);
Number_elem  = (Nx  ) * (Ny  );
fid          = fopen('rectangle.dmp','w');
fprintf(fid,'Number of nodes : %5.0f \n',Number_nodes);

for j = 1:Ny+1;
   for i = 1:Nx+1;
      node_id   = (j-1)*(Nx+1)+i;
      x         = (i-1)*hx;
      y         = (j-1)*hy;
      z         = 0;
      NODE(node_id,:) = [node_id x y z];
      fprintf(fid,'%6.0f  %12.6f  %12.6f  %12.6f \n',NODE(node_id,:));
   end
end

fprintf(fid,'Number of elements : %5.0f \n',Number_elem);

for j = 1:Ny;
   for i = 1:Nx;
      elem_id   = (j-1)*(Nx )+i;
      k1        = (j-1)*(Nx+1)+i;
      k2        = k1+1;
      k3        = k1+(Nx+1)+1;
      k4        = k3-1;
      ELEMENTS(elem_id,:) = [elem_id 4 k1 k2 k3 k4 H Vf Kxx Kxy Kyy];
      fprintf(fid,'%6.0f %4.0f %6.0f    ........  %15.4e \n',ELEMENTS(elem_id,:));
   end
end

fprintf(fid,'Resin Viscosity model NEWTON \n');
fprintf(fid,'Viscosity : %15.6f \n',mu);
fprintf(fid,'# Done ************************* \n');
fclose(fid);
%%%%%%%%%%%%%%%%%%%%%%%%%%%%%%%%%%%%%%%%%%%%%%%%%%%%%%%%%%%%%%%%%%%%%%%%%%%%%%
```

```
%%%%%%%%%% Matlab code for Example 8.21_mesh %%%%%%%%%%%%%%%%%%%%%%%%%%%%%%%%%%%%%%%
%%%%%%% FEM MESH GENERATOR FOR THE SEMICIRCLE MOLD
clear all, close all, clc

theta = [180:-3:0]*(pi/180);
R     = 0.25;
x1    = -0.325:0.025:-0.275;
z1    = zeros(size(x1));
x2    = R*cos(theta);
z2    = R*sin(theta);
x3    =  0.275:0.025: 0.325;
z3    = zeros(size(x3));
X     = [x1 x2 x3];
Z     = [z1 z2 z3];
H     = 0.005000; Vf  = 0.400000;
Kxx   = 5.3e-010; Kxy = 0;        Kyy = 5.3e-010;    mu  = 0.10;
Nx    = length(X);
Ny    = 13;
Number_nodes = Nx*Ny;
fid          = fopen('mesh.dmp','w+');
fprintf(fid,'Number of nodes : %5.0f \n',Number_nodes);

for j = 1:Ny;
  for i = 1:Nx;
  node_id   = (j-1)*(Nx)+i;
  x         = X(i);
  y         = (j-1)*(0.3600/(Ny-1));
  z         = Z(i);
  NODE(node_id,:) = [node_id x y z];
  fprintf(fid,'%6.0f  %12.6f  %12.6f  %12.6f \n',NODE(node_id,:));
  end
end
Nx = length(X) - 1;
Number_elem  = Nx*(Ny-1);
fprintf(fid,'Number of elements : %5.0f \n',Number_elem);
for j = 1:Ny-1;
  for i = 1:Nx;
  elem_id   = (j-1)*(Nx ) +i;
  k1        = (j-1)*(Nx+1)+i;
  k2        = k1+1;
  k3        = k1+(Nx+1)+1;
  k4        = k3-1;
  ELEMENTS(elem_id,:) = [elem_id 4 k1 k2 k3 k4 H Vf Kxx Kxy Kyy];
  fprintf(fid,'%6.0f %4.0f %6.0f %6.0f [illegible] %15.4e \n',ELEMENTS(elem_id,:));
  end
end
fprintf(fid,'Resin Viscosity model NEWTON \n');
fprintf(fid,'Viscosity : %15.6f \n',mu);
fprintf(fid,'# Done ************************** \n');
fclose(fid);
%%%%%%%%%%%%%%%%%%%%%%%%%%%%%%%%%%%%%%%%%%%%%%%%%%%%%%%%%%%%%%%%%%%%%%%%%%%%%%%%%%%%
```

```
%%%%%%%%%% Matlab code for Example 8.16_plot %%%%%%%%%%%%%%%%%%%%%%%
clc, close all, clear all
B     = load('semicircle_P_Q.txt');
t     = B(:,1);
P1    = B(:,2);
Q1    = B(:,3)*1.0e6;
P2    = B(:,4);
Q2    = B(:,5)*1.0e6;

figure(1)
plot(t,Q1,'k-','linewidth',2)
xlabel('Time, t [s]'), ylabel('Injection Flow Rate, Q_{in} [cc/s]')
grid, ylim([0 4])

A  = load('semicircle_allresults.txt');
Nx = 66;  Ny = 12;
for j = 1:Ny+1
    for i = 1:Nx+1
        X(i,j)=A(i+(Nx+1)*(j-1),2);
        Y(i,j)=A(i+(Nx+1)*(j-1),3);
        Z(i,j)=A(i+(Nx+1)*(j-1),4);
        P(i,j)=A(i+(Nx+1)*(j-1),5);
        Q(i,j)=A(i+(Nx+1)*(j-1),6);
        T(i,j)=A(i+(Nx+1)*(j-1),7);
        F(i,j)=A(i+(Nx+1)*(j-1),8);
    end
end

figure(2)
surf(X,Y,Z,T); colorbar; axis equal;
title('Flow Front Locations [s]')
xlabel('x [m]'); ylabel('y [m]'); zlabel('z [m]');

figure(3)
surf(X,Y,Z,P*0.001);colorbar;
axis equal;
title('Pressure distribution at the end of injection [kPa]')
xlabel('x [m]'); ylabel('y [m]'); zlabel('z [m]');
%%%%%%%%%%%%%%%%%%%%%%%%%%%%%%%%%%%%%%%%%%%%%%%%%%%%%%%%%%%%%%%%%%%%%
```

```
%%%%%%%%% Matlab code for Example 8.18_plot %%%%%%%%%%%%%%%%%%%%%%%%%%%%%
%%%%%% Injection Pressure Plotter
clear all, close all, clc,
Lx = 0.700; Nx = 35; hx = Lx / (Nx);
Ly = 0.200; Ny = 10; hy = Ly / (Ny);
A1 = load('scn_1_P_Q.txt');         t1 = A1(:,1); P1 = A1(:,2); Q1 = A1(:,3);
A2 = load('scn_2_P_Q.txt');         t2 = A2(:,1); P2 = A2(:,2); Q2 = A2(:,3);
A3 = load('scn_3_P_Q.txt');         t3 = A3(:,1); P3 = A3(:,2); Q3 = A3(:,3);
B1 = load('scn_1_allresults.txt'); T1 = B1(:,7);
B2 = load('scn_2_allresults.txt'); T2 = B2(:,7);
B3 = load('scn_3_allresults.txt'); T3 = B3(:,7);

figure(1)
plot(t1,0.001*P1,'k-',t2,0.001*P2,'k-.',t3,0.001*P3,'k--','linewidth',2)
xlabel('Time, t [s]'), ylabel('Injection Pressure, P_{inj} [kPa]'), grid
legend('K_B / K_A =   1','K_B / K_A =  10','K_B / K_A = 100',0)

figure(2)
plot(t1,1e6*Q1,'k-',t2,1e6*Q2,'k-.',t3,1e6*Q3,'k--','linewidth',2)
xlabel('Time, t [s]'), ylabel('Injection Flow Rate, Q_{inj} [cc/s]'), grid
legend('K_B / K_A =   1','K_B / K_A =  10','K_B / K_A = 100',0)

time_to_fill = [t1(end),t2(end),t3(end)]
P_inj_max    = [P1(end),P2(end),P3(end)] * 0.001

for j = 1:Ny+1
    for i = 1:Nx+1
        k = (j-1)*(Nx+1) + i;
      TIME1(i,j) = T1(k);
      TIME2(i,j) = T2(k);
      TIME3(i,j) = T3(k);
    end
end

TIME1_physical = TIME1'; TIME1_physical = TIME1_physical(end:-1:1,:);
TIME2_physical = TIME2'; TIME2_physical = TIME2_physical(end:-1:1,:);
TIME3_physical = TIME3'; TIME3_physical = TIME3_physical(end:-1:1,:);
[X,Y]          = meshgrid(0:hx:Lx, Ly:-hy:0);

figure(3)
subplot(311), contour(X,Y,TIME1_physical), hold on, shading flat
[aa,bb] = contourf(X,Y,TIME1_physical,[0:20:400]); clabel(aa,bb), colorbar
title('Flow Front Positions [s] for K_B / K_A =   1'), xlabel('x [m]')
ylabel('y [m]'), grid
subplot(312), contour(X,Y,TIME2_physical), hold on, shading flat
[aa,bb] = contourf(X,Y,TIME2_physical,[0:20:400]); clabel(aa,bb), colorbar
title('Flow Front Positions [s] for K_B / K_A = 10')
xlabel('x [m]'), ylabel('y [m]'), grid
subplot(313), contour(X,Y,TIME3_physical), hold on, shading flat
[aa,bb] = contourf(X,Y,TIME3_physical,[0:20:400]); clabel(aa,bb), colorbar
title('Flow Front Positions [s] for K_B / K_A = 100')
xlabel('x [m]'), ylabel('y [m]'), grid
%%%%%%%%%%%%%%%%%%%%%%%%%%%%%%%%%%%%%%%%%%%%%%%%%%%%%%%%%%%%%%%%%%%%%%%%%%%
```

```
%%%%%%%%%% Matlab code for Example 8.18_analytical %%%%%%%%%%%%%%
clear all, close all, clc
Lx         = 0.700;
Ly         = 0.200;
Lz         = 0.005;
VF         = [0.40 0.20 0.10];
Q          = 2.0e-6;

Vgate      = (0.020)*(0.010)*Lz;  % initially filled c.v. at gate
VA         = (Lx - 0.080)*(Ly-0.040)*Lz*(1-VF(1));
VC1        = (Ly)*(0.040)*Lz - Vgate;
VC2        = (Ly)*(0.040)*Lz;
VB         = (Lx - 0.080)*(0.040)*Lz*(1-VF);

VEMPTY     = VA + VC1 + VC2 + VB;
TIMEtoFILL = VEMPTY./Q

T_sim      = [225.4931 237.8988 244.0999]
Error      = (T_sim - TIMEtoFILL)./TIMEtoFILL * 100
%%%%%%%%%%%%%%%%%%%%%%%%%%%%%%%%%%%%%%%%%%%%%%%%%%%%%%%%%%%%%%%%%
```

```
%%%%%%%%% Matlab code for Example 8.19_plot %%%%%%%%%%%%%%%%%%%%%%%%%%%%%%
clear all, close all, clc,
Lx = 0.700; Nx = 35; hx = Lx / (Nx);
Ly = 0.200; Ny = 10; hy = Ly / (Ny);
A1 = load('scn_1_P_Q.txt');         t1 = A1(:,1); P1 = A1(:,2); Q1 = A1(:,3);
A2 = load('pmaxlimit_P_Q.txt');     t2 = A2(:,1); P2 = A2(:,2); Q2 = A2(:,3);
B1 = load('scn_1_allresults.txt');      T1 = B1(:,7);
B2 = load('pmaxlimit_allresults.txt'); T2 = B2(:,7);

figure(1)
subplot(121), plot(t1,0.001*P1,'k-',t2,0.001*P2,'bo','linewidth',2)
xlabel('Time, t [s]'), ylabel('Injection Pressure, P_{inj} [kPa]'), grid
legend('S1','S4',0)

subplot(122), plot(t1,1e6*Q1,'k-',t2,1e6*Q2,'bo','linewidth',2)
xlabel('Time, t [s]'), ylabel('Injection Flow Rate, Q_{inj} [cc/s]'), grid
legend('S1','S4',0)

time_to_fill = [t1(end),t2(end)]
P_inj_max    = [P1(end),P2(end)] * 0.001

for j = 1:Ny+1
    for i = 1:Nx+1
        k = (j-1)*(Nx+1) + i;
      TIME1(i,j) = T1(k);
      TIME2(i,j) = T2(k);
    end
end

TIME1_physical = TIME1'; TIME1_physical = TIME1_physical(end:-1:1,:);
TIME2_physical = TIME2'; TIME2_physical = TIME2_physical(end:-1:1,:);
[X,Y]          = meshgrid(0:hx:Lx, Ly:-hy:0);

figure(3)
subplot(211), contour(X,Y,TIME1_physical), hold on, shading flat
[aa,bb] = contourf(X,Y,TIME1_physical,[0:20:400]); clabel(aa,bb), colorbar
title('Flow Front Positions [s] for S1'), xlabel('x [m]')
ylabel('y [m]'), grid
subplot(212), contour(X,Y,TIME2_physical), hold on, shading flat
[aa,bb] = contourf(X,Y,TIME2_physical,[0:20:400]); clabel(aa,bb), colorbar
title('Flow Front Positions [s] for S4')
xlabel('x [m]'), ylabel('y [m]'), grid
%%%%%%%%%%%%%%%%%%%%%%%%%%%%%%%%%%%%%%%%%%%%%%%%%%%%%%%%%%%%%%%%%%%%%%%%%%%%%
```

```
%%%%%%%%% Matlab code for Example 8.20_plot %%%%%%%%%%%%%%%%%%%%%%%%%%%%%%%%%%%%%%%
clear all, close all, clc
Lx = 0.700; Nx = 35; hx = Lx / (Nx);
Ly = 0.200; Ny = 10; hy = Ly / (Ny);
A1 = load('scn_1_P_Q.txt');
t1 = A1(:,1); P1  = A1(:,2); Q1  = A1(:,3);
A2 = load('sequentialgating_P_Q.txt');
t2 = A2(:,1); P2a = A2(:,2); Q2a = A2(:,3);  P2b = A2(:,4); Q2b = A2(:,5);
B1 = load('scn_1_allresults.txt');            T1 = B1(:,7);
B2 = load('sequentialgating_allresults.txt'); T2 = B2(:,7);

figure(1)
subplot(121), plot(t1,0.001*P1,'k-',t2,0.001*P2a,'ro',t2,0.001*P2b,'bs','linewidth',2)
xlabel('Time, t [s]'), ylabel('Injection Pressure, P_{inj} [kPa]'), grid
legend('S1','S5, Gate 1','S5, Gate 2',0)

subplot(122), plot(t1,1e6*Q1,'k-',t2,1e6*Q2a,'ro',t2,1e6*Q2b,'bs','linewidth',2)
xlabel('Time, t [s]'), ylabel('Injection Flow Rate, Q_{inj} [cc/s]'), grid
legend('S1','S5, Gate 1','S5, Gate 2',0)

time_to_fill = [t1(end),t2(end)]
P_inj_max    = [P1(end),max([P2a(end),P2b(end)])] * 0.001

for j = 1:Ny+1
    for i = 1:Nx+1
        k = (j-1)*(Nx+1) + i;
      TIME1(i,j) = T1(k);
      TIME2(i,j) = T2(k);
    end
end

TIME1_physical = TIME1'; TIME1_physical = TIME1_physical(end:-1:1,:);
TIME2_physical = TIME2'; TIME2_physical = TIME2_physical(end:-1:1,:);
[X,Y]          = meshgrid(0:hx:Lx, Ly:-hy:0);

figure(3)
subplot(211), contour(X,Y,TIME1_physical), hold on, shading flat
[aa,bb] = contourf(X,Y,TIME1_physical,[0:20:400]); clabel(aa,bb), colorbar
title('Flow Front Positions [s] for S1'), xlabel('x [m]')
ylabel('y [m]'), grid
subplot(212), contour(X,Y,TIME2_physical), hold on, shading flat
[aa,bb] = contourf(X,Y,TIME2_physical,[0:20:400]); clabel(aa,bb), colorbar
title('Flow Front Positions [s] for S5')
xlabel('x [m]'), ylabel('y [m]'), grid
%%%%%%%%%%%%%%%%%%%%%%%%%%%%%%%%%%%%%%%%%%%%%%%%%%%%%%%%%%%%%%%%%%%%%%%%%%%%%%%%%%%%%
```

```
%%%%%%%%% Matlab code for Example 8.21 %%%%%%%%%%%%%%%%%%%%%%%%%%%%%%%%
%%%%%%% Permeability Measurement
clear all, close all, clc
A    = load('scn_1_P_Q.txt');
t    = A(:,1);
P    = A(:,2);

mu   = 0.10;     % viscosity, [Pa.s]
Vf   = 0.40;     % fiber volume fraction
Q    = 2.00e-6;  % flow rate (constant), [m^3/s]
Cw   = 0.200;    % width of mold, [m]
Ch   = 0.005;    % thickness of mold, [m]

tt   = t(find(t > 80 & t < 180));
PP   = P(find(t > 80 & t < 180));

c    = polyfit(tt,PP,1);
dPdt = c(1)
tfit = linspace(80,180,21);
Pfit = c(2) + dPdt * tfit;

Kfit = (mu/(1-Vf)) * (Q/(Cw*Ch))^2 * (1/dPdt)

figure(1)
plot(t,0.001*P,'k-',tfit,0.001*Pfit,'ro','linewidth',2)
xlabel('Time, t [s]'), ylabel('Injection Pressure, P_{inj} [kPa]'), grid
legend('Virtual Experiment','Linear Curve Fit',0)
%%%%%%%%%%%%%%%%%%%%%%%%%%%%%%%%%%%%%%%%%%%%%%%%%%%%%%%%%%%%%%%%%%%%%%%%%%%%%
```

```
%%%%%%%%%% Matlab code for Example 8.22_plot %%%%%%%%%%%%%%%%%%%%%%%%%%%%%%
clear all, close all, clc
mu     = 0.10;      % viscosity, [Pa.s]
Vf     = 0.55;      % fiber volume fraction
Pin    = 200000;    % constant injection pressure, [Pa]
Cw     = 0.200;     % width of mold, [m]
Ch     = 0.005;     % thickness of mold, [m]
xf     = [40    100    200    300    400    500     600    660  ]' * 0.001;
t      = [ 0     14.5  102.9  271.6  520.7  850.2  1260.0  1544.5]';
D      = [sqrt(t)];
RHS    = xf - 0.04;
A      = (D'*D)\(D'*RHS);
Kfit   = (A(1))^2*mu*(1-Vf) / (2*Pin)

tfit(1)  = 0;
xffit(1) = 0.040;
i        = 1;

while xffit(end) < 0.660
i        = i + 1;
tfit(i)  = tfit(i-1) + 0.1;
xffit(i) = sqrt(2*Kfit*Pin/(mu*(1-Vf)) * tfit(i)) + 0.04;
end

figure(1)
plot(t,xf*1000,'ro',tfit,xffit*1000,'b-','linewidth',2)
xlabel('Time, t [s]'), ylabel('Flow Front Position, x [mm]'), grid
legend('Experimental data','Curve fit',0)

%%% results from LIMS:
Lx = 0.700; Nx = 35; hx = Lx / (Nx);
Ly = 0.200; Ny = 10; hy = Ly / (Ny);
A = load('Kmeasurement_P_Q.txt');
tt= A(:,1);
P = A(:,2);
Q = A(:,3);
B = load('Kmeasurement_allresults.txt');
X = B(:,2);
T = B(:,7);

figure(2)
subplot(121), plot(t,xf*1000,'ro',tfit,xffit*1000,'b--', ...
                   T(1:36),X(1:36)*1000,'k-','linewidth',2)
xlabel('Time, t [s]'), ylabel('Flow Front Position, x_f [mm]'), grid
legend('Experimental data','Curve fit by Darcy Law','LIMS simulation',0)
subplot(122), plot(tt,1e6*Q,'k-','linewidth',2), ylim([-0.1 1])
xlabel('Time, t [s]'), ylabel('Injection Flow Rate, Q_{in} [cc/s]'), grid
legend('LIMS simulation',0)
```

```
for j = 1:Ny+1
    for i = 1:Nx+1
        k = (j-1)*(Nx+1) + i;
        TIME(i,j) = T(k);
    end
end

TIME_physical = TIME'; TIME_physical = TIME_physical(end:-1:1,:);
[X,Y]           = meshgrid(0:hx:Lx, Ly:-hy:0);
figure(3)
contour(X,Y,TIME_physical), hold on, shading flat
[aa,bb] = contourf(X,Y,TIME_physical,[0:200:2200]); clabel(aa,bb), colorbar
title('Flow Front Positions [s] by LIMS'),
xlabel('x [m]'), ylabel('y [m]'), grid
%%%%%%%%%%%%%%%%%%%%%%%%%%%%%%%%%%%%%%%%%%%%%%%%%%%%%%%%%%%%%%%%%%%%%%%%%%%%
```

```
%%%%%%%%%% Matlab code for Example 8.23_mesh %%%%%%%%%%%%%%%%%%%%%%%%%%%%%%%%%%%
%%%%%%%%%% FEM MESH GENERATOR FOR A HOLLOW DISK
clear all, close all, clc
Rin  = 0.070; Rout = 0.500; Nr   = 16;
Nt   = 60;    Dt   = 360/Nt;
R    = [0.050 0.060 logspace(log10(Rin),log10(Rout),Nr-2)];
T    = (pi/180)*linspace(Dt/2,360-Dt/2,Nt);
H   = 0.004000; Vf  = 0.300000;
Kxx = 3.0e-010; Kxy = 0;        Kyy = 3.0e-010;   mu  = 0.24;

Number_nodes =  Nr   * Nt;
Number_elem  = (Nr-1)* Nt;
fid          = fopen('disk.dmp','w');
fprintf(fid,'Number of nodes : %5.0f \n',Number_nodes);
for i = 1:Nr
   for j = 1:Nt
      node_id   = (i-1)*Nt + j;
      r         = R(i);
      t         = T(j);
      x         = r*cos(t);
      y         = r*sin(t);
      z         = 0;
      NODE(node_id,:) = [node_id x y z];
      fprintf(fid,'%6.0f  %12.6f  %12.6f  %12.6f \n',NODE(node_id,:));
   end
end
fprintf(fid,'Number of elements : %5.0f \n',Number_elem);
for i = 1:Nr-1
   for j = 1:Nt-1
      elem_id   = (i-1)* Nt + j;
      k1        = (i-1)* Nt + j;
      k2        = k1+Nt;
      k3        = k2+1;
      k4        = k1+1;
      ELEMENTS(elem_id,:) = [elem_id 4 k1 k2 k3 k4 H Vf Kxx Kxy Kyy];
      fprintf(fid,'%6.0f %4.0f %6.0f ............ %15.4e \n',ELEMENTS(elem_id,:));
   end
      j = Nt;
      elem_id   = (i-1)* Nt + j;
      k1        = (i-1)* Nt + j;
      k2        = k1+Nt;
      k3        = k2-Nt+1;
      k4        = k3-Nt;
      ELEMENTS(elem_id,:) = [elem_id 4 k1 k2 k3 k4 H Vf Kxx Kxy Kyy];
      fprintf(fid,'%6.0f %4.0f %6.0f ............ %15.4e \n',ELEMENTS(elem_id,:));
end
fprintf(fid,'Resin Viscosity model NEWTON \n');
fprintf(fid,'Viscosity : %15.6f \n',mu);
fprintf(fid,'# Done ************************* \n');
fclose(fid);
%%%%%%%%%%%%%%%%%%%%%%%%%%%%%%%%%%%%%%%%%%%%%%%%%%%%%%%%%%%%%%%%%%%%%%%%%%%%%%%%%%
```

```
%%%%%%%%% Matlab code for Example 8.23_plot %%%%%%%%%%%%%%%%%%%%%%
clear all, close all, clc,
mu   = 0.24;
Vf   = 0.30;
Pin  = 100000;
h    = 0.004;
K    = 3.0e-10;
R0   = 0.070;
Rf   = 0.500;

dt   = 0.01;
t(1) = 0;
p(1) = 0;
R(1) = R0;
i    = 1;

%%%% IF Pin = constant:
R = linspace(R0,Rf,10000);
t = (0.5*(R/R0).^2.*(log(R/R0)-0.5)+0.25)*((1-Vf)*mu*R0^2)/(Pin*K);
t = t(find(R <= Rf));
R = R(find(R <= Rf));
%%%%

A = load('disk_allresults.txt');
N = 121:60:901;
T = A(N,7);
X = A(N,2); Y = A(N,3); RR = sqrt(X.^2+Y.^2);

for i  = 1:size(RR)-1
RRR(i) = (RR(i)+RR(i+1))*0.5;
end
RRR(i+1) = RR(end);

figure(1)
plot(t,R*1000,'k-', T,RRR*1000,'bo','linewidth',2)
xlabel('Time, t [s]'), ylabel('Flow Front Location, x_{f} [mm]')
grid, legend('Analytical solution','LIMS simulation',0)
%%%%%%%%%%%%%%%%%%%%%%%%%%%%%%%%%%%%%%%%%%%%%%%%%%%%%%%%%%%%%%%%%%%%
```

```
%%%%%%%%%% Matlab code for Example 8.24 %%%%%%%%%%%%%%%%%%%%%%%%%%%%%%%%%%%
clc, close all, clear all
B1 = load('box_1_P_Q.txt'); t1 = B1(:,1); P1 = B1(:,2); Q1 = B1(:,3)*1.0e6;
B2 = load('box_2_P_Q.txt'); t2 = B2(:,1); P2 = B2(:,2); Q2 = B2(:,3)*1.0e6;

figure(1)
plot(t1,P1*0.001,'k-','linewidth',2), grid
xlabel('Time, t [s]'), ylabel('Injection Pressure, P_{in} [kPa]')

figure(2)
plot(t2,P2*0.001,'k-','linewidth',2), grid
xlabel('Time, t [s]'), ylabel('Injection Pressure, P_{in} [kPa]')
%%%%%%%%%%%%%%%%%%%%%%%%%%%%%%%%%%%%%%%%%%%%%%%%%%%%%%%%%%%%%%%%%%%%%%%%%%
```

Appendix B

Solution to Example 8.13 Using FDM

Example 8.13: L-Shaped Mold

Solution

As in Example 8.12, FDM will be applied here to solve the pressure distribution at the time when the mold is filled with resin. The mold cavity is discretized as shown in Figure B.1. The discretized differential equation and the boundary conditions are also shown in the same figure. The solution procedure of the previous example is applied here as well, with the following major differences:

- The discretized differential equation, Equation (8.110) is applied to all the internal points with a different and yet straightforward coding scheme here. Arrays of order $N_x \times N_y$) are assigned for the major variables of this problem, P, u and v; however, only the indexes within the L-shaped domain are considered and the others (in the lower right-corner of the rectangle) are assigned as zeros. The L-shaped domain is divided into two sub-domains, and the nested "for loops" are written for each sub-domain to iterate P values until a convergence is achieved (see the previous example for the details).

- While handling the boundary condition at the corner point, $(x, y) = (0.6, 0.5)$, one should be careful about implementing the "no-leakage" condition, $\partial P/\partial n$ where n is the normal vector to the mold wall at a point. That is singular; but it is realistic to consider the neighboring points on the wall, and use n as the direction making 45° CCW with the positive y axis. Then, the corresponding second order forward finite difference equation for this boundary condition is

$$P_{13,11} = \frac{1}{3}(4P_{12,12} - P_{11,13}) \tag{B.1}$$

  which involves the two neighbor nodes along n direction. If this boundary condition is not applied appropriately, as explained above, it will lead to incorrect mass balance and hence error in the pressure field calculation.

The MATLAB code, *example8_13.m*, is listed in the Appendix. Pressure and velocity distributions are plotted in Figures B.2 and B.3. The inlet and exit flow rates are calculated as $Q_{in} = 8.42\,\text{cc/s}$ and $Q_{vent} = 8.47\,\text{cc/s}$, thus the mass balance is satisfied within the accuracy of the numerical method at this level of mesh resolution. As the mesh gets finer, the numerical error gets smaller. The resin pressure at the point of interest is $P(0.3, 0.5) = 341{,}390\,\text{Pa}$. ■

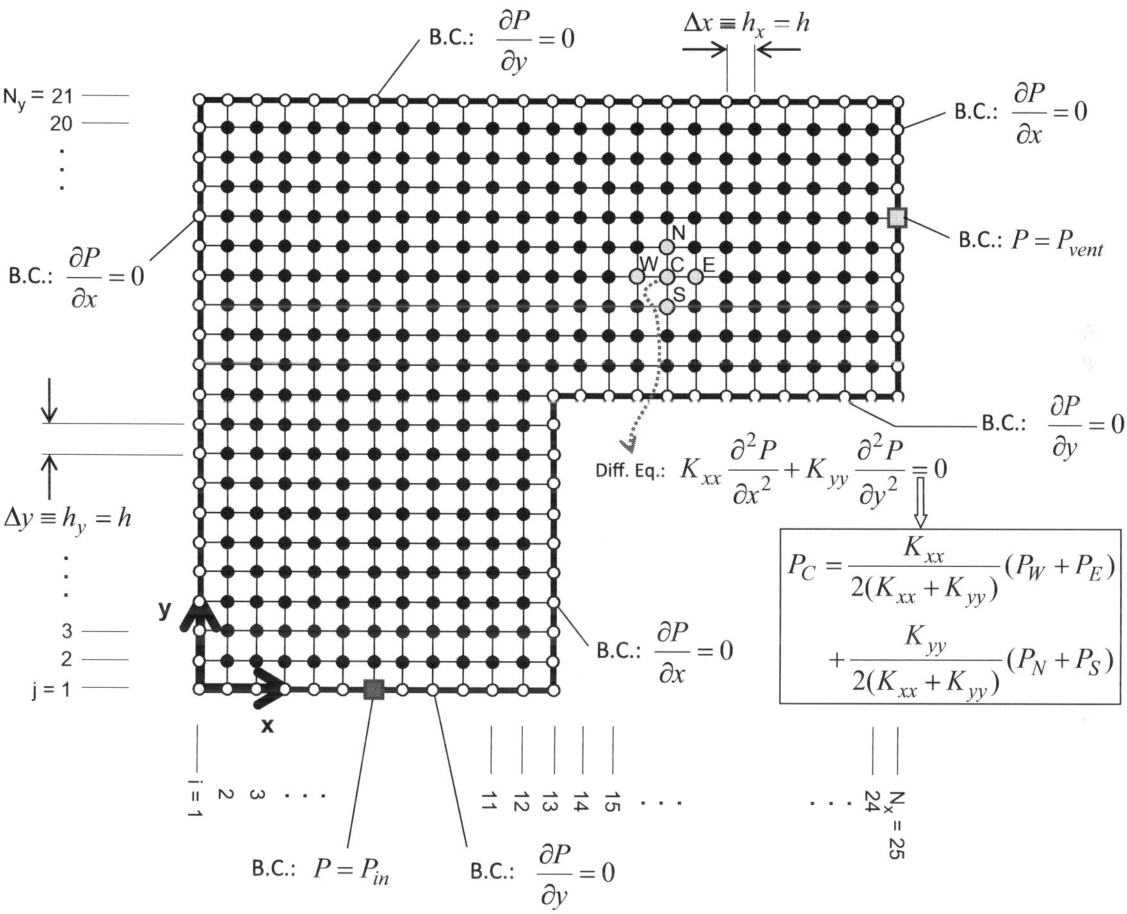

Figure B.1: Discretized domain, governing differential equation for P along with boundary conditions.

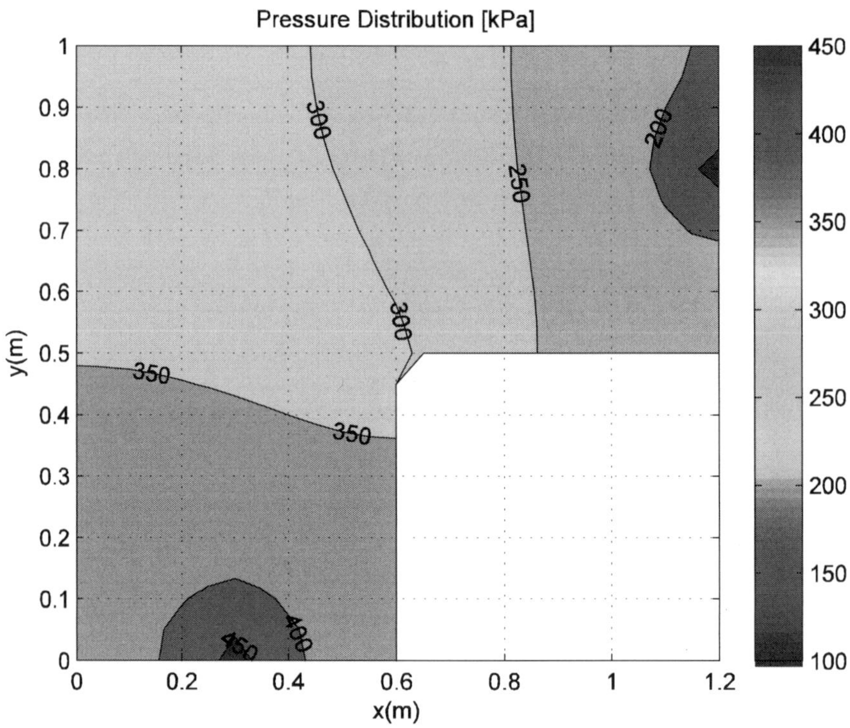

Figure B.2: Pressure distribution at the end of mold filling.

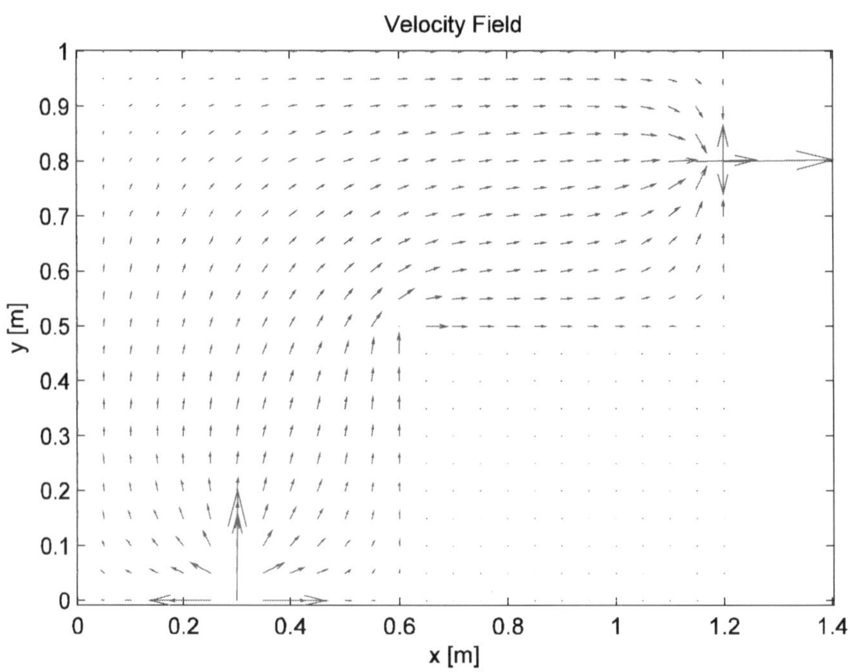

Figure B.3: Average velocity distribution at the end of mold filling.

Appendix C

Additional Examples with LIMS to Model Liquid Mold Filling

Example 8.18: Sequential Simulations of Scenarios S1-S3 of Example 8.15 with a Script File

To sequentially run the simulations of the three different scenarios and store the results in output files, a single LIMS script file will be written in LBASIC programming language. LBASIC is the combination of standard Basic and the functions and commands of LIMS. This script file will be called on the command window. The LIMS script file, "**scenarios.lb**" is listed below:

```
proc auto
defint i, is, igate
defdbl x, y, z, a, b
defstring base
dim Kratio(3), Vf(3)

let base = "rectanglemold"

let Kratio(1) = 1
let Kratio(2) = 10
let Kratio(3) = 100

let Vf(1) = 0.40
let Vf(2) = 0.20
let Vf(3) = 0.10

let igate = 181

for is=1 to 3
    read base+".dmp"
    settime 0.0
    let a = Kratio(is)
    let b = Vf(is)

    fiopen 1,"scn_"+str(is)+"_P_Q.txt"
    fiformat "%22.10f"

    for i=1 to 35
       SCALEPERM(i,a)
       SETTHICK i,0.005,b
    next i

    for i=316 to 350
       SCALEPERM(i,a)
       SETTHICK i,0.005,b
    next i

    for i=1 to 10
       SCALEPERM((i-1)*35+1,1000.)
       SCALEPERM((i-1)*35+2,1000.)
       SETTHICK  (i-1)*35+1,0.005,0
       SETTHICK  (i-1)*35+2,0.005,0
    next i

    for i=1 to 10
       SCALEPERM((i-1)*35+34,1000.)
       SCALEPERM((i-1)*35+35,1000.)
       SETTHICK  (i-1)*35+34,0.005,0
       SETTHICK  (i-1)*35+35,0.005,0
    next i

    do
       solve
       print "At time = ", socurrenttime, " ...
          # of empty nodes = ", sonumberempty
       fiprint 1,socurrenttime,       ...
                 sopressure(igate), ...
                 soflowrate(igate)
       fiprintnl 1
    loop while sonumberempty>0

    ficlose 1

    setouttype "tplt"
    write "scn_"+str(is)+"_tecplot"

    fiopen 1,"scn_"+str(is)+"_allresults.txt"
    fiformat "%22.10f"
    for i=1 to sonumbernodes
       let x = sonodecoord(i,1)
       let y = sonodecoord(i,2)
       let z = sonodecoord(i,3)
       fiprint 1,i,x,y,z,sopressure(i),    ...
                         soflowrate(i),    ...
                         sotimetofill(i), ...
                         sofillfactor(i)
       fiprintnl 1
    next i
    ficlose 1
next is
endproc
```

The major elements (commands, functions, etc.) of LBASIC are briefly described below:

- The content of the script file, which is given a procedure name "auto" here, will be called from the command window.

- All of the variables in the form of (1) integers, (2) real numbers, (3) strings (character groups) and (4) arrays must be be declared using (1) `defint`, (2) `defdbl`, (3) `defstring` and (4) `dim`, respectively.

- All of the assignments must start with `let`, for example, `let igate = 181`.

- `SCALEPERM(36,1000)` scales the previously assigned permeability tensor of element 36 with a factor of 1000.

- `SETTHICK 36,0.005,0` re-assigns the thickness and fiber volume fraction of element 36 to 0.005 m and 0, respectively.

- `solve` command solves (1) all the nodal pressures using FEM, and then (2) advances the flow front one time step using "control volume" solution method. However, recall that the time step is not fixed; it is determined by advancing the flow front until at least one more node's control volume is filled completely.

- The "for" loop, `for i=1 to 10 ... next i`, increments the integer variable `i` from 1 to 10.

- The "do" loop (a.k.a. "while" loop), `do ... loop while sonumberempty > 0` continues as long as the number of empty nodes, `sonumberempty` is greater than zero (i.e., until the mold is filled completely).

- `fiopen` and `ficlose` opens and closes a file, respectively, when storing the simulation results in an output file.

- `fiformat "%22.10f"` prints each variable with 22 characters of which 10 are decimal points.

- `socurrenttime` is the current time of the virtual mold filling.

- `sopressure(i)` is the pressure of the node `i`.

- `soflowrate(i)` is the net flow rate of the node `i`'s control volume.

- `sotimetofill(i)` is the time of flow front arrival to node `i` (i.e., at what time the `sofillfactor(i)` reaches 1 (100% saturation) in that node's control volume).

- `sonumberempty` is the number of emtpy or partially filled nodes (i.e., with a fill factor less than 1).

Executing the simulations: On the command window (see Figure C.1), type `load scenarios` to call the script file's name, `scenarios` and then start executing it by typing and entering the name of the procedure `auto` in that script file.

Storing the results: For scenario S1, the injection pressure and flow rate are stored in an output file, `scn_1_P_Q.txt` with time. When the mold cavity is filled completely, the final results (coordinates, pressure, flow rate, time of resin arrival and fill factor) of all nodes are stored in `scn_1_allresults.txt`. Similar output files are also generated for the other

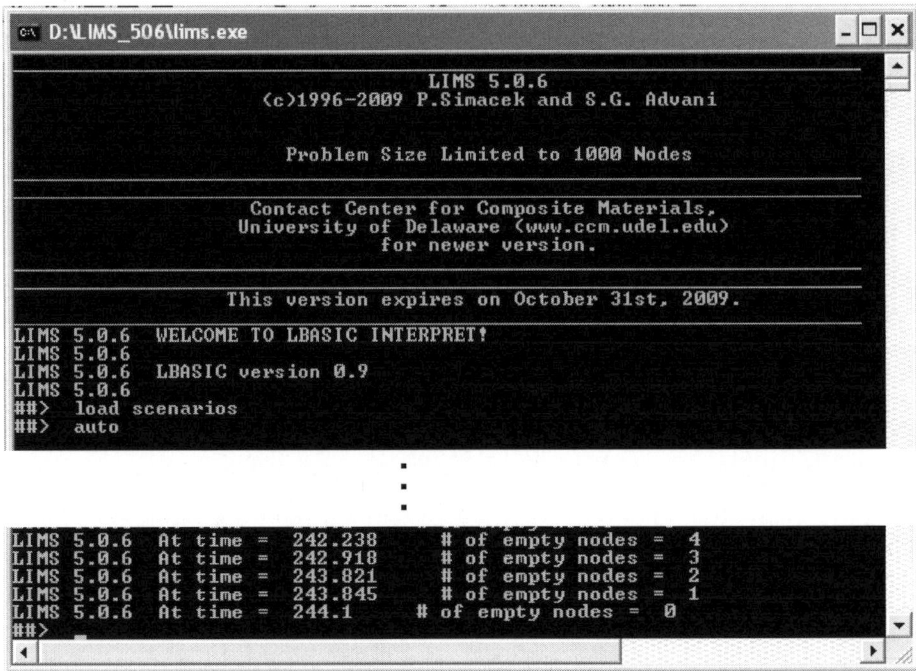

Figure C.1: Command window of LIMS program.

scenarios S2 and S3 by using the same script file. This is an advantage of LIMS script file programming; multiple simulations can be sequentially performed by varying a set of material and/or process parameters in a "for" loop, as done here. This way, an optimal design may be achieved for the mold and process window. Although we held the number of simulations to three here, in general many more simulations can be done by using a single script file.

Viewing the results: One can use any program that has a graphing feature such as Microsoft Excel or MATLAB. The MATLAB code, *example8_18_plot.m*, (listed in the Appendix) was used to plot (i) propagation of the flow front with time and (ii) injection pressure versus time for the three scenarios considered here (see Figure 8.79). ■

Example 8.19: Re-solve Example 8.15 with a Constraint of $P_{in} \leq 200\,\text{kPa}$

We will re-consider the mold cavity (shown in Figure 8.73) with the mesh (shown in Figures 8.75 and 8.76) and material and process parameters of Scenario S1 used in Example 8.15. However, this time we will illustrate how to implement a simple control action such as converting an injection gate's type from (1) constant flow-rate to (2) constant pressure once the injection pressure reaches a critical value, 200 kPa. The corresponding resin flow will be called **Scenario S4**. There are several reasons why this control action may be needed to avoid any or all of the followings: (1) fiber wash, (2) formation of microvoids, and (3) resin leakage from the mold seal or injection tubing.

Some RTM injection machines, such as Radius Engineering Inc.'s flow-rate controlled injectors shown in Figure 8.29(a) have the feature of automatically switching the injection condition from a constant flow rate to a constant pressure once the inlet resin pressure reaches the user's defined critical value. Thus, it is very beneficial to be able to simulate such a scenario using LIMS program if an engineer wants to design the process window under actual working conditions.

Solution

To simulate Scenario S4, the LIMS script file, "**pmaxlimit.lb**" was written. It is listed in the Appendix, and only major lines of the code are shown below:

```
do
solve
...
loop while sopressure(181) < 200000

setgate 181,1,200000

do
solve
...
loop while sonumberempty > 0
```

The boundary condition at the gate G1 is $Q_{in} = 2\,\text{cc/s}$ within the first "while" loop. Injection pressure, $P_{in}(t)$ increases with time and reaches a critical value, 200 kPa at $t = 121.5\,\text{s}$ (see Figure C.2 which was plotted by using the MATLAB code, *example8_19_plot.m*). However, since this critical time is not known beforehand, a "while" loop with a conditional expression, `sopressure(181) < 200000` was used instead of a "for" loop. At that time, the injection gate's boundary condition is converted to a constant pressure type by using the command `setgate 181,1,200000`, i.e., the gate at node 181 injects resin under constant 200 kPa. The following second loop ends when there is no empty node remaining (i.e., when the mold is filled completely)

It is not surprising to see that t_{fill} became longer (254.7 s) in S4 than 225.5 s in S1 because the injection pressure was kept constant at a lower level than the original Scenario 1. But the advantage of this application is to keep the maximum injection pressure at 200 kPa instead of 310.1 kPa in S1. Also notice that the flow front patterns are not affected (still 1D flow), but they are delayed. ∎

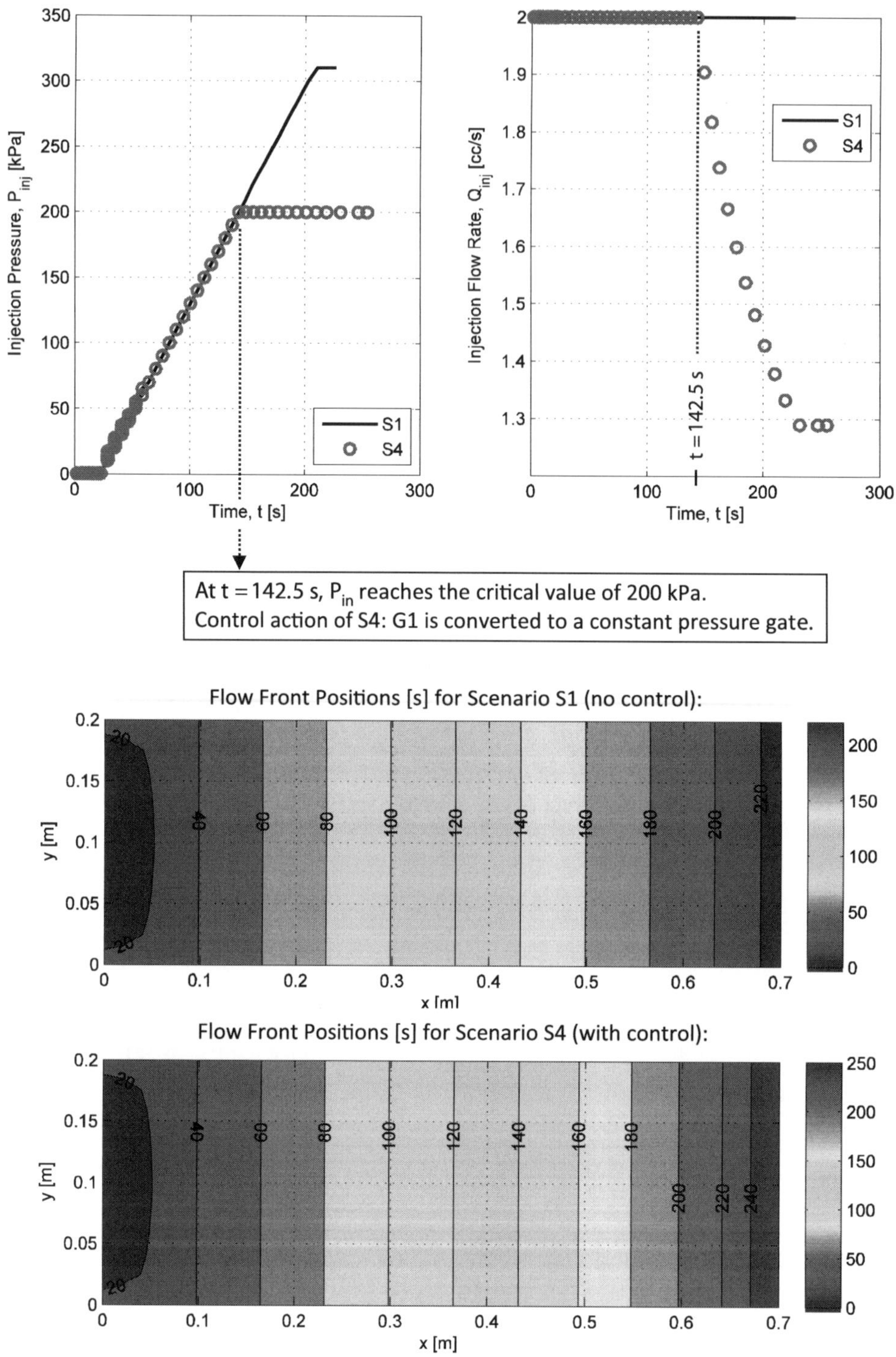

Figure C.2: Simulated injection pressure, flow rate and flow front propagation for Scenario S4 (in which there is a constraint of $P_{in} \leq 200\,\text{kPa}$) plotted together with Scenario S1 for comparison.

Example 8.20: Re-solve Example 8.15 with a Sequential Gating Control Scheme

We will re-consider the same mold cavity and the mesh used in Examples 15 and 19. This time, another control scheme will be applied for the purpose of reducing the maximum injection pressure needed. Initially, every parameter of Scenario S1 will be re-used until the resin arrives at the sensor which is located at $(x, y) = (0.36, 0.10)$ m. At that instant, gate G1 will be turned off and the second gate G2, located at $(x, y) = (0.34, 0.10)$ m, will be activated with $Q_{in} = 2$ cc/s, which is the same flow rate of the original gate. The corresponding resin flow will be called **Scenario S5**.

Solution

To simulate Scenario S5, the LIMS script file, "**sequentialgating.lb**" was written and is listed in the Appendix. Its major lines are shown below:

```
let igate1  = 181
let igate2  = 198
let isensor = 199
read "rectanglemold.dmp"
setgate igate1,2,0.000002
settime 0.0

   do
   solve
   ...
   loop while sofillfactor(isensor) < 0.5

setgate igate1,0,0
setgate igate2,2,0.000002

   do
   solve
   ...
   loop while sonumberempty > 0
```

The condition of the first loop is `sofillfactor(isensor) < 0.5`, i.e., the loop continues until the resin reaches the sensor node's (node 199) control volume and fills 50% of it. One can question why 50%. Practically, it does not make a big difference on timing of the control schedule as long as the critical fill factor is between 0^+ and 1^-. Equating it to 1 has a numerical issue since the fill factor may be 0.999 ... 999, and still the condition is not satisfied. Besides, some sensor types may be very sensitive and will be triggerred even when a very small amount of resin covers the sensor tip, while some sensor types may respond much later. Therefore, we consider a case where half-filled control volume is sufficient to trigger the sensor. The results of this Scenario S5 are shown in Figure C.3 along with the results of original S1. As planned, sequential gating (i.e., activating a gate after the flow front arrives there and closing the previous one in a multiple gate usage) helps to reduce the injection pressure (193.2 kPa in S5 instead of 310.1 kPa in S1) in constant flow-rate injection scenario, as in this example. If it were a constant pressure injection scenario, then it would allow to reduce the time to fill the mold, t_{fill}.

The pressure for G1 corresponds to the injection pressure at the gate G1 until the control action time, $t = 121.5$ s. After this time, the pressure in Figure C.3 is the resin's pressure at the gate's location within the mold. MATLAB code, *example8_20_plot.m*, was used for

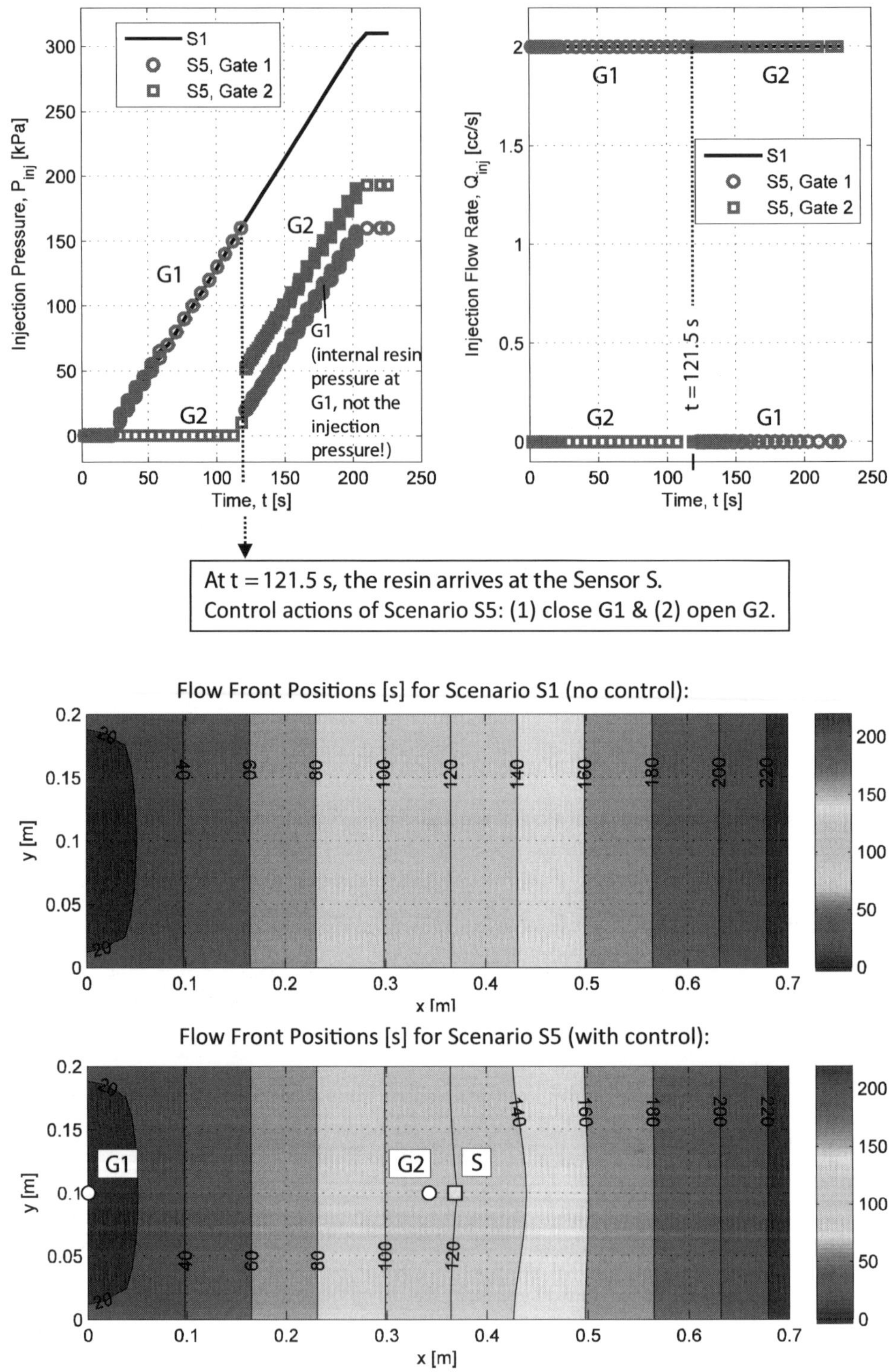

Figure C.3: Simulated injection pressure, flow rate and flow front propagation for Scenario S5 (in which a sequential gating is applied) plotted together with Scenario S1 for comparison.

plotting the LIMS' results. One can see that the flow rate becomes zero for $t > 121.5$ s since we closed it with the command, `setgate igate1,0,0`. The pattern of flow front propagation is almost not affected as seen in the same figure. ■

Example 8.21: 1D Permeability Measurement Experiment

Assume that the result (P_{in} versus t) of Scenario 1 in Example 8.15 was obtained during an actual experiment. By using this virtual experimental data, in the analytical formula of "1D permeability measurement experiment," calculate the permeability, K_{xx} of the fabric preform in the direction of flow. Then, verify that it is in agreement with $K_{xx} = 4.0 \times 10^{-10}\,\text{m}^2$ which had been used in the original LIMS simulation.

Solution

By using Darcy's Law for 1D flow, Equation (5.116) had been obtained to express the permeability in terms of material and process parameters and dP_{in}/dt as follows:

$$
\begin{aligned}
K_{xx,fit} &= \frac{\mu}{(1-V_f)}\left(\frac{Q}{C_w C_h}\right)^2 \frac{1}{\frac{dP_{in}}{dt}} \\
&= \frac{0.10\,\text{Pa.s}}{(1-(0.40))}\left(\frac{2\times 10^{-6}\,\text{m}^3/\text{s}}{(0.200\,\text{m})\,(0.005\,\text{m})}\right)^2 \frac{1}{1666.7\,\text{Pa/s}} \\
&= = 4.0000 \times 10^{-10}\,\text{m}^2
\end{aligned}
$$

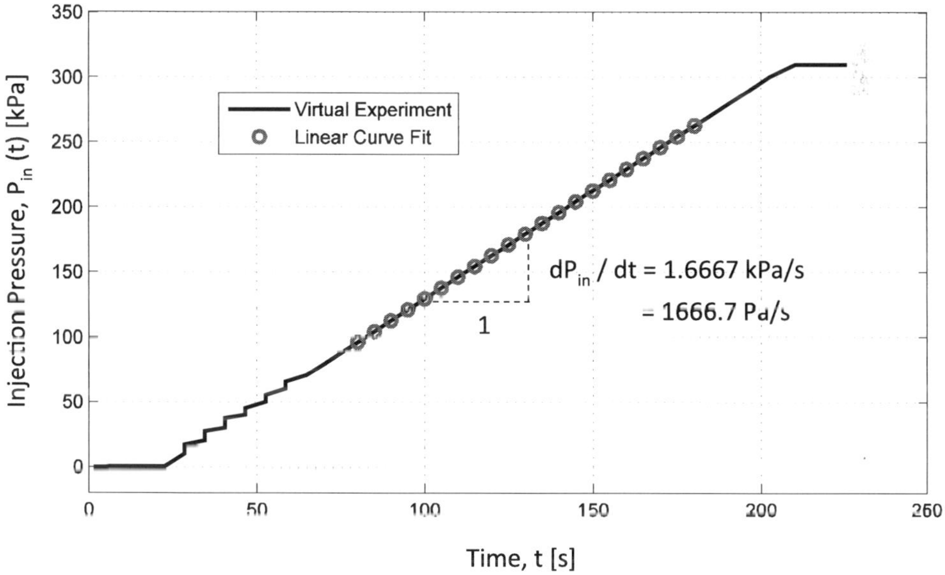

Figure C.4: Simulated injection pressure of S1 used as a virtual experimental data to calculate the permeability of the fabric preform.

where μ is the resin viscosity, V_f is the fiber volume fraction, Q is the flow rate, C_w and C_h are the width and thickness of the mold cavity. Calculation of dP_{in}/dt is shown in Figure C.4 which was done by using "polyfit" curve-fitting function of MATLAB (see the MATLAB code, *example8_21.m*, which is listed in the Appendix). The result, $K_{xx,fit} = 4.0000 \times 10^{-10}\,\text{m}^2$ agrees very accurately with the input data used in the original LIMS simulation. ■

Example 8.22: Lineal Injection under Constant Injection Pressure

A design engineer is given a task of manufacturing many composite parts using the same mold of Example 8.15, but with different fiber volume fractions (by compacting more layers in the mold cavity) to enhance the mechanical properties. In one of those RTM injections, the following parameters were used:

- Constant injection pressure, $P_{in} = 200\,\text{kPa}$ from a single gate, Gate 1
- Resin viscosity, $\mu = 0.10\,\text{Pa.s}$
- Fiber volume fraction, $V_f = 0.55$
- Fabric preform's permeability tensor, $\mathbf{K}$ is not known.

The experimental data for the resin flow position is given below:

x_f [mm]	40	100	200	300	400	500	600	660
t [s]	~ 0	14.5	102.9	271.6	520.7	850.2	1260.0	1544.5

where the fabric preform is placed in $40 \leq x \leq 660\,\text{mm}$ as shown in Figure 8.73.

Obtain a formula to calculate the permeability, K_{xx} of the fabric preform in the direction of flow by using the material and process parameters. Then, use this value in a LIMS simulation and verify your K_{xx}.

This example illustrates an unconventional experimental approach to measure the permeability of a fabric preform using constant pressure injection instead of constant flow rate injection. In some cases, this approach may be preferable since RTM injectors with constant pressure boundary condition are usually at a lower cost than injectors with flow-rate control and will be useful for VARTM process as well.

Solution

1D Darcy's Law is written as

$$
\begin{aligned}
u = (1 - V_f)u_f &= -\frac{K_{xx}}{\mu}\frac{dP}{dz} \\
(1 - V_f)\frac{dz_f}{dt} &= -\frac{K_{xx}}{\mu}\frac{0 - P_{in}}{z_f - 0}
\end{aligned}
\tag{C.1}
$$

where $z = x - 0.040\,\text{m}$ (i.e., the shifted longitudional axis starting from the edge of the preform instead of the empty pool) and z_f is the flow front velocity in z. Equation (C.1) is re-expressed as

$$z_f \, dz_f = \frac{K_{xx} P_{in}}{\mu(1 - V_f)} \, dt. \tag{C.2}$$

If the integration of both sides is taken, the following is obtained:

$$\frac{1}{2} z_f^2 = \frac{K_{xx} P_{in}}{\mu(1 - V_f)} t + C \tag{C.3}$$

where C is determined as zero by using the initial condition, $z_f(0) = 0$. Thus, the flow front is expected to propagate with time as a square-root function:

$$z_f(t) = \sqrt{\frac{2K_{xx} P_{in}}{\mu(1 - V_f)} t} \qquad \Longrightarrow \qquad x_f(t) = \sqrt{\frac{2K_{xx} P_{in}}{\mu(1 - V_f)} t} + 0.040\,\text{m} \tag{C.4}$$

and the time to fill the preform is given as

$$t_{fill} = \frac{L^2 \mu(1 - V_f)}{2K_{xx} P_{in}} \tag{C.5}$$

where L is the length of the preform (which is 0.62 m in this example).

It is easier to use Equation (C.5) to calculate K_{xx} as there is one unknown and one equation:

$$1544.5\,\text{s} = \frac{(0.62\,\text{m})^2\,(0.10\,\text{Pa.s})(1 - 0.55)}{2K_{xx}(200,000\,\text{Pa})} \qquad \Longrightarrow \qquad K_{xx} = 2.800 \times 10^{-11}\,\text{m}^2. \tag{C.6}$$

Notice that, this way of solving K_{xx} involves the data which represents only the end of the mold filling and not the earlier stage. Thus, if significant experimental error is expected (such as deviation from 1D flow or DAQ inaccuracy), then the calculated permability may be very inaccurate. A better way of solution is done by curve-fitting to $x_f(t)$ using the expected analytical formula, Equation (C.4). A MATLAB code was written by using the least square method on the "overdetermined system of equations" [277], as previously detailed in Example 5.9. That method also resulted in $K_{xx} = 2.800 \times 10^{-11}\,\text{m}^2$ which tells us that there is no significant experimental error in $x_f(t)$, i.e., the flow was 1D so that Equation (C.4) was applicable.

To simulate the mold filling, the LIMS script file, "**Kmeasurement.lb**" was written and is listed in the Appendix. Its major lines are shown below:

```
let igate = 181
setgate 181,1,200000
let f = 2.800e-11/4.000e-10

for i=1 to 350
    SCALEPERM(i,f)
    SETTHICK  i,0.005,0.55
next i
```

Notice that the original input file, "rectanglemold.dmp" is used, but the permeability and fiber volume fraction of the preform are adjusted in the script file (see the "for" loop above). The gate at node 181 is also changed to constant pressure injection type (type 2 of LIMS).

To run the script file's content, type `load Kmeasurement` and then `auto` on the command window of LIMS. It will generate output files to be viewed by the MATLAB code, *example8_21_plot.m* (listed in the Appendix). The simulated $x_f(t)$, $Q_{in}(t)$ and the locations of flow front are shown in Figure C.5 together with the experimental data and the curve fit data by using 1D Darcy's Law. The simulated result agrees well with the experimental and curve-fit data. The slight deviation is due to use of a coarse mesh. That means, in the post-processing of the results, the "resin arrival time to a node" was assumed to be equivalent to the instant when the control volume centered at that node has a "full" saturation (fill factor = 1) (see Figure C.6).

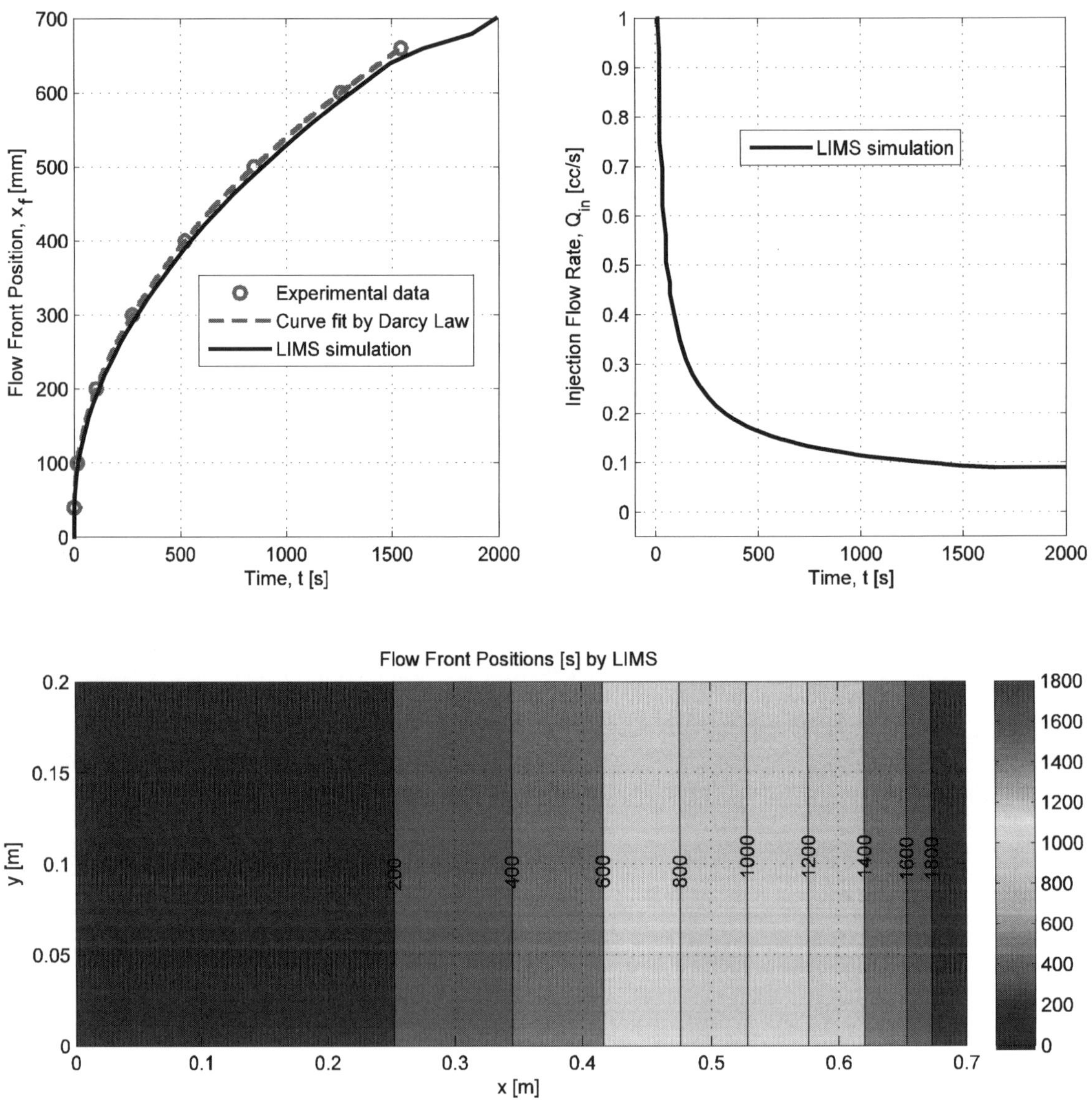

Figure C.5: Simulated flow front propagation by using calculated K_{xx} in LIMS program together with the experimental data to verify the value of K_{xx}.

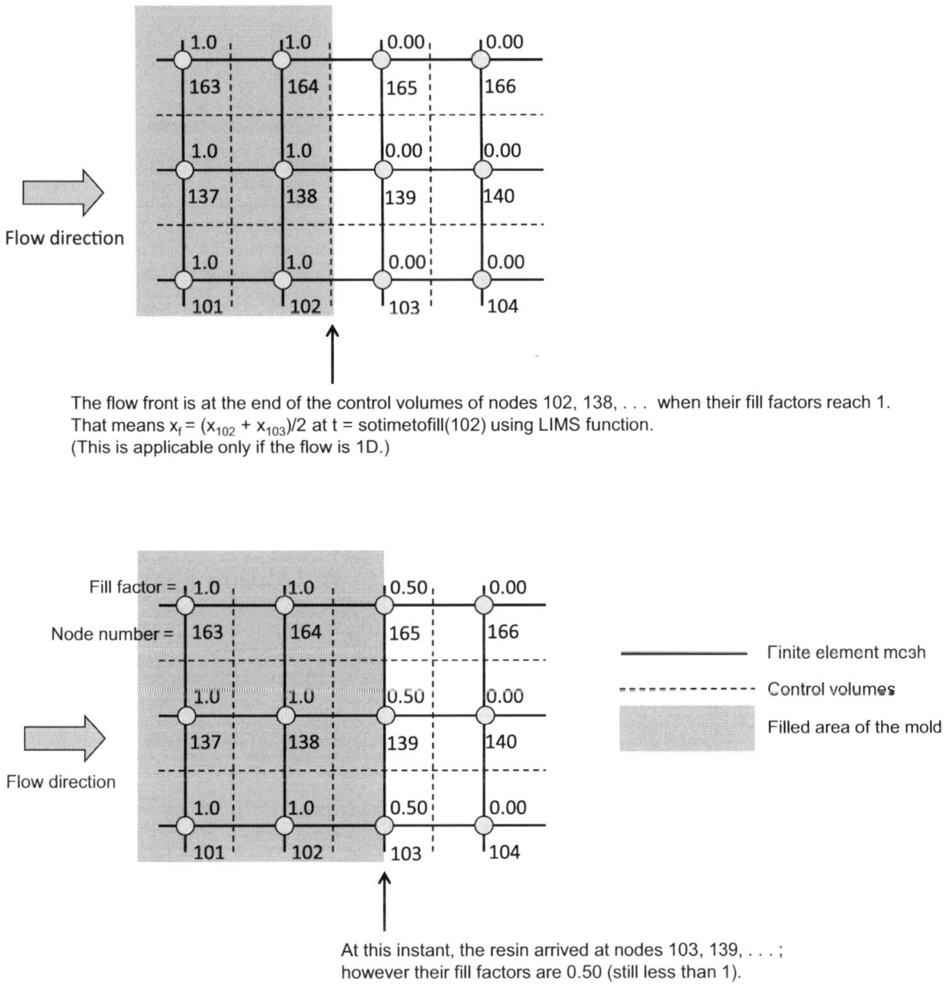

Figure C.6: The figure at the top explains how the flow front was tracked in Figure C.5.

Thus, with this definition, the flow front not only arrived at the central node, but also filled some extra volume nearby. When the mesh is coarse, this definition of resin arrival may cause significant deviation from the analytic solution. However, in the next example this issue will be resolved with a different approach.

Also notice that the fabric is filled completely at $t = 1656.2$ s, but the remaining empty ventilation pool is filled completely at $t = 1988.5$ s. This clearly indicates that, toward the end of mold filling, the flow front slowed down very significantly due to smaller pressure gradient available. This slow propagation is also seen in the bottom plot of Figure C.5. ■

Example 8.23: Radial Flow in a Hollow Disk

(a) Simulate the radial resin flow in the hollow disk (shown in Figure C.7) by using LIMS.
(b) Compare the simulated propagation of the flow front, $R(t)$ with the analytical solution of Darcy's Law.

Use the following geometry, material and process parameters:

- Resin viscosity, $\mu = 0.24\,\text{Pa.s}$
- Part thickness, $h = 4\,\text{mm}$
- Injection presure, $P_{in} = 100{,}000\,\text{Pa}$
- Different regions in the mold cavity:
  - Inner empty pool: $0.05 \leq r < 0.07\,\text{m}$; $V_f = 0$
  - Fabric preform: $0.07 \leq r \leq 0.50\,\text{m}$; $V_f = 0.30$ and $K_{xx} = K_{yy} = 3.0 \times 10^{-10}\,\text{m}^2$ and $K_{xy} = K_{yx} = 0$. Thus $K_{rr} = 3.0 \times 10^{-10}\,\text{m}^2$ due to isotropy.

It is suggested that you use $K_{rr} = 3.0 \times 10^{-7}\,\text{m}^2$ (high pemeability) in the empty pool.

Solution

(a) Simulation using LIMS: The MATLAB code, *example8_23_mesh.m* (listed in the Appendix), generates the nodes and elements of the mesh. By editing the rest manually, the LIMS input file, "**diskmold.dmp**" is listed below:

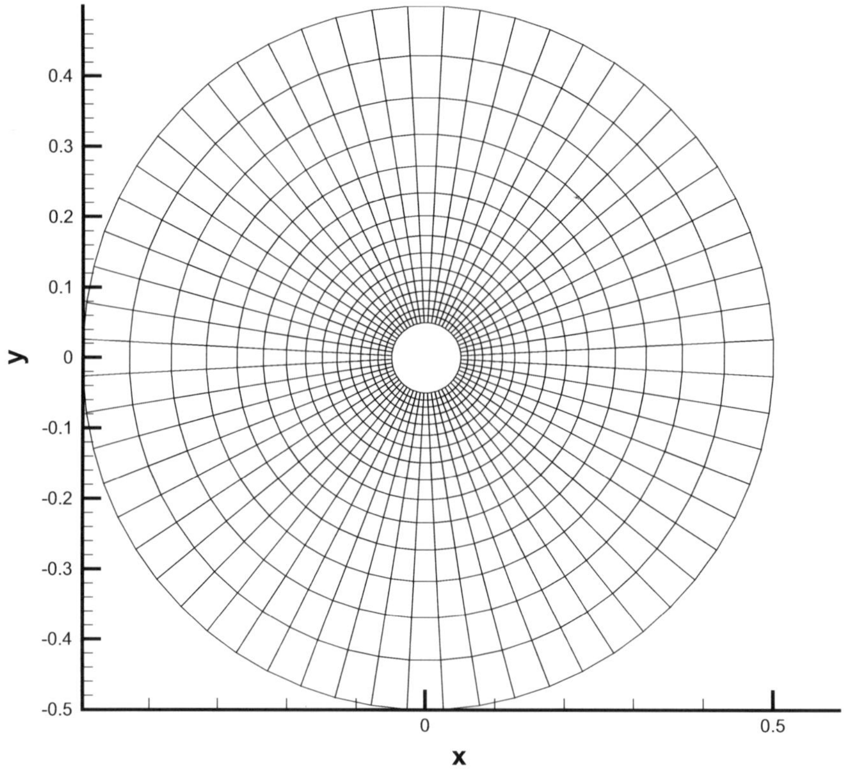

Figure C.7: Mesh for the mold cavity. The thickness is 4 mm everywhere. The inlet hole has a diameter of 10 mm; the fabric preform has an inner diameter of 0.14 m. That means, between $0.05 \leq r < 0.07\,\text{m}$, there is resin pool to achieve a radial injection.

```
#Lims dump File (.dmp) generated by LimsUI

Number of nodes : 960
 Index        x              y              z
====================================================
     1      0.049931       0.002617       0.000000
     2      0.049384       0.007822       0.000000
     3      0.048296       0.012941       0.000000
......................................
   958      0.482963      -0.129410       0.000000
   959      0.493844      -0.078217       0.000000
   960      0.499315      -0.026168       0.000000
Number of elements : 900
  Index  NNOD  N1    N2    N3    (N4)  ...  h       Vf         Kxx       Kxy    Kyy    ...
===========================================================================================
     1    4     1    61    62     2        0.004   0.300000   3e-010    0     3e-010
     2    4     2    62    63     3        0.004   0.300000   3e-010    0     3e-010
     3    4     3    63    64     4        0.004   0.300000   3e-010    0     3e-010
...........................................................................................
   898    4   898   958   959   899        0.004   0.300000   3e-010    0     3e-010
   899    4   899   959   960   900        0.004   0.300000   3e-010    0     3e-010
   900    4   900   960   901   841        0.004   0.300000   3e-010    0     3e-010
Resin Viscosity model NEWTON
Viscosity :             0.24
```

In contrast with the previous example, the locations and boundary conditions of the injection gates will be assigned in the script file, "**diskmold.lb**" (listed in the Apendix). The major lines of the script file are given below:

```
for i = 1 to 60
SETGATE i,1,100000
next i

for i = 1 to 120
SCALEPERM(i,1000)
SETTHICK i,0.004,0
next i

let igate1 = 1
```

The command `SETGATE i,1,100000` assigns the constant pressure boundary condition at the inner edge (at Nodes 1 through 60). The commands `SCALEPERM(i,1000)` and `SETTHICK i,0.004,0` assign the permeability and fiber volume fraction of the empty pool.

On the command window, type `load diskmold` to call the script file's name, and then start executing it by typing `auto`.

As briefly discussed in the previous example, LIMS calculates at what time each control volume is filled 100%. If `socurrenttime(i)` is treated as the resin arrival time to the node, it may mislead the user especially when the mesh is coarse, as in this example. However, it will be handled as follows:

- The flow front will be treated at the edge of a control volume, not at the central nodal point. This was implemented in the MATLAB code, *example8_23_plot.m* (listed in the Appendix), as follows:

$$R(t) = (r_i + r_{i+1})/2 \quad \text{at } t = t_i$$

where t_i is the time at which the fill factor $= 1$ at the node `i`, r_i and r_{i+1} are the radial positions of the node and the neighboring node, respectively, and $(r_i + r_{i+1})/2$ is where the control volume ends.

- However, the above approach is applicable only if the flow front is expected to propagate in a 1D radial flow, otherwise we may not know which edge (or point) on the control volume will be filled at the latest. In fact, one does not need to worry about this issue as long as it is remembered that the resin fills the "control volume" completely at $t =$`socurrenttime(i)`, and that value of the time is not exactly equal to the resin arrival time to the "nodal point" itself.

(b) Analytical Solution: For a radial flow under constant injection pressure, P_{in} the flow front position, $R(t)$ is given as [123]

$$\frac{1}{2}\left(\frac{R(t)}{R_o}\right)^2\left[\ln\left(\frac{R(t)}{R_o}\right) - \frac{1}{2}\right] = \frac{P_{in}K_{rr}\,t}{(1-V_f)\mu R_o^2} - \frac{1}{4} \tag{C.7}$$

which is not in a closed form. However, we will handle plotting of $R(t)$ in a simple approach: instead of treating t as the independent variable and R as the dependent one, we will treat them vice versa, since $R(t)$ is a single-valued function. Thus,

$$t = \frac{(1-V_f)\mu R_o^2}{P_{in}K_{rr}}\left\{\frac{1}{2}\left(\frac{R(t)}{R_o}\right)^2\left[\ln\left(\frac{R(t)}{R_o}\right) - \frac{1}{2}\right] + \frac{1}{4}\right\}. \tag{C.8}$$

Here $R_o = 0.070\,\text{m}$ is the inner radius of the preform.

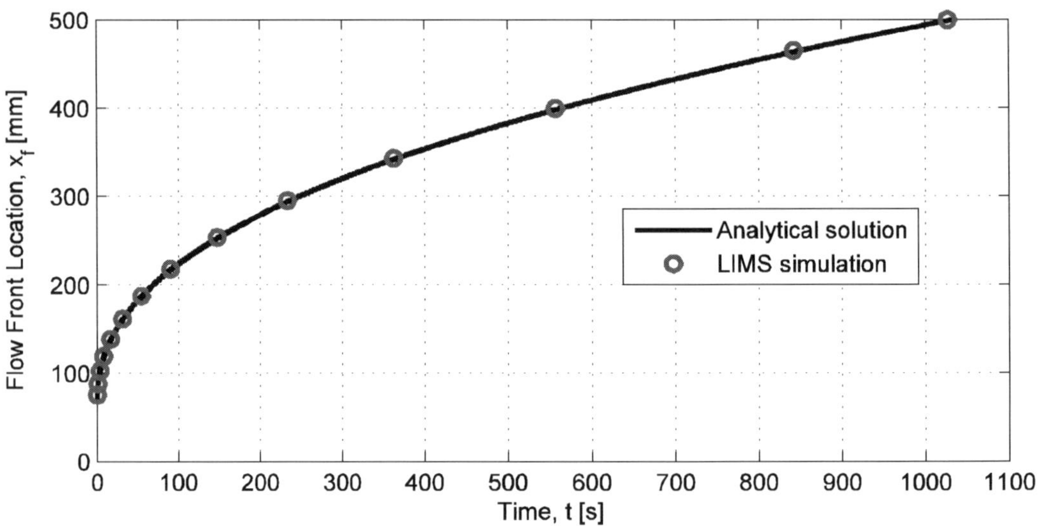

Figure C.8: Simulated flow front locations by using LIMS along with the analytical solution.

Both the simulated and the analytical solutions (Equation (C.8)) of the flow front position, $R(t)$, are plotted in Figure C.8 by using a post-processing MATLAB code, *example8_23_plot.m* (listed in the Appendix). There is a good agreement between the two results. ■

Example 8.24: Box Mold in RTM Process

An open box with dimensions of $L_x \times L_y \times L_z = 140\,\text{mm} \times 203\,\text{mm} \times 102\,\text{mm}$ will be manufactured by RTM. The wall thickness of all the five surfaces is 3.175 mm. For mold filing simulations in LIMS, a FEM mesh was prepared using PATRAN program (see Figure C.9).

The material, geometry and process parameters are listed below:

- Resin viscosity, $\mu = 0.10\,\text{Pa.s}$
- Part thickness, $h - 3.175\,\text{mm}$
- Fiber volume fraction, $V_f = 0.35$
- Fabric preform's permeability tensor, $K_{xx} = K_{yy} = 1.80 \times 10^{-10}\,\text{m}^2$ and $K_{xy} = K_{yx} = 0$.

Simulate the resin flow for the following two different choices of injection gates:

- Scenario **S1**: The injection gate is centered at the center of the surface, $y = 0$ at the two neighbor nodes, Nodes 77 and 78 located at $(x, y, z) = (63.5, 0, 50.8)$ and $(x, y, z) = (76.2, 0, 50.8)$ in [mm], respectively. That means the resin enters into the mold through a rectangular hole with a total size of two conrol volumes around those nodes.
- Scenario **S2**: Two separate injection gates (Nodes 71 and 398) are centered at the two vertical edges as shown in Figure C.9.

In both S1 and S2, each gate injects resin under constant flow rate, $Q_{in} = 24\,\text{cc/min} = 0.4\,\text{cc/s}$.

Simulate the resin flow using LIMS and plot (i) the locations of the flow front at different times, and (ii) injection pressure versus time.

Solution

The input file, "**boxmold.dmp**" and the script file, "**boxmold.lb**" were edited and are listed in the Apendix. The major lines of the script file are given below:

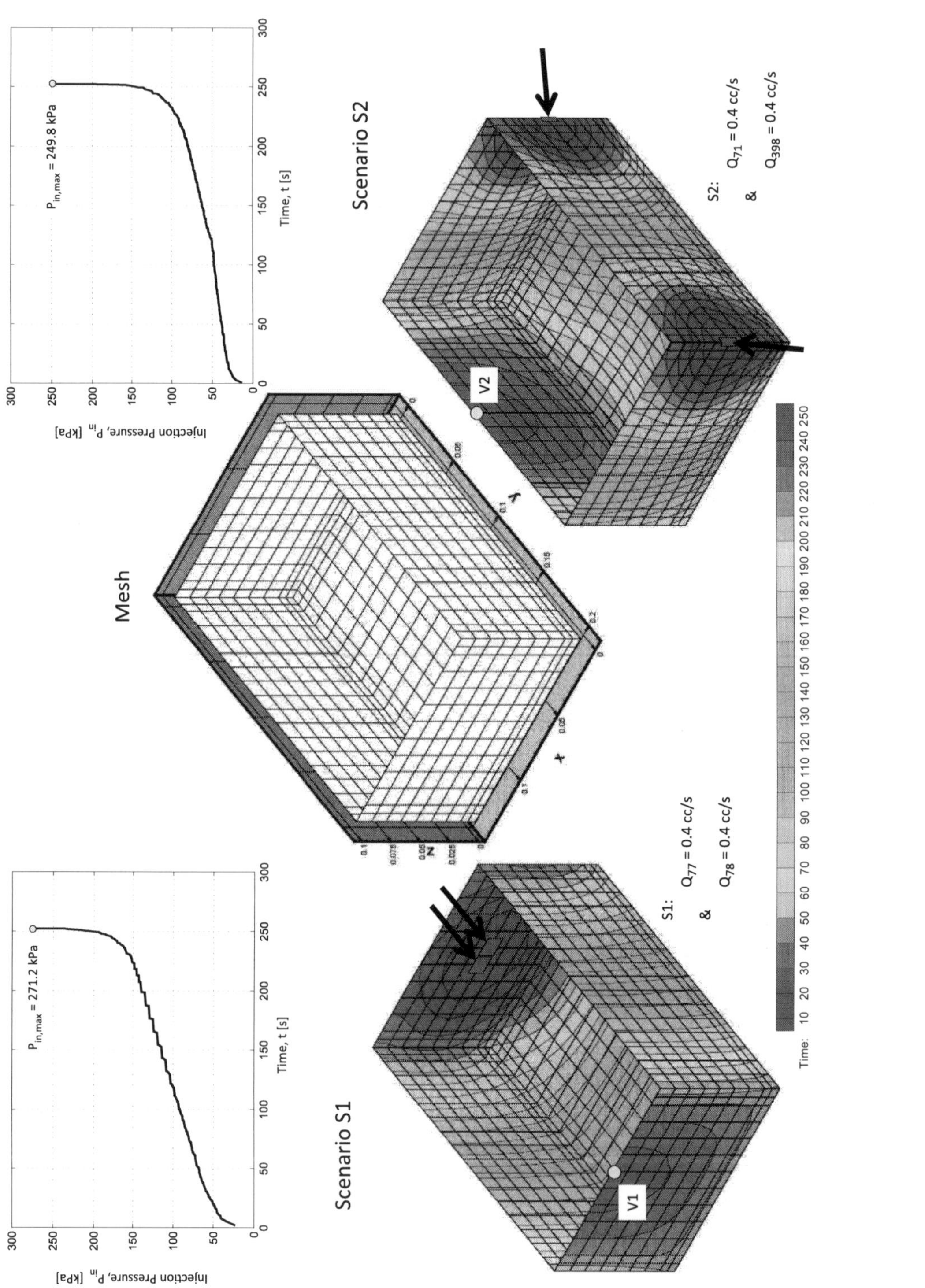

Figure C.9: Mesh and two different mold filling simulations for alternative locations of injection gates.

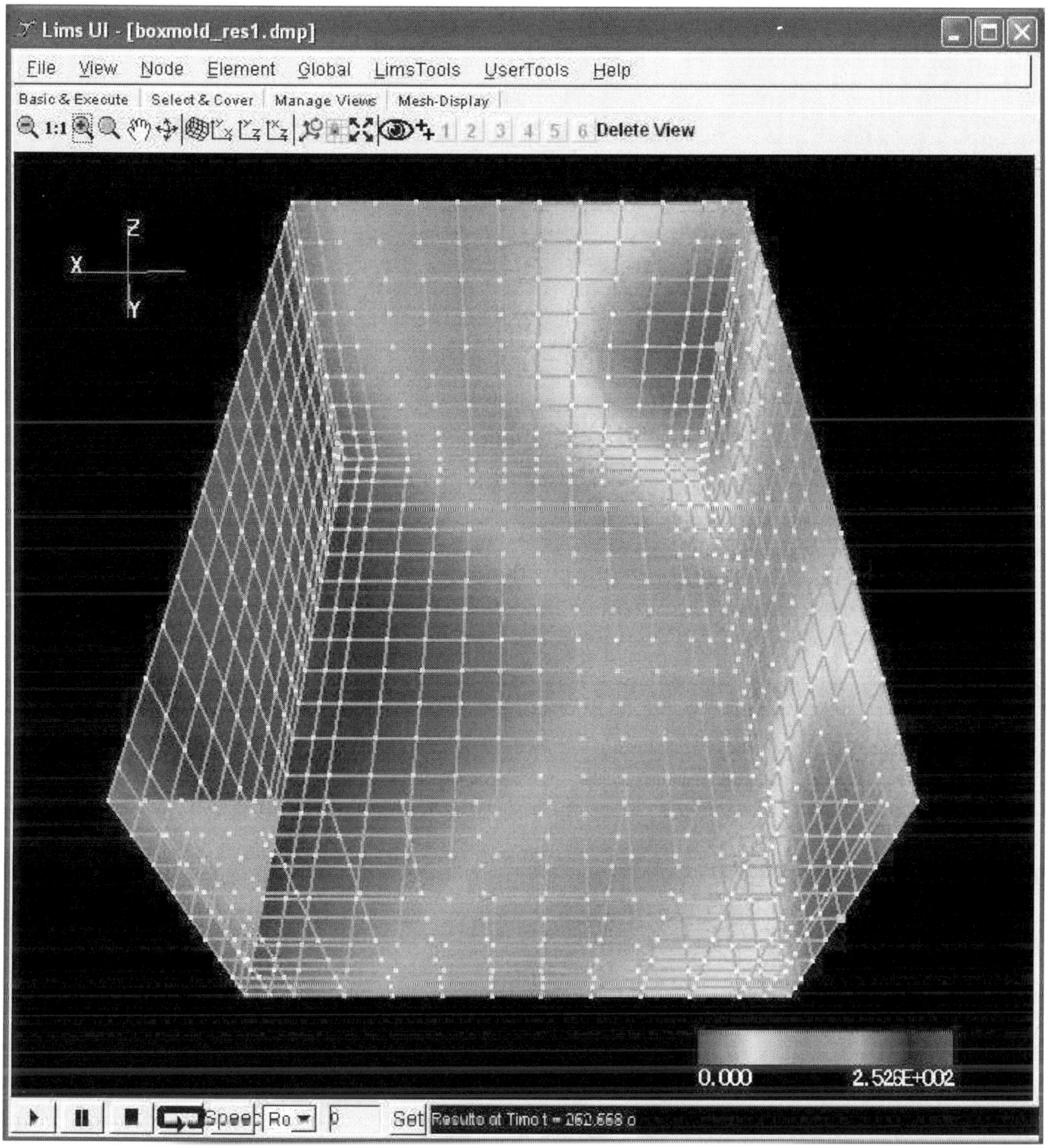

Figure C.10: Mold filling simulation for the scenario S2 using LimsUI.

```
if iscenario=1 then
   let igate1 = soclosestnode(0.0635,0.0000,0.0508,100)
   let igate2 = soclosestnode(0.0762,0.0000,0.0508,100)
endif

if iscenario=2 then
   let igate1 = soclosestnode(0.0000,0.0000,0.0508,100)
   let igate2 = soclosestnode(0.0000,0.2032,0.0508,100)
endif

setgate igate1,2,0.0000004
setgate igate2,2,0.0000004
```

Both scenarios, S1 and S2 were simulated sequentially using one single script file, and their results were recorded in separate output files. The function `soclosestnode` determines the node number of a point. Its first three arguments are the x, y and z components of the point, and the fourth one is related with a tolerance of the convergence used in the numerical calculations, and it should be larger than the minimum distance between the nodes (here it was taken as 100).

The simulated $P_{in}(t)$ and the locations of flow front are shown in Figure C.9.

Alternatively, the simulation and post-processing can be done by using a single program, LimsUI as illustrated in Chapter 8. The result is shown for one of the scenario in Figure C.10.

The maximum injection pressures are 271.2 and 249.8 kPa, respectively, for the scenarios S1 and S2, respectively. The corresponding last points to be filled are shown in the same figure. The ventilation ports should be located at these points on the mold, otherwise voids (dry spots) may be formed as air entraps in the mold cavity.

In this example, uniform material properties (h, V_f and $[\mathbf{K}]$) for the fabric preform were used everywhere in the mold. However, as the preform is being draped over the tool surface, some of the fiber tows may deform very significantly. Especially along the corners, the tows may be more compacted than the adjacent sections; and the empty spaces will form a different structure than the less deformed regions. As studied in [425–427], one can get an experimentally measured distribution of tow spacing and predict what the corresponding permeability is by using experimental or predictive model results. Along these corners, lower and upper bounds of the permeability range can be used in different mold filling scenarios in LIMS, as similarly done in Example 8.15 for a rectangular mold. ■

Bibliography

[1] S. Kueh. Investigation into the effects of applying pressure during compression molding. Mechanical Engineering Department, University of Delaware, Newark, Report 99.1, 1999.

[2] M.R. Barone and D.A. Caulk. A model for the flow of a chopped fiber reinforced polymer composite in compression molding. *Journal of Applied Mechanics*, 53(2):361–371, 1986.

[3] M.R. Barone and D.A. Caulk. The Effect of Deformation and Thermoset Cure on Heat Conduction in a Chopped-Fiber Reinforced Polyester during Compression Molding. *Int. J. Heat Mass Transfer*, 22:1021–1032, 1979.

[4] S. Bickerton. Compression molding with partial slip boundary condition. Mechanical Engineering Department, University of Delaware, Report 98.1, 1998.

[5] S. Middleman. *Fundamentals of Polymer Processing.* McGraw Hill, New York, 1977.

[6] P.J. Hepola. *Thermoplastic pultrusion with on-line dry powder impregnation of fibers.* PhD thesis, Mechanical Engineering Dept, University of Delaware, Newark, 1993.

[7] S. G. Advani, editor. *Flow and Rheology in Polymer Composites Manufacturing.* Elsevier, Amsterdam; New York, 1994.

[8] T.G. Gutowski. *Advanced Composites Manufacturing.* Wiley, New York, 1997.

[9] R.D. Dave and A.C. Loos, editors. *Processing of Composites.* Hanser Publishers, Munich, 2000.

[10] M. Schwartz. *Composite Materials II: Processing, Fabrication and Applications.* Prentice Hall, Upper Saddle River, N.J., 1996.

[11] T. Astrom. *Manufacturing of Polymer Composites.* Chapman and Hall, London, 1997.

[12] T.S. Chou, R.L. McCullough, and R.B. Pipes. Composites. *Scientific America*, 254(18):193–203, 1986.

[13] S. Shuler and S.G. Advani. Transverse squeeze flow of concentrated aligned fibers in viscous fluids. *Journal of Non-Newtonian Fluid Mechanics*, 65:47–74, 1996.

[14] H. Domininghaus. *Die Kunststoffe und ihre Eigenschaften.* VDI-Verlag, 4th Edition, Dusseldorf, 1992.

[15] A.B. Strong, editor. *Fundamentals of Composites Manufacturing*. Society of Manufacturing Engineers, 2007.

[16] H. Stadtfeld. Thermoplastic pultrusion. Mechanical Engineering Department, University of Delaware, Newark, Report 99.8, 1999.

[17] M. Pahl, W. Gleißle, and H.-M. Laun. *Praktische Rheologie der Kunststoffe und Elastomere*. VDI-Verlag Düsseldorf, 4th Edition, 1995.

[18] K. van Rijswijk and Bersee H.E.N. Reactive processing of textile fiber-reinforced thermoplastic composites - an overview. *Composites: Part A*, 38(3):666–681, 2007.

[19] G.I. Williams and R.P. Wool. Composites from natural fibers and soy oil resins. *Applied Composite Materials*, 7:421–432, 2000.

[20] www.rajasthancrafts.com/products/index.php?cat=60.

[21] http://www.pacific-home.com/shop/living-room/sofas-sectionals/slipcovered/maxim sofa.

[22] www.repar.com.tw/images/33003.jpg.

[23] D.W. Becker. *International Encyclopedia of Composites*, volume 3. Edited by Lee, S. M., VCH Publishers, Inc., New York, 1990.

[24] A. Reinhardt. *Injection Molding of Long Fiber Reinforced Thermoplastics*. PhD thesis, Institute for Composite Materials Ltd., University of Kaiserslautern, Erwin-Schroedinger-Strasse, 67663 Kaiserslautern, Germany, 2002.

[25] J.M. Lawrence, E.M. Sozer, H. Stadtfeld, P. Simacek, G. Estrada, R. Don, and S.G. Advani. Use of sensors and simulations for process control to manufacture a feature based composite with resin transfer molding process. In *Proceedings of the Society for the Advancement of the Material and Process Engineering (SAMPE)*, pages 146–155, 2000.

[26] C.-C. Lee, F. Folgar, and C.L. Tucker III. Simulation of compression molding for fiber-reinforced thermosetting polymers. *Journal of Engineering for Industry*, 106:114–125, 1984.

[27] S. Shuler and S.G. Advani. Transverse squeeze flow of concentrated aligned fibers in viscous fluids. *Journal of Non-Newtonian Fluid Mechanics*, (65):47–74, 1996.

[28] S.G. Advani, S. Shuler, and T. Creasy. *Sheet Forming of Composites, Chapter 8: Rheology of Long Fiber Composites in Sheet Forming*. Edited by Debes Bhattacharrya, Elsevier Publishers, Amsterdam, 1997.

[29] S.G. Advani and T.S. Creasy. *Advances in the Flow and Rheology of Non-Newtonian Fluids, Chapter: Rheology of Long Discontinuous Fiber Thermoplastic Composites*. Edited by D. A. Siginer and D.De Kee and R.P. Chhabra, Elsevier Publishers, 1999.

[30] W. Vanwest, R.B. Pipes, and S.G. Advani. Consolidation mechanics in co-mingled fabrics. *Polymer Composites*, (12):417–427, 1991.

[31] B. T. Astrom, R. B. Pipes, and S. G. Advani. On flow through aligned fiber beds and its application to composites processing. *J. Comp. Mater.*, 26(9):1351–1373, 1992.

[32] J.D. Muzzy, X. Wu, and J.S. Colton. Thermoforming of high performance thermoplastic composites. In *ANTEC*, pages 171–177, 1989.

[33] D.L. Denton. In *Proceeding SPI RP/C Inst.*, volume 16, 1981.

[34] R. Aluru, M. Keefe, and S.G. Advani. Simulation of injection molding into rapid-prototyped molds. *Rapid Prototyping Journal*, 17(1), 2001.

[35] A. Whelan. *Injection Moulding Machines.* Elsevier, London, 1984.

[36] F. Johannaber. *Injection Molding Machines: User's Guide.* Hanser, Munich, Germany, 1983.

[37] T. Whelan and J. Goff. *Injection Molding of Thermoplastics Materials.* Van Nostrand Reinhold, New York, 1990.

[38] H.D. Keith and F.J. Padden Jr. Spherulitic morphology in polyethylene and isotactic polystyrene: Influence of diffusion of segregated species. *Journal of Polymer Science, Part B: Polymer Physics*, 25(11):2371–2392, 1987.

[39] S. Ranganathan and S.G. Advani. Characterization of orientation clustering in short fiber composites. *J. of Polymer Science: Part B Polymer Physics*, 28:2651–2672, 1990.

[40] D.J. Coyle, J.W. Blake, and C.W. Macosko. Kinematics of fountain flow in mold-filling. *AIChE Journal*, 33(7):1168–1177, 1987.

[41] J.R. Wood, Q. Zhao, and H.D. Wagner. Orientation of carbon nanotubes in polymers and its detection by raman spectroscopy. *Composites Part A: Applied Science and Manufacturing*, 32(3-4, sp. iss. SI):391–399, 2001.

[42] C.A. Cooper, D. Ravich, D. Lips, J. Mayer, and H.D. Wagner. Distribution and alignment of carbon nanotubes and nanofibrils in a polymer matrix. *Composites Science and Technology*, 62(7-8):1105–1112, 2002.

[43] Z.H. Fan and S.G. Advani. Characterization of orientation state of carbon nanotubes in shear flow. *Polymer*, 46(14):5232–5240, June 27 2005.

[44] W.Z. Tang and S.G. Advani. Nonequilibrium molecular dynamics simulation to describe the rotation of rigid, low aspect ratio carbon nanotubes in simple shear flow. *Journal of Chemical Physics*, 126(14), April 14 2007.

[45] E. Camponeschi, B. Florkowski, R. Vance, G. Garrett, H. Garmestani, and R. Tannenbaum. Uniform directional alignment of single-walled carbon nanotubes in viscous polymer flow. *LANGMUIR*, 22(4):1858–1862, February 14 2006.

[46] W.K.A. Ma, F. Chinesta, A. Ammar, and M. Mackley. Rheological modeling of carbon nanotube aggregate suspensions. *Journal of Rheology*, 52(6):1311–1330, November-December 2008.

[47] W.Z. Tang and S.G. Advani. Dynamic simulation of carbon nanotubes in simple shear flow. *CMES-Computer Modeling in Engineering and Sciences*, 25(3):149–164, March 2008.

[48] Z.H. Fan and S.G. Advani. Rheology of multiwall carbon nanotube suspensions. *Journal of Rheology*, 51(4):585–604, July-August 2007.

[49] M.J. Stevens. *Extruder principles and operation.* Elsevier Applied Science, London; New York, 1985.

[50] plastic (thermoplastic and thermosetting resins). Encyclopedia Britannica, http://search.eb.com/eb/article?eu=115139&hook=625187.

[51] S. Kalpakjian and S.R. Schmid. *Manufacturing Engineering and Technology.* Prentice Hall, Upper Saddle River, N.J., 5th edition, 2006.

[52] P.N. Richardson. *Introduction to Extrusion.* Society of Plastics Engineers, Greenwich, Conn., 1974.

[53] Z. Tadmor and I. Klein. *Engineering Principles of Plasticating Extrusion*, volume c1970. Krieger, Malabar, Fla, 1970.

[54] L.P.B.M. Janssen. *Twin screw extrusion.* Elsevier Scientific Pub. Co., Amsterdam; New York, 1978.

[55] C.L. Tucker. Compression molding of reinforced plastics. In Avraam I. Isayev, editor, *Injection and Compression Molding Fundamentals.* M. Dekker, New York, 1987.

[56] M.L. Stuart. Molding, compression. In *International Encyclopedia of Composites*, volume 5, pages 458–465. VCH Publishers, Inc., New York, NY, 1990.

[57] Ford to make millionth truck with SMC beam. *Composites Technology*, 1(3), September/October 1995.

[58] D. Nichols. *Adv. Mater. Process.*, 146(4):29–32, 1994.

[59] M.D. Bigg. *Plast. Eng.*, 46(10):33–34, 1990.

[60] U.K. Vaidya and K.K. Chawla. Processing of fibre reinforced thermoplastic composites. *International Materials Reviews*, 53(4):185, 2008.

[61] C.F. Johnson. Compression molding. In T. Reinhart, editor, *Engineered Materials Handbook*, volume 1: Composites. ASM International, 1987.

[62] A. Arefmanesh, S.G. Advani, and E.E. Michaelides. A parametric study of bubble growth in low pressure foam molding. *Polymer Engineering and Science*, 30:1330–1338, 1990.

[63] A. Arefmanesh, S.G. Advani, and E.E. Michaelides. An accurate solution technique for diffusion induced growth of a gas bubble in a viscous fluid. *International J. of Heat and Mass Transfer*, 35:1711–1722, 1992.

[64] J. Ward, S. Panigrahi, L.G. Tabil, W.J. Crerar, and T. Powell. Rotational molding of flax fiber reinforced thermoplastics. *ASAE Meeting Presentation*, Paper No: MBSK 02-209.

[65] S.F. Shuler. *Rheology and forming of long fiber reinforced thermoplastic composite materials.* PhD thesis, Mechanical Engineering Department, University of Delaware, Newark, DE, 1995.

[66] S.G. Pantelakis and E.A. Baxevani. *Composites Part A: Applied Science and Manufacturing*, 33A:459–470, 2002.

[67] M. Howald and L.S. Meyer. Shaft for fishing rods. U.S. Patent 2,571,717, 1951.

[68] P.J. Hepola. Thermoplastic pultrusion with on-line dry powder impregnation of fibers. Center for Composite Materials, Report 93-40, University of Delaware, 1993.

[69] Shakespeare Composite Structures. 19845 U.S. Highway 76, Newberry, SC 29108, www.shakespeare-ce.com.

[70] D. Heider. *Model-based control incorporating neural network optimization of the automated thermoplastic tow-placement process.* PhD thesis, Department of Electrical Engineering, University of Delaware, Newark, 1998.

[71] P.G. deGennes. Reptation of a polymer chain in the presence of fixed obstacles. *J. Chem. Phys.*, (55):572–579, 1971.

[72] W. Lee and G.S. Springer. A model of the manufacturing process of thermoplastic matrix composites. *J. of Composite Materials*, 21:1017–1055, 1987.

[73] J.D. Muzzy. Processing of advanced thermoplastic composites. In T. Gutowski, editor, *Proceedings of Manufacturing Science of Composites, ASME*, pages 27–39, 1988.

[74] R. Chowdhury and S.G. Advani. An experimental investigation of consolidation in filament winding. *Composites Manufacturing*, 2:97–104, 1991.

[75] M.A. Lamontia, M.B. Gruber, and B.J. Jenson. Optimal thermoplastic composite material for low cost fabrication of large composite aerospace structure using NASA resins or poss nanoparticle modifications. In *27th SAMPE EUROPE, Paris*, March 27-29 2006.

[76] J. Tierney and J.W.Jr. Gillespie. Modeling of in situ strength development for the thermoplastic composite tow placement process. *Journal of Composite Materials*, pages 1487–1506, 2006.

[77] C. Nicodeau, J. Cinquin, G. Regnier, and J. Verdu. In-situ consolidation process optimization for thermoplastic matrix composites. In *SAMPE 2006, Long Beach, CA*, April 30 - May 4 2006.

[78] C. Dillon, P. Mallon, and M. Monaghan. The autoclave processing of composites. In T.G. Gutowski, editor, *Advanced Composites Manufacturing*, chapter 6. John Wiley and Sons, New York, 1997.

[79] *Liquid moulding technologies: resin transfer moulding, structural reaction injection moulding, and related processing techniques.* SAE International and Woodhead Pub Ltd., Warrendale, PA; Cambridge, England, 1997.

[80] R.S. Parnas. *Liquid Composite Molding.* Hanser Publishers, Munich, Germany, 2000.

[81] S. Bickerton. *Modeling and control of flow during impregnation of heterogeneous porous media, with application to composite mold filling processes.* PhD thesis, University of Delaware, Newark, DE, 1999.

[82] E.F. Gillio. Co-injection resin transfer molding of hybrid composites. Master's thesis, University of Delaware, Newark, DE, 1998.

[83] D.A. Eckel. An all fiber-reinforced-polymer-composite bridge: design, analysis, fabrication, full-scale experimental structural validation, construction and erection. Department of Civil and Environmental Engineering, University of Delaware, 2001.

[84] Hardcore Composites. GFRP composite bridge deck off Delaware rte 896, http://www.hardcorecomposites.com.

[85] REpower Systems AG. http://www.repower.de/index.php?id=81&l=1.

[86] P. Simacek and S.G. Advani. Simulating three-dimensional flow in compression resin transfer molding process. *Revue Européenne des Elements Finis*, 14(6-7):777–802, 2005.

[87] E.F. Gillio, S.G. Advani, B.K. Fink, and J.W. Gillespie Jr. Investigation of the role of transverse flow in co-injection resin transfer molding. *Polymer Composites*, 19:738–749, 1998.

[88] R. Mathur. Thermoset filament winding. University of Delaware, Report 99.2, 1999.

[89] S. Scarborough. Modeling of filament winding process. Department of Mechanical Engineering, University of Delaware, Report 0.03, 2000.

[90] R. Dave, J.L. Kardos, and M.P. Dudukovic. A model for resin flow during composite processing: Part 1 - General Mathematical development. *Polymer Composites*, 8(1):29–38, 1987.

[91] R.H. Sabersky, A.J. Acosta, E.G. Hauptmann, and E.M. Gates. *Fluid Flow, A First Course in Fluid Mechanics.* Prentice Hall, New Jersey, 1999.

[92] R.B. Bird, R.C. Armstrong, and O. Hassager. Fluid mechanics. In *Dynamic of Polymeric Liquids*, volume 1. John Wiley and Sons, New York, 1987.

[93] S.M. Richardson. *Fluid Mechanics.* Hemisphere Publishing Corporation, New York, 1989.

[94] M.N. Ozisik. *Heat Conduction.* John Wiley and Sons, Inc., New York, 1993.

[95] M. Kamal and M.E. Ryan. *Fundamentals of Computer Modeling for Polymer Processing.* Edited by C. L. Tucker III, Hanser Publishers, New York, 1989.

[96] T.G. Gutowski. *Advanced Composite Manufacturing, Chapter 1: A Brief Introduction to Composite Materials and Manufacturing Processes.* Edited by T. G. Gutowski, John Wiley and Sons, New York, 1997.

[97] S. Barsky and G.W. Slater. Nonequilibrium molecular dynamics simulation of the time-dependent orientational coupling between long and short chains in a bimodal polymer melt upon uniaxial stretching. *Macromolecules*, 32(19):6348–6358, 1999.

[98] W. Ostwald and et al. Zut theorie der liesegang'schen ringe. *Colloid and Polymer Science*, 36:380–390, 1925.

[99] R.I. Tanner. *Engineering Rheology.* Oxford University Press, New York, second edition, 2000.

[100] T.W. Spriggs and R.B. Bird. Some nonlinear viscoelastic models with inclusion of results from molecular theory. *Ind. Eng. Chem. Fundam.*, 4:182–186, 1965.

[101] P.J. Carreau. Rheological equations from molecular network theories. *Trans. Soc. Rheol.*, 16(1):99–127, 1972.

[102] P.J. Carreau, D. De Kee, and M. Daroux. An analysis of the viscous behaviour of polymeric systems. *Canadian Journal of Chemical Engineering*, 57:135–140, 1979.

[103] J.M. Kenny. Application of Modeling to the Control and Optimization of Composite Processing. *Composite Structures*, 27:129–139, 1994.

[104] M.L. Williams, R.F. Landel, and J.D. Ferry. The temperature dependence of relaxation mechanisms in amorphous polymers and other glass-forming liquids. *J. Am. Chem. Soc.*, 77:3701, 1955.

[105] W.M. Sanford and R.L. McCullough. Free-Volume Approach to Thermoset Cure. *Journal of Polymer Science: Part B: Polymer Physics*, 28:973–1000, 1990.

[106] J.B. Enns and J.K. Gillham. Time-Temperature-Transformation (TTT) Cure Diagram: Modeling the Cure Behavior of Thermosets. *Journal of Applied Polymer Science*, 28:2567–2591, 1983.

[107] G. Wisanrakkit, J.K. Gillham, and J.B. Enns. The Glass Transition Temperature (Tg) as a Parameter for Monitoring the Cure of an Amine/Epoxy System at Constant Heating Rates. *Journal of Applied Polymer Science*, 41:1895–1912, 1990.

[108] D.R. Miller and C.W. Macosko. Average Property Realtions for Nonlinear Polymerization with Unequal Reactivity. *Macromolecules*, 11(4):656–662, 1978.

[109] C.W. Macosko and D.R. Miller. A new derivation of average molecular weights of nonlinear polymers. *Macromolecules*, 9(199), 1976.

[110] P.R. Ciriscioli, Q. Wang, and G.S. Springer. Autoclave Curing - Comparisons of Model and Test Results. *Journal of Composite Materials*, 26(1):90–102, 1992.

[111] J.T. Tzeng and A.C. Loos. A Cure Analysis for Axisymmetric Composites. *Composites Manufacturing*, 4(3):157–165, 1993.

[112] T.E. Twardowski, S.E. Lin, and P.H. Geil. Curing in Thick Composite Laminates: Experiment and Simulation. *Journal of Composite Materials*, 27(3):216–250, 1993.

[113] L.N. Hjellming and J.S. Walker. Thermal Curing Cycles for Composite Cylinders with Thick Walls and Thermoset Resins. *Journal of Composite Materials*, 23:1048–1064, 1989.

[114] W.I. Lee, A.C. Loos, and G.S. Springer. Heat of Reaction, Degree of Cure, and Viscosity of Hercules 3501-6 Resin. *Journal of Composite Materials*, 16:510–520, 1982.

[115] W.B. Young, K. Han, L.H. Fong, and L.J. Lee. Flow simulation in molds with preplaced fiber mats. *Polymer Composites*, 12(6):391–403, December 1991.

[116] M.R. Dusi, W.I. Lee, P.R. Ciriscioli, and G.S. Springer. Cure Kinetics and Viscosity of Fiberite 976 Resin. *Journal of Composite Materials*, 21:243–261, 1987.

[117] A.M. Stolin, A.G. Merzhanov, and A.Y. Malkin. Non-Isothermal Phenomena in Polymer Engineering and Science: A Review. Part II: Non-Isothermal Phenomena in Polymer Deformation. *Polymer Engineering and Science*, 19(15):1074–1080, 1979.

[118] J.M. Castro and C.W. Macosko. Kinetics and Rheology of Typical Polyurethane Reaction Molding Systems. *Soc. Plast. Eng. Tech. Papers*, 26:434, 1980.

[119] J.M. Kenny, A. Maffezzoli, and L. Nicolais. A Model for the Thermal and Chemorheology Behavior of Thermosets. (II): Unsaturated Polyester Based Composites. *Composites Science and Technology*, 38:339–358, 1990.

[120] D.S. Kim and S.C. Kim. Rubber Modified Epoxy-Resin. I: Cure Kinetics and Chemorheology. *Polymer Engineering and Science*, 34:625–631, 1994.

[121] D.-S. Lee and C.D. Han. A Chemorheological Model for the Cure of Unsaturated Polyester Resin. *Polymer Engineering and Science*, 27(13):955–963, 1987.

[122] D.J. Michaud. Investigation of Curing Behavior in Thick Thermoset Composites Manufactured by Resin Transfer Molding. Master's thesis, Chemical Engineering Department, University of Delaware, Newark, Spring 1996.

[123] S.G. Advani, M.V. Bruschke, and R.S. Parnas. Resin Transfer Molding Flow Phenomena in Polymeric Composites. In Suresh G. Advani, editor, *Flow and Rheology in Polymer Composite Manufacturing*, volume 10 of *Composite Materials Series*, chapter 12, pages 465–515. Elsevier Science, Amsterdam, 1994.

[124] Y.A. Tajima and D. Crozier. Thermokinetic modeling of an epoxy resin I. chemoviscosity. *Polymer Engineering and Science*, 23:186–274, 1983.

[125] T.H. Hou. Autoclave processing for composite material fabrication. 1. An analysis of resin flows and fiber compactions for thin laminate. In *SPE ANTEC Tech. Papers*, volume 31, page 1253, 1985.

[126] Dienes G. and H. Klemm. Theory and application of the parallel plate plastometer. *J. of Applied Phys.*, 17:459–471, 1946.

[127] S. Oka. *Rheology, Chapter 2*. Edited by F. R. Eirich, Academic Press, New York, 1960.

[128] P.J. Leider and R.B. Bird. University of Wisconsin Rheology Center Report Number 22, 1973.

[129] M. McClelland and B. Finlayson. Squeezing flow of highly viscous polymers. *J. Rheol.*, 32(2):101–133, 1988.

[130] P. Leider and R. Bird. Squeezing flow between parallel disks. I: Theoretical analysis. *Ind. Eng. Chem. Fundam.*, 13(4):336–341, 1974.

[131] P. Leider. Squeezing flow between parallel disks. II: Experimental results. *Ind. Eng. Chem. Fundam.*, 13(4):342–346, 1974.

[132] Z. Tadmor and C.G. Gogos. *Principles of Polymer Processing.* John Wiley and Sons, New York, 1979.

[133] G. Brindley, J. Davies, and K. Walters. Elastico-viscous squeeze films. part I. *JNNFM*, 1:19–37, 1976.

[134] D.M. Binding, J.M. Davies, and K. Walters. Elastico-viscous squeeze films. part II, superimposed rotation. *JNNFM*, pages 259–275, 1976.

[135] M. McClelland and B. Finlayson. Squeezing flow of elastic liquids. *JNNFM*, 13:181–201, 1983.

[136] R. Grimm. Squeezing flows of polymeric liquids. *AIChE Journal*, 24(3):427–439, 1978.

[137] N. Phan-Thien and R. Tanner. Viscoelastic squeeze -film flows- maxwell fluids. *J. Fluid Mech.*, 129:265–281, 1983.

[138] P. Shrodikar and S. Middleman. Lubrication flows in viscoelastic liquids. i. squeezing flow between approacing parallel rigid planes. *J. Rheol.*, 26(1):1–17, 1982.

[139] M.R. Barone and D.A. Caulk. Kinematics of flow in sheet molding compounds. *Polymer Composites*, 6(2):105–109, 1985.

[140] R. Balasubramanyam, R.S. Jones, and A.B. Wheeler. Modeling transverse flows of reinforced thermoplastic materials. *Composites*, 20(1), 1989.

[141] J.A. Barnes and F.N. Cogswell. Transverse flow processes in continuous fibre-reinforced thermoplastic composites. *Composites*, 20(1):38–42, 1989.

[142] S. Shuler and S.G. Advani. Transverse squeeze flow of concentrated aligned fibers in viscous fluids. *Journal of Non-Newtonian Fluid Mechanics*, (65):47–74, 1996.

[143] C.M. O'Bradaigh. Sheet forming of composite materials. *Composite Materials Series*, 10:517, 1994.

[144] B. Hull, T. Rogers, and A. Spencer. A linear stability analysis of flows of transversely isotropic non-newtonian fluids. *Journal of Non-Newtonian Fluid Mechanics*, 43:325–349, 1992.

[145] R.B. Pipes, D.W. Coffin, P. Simacek, S.F. Shuler, and R.K. Okine. Rheological behavior of long collimated fibers. In S. G. Advani, editor, *Flow and Rheology in Polymeric Composites Processing*, chapter 4, pages 85–125. Elsevier, Amsterdam; New York, 1994.

[146] E. Seiler, M. Welz, and H.H. Wald. *Kunstst. Germ. Plast.*, 72(6):10–14, 1982.

[147] C.L. Tucker III and S.G. Advani. Processing of short-fiber systems. In Suresh G. Advani, editor, *Flow and Rheology in Polymeric Composites Manufacturing*, chapter 6. Elsevier Publishers, Amsterdam, 1994.

[148] W.C. Jackson, S.G. Advani, and C.L. Tucker III. Predicting the orientation of short fibers in thin compression moldings. *J. Composite Materials*, 20:539, 1986.

[149] S.G. Advani. *Prediction of fiber orientation during processing of short fiber composites.* PhD thesis, University of Illinois, Urbana, 1987.

[150] S.G. Advani and C.L. Tucker III. The use of tensors to describe and predict fiber orientation in short fiber composites. *J. of Rheology*, 31:751–784, 1987.

[151] C.L. Tucker III. Predicting fiber orientation in short fiber composites. In T. Gutowski, editor, *Manufacturing Science of Composites, Manufacturing International Conference Proceedings*. 1988.

[152] C.L. Tucker. Flow regimes for fiber suspensions in narrow gaps. *J. Non-Newtonian Fluid Mech*, 39:239–268, 1991.

[153] H. Giesekus. Elasto-viskose flussigkeiten, fur die in stationaren schichtstromungen samtliche normalspannungskomponenten verschieden gross siend. *Rheol. Acta*, 2:50–62, 1962.

[154] G.G. Lipscomb, M.M. Denn, D.U. Hur, and D.V. Boger. The flow of fiber suspensions in complex geometries. *J. Non-Newtonian Fluid Mech.*, 26:297–325, 1988.

[155] J.M. Burgers. On the motion of small particles of elongated form, suspended in a viscous liquid. *Verhandelingen Koninkl. Akad. Wetenschap*, 16:8–184, 1938.

[156] G.K. Batchelor. The stress generated in a non-dilute suspension of elongated particles by pure straining motion. *Journal of Fluid Mechanics*, 46:813–829, 1971.

[157] J.G. Evans. The flow of a suspension of force-free rigid rods in a newtonian fluid. Master's thesis, Cambridge University, 1975.

[158] J.G. Evans. *Theoretical Rheology, Chapter: The Effect of Non-Newtonian Properties of a Suspension of Rod-like Particles on Flow Fields.* Edited by J. F. Hutton and J. R. A. Pearson and K. Walters, Halstead Press, New York, 1975.

[159] E.S.G. Shaqfeh and G.H. Fredrickson. The hydrodynamic stress in a suspension of rods. *Phys. Fluids A*, 2:7–24, 1990.

[160] S.M. Dinh and R.C. Armstrong. A rheological equation of state for semiconcentrated fiber suspensions. *Journal of Rheololgy*, 28(3):207 227, 1984.

[161] V. Calado and S.G. Advani. Theromoset resin cure kinetics and rheology. In A. Loos and Raju Dave, editors, *Transport Processes in Composites*, pages 31–105. 2000.

[162] H. Dannesburg. *Soc. Plast. Eng. J.*, 15:875, 1959.

[163] W. Fisch, W. Hoffman, and R. Schmid. *J. Appl. Polym. Sci.*, 13:295, 1969.

[164] P.E. Willard. Determination of cure of diallyl phthalate using differential scanning calorimetry. *Polymer Engineering and Science*, 12(2):120–124, 1972.

[165] J.A. Aukward, R.W. Warfield, and M.C. Petree. Change in electrical resistivity of some high polymers during isothermal polymerization. *Journal of Polymer Science*, 27:199–205, 1958.

[166] S. Sourour and M.R. Kamal. Differential Scanning Calorimetry of Epoxy Cure: Isothermal Cure Kinetics. *Thermochimica Acta*, 14(41):41–59, 1976.

[167] J.P. Creedon. Thermal methods for determination of degree of cure of thermosets. *Anal. Calorim.*, 2:185–199, 1970.

[168] R.B. Prime. In E. A. Turi, editor, *Thermal Characterization of Polymeric Materials*, pages 435–569. Academic Press, New York, second edition, 1981.

[169] E.M. Barrall II and J.F. Johnson. Differential Scanning Calorimetry Theory and Applications. In Jr. Philip E. Slade and Lloyd T. Jenkis, editors, *Thermal Characterization Techniques*, volume 2, chapter 1, pages 1–39. Marcel Dekker, Inc., 1970.

[170] M.J. Richardson. Quantitative Differential Scanning Calorimetry. In J. V. Dawkins, editor, *Developments in Polymer Characterization*, volume 1, chapter 7, pages 205–244. Applied Science Publishers Ltd, 1978.

[171] K.E.J. Barret. Determination of Rates of Thermal Decomposition of Polymerization Initiators with a Differential Scanning Calorimeter. *Journal of Applied Polymer Science*, 11:1617–1626, 1967.

[172] J.-M. Vergnaud and J. Bouzon. *Cure of Thermosetting Resins - Modelling and Experiments*. Springer-Verlag, Germany, 1992.

[173] R.B. Prime. Thermosetting. In Edith A. Turi, editor, *Thermal Characterization of Polymeric Materials*, volume 2, pages 435–569. Academic Press, Inc., 1981.

[174] B. Miller. Polymerization behavior of silver-filled epoxy resins by resistivity measurements. *Journal of Applied Polymer Science*, 10:217–228, 1966.

[175] R.A. Fava. Differential Scanning of Calorimetry of Epoxy Resins. *Polymer*, 9:137–151, 1968.

[176] S.R. White and H.T. Hahn. Cure Cycle Optimization for the Reduction of Processing-Induced Residual Stresses in Composite Materials. *Journal of Composite Materials*, 27(14):1352–1378, 1993.

[177] R.G.C. Arridge and J.H. Speake. Mechanical Relaxation Studies of the Cure of Epoxy Resins: 1. Measurement of Cure. *Polymer*, 13:443–449, 1972.

[178] C.C. Riccardi, H.E. Adabbo, and R.J.J. Williams. Curing Reaction of Epoxy Resins with Diamines. *Journal of Applied Polymer Science*, 29:2481–2492, 1984.

[179] J.R. MacCallum and J. Tanner. Derivation of rate equations used in thermogravimetry. *Nature*, 225:1127, 1970.

[180] J. Sestak and J. Kratochvil. Rational approach to thermodynamic processes and constitutive equations in isothermal and non isothermal kinetics. *J. Thermal Anal.*, 5:192–201, 1973.

[181] A. Dutta and M.E. Ryan. The Relantionship Between Isothermal and Non-Isothermal Kinetics for Thermoset Characterization. *Thermochimica Acta*, 33:87–92, 1979.

[182] T. Ozawa. Non-Isothermal Kinetics and Generalized Time. *Thermochimica Acta*, 100:109–118, 1986.

[183] A. Letton and P.B. Chiou. Development of a Cure Model for Determining Optimum Cure Cycles in Kinetically Complex Systems. In A. A. Collyer and L. A. Utracki, editors, *Polymer Rheology and Processing*, chapter 16, pages 431–458. Elsevier, 1990.

[184] R.B. Prime. Dynamic Cure Analysis of Thermosetting Polymers. In Roger S. Porter and Julian F. Johnson, editors, *Analytical Calorimetry*, volume 2, pages 201–210. Plenum Press, 1970.

[185] R.B. Prime. Differential Scanning Calorimetry of the Epoxy Cure Reaction. *Polymer Engineering and Science*, 13(5):365–371, 1973.

[186] J. Kratochvil and J. Sesta. The Role of Constitutive Equation in Chemical Kinetics. *Thermochimica Acta*, 7:330–332, 1973.

[187] A.L. Draper. In *Proc. Toronto Symp. Therm. Anal.*, volume 3rd, page 63, 1970.

[188] E.L. Simmons and W.W. Wendlandt. Non-Isothermal Rate Equations. *Thermochimica Acta*, 3:498–500, 1972.

[189] R.A.W. Hill. Rate equations in thermogravimetry. *Nature*, 227:703, 1970.

[190] V.M. Gorbatchev and V.A. Logvinenko. The correlation between isothermal and non-isothermal kinetics in thermogravimetry. *Journal of Thermal Anal.*, 4:475–477, 1972.

[191] J.D. Nam and J.C. Seferis. Application of the Kinetic Composite Methodology to Autocatalytic-Type Thermoset Prepreg Cures. *Journal of Applied Polymer Science*, 50:1555–1564, 1993.

[192] J.P. Creedon. Thermal Methods for Determination of Degree of Cure of Thermosets. *Anal. Calorim.*, 2:185–199, 1970.

[193] K. Horie, M. Sawada, I. Mita, and H. Kambe. Calorimetric Investigation of Polymerization Reactions. III. Curing Reaction of Epoxides with Amines. *Journal of Polymer Science: Part A-1*, 8:1357–1472, 1970.

[194] A. Lewis. Torsional Braid Analysis. *Polymer Engineering and Science*, 19(10):662–663, 1979.

[195] J.K. Gillham. Formation and Properties of Network Polymeric Materials. *Polymer Engineering and Science*, 19(10):676–682, 1979.

[196] J. Heijboer. The Torsion Pendulum in the Investigation of Polymers. *Polymer Engineering and Science*, 19(10):664–675, 1979.

[197] J.K. Gillham. Torsional Braid Analysis (TBA) of Polymers. In J. V. Dawkins, editor, *Developments in Polymer Characterization*, volume 3, chapter 5, pages 159–227. Applied Science Publishers Ltd, London, 1978.

[198] P.G. Babayevsky and J.K. Gillham. Epoxy Thermosetting Systems: Dynamic Mechanical Analysis of the Reactions of Aromatic Diamines with the Diglycidyl Ether of Bisphenol A. *Journal of Applied Polymer Science*, 17:2067–2088, 1973.

[199] J.K. Gillham and M.B. Roller. Advances in Instrumentation and Technique of Torsional Pendulum and Torsional Braid Analysis. *Polymer Engineering and Science*, 11(4):295–304, 1971.

[200] J.K. Gillham. Torsional Braid Analysis: A Semimicro Thermomechanical Approach to Polymer Characterization. In Jr. Philip E. Slade and Lloyd T. Jenkis, editors, *Thermal Characterization Techniques*, volume 2, chapter 4, pages 225–292. Marcel Dekker, Inc., New York, N.Y., 1970.

[201] C.D. Wingard and C.L. Beatty. Crosslinking of an Epoxy with a Mixed Amine as a Function of Stoichiometry. I. Cure Kinetics via Dynamic Mechanical Spectroscopy. *Journal of Applied Polymer Science*, 40:1981–2005, 1990.

[202] C.D. Wingard and C.L. Beatty. Crosslinking of an Epoxy with a Mixed Amine as a Function of Stoichiometry. II. Final Properties via Dynamic Mechanical Spectroscopy. *Journal of Applied Polymer Science*, 41:2539–2554, 1990.

[203] F.G. Mussati and C.W. Macosko. Rheology of Network Forming Systems. *Polymer Engineering and Science*, 13(3):236–240, 1973.

[204] F.M. White. *Viscous Fluid Flow*. McGraw-Hill, New York, N.Y., 1974.

[205] A. Shimazaki. Viscoelastic Changes of Epoxy Resin-Acid Anhydride System during Curing. *Journal of Applied Polymer Science*, 12:2013–2021, 1968.

[206] M.A. Acitelli, R.B. Prime, and E. Sacher. Kinetics of epoxy cure: (1) The system bisphenol-A Diglycidyl Ether/m-Phenylene Diamine. *Polymer*, 12:335–343, 1971.

[207] Y. Deng and G.C. Martin. Modeling Diffusion during Thermoset Cure: An Approach Based on Dielectric Analysis. *Macromolecules*, 27:5141–5146, 1994.

[208] G.S. Learmonth and G. Pritchard. Dielectric Relaxation and Crosslinking in Unsaturated Polyester Resins. *Journal of Applied Polymer Science*, 13:2119–2127, 1969.

[209] G.M. Maistros and I.K. Partridge. Dielectric Monitoring of Cure in a Commerical Carbon-Fibre Composite. *Composites Science and Technology*, 53:355–359, 1995.

[210] R.J. Morgan and E.T. Mones. The Cure Reactions, Network Structure, and Mechanical Response of Diaminodiphenyl Sulphone-Cured Tetraglycidyl 4,4' Diaminodiphenyl Methane Epoxies. *Journal of Applied Polymer Science*, 33:999–1020, 1987.

[211] J.F. Stevenson. Free Radical Polymerization Models for Simulating Reactive Processing. *Polymer Engineering and Science*, 26:746–759, 1986.

[212] V. Bellenger, J. Verdu, J. Francillette, P. Hoarau, and E. Morel. Infra-Red Study of Hydrogen Bonding in Amine-Crosslinked Epoxies. *Polymer*, 28:1079–1086, 1987.

[213] A. Sabra, T.M. Lam, J.P. Pascault, M.F. Grenier-Loustalot, and P. Grenier. Characterization and Behaviour of Epoxy-Based Diaminodiphenylsulphone Networks. *Polymer*, 28:1030–1036, 1987.

[214] A.J. Attias and B.J. Bloch. Chemical Structure of Networks Resulting from Curing of N,N-Diglycidylaniline - Type Resins with Aromatic Amines. IV. Characterization of TGDDM/DDS and TGDDM/DDM Networks by High-Resolution Solid State ^{13}C NMR. *Journal of Applied Polymer Science: Part A: Polymer Chemistry*, 28:3445–3466, 1990.

[215] M.G. Rogers. The Structure of Epoxy Resins using NMR and GPC Techniques. *Journal of Applied Polymer Science*, 16:1953–1958, 1972.

[216] E. Mertzel and J.L. Koenig. Application of FT-IR and NMR to Epoxy Resins. In K. Dusek, editor, *Advances in Polymer Science - Epoxy Resins and Composites II*, volume 75, chapter 3, pages 73–112. Springer-Verlag, Berlin, 1986.

[217] T.F. Saunders, M.F. Levy, and J.F. Serino. Mechanism of the Tertiary Amine-Catalyzed Dicyandiamine Cure of Epoxy Resins. *Journal Polymer Science A1*, 5:1609–1617, 1967.

[218] H. Batzer and S.A. Zahir. Studies in the Molecular Weight Distribution of Epoxide Resins. II-Chain Branching in Epoxide Resins. *Journal of Applied Polymer Science*, 19:601–607, 1975.

[219] N.A. St. John and G.A. George. Cure Kinetics and Mechanisms of a Tetraglycidyl-4,4'-diaminodiphenylmethane/Diaminodiphenylsulphone Epoxy Resin using Near i.r. Spectroscopy. *Polymer*, 33(13):2679–2688, 1992.

[220] B.-G. Min, Z.H. Stachurski, J.H. Hodgkin, and G.R. Heath. Quantitative Analysis of the Cure Reaction of DGEBA/DDS Epoxy Resins without and with Thermoplastic Polysulphone Modifier using Near Infra-Red Spectrocospy. *Polymer*, 34(17):3620–3627, 1993.

[221] C.J. de Bakker, N.A. St. John, and G.A. George. Simultaneous Differential Scanning Calorimetry and Near-Infra-Red Analysis of the Curing of Tetraglycidyldiaminodiphenylmethane with Diaminodiphenylsulphone. *Polymer*, 34(4):716–725, 1993.

[222] L. Xu, J.H. Fu, and J.R. Schlup. In Situ Near-Infrared Spectroscopic Investigation of Epoxy Resin-Aromatic Amine Cure Mechanisms. *Journal of American Chemistry Society*, 116:2821–2826, 1994.

[223] P. Banks and R.H. Peters. Polymerization and Crosslinking of Epoxides: Base-Catalyzed Polymerization of Phenyl Glycidyl Ether. *Journal of Polymer Science: Part A-1*, 8:2595–2610, 1970.

[224] W.M. Sanford. *Cure Behavior of Thermosetting Resin Composites.* PhD thesis, University of Delaware, Newark, DE, 1987.

[225] G.R. Palmese and V.M. Karbhari. Effects of Sizings on Microscopic Flow in Resin Transfer Molding. *Polymer Composites*, 16(4):313–318, 1995.

[226] K.-W. Lem and C.D. Han. Chemorheology of Thermosetting Resins. II. Effect of Particulates on the Chemorheology and Curing kinetics of unsaturated polyester resin. *Journal of Applied Polymer Science*, 28:3185–3206, 1983.

[227] A. Dutta and M.E. Ryan. Effect of Fillers on Kinetics of Epoxy Cure. *Journal of Applied Polymer Science*, 24:635–649, 1979.

[228] S.H. McGee. Curing Characteristics of Particulate-Filled Thermosets. *Polymer Engineering and Science*, 22(8):484–491, 1982.

[229] J.H. Lee and J.W. Lee. Kinetic Parameters Estimation for Cure Reaction of Epoxy Based Vinyl Ester Resin. *Polymer Engineering and Science*, 34(9):742–749, 1994.

[230] K. Han, C.H. Wu, and L.J. Lee. Characterization and simulation of resin transfer molding - race tracking and dry spot formation. In *Proceedings of Ninth Annual ASM-ESD Advanced Composite Conference*, Michigan, 1993.

[231] J. Mijovic and H.T. Wang. *Sampe Journal*, 24:42–55, 1988.

[232] L. Mascia. *Thermoplastics: Materials Engineering*. Applied Science Pub. Ltd., London, 1982.

[233] P.C. Painter and M.M. Coleman. *Fundamentals of Polymer Science*. Technomic Publishing Co., Inc., Lancaster, PA, 1997.

[234] J.W.S. Hearle. *Polymers and Their Properties, Volume 1: Fundamentals of Structure and Mechanics*. Ellis Horwood limited, Chichestor, England, 1982.

[235] R. Phillips and J.-A.E. Manson. Prediction and analysis of non-isothermal crystallization of polymers. *Polymer*, 1997.

[236] A. Wasaik. Kinetics of polymer crystallization in non-isothermal conditions. *Macromolecular Chemistry*, (2):211–245, 1991.

[237] H.D. Keith and F.J. Padden. Spherulitic crystallization from the melt. i. fractionation and impurity segregation and their influence on crystalline morphology. *Journal of Applied Physics*, 35(4):1270, 1964.

[238] A. Benard. *Transport Phenomena and Microstructure Development during Solidification of Semicrystalline Polymers and Composites*. PhD thesis, Mechanical Engineering Department, University of Delaware, Newark, 1995.

[239] A. Benard and S.G. Advani. A cell model to describe the spherullitic growth in semi-crystalline polymers. *Polymer Engineering and Science*, 36(4):520–534, 1996.

[240] A. Benard and S.G. Advani. Energy equation and crystallization kinetics of semi-crystalline polymers: Regimes of coupling. *International J. of Heat and Mass Transfer*, 38:819–832, 1995.

[241] J.D. Hoffman and J.I. Lauritzen. Crystallization of bulk polymers with chain folding: Theory of lamellar spherulites. *Journal of Research of the National Bureau of Standards, A. Physics and Chemistry*, 65A(4):297–336, 1961.

[242] J.D. Hoffman, G.T. Davis, and J.I. Lauritzen. Theory of the crystallization of linear polymers with chain folding. In N. B. Hannay, editor, *Treatise on Solid State Chemistry*, volume 3, pages 497–614. New York, Plenum, 1976.

[243] P. Supaphol and J.E. Spruiell. Non-isothermal bulk crystallization studies of high density polyethylene using light depolarizing microscopy. *Journal of Polymer Science: Part B: Polymer Physics*, 36:681–692, 1998.

[244] Z. Ding and J.E. Spruiell. An experimental method for studying non-isothermal crystallization of polymers at very high cooling rates. *Journal of Polymer Science: Part B: Polymer Physics*, 34:2783–2804, 1996.

[245] H. Darcy. *Les fontaines publiques de la ville de Dijon*. Dalmont, Paris, 1856.

[246] S.G. Advani, M.V. Bruschke, and R. Parnas. Resin transfer molding. In Suresh G. Advani, editor, *Flow and Rheology in Polymeric Composites Manufacturing*, chapter 12, pages 465–516. Elsevier Publishers, Amsterdam, 1994.

[247] B.R. Gebart. Permeability of unidirectional reinforcements for RTM. *Journal of Composite Materials*, 26(8):1100–1133, 1992.

[248] M.V. Bruschke and S.G. Advani. Flow of generalized newtonian fluids across a periodic array of cylinders. *J. Rheology*, 37(3):479–498, 1993.

[249] S. Ranganathan, F.R. Phelan Jr., and S.G. Advani. Generalized model for the transverse fluid permeability in unidirectional fibrous media. *Polymer Composites*, 17(2):222–230, April 1996.

[250] P.B. Nedanov and S.G. Advani. Numerical computation of the fiber preform permeability tensor by the homogenization method. *Polymer Composites*, 23:758–770, 2002.

[251] P. Simacek and S.G. Advani. Permeability model for woven fabrics. *Polymer Composites*, 17(6):887–899, December 1996.

[252] S. Rangatnathan, R.G. Easterling, S.G. Advani, and F.R. Phelan Jr. Effect of microstructure variations on the permeability of preform materials. *Polymers and Polymer Composites*, 6(2):63–73, 1998.

[253] T.D. Papathanasiou. Structure-oriented micromechanical model for viscous flow through square arrays of fibre clusters. *Composites Science and Technology*, 5(9):1055–1069, 1996.

[254] S.G. Advani and Z. Dimitrovova. Role of capillary driven flow in composites manufacturing. In Stanley Hartland, editor, *Surface and Interfacial Tension: Measurement, Theory and Applications*, volume 119, pages 263–312. Marcel and Dekker, Surfactant Science Series, 2004.

[255] P.B. Nedanov and S.G. Advani. A method to determine 3d permeability of fibrous reinforcements. *Journal of Composite Materials*, 36(2):241–254, 2002.

[256] G. Estrada. Role of binders and compaction on permeability of fabrics. Master's thesis, Mechanical Engineering Department, University of Delaware, Newark, 2002.

[257] P.B. Nedanov, S.G. Advani, S.W. Walsh, and W. O Ballata. Determination of the permeability tensor of fibrous reinforcements for VARTM. In *ASME*, 1999.

[258] S. Walsh. In-situ sensors method and device. U.S. Patent 5,210,499, May 1993.

[259] Y. Luo, I. Verpoest, K. Hoes, M. Vanheule, H. Sol, and A. Cardon. Permeability measurement of textile reinforcements with several test fluids. *Composites - Part A: Applied Science and Manufacturing*, 32(10):1497–1504, 2001.

[260] K.M. Pillai and S.G. Advani. Model for unsaturated flow in woven fiber preforms during mold filling in resin transfer molding. *Journal of Composite Materials*, 32(19):1753–1783, 1998.

[261] F. Zhou, J. Alms, and S.G. Advani. A closed form solution for flow in dual scale fibrous porous media under constant injection pressure conditions. *Composites Science and Technology*, 68(3-4):699–708, 2008.

[262] F. Zhou, N. Kuentzer, P. Simacek, S.G. Advani, and S. Walsh. Analytic characterization of the permeability of dual-scale fibrous porous media. *Composites Science and Technology*, 66(15):2795–2803, 2006.

[263] A. Belloli. Automated fiber preform permeability measurement station for VARTM process, diploma thesis, ETH, Zurich. 2001.

[264] Z. Cai and T. Gutowski. 3-d deformation behavior of a lubricated fiber bundle. *Journal of Composite Materials*, 26(8):1207–1237, 1992.

[265] B. Chen, A.H.-D. Cheng, and T.-W. Chou. Nonlinear compaction model for fibrous preforms. *Composites - Part A: Applied Science and Manufacturing*, 32(5):701–707, 2001.

[266] B. Chen and T.-W. Chou. Compaction of woven-fabric preforms: Nesting and multi-layer deformation. *Composites Science and Technology*, 60(12-13):2223–2231, 2000.

[267] B. Chen and T.-W. Chou. *Composites Science and Technology*, 59(10):1519–1526, 1999.

[268] E.M. Sozer, B. Chen, P.J. Graham, T.-W. Chou, and S.G. Advani. Characterization and prediction of compaction force and preform permeability of woven fabrics during the resin transfer molding process. In *Proceedings of 5th International Conference on Flow Processes in Composite Materials*, pages 25–36, July 1999.

[269] C.L. Tucker III, editor. *Fundamentals of Computer Modeling for Polymer Processing.* Hanser Publishers, New York, 1989.

[270] M.M. Denn. *Process Modeling.* Longman, New York, 1986.

[271] C.-C. Lee and J.M. Castro. Model simplifications. In C. L. III Tucker, editor, *Fundamentals of computer modeling for polymer processing*, volume 3. Hanser Publishers, Munich, Germany, 1989.

[272] V.L. Streeter, E.B. Wylie, and K.W. Bedford. *Fluid Mechanics.* WCB/McGraw Hill, Boston, 9 edition, 1998.

[273] C.F. Gerald and P.O. Wheatley. *Applied Numerical Analysis.* Pearson, Boston, 2004.

[274] F.P. Incropera and D.P. DeWitt. *Fundamentals of Heat Transfer.* Wiley, New York, 1981.

[275] O. Reynolds. On the theory of lubrication and its application to mr. beauchamps tower's experiments. *Phil. Trans. Royal Society*, 177:157–234, 1886.

[276] M.V. Bruschke and S.G. Advani. A numerical approach to model non-isothermal viscous flow through fibrous media with free surfaces. *International Journal for Numerical Methods in Fluids*, 19:575–603, 1994.

[277] G.W. Recktenwald. *Introduction to Numerical Methods and MATLAB: Implementations and Applications.* Addison-Wesley Longman Publishing Co., Inc., Boston, 2000.

[278] N.C. Correia, F. Robitaille, A.C. Long, C.D. Rudd, P. Simacek, and S.G. Advani. Analysis of the vacuum infusion molding process: I. analytical formulation. *Composites Part A: Applied Science and Manufacturing*, 26:1645–1656, 2005.

[279] B. Yenilmez, M. Senan, and E.M. Sozer. Variation of part thickness and compaction pressure in vacuum infusion process. *Composites Science and Technology*, 69(11-12):1710–1719, 2009.

[280] A.W. Chan and S. Hwang. Anisotropic in-plane permeability of fabric media. *Polymer Engineering and Science*, 31(16):1233–1239, 1991.

[281] J.R. Weitzenbock, R.A. Shenoi, and P.A. Wilson. Analysis of the vacuum infusion molding process: I. analytical formulation. *Composites Part A: Applied Science and Manufacturing*, 30:797–813, 1999.

[282] *Fundamentals of Computer Modeling for Polymer Processing.* Edited by Tucker III, C.L., Hanser Publications, New York, 1989.

[283] P. Simacek and S.G. Advani. A numerical model to predict fiber tow saturation during liquid composite molding. *Composites Science and Technology*, 63:1725–1736, 2003.

[284] F. Zhou, N. Kuentzer, P. Simacek, S.G. Advani, and S. Walsh. Analytic characterization of the permeability of dual-scale fibrous porous media. *Composites Science and Technology*, 66(15):2795–2803, 2006.

[285] N. Kuentzer, P. Simacek, S.G. Advani, and S. Walsh. Permeability characterization of dual scale fibrous porous media. *Composites Part A: Applied Science and Manufacturing*, 37:2057–2068, 2006.

[286] S. Ranganathan. *Mechanics of Fiber-Fiber Interactions during the Flow of Non-Dilute Fiber Suspensions.* PhD thesis, Mechanical Engineering Department, University of Delaware, Newark, 1992.

[287] S.K. Gupte and S.G. Advani. Role of process modeling to manufacture orthopaedic implants from short fiber composites. *Polymer Composites*, 15:7–17, 1994.

[288] A. Ahmed and A.N. Alexandrou. Compression molding using a generalized Eulerian-Lagrangian formulation with automatic remeshing. *Advances in Polymer Technology*, 11(3):203–211, 1992.

[289] W.I. Childs. Molding, compression. In S. M. Lee, editor, *International Encyclopedia of Composites*, volume 3, pages 458–465. VCH Publishers, New York, 1990.

[290] R. Liebold. Compression moulding and processing of thermosets. *Kunststoffe Plast Europe*, 87(11):50–52, 1997.

[291] E.W. Liang and C.L. Tucker III. Finite element method for flow in compression molding of thin and thick parts. In *ASME MD Use of Plastics and Plastic Composites: Materials and Mechanics Issues*, volume 46, pages 569–585, 1993.

[292] M.A. Dweib and C.M. O'Bradaigh. Anisotropic modeling of isothermal squeezing flow of glass-mat reinforced thermoplastics (GMT). *Polymer Composites*, 19(5):588–599, 1998.

[293] M.R. Barone and D.A. Caulk. Mechanics of compression molding. In *Proceedings of Manufacturing International*, volume 4, pages 63–77, 1988.

[294] G. Kotsikos and A.G. Gibson. Investigation of the squeeze flow behaviour of sheet moulding compounds (SMC). *Composites: Part A*, 29A(12):1569–1577, 1998.

[295] R. Ducloux, J.F. Agassant, and M. Vincent. Non-isotropic model for compression molding of long glass fiber reinforced thermoplastic sheets. In *ANTEC '91*, volume 49, pages 2062–2064, 1991.

[296] R.C. Harper and J.H. Pugh. Thermoforming of thermoplastics matrix composites. In S. M. Lee, editor, *International Encyclopedia of Composites*, volume 5, pages 496–530. VCH Publishers, New York, 1990.

[297] S.C. Chen, Y.C. Chen, and N.T. Cheng. Simulation of injection-compression mold-filling process. *International Communications in Heat and Mass Transfer*, 25(7):907–917, 1998.

[298] C.L. Tucker III and F. Folgar. A model of compression mold filling. *Polymer Engineering and Science*, 23(2):69–73, 1983.

[299] C.A. Heiber and S.F. Shen. A finite element/finite difference simulation of the injection mold filling process. *Journal of Non-Newtonian Fluid Mechanics*, 17(1):1–32, 1980.

[300] T.A. Osswald and C.L. Tucker. An automated simulation of compression mold filling for complex parts. In *Proceedings of 43rd SPE ANTEC*, pages 169–172, 1985.

[301] T.A. Osswald and C.L. Tucker. A Boundary Element Simulation of Compression Mold Filling. *Polymer Engineering and Science*, 28(7):413–420, 1988.

[302] T.A. Osswald and C.L. Tucker III. Compression mold filling simulation for non-planar parts. *International Polymer Processing*, 5(2):79–87, 1990.

[303] Y.-E. Yoo, J.S. Park, S.T. Lim, and W.I. Lee. Numerical study on the mold filling process in compression molding. *42nd International SAMPE Symposium and Exhibition*, 42(2):1439–1449, 1997.

[304] C.L. Tucker III. Compressing molding of reinforced thermoset polymers. In Avraam I. Isayev, editor, *Injection and Compression Molding Fundamentals*. M. Dekker, New York, 1987.

[305] C.L. Tucker III. Flow regimes for fiber suspensions in narrow gaps. *Journal of Non-Newtonian Fluid Mechanics*, 39(3):239–268, 1991.

[306] http://www.moldflow.com/stp/deutsch/products/mpi/mpi_fiber.htm

[307] H.H. Chiang, C.A. Hieber, and K.K. Wang. A unified simulation of the filling and post-filling stages in the injection-molding, part i: Formulation; part ii: Experimental verification. *Polymer Engineering and Science*, 31(2):116–139, 1991.

[308] S. Ranganathan and S.G. Advani. A simultaneous solution for flow and fiber orientation in axisymmetric radial flow. *J. of Non-Newtonian Fluid Mechanics*, 43:107–136, 1993.

[309] K.K. Wang. Numerical simulation of injection molding process. In Tim Gutowski, editor, *Manufacturing International, ASME Conference Proceedings.* 1988.

[310] A. Larsen. Injection molding of short fiber reinforced thermoplastics in a center-gated mold. *Polymer Composites*, 21(1):51–64, 2000.

[311] V. Verieye and F. Dupret. Prediction of fiber orientation in complex injection molded parts. In *Developments in Non-Newtonian Flows, Proceedings of the 1993 ASME Winter Annual Meeting*, volume 175, pages 139–163, New Orleans, LA, Nov 28 - Dec 3 1993. American Society of Mechanical Engineers, Applied Mechanics Division, AMD.

[312] J.S. Cintra Jr. and C.L. Tucker III. Orthotropic closure approximations for flow-induced fiber orientation. *Journal of Rheology*, 39(6):1095, 1995.

[313] G.L. Hand. *Journal of Fluid Mechanics*, 13:33, 1962.

[314] L.G. Leal and E.J. Hinch. *Rheol. Acta*, 12:127, 1973.

[315] S.G. Advani and C.L. Tucker III. Closure approximations for three dimensional structural tensors. *J. of Rheology*, 34:367–386, 1990.

[316] E.G. Kim, J.K. Park, and S.H. Jo. A study on fiber orientation during the injection molding of fiber-reinforced polymeric composites (comparison between image processing results and numerical simulation). *Journal of Materials Processing Technology*, 111:225–232, 2001.

[317] S. Patcharaphun, B. Zhang, and G. Mennig. Simulation of three-dimensional fiber orientation in weldline areas during push-pull-processing. *Journal of Reinforced Plastics and Composites*, 26(10):977–985, 2007.

[318] D.A. Cianelli, J.E. Travis, and R.S. Bailey. How to process long-fiber reinforced thermoplastics. *Plastics Technology*, 34(4):83–85, 87, 89, 1988.

[319] F.N. Cogswell and D.C. Leach. Continuous fiber reinforced thermoplastics: A change in the rules for composite technology. *Plastics and Rubber Processing and Applications*, 4(3):271–276, 1984.

[320] C.M. O'Bradaigh. Sheet forming of composite materials. In S.G. Advani, editor, *Flow and Rheology in Polymeric Composites Manufacturing*, chapter 13. Elsevier, Amsterdam; New York, 1994.

[321] S.G. Advani, T.S. Creasy, and S.F. Shuler. Rheology of long fiber-reinforced composites in sheet forming. In D. Bhattacharrya, editor, *Composite Sheet Forming*, chapter 8, pages 323–369. Elsevier Science B.V., 1997.

[322] M.R. Monaghan, C.M. O'Bradaigh, P.J. Mallon, and R.B. Pipes. Effect of diaphragm stiffness on the quality of diaphragm formed thermoplastic composite components. In *Proceedings of the National SAMPE Symposium and Exhibition, Advanced Materials: the Challenge for the Next Decade. Part 1*, volume 35, pages 810–824, Anaheim, CA, April 1990.

[323] A.J Smiley and R.B. Pipes. Simulation of the diaphragm forming of carbon fiber/thermoplastic composite laminates. In *Proceedings of the American Society for Composites, Second Technical Conference*, pages 218–221, Newark, DE, 1987.

[324] T.S. Creasy. *Elongational flow of long discontinuous fiber composites.* PhD thesis, Department of Mechanical Engineering, University of Delaware, Newark, DE, 1997.

[325] MATLAB, The Language of Technical Computing, Version 6.1, The MathWorks, Inc., http://www.mathworks.com.

[326] Maple V, Waterloo Maple Inc., http://www.maplesoft.com/flash/index.html.

[327] S. Ranganathan, S.G. Advani, and M.A. Lamontia. A non-isothermal process model for consolidation and void reduction during in-situ tow placement of thermoplastic composites. *Journal of Composite Materials*, 29:1040 1062, 1995.

[328] R. Pitchumani, R.C. Don, J.W. Gillespie Jr., and S. Ranganathan. Analysis of on-line consolidation during thermoplastic tow-placement process. In *Thermal Processing of Materials: Thermo-Mechanics, Controls, and Composites, Proceedings of the 1994 International Mechanical Engineering Congress and Exposition*, volume 289, pages 223–234, Chicago, Nov 1994. ASME, Heat Transfer Division.

[329] M.-C. Li and A.C. Loos. Effects of processing on interply bond strength of thermoplastic composites. *Journal of Reinforced Plastics and Composites*, 11(10):1142–1162, 1992.

[330] M.N. Ghasemi Nejhad. Issues related to processability during the manufacture of thermoplastic composites using on-line consolidation techniques. *Journal of Thermoplastic Composite Materials*, 6(2):130–146, 1993.

[331] S. Ranganathan and S.G. Advani. Consolidation process window in thermoplastic composite materials. Center for Composite Materials, University of Delaware, Report.

[332] R. Phillips, D.A. Akyuz, and J.-A.E. Manson. Prediction of the consolidation of woven fiber-reinforced thermoplastic composites. part i. isothermal case. *Composites - Part A: Applied Science and Manufacturing*, 29(4):395–402, 1998.

[333] R. Pitchumani, S. Ranganathan, R.C. Don, J.W. Jr. Gillespie, and M.A. Lamontia. Analysis of transport phenomena governing interfacial bonding and void dynamics during thermoplastic tow-placement. *International Journal of Heat and Mass Transfer*, 39(9):1883–1897, 1996.

[334] R.P. Wool. Theory of crack healing in polymers. *Journal of Applied Physics*, 52(10):5953–5963, 1981.

[335] S.C. Mantell, Q. Wang, and G.S. Springer. Processing thermoplastic composites in a press and by tape laying - experimental results. *Journal of Composite Materials*, 26(16):2378–2401, 1992.

[336] S.C. Mantell and G.S. Springer. Filament winding process models. *Composite Structures*, 27(1-2):141–147, 1994.

[337] A. Arefmanesh, S.G. Advani, and E.E. Michaelides. A parametric study of bubble growth in low pressure foam molding. *Polymer Engineering and Science*, (30):1330–1338, 1990.

[338] A. Arefmanesh and S.G. Advani. Non-isothermal bubble growth in polymeric foams. *Polymer Engineering and Science*, (35):252–260, 1995.

[339] S. Roychowdhury, J.W. Gillespie Jr., and S.G. Advani. Volatile-induced void formation in amorphous thermoplastic polymeric materials: I. modeling and parametric studies. *Journal of Composite Materials*, 35:340–366, 2001.

[340] T. Taricco. Autoclave cure systems. In *Engineered Materials Handbook, Composites*, volume 1. 1987.

[341] T.G. Gutowski and G. Dillon. Elastic deformation of fiber bundles. In T. G. Gutowski, editor, *Advanced Composite Manufacturing*, chapter 4. Wiley, 1997.

[342] T.G. v, T. Morigaki, and Z. Cai. The consolidation of laminate composites. *J. Compos. Mater.*, 21:172–188, 1987.

[343] V. Antonucci, M. Giordano, K.-T. Hsiao, and S.G. Advani. Methodology to reduce thermal gradients due to the exothermic reactions. *International Journal of Heat and Mass Transfer*, 45(8):1675–1684, 2002.

[344] D.J. Michaud. *Simulation-based design optimization and control of thick composite laminates manufactured by resin transfer molding*. PhD thesis, University of Delaware, Newark, 2000.

[345] E. Watkins, S.C. Mantell, and P.R. Ciriscioli. Optimization of autoclave curing of composites: an expert system approach. In *American Society of Mechanical Engineers, Materials Division*, volume 69. 1995.

[346] S. Bickerton, P. Simacek, S.E. Guglielmi, and S.G. Advani. Investigation of draping and its effects on the mold filling process during manufacturing of a compound curved composite part. *Composites Part A*, 28A:801–816, 1997.

[347] C.-H. Shih and L.J. Lee. Tackification of textile fiber preforms in resin transfer molding. *Journal of Composite Materials*, 35(21):1954–1981, 2001.

[348] G. Estrada and S.G. Advani. Experimental characterization of the influence of tackifier material on preform permeability. *J. of Composite Materials*, 36(19):2297–2310, 2002.

[349] P. Bhat, J. Merotte, P. Simacek, and S.G. Advani. Process analysis of compression resin transfer molding. *Composites Part A-Applied Science and Manufacturing*, 40(4):431–441, 2009.

[350] K.M. Pillai, C.L. Tucker, and F.R. Phelan. Numerical simulation of injection/compression liquid composite molding. part 2: Preform compression. *Composites - Part A: Applied Science and Manufacturing*, 32(2):207–220, 2001.

[351] W.H. Seeman III. Plastic transfer molding techniques for the production of fiber reinforced plastic structures. US Patent 4,902,215, February 1990.

[352] S.M. Walsh, E.J. Rigas, W.A. Spurgeon, W.N. Roy, D. Heider, and J. Gillespie. A non-contact distribution scheme for promoting and controlling resin flow for VARTM processes. In *International SAMPE Technical Conference*, pages 284–293, Boston, MA, Nov. 5-9 2000.

[353] R. Mathur, D. Heider, S.G. Advani, S. Walsh, E.J. Rigas, K. Tackitt, M. Allende, B.K. Fink, and J.W. Gillespie Jr. Experimentation and modeling for the investigation of fast remotely actuated channeling in the VARTM process. In *Proceeding of ASME (2001) Winter Annual Meeting*, 2001.

[354] N. Vernin, M. Chohra, J. Alms, M. Fuqua, S. Kasprzak, S.G. Advani, and et al. Vacuum assisted resin transfer molding techniques with flow flooding chamber. Invention Patent, UD 06-01 (131 * 414), Office of Naval Research (ONR), Grant N00014-04-1-0891.

[355] J. Alms and S.G. Advani. Simulation and experimental validation of flow flooding chamber method of resin delivery in liquid composite molding. *Composites: Part A*, 38:2131 2141, 2007.

[356] C.D. Rudd. *Composites for automotive applications.* Rapra Technology Ltd., Shawbury, Shrewsbury, Shropshire, U.K., 2000.

[357] L.J. Lee. Liquid composite molding. In T. G. Gutowski, editor, *Advanced Composites Manufacturing*, chapter 10. John Wiley and Sons, New York, 1997.

[358] LLC Weiler Rubber Technologies. Inflatable Bladders, http://www.weilertechnologies.com/bladders.html.

[359] E.O. Ayorinde, R.F. Gibson, and F. Song. Vibration-assisted liquid composite molding simulation and analysis. In *Proceedings of the 1996 ASME International Mechanical Engineering Congress and Exposition: Development, Characterization Processing, and Mechanical Behavior*, volume 74, pages 57–58, Atlanta, GA, Nov. 17-22 1996.

[360] Radius Engineering, Inc., 3474 So. East, Salt Lake City, Utah 84109. Web page: www.radiusengineering.com. *Radius FloWare TM 2500cc RTM Injector Operation & Maintenance Manual Version: 1.4.99*, 1999.

[361] Composite Integration. http://www.composite-integration.co.uk.

[362] Magnum Venus Plastech Limited. http://www.plastech.co.uk.

[363] Liquid Control Corp., 7576 Freedom Ave. N. W., P.O. Box 2747, North Canton, Ohio 44720. Web page: www.liquidcontrol.com. *Meter, Mix and Dispense Equipment Engineering Handbook*, 1996.

[364] S. Bickerton and S.G. Advani. Characterization of corner and edge permeabilities during mold filling in resin transfer molding. In *Proceedings of the ASME AMD-MD Summer Meeting*, volume 56, pages 143–150, Los Angeles, CA, June 1995.

[365] S. Bickerton and S.G. Advani. Characterization of racetracking in liquid composite molding processes. *Composites Science and Technology*, 59:2215 2229, 1999.

[366] S. Bickerton and S.G. Advani. Experimental investigation and flow visualization of the resin transfer molding process in a non-planar geometry. *Composites Science and Technology*, 57(1):23–33, 1997.

[367] S. Bickerton, S.G. Advani, R.V. Mohan, and D.R. Shires. Experimental analysis and numerical modeling of flow channel effects in resin transfer molding. *Polymer Composites*, pages 134–153, 2000.

[368] S. Bickerton, P. Simacek, K.M. Pillai, J. Mogavero, and S.G. Advani. Important mold filling issues in liquid composite molding processes: modeling and experiments. In *Annual Technical Conference - ANTEC, Materials, Proceedings of the 55th Annual Technical Conference*, volume 2, pages 2411–2423, Apr. 27-May 2 1997.

[369] E.M. Sozer. Effect of preform non-uniformity on mold filling in RTM process. In *33rd International SAMPE Technical Conference - Advancing Affordable Materials Technology*, pages 176–189, Seattle, WA, Nov. 5-8 2001.

[370] K. Han, B. Lee, and B. Rice. Permeability measurements of fiber preforms and applications. In *Mechanical Behavior of Advanced Materials, Proceedings of the 1998 ASME International Mechanical Engineering Congress and Exposition*, page 275, Anaheim, CA, Nov. 15-20 1998.

[371] J.M. Lawrence. *Methodologies for resin flow prediction and manipulation in liquid composite molding processes.* PhD thesis, University of Delaware, 2005.

[372] S.G. Advani and E.M. Sozer. *Comprehensive Composite Materials, Chapter 24: Resin impregnation in liquid molding processes.* Edited by A. Kelly, Elsevier Science.

[373] L.L. Fong and S.G. Advani. Preforming analysis of thermoformable fiber mats: Preforming effect on mold filling. *Journal of Reinforced Plastics and Composites*, 13(7):637–663, 1994.

[374] J. Ni, Y. Zhao, L.J. Lee, and S. Nakamura. Analysis of two-regional flow in liquid composite molding. *Polymer Composites*, 18(2):254–269, April 1997.

[375] E.M. Sozer, S. Bickerton, and S.G. Advani. On-line strategic control of liquid composite mold filling process. *Composites part A: Applied Science and Manufacturing*, (31):1383–1394, 2000.

[376] K.T. Hsiao, R. Mathur, J.W. Gillespie, B.K. Fink, and S.G. Advani. A closed form solution for the vacuum assisted resin transfer molding process. *Journal of Manufacturing Science*, (122):463–472, 2000.

[377] M.J. Tari, J.-P. Imbert, M.Y. Lin, A.S. Lavine, and H.T. Hahn. Analysis of resin transfer molding with high permeability layers. *Journal of Manufacturing Science and Engineering*, 120(3):609–616, 1998.

[378] F. Trochu, J.F. Boudreault, D.M. Gao, and R. Gauvin. Three-dimensional flow simulations for the resin transfer molding process. *Materials and Manufacturing Processes*, 10(1):21–26, 1995.

[379] S.G. Advani and P. Simacek. Modeling and simulation of flow, heat transfer and cure. In Teresa Krukenberg and Rowan Paton, editors, *Resin Transfer Molding For Aerospace Structures*, pages 225–281. Kluwer Academic Publishers, the Netherlands, 1999.

[380] P. Simacek, E.M. Sozer, and S.G. Advani. Numerical simulations of mold filling for design and control of RTM process. In *ANTEC, Proceedings of the 56th Annual Technical Conference*, pages 2327–2335, Atlanta, GA, Apr. 26-30 1998.

[381] J. Luo, Z. Liang, C. Zhang, and B. Wang. Optimum tooling design for resin transfer molding with virtual manufacturing and artificial intelligence. *Composites - Part A: Applied Science and Manufacturing*, 32(6):877–888, 2001.

[382] R.S. Parnas, J.G. Howard, T.L. Luce, and S.G. Advani. Permeability Characterization, Part 1: A Proposed Standard Reference Material for Permeability. *Polymer Composites*, 16(6):430–446, 1996.

[383] S.H. Ahn, W.I. Lee, and G.S. Springer. Measurement of the three-dimensional permeability of fiber preforms using embedded fiber optic sensors. *Journal of Composite Materials*, 29(6):714–733, 1995.

[384] B. Ballata, S. Walsh, and S.G. Advani. Measurement of the transverse permeability of fiber preforms. *Journal of Reinforced Plastics and Composites*, 18:1450–1464, 1999.

[385] Nirajkumar Patel. *Micro Scale Flow Behavior, Fiber Wetting and Void Formation in Liquid Composite Molding.* PhD thesis, Ohio State University, 1997.

[386] N.C. Correia, F. Robitaille, A.C. Long, C.D. Rudd, P. Simacek, and S.G Advani. Use of resin transfer molding simulation to predict flow, saturation, and compaction in the VARTM process. *Journal of Fluids Engineering*, 126(2):210–215, 2004.

[387] M.V. Bruschke and S.G. Advani. A finite element/control volume approach to mold filling in anisotropic porous media. *Polymer Composites*, 11(6):398–405, December 1990.

[388] J.P. Coulter and S.I. Guceri. Resin impregnation during the manufacturing of composite materials subject to prescribed injection rate. *Journal of Reinforced Plastics and Composites*, 7(3):200–219, 1988.

[389] R. Mathur, D. Heider, C. Hoffmann, J.W. Gillespie Jr., S.G. Advani, and B.K. Fink. Flow front measurements and model validation in the vacuum assisted resin transfer molding process. *Polymer Composites*, 22(4):477–490, 2001.

[390] A.C. Loos, J. Sayre, R. McGrane, and B. Grimsley. VARTM process model development. In *Proceedings of the International SAMPE Symposium and Exhibition, Materials and Processes Odyssey*, volume 46, pages 1049–1060, Long Beach, CA, May 2001.

[391] B.W. Grimsley, P. Hubert, X. Song, R.J. Cano, Alfred C. Loos, and R. Byron Pipes. Flow and compaction during the vacuum assisted resin transfer molding process. In *International SAMPE Technical Conference - Advancing Affordable Materials Technology*, volume 33, pages 140–153, Seattle, WA, Nov. 5-8 2001.

[392] M.K. Kang, W.I. Lee, and H.T. Hahn. Analysis of vacuum bag resin transfer molding process. *Composites, Part A: Applied Science and Manufacturing*, 32(11):1553–1560, 2001.

[393] B. Liu, S. Bickerton, and S.G. Advani. Modeling and simulation of resin transfer molding (RTM) gate control, venting and dry spot prediction. *Composites: Part A*, 27(2):135–141, 1996.

[394] P. Simacek and S.G. Advani. *Liquid injection molding simulation - LIMS User's Manual - version 5.0.* University of Delaware, Newark, 2009.

[395] C. Binetruy, V. Hilaire, and J. Pabiot. The interactions between flows occurring inside and outside fabric tows during RTM. *Composites Science and Technology*, 57:587–596, 1997.

[396] N. Patel, V. Rohatgi, and J.L. Lee. Micro scale flow behavior and void formation mechanism during impregnation through a unidirectional stitched fiberglass mat. *Polymer Engineering and Science*, 35(10):837–851, May 1995.

[397] F.R. Jr. Phelan, Y. Leung, and R.S. Parnas. Modeling of microscale flow in unidirectional fibrous porous media. *Journal of Thermoplastic Composite Materials*, 7(3):208–218, 1994.

[398] S. Amico and C. Lekakou. Mathematical modeling of capillary micro-flow through woven fabrics. *Composites, Part A: Applied Science and Manufacturing*, 31(12):1331–1344, 2000.

[399] K.M. Pillai and S.G. Advani. Numerical simulation of unsaturated flow in woven fiber preforms during resin transfer molding process. *Polymer Composites*, 19(1):71–80, 1998.

[400] J. Slade, K. Pillai, and S.G. Advani. Investigation of unsaturated flow in woven, braided and stitched fiber mats during mold-filling in resin transfer molding. *Polymer Composites*, 22:491–505, 2001.

[401] M. Kaviany. *Principles of Heat Transfer in Porous Media*. Springer, New York, 1995.

[402] H.-T. Chiu, B. Yu, S.C. Chen, and L.J. Lee. Heat transfer during flow and resin reaction through fiber reinforcement. *Chemical Engineering Science*, 55(17):3365–3376, 2000.

[403] C.L. Tucker III and R.B. Dessenberger. Governing equations for flow and heat transfer in stationary fiber beds. In S. G. Advani, editor, *Flow and Rheology in Polymer Composites Manufacturing*, chapter 8. Elsevier, 1994.

[404] S.G. Advani and K.-T. Hsiao. Heat transfer during mold filling in liquid composite manufacturing. In Kambiz Vafai, editor, *Handbook of Porous Media*, chapter 19, pages 845–891. Marcel Dekker, Inc., New York, NY, 2000.

[405] K.-T. Hsiao, H. Laudorn, and S.G. Advani. Experimental investigation of heat dispersion due to impregnation of viscous fluids in heated fibrous porous during composites processing. *ASME J. of Heat Transfer*, 123:178–187, 2001.

[406] R.B. Dessenberger and C.L. Tucker III. Thermal Dispersion in Resin Transfer Molding. *Polymer Composites*, 16(6):495–506, 1995.

[407] K.-T. Hsiao and S.G. Advani. A method to predict microscopic temperature distribution inside a periodic unit cell of non-isothermal flow in porous media. *Journal of Porous Media*, 5:69–86, 2002.

[408] H.-T. Chiu, S.-C. Chen, and L.J. Lee. Analysis of heat transfer and resin reaction. In *Proceedings of the 55th Annual Technical Conference, ANTEC, Part 2*, pages 2424–2429, Toronto, Canada, Apr. 27-May 2 1997.

[409] R.B. Dessenberger and C.L. Tucker III. *Thermal Dispersion in Resin Transfer Molding*. ASME, 1994.

[410] M.R. Kamal, S. Sourour, and M. Ryan. Integrated thermo-rheological analysis of the cure of thermosets. *SPE Tech. Pap.*, 18:187–191, 1973.

[411] M.R. Kamal and S. Sourour. Kinetics and thermal characterization of thermoset cure. *Polym. Eng. & Sci.*, 13:59–64, 1973.

[412] C.A. Hieber and S.F. Shen. A finite element/finite difference simulation of the injection mold filling process. *J. Non-Newtonian Fluid Mech.*, 7(1):1–31, 1980.

[413] S.F. Shen. Simulation of polymer flows in the injection molding process. *Int. J. Num. Meth. Fluids*, 4:171–183, 1984.

[414] B. Liu and S.G. Advani. Operator splitting scheme for 3-D temperature solution based on 2-D flow approximation. *Computational Mechanics*, 38:74–82, 1995.

[415] P. Simacek and S.G. Advani. Desirable features in mold filling simulations for liquid molding processes. *Polymer Composites*, 25:355–367, 2004.

[416] P. Simacek, E.M. Sozer, and S.G. Advani. *User Manual for DRAPE 1.1 and LIMS 4.0 (Liquid Molding Simulation)*. Center of Composite Materials, University of Delaware, Newark, 1998.

[417] E.M. Sozer, P. Simacekek, and S.G. Advani. Draping and filling of long axisymmetric molds in RTM process. In *Proceedings of the 8th Japan-U.S. Conference on Composite Materials*, pages 73–84, September 1998.

[418] B. Yenilmez and E.M. Sozer. Compaction of e-glass fabric preforms in the vacuum infusion process, A: Characterization experiments. *Composites Part A: Applied Science and Manufacturing*, 40(4):499–510, 2009.

[419] http://www.dogma.org.uk/vtt/process/processes/filament_winding.htm.

[420] S.Y. Lee and G.S. Springer. Filament winding cylinders I: Process model. *J. Comp. Mat.*, 24(12):1270–1298, 1990.

[421] E.P. Calius, S.Y. Lee, and G.S. Springer. Filament winding cylinders II: Validation of process model. *J. Comp. Mat.*, 24(12):1299–1343, 1990.

[422] S.Y. Lee and G.S. Springer. Filament winding cylinders III: Selection of the process variables. *J. Comp. Mat.*, 24(12):1343–1366, 1990.

[423] W.I. Lee, A.C. Loos, and G.S. Springer. Heat of reaction, degree of cure and viscosity of hercules 3501-6 resin. *J. Comp. Mat.*, 16, 1982.

[424] S. Mantell and D. Cohen. Filament winding. In R. Dave and A. Loos, editors, *Processing of Composites*, chapter 13. Hanser, Munich, 1999.

[425] S. Bickerton, E.M. Sozer, P. Simacek, and S.G. Advani. Fabric structure and mold curvature effects on preform permeability and mold filling in the RTM process, part II: Predictions and comparisons with experiments. *Composites Part A: Applied Science and Manufacturing*, 31(5):439–458, 2000.

[426] S. Bickerton, E.M. Sozer, P.J. Graham, and S.G. Advani. Fabric structure and mold curvature effects on preform permeability and mold filling in the RTM process, part I: Experiments. *Composites Part A: Applied Science and Manufacturing*, 31(5):423–438, 2000.

[427] N. Laval, E.M. Sozer, and S.G. Advani. Characterisation of permeability around a 90° corner. *Advanced Composites Letters*, 9(1):17–24, 2000.

Index

An environmentally friendly book printed and bound in England by www.printondemand-worldwide.com

PEFC Certified

This product is from sustainably managed forests and controlled sources

www.pefc.org

PEFC/16-33-415

MIX

Paper from responsible sources

FSC® C004959

www.fsc.org

This book is made entirely of sustainable materials; FSC paper for the cover and PEFC paper for the text pages.

#0207 - 300913 - C0 - 254/178/33 [35] - CB